BAYESIAN SIGNAL PROCESSING

Wiley Series on Adaptive and Cognitive Dynamic Systems

Editor: Simon Haykin

A complete list of titles in this series appears at the end of this volume.

BAYESIAN SIGNAL PROCESSING

Classical, Modern, and Particle Filtering Methods

SECOND EDITION

James V. Candy

Lawrence Livermore National Laboratory
University of California, Santa Barbara

IEEE
IEEE PRESS
WILEY

Published by John Wiley & Sons, Inc., Hoboken, New Jersey.
Published simultaneously in Canada.

Library of Congress Cataloging-in-Publication Data

Names: Candy, James V., author.
Title: Bayesian signal processing : classical, modern, and particle filtering methods / James V. Candy.
Description: Second edition. | Hoboken, New Jersey : John Wiley & Sons Inc., [2016] | Includes index.
Identifiers: LCCN 2016019012 | ISBN 9781119125457 (cloth) | ISBN 9781119125488 (epub)
Subjects: LCSH: Signal processing–Mathematics. | Bayesian statistical decision theory.
Classification: LCC TK5102.9 .C3187 2016 | DDC 621.382/201519542–dc23
LC record available at https://lccn.loc.gov/2016019012

10 9 8 7 6 5 4 3 2 1

"…no mere man has ever seen, heard or even imagined what wonderful things God has ready for those who love the Lord" (1 Cor. 2:9)

CONTENTS

PREFACE TO SECOND EDITION

The second edition of *Bayesian Signal Processing* incorporates a chapter on "Sequential Bayesian Detection" (Chapter 10) and a section on "Ensemble Kalman Filters" (Section 6.7), as well as an expansion of case studies in the final chapter (Chapter 11). These new "physics-based" studies detail Bayesian approaches to problem solving in real-world applications incorporating detailed particle filter designs, adaptive particle filters, and sequential Bayesian detection. In addition to these major developments, a variety of sections are expanded to "fill in the gaps" of the first edition. Here, metrics for particle filter (PF) designs with emphasis on classical "sanity tests," introduced earlier in model-based processors, led to *ensemble techniques* as a basic requirement for performance analysis. Next, the expansion of *information theory metrics* (Kullback–Leibler divergence (KD), Hellinger distance (HD)) and their application to PF designs is discussed. These "fill-in-the-gap" expansions provide a more cohesive discussion with examples and applications enabling the comprehension of these alternative approaches to solving estimation/detection problems.

Detection theory, and more specifically sequential detection theory, is closely coupled to sequential estimation techniques presented in this text and is often the primary reason for constructing the estimators in the first place [1]–[14]. Sequential techniques find application in many technical application areas such as radar, sonar (detection/tracking), biomedical (anomaly detection/localization), speech (recognition/tracking), communications (real-time/obstructed environments), the sciences (e.g., seismology (earthquakes), structures (vibrations), materials (additive manufacturing/threat detection), radiation (threat detection, etc.), and of course, a huge variety of military applications [3], [7]. By incorporating a new chapter on sequential detection techniques primarily aimed at the binary decision problem, we enable the extension of these estimation methods to an entire class of problems especially when a physical model is available that can be incorporated into the algorithm [4], [6]. This new chapter, in itself, will provide wider application, since sequential detection is such a natural extension to sequential estimation and vice versa.

The ensemble Kalman Filter (EnKF) addition to the second edition is an area that has been neglected in most non-specialized texts. The EnKF is basically a little known hybrid in the engineering area, but well-known in the sciences. It is a hybrid of a regression-based processor (e.g., unscented Kalman filter (UKF)) and a particle-like (PF) "sampling" filter. The EnKF is well known in science areas where large-scale computations are required such as seismology, energy systems (wind, ocean waves, etc.), weather prediction, climatology (global warming), computational biology, large structures (vibrations), and more because of its computational efficiency for very large-scale computational problems (super-computer applications). Here, the coupling of model-based Bayesian techniques to these large-scale problems is unique.

With this in mind, let us consider the construct of the new chapter entitled "Sequential Bayesian Detection." Here, we develop the Bayesian approach to decision theory primarily aimed at a coupling of sequential Bayesian estimation to sequential decision-making. We start with the binary decision problem for multi-channel measurements and develop the usual Bayesian solutions based on probability-of-error minimization leading to the well-known Bayes' risk criterion. Next, the Neyman–Pearson detection approach (maximize detection probability for fixed false-alarm probability) is developed and compared to the classical Bayesian schemes illustrating their similarity and differences. Once these "batch schemes" are developed, we introduce the Wald sequential approach to solving these problems in pseudo real time [3], [7]. Once developed, we then investigate a variety of performance criteria based on the receiver operating characteristic (ROC) curve and its variants that provide the foundations for classical analysis [9], [10]. Other metrics (e.g., area-under-curve, and so on) associated with the ROC curve are introduced and applied as well. With the sequential detection theory developed, we investigate the basic linear Gaussian case and demonstrate that a sequential scheme easily follows when coupled to the model-based (Kalman) processor. Next, we generalize this approach to nonlinear models and again under Gaussian-like approximations develop the sequential detection scheme [7]. Finally, we remove the Gaussian assumptions and show how, using an MCMC (particle filter), sequential detection schemes can be developed and applied. A variety of applications are included in case studies on anomaly/change detection.

Finally, sequential detection enables the inclusion of more relevant case studies (Chapter 11) in ocean acoustics and physics-based radiation detection as well as X-ray threat material detection offering a completely different perspective on classical problem solving incorporating these physics-based approaches from the sequential Bayesian framework.

James V. Candy

Danville, CA

REFERENCES

1. S. Kay, *Fundamentals of Statistical Signal Processing: Detection Theory* (Englewood Cliffs, NJ: Prentice-Hall, 1998).
2. C. Therrien, *Decision, Estimation, and Classification: An Introduction to Pattern Recognition and Related Topics* (New York: John Wiley & Sons, 1989).
3. J. Melsa and D. Cohn, *Detection and Estimation Theory* (New York: McGraw-Hill, 1978).
4. M. Basseville and I. Nikiforov, *Detection of Abrupt Changes: Theory and Application* (Englewood Cliffs, NJ: Prentice-Hall, 1993).
5. F. Gustafasson, *Adaptive Filtering and Change Detection* (Hoboken, NJ: John Wiley & Sons, 2000).
6. K. Burnham and D. Anderson, *Model Selection and Multimodal Inference* (New York: Springer, 1998).
7. A. Sage and J. Melsa, *Estimation Theory with Applications to Communications and Control* (New York: McGraw-Hill, 1971).
8. L. Scharf, *Statistical Signal Processing: Detection, Estimation, and Time Series Analysis* (Reading, MA: Addison-Wesley, 1990).
9. H. Van Trees, *Detection, Estimation and Modulation Theory, Part 1* (New York: John Wiley & Sons, 1968).
10. R. Duda, P. Hart, and D. Stork, *Pattern Classification*, 2nd Ed. (Hoboken, NJ: John Wiley & Sons, 2001).
11. A. Gelman, J. Carlin, H. Stern, and D. Rubin, *Bayesian Data Analysis* (New York: Chapman & Hall, 2004).
12. J. Hancock and P. Wintz, *Signal Detection Theory* (New York: McGraw-Hill, 1966).
13. D. Middleton, *Introduction to Statistical Communication Theory* (New York: McGraw-Hill, 1960).
14. S. Kassam, *Signal Detection in Non-Gaussian Noise* (New York: Springer-Verlag, 1988).

PREFACE TO FIRST EDITION

In the real world, systems designed to extract signals from noisy measurements are plagued by errors evolving from constraints of the sensors employed, random disturbances and noise, and probably, most common, the lack of precise knowledge of the underlying physical phenomenology generating the process in the first place! Methods capable of extracting the desired signal from hostile environments require approaches that capture all of the a priori information available and incorporate them into a processing scheme. This approach is typically model-based [1], employing mathematical representations of the component processes involved. However, the actual implementation providing the algorithm evolves from the realm of statistical signal processing using a Bayesian approach based on Bayes' rule. Statistical signal processing is focused on the development of processors capable of extracting the desired information from noisy, uncertain measurement data. This is a text that develops the "Bayesian approach" to statistical signal processing for a variety of useful model sets. It features the next generation of processors which have recently been enabled with the advent of high-speed/high-throughput computers. The emphasis is on nonlinear/non-Gaussian problems, but classical techniques are included as special cases to enable the reader familiar with such methods to draw a parallel between the approaches. The common ground is the model sets. Here, the state–space approach is emphasized because of its inherent applicability to a wide variety of problems both linear and nonlinear as well as time invariant and time-varying problems including what has become popularly termed "physics-based" models. This text brings the reader from the classical methods of model-based signal processing including Kalman filtering for linear, linearized and approximate nonlinear processors as well as the recently developed unscented or sigma-point filters to the next generation of processors that will clearly dominate the future of model-based signal processing for years to come. It presents a unique viewpoint of signal processing from the Bayesian perspective in contrast to the pure statistical approach found in many textbooks. Although designed primarily as a graduate textbook, it will prove very useful to the practicing signal processing professional or scientist, since a wide variety of applications are included to demonstrate the applicability of the Bayesian approach to real-world problems. The prerequisites for such a text is a melding of undergraduate work in linear algebra, random processes, linear systems, and digital signal processing as well as a minimal background in model-based signal processing illustrated in the recent text [1]. It is unique in the sense that few texts cover the breadth of its topics, whereas, the underlying theme of this text is the Bayesian approach that is uniformly developed and followed throughout in the algorithms, examples, applications, and case studies. It is this theme coupled with the hierarchy of physics-based models developed that contribute to its uniqueness. This text has evolved from three previous texts, Candy [1–3] coupled with a wealth of practical applications to real-world Bayesian problems.

The Bayesian approach has existed in statistical physics for a long time and can be traced back to the 1940s with the evolution of the Manhattan project and the work of such prominent scientists as Ulam, von Neumann, Metropolis, Fermi, Feynman, and Teller. Here the idea of Monte Carlo (MC) techniques to solve complex integrals evolved [4]. Since its birth, Monte Carlo related methods have been the mainstay of many complex statistical computations. Many applications have evolved from this method in such areas as physics, biology, chemistry, computer science, economics/finance, material science, statistics and more recently in engineering. Thus, statisticians have known for a long time about these methods, but their practicalities have not really evolved as a working tool until the advent of high-speed super computers around the 1980s. In signal processing, it is hard to pinpoint the actual initial starting point but clearly the work of Handschin and Mayne in the late 1960s and early 1970s [5, 6] was the initial evolution of Monte Carlo techniques for signal processing and control. However, from the real-time perspective, it is probably the development of the sequential Bayesian processor made practical by the work of Gordon, Salmond, and Smith in 1993 [7] enabling the evolution and the explosion of the Bayesian sequential processor that is currently being researched today. To put this text in perspective, we must discuss the current signal processing texts available on Bayesian processing. Since its evolution much has been published in the statistical literature on Bayesian techniques for statistical estimation; however, the earliest texts are probably those of Harvey [8], Kitigawa and Gersch [9], and West [10] which emphasize the Bayesian model-based approach incorporating dynamic linear or nonlinear models into the processing scheme for additive Gaussian noise sources leading to the classical approximate (Kalman) filtering solutions. These works extend those results to nonGaussian problems using Monte Carlo techniques for eventual solution laying the foundation for works to follow. Statistical MC techniques were also available, but not as accessible to the signal processor due to statistical jargon and abstractness of the discussions. Many of these texts have evolved during the 1990s such as Gilks [11], Robert [12], Tanner [13], Tanizaki [14], with the more up-to-date expositions evolving in the late 1990s and currently such as Liu [4], Ruanaidh [15], Haykin [16], Doucet [17], Ristic [18], and Cappe [19]. Also during the last period a sequence of tutorials and special IEEE issues evolved exposing the MC methods to the signal processing community such as Godsill [20], Arulampalam [21], Djuric [22], Haykin [23], Doucet [24], Candy [25], as well as a wealth of signal processing papers (see references for details). Perhaps the most complete textbook from the statistical researcher's perspective is that of Cappe [19]. In this text, much of the statistical MC sampling theory is developed along with all of the detailed mathematics—ideal for an evolving researcher. But what about the entry level person—the engineer, the experimentalist, and the practitioner? This is what is lacking in all of this literature. Questions like, how do the MC methods relate to the usual approximate Kalman methods? How does one incorporate models (model-based methods) into a Bayesian processor? How does one judge performance compared with classical methods? These are all basically pragmatic questions that the proposed text will answer in a lucid manner through coupling the theory to real-world examples and applications. Thus, the goal of this text is to provide a bridge for the practitioners with enough theory and applications to provide the basic background to comprehend the Bayesian framework and enable the application of these powerful techniques to real-world problem solving. Next, let us discuss the structure of the proposed text in more detail to understand its composition and approach.

We first introduce the basic ideas and motivate the need for such processing while showing that they clearly represent the next generation of processors. We discuss potential application areas and motivate the requirement for such a generalization. That is, we discuss how the simulation-based approach to Bayesian processor design provides a much needed capability, while well known in the statistical community, not very well known (until recently) in the signal processing community. After introducing the basic concepts in Chapter 1, we begin with the basic Bayesian processors in Chapter 2. We start with the Bayesian "batch" processor and establish its construction by developing the fundamental mathematics required. Next we discuss the well-known maximum likelihood (ML) and minimum (error) variance (MV) or equivalently minimum mean-squared error (MMSE) processors. We illustrate the similarity and differences between the schemes. Next, we launch into sequential Bayesian processing schemes which forms the foundation of the text. By examining the "full" posterior distribution in both dynamic variables of interest as well as the full data set, we are able to construct the sequential Bayesian approach and focus on the usual *filtered* or *filtering* distribution case of highest interest demonstrating the fundamental prediction/update recursions inherent in the sequential Bayesian structure. Once establishing the general Bayesian sequential processor (BSP), the schemes that follow are detailed depending on the assumed distribution with a variety of model sets.

We briefly review simulation-based methods starting with sampling methods, progressing to Monte Carlo approaches leading to the basic iterative methods of sampling using the Metropolis, Metropolis-Hastings, Gibb's, and slice samplers. Since one of the major motivations of recursive or sequential Bayesian processing is to provide a real-time or pseudo real-time processor, we investigate the idea of *importance sampling* as well as *sequential importance sampling* techniques leading to the generic Bayesian sequential importance sampling algorithm. Here we show the solution can be applied, once the importance sampling distribution is defined.

In order to be useful, Bayesian processing techniques must be specified through a set of models that represent the underlying phenomenology driving the particular application. For example, in radar processing, we must investigate the propagation models, tracking models, geometric models, and so forth. In Chapter 4, we develop the state–space approach to signal modeling which forms the basis of many applications such as speech, radar, sonar, acoustics, geophysics, communications, control, etc. Here, we investigate continuous, sampled-data and discrete state–space signals and systems. We also discuss the underlying systems theory and extend the model-set to include the stochastic case with noise driving both process and measurements leading the well-known Gauss–Markov (GM) representation which forms the starting point for the *classical* Bayesian processors to follow. We also discuss the equivalence of the state–space model to a variety of time series (ARMA, AR, MA, etc.) representations as well as the common engineering model sets (transfer functions, all-pole, all-zero, pole-zero, etc.). This discussion clearly demonstrates why the state–space model with its inherent generality is capable of capturing the essence of a broad variety of signal processing representations. Finally, we extend these ideas to *nonlinear* state–space models leading to "approximate" Gauss-Markov representation evolving from nonlinear, perturbed and linearized systems.

In the next chapter, we develop *classical* Bayesian processors by first motivating the Bayesian approach to the state–space where the required conditional distributions use the embedded state–space representation. Starting with the linear, time-varying, state–space models, we show that the "optimum" classical Bayesian processor under multivariate Gaussian assumptions leads to minimum (error) variance (MV) or equivalently minimum mean-squared error (MMSE), which is the much heralded *Kalman filter* of control theory [1]. That is, simply substituting the underlying Gauss-Markov model into the required conditional distributions leads directly to the BSP or Kalman filter in this case. These results are then extended to the *nonlinear* state–space representation which are linearized using a known reference trajectory through perturbation theory and Taylor-series expansions. Starting with the linearized or approximate GM model of Chapter 4, we again calculate the required Bayesian sequential processor from the conditionals which lead to the "linearized" BSP (or linearized Kalman filter) algorithm. Once this processor is developed, it is shown that the "extended" Bayesian processor follows directly by linearizing about the most currently available estimate rather than the reference trajectory. The extended Bayesian processor (XBP) or equivalently extended Kalman filter (EKF) of nonlinear processing theory evolves quite naturally from the Bayesian perspective, again following the usual development by defining the required conditionals, making nonlinear approximations and developing the posterior distributions under multivariate Gaussian assumptions. Next, we briefly investigate an iterative version of the XBP processor, again from the Bayesian perspective which leads directly to the iterative version of the extended Bayesian processor (IX-BP) algorithm—an effective tool when nonlinear measurements dominate the uncertain measurements required.

Chapter 6 focuses on *statistical linearization* methods leading to the *modern* unscented Bayesian processor (UBP) or equivalently sigma-point Bayesian processor (SPBP). Here we show how statistical linearization techniques can be used to transform the underlying probability distribution using the sigma-point or unscented nonlinear transformation technique (linear regression) leading to the *unscented* Bayesian processor or equivalently the unscented Kalman filter (UKF). Besides developing the fundamental theory and algorithm, we demonstrate its performance on a variety of example problems. We also briefly discuss the Gaussian–Hermite quadrature (G-H) and Gaussian sum (G-S) techniques for completeness.

We reach the heart of the *particle filtering* methods in Chapter 7, where we discuss the Bayesian approach to the state–space. Here the ideas of Bayesian and model-based processors are combined through the development of Bayesian state–space particle filters. Initially, it is shown how the state–space models of Chapter 4 are incorporated into the conditional probability distributions required to construct the sequential Bayesian processors through importance sampling constructs. After investigating a variety of importance proposal distributions, the basic set of state-space particle filters (SSPF) are developed and illustrated through a set of example problems and simulations. The techniques including the Bootstrap, auxiliary, regularized MCMC and linearized particle filters are developed and investigated when applied to the set of example problems used to evaluate algorithm performance.

In Chapter 8, the important joint Bayesian SSPF are investigated by first developing the joint filter popularly known as the parametrically adaptive processor [1]. Here both states and static as well as dynamic parameters are developed as solutions to this joint estimation problem. The performance of these processors are compared to classical and modern processors through example problems.

In Chapter 9, the hidden Markov models (HMM) are developed for event-related problems (e.g., Poisson point processes). This chapter is important in order to place purely discrete processes into perspective. HMM evolve for any type of memoryless, counting processes and become important in financial applications, communications, biometrics, as well as radiation detection. Here we briefly develop the fundamental ideas and discuss them in depth to develop a set of techniques used by the practitioner while applying them to engineering problems of interest.

In the final chapter, we investigate a set of physics-based applications focusing on the Bayesian approach to solving real-world problems. By progressing through a step-by-step development of the processors, we see explicitly how to develop and analyze the performance of such Bayesian processors. We start with a practical laser alignment problem followed by a broadband estimation problem in ocean acoustics. Next, the solid-state microelectromechanical (MEM) sensor problem for biothreat detection is investigated followed by a discrete radiation detection problem based on counting statistics. All of these methods invoke Bayesian techniques to solve the particular problems of interest enabling the practitioner the opportunity to track "real-world" Bayesian model-based solutions.

The place of such a text in the signal processing textbook community can best be explained by tracing the technical ingredients that comprise its contents. It can be argued that it evolves from the digital signal processing area primarily from those texts that deal with random or statistical signal processing or possibly more succinctly "signals contaminated with noise." The texts by Kay [26–28], Therrien [29], and Brown [30] all provide the basic background information in much more detail than this text, so there is little overlap at the detailed level with them.

This text, however, possesses enough theory for the graduate or advanced graduate student to develop a fundamental basis to go onto more rigorous texts like Jazwinski [31], Sage [32], Gelb [33], Anderson [34], Maybeck [35], Bozic [36], Kailath [37, 38], and more recently, Mendel [39], Grewel [40], Bar-Shalom [41], and Simon [42]. These texts are rigorous and tend to focus on Kalman filtering techniques ranging from continuous to discrete with a wealth of detail on all of their variations. The Bayesian approach discussed in this text certainly includes the state–space models as one of its model classes (probably the most versatile), but the emphasis is on various classes of models and how they may be used to solve a wide variety of signal processing problems. Some of the more recent texts about the same technical level, but again, with a different focus are Widrow [43], Orfanidis [44], Sharf [45], Haykin [46], Hayes [47], Brown [30], and Stoica [48]. Again, the focus of these texts is not the Bayesian approach but the narrow set of specific models and the development of a variety of algorithms to estimate these sets. The system identification literature and texts therein also provide some overlap with this text, but again the approach is focused on estimating a model from noisy data sets and not really aimed at developing a Bayesian solution to a particular signal processing problem. The texts in this area are Ljung [49, 50], Goodwin [51], Norton [52], and Soderstrom [53].

The recent particle filtering texts of Ristic [18] and Cappe [19] are useful as references to accompany this text, especially if more details are required on the tracking problem and the fundamental theorems governing statistical properties and convergence proofs. That is, Ristic's text provides an introduction that closely follows the 2002 tutorial paper by Arulampalam [21] but provides little of the foundational material necessary to comprehend this approach. It focuses primarily on the tracking problem. Cappe's text is at a much more detailed technical level and is written for researchers in this area not specifically aimed at the practitioner's viewpoint. The proposed text combines the foundational material, some theory along with the practice and application of PF to real-world applications and examples.

The approach we take is to introduce the basic idea of Bayesian signal processing and show where it fits in terms of signal processing. It is argued that BSP is a natural way to solve basic processing problems. The more a priori information we know about data and its evolution, the more information we can incorporate into the processor in the form of mathematical models to improve its overall performance. This is the theme and structure that echoes throughout the text. Current applications (e.g., structures, tracking, equalization, biomedical) and simple examples are incorporated to motivate the signal processor. Examples are discussed to motivate all of the models and prepare the reader for further developments in subsequent chapters. In each case, the processor, along with accompanying simulations, is discussed and applied to various data sets demonstrating the applicability and power of the Bayesian approach. The proposed text is linked to the MATLAB (signal processing standard software) software package providing notes at the end of each chapter.

In summary, this Bayesian signal processing text will provide a much needed "down-to-earth" exposition of modern MC techniques. It is coupled with well-known signal processing model sets along with examples and problems that can be used to solve many real-world problems by practicing engineers and scientists along with entry-level graduate students as well as advanced undergraduates and post-doctorates requiring a solid introduction to the "next generation" of model-based signal processing techniques.

James V. Candy

Danville, CA

REFERENCES

1. J. Candy, *Model-Based Signal Processing* (Hoboken, NJ: John Wiley & Sons/IEEE Press, 2006).
2. J. Candy, *Signal Processing: The Model-Based Approach* (New York: McGraw-Hill, 1986).
3. J. Candy, *Signal Processing: The Modern Approach* (New York: McGraw-Hill, 1988).
4. J. Liu, *Monte Carlo Strategies in Scientific Computing* (New York: Springer-Verlag, 2001).
5. J. Handschin and D. Mayne, "Monte Carlo techniques to estimate the conditional expectation in multi-stage nonlinear filtering," *Intl. J. Control*, **9**, 547–559, 1969.
6. J. Handschin, "Monte Carlo techniques for prediction and filtering of nonlinear stochastic processes," *Automatica*, **6**, 555–563, 1970.
7. N. Gordon, D. Salmond, and A. F. M. Smith, "Novel approach to nonlinear/non-Gaussian Bayesian state estimation," *IEEE Proc. F*, **140**, 107–113, 1993.
8. A. Harvey, *Forecasting, Structural Time Series Models and the Kalman Filter* (Cambridge, UK: Cambridge University Press, 1989).
9. G. Kitagawa and W. Gersch, *Smoothness Priors Analysis of Time Series* (New York: Springer-Verlag, 1996).
10. M. West and J. Harrison, *Bayesian Forecasting and Dynamic Models*, 2nd Ed. (New York: Springer-Verlag, 1997).
11. W. Gilks, S. Richardson, and D. Spiegelhalter, *Markov Chain Monte Carlo in Practice* (New York: Chapman & Hall/CRC Press, 1996).
12. C. Robert and G. Casella, *Monte Carlo Statistical Methods* (New York: Springer, 1999).
13. M. Tanner, *Tools for Statistical Inference: Methods for the Exploration of Posterior Distributions and Likelihood Functions*, 2nd Ed. (New York: Springer-Verlag, 1993).
14. H. Tanizaki, "Nonlinear filters: estimation and applications," in *Lecture Notes in Economics and Mathematical Systems*, No. 400, (New York: Springer, 1993).
15. J. Ruanaidh and W. Fitzgerald, *Numerical Bayesian Methods Applied to Signal Processing* (New York: Springer-Verlag, 1996).
16. S. Haykin, *Kalman Filtering and Neural Networks* (Hoboken, NJ: John Wiley & Sons: 2001).
17. A. Doucet, N. de Freitas, and N. Gordon, *Sequential Monte Carlo Methods in Practice* (New York: Springer-Verlag, 2001).
18. B. Ristic, S. Arulampalam, and N. Gordon, *Beyond the Kalman Filter: Particle Filters for Tracking Applications* (Boston, MA: Artech House, 2004).

19. O. Cappe, E. Moulines, and T. Ryden, *Inference in Hidden Markov Models* (New York: Springer-Verlag, 2005).
20. S. Godsill and P. Djuric, "Special Issue: Monte Carlo methods for statistical signal processing," *IEEE Trans. Signal Proc.*, **50**, 173–499, 2002.
21. M. Arulampalam, S. Maskell, N. Gordon, and T. Clapp "A tutorial on particle filters for online nonlinear/non-gaussian Bayesian tracking," *IEEE Trans. Signal Proc.*, **50**, 2, 174–188, 2002.
22. P. Djuric, J. Kotecha, J. Zhang, Y. Huang, T. Ghirmai, M. Bugallo, and J. Miguez, "Particle filtering," *IEEE Signal Proc. Mag.***20**, 5, 19–38, 2003.
23. S. Haykin and N. de Freitas, "Special Issue: sequential state estimation: from Kalman filters to particle filters," *Proc. IEEE*, **92**, 3, 399–574, 2004.
24. A. Doucet and X. Wang, "Monte Carlo methods for signal processing," *IEEE Signal Proc. Mag.*, **24**, 5, 152–170, 2005.
25. J. Candy, "Bootstrap particle filtering for passive synthetic aperture in an uncertain ocean environment," *IEEE Signal Proc. Mag.*, **24**, 4, 63–75, 2007.
26. S. Kay, *Modern Spectral Estimation: Theory and Applications* (Englewood Cliffs, NJ: Prentice-Hall, 1988).
27. S. Kay, *Fundamentals of Statistical Signal Processing: Estimation Theory* (Englewood Cliffs, NJ: Prentice-Hall, 1993).
28. S. Kay, *Fundamentals of Statistical Signal Processing: Detection Theory* (Englewood Cliffs, NJ: Prentice-Hall, 1998).
29. C. Therrian, *Random Signal Processing: Detection Theory* (Englewood Cliffs, NJ: Prentice-Hall, 1991).
30. R. Brown and P. Hwang, *Introduction to Random Signals and Applied Kalman Filtering* (New York: John Wiley & Sons, 1997).
31. A. Jazwinski, *Stochastic Processes and Filtering Theory* (New York: Academic Press, 1970).
32. A. Sage and J. Melsa, *Estimation Theory with Applications to Communcations and Control* (New York: McGraw-Hill, 1971).
33. A. Gelb, *Applied Optimal Estimation* (Cambridge, MA: MIT Press, 1974).
34. B. Anderson and J. Moore, *Optimum Filtering* (Englewood Cliffs, NJ: Prentice-Hall, 1979).
35. P. Maybeck, *Stochastic Models, Estimation and Control* (New York: Academic Press, 1979).
36. M. S. Bozic, *Linear Estimation Theory* (Englewood Cliffs, NJ: Prentice-Hall, 1979).
37. T. Kailath, *Lectures on Kalman and Wiener Filtering Theory* (New York: Springer-Verlag, 1981).
38. T. Kailath, A. Sayed, and B. Hassibi, *Linear Estimation* (Englewood Cliffs, NJ: Prentice-Hall, 2000).
39. J. Mendel, *Lessons in Estimation Theory for Signal Processing, Communications, and Control* (Englewood Cliffs, NJ: Prentice-Hall, 1995).
40. M. Grewal and A. Andrews, *Kalman Filtering: Theory and Practice* (Englewood Cliffs, NJ: Prentice-Hall, 1993).
41. Y. Bar-Shalom and X. Li, *Estimation and Tracking: Principles, Techniques and Software* (Boston, MA: Artech House, 1993).

42. D. Simon, *Optimal State Estimation: Kalman, H_∞ and Nonlinear Approaches* (Hoboken, NJ: John Wiley & Sons, 2006).
43. B. Widrow and S. Stearns, *Adaptive Signal Processing* (Englewood Cliffs, NJ: Prentice-Hall, 1985).
44. S. Orfanidis, *Optimal Signal Processing* (New York: MacMillan, 1988).
45. L. Sharf, *Statistical Signal Processing: Detection, Estimation and Time Series Analysis* (Reading, MA: Addison-Wesley, 1991).
46. S. Haykin, *Adaptive Filter Theory* (Englewood Cliffs, NJ: Prentice-Hall, 1993).
47. M. Hayes, *Statistical Digital Signal Processing and Modeling* (Hoboken, NJ: John Wiley & Sons, 1996).
48. P. Stoica and R. Moses, *Introduction to Spectral Analysis* (Englewood Cliffs, NJ: Prentice-Hall, 1997).
49. L. Ljung, *System Identification: Theory for the User* (Englewood Cliffs, NJ: Prentice-Hall, 1987).
50. L. Ljung and T.Soderstrom, *Theory and Practice of Recursive Identification* (Cambridge, MA: MIT Press, 1983).
51. G. Goodwin and K. S. Sin, *Adaptive Filtering, Prediction and Control* (Englewood Cliffs, NJ: Prentice-Hall, 1984).
52. J. Norton, *An Introduction to Identification* (New York: Academic Press, 1986).
53. T. Soderstrom and P. Stoica, *System Identification* (New York: Academic Press, 1989).

ACKNOWLEDGMENTS

The unwavering support and encouragement by my wife is the motivational drive needed to undertake this endeavor, thank you with love to Patricia. My family and extended family and friends having endured the many regrets, thanks folks. Of course, the constant support of my great colleagues and friends, especially Drs. M. Horsley, E. Sullivan, Prof. H. Medeiros, Mr. John Breneman, and Mr. B. Beachamp who carefully reviewed the manuscript and suggested many improvements cannot go without a strong acknowledgement.

LIST OF ABBREVIATIONS

ABSP	adaptive Bayesian processor
ACC	accuracy
ACS	attenuation coefficient sequence
ADC	analog-to-digital conversion
AIC	Akaike information criterion
ALE	adaptive line enhancer
AMBP	adaptive model-based processor
AR	autoregressive (model)
ARMA	autoregressive moving average (model)
ARMAX	autoregressive moving average exogenous input (model)
AROC	average receiver operating characteristic (curve)
ARX	autoregressive exogenous input (model)
ASIR	auxiliary sequential importance resampling
AUC	area-under-curve (ROC curve)
BP	Bayesian processing
B-R	Bayes' risk (detector)
BSP	Bayesian signal processing
BW	bandwidth
CD	central difference
CDF	cumulative distribution
CM	conditional mean
CRLB	Cramer–Rao lower bound
C-Sq	Chi-squared (distribution or test)
CT	continuous-time
CTD	concentration-temperature-density (measurement)
EKF	extended Kalman filter
EM	expectation-maximization
EnBP	ensemble Bayesian processor
EnKF	ensemble Kalman filter
FN	false negatives
FP	false positives

FPE	final prediction error
G-H	Gauss–Hermite
GLRT	generalized likelihood-ratio test
G-M	Gaussian mixture
GM	Gauss–Markov
G-S	Gaussian sum
HD	Hellinger distance
HMM	hidden Markov model
HPR	high probability region
IEKF	iterated extended Kalman filter
i.i.d.	independent-identically distributed (samples)
IXBP	iterated extended Bayesian processor
KD	Kullbach divergence
KL	Kullbach–Leibler
K-S	Kolmogorov–Smirnov statistical test
KSP	Kalman–Szego–Popov (equations)
LBP	linear Bayesian processor
LD	lower-diagonal (matrix) decomposition
LKF	linear Kalman filter
LMS	least mean square
LS	least-squares
LTI	linear time-invariant (system)
LWR	Levinson–Wiggins–Robinson (algorithm)
LZ-BP	linearized Bayesian processor
LZKF	linearized Kalman filter
LZ-MBP	linearized model-based processor
MA	moving average (model)
MAICE	minimum Akaike information criterion
MAP	maximum a posteriori
MATLAB®	mathematical software package
MBMFP	model-based matched-field processor
MBP	model-based processor
MBSP	model-based signal processing
MC	Monte Carlo
MCEM	Monte Carlo expectation-maximization
MCMC	Markov chain Monte Carlo
MDL	minimum data length (description)
MEM	micro-electro-mechanical
M-H	Metropolis–Hastings (sampling)
MIMO	multiple input/multiple output (system)
MinE	minimum probability-of-error
ML	maximum likelihood
MMSE	minimum mean-squared error
MSE	mean-squared error
MV	minimum variance

NETLAB	neural network software package
NMSE	normalized mean-squared error
N-P	Neyman–Pearson (detector)
NPV	negative predictive value
ODP	optimal decision (threshold) point
PDF	probability density function (continuous)
P-E	probability-of-error (detector)
PEM	prediction error method
PF	particle filter
PHS	pulse height spectrum
PMF	probability mass function (discrete)
PPV	positive predictive value
PRTools	pattern recognition software package
PSD	power spectral density
RC	resistor capacitor (circuit)
REBEL	recursive Bayesian estimation library
RLC	resistor–inductor–capacitor (circuit)
RLS	recursive least-squares
RMS	root mean squared
RMSE	root minimum mean-squared error
ROC	receiver operating characteristic (curve)
RPE	recursive prediction error
RPEM	recursive prediction error method
RPF	regularized particle filter
SBP	sequential Bayesian processor
SENS	sensitivity
SIG®	signal processing software package
SIR	sequential importance sampling–resampling
SIS	sequential importance sampling
SMC	sequential Markov chain
SNR	signal-to-noise ratio
SPBP	sigma-point Bayesian processor
SPEC	specificity
SPKF	sigma-point Kalman filter
SPRT	sequential probability ratio test
SPT	sigma-point transformation
SRBP	square root Bayesian processor
SSIS	sequential sampling importance sampling
SSP	state-space processor
SSPACK®	state-space systems software package
SSPF	state-space particle filter
SSQE	sum-squared error
SVD	singular value (matrix) decomposition
TN	true negatives
TNR	true negative rate

TP	true positives (detection)
TPR	true positive rate
T/R	time-reversal
UBP	unscented Bayesian processor
UBSP	unscented Bayesian signal processor
UD	upper-diagonal matrix decomposition
UKF	unscented Kalman filter
UMBP	unscented model-based processor
UT	unscented transform
WSSR	weighted sum-squared residual statistical test
W-Test	whiteness test
XBP	extended Bayesian processor
Z	Z-transform
Z-M	zero-mean statistical test

1

INTRODUCTION

1.1 INTRODUCTION

In this chapter we motivate the philosophy of Bayesian processing from a probabilistic perspective. We show the coupling between model-based signal processing (MBSP) incorporating the a priori knowledge of the underlying processes and the Bayesian framework for specifying the distribution required to develop the processors. The idea of the sampling approach evolving from Monte Carlo (MC) and Markov chain Monte Carlo (MCMC) methods is introduced as a powerful methodology for simulating the behavior of complex dynamic processes and extracting the embedded information required. The main idea is to present the proper perspective for the subsequent chapters and construct a solid foundation for solving signal processing problems.

1.2 BAYESIAN SIGNAL PROCESSING

The development of Bayesian signal processing has evolved in a manner proportional to the evolution of high performance/high throughput computers. This evolution has led from theoretically appealing methods to pragmatic implementations capable of providing reasonable solutions for nonlinear and highly multi-modal (multiple distribution peaks) problems. In order to fully comprehend the Bayesian perspective, especially for signal processing applications, we must be able to separate our thinking and in a sense think more abstractly about probability distributions without worrying about how these representations can be "applied" to realistic processing problems.

Bayesian Signal Processing: Classical, Modern, and Particle Filtering Methods, Second Edition. James V. Candy.

Our motivation is to first present the Bayesian approach from a statistical viewpoint and then couple it to useful signal processing implementations following the well-known model-based approach [1, 2]. Here we show that when we constrain the Bayesian distributions in estimation to Markovian representations using primarily state–space models, we can construct sequential processors capable of "pseudo real-time" operations that can easily be utilized in many physical applications. Bayes' rule provides the foundation of all Bayesian estimation techniques. We show how it can be used to both theoretically develop processing techniques based on a specific distribution (e.g., Poisson, Gaussian, etc.) and then investigate properties of such processors relative to some of the most well-known approaches discussed throughout texts in the field.

Bayesian signal processing is concerned with the estimation of the underlying probability distribution of a random signal in order to perform statistical inferences [3]. These inferences enable the extraction of the signal from noisy uncertain measurement data. For instance, consider the problem of extracting the random variate, say X, from the noisy data, Y. The Bayesian approach is to first *estimate* the underlying conditional probability distribution, $\Pr(X|Y)$, and then perform the associated inferences to extract $\hat{X}$, that is,

$$\hat{\Pr}(X|Y) \Rightarrow \hat{X} = \arg\max_{X} \hat{\Pr}(X|Y)$$

where the caret, $\hat{X}$ denotes an estimate of X. This concept of estimating the underlying distribution and using it to extract a signal estimate provides the foundation of Bayesian signal processing developed in this text.

Let us investigate this idea in more detail. We start with the previous problem of trying to estimate the random parameter, X, from noisy data $Y = y$. Then the associated conditional distribution $\Pr(X|Y = y)$ is called the *posterior distribution* because the estimate is conditioned "after (*post*) the measurements" have been acquired. Estimators based on this a posteriori distribution are usually called *Bayesian* because they are constructed from *Bayes' rule*, since $\Pr(X|Y)$ is difficult to obtain directly. That is,

$$\Pr(X|Y) = \frac{\Pr(Y|X) \times \Pr(X)}{\Pr(Y)} \tag{1.1}$$

where $\Pr(X)$ is called the *prior distribution* (before measurement), $\Pr(Y|X)$ is called the *likelihood* (more likely to be true), and $\Pr(Y)$ is called the *evidence* (scales the posterior to assure its integral is unity). Bayesian methods view the sought-after parameter as random possessing a "known" a priori distribution. As measurements are made, the *prior* is transformed to the *posterior distribution* function adjusting the parameter estimates. Thus, the result of increasing the number of measurements is to improve the a posteriori distribution resulting in a sharper peak closer to the true parameter as shown in Fig. 1.1.

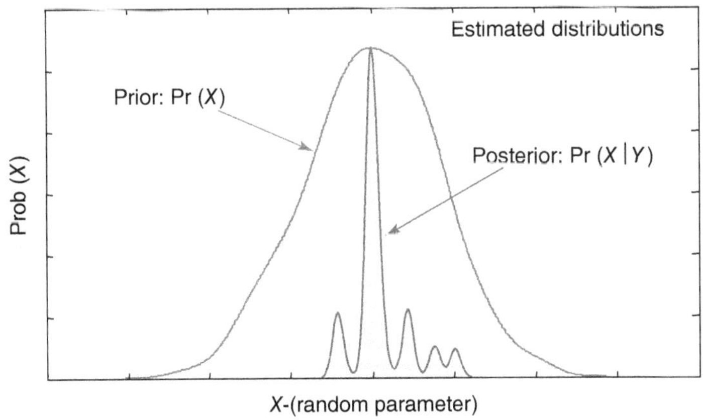

FIGURE 1.1 Bayesian estimation of the random variate X transforming the prior, $\Pr(X)$ to the posterior, $\Pr(X|Y)$ using Bayes' rule.

When the variates of interest are *dynamic*, they are functions of time and therefore, $X_t \rightarrow X$ and $Y_t \rightarrow Y$. Bayes' rule for the joint dynamic distribution is

$$\Pr(X_t|Y_t) = \frac{\Pr(Y_t|X_t) \times \Pr(X_t)}{\Pr(Y_t)} \tag{1.2}$$

In Bayesian theory, the *posterior* defined by $\Pr(X_t|Y_t)$ is decomposed in terms of the *prior* $\Pr(X_t)$, its *likelihood* $\Pr(Y_t|X_t)$, and the *evidence* or normalizing factor, $\Pr(Y_t)$. Bayesian signal processing in this dynamic case follows the identical path, that is,

$$\hat{\Pr}(X_t|Y_t) \Rightarrow \hat{X}_t = \arg\max_{X_t} \hat{\Pr}(X_t|Y_t)$$

So we begin to see the versatility of the Bayesian approach to random signal processing. Once the posterior distribution is determined, all statistical inferences or estimates are made. For instance, suppose we would like to obtain the prediction distribution. Then it can be obtained as

$$\Pr(X_{t+1}|Y_t) = \int \Pr(X_{t+1}|X_t, Y_t) \times \Pr(X_t|Y_t)\, dX_t$$

and a point estimate might be the conditional mean of this distribution, that is,

$$E\{X_{t+1}|Y_t\} = \int X_{t+1} \Pr(X_{t+1}|Y_t)\, dX_{t+1}$$

This relation shows how information that can be estimated from the extracted distribution is applied in the estimation context by performing statistical inferences.

Again, even though the Bayesian signal processing concept is simple, conceptually, the real problem to be addressed is that of evaluating the integrals which is very difficult because they are only analytically tractable for a small class of priors and likelihood distributions. The large dimensionality of the integrals causes numerical integration techniques to break down, which leads to the approximations we discuss subsequently for stabilization. Next let us consider the various approaches taken to solve the probability distribution estimation problems using non-parametric or parametric representations. This will eventually lead to the model-based approach [4].

1.3 SIMULATION-BASED APPROACH TO BAYESIAN PROCESSING

The *simulation-based approach* to Bayesian processing is founded on MC methods that are stochastic computational techniques capable of efficiently simulating highly complex systems. Historically motivated by games of chance and encouraged by the development of the first electronic computer (ENIAC), the MC approach was conceived by Ulam (1945), developed by Ulam, Metropolis and von Neumann (1947) and coined by Metropolis (1949) [5–9]. The method evolved in the mid-1940s during the Manhattan project by scientists investigating calculations for atomic weapon designs [10]. It evolved further from such areas as computational physics, biology, chemistry, mathematics, engineering, materials and finance to name a few. *Monte Carlo* methods offer an alternative approach to solving classical numerical integration and optimization problems. Inherently, as the dimensionality of the problem increases, classical methods are prone to failure while MC methods tend to increase their efficiency by reducing the error—an extremely attractive property. For example, in the case of classical grid-based numerical integration or optimization problems, as the number of grid points increase along with the number of problem-defining vector components, there is an accompanying exponential increase in computational time [10–15]. The stochastic MC approach of selecting random samples and averaging over a large number of points actually reduces the computational error by the Law of Large Numbers irrespective of the problem dimensionality. It utilizes Markov chain theory as its underlying foundation establishing the concept that through random sampling the resulting "empirical" distribution converges to the desired posterior called the stationary or invariant distribution of the chain. MCMC techniques are based on sampling from probability distributions based on a Markov chain, which is a stochastic system governed by a transition probability, having the desired posterior distribution as its invariant distribution. Under certain assumptions the chain converges to the desired posterior through proper random sampling as the number of samples become large—a crucial property (see Ref. [10] for details). Thus, the Monte Carlo approach has evolved over a long time period and is well understood by scientists and statisticians, but it must evolve even further to be useful for signal processors to become an effective tool in their problem solving repertoire.

Perhaps the best way to visualize the MC methods follows directly from the example of Frenkel [11]. Suppose that a reasonable estimate of the depth of the Mississippi river is required. Using numerical quadrature techniques the integrand value is

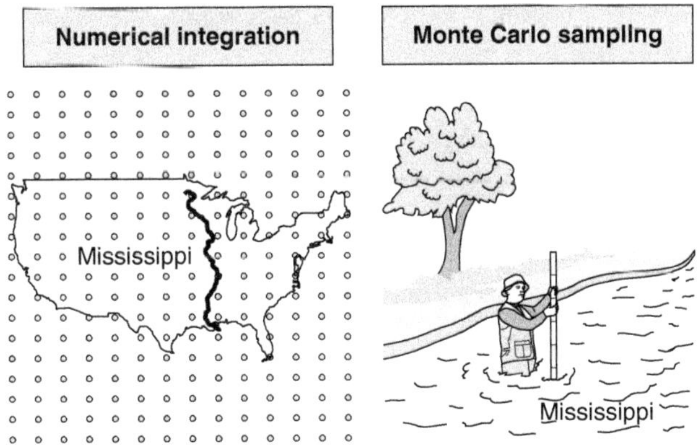

FIGURE 1.2 Monte Carlo sampling compared with numerical grid-based integration for depth of Mississippi estimation.

measured at prespecified grid points. We also note that the grid points may not be in regions of interest and, in fact, the integrand may vanish as shown in Fig. 1.2. On the other hand, the surveyor is in the Mississippi and performing a (random) walk within the river measuring the depth of the river directly. In this sampling approach measurements are accepted as long as the surveyor is in the river and rejected if outside. Here the "average" depth is simply the sample average of the measurements much the same as a sampling technique might perform. So we see that a refinement of the brute force integration approach is to use random points or samples that "most likely" come from regions of high contribution to the integral rather than from low regions.

Modern MC techniques such as in numerical integration seek to select random samples in high regions of concentration of the integrand by drawing samples from a proposed function very similar to the integrand. These methods lead to the well-known importance sampling approaches (see Chapter 3). Besides numerical integration problems that are very important in statistical signal processing for extracting signals/parameters of interest, numerical optimization techniques (e.g., genetic algorithms and simulated annealing) benefit directly from sampling technology. This important discovery has evolved ever since and become even more important with the recent development of high-speed/high-throughput computers.

Consider the following simple example of estimating the area of a circle to illustrate the MC approach.

Example 1.1

Define a sample space bounded by a square circumscribing (same center) a circle of radius r. Draw uniform random samples say $z := (X, Y)$ such that $z \sim \mathcal{U}(-r, +r)$; therefore, the number of random samples drawn from within the circle of radius r to the number of total samples drawn (bounded by the square) defines the probability

$$\Pr(Z = z) = \frac{\text{No. of circle samples}}{\text{Total no. of (square) samples}}$$

From geometry we know that the probability is simply the ratio of the two areas (circle-to-square), that is,

$$\Pr(Z = z) = \frac{\pi r^2}{4r^2} = \pi/4$$

Let $r = 1$, then a simple computer code can be written that

- draws the X,Y-coordinates from $z \sim \mathcal{U}(-1, +1)$;
- calculates the range function, $\rho = \sqrt{X^2 + Y^2}$;
- counts the number of samples that are less than or equal to ρ;
- estimates the probability, $\hat{\Pr}(Z = z)$.

The area is determined by multiplying the estimated probability by the area of the square. The resulting sample scatter plot is shown in Fig. 1.3 for a 10,000 sample realization resulting in $\pi \approx 3.130$. As the number of samples increase the estimate of the area (π) gets better and better demonstrating the MC approach. △△△

In signal processing, we are usually interested in some statistical measure of a random signal or parameter usually expressed in terms of its moments [16–23]. For example, suppose we have some signal function, say $f(X)$, with respect to some

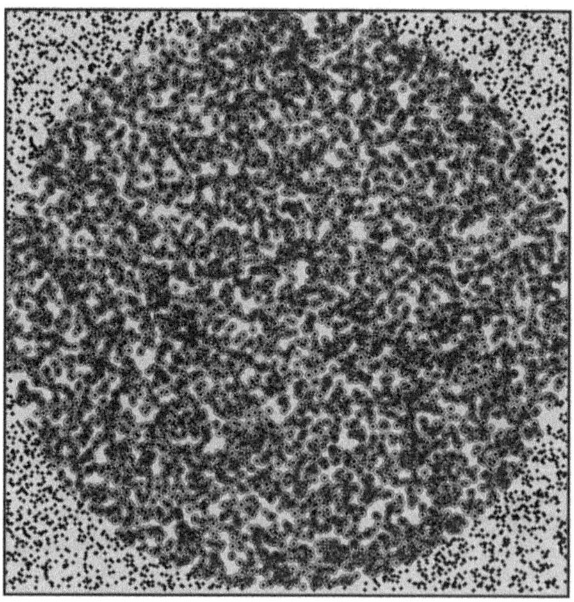

FIGURE 1.3 Area of a circle of unit radius using a Monte Carlo approach (area is estimated as 3.130 using 10, 000 samples).

underlying probabilistic distribution $\Pr(X)$. Then a typical measure to seek is its performance "on the average" which is characterized by the expectation

$$E_X\{f(X)\} = \int f(X)\Pr(X)\,dX \tag{1.3}$$

Instead of attempting to use direct numerical integration techniques, stochastic sampling techniques or *Monte Carlo integration* is an alternative. As mentioned, the *key idea* embedded in the MC approach is to represent the required distribution as a set of random *samples* rather than a specific analytic function (e.g., Gaussian). As the number of samples becomes large, they provide an equivalent (empirical) representation of the distribution enabling moments to be estimated directly (inference).

MC integration draws samples from the required distribution and then forms sample averages to approximate the sought after distributions. That is, MC integration evaluates integrals by drawing samples, $\{X(i)\}$ from the designated distribution $\Pr(X)$. Assuming perfect sampling, this produces the estimated or *empirical distribution* given by

$$\hat{\Pr}(X) \approx \frac{1}{N}\sum_{i=1}^{N} \delta(X - X(i))$$

which is a probability mass distribution with weights $\frac{1}{N}$ and random variable or sample $X(i)$. Substituting the empirical distribution into the integral gives

$$E_X\{f(X)\} = \int f(X)\hat{\Pr}(X)\,dX \approx \frac{1}{N}\sum_{i=1}^{N} f(X(i)) \equiv \overline{f} \tag{1.4}$$

which follows directly from the sifting property of the delta or impulse function. Here $\overline{f}$ is said to be a *Monte Carlo estimate* of $E_X\{f(X)\}$.

As stated previously, scientists (Ulam, von Neumann, Metropolis, Fermi, Teller, etc. [7]) created statistical sampling-based or equivalently *simulation-based* methods for solving problems efficiently (e.g., neutron diffusion or eigenvalues of the Schrodinger relation). The MC approach to problem solving is a class of stochastic computations to simulate the dynamics of physical or mathematical systems capturing their inherent uncertainties. The MC *method* is a powerful means for generating random samples used in estimating conditional and marginal probability distributions required for statistical estimation and therefore signal processing. It offers an alternative numerical approach to find solutions to mathematical problems that cannot easily be solved by integral calculus or other numerical methods. As mentioned, the efficiency of the MC method increases (relative to other approaches) as the problem dimensionality increases. It is useful for investigating systems with a large number of degrees of freedom (e.g., energy transport, materials, cells, genetics) especially for systems with input uncertainty [5].

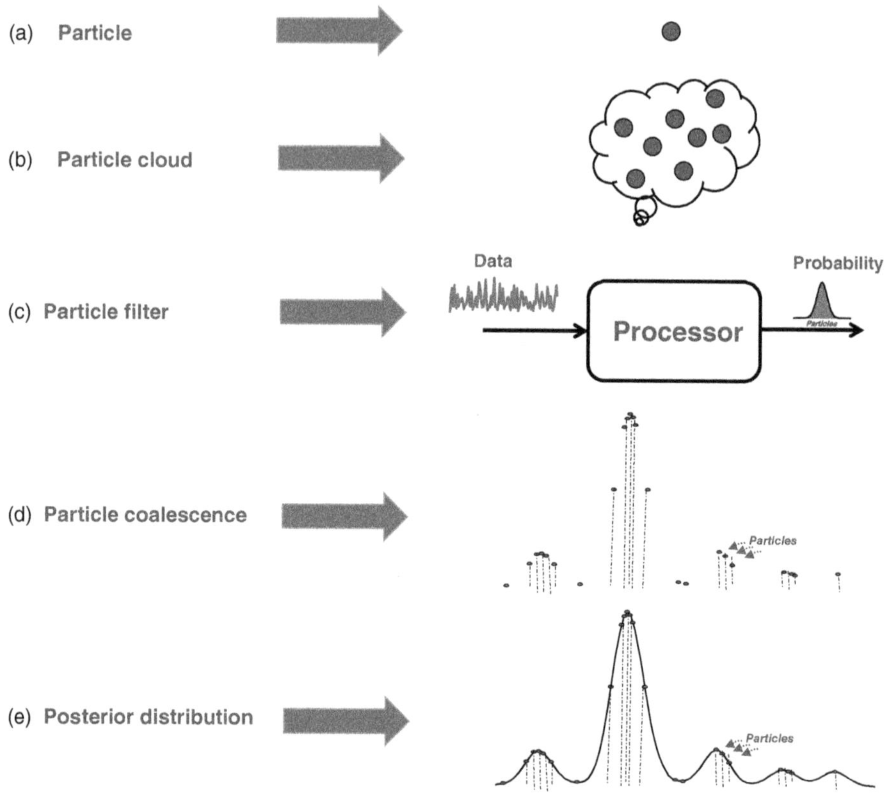

FIGURE 1.4 Particle filtering: (a) particle; (b) particle cloud; (c) particle filter processor; (d) particle coalescence; (e) posterior distribution.

1.3.1 Bayesian Particle Filter

Before we leave this section, let us introduce the evolution of these ideas to a sequential version of the sampling approach that is very important in the signal processing area—the *particle filter*. Referring to Fig. 1.4, we start with a single particle or equivalently a discrete *random sample* and realize that it is not sufficient, so the particle can migrate into a group or so-called *cloud* or *swarm* of particles that can be utilized to develop a nonparametric characterization of a probability mass function. The key, of course, is to select particles to drive a model of the underlying process that generates the parameters of interest. The particle filter itself is a *processor* that has *data* on input and produces an estimate of the corresponding *posterior distribution* on output as depicted in Fig. 1.4 c. These random samples or particles are actually the "location" parameters along with their associated weights that gather or *coalesce* in the regions of *highest probabilities* (mean, mode, etc.) providing the nonparametric estimate of the *empirical* posterior distribution. With these particles generated, the appropriate weights based on Bayes' rule are estimated from the data and lead to the desired result as illustrated in Fig. 1.4 d. The resulting posterior can be observed through probability mass function estimation (histogram, kernel density, etc.) as shown in Fig. 1.4 e. Once the posterior is estimated, then inferences can be performed to extract a wealth of descriptive statistics characterizing the underlying process.

These concepts have recently evolved to the signal processing area and are of high interest in nonlinear estimation problems especially in model-based signal processing applications [16] as discussed next.

1.4 BAYESIAN MODEL-BASED SIGNAL PROCESSING

The estimation of probability distributions required to implement Bayesian processors is at the heart of this approach. How are these distributions obtained from data or simulations? Nonparametric methods of distribution estimation ranging from simple histogram estimators to sophisticated kernel smoothing techniques rooted in classification theory [3] offer reasonable approaches when data are available. However, these approaches usually do not take advantage of prior knowledge about the underlying physical phenomenology generating the data. An alternative is to parameterize the required distributions by prior knowledge of their actual form (e.g., exponential, Poisson, etc.) and fit their parameters from data using optimization techniques [3]. Perhaps the ideal realization is the parameterization of the evolution dynamics associated with the physical phenomenology using underlying mathematical representation of the process combined with the data samples. This idea provides the essence of the model-based approach to signal processing which (as we shall see) when combined with the Bayesian processors provide a formidable tool to attack a wide variety of complex processing problems in a unified manner. An alternative view of the underlying processing problem is to decompose it into a set of steps that capture the strategic essence of the processing scheme. Inherently, we believe that the more a priori knowledge about the measurement and its underlying phenomenology we can incorporate into the processor, the better we can expect the processor to perform—as long as the information that is included is correct! One strategy called the model-based approach provides the essence of model-based signal processing [1].

Simply stated, the *model-based approach* is "incorporating mathematical models of both physical phenomenology and the measurement process (including noise) into the processor to extract the desired information." This approach provides a mechanism to incorporate knowledge of the underlying physics or dynamics in the form of mathematical process models along with measurement system models and accompanying noise as well as model uncertainties directly into the resulting processor. In this way the model-based processor (MBP) enables the interpretation of results directly in terms of the problem physics. It is actually a modeler's tool enabling the incorporation of any a priori information about the problem to extract the desired information. The fidelity of the model incorporated into the processor determines the complexity of the model-based processor with the ultimate goal of increasing the inherent signal-to-noise ratio (SNR). These models can range from simple, implicit, non-physical representation of the measurement data such as the Fourier or wavelet transforms to parametric black-box models used for data prediction, to lumped mathematical representation characterized by ordinary differential equations, to distributed representations characterized by partial differential equation models to capture the underlying physics of the process under investigation. The dominating factor of which model is the most appropriate is usually determined by how severe the measurements are contaminated with noise and the underlying uncertainties. If the SNR of the measurements is high, then simple non-physical techniques can be used to extract the desired information; however, for low SNR measurements more and more of the physics and instrumentation must be incorporated for the extraction. For instance, consider the example of detecting the presence of a particular species in a test solution using a microcantilever sensor measurement system [4].

Example 1.2

The model-based processing problem is characterized in Fig. 1.5 representing the process of estimating the presence of a particular species of material in solution using the multichannel microcantilever sensor system. Here the microcantilever sensor is pre-conditioned by depositing attractor material on its levers to attract molecules of the target species. Once calibrated, the test solution flows along the levers with the target molecules attracted and deposited on each "tuned" microcantilever creating a deflection that is proportional to the concentration. This deflection is measured using a laser interferometric technique and digitized for processing. The process model is derived directly from the fluidics, while the measurement model evolves from the dynamics of the microcantilever structure. The resulting processor is depicted in Fig. 1.6, where we note the mathematical models of both the process dynamics and

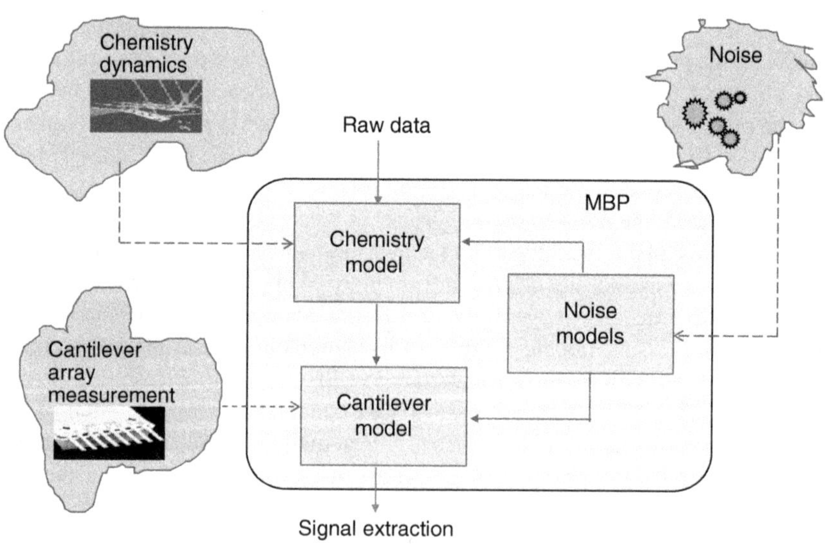

FIGURE 1.5 Model-based approach to signal processing: process (chemistry and physics), measurement (microcantilever sensor array), and noise (Gaussian) representations.

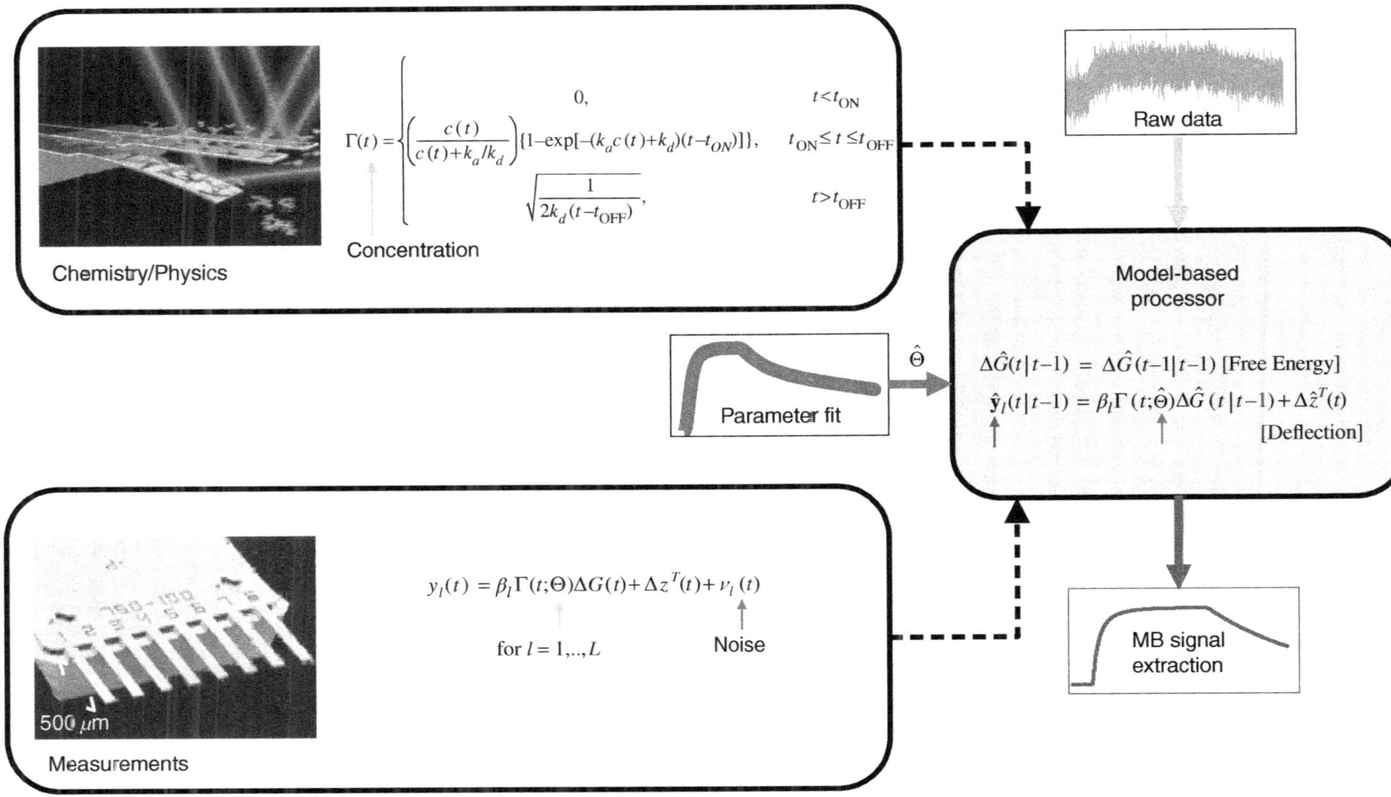

FIGURE 1.6 Model-based processor representation of species detection problem: process (concentration model), measurement (microcantilever sensor array), raw data, parameter estimator (coefficients), and model-based processor (enhancement).

microcantilever measurement system. Since parameters Θ of the model are unknown a priori calibration data are used to estimate them directly and then they are employed in the MBP to provide the enhanced signal estimate shown in the figure. Even though nonlinear and non-Gaussian, the processor appears to yield reasonable estimates. See Ref. [4] for details. △△△

The above example demonstrates that incorporating reasonable mathematical models of the underlying phenomenology can lead to improved processing capability; however, even further advantages can be realized by combining the MBP concepts in conjunction with Bayesian constructs to generalize solutions.

Combining Bayesian and model-based signal processing can be considered a parametric representation of the required distributions using mathematical models of the underlying physical phenomenology and measurement (sensor) system. Certainly, if we assume the distribution is Gaussian and further constrain the processes to be Markovian (only depending on the previous sample), then the multivariate Gaussian can be completely characterized using state–space models resulting in the well-known Kalman filter in the linear model case [2].

Since we are primarily concerned with pseudo real-time techniques in this text, we introduce the notion of a recursive form leading to the idea of sequential processing techniques. That is, we investigate "recursive" or equivalently "sequential" solutions to the estimation problem. Recursive estimation techniques evolved quite naturally during the advent of the digital computer in the late fifties, since both are sequential processes. It is important to realize the recursive solution is identical to the batch solution after it converges, so there is *no gain* in estimator performance properties; however, the number of computations is significantly less than the equivalent batch technique. It is also important to realize that the recursive approach provides the underlying theoretical and pragmatic basis of all *adaptive estimation* techniques; thus, they are important in their own right [2]!

Many processors can be placed in a recursive form with various subtleties emerging in the calculation of the current estimate ($\hat{X}_{old}$). The standard technique employed is based on correcting or updating the current estimate as a new measurement data sample becomes available. The estimates generally take the *recursive form*:

$$\hat{X}_{new} = \hat{X}_{old} + KE_{new} \tag{1.5}$$

where

$$E_{new} = Y - \hat{Y}_{old} = Y - C\hat{X}_{old}$$

Here we see that the new estimate is obtained by correcting the old estimate with a K-weighted error. The error term E_{new} is the new information or innovation—the difference between the actual and the predicted measurement ($\hat{Y}_{old}$) based on the old estimate ($\hat{X}_{old}$). The computation of the weight matrix K depends on the criterion used (e.g., mean-squared error and absolute error).

Consider the following example, which shows how to recursively estimate the sample mean.

Example 1.3

The sample mean estimator can easily be put in recursive form. The estimator is given by

$$\hat{X}(N) = \frac{1}{N} \sum_{t=1}^{N} y(t)$$

Extracting the *Nth* term from the sum, we obtain

$$\hat{X}(N) = \frac{1}{N} y(N) + \frac{1}{N} \sum_{t=1}^{N-1} y(t)$$

Identify $\hat{X}(N-1)$ from the last term,

$$\hat{X}(N) = \frac{1}{N} y(N) + \frac{N-1}{N} \hat{X}(N-1)$$

The recursive form is given by

$$\underbrace{\hat{X}(N)}_{New} = \underbrace{\hat{X}(N-1)}_{Old} + \underbrace{\frac{1}{N}}_{WT} \underbrace{[y(N) - \hat{X}(N-1)]}_{Error}$$

This procedure to develop the "recursive form" is very important and can be applied to a multitude of processors. Note the steps in determining the form.

1. Remove the *N*th term from the summation;
2. Identify the previous estimate in terms of the $N - 1$ remaining terms; and
3. Perform the algebra to determine the gain factor and place the estimator in the recursive form of Eq. 1.5 for a *scalar* measurement.

△△△

1.5 NOTATION AND TERMINOLOGY

The notation used throughout this text is standard in the literature. Where necessary, vectors are represented by boldface, lowercase, **x**, and matrices by boldface, uppercase, **A**. We denote the real part of a signal by *Re x* and its imaginary part by *Im x*. We define the notation $\underline{N}$ to be a shorthand way of writing $1, 2, \ldots, N$. It will be used in matrices, $A(\underline{N})$ to mean there are *N*-columns of *A*. As mentioned previously, estimators are annotated by the caret, such as $\hat{x}$. We also define partial derivatives at the component level by $\frac{\partial}{\partial \theta_i}$, the N_θ- gradient vector by ∇_θ and higher order partials by ∇^2_θ.

The most difficult notational problem will be with the "time" indices. Since this text is predominantly discrete-time, we will use the usual time symbol, *t* to mean a discrete-time index, that is, $t \in \mathcal{I}$ for $\mathcal{I}$ the set of integers. However, and hopefully not too confusing, t will also be used for continuous-time, that is, $t \in \mathcal{R}$ for $\mathcal{R}$ the set of real numbers denoting the continuum. When used as a continuous-time variable, $t \in \mathcal{R}$ it will be represented as a subscript to distinguish it, that is, x_t. This approach of choosing $t \in \mathcal{I}$ primarily follows the system identification literature and for the ease of recognizing discrete-time variable in transform relations (e.g., discrete Fourier transform). The rule-of-thumb is therefore to "interpret t as a discrete-time index unless noted by a subscript as continuous in the text." With this in mind we will define a variety of discrete estimator notations as $\hat{x}(t|t-1)$ to mean the estimate at time (discrete) t based upon all of the previous data up to $t-1$. We will define these symbols prior to their use within the text to assure no misunderstanding of its meaning.

With a slight abuse of notation, we will use the terminology distribution of X, $\Pr(X)$ in general, so as not to have to differentiate between density for continuous random variables or processes and mass for discrete variates. It will be obvious from the context which is meant. In some cases, we will be required to make the distinction between cumulative distribution function (CDF) and density (PDF) or mass (PMF) functions. Here we use the uppercase notation, $\mathrm{P}_X(x)$ for the CDF and lower case $\mathrm{p}_X(x)$ for the PDF or PMF.

Subsequently, we will also need to express a discrete PMF as a continuous PDF using impulse or delta functions as "samplers" much the same as in signal processing when we assume there exists an impulse sampler that leads to the well-known Nyquist sampling theorem [2]. Thus, corresponding to a discrete PMF we can define a continuous PDF through the concept of an *impulse sampler*, that is, given a discrete PMF defined by

$$\mathrm{p}_X(x) \approx \mathrm{p}(X = x_i) = \sum_i p_i \delta(x - x_i) \tag{1.6}$$

then we define the *equivalent continuous* PDF as $\mathrm{p}_X(x)$. Moments follow from the usual definitions associated with a continuous PDF, for instance, consider the definition of the expectation or mean. Substituting the equivalent PDF and utilizing the sifting property of the impulse function gives

$$E\{x\} = \int_{-\infty}^{\infty} x \mathrm{p}_X(x)\, dx = \int_{-\infty}^{\infty} x \left(\sum_i p_i\, \delta(x - x_i) \right) dx = \sum_i x_i p_i \tag{1.7}$$

which is precisely the mean of the discrete PMF.

Also, as mentioned, we will use the symbol $\sim$ to mean "distributed according to" as in $x \sim \mathcal{N}(m, v)$ defining the random variable x as Gaussian distributed with mean m and variance v. We may also use the extended notation: $\mathcal{N}(x : m, v)$ to include the random variable x as well. When *sampling* we use the *non-conventional* right arrow "action" notation $\rightarrow$ to mean "draw a sample from" a particular distribution such as $x_i \rightarrow \Pr(x)$—this again will be clear from the context. When *resampling*, that is, replacing samples with new ones we use the "block" right arrow such as $x_j \Rightarrow x_i$ meaning new sample x_j replaces current sample x_i.

Finally in a discrete (finite) probabilistic representation, we define a purely discrete variate as $x_k(t) := \Pr(x(t) = \mathcal{X}_k)$ meaning that x can only take on values (integers) k from a known set $\mathcal{X} = \{\mathcal{X}_1, \ldots, \mathcal{X}_k, \ldots, \mathcal{X}_N\}$ at time t. We also use the symbol, △△△ to mark the end of an Example.

MATLAB NOTES

MATLAB is a command-oriented vector-matrix package with a simple yet effective command language featuring a wide variety of embedded C language constructs making it ideal for signal processing applications and graphics. Most of the algorithms we have applied to the examples and problems in this text are MATLAB-based in solution ranging from simple simulations to complex applications. We will develop these notes primarily as a summary to point out to the reader many of the existing commands that already perform the signal processing operations discussed in the presented chapter and throughout the text.

REFERENCES

1. J. Candy, *Signal Processing: The Model-Based Approach* (New York: McGraw-Hill, 1986).
2. J. Candy, *Model-Based Signal Processing* (Hoboken, NJ: John Wiley & Sons, Inc./IEEE Press, 2006).
3. R. Duda, P. Hart, and D. Stork, *Pattern Classification* (Hoboken, NJ: John Wiley & Sons, Inc./IEEE Press, 2001).
4. J. Tringe, D. Clague, J. Candy, and C. Lee, "Model-based signal processing of multichannel cantilever arrays," *IEEE J. Micromech. Syst.*, **15**, 5, 1371–1391, 2006.
5. S. Ulam, R. Richtmyer, and J. von Neumann, "Statistical methods in neutron diffusion," *Los Alamos Scientific Laboratory Report*, LAMS-551, 1947.
6. N. Metropolis and S. Ulam, "The Monte Carlo method," *J. American Stat. Assoc.*, **44**, 335–341, 1949.
7. N. Metropolis, A. Rosenbluth, M. Rosenbluth, A. Teller, and E. Teller, "Equations of state calculations by fast computing," *J. Chem. Phys.*, **21**, 6, 1087–1091, 1953.
8. W. Hastings, "Monte Carlo sampling methods using Markov chains and their applications," *Biometrika*, **57**, 1, 97–109, 1970.
9. N. Metropolis, "The beginning of the Monte Carlo method," *Los Alamos Sci.*, Special Issue, 125–130, 1987.
10. J. Liu, *Monte Carlo Strategies in Scientific Computing* (New York: Springer-Verlag, 2001).
11. D. Frenkel, "Introduction to Monte Carlo methods," in *Computational Soft Matter: From Synthetic Polymers to Proteins*, edited by N. Attig, K. Binder, H. Grubmuller, and K. Kremer (John von Neumann Institute for Computing, 2004), NIC Series, Vol. 23, pp. 29–60.
12. C. Robert and G. Casella, *Monte Carlo Statistical Methods* (New York: Springer, 1999).

13. M. Tanner, *Tools for Statistical Inference: Methods for the Exploration of Posterior Distributions and Likelihood Functions*, 2nd Ed. (New York: Springer-Verlag, 1993).
14. J. Ruanaidh and W. Fitzgerald, *Numerical Bayesian Methods Applied to Signal Processing* (New York: Springer-Verlag, 1996).
15. W. Gilks, S. Richardson, and D. Spiegelhalter, *Markov Chain Monte Carlo in Practice* (New York: Chapman & Hall/CRC Press, 1996).
16. A. Doucet, N. de Freitas, and N. Gordon, *Sequential Monte Carlo Methods in Practice* (New York: Springer-Verlag, 2001).
17. B. Ristic, S. Arulampalam, and N. Gordon, *Beyond the Kalman Filter: Particle Filters for Tracking Applications* (Boston: Artech House, 2004).
18. O. Cappe, E. Moulines, and T. Ryden, *Inference in Hidden Markov Models* (New York: Springer-Verlag, 2005).
19. S. Godsill and P. Djuric, "Special issue: Monte Carlo methods for statistical signal processing." *IEEE Trans. Signal Proc.*, **50**, 173–499, 2002.
20. P. Djuric, J. Kotecha, J. Zhang, Y. Huang, T. Ghirmai, M. Bugallo, and J. Miguez, "Particle filtering," *IEEE Signal Proc. Mag.*, **20**, 5, 19–38, 2003.
21. S. Haykin and N. de Freitas, "Special issue: sequential state estimation: from Kalman filters to particle filters." *Proc. IEEE*, **92**, 3, 399–574, 2004.
22. A. Doucet and X. Wang, "Monte Carlo methods for signal processing," *IEEE Signal Proc. Mag.*, **24**, 5, 152–170, 2005.
23. J. Candy, "Bootstrap particle filtering for passive synthetic aperture in an uncertain ocean environment." *IEEE Signal Proc. Mag.*, **24**, 4, 73–85, 2007.

PROBLEMS

1.1 Estimate the number of times a needle when dropped between two parallel lines intersects a line. One way to accomplish this is experimentally by setting up the experiment and doing it—this is the famous Buffon's needle experiment performed in 1725.

(*a*) Set up the experiment and perform the measurements for 100 samples. Estimate the underlying probabilities.

(*b*) Analyze the experiment using a "closed form" approach.

(*c*) How do your answers compare?

Note that this is one of the first Monte Carlo approaches to problem solving.

1.2 Suppose we have three loaded dice with the following six "face" probabilities (each):

$$D1 = \left\{\frac{1}{12}, \frac{1}{6}, \frac{1}{12}, \frac{1}{3}, \frac{1}{6}, \frac{1}{6}\right\}$$

$$D2 = \left\{\frac{1}{6}, \frac{1}{6}, \frac{1}{6}, \frac{1}{12}, \frac{1}{12}, \frac{1}{3}\right\}$$

$$D3 = \left\{\frac{1}{6}, \frac{1}{6}, \frac{1}{6}, \frac{1}{12}, \frac{1}{12}, \frac{1}{3}\right\}$$

Applying Bayes' rule, answer the following questions:

(*a*) Selecting a die at random from the three, what is the probability of rolling a 6?

(*b*) What is the probability that die two ($D = D2$) was selected, if a six ($R = 6$) is rolled with the chosen die?

1.3 A binary communication transmitter (T) sends either a 0 or a 1 through a channel to a receiver (R) with the following probabilities for each as:

$$\Pr(T_1) = 0.6 \quad \Pr(T_0) = 0.4$$

$$\Pr(R_1|T_1) = 0.9 \quad \Pr(R_0|T_1) = 0.1$$

$$\Pr(R_1|T_0) = 0.1 \quad \Pr(R_0|T_0) = 0.9$$

(*a*) What is the probability that R_1 is received?

(*b*) What is the probability that R_0 is received?

(*c*) What is the probability that the true transmitted signal was a 1, when a 1 was received?

(*d*) What is the probability that the true transmitted signal was a 0, when a 0 was received?

(*e*) What is the probability that the true transmitted signal was a 1, when a 0 was received?

(*f*) What is the probability that the true transmitted signal was a 0, when a 1 was received?

(*g*) Draw a probabilistic directed graph with nodes being the transmitters and receivers and links being the corresponding prior and conditional probabilities?

1.4 We are asked to estimate the displacement of large vehicles (semi-trailers) when parked on the shoulder of a freeway and subjected to wind gusts created by passing vehicles. We measure the displacement of the vehicle by placing an accelerometer on the trailer. The accelerometer has inherent inaccuracies which is modeled as

$$y = K_a x + n$$

with y, x, n the measured and actual displacement and white measurement noise of variance R_{nn} and K_a the instrument gain. The dynamics of the vehicle can be modeled by a simple mass-spring-damper.

(*a*) Construct and identify the measurement model of this system.

(*b*) Construct and identify the process model and model-based estimator for this problem.

1.5 Think of measuring the temperature of a liquid in a beaker heated by a burner. Suppose we use a thermometer immersed in the liquid and periodically observe the temperature and record it.

(*a*) Construct a measurement model assuming that the thermometer is linearly related to the temperature, that is, $y(t) = k\,\triangle T(t)$. Also model the uncertainty of the visual measurement as a random sequence $v(t)$ with variance R_{vv}.

(*b*) Suppose we model the heat transferred to the liquid from the burner as

$$Q(t) = CA\,\triangle T(t)$$

where C is the coefficient of thermal conductivity, A is the cross-sectional area, and $\triangle T(t)$ is the temperature gradient with assumed random uncertainty $w(t)$ and variance R_{ww}. Using this process model and the models developed above, identify the model-based processor representation.

1.6 We are given an RLC series circuit driven by a noisy voltage source $V_{in}(t)$ and we use a measurement instrument that linearly amplifies by K and measures the corresponding output voltage. We know that the input voltage is contaminated by an additive noise source $w(t)$ with covariance R_{ww} and the measured output voltage is similarly contaminated with noise source $v(t)$ with R_{vv}.

(*a*) Determine the model for the measured output voltage $V_{out}(t)$ (measurement model).

(*b*) Determine a model for the circuit (process model).

(*c*) Identify the general model-based processor structures. In each scheme, specify the models for the process, measurement, and noise.

1.7 A communications satellite is placed into orbit and must be maneuvered using thrusters to orientate its antennas. Restricting the problem to the single axis perpendicular to the page, the equations of motion are

$$J\frac{d^2\theta}{dt^2} = T_c + T_d$$

where J is the moment of inertia of the satellite about its center of mass, T_c is the thruster control torque, T_d is the disturbance torque, and θ is the angle of the satellite axis with respect to the inertial reference (no angular acceleration) A. Develop signal and noise models for this problem and identify each model-based processor component.

1.8 Consider a process described by a set of linear differential equations

$$\frac{d^2c}{dt^2} + \frac{dc}{dt} + c = Km$$

The process is to be controlled by a proportional-integral-derivative (PID) control law governed by the equation

$$m = K_p \left(e + \frac{1}{T_i} \int e\, dt + T_d \frac{de}{dt} \right)$$

and the controller reference signal r is given by

$$r = e + c$$

Suppose the reference is subjected to a disturbance signal and the measurement sensor, which is contaminated with additive noise, measures the "square" of the output. Develop the model-based signal and noise models for this problem.

1.9 The elevation of a tracking telescope is controlled by a DC motor. It has a moment of inertia J and damping B due to friction, the equation of motion is given by

$$J\frac{d^2\theta}{dt^2} + B\frac{d\theta}{dt} = T_m + T_d$$

where T_m and T_d are the motor and disturbance torques and θ is the elevation angle. Assume a sensor transforms the telescope elevation into a proportional voltage that is contaminated with noise. Develop the signal and noise models for the telescope and identify all of the model-based processor components.

1.10 Suppose we have a two-measurement system given by

$$y = \begin{bmatrix} 3 \\ 4 \end{bmatrix} + v$$

where $R_{vv} = \text{diag}[1, 0.1]$.

(*a*) What is the batch least-squares estimate ($W = I$) of the parameter x, if $y = [7 \quad 21]'$?

(*b*) What is the batch weighted least-squares estimate of the parameter x with W selected for minimum variance estimation?

1.11 Calculate the batch and sequential least-squares estimate of the parameter vector x based on two measurements $y(1)$ and $y(2)$ where

$$y(1) = C(1)x + v(1) = \begin{bmatrix} 2 \\ 1 \end{bmatrix}$$

$$y(2) = c'x + v(2) = 4$$

$$C = \begin{bmatrix} 1 & 1 \\ 0 & 1 \end{bmatrix}, \quad c'(1) = [1 \quad 2], \quad W = I$$

2

BAYESIAN ESTIMATION

2.1 INTRODUCTION

In this chapter we motivate the idea of Bayesian estimation from probabilistic perspective, that is, we perform the required estimation using the underlying densities or mass functions. We start with the "batch" approach and evolve to the Bayesian sequential techniques. We discuss the most popular formulations: maximum a posteriori (MAP), maximum likelihood (ML), minimum variance (MV) or equivalently minimum mean-squared error (MMSE) and least-squares (LS) methods. Bayesian sequential techniques are then developed. The main idea is to develop the proper perspective for the subsequent chapters and construct a solid foundation for the techniques to follow.

2.2 BATCH BAYESIAN ESTIMATION

Suppose we are trying to estimate a random parameter X from data $Y = y$. Then the associated conditional density $\Pr(X|Y = y)$ is called the *posterior density* because the estimate is conditioned "after (*post*) the measurements" have been acquired. Estimators based on this a posteriori density are usually called *Bayesian* because they are constructed from Bayes' theorem, since $\Pr(X|Y)$ is difficult to obtain directly. That is, *Bayes' rule* is defined,

$$\Pr(X|Y) := \Pr(Y|X)\frac{\Pr(X)}{\Pr(Y)} \tag{2.1}$$

Bayesian Signal Processing: Classical, Modern, and Particle Filtering Methods, Second Edition. James V. Candy.

where $\Pr(X)$ is called the *prior density* (before measurement), $\Pr(Y|X)$ is called the likelihood (more likely to be true) and $\Pr(Y)$ is called the *evidence* (normalizes the posterior to assure its integral is unity). Bayesian methods view the sought-after parameter as random possessing a "known" a priori density. As measurements are made, the *prior* is converted to the *posterior density* function adjusting the parameter estimates. Thus, the result of increasing the number of measurements is to improve the a posteriori density resulting in a sharper peak closer to the true parameter as depicted in Fig. 1.1.

To solve the estimation problem, the first step requires the determination of the a posteriori density. A logical solution to this problem leads us to find the "most probable" value of $\Pr(X|Y)$—its *maximum* [1]. The *MAP estimate* is the value of x that maximizes the posterior density, that is,

$$\max_X \Pr(X|Y) \tag{2.2}$$

The optimization is carried out in the usual manner by differentiating, setting the result to zero and solving to obtain the *MAP equation*

$$\nabla_X \Pr(X|Y)\Big|_{X=\hat{X}_{\rm MAP}} = 0 \tag{2.3}$$

with the *gradient vector* $\nabla_X \in R^{N_X \times 1}$ defined by

$$\nabla_X := \left[\frac{\partial}{\partial X_1} \cdots \frac{\partial}{\partial X_{N_X}}\right]' \tag{2.4}$$

Because many problems are based on the exponential class of densities, the $\ln \Pr(X|Y)$ is considered instead. Since the logarithm is a monotonic function, the maximum of $\Pr(X|Y)$ and $\ln \Pr(X|Y)$ occur at the same value of X. Therefore, the logarithmic *MAP equation* is

$$\nabla_X \ln \Pr(X|Y)\Big|_{X=\hat{X}_{\rm MAP}} = 0 \tag{2.5}$$

Now, if we apply Bayes' rule to Eq. 2.5, then

$$\ln \Pr(X|Y) = \ln \Pr(Y|X) + \ln \Pr(X) - \ln \Pr(Y) \tag{2.6}$$

Since $\Pr(Y)$ is not a function of the parameter X, the MAP equation can be written succinctly as

$$\nabla_X \ln \Pr(X|Y)\Big|_{X=\hat{X}_{\rm MAP}} = \nabla_X(\ln \Pr(Y|X) + \ln \Pr(X))\Big|_{X=\hat{X}_{\rm MAP}} = 0 \tag{2.7}$$

With a variety of estimators available, we must construct some way of ascertaining performance. The quality of an estimator is usually measured in terms of its *estimation error*,

$$\tilde{X} = X - \hat{X} \tag{2.8}$$

A common measure of estimator quality is called the Cramer–Rao lower bound (CRLB). The CRLB offers a means of assessing estimator quality prior to processing the measured data. We restrict discussion of the CRLB to the case of unbiased estimates $\hat{X}$ of a "nonrandom" parameter X. The bound is easily extended to more complex cases for biased estimates as well as random parameters [2, 3]. The *Cramer–Rao lower bound*[1] for any unbiased estimate $\hat{X}$ of X based on the measurement Y is given by

$$R_{\tilde{X}|X} = \text{cov}(X - \hat{X}(Y)|X = x) \geq \mathcal{I}^{-1} \tag{2.9}$$

where $\mathcal{I}$, the $N_X \times N_X$ *information matrix*, is defined by

$$\mathcal{I} := -E_y\{\nabla_X(\nabla_X \ln \Pr(Y|X))'\} \tag{2.10}$$

with the gradient vector defined above. Any estimator satisfying the CRLB with equality is called *efficient*. The bound is easily calculated using the *chain rule* from vector calculus [4] defined by

$$\nabla_X(a'b) := (\nabla_X a')b + (\nabla_X b')a \quad a, b \in R^{N_X \times 1} \tag{2.11}$$

where a, b are functions of X. Consider the following example illustrating the calculation of the CRLB.

Example 2.1

Suppose we would like to estimate a nonrandom but unknown parameter X from a measurement y contaminated by additive Gaussian noise, that is,

$$y = X + v$$

where $v \sim \mathcal{N}(0, R_{vv})$ and X is unknown. Thus, we have that

$$E\{Y|X\} = E\{X + v|X\} = X$$

and

$$\text{var}(Y|X) = E\{(y - E\{Y|X\})^2|X\} = E\{v^2|X\} = R_{vv}$$

[1] We choose the matrix-vector version, since parameter estimators are typically vector estimates.

which gives

$$\Pr(Y|X) \sim \mathcal{N}(X, R_{vv})$$

and therefore

$$\ln \Pr(Y|X) = -\frac{1}{2}\ln(2\pi R_{vv}) - \frac{1}{2}\frac{(y-X)^2}{R_{vv}}$$

Differentiating according to the chain rule of Eq. 2.11 and taking the expectation, we obtain

$$\mathcal{I} = -E\left\{\frac{\partial^2}{\partial X^2}\ln \Pr(Y|X)\right\} = -E\left\{\frac{\partial}{\partial X}\frac{(y-X)}{R_{vv}}\right\} = \frac{1}{R_{vv}}$$

and therefore the CRLB is

$$R_{\tilde{X}|X} \geq R_{vv} \qquad \triangle\triangle\triangle$$

The utility of the CRLB is twofold: (1) it enables us to measure estimator quality because it indicates the "best" (minimum error covariance) that any estimator can achieve, and (2) it allows us to decide whether or not a designed estimator is efficient, that is, any estimator achieving the CRLB with equality is *efficient*—a desirable statistical property.

In summary, the properties of an estimator can be calculated prior to estimation (in some cases), and these properties can be used to answer the question "how well does this estimator perform?" Next, we consider the case when the parameter X is not random, leading to the maximum likelihood estimator.

2.3 BATCH MAXIMUM LIKELIHOOD ESTIMATION

In contrast to the Bayesian approach, the likelihood method views the parameter as deterministic but *unknown*. We include it in the Bayesian discussion because as we will show, both estimators are in fact intimately linked. Maximum likelihood produces the "best" estimate as the value which maximizes the probability of the measurements given that the parameter value is "most likely" true. In the estimation problem, the measurement data are given along with the underlying structure of the probability density function (as in the Bayesian case), but the parameters of the density are unknown and must be determined from the measurements; therefore, the maximum likelihood estimate can be considered heuristically as that value of the parameter that best "explains" the measured data giving the most likely estimation.

More formally, let $\mathbf{X}$ be a vector of unknown parameters, $\mathbf{X} \in R^{N_X \times 1}$ and the corresponding set of *N-conditionally independent* measurements, $Y(N) := \{\mathbf{y}(1) \cdots \mathbf{y}(N)\}$ for $\mathbf{y} \in R^{N_y \times 1}$. The *likelihood* of X given the measurements is defined to be proportional to the value of the probability density of the measurements given the parameters, that is,

$$\mathcal{L}(Y(N);X) \propto \Pr(Y(N)|X) = \Pr(\mathbf{y}(1) \cdots \mathbf{y}(N)|X) = \prod_{i=1}^{N} \Pr(\mathbf{y}(i)|X) \quad (2.12)$$

where $\mathcal{L}$ is the likelihood function and $\Pr(Y|X)$ is the joint probability density functions of the measurements given the unknown parameters. This expression indicates the usefulness of the likelihood function in the sense that in many applications, measurements are available and are assumed drawn as a sample from a "known" or assumed known probability density function with unknown parameters (e.g., Poisson with unknown mean). Once we have the measurements (given) and the likelihood function, we would like to find the best estimate of the parameters. If we search through the parameter space over various values of X, say X_i, then we select the value of $\hat{X}$ that most likely specifies the underlying probability function that the measurement sample was drawn from, that is, suppose we have two estimates, $\hat{X}_i$ and $\hat{X}_j$ for which

$$\Pr(Y|X_i) > \Pr(Y|X_j) \quad (2.13)$$

Thus, it is "more likely" that the $Y(N)$ were drawn for parameter value $\hat{X}_i$ than $\hat{X}_j$, since, equivalently, $\mathcal{L}(Y;\hat{X}_i) > \mathcal{L}(Y;\hat{X}_j)$. Searching over all X and selecting that value of X that is maximum (most probable) leads to the maximum likelihood estimate (ML) given by

$$\hat{X}_{\mathrm{ML}}(Y) = \arg\max_{X} \Pr(Y|X) \quad (2.14)$$

As noted previously, many problems are characterized by the class of exponential densities for the Bayesian estimator making it more convenient to use the natural logarithm function; therefore, we define the *log-likelihood function* as

$$\Lambda(Y(N)|X) := \ln\mathcal{L}(Y;X) = \ln\Pr(Y(N)|X) \quad (2.15)$$

Since the logarithm is monotonic, it preserves the maximum of the likelihood providing the identical result,

$$\hat{X}_{\mathrm{ML}}(Y) = \arg\max_{X} \ln \Pr(Y|X) \quad (2.16)$$

What makes the ML estimator popular is the fact that it enjoys some very desirable properties that we list without proof (see Ref. [2] for details).

1. ML estimates are *consistent*.
2. ML estimates are *asymptotically efficient* with $R_{\tilde{X}|X} = \mathcal{I}^{-1}$.

3. ML estimates are *asymptotically Gaussian* with $\mathcal{N}(X, R_{\tilde{X}\tilde{X}})$.
4. ML estimates are *invariant*, that is, if $\hat{X}_{\text{ML}}$, then any function of the ML estimate is the ML estimate of the function, $\hat{f}_{\text{ML}} = f(\hat{X}_{\text{ML}})$.
5. ML estimates of the *sufficient statistic* are equivalent to the ML estimates over the original data.

These properties are asymptotic and therefore imply that a large amount of data must be available for processing.

Mathematically, the relationship between the MAP and ML estimators is clear even though philosophically they are quite different in construct. If we take the MAP relation of Eq. 2.6 and ignore the a priori distribution $\Pr(X)$ (assume X is unknown but deterministic), then the ML estimator is only a special case of MAP. Using the same arguments as before, we use the $\ln \Pr(X|Y)$ instead of $\Pr(X|Y)$. We obtain the ML estimate by solving the *log-likelihood equation* and checking for the existence of a maximum; that is,

$$\nabla_X \ln \Pr(X|Y)\Big|_{X=\hat{X}_{\text{ML}}} = 0 \tag{2.17}$$

Of course, to check for a maximum we have that $\nabla_X(\nabla_X \ln \Pr(X|Y)') < 0$. Again applying Bayes' rule as in Eq. 2.1 and ignoring $\Pr(X)$, we have

$$\nabla_X \ln \Pr(X|Y) = \nabla_X \ln \Pr(Y|X)\Big|_{X=\hat{X}_{\text{ML}}} = 0. \tag{2.18}$$

Consider the following example to demonstrate this relationship between MAP and ML.

Example 2.2

Consider estimating an unknown *constant*, from a noisy measurement as in the previous example. Further, assume that the noise is an independent Gaussian random variable such that $\upsilon \sim N(0, R_{\upsilon\upsilon})$ and the measurement model is given by

$$y = X + \upsilon \tag{2.19}$$

First, we assume no a priori statistical knowledge of X just that it is an unknown, nonrandom constant. Thus, we require the ML estimate, since no prior information is assumed about $\Pr(X)$. The associated conditional density is

$$\Pr(Y|X) = \frac{1}{\sqrt{2\pi R_{\upsilon\upsilon}}} e^{-\frac{1}{2}\frac{(y-X)^2}{R_{\upsilon\upsilon}}} \tag{2.20}$$

The ML estimate of X is found by solving the log-likelihood equation

$$\nabla_X \ln \Pr(Y|X)\Big|_{X=\hat{X}_{\text{ML}}} = 0 \tag{2.21}$$

or

$$\frac{\partial}{\partial X} \ln \Pr(Y|X) = \frac{\partial}{\partial X}\left\{-\frac{1}{2}\ln 2\pi R_{vv} - \frac{1}{2R_{vv}}(y-X)^2\right\}$$
$$= \frac{1}{R_{vv}}(y-X)$$

Setting this expression to zero and solving for X, we obtain

$$\hat{X}_{\mathrm{ML}} = y \tag{2.22}$$

The best estimate of X in a maximum likelihood sense is the raw data y. The corresponding error variance is easily calculated (as before),

$$R_{\tilde{X}|X} = R_{vv} \tag{2.23}$$

Next, we model X as a random variable with Gaussian prior, that is, $X \sim N(\overline{X}, R_{XX})$ and desire the MAP estimate. The MAP equation is

$$\mathcal{J}_{\mathrm{MAP}} = \nabla_X(\ln \Pr(Y|X) + \ln \Pr(X))$$
$$\mathcal{J}_{\mathrm{MAP}} = \frac{\partial}{\partial X}\left\{-\frac{1}{2}\ln 2\pi R_{vv} - \frac{1}{2R_{vv}}(y-X)^2 - \frac{1}{2}\ln 2\pi R_{XX} - \frac{\frac{1}{2}(X-\overline{X})^2}{R_{XX}}\right\}$$

or

$$\mathcal{J}_{\mathrm{MAP}} = \frac{1}{R_{vv}}(y-X) - \frac{1}{R_{XX}}(X-\overline{X})$$

Setting this expression to zero and solving for $X = \hat{X}_{\mathrm{MAP}}$, we obtain

$$\hat{X}_{\mathrm{MAP}} = \frac{y + \frac{R_{vv}}{R_{XX}}\overline{X}}{1 + \frac{R_{vv}}{R_{XX}}}$$

It can be shown from Eq. 2.9 that the corresponding error variance is

$$R_{\tilde{X}|X} = \frac{R_{vv}}{1 + \frac{R_{vv}}{R_{XX}}}$$

Examining the results of this example, we see that when the parameter variance is large ($R_{XX} \gg R_{vv}$), the MAP and ML estimates perform equivalently. However, when the variance is small, the MAP estimator performs better because the corresponding error variance is smaller. △△△

The main point to note is that the MAP estimate provides a mechanism to incorporate the a priori information, while the ML estimate does not. Therefore, for some problems, MAP is the efficient estimator. In the above example, if X were actually Gaussian, then the ML solution, which models X as an unknown parameter, is not an efficient estimator, while the MAP solution that incorporates this information by using the prior $\Pr(X)$ is efficient.

This completes the introduction to batch Bayesian estimation using the MAP and ML estimation. Next, we consider a very popular approach to solving ML estimation problems.

2.3.1 Expectation-Maximization Approach to Maximum Likelihood

Solving maximum likelihood parameter estimation problems is a formidable task especially when the underlying probability distribution functions are unknown. Therefore, we must resort to numerical approaches that will successfully converge to the parameters of interest. Expectation–maximization (EM) is a general method of determining the ML estimate of parameters of the underlying distribution from a given set of data which is incomplete, that is, having "missing" (data) values [5]. Missing data could be considered a misnomer; however, if we include "hidden variables" (not directly measured) as missing, then a wide variety of state/parameter estimation problems can be incorporated into these problem classes. Probably the most popular applications of the EM technique occur in tomographic image reconstruction, pattern recognition, communications and the training of hidden Markov models (see Chapter 9) for speech recognition [6, 7].

The EM technique produces maximum likelihood parameter estimates in two steps: an expectation step followed by a maximization step. The expectation step with respect to the unknown parameters uses the most recently available parameter estimate conditioned on the measurements, while the maximization step provides an updated estimate of the parameters.

To be more precise, we mathematically formulate the general "missing data" problem by first defining the unknown parameters or variables to be estimated as $\theta \in \mathcal{R}^{N_\theta \times 1}$ with $\theta \in \Theta$, the parameter space. We further define three distinct spaces for our problem: (1) the *complete* data space, $\mathcal{Z}$; (2) the *incomplete* data space, $\mathcal{Y}$; and (3) the *missing* data space, $\mathcal{X}$, where the complete space is the union: $\mathcal{Z} = (\mathcal{X}, \mathcal{Y})$. Analogously, we define the corresponding complete, incomplete, and missing/hidden vectors as: $\mathbf{z} \in \mathcal{R}^{N_z \times 1}$, $\mathbf{y} \in \mathcal{R}^{N_y \times 1}$, and $\mathbf{x} \in \mathcal{R}^{N_x \times 1}$, respectively.

With this in mind, we can now define the underlying distributions as the joint or *complete* distribution along with its Bayesian decompositions as

$$\Pr(\mathbf{z}|\theta) = \Pr(\mathbf{x}, \mathbf{y}|\theta) = \Pr(\mathbf{x}|\mathbf{y}, \theta) \times \Pr(\mathbf{y}|\theta) = \Pr(\mathbf{y}|\mathbf{x}, \theta) \times \Pr(\mathbf{x}|\theta) \quad (2.24)$$

or taking logarithms, we have the *complete* (data) log-likelihood

$$\Lambda_{CD}(\mathbf{z}|\theta) = \ln \Pr(\mathbf{z}|\theta) = \ln \Pr(\mathbf{x}, \mathbf{y}|\theta) = \ln \Pr(\mathbf{x}|\mathbf{y}, \theta) + \Lambda_{ID}(\mathbf{y}|\theta) \quad (2.25)$$

where $\Lambda_{ID}(\mathbf{y}|\theta) = \ln \Pr(\mathbf{y}|\theta)$ is the corresponding *incomplete* (data) log-likelihood and the missing data is *random* with $\mathbf{x} \sim \Pr(\mathbf{x})$. Since $\mathbf{x}$ is random, then so is $\Lambda_{CD}(\mathbf{z}|\theta)$ with $\mathbf{y}$ and θ assumed fixed (constant). Thus, the basic maximum likelihood parameter estimation problem for the complete data is to find the value of the parameter such that

$$\hat{\theta}_i = \arg\max_{\theta} \Lambda_{CD}(\mathbf{z}|\theta)$$

given the measured or observed (incomplete) data $\mathbf{y}$ and the previous parameter estimate $\theta = \hat{\theta}$. However, since Λ_{CD} is random, we must search for the parameter vector that maximizes its expectation (over $\mathbf{x}$), that is, we seek

$$\hat{\theta}_i = \arg\max_{\theta} E_{\mathbf{x}}\{\Lambda_{CD}(\mathbf{z}|\theta)\} \tag{2.26}$$

given the measured data $\mathbf{y}$ and the current parameter estimate θ. Multiplying both sides of Eq. 2.25 by the underlying marginal posterior distribution of the missing data, $\Pr(\mathbf{x}|\mathbf{y},\theta)$ and summing over $\mathbf{x}$, we obtain

$$\begin{aligned}\sum_x \Lambda_{CD}(\mathbf{z}|\theta) \times \Pr(\mathbf{x}|\mathbf{y},\theta) &= \sum_x \ln \Pr(\mathbf{z}|\theta) \times \Pr(\mathbf{x}|\mathbf{y},\theta)\\ &= \sum_x \ln \Pr(\mathbf{x}|\mathbf{y},\theta) \times \Pr(\mathbf{x}|\mathbf{y},\theta)\\ &\quad + \sum_x \Lambda_{ID}(\mathbf{y}|\theta) \times \Pr(\mathbf{x}|\mathbf{y},\theta)\end{aligned}$$

Using the definition of the conditional expectation and recognizing that the last term is not a function of the random vector $\mathbf{x}$, we obtain

$$E_{\mathbf{x}}\{\Lambda_{CD}(\mathbf{z}|\theta)|\mathbf{y},\hat{\theta}\} = E_{\mathbf{x}}\{\ln \Pr(\mathbf{x}|\mathbf{y},\hat{\theta})\} + \Lambda_{ID}(\mathbf{y}|\hat{\theta}) \tag{2.27}$$

Since we do not know the complete data, we cannot calculate the exact log-likelihood for this problem. But, given the measured data $\mathbf{y}$, we can estimate the posterior probability for the missing (data) variables $\mathbf{x}$. For each $\mathbf{x}$, there exists a $\hat{\theta}$, and therefore we can calculate an expected value of the complete log-likelihood.

The basic EM principle is to find the θ that maximizes $\Pr(\mathbf{z}|\theta)$ using the available data $\mathbf{y}$ and current parameter estimate. Let $\hat{\theta}_{i-1}$ be the current parameter estimate, then the complete log-likelihood is given by the expectation step:

$$\text{E-step:} \quad Q(\theta,\hat{\theta}_{i-1}) := E_x\{\Lambda_{CD}(\mathbf{z}|\theta)|\mathbf{y},\hat{\theta}_{i-1}\} \tag{2.28}$$

for θ the new parameter vector to be optimized in the next step. In this expression the $\mathbf{y}$ and $\hat{\theta}$ are assumed fixed (constant) and $\mathbf{x}$ is the random vector such that $\mathbf{x} \sim \Pr(\mathbf{x}|\mathbf{y},\hat{\theta}_{i-1})$ so that

$$E_x\{\Lambda_{CD}(\mathbf{z}|\theta)|\mathbf{y},\hat{\theta}_{i-1}\} = \sum_{\mathbf{x}} \ln \Pr(\mathbf{x}|\mathbf{y},\theta)\Pr(\mathbf{x}|\mathbf{y},\hat{\theta}_{i-1}) \tag{2.29}$$

where $\Pr(\mathbf{x}|\mathbf{y}, \hat{\theta}_{i-1})$ is the marginal of the missing data (hidden variable) based on the measured data and current parameter estimate (as shown).

The maximization step is to find the parameter vector update, $\hat{\theta}_i$ that will maximize the computed expectation, that is,

$$\text{M-step:} \quad \hat{\theta}_i = \arg\max_{\theta} Q(\theta, \hat{\theta}_{i-1}) \tag{2.30}$$

Each iteration is guaranteed to *increase* the log-likelihood eventually converging to a local maximum at an exponential rate [8, 9]. Different forms of the EM have evolved with the "generalized" EM (GEM) method by finding an alternative (simpler) expectation function and using the updated parameter vector such that $Q(\hat{\theta}_i, \hat{\theta}_{i-1}) > Q(\theta, \hat{\theta}_{i-1})$ which is also guaranteed to converge. The Monte Carlo EM (MCEM) is another form that is used when a closed form distribution for the E-step is replaced by simulation-based methods (sampling) [10]. Consider the following example from Snyder [11].

Example 2.3

Suppose we would like to estimate the rate or intensity parameter λ_s of a signal contaminated in Poisson noise with known mean λ_v. We measure the counts from a photodetector characterized by

$$y_n = s_n + v_n$$

where y_n is the observation with Poisson distribution

$$y_n \sim \mathcal{P}(\lambda_y) = (\lambda_y)^{y_n} e^{-\lambda_y}/y_n!$$

The respective signal and noise counts during the measurement period are independent of each other with $s_n \sim \mathcal{P}(\lambda_s)$ and $v_n \sim \mathcal{P}(\lambda_v)$. We have that the log-likelihood of the incomplete data is

$$\Lambda_{ID}(y_n|\lambda_y) = \ln \mathcal{P}(y_n|\lambda_y) = y_n \ln \lambda_y - \lambda_y - \ln y_n!$$

Differentiating Λ_{ID} with respect to λ_s, setting to the result to zero and solving for the rate parameter, that is,

$$\frac{d}{d\lambda_s}(y_n \ln(\lambda_s + \lambda_v) - (\lambda_s + \lambda_v) - \ln y_n!) = \frac{y_n}{\lambda_s + \lambda_v} - 1 = 0$$

which yields the maximum likelihood parameter estimate

$$\hat{\lambda}_s = y_n - \lambda_v > 0$$

Here we have used the fact that the superposition of Poisson processes is Poisson with intensity parameter, $\lambda_y = \lambda_s + \lambda_v$. Next, let us investigate the EM approach to this problem. Here the complete data space is $z_n = (s_n, v_n)$ and y_n is the incomplete (observed) data. Therefore, we have that the complete log-likelihood is

$$\Lambda_{CD}(z_n|\lambda_s) = -(\lambda_s + \lambda_v) + s_n \ln \lambda_s + v_n \ln \lambda_v - \ln s_n! - \ln v_n!$$

because s_n and v_n are independent. Thus, the expectation step is given by

$$\text{E-step:} \quad Q(\lambda_s|\hat{\lambda}_s(i-1)) = -\lambda_s + \hat{s}_n(i-1)\ln \lambda_s - \lambda_v + \hat{v}_n(i-1)\ln \lambda_v$$

with

$$\hat{s}_n(i-1) = E\{s_n|y_n, \hat{\lambda}_s(i-1)\} = \frac{\hat{\lambda}_s(i-1)}{\hat{\lambda}_s(i-1) + \lambda_v} y_n$$

and

$$\hat{v}_n(i-1) = E\{v_n|y_n, \hat{\lambda}_v(i-1)\} = y_n - \hat{s}_n(i-1)$$

Since the maximization step does not depend on λ_v or $\hat{v}_n(i-1)$, we have

$$\hat{\lambda}_s(i) = \arg\max_{\lambda_s}(-\lambda_s + \hat{s}_n(i-1)\ln \lambda_s)$$

giving

$$\text{M-step:} \quad \hat{\lambda}_s(i) = \hat{s}_n(i-1)$$

which completes the EM algorithm.

We simulated a sequence of Poisson counts for 500 samples composed of the additive signal ($\lambda_s = 14$) and noise ($\lambda_N = 3.5$). The estimated signal intensity at each measurement is shown in Fig. 2.1a along with the estimated probability mass functions in 2.1b of the true signal, EM-estimated signal, measurement and noise. Here we see the estimated signal PDF closely matches the true PDF and the average intensity is very close to the true signal intensity at $\lambda_s = 14.02$. △△△

In summary, the EM approach is a two-step, iterative technique to solve the maximum likelihood parameter estimation problem when missing data or hidden variables (states) are present. Clearly, we have just discussed the generic form of the EM approach, the actual algorithm is problem-specific, based on the knowledge of the underlying distributions; however, we can consider a class of distributions to develop a more specific algorithm such as the so-called *exponential family* [12].

2.3.2 EM for Exponential Family of Distributions

In this subsection we apply the EM technique to the exponential class of distribution functions, many of which are well-known forms like the Gaussian, Poisson,

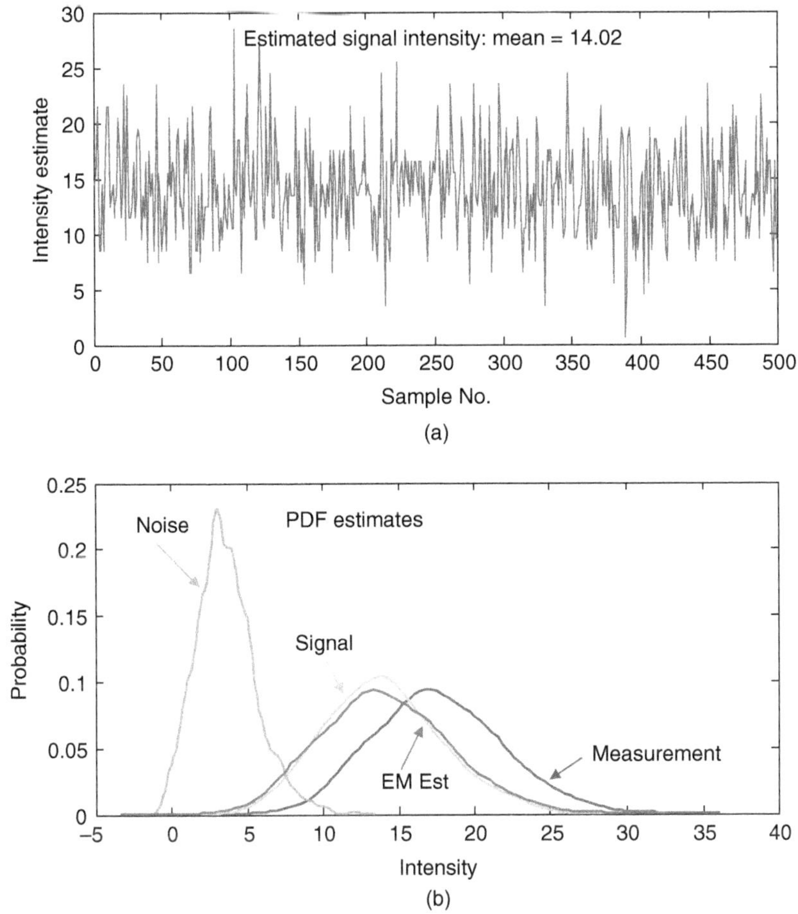

FIGURE 2.1 EM estimation for Poisson intensity: (a) signal intensity estimate (mean = 14.02); (b) PDF estimation of Poisson processes: noise, signal, EM estimate, measurement.

exponential, Raleigh, binomial, and more. The exponential family is defined by the generic form:

$$\Pr(\mathbf{z}|\theta) = \mathbf{b}(\mathbf{z}) \exp(\mathbf{c}'(\theta)\mathbf{s}(\mathbf{z}))/a(\theta) \tag{2.31}$$

with $\theta \in \mathcal{R}^{N_\theta}$, the parameters defining the class [5, 13, 14] and $\mathbf{s}(\mathbf{z})$, the *sufficient* statistic providing all of the information necessary about the underlying process for estimation with $\mathbf{s}, \mathbf{c} \in \mathcal{R}^{N_\theta \times 1}$. Since the complete log-likelihood can be written as

$$\Lambda_{CD}(\mathbf{z}|\theta) = \ln \Pr(\mathbf{z}|\theta) = \ln \mathbf{b}(\mathbf{z}) + \mathbf{c}'(\theta)\mathbf{s}(\mathbf{z}) - \ln a(\theta) \tag{2.32}$$

taking the conditional expectations, we obtain the expectation step as:

$$\text{E-step:} \quad Q(\theta, \hat{\theta}_{i-1}) = E_{\mathbf{x}}\{\ln \mathbf{b}(\mathbf{z})|\mathbf{y}, \theta) + \mathbf{c}'(\theta)E\{\mathbf{s}(\mathbf{z})|\mathbf{y}, \hat{\theta}_{i-1}\} - \ln a(\theta) \tag{2.33}$$

Defining $\hat{\mathbf{s}}_i := E\{\mathbf{s}(\mathbf{z})|\mathbf{y}, \hat{\theta}_{i-1}\}$, the maximization step with respect to θ, is performed on

$$E\{\ln \mathbf{b}(\mathbf{z})|\mathbf{y}, \theta) + \mathbf{c}'(\theta)\hat{\mathbf{s}}_i - \ln a(\theta) \tag{2.34}$$

Since the first term is not a function of θ, it need not be included. The EM technique for the exponential family is:

$$\begin{aligned} &\text{E-step:} \quad \hat{\mathbf{s}}_i \\ &\text{M-step:} \quad \hat{\theta}_i = \arg\max_{\theta} \; \mathbf{c}'(\theta)\hat{\mathbf{s}}_i - \ln a(\theta) \end{aligned} \tag{2.35}$$

completing the method.

For instance, in the previous example, we have $\mathbf{b}(\mathbf{z}) = (\lambda_y)^{y_n}$; $\exp(\mathbf{c}'\mathbf{s}) = e^{-\lambda_y}$ and $a(\theta) = y_n!$. Consider the following example of producing an image by estimating the intensity of a Poisson process discussed in McLachlan [15].

Example 2.4

Photons emitted from a radioactive tracer are injected into tissue in order to create a single photon emission computed tomography (SPECT) image for medical diagnosis [11, 13–15]. The basic idea is to create an image that represents the photon emitted, counted by photon detectors and displayed as an image based on pixel counts. For simplicity, we assume all photons are counted by the instrument. The emissions are assumed to be Poisson-distributed with unknown intensities or *emission densities* corresponding to the Poisson rate parameters λ. Thus, the basic problem is:

GIVEN a set of incomplete measurements (counts), $\{y_n\}$, **FIND** the maximum likelihood estimate of the unknown emission densities, λ, (rate parameter vector) assuming that each of these component intensities is constant within a pixel, that is, λ_m is constant for $m = 1, \ldots, M_p$.

Let $y_n; n = 1, \ldots, N_d$ be the number of counts measured by the nth detector, so mathematically the problem is to estimate the M_p-intensity vector of emission densities, $\hat{\lambda}$, from the N_d-measurement vector $\mathbf{y}$ assuming that the counts are conditionally independent such that $\Pr(y_n|\lambda_y(n))$ is Poisson, that is,

$$y_n \sim \mathcal{P}(\lambda_y(n)) = \frac{(\lambda_y(n))^{y_n} e^{-\lambda_y(n)}}{y_n!}$$

For imaging considerations, it is assumed that individual photons are emitted within a given pixel and detected; therefore, we define x_{nm} as the unobservable counts (missing data), assuming λ is known and conditionally independent of x, such that,

$$x_{mn} \sim \mathcal{P}(\lambda_x(m,n)) = \frac{(\lambda_x(m,n))^{x_{nm}} e^{-\lambda_x(m,n)}}{x_{nm}!}$$

where $\lambda_x(m,n) = \lambda_m p_{mn}$, the emission density corresponding to the number of photons emitted in the mth pixel detected by the nth detector with emission probability p_{mn}—a marked Poisson process [11, 16, 17].

The photon counter measurement at the nth detector is the superposition of the photons emitted by the radionuclide at the mth pixel; therefore, we have

$$y_n = \sum_{m=1}^{M_p} x_{mn}$$

which is the sum of Poisson variates, so that

$$\lambda_y(n) = \sum_{m=1}^{M_p} \lambda_x(m,n) = \sum_{m=1}^{M_p} \lambda_m p_{mn}$$

It has been shown [15] that the conditional probability of the missing data is binomial with parameter y_n and probability $\Pr(\lambda_m p_{mn})$, that is,

$$x_{mn} \sim \mathcal{B}(y_n, \Pr(\lambda_m p_{mn})) \quad \text{with } \Pr(\lambda_m p_{mn}) = \lambda_m p_{mn} \bigg/ \sum_{j=1}^{M_p} \lambda_i p_{jn}$$

The EM approach for this problem is straightforward, since the underlying distributions are all in the exponential family [15]. The sufficient statistic is the data x_{mn} which can be estimated. That is, since x_{mn} is binomial, the conditional mean (Q-step) is given by

$$\hat{x}_{mn}(i-1) := E\{x_{mn}|y_n, \hat{\lambda}_m(i-1)\} = y_n \left(\hat{\lambda}_x(m,n) \bigg/ \sum_{j-1}^{M_p} \hat{\lambda}_x(j,n) \right)$$

where $\hat{\lambda}_x(m,n) = \hat{\lambda}_m(i-1)p_{mn}$. Next, define the likelihood based on $\mathbf{z} = (\mathbf{x}, \mathbf{y})$ assuming λ is known and conditionally independent. Therefore, the complete likelihood is given by

$$\mathcal{L}(\mathbf{x}, \mathbf{y}|\lambda) = \prod_{m,n} e^{-\lambda_x(m,n)} (\lambda_x(m,n))^{x_{mn}} / x_{mn}!$$

and

$$\Lambda(\mathbf{x}, \mathbf{y} | \lambda) = \sum_{m,n} -\lambda_x(m, n) + x_{nm} \ln \lambda_x(m, n) - \ln x_{nm}!$$

or substituting for $\lambda_x(m, n)$ and expanding, we have

$$\Lambda(\mathbf{x}, \mathbf{y} | \lambda) = \sum_{m=1}^{M_p} \sum_{n=1}^{N_d} -\lambda_m p_{nm} + x_{nm} \ln(\lambda_m p_{nm}) - \ln x_{nm}!$$

Differentiating this expression with respect to λ_m, setting to the result zero and solving gives the maximization step as

$$\hat{\lambda}_m(i) = \sum_{n=1}^{N_d} \hat{x}_{mn}(i-1)$$

Summarizing the E- and M-steps for this problem,

$$\begin{aligned} \text{E-step:} \quad Q(\lambda_m, \hat{\lambda}_m(i-1) &= E_x\{\Lambda(\mathbf{x}, \mathbf{y}) | \mathbf{y}, \hat{\lambda}_m(i-1))\} \\ &= y_n \hat{\lambda}_m(i-1) p_{mn} \Big/ \sum_{j=1}^{M_p} \hat{\lambda}_j(i-1) p_{jn} \end{aligned}$$

$$\text{M-step: } \hat{\lambda}_m(i) = \hat{\lambda}_m(i-1) \sum_{n=1}^{N_d} \left(\frac{y_n p_{mn}}{\sum_{j=1}^{M_p} \hat{\lambda}_j(i-1) p_{jn}} \right)$$

More details can be found in Refs. [11], [13], and [15]. △△△

This completes our discussion of the EM approach to parameter estimation problems with incomplete data/parameters. Next, we briefly describe one of the most popular approaches to the estimation problem—MV or equivalently MMSE estimation.

2.4 BATCH MINIMUM VARIANCE ESTIMATION

To complete this discussion evolving from the batch Bayesian perspective, we discuss the development of the minimum (error) variance (MV) or equivalently MMSE estimator. The general techniques developed in this section can be applied to develop various model-based processors [1].

Suppose we are asked to obtain an estimate of a N_x-parameter vector $\mathbf{X}$ from a set of noisy measurements characterized by the N_y-measurement vector $\mathbf{Y}$. We would like to construct an estimate that is best or optimal in some sense. The most natural criterion to consider is one that minimizes the error between the true parameter and its estimate based on the measured data. The *error variance criterion* is defined by

$$J(X) = E_X\{[X - \hat{X}(Y)]'[X - \hat{X}(Y)]|Y\} \tag{2.36}$$

where

X is the true random N_x-vector
Y is the measured random N_y-vector (data)
$\hat{X}$ is the estimate of X given Y

whose minimization leads to the *minimum variance estimator* [1]. Thus, if we minimize $J(X)$ using the chain rule of Eq. 2.11, then we have

$$\begin{aligned}\nabla_{\hat{X}} J(X) &= E_x\{\nabla_{\hat{X}}(X - \hat{X}(Y))'(X - \hat{X}(Y))|Y\} \\ &- E_x\{-(X - \hat{X}(Y)) - (X - \hat{X}(Y))|Y\} \\ &= -2[E_x\{X - \hat{X}(Y)|Y\}]\end{aligned}$$

performing the conditional expectation operation gives

$$\nabla_{\hat{X}} J(X) = -2[E_x\{X|Y\} - \hat{X}(Y)] \tag{2.37}$$

and setting this equation to zero and solving yields the minimum variance estimate as

$$\hat{X}_{MV} = \hat{X}(Y) = E_x\{X|Y\} \tag{2.38}$$

We check for the global minimum by examining the sign (positive) of the second derivative, and since

$$\nabla_{\hat{X}}(\nabla_{\hat{X}} J(X))' = 2I \tag{2.39}$$

a unique minimum exists. Thus, the minimum variance estimator is the *conditional mean*. The associated error variance is determined by substituting for $\hat{X}(Y)$ into Eq. 2.36 giving

$$\begin{aligned}J_{MV} = R_{\tilde{X}|Y} &= E_x\{(X - E\{X|Y\})'(X \quad E\{X|Y\})|Y\} \\ &= E_x\{X'X|Y\} - E_x^2\{X|Y\}\end{aligned} \tag{2.40}$$

This estimator is linear, unconditionally and conditionally unbiased and possesses general orthogonality properties. To see this, we investigate the estimation error defined by

$$\tilde{X} = X - \hat{X}(Y) \tag{2.41}$$

Taking expectations, we have the fact that the estimator is unconditionally unbiased,

$$E_x\{\tilde{X}\} = E_x\{X - \hat{X}(Y)\} = E_x\{X\} - E_x\{E_x\{X|Y\}\} = E_x\{X\} - E_x\{X\} = 0 \quad (2.42)$$

as well as conditionally unbiased, since

$$\begin{aligned} E_x\{\tilde{X}|Y\} &= E_x\{X - \hat{X}(Y)|Y\} = E_x\{X|Y\} - E_x\{E_x\{X|Y\}|Y\} \\ &= E_x\{X|Y\} - E_x\{X|Y\} = 0 \end{aligned} \quad (2.43)$$

Another important property of the minimum variance estimator is that the estimation error is orthogonal to *any* function, say $f(\cdot)$, of the data vector Y [18], that is,

$$E_{xY}\{f(Y)\tilde{X}'\} = 0 \quad (2.44)$$

Also,

$$E_x\{f(Y)\tilde{X}'|Y\} = 0 \quad (2.45)$$

This is the well-known orthogonality condition. To see this we substitute for the error

$$\begin{aligned} E_x\{f(Y)\tilde{X}'|Y\} &= E_x\{f(Y)(X - \hat{X}(Y))'|Y\} \\ &= f(Y)E_x\{(X - \hat{X}(Y))'|Y\} \\ &= f(Y)(E_x\{X|Y\} - \hat{X}(Y)') = 0 \end{aligned}$$

Taking the expectation over Y proves the unconditional result as well. Thus, the estimation error is orthogonal to *any* function of Y, a fact that is used in proofs throughout the literature for both linear and nonlinear estimators.

Let us now investigate the special case of the *linear* minimum variance estimator. The *estimation error* is *orthogonal* to all past data Y, that is,

$$E_{XY}\{Y\tilde{X}'\} = 0 \quad (2.46)$$

or as before

$$E_X\{Y\tilde{X}'|Y\} = 0 \quad (2.47)$$

This is the well-known minimum variance estimator that results in the linear case [1]. For a linear function of the parameter, we have that

$$y = CX + v \quad (2.48)$$

where $y, v, \in R^{N_y \times 1}, X \in R^{N_x \times 1}, C \in R^{N_y \times N_x}$ and v is zero-mean, white with R_{vv}. The mean-squared error criterion

$$J(X) = E\{\tilde{X}'\tilde{X}\} \quad (2.49)$$

is minimized to determine the estimate. The minimization results in the orthogonality condition of Eq. 2.47 which we write as

$$E\{y\tilde{X}'\} = E\{yX'\} - E\{y\hat{X}'_{MV}\} = 0 \tag{2.50}$$

for $\hat{X}_{MV} = K_{MV}y$, a linear function of the data vector. Substituting for y and $\hat{X}$ in this equation,

$$K_{MV} = R_{XX}C'(CR_{XX}C' + R_{vv})^{-1} \tag{2.51}$$

where R_{XX} is the covariance matrix of X. The corresponding quality is obtained as

$$R_{\tilde{X}\tilde{X}} = (R_{XX}^{-1} + C'R_{vv}^{-1}C)^{-1} \tag{2.52}$$

It is also interesting to note that the fundamental Wiener result [1] is easily obtained from the orthogonality condition of Eq. 2.47, that is,

$$E\{y\tilde{X}'\} = E\{yX'\} - E\{yy'\}K'_{MV} = R_{yX} - R_{yy}K'_{MV} = 0 \tag{2.53}$$

which is called the *discrete Wiener equation*. Solving for K_{MV}, we obtain the Wiener solution for a linear (batch) estimation scheme, that is,

$$K_{MV} = R_{Xy}R_{yy}^{-1} \tag{2.54}$$

Note that *least-squares estimation* is similar to that of minimum variance except that no statistical information (expectation removed) is assumed known about the process, that is, the least-squares estimator, $\hat{y}_{LS}$ minimizes the sum-squared error criterion

$$\min_{\hat{y}} J = \tilde{y}'\tilde{y} = \sum_i \tilde{y}_i^2 \tag{2.55}$$

for $\tilde{y} = y - \hat{y}_{LS}$.

This completes the introduction to *batch* minimum variance, maximum a posteriori and maximum likelihood estimation. Next, we consider the sequential problem.

2.5 SEQUENTIAL BAYESIAN ESTIMATION

Modern statistical signal processing techniques evolve directly from a Bayesian perspective, that is, they are cast into a probabilistic framework using Bayes' theorem as the fundamental construct. More specifically, the information about the random signal, $x(t)$, required to solve a vast majority of estimation/processing problems is incorporated in the underlying probability distribution generating the process. For instance, the usual signal enhancement problem is concerned with providing the "best" (in some sense) estimate of the signal at time t based on all of the data available at that time. The filtering distribution provides that information directly in terms of its underlying statistics. That is, by calculating the statistics of the process directly from the filtering distribution, the enhanced signal can be extracted using a variety of estimators like MAP, ML or MMSE, accompanied by a set of performance statistics such as error covariances and bounds. Sequential methods to calculate these distributions become extremely important in pragmatic problems where implementation and speed are an issue. Therefore from an engineering perspective, they are our primary focus.

The roots of this theory are based on Bayesian estimation—in fact, sequential Bayesian estimation. We will see that many of our well-known techniques are easily cast into this unifying framework, especially in the nonlinear signal processing area. The Bayesian algorithms that provide posterior distribution estimates are optimal; however, they are impossible to implement directly because they require integrations or summations with an infinite number of terms. We will develop the optimal Bayesian algorithms mathematically and perform the required calculations in a sequential manner, but it must be realized that only under certain restricted circumstances can these actually be realized (e.g., the linear Gaussian case). Starting with Bayes' theorem, it is possible to show how this leads directly to a recursive or sequential estimation framework that is the foundation of these new approaches.

We cast this discussion into a dynamic variable/parameter structure by defining the "unobserved" signal or equivalently "hidden" variables as the set of N_x-vectors, $\{x(t)\}, t = 0, \ldots, N$. On the other hand, we define the observables or equivalent measurements as the set of N_y-vectors, $\{y(t)\}, t = 0, \ldots, N$ considered to be *conditionally independent* of the signal variables. The goal in recursive Bayesian estimation is to sequentially (in-time) estimate the joint *posterior* distribution, $\Pr(x(0), \ldots, x(N); y(0), \ldots, y(N))$. Once the posterior is estimated, many of the interesting statistics characterizing the process under investigation can be exploited to extract meaningful information.

We start by defining two sets of random (vector) processes: $X_t := \{x(0), \ldots, x(t)\}$ and $Y_t := \{y(0), \ldots, y(t)\}$, as before. Here we can consider X_t to be the set of dynamic random variables or parameters of interest and Y_t as the set of measurements or observations of the desired process as before.[2] In any case, we start with Bayes' theorem for the joint posterior distribution as

$$\Pr(X_t|Y_t) = \Pr(Y_t|X_t) \times \frac{\Pr(X_t)}{\Pr(Y_t)} \tag{2.56}$$

In Bayesian theory (as before), the *posterior* defined by $\Pr(X_t|Y_t)$ is decomposed in terms of the *prior* $\Pr(X_t)$, its *likelihood* $\Pr(Y_t|X_t)$ and the *evidence* or normalizing factor $\Pr(Y_t)$. Each has a particular significance in this construct which we shall discuss subsequently.

[2] In *Kalman filtering theory*, the X_t are considered the states or hidden variables not necessarily observable directly, while the Y_t are observed or measured directly.

We can replace the evidence by realizing that it is the *total* probability given by

$$\Pr(Y_t) = \int \Pr(Y_t|X_t)\Pr(X_t)\, dX_t \tag{2.57}$$

which is integrated over the entire X_t-dimensional parameter space. Substituting this into Eq. 2.56 we obtain

$$\Pr(X_t|Y_t) = \frac{\Pr(Y_t|X_t) \times \Pr(X_t)}{\int \Pr(Y_t|X_t)\Pr(X_t)\, dX_t} \tag{2.58}$$

Once the posterior distribution is determined, all statistical inferences or estimates are made by integration or equivalently summation. For example, suppose we would like to obtain a prediction distribution, then it is obtained as

$$\Pr(X_{t+1}|Y_t) = \int \Pr(X_{t+1}|X_t, Y_t) \times \Pr(X_t|Y_t)\, dX_t$$

and a point estimate might be the conditional mean of this distribution, that is,

$$E\{X_{t+1}|Y_t\} = \int X_{t+1}\Pr(X_{t+1}|Y_t)\, dX_{t+1}$$

while the variance of the distribution can be calculated similarly and used for performance evaluation. This calculation illustrates how information that can be estimated from the extracted distribution is applied in the estimation context.

Again, even though simple conceptually, the real problem to be addressed is that of evaluating these integrals which is very difficult because they are only analytically tractable for a small class of priors and likelihood distributions. The large dimensionality of the integrals cause numerical integration techniques to break down which leads to the approximations we discuss subsequently for stabilization. Let us investigate further to consider posing problems of signal processing interest. First, we review some of the basics.

Many of the derivations and algorithms to follow are based on the simple, but sometimes not so obvious, manipulations of Bayes' theorem; therefore, we develop these variants here first. Starting with a joint distribution of variables indexed by t, we apply Bayes' rule[3] in steps to expose some of the manipulations:

$$\Pr(y(t), y(t-1), y(t-2)) = \Pr(y(t), y(t-1)|y(t-2)) \times \Pr(y(t-2)) \tag{2.59}$$

Applying the rule again to the first term in this expression gives

$$\Pr(y(t), y(t-1)|y(t-2)) = \Pr(y(t)|y(t-1), y(t-2)) \times \Pr(y(t-1)|y(t-2)) \tag{2.60}$$

[3] In set notation, we have (simply) that

$$\Pr(ABC) = \Pr(AB|C) \times \Pr(C) = [\Pr(A|BC) \times \Pr(B|C)] \times \Pr(C)$$

Combining these results, we have

$$\begin{aligned}\Pr(y(t), y(t-1), y(t-2)) &= \Pr(y(t)|y(t-1), y(t-2)) \\ &\quad \times \Pr(y(t-1)|y(t-2)) \times \Pr(y(t-2))\end{aligned} \tag{2.61}$$

Additionally, if $\{y(t)\}$ is considered first-order Markov, then Eq. 2.61 simplifies even further to

$$\begin{aligned}\Pr(y(t), y(t-1), y(t-2)) &= \Pr(y(t)|y(t-1)) \times \Pr(y(t-1)|y(t-2)) \\ &\quad \times \Pr(y(t-2))\end{aligned} \tag{2.62}$$

Generalizing these expansions, we can obtain the *chain rule of probability* which states that the joint distribution Y_t can be decomposed using Bayes' rule as

$$\Pr(Y_t) = \Pr(y(t)|Y_{t-1}) \times \Pr(Y_{t-1}) = \Pr(y(t)|Y_{t-1}) \times \Pr(y(t-1)|Y_{t-2}) \times \Pr(Y_{t-2})$$

or expanding this expression, we obtain

$$\begin{aligned}\Pr(Y_t) &= \prod_{k=0}^{t} \Pr(y(t-k)|Y_{t-k-1}) \\ &= \Pr(y(t)|Y_{t-1}) \times \cdots \times \Pr(y(1)|Y_0) \times \Pr(y(0))\end{aligned} \tag{2.63}$$

Further assuming that the process is *Markov*,

$$\Pr(y(t)|Y_{t-1}) = \Pr(y(t)|y(t-1))$$

and the chain rule simplifies even further to

$$\Pr(Y_t) = \prod_{k=0}^{t} \Pr(y(t-k)|Y_{t-k-1}) = \prod_{k=0}^{t} \Pr(y(t-k)|y(t-k-1)) \tag{2.64}$$

Marginal distributions are used throughout these derivations and they follow directly from the *Chapman–Kolmogorov* evolution equation [17], that is,

$$\Pr(y(t)|y(t-2)) = \int \Pr(y(t)|y(t-1)) \times \Pr(y(t-1)|y(t-2))\, dy(t-1) \tag{2.65}$$

Before we close this section, we must mention that the derivations to follow rely on the following two fundamental *assumptions*:

- the dynamic variable $x(t)$ is Markov; and
- the data $y(t)$ are conditionally independent of the *past* dynamic variables, $\{x(t-k)\}\forall k > 0$.

Next, we proceed with the development of generic Bayesian processors.

2.5.1 Joint Posterior Estimation

With this information in mind, we develop the sequential Bayesian solution to the joint posterior estimation problem. Starting with Eq. 2.56, we decompose this expression by first extracting the tth term from the joint distributions, that is,

$$\Pr(Y_t|X_t) = \Pr(y(t), Y_{t-1}|x(t), X_{t-1})$$

and applying Bayes' rule to obtain

$$\Pr(Y_t|X_t) = \Pr(y(t)|Y_{t-1}, x(t), X_{t-1}) \times \Pr(Y_{t-1}|x(t), X_{t-1}) \tag{2.66}$$

The data at time t, $y(t)$, is assumed independent of X_{t-1} and Y_{t-1}; therefore, the first term of Eq. 2.66 simplifies to

$$\Pr(y(t)|x(t), Y_{t-1}, X_{t-1}) \longrightarrow \Pr(y(t)|x(t)) \tag{2.67}$$

The second term also simplifies based on the independence of the past data and $x(t)$ to give

$$\Pr(Y_{t-1}|x(t), X_{t-1}) \longrightarrow \Pr(Y_{t-1}|X_{t-1}) \tag{2.68}$$

We now have the final expression for the *likelihood* as

$$\Pr(Y_t|X_t) = \Pr(y(t)|x(t)) \times \Pr(Y_{t-1}|X_{t-1}) \tag{2.69}$$

Similarly, for the *prior*, extracting the tth term and applying the rule, we obtain

$$\Pr(X_t) = \Pr(x(t), X_{t-1}) = \Pr(x(t)|X_{t-1}) \times \Pr(X_{t-1})$$

Assuming $x(t)$ is a first-order Markov process, this expression simplifies to

$$\Pr(X_t) = \Pr(x(t)|x(t-1)) \times \Pr(X_{t-1}) \tag{2.70}$$

Finally the *evidence* is obtained in a similar manner to give

$$\Pr(Y_t) = \Pr(y(t), Y_{t-1}) = \Pr(y(t)|Y_{t-1}) \times \Pr(Y_{t-1}) \tag{2.71}$$

Therefore, substituting these results into Eq. 2.56, we obtain

$$\Pr(X_t|Y_t) = \frac{\overbrace{[\Pr(Y_{t-1}|X_{t-1}) \times \Pr(y(t)|x(t))]}^{\Pr(Y_t|X_t)}\;\overbrace{[\Pr(x(t)|x(t-1)) \times \Pr(X_{t-1})]}^{\Pr(X_t)}}{\underbrace{\Pr(y(t)|Y_{t-1}) \times \Pr(Y_{t-1})}_{\Pr(Y_t)}} \tag{2.72}$$

but the *posterior* at the previous time, $t-1$, is given by

$$\Pr(X_{t-1}|Y_{t-1}) = \frac{\Pr(Y_{t-1}|X_{t-1}) \times \Pr(X_{t-1})}{\Pr(Y_{t-1})} \tag{2.73}$$

Identifying the terms on the right-hand side of Eq. 2.72 and grouping them together enables the joint *sequential Bayesian posterior estimator* to be expressed as

$$\Pr(X_t|Y_t) = \left[\frac{\Pr(y(t)|x(t)) \times \Pr(x(t)|x(t-1))}{\Pr(y(t)|Y_{t-1})}\right] \times \Pr(X_{t-1}|Y_{t-1}) \tag{2.74}$$

This result is satisfying in the sense that we need only know the *joint* posterior distribution at the previous stage, $t-1$, scaled by a weighting function to sequentially propagate the posterior to the next stage, that is,

$$\overbrace{\Pr(X_t|Y_t)}^{New} = \overbrace{\mathcal{W}(t,t-1)}^{Weight} \times \overbrace{\Pr(X_{t-1}|Y_{t-1})}^{Old} \tag{2.75}$$

where the *weight* is defined by

$$\mathcal{W}(t,t-1) := \left[\frac{\Pr(y(t)|x(t)) \times \Pr(x(t)|x(t-1))}{\Pr(y(t)|Y_{t-1})}\right]$$

The sequential Bayesian processor is shown diagrammatically in Fig. 2.2. Even though this expression provides the full joint posterior solution, it is not physically realizable unless the distributions are known in closed form and the underlying multiple integrals or sums can be analytically determined. In fact, a more useful solution is the *marginal* posterior distribution.

2.5.2 Filtering Posterior Estimation

In this subsection, we develop a more realizable Bayesian processor for the posterior distribution of the random $x(t)$. Instead of requiring that the posterior possess the

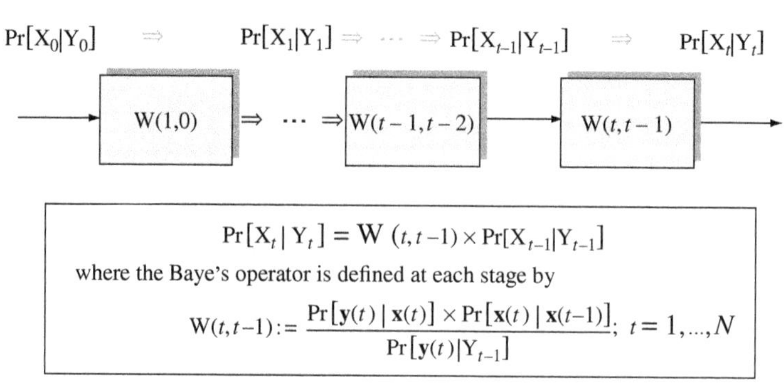

FIGURE 2.2 Sequential Bayesian processor for joint posterior distribution.

entire set of dynamic variables, X_t, we need only restrict our attention to the current variable at time t. That is, for signal enhancement, we would like to estimate $\Pr(x(t)|Y_t)$ where we restrict our attention to the value of the parameter based on all of the available measurements at time t. We start with the prediction recursion "marginalizing" the joint posterior distribution

$$\Pr(x(t)|Y_{t-1}) = \int \Pr(x(t), x(t-1)|Y_{t-1})\, dx(t-1)$$

Applying Bayes' rule to the integrand yields

$$\Pr(x(t), x(t-1)|Y_{t-1}) = \Pr(x(t)|x(t-1), Y_{t-1}) \times \Pr(x(t-1)|Y_{t-1})$$

or

$$\Pr(x(t), x(t-1)|Y_{t-1}) = \Pr(x(t)|x(t-1)) \times \Pr(x(t-1)|Y_{t-1}) \qquad (2.76)$$

where the final expression evolves from the first-order Markovian assumption. Applying the *Chapman–Kolmogorov* equation, the *prediction recursion* can be written as

$$\Pr(x(t)|Y_{t-1}) = \int \Pr(x(t)|x(t-1)) \times \Pr(x(t-1)|Y_{t-1})\, dx(t-1) \qquad (2.77)$$

Examining the update or correction equation based on the filtering relation and applying Bayes' rule, we have

$$\Pr(x(t)|Y_t) = \frac{\Pr(x(t), Y_t)}{\Pr(Y_t)} = \frac{\Pr(x(t), y(t), Y_{t-1})}{\Pr(y(t), Y_{t-1})}$$

Again applying the rule to this relation gives

$$\Pr(x(t)|Y_t) = \frac{\Pr(y(t)|x(t), Y_{t-1}) \times \Pr(x(t)|Y_{t-1}) \times \Pr(Y_{t-1})}{\Pr(y(t)|Y_{t-1}) \times \Pr(Y_{t-1})} \qquad (2.78)$$

Canceling like terms in numerator and denominator and applying the Markov assumption, we obtain the final expression for the *update recursion*[4] as

$$\overbrace{\Pr(x(t)|Y_t)}^{Posterior} = \frac{\overbrace{\Pr(y(t)|x(t))}^{Likelihood} \times \overbrace{\Pr(x(t)|Y_{t-1})}^{Prior}}{\underbrace{\Pr(y(t)|Y_{t-1})}_{Evidence}} \qquad (2.79)$$

[4] Note that this expression precisely satisfies Bayes' rule as illustrated in the equation.

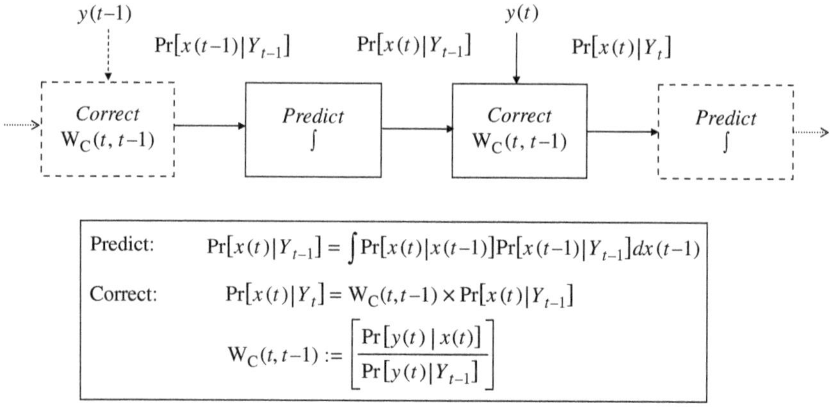

FIGURE 2.3 Sequential Bayesian processor for filtering posterior distribution.

where we can consider the update or filtering distribution as a weighting of the prediction distribution as in the full joint case above, that is,

$$\overbrace{\Pr(x(t)|Y_t)}^{Update} = \overbrace{\mathcal{W}_c(t, t-1)}^{Weight} \times \overbrace{\Pr(x(t)|Y_{t-1})}^{Prediction} \tag{2.80}$$

where the *weight* in this case is defined by

$$\mathcal{W}_c(t, t-1) := \frac{\Pr(y(t)|x(t))}{\Pr(y(t)|Y_{t-1})}$$

The resulting processor is shown diagrammatically in Fig. 2.3.

We summarize the sequential Bayesian processor in Table 2.1.

TABLE 2.1 Sequential Bayesian (Filtering) Processor

	Prediction	
$\Pr(x(t)\|Y_{t-1}) =$	$\int \Pr(x(t)\|x(t-1)) \times \Pr(x(t-1)\|Y_{t-1})\, dx(t-1)$	[Prediction]
where	$\Pr(x(t)\|x(t-1))$	[Transition probabliity]
	Correction/Update	
$\Pr(x(t)\|Y_t) =$	$\Pr(y(t)\|x(t)) \times \Pr(x(t)\|Y_{t-1})/\Pr(y(t)\|Y_{t-1})$	[Posterior]
where	$\Pr(y(t)\|x(t))$	[Likelihood]
	Initial Conditions	
	$\bar{x}(0) \quad \bar{P}(0) \quad \Pr(x(0)\|Y_0)$	

2.5.3 Likelihood Estimation

For a variety of applications, the "data" or "measurement" likelihood is a critical distribution function that is required. For instance, in the sequential detection of both Gaussian and non-Gaussian processes, the data likelihood is the essential ingredient required to make a timely decision (see Chapter 10 for more details).

We note from Bayes' rule for the filtering posterior of Eq. 2.79 that the data likelihood is given by the Chapman–Kolmogorov relation

$$\Pr(y(t)|Y_{t-1}) = \int \underbrace{\Pr(y(t)|x(t))}_{\textit{Likelihood}} \times \underbrace{\Pr(x(t)|Y_{t-1})}_{\textit{Prediction}}\, dx(t) \tag{2.81}$$

Substituting for the prediction distribution of Eq. 2.77, we can also express the likelihood in terms of the filtering distribution as

$$\Pr(y(t)|Y_{t-1}) = \int\int \underbrace{\Pr(y(t)|x(t))}_{\textit{Likelihood}} \times \underbrace{\Pr(x(t)|x(t-1))}_{\textit{Transition}} \times \underbrace{\Pr(x(t-1)|Y_{t-1})}_{\textit{Filtered}} \times dx(t-1)dx(t) \tag{2.82}$$

This completes the description of the data likelihood distribution that will prove critical in developments to follow in subsequent chapters.

2.6 SUMMARY

In this chapter, we have developed the idea of statistical signal processing from the Bayesian perspective. We started with the foundations of Bayesian processing by developing the "batch" solutions to the dynamic variable (state) or parameter estimation problem. These discussions led directly to the generic Bayesian processor—the maximum a posteriori solution evolving from Bayes' theorem. We showed the relationship of MAP to the popular maximum likelihood estimator demonstrating the ML is actually a special case when the dynamic variable is assumed deterministic but unknown and therefore not random. We also included the minimum variance estimator (MV, MMSE) and mentioned the least-squares (LS) approach. After discussing some of the basic statistical operations required (chain rule, performance (CRLB), etc.), we developed the idea of Bayesian sequential or recursive processing. We started with the full or joint posterior distribution assuming all of the information about the dynamic variable and observations was available. The solution to this problem led to a sequential Bayesian processor. Lastly, we followed with the development of a solution to the more practical marginal posterior or "filtering" distribution and illustrated the similarity to the previous processor.

MATLAB NOTES

MATLAB is a command oriented vector-matrix package with a simple yet effective command language featuring a wide variety of embedded *C* language constructs making it ideal for statistical operations and graphics. Least-squares problems are solved with the pseudo-inverse (**pinv**). When the covariance is known (minimum variance) the (**lscov**) command can be applied. Individual linear algebraic techniques including the singular-value decomposition, QR-decomposition (Gram–Schmidt) and eigenvalue/vector-decomposition techniques (**svd**, **qr**, **eig**, etc.) are available. The *Statistics* toolbox offers a wide variety of commands to perform estimation. For instance, "fit" tools are available to perform parameter estimation for a variety of distributions: exponential (**expfit**), Gaussian or normal (**normfit**), Poisson (**poissfit**), etc., as well as the generic maximum likelihood estimation (**mle**), as well as specific likelihood estimator for negative Gaussian/normal (**normlike**), and negative exponential (**explike**).

REFERENCES

1. A. Sage and J. Melsa, *Estimation Theory with Applications to Communications and Control* (New York: McGraw-Hill, 1971).
2. H. Van Trees, *Detection, Estimation and Modulation Theory* (New York: John Wiley & Sons, Inc., 1968).
3. J. Candy, "The Cramer– Rao bound: A measure of estimator quality," *Lawrence Livermore National Laboratory Report*, UCID-17660, 1977.
4. J. Candy, *Model-Based Signal Processing* (Hoboken, NJ: John Wiley & Sons/IEEE Press, 2006).
5. A. Dempster, N. Laird, and D. Rubin, "Maximum likelihood from incomplete data via the EM algorithm," *J. R. Statist. Soc.B*, **39**, 1, 1–38, 1977.
6. R. Duda and P. Hart, *Pattern Classsification* (New York: John Wiley & Sons, Inc., 2000).
7. I. Nabney, *NETLAB: Algorithms for Pattern Recognition* (New York: Springer, 2002).
8. C. Wu, "On the convergence properties of the EM algorithm," *Ann. Statist.*, **11**, 1, 95–103, 1983.
9. R. Redner and H. Walker, "Mixture densities, maximum likelihood and the EM algorithm," *SIAM Rev.*, **26**, 2, 195–239, 1984.
10. W. Gilks, S. Richardson, and D. Spiegelhalter, *Markov Chain Monte Carlo in Practice* (New York: Chapman & Hall, 1996).
11. D. Snyder and M. Miller, *Random Point Processes in Time and Space*, 2nd Ed. (New York: Springer, 1991).
12. M. Tanner, *Tools for Statistical Inference*, 2nd Ed. (New York: Springer-Verlag, 1993).
13. T. Moon, "The expectation-maximization algorithm," *IEEE Sig. Proc. Mag.*, **13**, 6, 47–60, 1996.

14. J. Bilmes, "A gentle tutorial of the EM algorithm and its application to parameter estimation for Gaussian mixture and hidden Markov models," *U. C. Berkeley Report*, TR-97-021, 1998.

15. G. McLachlan and T. Krishnan, *The EM Algorithm and Extensions* (New York: John Wiley & Sons, Inc., 1997).

16. A. Gelman, J. Carlin, H. Stern and D. Rubin, *Bayesian Data Analysis*, 2nd Ed. (New York: Chapman & Hall, 2004).

17. A. Papoulis and S. Pillai, *Probability, Random Variables, and Stochastic Processes*, 4th Ed. (New York: McGraw-Hill, 2002).

18. I. Rhodes, "A tutorial introduction to estimation and filtering," *IEEE Trans. Autom. Contr.*, **16**, 6, 688–706, 1971.

PROBLEMS

2.1 Derive the following properties of conditional expectations:

(*a*) $E_x\{X|Y\} = E\{X\}$, if X and Y are independent.

(*b*) $E\{X\} = E_y\{E\{X|Y\}\}$.

(*c*) $E_x\{g(Y)\,X\} = E_y\{g(Y)\,E\{X|Y\}\}$.

(*d*) $E_{xy}\{g(Y)\,X\} = E_y\{g(Y)\,E\{X|Y\}\}$.

(*e*) $E_x\{c|Y\} = c$.

(*f*) $E_x\{g(Y)|Y\} = g(Y)$.

(*g*) $E_{xy}\{cX + dY|Z\} = cE\{X|Z\} + dE\{Y|Z\}$.

2.2 Verify the following properties:

(*a*) $\nabla_x(a'b) = (\nabla_x a')b + (\nabla_x b')a$, for $a, b \in R^n$ and functions of x.

(*b*) $\nabla_x(b'x) = b$.

(*c*) $\nabla_x(x'C) = C, C \in R^{n\times m}$.

(*d*) $\nabla_x(x') = I$.

(*e*) $\nabla_x(x'x) = 2x$.

(*f*) $\nabla_x(x'Ax) = Ax + A'x$, for A not a function of x.

2.3 Show that for any unbiased estimate of $\hat{\mathbf{x}}(y)$ of the parameter vector $\mathbf{x}$ the *Cramer–Rao bound* is

$$\text{Cov}(\tilde{x}|x) \geq \mathcal{I}^{-1} \quad \text{for } \tilde{x} - x = \hat{x}(y) \text{ and } \mathbf{x} \in \mathcal{R}^n, \mathcal{I} \in \mathcal{R}^{n\times n}$$

where the *information matrix* is defined by

$$\mathcal{I} := -E_y\{\nabla_x(\nabla_x \ln Pr(Y|X))'\}$$

2.4 The sample mean, $\overline{\theta} = 1/N \sum_{t=1}^{N} \theta(t)$, is a very important statistic in data processing.

(*a*) Show that the sample mean is an *unbiased* estimator of $\theta \sim \mathcal{E}xp(1/\theta)$.

(*b*) Estimate the corresponding variance of $\overline{\theta}$.

(*c*) Show that the sample mean is a *consistent* estimator of θ.

(*d*) Construct the two standard deviation *confidence interval* about the sample mean estimator. (*Hint*: Let $\overline{\theta} \sim N(\theta, \sigma^2), \sigma^2$ known.)

(*e*) Show that $\overline{\theta}$ is a *minimum variance* estimator. (*Hint*: Show it satisfies the *Cramer–Rao bound* with equality).

(*f*) If $\hat{\theta}$ is an estimator or θ, calculate the corresponding *mean-squared error.*

2.5 Let $x(t)$ be an unknown dynamic parameter obtained from the linear measurement system

$$y(t) = C(t)x(t) + v(t)$$

where $v \sim \mathcal{N}(0, R_{vv}(t))$ with $y, v \in \mathcal{R}^p$ and x is governed by the state transition mechanism

$$x(t+1) = \Phi(t+1, t)x(t) \quad \text{for } \Phi \in \mathcal{R}^{n\times n}$$

Calculate the Cramer–Rao bound for the case of estimating the initial state $x(0)$.

2.6 Suppose we have the following signal in additive Gaussian noise:

$$y = x + n \quad \text{with } n \sim N(0, R_{nn})$$

(*a*) Find the Cramer–Rao bound if x is assumed unknown.

(*b*) Find the Cramer–Rao bound if $p(x) = xe^{-x^2/R_{nn}},\ x \geq 0$.

2.7 Suppose we have two jointly distributed random vectors x and y with known means and variances. Find the linear minimum variance estimator. That is, find

$$\hat{x}_{MV} = \hat{A}y + \hat{b}$$

and the corresponding error covariance $\operatorname{cov}\tilde{x}, \tilde{x} = x - \hat{x}$.

2.8 We would like to estimate a vector of unknown parameters $p\epsilon R^n$ from a sequence of N-noisy measurements given by

$$y = Cp + n$$

where $C\epsilon R^{p\times n}, y, n\epsilon R^p$, and v is zero-mean and white with covariance R_{nn}.

(*a*) Find the minimum variance estimate $\hat{p}_{MV}$.

(*b*) Find the corresponding quality $\operatorname{cov}(p - \hat{p}_{MV})$.

2.9 Suppose we have N samples $\{x(0), \ldots, x(N-1)\}$ of a process $x(t)$ to be estimated by N complex sinusoids of arbitrary frequencies $\{f_0, \ldots, f_{n-1}\}$. Then

$$x(k \triangle T) = \sum_{m=0}^{N-1} a_m \exp(j2\pi f_m k \triangle T) \quad \text{for } k = 0, \ldots, N-1$$

Find the least-squares estimate $\hat{a}_{LS}$ of $\{a_m\}$.

2.10 Suppose we are given a measurement modeled by

$$y(t) = s + n(t)$$

where s is random and zero-mean with variance $\sigma_s^2 = 4$, and n is zero-mean and white with a unit variance. Find the two-weight minimum variance estimate of s, that is,

$$\hat{s}_{MV} = \sum_{i=1}^{2} w_i y(i)$$

that minimizes

$$J = E\{(s - \hat{s})^2\}$$

2.11 Find the maximum likelihood and maximum a posteriori estimates of the parameter x :

$$p(x) = \alpha e^{-\alpha x} \quad x \geq 0, \ \ \alpha \geq 0$$

and

$$p(y|x) = x e^{-xy} \quad x \geq 0, \ \ y \geq 0$$

2.12 Suppose we have two classes of objects, black and white, with shape subclasses, circle and square, and we define the random variables as:

$$\begin{aligned} X_1 &= \text{number of black circular objects} \\ X_2 &= \text{number of black square objects} \\ X_3 &= \text{number of white objects} \end{aligned}$$

Assume that these objects are trinomially distributed such that

$$\Pr(\mathbf{x}|\Theta) = \left(\frac{(x_1 + x_2 + x_3)!}{x_1!x_2!x_3!}\right)\left(\frac{1}{4}\right)^{x_1}\left(\frac{1}{4} + \frac{\Theta}{4}\right)^{x_2}\left(\frac{1}{2} - \frac{\Theta}{4}\right)^{x_3}$$

Suppose a person with blurred vision cannot distinguish shape (circle or square), but can distinguish between black and white objects. In a given batch of objects, the number of objects detected by this person is: $\mathbf{y} = [y_1 y_2]'$ with

$$y_1 = x_1 + x_2$$
$$y_2 = x_3$$

(*a*) What is the probability, $\Pr(y_1|\Theta)$?

(*b*) What is the expression for the E-step of the EM algorithm assuming $\hat{\Theta}_k$ is the current parameter estimate, that is, find the expression for $E\{x_1|y_1, \hat{\Theta}_k\}$ with x_3 known?

(*c*) What is the corresponding M-step?

(*d*) Take the EM solution (a)–(c) above based on 100 samples with $y_1 = 100$ and start iterating with $\Theta_0 = 0$ for 10-steps; what is your final estimate of the parameter $\hat{\Theta}_{10}$? ($\Theta_{true} = 0.5$) for simulation $x_1 = 25, x_2 = 38$.

2.13 Suppose we have a bimodal distribution consisting of a Gaussian mixture with respective means, variances and mixing coefficients: $\{\mu_1, \sigma_1^2, p_1\}$ and $\{\mu_2, \sigma_2^2, p_2\}$ such that $\mathrm{p}_X(x) = \sum_{i=1}^{2} p_i \mathcal{N}(\mu_i, \sigma_i^2)$ with $p_2 = 1 - p_1$. We would like to fit the parameters $\Theta = (p_1, \mu_1, \sigma_1^2, \mu_2, \sigma_2^2)$ to data. Develop the EM algorithm for this problem.

2.14 Suppose we are given a measurement system

$$y(t) = x(t) + v(t) \quad t = 1, \ldots, N$$

where $v(t) \sim N(0, 1)$.

(*a*) Find the maximum likelihood estimate of $x(t)$, that is, $\hat{x}_{\mathrm{ML}}(t)$, for $t = 1, \ldots, N$.

(*b*) Find the maximum a posteriori estimate of $x(t)$ that is, $\hat{x}_{\mathrm{MAP}}(t)$, if $p(x) = e^{-x}$.

2.15 Suppose we have a simple AM receiver with signal

$$y(t) = \Theta s(t) + v(t) \quad t = 1, \ldots, N$$

where Θ is a random amplitude, s is the known carrier, and $v \sim N(0, R_{vv})$.

(*a*) Find the maximum likelihood estimate $\hat{\Theta}_{\mathrm{ML}}$.

(*b*) Assume $\Theta \sim N(\Theta_0, R_{\Theta\Theta})$. Find the maximum a posteriori estimate $\hat{\Theta}_{\mathrm{MAP}}$.

(*c*) Assume Θ is Rayleigh-distributed (a common assumption). Find $\hat{\Theta}_{\mathrm{MAP}}$.

2.16 We would like to estimate a signal from a noisy measurement

$$y = s + v$$

where $\upsilon \sim N(0, 3)$ and s is Rayleigh-distributed.

$$p(s) = se^{-s^2/2}$$

with

$$p(y|s) = \frac{1}{\sqrt{6\pi}} e^{-\frac{(y-s)^2}{6}}$$

(*a*) Find the maximum likelihood estimate.
(*b*) Find the maximum a posteriori estimate.
(*c*) Calculate the Cramer–Rao bound (ignoring $p(s)$).

2.17 Assume that a planet travels an elliptical orbit centered about a known point. Suppose we make M observations.

(*a*) Find the best estimate of the orbit and the corresponding quality, that is, for the elliptical orbit (Gaussian problem)

$$\beta_1 x^2 + \beta_2 y^2 = 1$$

Find $\hat{\beta} = [\hat{\beta}_1 \hat{\beta}_2]'$.

(*b*) Suppose we are given the following measurements: $(x, y) = \{(2, 2), (0, 2), (-1, 1)$, and $(-1, 2)\}$, find $\hat{\beta}$ and $\hat{J}$, the cost function.

2.18 Find the parameters, β_1, β_2, and β_3 such that

$$f(t) = \beta_1 t + \beta_2 t^2 + \beta_3 \sin t$$

fits $(t, f(t))$: $= \{(0, 1), (\pi/4, 2), (\pi/2, 3)$, and $(\pi, 4)\}$ with corresponding quality estimate $\hat{J}$.

3

SIMULATION-BASED BAYESIAN METHODS

3.1 INTRODUCTION

In this chapter, we investigate the idea of Bayesian estimation [1–13] using approximate sampling methods to obtain the desired solutions. We first motivate the simulation-based Bayesian processors and then review much of the basics required for comprehension of this powerful methodology. Next, we develop the idea of simulation-based solutions using the *Monte Carlo (MC)* approach [14–21] and introduce *importance sampling* as a mechanism to implement this methodology from a generic perspective [22–28]. Finally, we consider the class of iterative processors founded on Markov chain concepts leading to efficient techniques such as the foundational Metropolis–Hastings approach and the Gibbs sampler [29–37].

Starting from Bayes' rule and making assertions about the underlying probability distributions enables us to develop reasonable approaches to design approximate Bayesian processors. Given "explicit" distributions, it is possible to develop analytic expressions for the desired posterior distribution. Once the posterior is estimated, the Bayesian approach allows us to make inferences based on this distribution and its associated statistics (e.g., mode, mean, median). For instance, in the case of a *linear* Gauss–Markov (GM) model, calculation of the posterior distribution leads to the optimal minimum variance solution [11]. But again this was based completely on the assertion that the dynamic processes were strictly constrained to Gaussian distributions and a *linear* GM model therefore leading to a Gaussian posterior. When the dynamics become nonlinear, approximate methods evolve based on " linearization"

Bayesian Signal Processing: Classical, Modern, and Particle Filtering Methods, Second Edition. James V. Candy.

techniques either model-based (Taylor series transformations) or statistical (sigma-point transformations). In both cases, these are fundamentally approximations that are constrained to unimodal distributions. What happens when both the dynamics and statistics are nonstationary and non-Gaussian? Clearly, these approaches can be applied, but with little hope of success under most conditions. Therefore, we must resort to other less conventional (in signal processing) ways of attacking this class of problems that have dominated the science and engineering literature for a long time [38–61]. This question then leads us directly to statistical simulation-based techniques invoking random sampling theory and the Monte Carlo (MC) method—well known in statistics for a long time [14]. This method is essentially a collection of techniques to estimate statistics based on random sampling and simulation. It can be thought of simply as "performing estimation through sampling." The *goal* of Bayesian techniques using MC methods is to generate a set of independent samples from the target posterior distribution with enough samples to perform accurate inferences [21]. Monte Carlo techniques have been applied to a large variety of applications in science and engineering [56, 57, 61]. In this chapter, we start out with the simple idea of random sampling with the underlying motive of developing Monte Carlo simulation techniques to solve nonlinear/non-Gaussian signal processing problems.

MC methods involve techniques to estimate the posterior distribution of interest using either numerical integration-based methods (when possible) or sample-based simulation methods which attempt to produce *independent identically distributed* (*i.i.d.*) samples from a targeted posterior distribution and use them to make statistical inferences. Following Smith [18], we develop this idea further starting out with Bayes' rule for the variable or parameter X and the corresponding data Y, that is, the posterior distribution is given by

$$\Pr(X|Y) = \frac{\Pr(Y|X) \times \Pr(X)}{\Pr(Y)} = \frac{\Pr(Y|X) \times \Pr(X)}{\int \Pr(Y|X) \times \Pr(X)\, dX} \tag{3.1}$$

with the usual definitions of likelihood, prior, and evidence (normalization) distributions. Once the posterior is estimated, the inferences follow immediately. For instance, the conditional mean is given by

$$E\{X|Y\} = \int X \Pr(X|Y)\, dX \tag{3.2}$$

In the continuous (random variable) case, the explicit evaluation of these integrals is required, yet rarely possible, leading to sophisticated numerical or analytical approximation techniques to arrive at a solution. These methods rapidly become intractable or computationally intensive to be of any real value in practice. Thus an alternative method is necessary.

Just as we can use the distribution to generate a set of random samples, the samples can be used to generate (approximately) the underlying distribution (e.g., histogram). The idea of using samples generated from the prior through the likelihood to obtain the posterior parallels that of using analytic distributions. For example, ignoring the evidence we use the *method of composition* [16] to sample

$$\Pr(X|Y) \propto \Pr(Y|X) \times \Pr(X) \tag{3.3}$$

so that we can

- Generate samples from the prior: $X_i \rightarrow \Pr(X); i = 1, \ldots, N$;
- Calculate the likelihood: $\Pr(Y|X_i)$; and
- Estimate the posterior: $\hat{\Pr}(X|Y) \approx \Pr(Y|X_i) \times \Pr(X_i)$.

Unfortunately, this methodology is not realistic unless we can answer the following questions: (1) How do we generate samples from distributions for simulations? and (2) How do we guarantee that the samples generated are *i.i.d.* from the targeted posterior distribution?

However, before we develop the idea of sampling and simulation-based methods, we must have some mechanism to evaluate the performance of these samplers. In the next section, we briefly describe techniques to estimate the underlying probability distributions from data samples.

3.2 PROBABILITY DENSITY FUNCTION ESTIMATION

One of the requirements of Bayesian signal processing is to generate samples from an estimated posterior distribution. If the posterior is of a known closed form (e.g., Gaussian), then it is uniquely characterized by its particular parameters (e.g., mean and variance) that can be "fit" to the samples directly using parameter estimation techniques. But, if the probability density[1] is too complex or cannot readily be represented in closed form, then *nonparametric* techniques must be employed to perform the estimation.

Since most parametric forms rarely fit the underlying posterior distribution encountered in the real world, especially if they are multimodal (multiple peaks) distributions, we investigate the so-called *kernel* (smoothing) method of probability density function (PDF) estimation. The basic idea is to fit a simple model individually at each target sample (random variate) location, say $\hat{x}$, using the observations close to the target $\mathrm{p}(\hat{x})$. This estimate can be accomplished by using a *weighting function* or equivalently *kernel* function $\mathcal{K}(\hat{x}; x_i)$ that assigns a weight to x_i based on its proximity to $\hat{x}$ [4]. Different kernels lead to different estimators. For instance, the classical *histogram* can be considered a rectangular or box kernel PDF estimator [5], while

[1] In this section, we introduce more conventional notation for both the cumulative and density or mass functions, since it is required. The CDF of the random variable X with realization x is defined by $\mathrm{P}_X(x)$, while the PDF or PMF is $\mathrm{p}_X(x)$. In instances where it is obvious, we do not use the subscript X to identify the random variable.

the fundamental Parzen-window multidimensional estimator is based on a hypercube kernel [3]. Some of the popular kernel smoothing functions include the triangle, Gaussian, Epanechnikov, and others [4]. In any case, kernel density estimation techniques [6] provide the basic methodology required to estimate the underlying PDF from random data samples and subsequently evaluate sampler performance—one of the objectives of this chapter.

Theoretically, the underlying principle of distribution estimation is constrained by the properties of the PDF itself, that is,

- $\text{p}_X(x) \geq 0 \, \forall x \in \mathcal{R}_X$; and
- $\int_{\mathcal{R}_x} \text{p}_X(x)\, dx = 1$.

Formally, the probability that the random variate x will fall within the interval $x_i \leq X \leq x_j$ for $i < j$ is given by [3]

$$\Pr(x_i < X < x_j) = \int_{x_i}^{x_j} \text{p}_X(x)\, dx \approx \text{p}_X(x) \times \Delta_x \quad \text{for } \Delta_x \ll 1(\text{small}) \tag{3.4}$$

which is simply the *area* under the PDF in the interval (x_i, x_j). Thus, the probability that X is in a small interval of length Δ_x is proportional to its PDF $\text{p}_X(x)$.

Therefore, it follows that the probability at a point, say $\hat{x}$, can be approximated by allowing $x_i = x$ and $x_j = x + \Delta_x$, then

$$\Pr(x < \hat{x} < x + \Delta_x) = \int_{x}^{x+\Delta_x} \text{p}_X(x) dx \quad \text{for } x \in \mathcal{R}_x \tag{3.5}$$

demonstrating that

$$\text{p}_X(x) = \lim_{\Delta_x \to 0} \frac{\Pr(x < \hat{x} < x + \Delta_x)}{\Delta_x} \tag{3.6}$$

Consider a random draw of N *i.i.d.* (total) samples according to $\text{p}_X(x)$ above. The probability that n_N of these reside within $\mathcal{R}_x$ is binomial and can be approximated by the *frequency ratio* n_N/N leading to a relative frequency interpretation of the probability. That is, it follows that

$$\text{p}_X(x) \times \Delta_x = \frac{n_N}{N} \tag{3.7}$$

But this is just the well-known [1] *histogram* estimator given by

$$\hat{\text{p}}_X(x) := \hat{\text{p}}_X(\hat{x}) = \frac{1}{\Delta_x} \times \frac{n_N}{N} \quad |x - \hat{x}| \leq \frac{\Delta_x}{2} \tag{3.8}$$

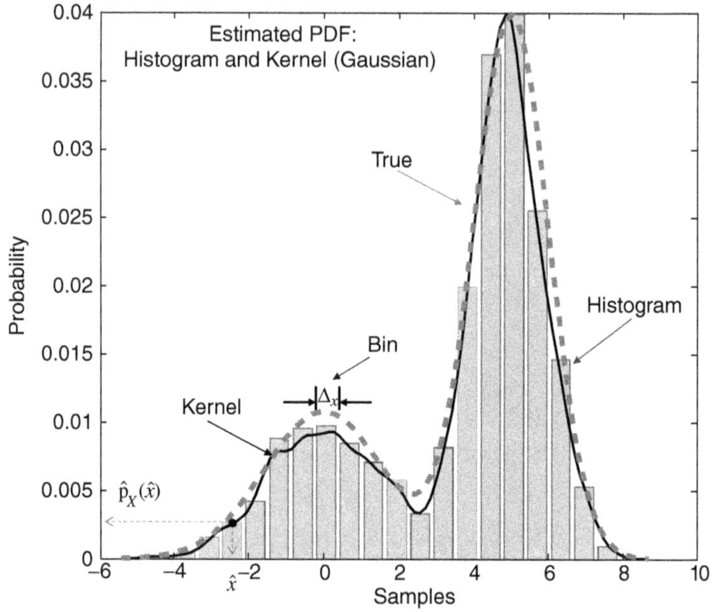

FIGURE 3.1 Estimated PDF using both histogram and kernel (Gaussian window $\mathcal{N}(0,10)$) methods for random samples generated from a Gaussian mixture $\mathcal{N}_1(0, 2.5)$ and $\mathcal{N}_2(5,1)$ and mixing coefficients $P_1 = 0.3$ and $P_2 = 0.7$.

Here $\hat{x}$ is located at the midpoint of the interval (bin) $\hat{x} - \frac{\Delta_x}{2} \leq x \leq \hat{x} + \frac{\Delta_x}{2}$ and the corresponding probability is assumed *constant* throughout the bin as shown in Fig. 3.1. If

$$\mathrm{p}_N(x) \approx \frac{n_N/N}{\Delta_x} \tag{3.9}$$

then as $N \to \infty$, histogram estimator converges to the true PDF, that is, $\mathrm{p}_N(x) \to \mathrm{p}_X(x)$ and the following three conditions must hold

1. $\lim_{N\to\infty} \Delta_x \to 0$;
2. $\lim_{N\to\infty} n_N \to \infty$; and
3. $\lim_{N\to\infty} \frac{n_N}{N} \to 0$.

The first condition must hold to achieve the limit in Eq. 3.6, while the second condition indicates how n_N is to grow to guarantee convergence. Thus, at all nonzero points of the PDF by fixing the size of the interval, the probability of samples falling within this interval is finite and therefore $n_N \approx \Pr(x) \times N$ and $n_N \to \infty$ as $N \to \infty$. The final condition arises because as the interval size shrinks $\Delta_x \to 0$, the corresponding probability $\Pr(x) \to 0$. These conditions indicate the manner in which the parameters should be chosen: *N large*, Δ_x*small* with n_N *large*. Satisfaction of all of these conditions imply convergence to $\mathrm{p}_X(x)$ (see Refs. [3] and [5] for more details).

There are two common ways of ensuring convergence: (1) "shrink" the region by selecting the interval as a function of the number of samples such as $\Delta_x = \frac{1}{\sqrt{N}}$ (Parzen-window method)[2]; and (2) "increase" the number of samples within the interval which can be done as $n_N = \sqrt{N}$. Here the local region grows until it encloses n_N neighbors of x (nearest-neighbor method). Both methods converge in probability. This is the principle theoretical result in PDF estimation [3, 5, 6]. Next, consider the kernel density estimator as a "generalized" histogram.

Let $x_1, \ldots, x_N$ be a set of *i.i.d.* samples of the random variate x then the *kernel density* estimator approximation of the underlying probability density *at* x is given by

$$\hat{\mathrm{p}}_X(x) = \frac{1}{\Delta_x}\left(\frac{1}{N}\sum_{i=1}^{N}\mathcal{K}\left(\frac{x - x_i}{\Delta_x}\right)\right) = \frac{1}{N\,\Delta_x}\sum_{i=1}^{N}\mathcal{K}\left(\frac{x - x_i}{\Delta_x}\right) \tag{3.10}$$

where $\mathcal{K}$ is defined as the kernel density (nonnegative and integrates to unity) with Δ_x the corresponding smoothing parameter or *bandwidth* used to "tune" the estimate. The classic bias–variance tradeoff arises here. That is, smaller Δ_x (high resolution) yields smaller bias, but higher variance. However, a larger Δ_x (low resolution) gives a larger bias with smaller variance [5]. We demonstrate the application of kernel techniques in the following example of estimating a multimodal distribution.

Example 3.1

Suppose we would like to estimate the PDF of a set of 1000 samples generated from a mixture of two Gaussian densities such that $\mathcal{N}(\mu_x, \sigma_x^2) \to \mathcal{N}_1(0, 2.5)$ and $\mathcal{N}_2(5, 1)$ with mixing coefficients $P_1 = 0.3$ and $P_2 = 0.7$. We use both the histogram and kernel density estimators. For the histogram we choose $N = 20$ bins to illustrate (graphically) the bin estimates, while for the kernel density estimator we choose a Gaussian window specified by $K(x) = \frac{1}{\sqrt{20\pi}} e^{\frac{-x^2}{20}}$ with an optimal bandwidth of $\Delta_x = \sigma_x \times \left(\frac{4}{3N}\right)^{\frac{1}{5}}$ [3].

The results are shown in Fig. 3.1 for 100 bins or points where we see the estimated PDF using both methods along with the "true" Gaussian density. Clearly the Gaussian kernel estimator provides a much "smoother" estimate of the true density. A smoother histogram estimate can be obtained by decreasing the bin-width. Both techniques capture the general shape of the bimodal distribution. △△△

[2] It has been shown that the "optimal" bin (interval) size can be obtained by searching for the bin that minimizes the cost function $C(\Delta_x) = \frac{(2\bar{x} - \overline{\sigma_x^2})}{\Delta_x^2}$ [7].

This completes the section on PDF estimation techniques. Note such details as selecting the appropriate bin-size for the histogram and the bandwidth for the kernel density estimator as well as its structure (Gaussian, Box, etc.) is discussed in detail in Refs. [3] and [5–7]. These relations enable "tuning" of the estimators for improved performance.

Next, we consider the simulation-based methods in an effort to answer the questions posed previously: (1) How do we generate samples from distributions for simulations? and (2) How do we guarantee that the samples generated are *i.i.d.* from the targeted posterior distribution?

3.3 SAMPLING THEORY

The generation of random samples from a known distribution is essential for simulation. If the distribution is standard and has a closed analytic form (e.g., Gaussian, exponential, Rayleigh), then it is usually possible to perform this simulation easily. This method is called the *direct method* because it evolves "directly" from the analytic form. An alternative is the " inversion" method which is based on uniform variates and is usually preferred. Thus, the uniform distribution becomes an extremely important distribution in sampling because of its inherent simplicity and the ease with which samples can be generated [15].

Many numerical sample generators evolve by first generating a set of random numbers that are uniformly distributed on the interval $[a, b]$ such that all of the subintervals are of equal length and have "equal probability." Thus, X is said to be a *uniform random variable*

$$X \sim \mathcal{U}(a, b) \tag{3.11}$$

that has a *cumulative distribution function* (*CDF*) defined by

$$\mathrm{P}_X(x) = \Pr(X \leq x) = \begin{cases} 0 & x < a \\ \frac{x-a}{b-a} & a \leq x < b \\ 1 & x \geq b \end{cases} \tag{3.12}$$

along with the corresponding probability density

$$\mathrm{p}_X(x) = \frac{1}{b-a} \quad a \leq x \leq b \tag{3.13}$$

with mean and variance $\mu_x = \frac{b+a}{2}$ and $\sigma_x^2 = \frac{(b-a)^2}{12}$.

Because of its simplicity and ease of numerical generation, the uniform variate is the starting point of many simulation schemes. For instance, to generate a binomial random number (e.g., coin tossing with Pr(heads) $= p$), we first generate uniform random variates and then count the number of times they are greater than p. The result is a set of binomial variates with trial parameter N and success (heads) probability p.

Since statistical sampling techniques are based on the generation of random samples from a given distribution, we must understand the relationship between an *input* PDF, $\mathrm{p}_X(x)$ and an *output* PDF, $\mathrm{p}_Y(y)$ when there exists a relationship between the random variables x and y defined by $y = y(x)$. The problem is

GIVEN x and $\mathrm{p}_X(x)$ **FIND** the output PDF, $\mathrm{p}_Y(y)$ defining the probability of y.

When we have functions of random variables with known analytic distributions, then the usual *transformation method applies*. That is, given the known distribution $\mathrm{P}_X(x)$ or density $\mathrm{p}_X(x)$ and the monotonic, one-to-one, invertible transformation $y = T(x)$, then the distribution or density of y is found by the using the transformation relation [1]

$$\mathrm{p}_Y(y) = \mathrm{p}_X(x) \times \left|\frac{\partial y}{\partial x}\right|^{-1} = \mathrm{p}_X(x = T^{-1}(y)) \times \left|\frac{\partial x}{\partial y}\right| \tag{3.14}$$

where $\left|\frac{\partial y}{\partial x}\right|$ is the Jacobian of the transformation and T^{-1} is its inverse. The derivation of this relationship follows directly from the definitions of CDF [2]. Although simple, this relation establishes one of the basic concepts in random sampling.

Perhaps the most fundamental relation is obtained by applying the transformation directly to the CDF defined by

$$y(x) = \mathrm{P}_X(x) = \int_{-\infty}^{\infty} \mathrm{p}_X(\alpha)d\alpha \tag{3.15}$$

That is, applying the transformation of Eq. 3.14 and taking the required derivatives gives

$$\frac{dy}{dx} = \frac{d\mathrm{P}_X(x)}{dx} = \mathrm{p}_X(x) \tag{3.16}$$

which follows from the fundamental theorem of calculus. Therefore, we have

$$\mathrm{p}_Y(y) = y(x) - \frac{\mathrm{p}_X(x)}{|\mathrm{p}_X(x)|} = 1 \tag{3.17}$$

yielding a uniformly distributed random variable $y \sim \mathcal{U}(0, 1)$, that is,

$$\mathrm{p}_Y(y) = y(x) = 1 \quad \text{for } 0 \leq y \leq 1 \tag{3.18}$$

Thus, the CDF of any random variable is *always* uniformly distributed on the interval $[0, 1]$ independently of $\mathrm{p}_X(x)$! This important result enables us to "sample" from any arbitrary $\mathrm{p}_X(x)$, since the random variable x can be sampled by first generating samples from $y \sim \mathcal{U}(0,1)$ and then performing the inversion $x = \mathrm{P}_X^{-1}(y)$. This is called the *inversion method* in sampling theory that simply entails generating random samples from $\mathcal{U}(0,1)$ and determining the samples x by inversion of its (known) CDF. Note that the CDF can be known analytically or just in tabular form to perform the inversion. We discuss this theorem more formally in the next section. First, consider the following example to illustrate this idea.

Example 3.2

Suppose we have a uniform random variable $x \sim \mathcal{U}(0,1)$ and a transformation $y = T(x) = -\frac{1}{\lambda}\ln x$. We would like to know the analytic form of the density $\mathrm{p}_Y(y)$. Since x is uniform, the density is $\mathrm{p}_X(x) = \frac{1}{b-a} = 1$, the inverse transformation is $x = T^{-1}(y) = e^{-\lambda y}$; therefore, taking the derivative and substituting its absolute value into Eq. 3.14 gives

$$\mathrm{p}_Y(y) = (1) \times |-\lambda e^{-\lambda y}| = \lambda e^{-\lambda y}$$

an exponential distribution with parameter λ. △△△

Transformations of *discrete* random variables are much simpler where we use the probability mass and discrete CDFs in place of their continuous counterparts. In this case the transformation is still continuous with the identical conditions (invertible, etc.), that is,

$$y_i = T(x_i) \quad i = 1, \ldots, N \tag{3.19}$$

and therefore we have the discrete CDF as

$$\mathrm{P}_Y(y) = \Pr(Y \le y_i) = \Pr(X \le x_i) = \Pr(X \le x_i = T^{-1}(y_i)); \quad i = 1, \ldots, N \tag{3.20}$$

3.3.1 Uniform Sampling Method

As noted, the uniform distribution plays a vital role in simulating random variates and is applied heavily in sample-based simulation schemes. We formally present two fundamental theorems and a corollary [1] to justify the theoretical foundation of this approach and motivate their use through some simple examples.

Uniform Transformation Theorem Given a random variable X with distribution $\mathrm{P}_X(x)$, there *exists* a unique, monotonic transformation $T(X)$ such that the random variable $U = T(X)$ is distributed as $\mathcal{U}(0,1)$, that is, if $T(X)$ is selected as

$$T(X) = U$$

then

$$U \sim \mathcal{U}(0, 1)$$

The proof of this theorem follows directly [1] from the properties of the CDF. Suppose x is arbitrary and $u = \mathrm{P}_X(x)$. Then with a monotonic $\mathrm{P}_X(x)$, we have $U \leq u$ iff $X \leq x$ and therefore it follows that

$$\mathrm{P}_U(u) = \Pr(U \leq u) = \Pr(X \leq x) = \mathrm{P}_X(x) = u$$

giving the desired result. This is a strong result stating that *any* CDF can be represented in terms of a set of uniform random variates as discussed before.

The second theorem [1] provides the basis of most sample-based simulations and is reliant on the existence of the "inverse" of the CDF.

Inverse Transformation Theorem Given a uniform random variable $U \sim \mathcal{U}(0, 1)$ and a distribution (target) $\mathrm{P}_X(x)$, there *exists* a unique, monotonic transformation $T(U)$, such that the random variable $X = T(U)$ is distributed as $\mathrm{P}_X(x)$, that is, if $T(U)$ is selected as

$$T(U) = X = \mathrm{P}_X^{-1}(U)$$

then

$$X \sim \mathrm{P}_X(x)$$

Again the proof of this theorem follows directly from the properties of the CDF [1]. From the Uniform Transformation Theorem above, we know that U is uniform and $\mathrm{P}_X(X)$ is arbitrary; therefore, $U = \mathrm{P}_X(X) \longrightarrow X = \mathrm{P}_X^{-1}(U)$ and it follows that

$$\mathrm{P}_X(X \leq x) = \mathrm{P}_X(X \leq \mathrm{P}_X^{-1}(U)) = \mathrm{P}_X(\mathrm{P}_X^{-1}(U)) = u = \mathrm{P}_X(X)$$

providing the proof. We illustrate this technique with the following example.

Example 3.3

Suppose we would like to generate a random variable X exponentially distributed with parameter λ, $X \sim \mathrm{Exp}(\lambda)$. From the Uniform Transformation Theorem, we have

$$u = \mathrm{P}_X(x) = 1 - e^{-\lambda x}$$

and

$$x = \mathrm{P}_X^{-1}(U) = -\frac{1}{\lambda}\ln(1 - u)$$

To generate exponential random variables

- Generate uniform samples: $u_i \longrightarrow \mathcal{U}(0, 1)$
- Transform to exponential: $x_i = -\frac{1}{\lambda}\ln(1 - u_i)$

Also since $\ln(1 - u_i) \sim \mathcal{U}(0, 1)$, then the exponential variates are generated more efficiently from $x_i = -\frac{1}{\lambda}\ln(u_i)$. △△△

To further illustrate the *inverse CDF method*, we apply it to a discrete problem that occurs quite frequently in nonlinear processing.

Example 3.4

Suppose we have estimated a discrete empirical distribution representing a posterior density given by

$$\hat{p}_X(x) = \sum_{i=1}^{N} W_i \delta(x - x_i)$$

with corresponding CDF

$$\hat{P}_X(x) = \sum_{i=1}^{N} W_i \mu(x - x_i)$$

where $\mu(x)$ is a unit step function, $W_i := \Pr(X = x_i)$ and of course, $\sum_{i=1}^{N} W_i = 1$. The CDF is shown in Fig. 3.2a. Using the inverse CDF method, we can generate

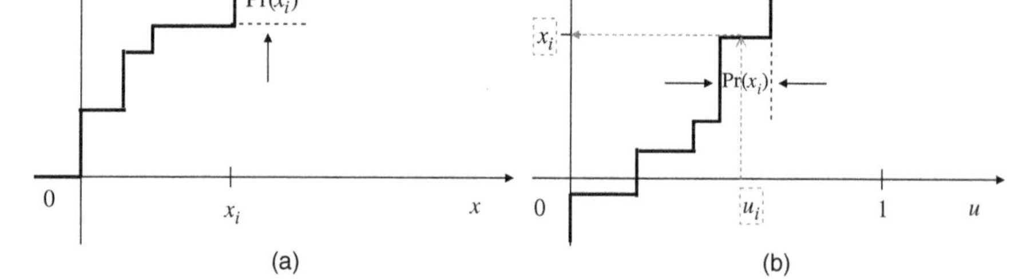

FIGURE 3.2 Empirical distribution: (a) estimated CDF; (b) estimated inverse cumulative distribution function (ICDF).

realizations of $X = x_i$ by

- Generate uniform samples: $u_k \sim \mathcal{U}(0, 1)$
- Simulate samples: $x_i = \mathrm{P}_X^{-1}(u)$ or

$$x_i = x_k \quad \text{for } \mathrm{P}_X(x_{k-1}) \leq u_k < \mathrm{P}_X(x_k)$$

or using the empirical distribution

$$x_i = x_k \quad \text{for } \sum_{j=1}^{k-1} W_j \leq u_k < \sum_{j=1}^{k} W_j$$

This transformation is illustrated in Fig. 3.2b. △△△

Summarizing, we first generate a uniform random variate u_i and then "bracket" its probability to determine the desired random variable x_i. So we see from these examples that using the inverse CDF method enables us to generate continuous and discrete random variates from both the analytic function as well as discrete samples from an empirical distribution. We will apply this technique frequently in the simulation-based processor. This approach is also used heavily in the resampling algorithms to follow (e.g., uniform, systematic). Numerical methods can also be used to perform the inversion by solving $\mathrm{P}_X(X) - U = 0$ for X [5, 15].

Before we close this section, let us note that the above theorems can be generalized to simulate any distribution from any other distribution using the uniform variates as an intermediate step leading to the following corollary.

Corollary Given a distribution $\mathrm{P}_X(X)$ and a corresponding target distribution $\mathrm{P}_Y(Y)$, there *exists* a unique, monotonic transformation $T(X)$, such that the random variable $Y = T(X)$ is distributed as $\mathrm{P}_Y(y)$, that is, if $T(X)$ is selected as

$$T(X) = Y = \mathrm{P}_Y^{-1}(\mathrm{P}_X(x))$$

then

$$\mathrm{P}_Y(Y \leq y) = \mathrm{P}_Y(y)$$

The proof of this corollary follows by applying the results of the previous theorems with $U = \mathrm{P}_X(x)$ and $Y = \mathrm{P}_Y^{-1}(U)$ for $U \sim \mathcal{U}(0, 1)$. Consider the following example to demonstrate this approach to simulation.

Example 3.5

Consider a random variable X distributed as $X \sim \mathrm{P}_X(x) = 1 - e^{-x} - x\,e^{-x}$. We would like to generate a random variable Y exponentially distributed with parameter λ, $Y \sim \mathrm{Exp}(\lambda)$. Then from the Corollary, we have

$$u = \mathrm{P}_Y(y) = 1 - e^{-\lambda y} \quad \text{and} \quad y = \mathrm{P}_Y^{-1}(U) = -\frac{1}{\lambda}\ln(1-u)$$

Thus, it follows that

$$\mathrm{P}_Y^{-1}(\mathrm{P}_X(x)) = -\frac{1}{\lambda}\ln(1 - \mathrm{P}_X(x)) = -\frac{1}{\lambda}\ln(e^{-x} + x\,e^{-x})$$

Therefore, we have

$$y_i = -\frac{1}{\lambda}\ln(e^{-x_i} + x_i\,e^{-x_i})$$

Sampling from the uniform (as before), we first generate $\{x_i\}$ and then the desired set of random variates $\{y_i\}$. △△△

We complete this subsection by stating the *Golden Rule of Sampling* [22] given by

- Generate uniform samples: $u_i \to \mathcal{U}(0,1); i = 1, \ldots, N$;
- Define the transformation: $u_i = \mathrm{P}_X(x_i)$; and
- Apply the inverse transformation: $x_i = \mathrm{P}_X^{-1}(u_i)$.

Next, we investigate a more general approach.

3.3.2 Rejection Sampling Method

In order to use the uniform sampling simulation methods of the previous section, we require the inverse CDF, which is not a simple task, even when the distribution is known in closed form. The rejection sampling method offers an alternative that not only eliminates the inversion problem, but also becomes an integral part of many of the sophisticated sampling algorithms to follow because of its simplicity and generality. In principle, the rejection method can be applied to *any* distribution with a density given up to a normalization constant [32].

The basic sampling problem in this context is that we are trying to generate samples from a density (or distribution) that is capable of being evaluated and we have a function

$$\mathrm{p}_X(x) = c \times \Pr(X)$$

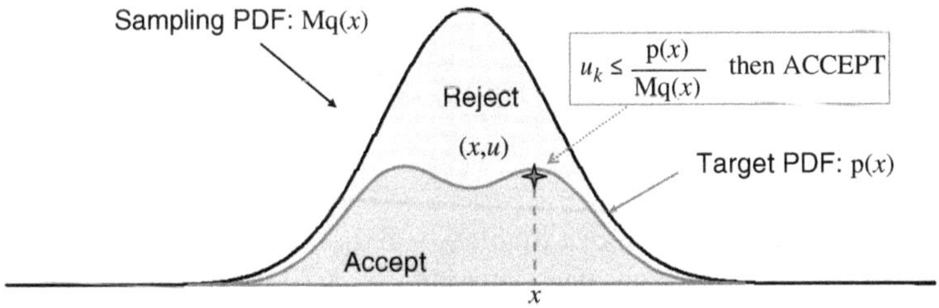

FIGURE 3.3 Rejection sampling method.

where Pr(X) is the target distribution and $p_X(x)$ is related and computable up to the normalization constant c which is *not* known. Suppose we select a sampling distribution, say q(x), such that there exists a " covering" constant M with

$$p_X(x) \leq Mq(x) \quad \forall x$$

The *rejection sampling method* is illustrated in Fig. 3.3 (c =1) and summarized as

- Generate a sample: $x_k \longrightarrow q(x)$
- Generate a uniform sample: $u_k \longrightarrow \mathcal{U}(0, 1)$
- ACCEPT the sample: $x_i = x_k$if $u_k \leq \frac{p_X(x)}{M \times q(x)}$
- Otherwise, REJECT the sample and generate the next trial sample: x_i

The proof of this method is given in Liu [32] using an *indicator* function $\mathcal{I}(x)$ defined by

$$\mathcal{I}(x) = \begin{cases} 1 & \text{if } x \text{ accepted} \\ 0 & \text{if } x \text{ rejected} \end{cases} \tag{3.21}$$

Proceeding

$$\Pr(\mathcal{I}(x) = 1) = \int \Pr(\mathcal{I}(x) = 1 | X = x)\, q(x)\, dx = \int \left(\frac{c\Pr(x)}{Mq(x)} \right) q(x) dx = \frac{c}{M}$$

and

$$\Pr(x | \mathcal{I}(x) = 1) = \frac{\left(\frac{c\Pr(x)}{Mq(x)} \right) q(x)}{\Pr(\mathcal{I}(x) = 1)} = \Pr(x)$$

which shows that the acceptance of the sample corresponds to sampling directly from the target distribution Pr(x).

The expected number of samples to accept an $x \sim \Pr(x)$ is M and therefore the key to using this approach is to select a good proposal distribution q(x) with a low M—a nontrivial problem. Note that Example 1.1 in which the area of a circle was estimated using sampling methods is a simple geometric illustration of this methodology. The following example from Papoulis [1] demonstrates the method in an analytic form.

Example 3.6

Suppose we are given a random variable $x \sim \text{Exp}(1)$ and we would like to simulate the random variable with truncated Gaussian $y \sim \mathcal{N}(0, 1)$, that is, we have

$$\text{p}(x) = \frac{2}{\sqrt{2\pi}} e^{\frac{-x^2}{2}} \times \mu(x) \quad \text{and} \quad \text{q}(x) = e^{-x}$$

where $\mu(x)$ is a unit step function as before, then for $x > 0$, we have

$$\frac{\text{p}(x)}{\text{q}(x)} = \sqrt{\frac{2e}{\pi}} e^{\frac{-(x-1)^2}{2}} \times e^1 \leq \sqrt{\frac{2e}{\pi}}$$

Setting $M = \sqrt{\frac{\pi}{2e}}$, we have the acceptance/rejection sampling rule:

$$x_i = x_k \quad \text{if } u_k < e^{-\frac{(x_k-1)^2}{2}} \quad \text{(Accept)}$$

△△△

This completes the background on sampling theory; next we consider the Monte Carlo approach.

3.4 MONTE CARLO APPROACH

The Monte Carlo approach to solving Bayesian estimation problems is to replace complex analytic or unknown probability distributions with sample-based representations to solve a variety of "unsolvable" problems in inference (integration, normalization, marginalization, expectation, etc.), optimization (parameter, maximum a-posteriori (MAP), maximum likelihood (ML) estimation), statistical mechanics (Boltzmann equation), and nuclear physics (fission, diffusion, etc.) [32, 35]. The key to the MC approach is to generate independent random samples from a probability distribution $\Pr(X)$ usually known only up to a normalizing constant (evidence) [32]. Typically, generating independent samples from this distribution is not feasible implying sample dependencies or using a proposal distribution $\text{q}(X)$, that is, similar to but not the exact target distribution $\Pr(X)$. *Independent* (*i.i.d.*) samples are important because both strong and weak Laws of Large Numbers (mean converges in distribution to population mean) ensure that the inferences (e.g., mean, variance) can be made as accurate as desired by increasing the number of samples. However, the samples can be *dependent* and still properly reflect the probability of the target distribution, opening the possibility of Markov chain methods (see Section 3.4.1).

The rejection method, just discussed, [22], importance sampling [16], and sampling-importance-resampling [26] are methodologies that do employ a proposal distribution. The Metropolis technique and its variants provide the foundation for this approach using Markov chain concepts generating dependent samples from a chain with $\Pr(X)$ as its invariant or stationary distribution. In this section, we develop the idea of the MC simulation-based approach to Bayesian estimation using *iterative* rather than *sequential* Monte Carlo techniques.

Theoretically, the Monte Carlo approach to sampling is based on the following principles [10, 27, 29, 30, 35]. Here N *i.i.d.* samples $\{X(i)\}_{i=1}^{N}$ are drawn from the *target density* $p(X)$ to produce an estimate of the *empirical distribution* (density)

$$\hat{\Pr}(X) = \hat{p}_N(X) = \frac{1}{N}\sum_{i=1}^{N} \delta(X - X(i)) \tag{3.22}$$

which can be used to approximate integrals (PDFs, areas, etc.) with sums, that is,

$$I_N(f) = \frac{1}{N}\sum_{i=1}^{N} f(X)\,\delta(X - X(i)) = \frac{1}{N}\sum_{i=1}^{N} f(X(i)) \underset{N\to\infty}{\overset{\text{a.s.}}{\longrightarrow}} I(f) = \int f(X)\,p(X)\,dX \tag{3.23}$$

Here $I_N(f)$ is *unbiased* and *converges* (almost surely) to $I(f)$ according to the *strong Law of Large Numbers*. Its corresponding *variance* is bounded ($\sigma_f^2 < \infty$) and

$$\text{var}(I_N(f)) = \frac{\sigma_f^2}{N} \tag{3.24}$$

A *central limit theorem* argument shows that the error converges in-distribution as

$$\sqrt{N}(I_N(f) - I(f)) \overset{N\to\infty}{\longrightarrow} \mathcal{N}(0, \sigma_f^2) \tag{3.25}$$

The main advantage of MC over deterministic integration is that it positions its samples over regions of *high probability*.

In signal processing, we are usually interested in some statistical measure of a random signal or parameter expressed in terms of its moments. Let us take a slightly more detailed look at just how MC concepts can be applied with this perspective in mind. Suppose we have some signal function, say $f(X)$, with respect to some underlying probabilistic distribution $\Pr(X)$. Then a typical measure to seek is its performance "on the average" which is characterized by the expectation

$$E_X\{f(X)\} = \int f(X)\Pr(X)\,dX \tag{3.26}$$

From the Bayesian perspective, the embedded distribution can be thought of as the "posterior" of the signals/parameters. The Bayesian approach must integrate over high-dimensional probability distributions to make inferences about parameters or predictions about signals. Unless the integral is analytically tractable, the usual method of evaluation is through numerical integration (deterministic) techniques. Unfortunately, the number of points to evaluate both $f(\cdot)$ and $\Pr(\cdot)$ increases exponentially with the dimensionality of the signal/parameter space. Also it is not possible to evaluate this integral over the entire space in practice; therefore, we concentrate on specific regions where the integrand is dense (not null). Instead of attempting to use numerical integration techniques, stochastic sampling techniques known as *Monte Carlo (MC) integration* have evolved as an alternative (see Fig. 1.2 for concept). As mentioned, the *key idea* embedded in the MC approach is to represent the required distribution as a set of random *samples* rather than a specific analytic function (e.g., Gaussian). As the number of samples becomes large, they provide an equivalent representation of the distribution enabling moments to be estimated directly. A functional estimate of the distribution could also be fit to the samples, if desired.

Integration has been used throughout statistics to evaluate probabilities and expectations. However, with Monte Carlo techniques the process is reversed and expectations are used to calculate integrals [1, 16, 28, 32]. Suppose we are asked to evaluate a multidimensional integral

$$I = \int_X g(x)\, dx \tag{3.27}$$

Then an MC approach would be to factorize the integrand as

$$g(x) \longrightarrow f(x)p(x) \ni p(x) \geq 0 \quad \text{and} \quad \int p(x)\, dx = 1$$

where $p(x)$ is interpreted as a probability distribution in which samples can be drawn. This is the foundation of sampling techniques based on MC integration. Monte Carlo approaches draw samples from the required distribution and then form sample averages to approximate the sought after distributions, that is, they map integrals to discrete sums. Thus, MC integration evaluates Eq. 3.26 by drawing samples $\{X(i)\}$ from $\Pr(X)$. Assuming perfect sampling, this produces the estimated or *empirical distribution* given by

$$\hat{\Pr}(X) \approx \frac{1}{N}\sum_{i=1}^{N} \delta(X - X(i)) \stackrel{N\to\infty}{\longrightarrow} \mathcal{N}(0, \sigma_f^2)$$

which is a probability distribution of mass or weights $\frac{1}{N}$ and random variable or location $X(i)$. Substituting the empirical distribution into the integral gives

$$E_X\{f(X)\} = \int f(X)\hat{\Pr}(X)\, dX \approx \frac{1}{N}\sum_{i=1}^{N} f(X(i)) \equiv \bar{f} \tag{3.28}$$

which follows directly from the sifting property of the delta function. Here $\bar{f}$ is said to be a *Monte Carlo estimate* of $E_X\{f(X)\}$. Clearly, it is *unbiased*, since

$$E\{\bar{f}\} = \frac{1}{N}\sum_{i=1}^{N} E\{f(X(i))\} = E_X\{f(X)\}$$

with *variance* given by

$$\mathrm{var}(\bar{f}) = \frac{1}{N}\int [f(X) - E_X\{f(X)\}]^2 \Pr(X)\, dX$$

Additionally, if the variance is *finite*, then the *central limit theorem* holds and the error in estimation converges to a zero-mean Gaussian with $\mathcal{N}(0, \mathrm{var}(\bar{f}))$. Consider the following examples to illustrate these concepts. First we discuss an analytic case, then a numerical case to solidify both MC approaches.

Example 3.7

Suppose we would like to solve for the integral

$$I = \int_0^1 f(x)\, dx$$

using the MC approach. Let u be defined as a random variate drawn from a uniform distribution $u \sim \mathcal{U}(0, 1)$, then we can express the integral in terms of an expectation as

$$I = E\{f(u)\} = \int_0^1 f(u) \Pr(u)\, du$$

Generating a set of *i.i.d.* uniform samples $u_1, \ldots, u_N$, the corresponding set of functions $\{f(u_i)\}$ are also *i.i.d.* with mean I as defined above. Therefore, the uniform sampling distribution is given by (equally likely)

$$\Pr(u = u_i) = \frac{1}{N}\sum_{i=1}^{N} \delta(u - u_i) \quad \text{for } 0 \le u_i \le 1$$

Substituting this expression into the expectation relation, we obtain

$$I = E\{f(u)\} = \int_0^1 f(u) \left[\frac{1}{N}\sum_{i=1}^{N} \delta(u - u_i)\right] du = \frac{1}{N}\sum_{i=1}^{N} f(u_i)$$

From the *Law of Large Numbers*[3] as $N \rightarrow \infty$, it follows that

$$\sum_{i=1}^{N} f(u_i) \longrightarrow E\{f(u)\} = I$$

Therefore, we can approximate the integral by generating a large number of random variables drawn from a uniform sampling distribution, transform them according to some functional relationship, and calculate the sample mean to approximate the desired integral. △△△

This simple example illustrates the basic MC concept that will be applied throughout this text. Consider another example that illustrates this approach further by developing a simulation-based solution to a familiar statistical estimation problem.

Example 3.8

Suppose we have a Gaussian random variable and we would like to estimate its mean and variance. Knowledge that it is Gaussian enables us to write the closed form expression for the distribution

$$x \sim \mathcal{N}(m_x, \sigma_x^2) \quad \text{or} \quad \Pr(x) = \frac{1}{\sqrt{2\pi\sigma_x^2}} \exp\left\{ \frac{-(x - m_x)^2}{2\sigma_x^2} \right\}$$

and the mean and the variance can be calculated analytically as

$$m_x = \int x \Pr(x)\, dx$$

$$\sigma_x^2 = \int (x - m_x)^2 \Pr(x) dx$$

In contrast, the Monte Carlo approach is to generate N samples from a Gaussian. Assuming perfect sampling $x_i \longrightarrow \mathcal{N}(m_x, \sigma_x^2)$ and we have that

$$\hat{\Pr}(x) \approx \frac{1}{N} \sum_{i=1}^{N} \delta(x - x_i)$$

Now the mean and variance can be estimated from the samples directly using

$$\hat{m}_x = \int x \hat{\Pr}(x) dx = \int x \left(\frac{1}{N} \sum_{i=1}^{N} \delta(x - x_i) \right) dx = \frac{1}{N} \sum_{i=1}^{N} x_i$$

[3] The Law of Large Numbers states simply that the mean converges with probability one to the population mean as the number of samples become large.

with variance

$$\hat{\sigma}_x^2 = \int (x - m_x)^2 \,\hat{\Pr}(x)\, dx = \int (x - m_x)^2 \left(\frac{1}{N} \sum_{i=1}^{N} \delta(x - x_i) \right) dx = \frac{1}{N} \sum_{i=1}^{N} (x_i - \hat{m}_x)^2$$

Summarizing the MC approach for this example, we must

- Generate N samples from a Gaussian: $x_i \longrightarrow \mathcal{N}(m_x, \sigma_x^2)$;
- Estimate the desired statistics of the distribution from its samples as $\hat{m}_x$ and $\hat{\sigma}_x^2$.

We performed a simulation in MATLAB to generate Gaussian variates with $m_x = 2$ and $\sigma_x^2 = 4$. The results are shown in Fig. 3.4. In Fig. 3.4a, we see a simulation for $N = 1000$ samples (+) with corresponding estimated mean (solid line) and upper and lower 95%-confidence limits about the mean ($m_x \pm 1.96\sigma_x$). The sample mean and variance for this MC realization were at $\hat{m}_x = 1.97$ and $\hat{\sigma}_x^2 = 4.01$. The distribution was estimated using a histogram with 100 bins and is shown in Fig. 3.4b with a near-perfect MC solution using $N = 10^6$ samples shown in the inset. △△△

So we see from this example how MC methods can be used to approximate (estimate) distributions and their associated statistics from simulated samples.

These methods are acceptable as long as high accuracy is not required. Monte Carlo techniques tend to perform a "divide and conquer" approach to integration by breaking the integral up into distinct regions around the integrand consisting of strong local peaks at known locations. They are typically very time consuming methods, of the order of tens and hundreds of thousands of points, and have only recently gained notable prominence in the signal processing literature due to the major advances in fast computers [28]. Next, let us consider the extension of Monte Carlo techniques using Markov chains.

3.4.1 Markov Chains

In MC integration, the population mean of $f(X)$ is estimated by the sample mean. When the samples $\{X_i(t) = \mathcal{X}_i\}$ are independent, the *Law of Large Numbers* ensures that the approximation can be made as accurately as required by increasing the number of samples N. Generally, however, drawing samples from $\Pr(X)$ is not feasible especially when it is a nonstandard distribution. However, dependent samples can be generated by any process that draws samples throughout the support (range) of the distribution. One efficient technique to accomplish this sampling is through a Markov chain having $\Pr(X)$ as its unique stationary or *invariant distribution*—this methodology is termed *Markov chain Monte Carlo (MCMC)*. This technique is basically MC integration where the random samples are produced using a Markov chain. Recall that a *Markov chain* is a discrete random process possessing the property

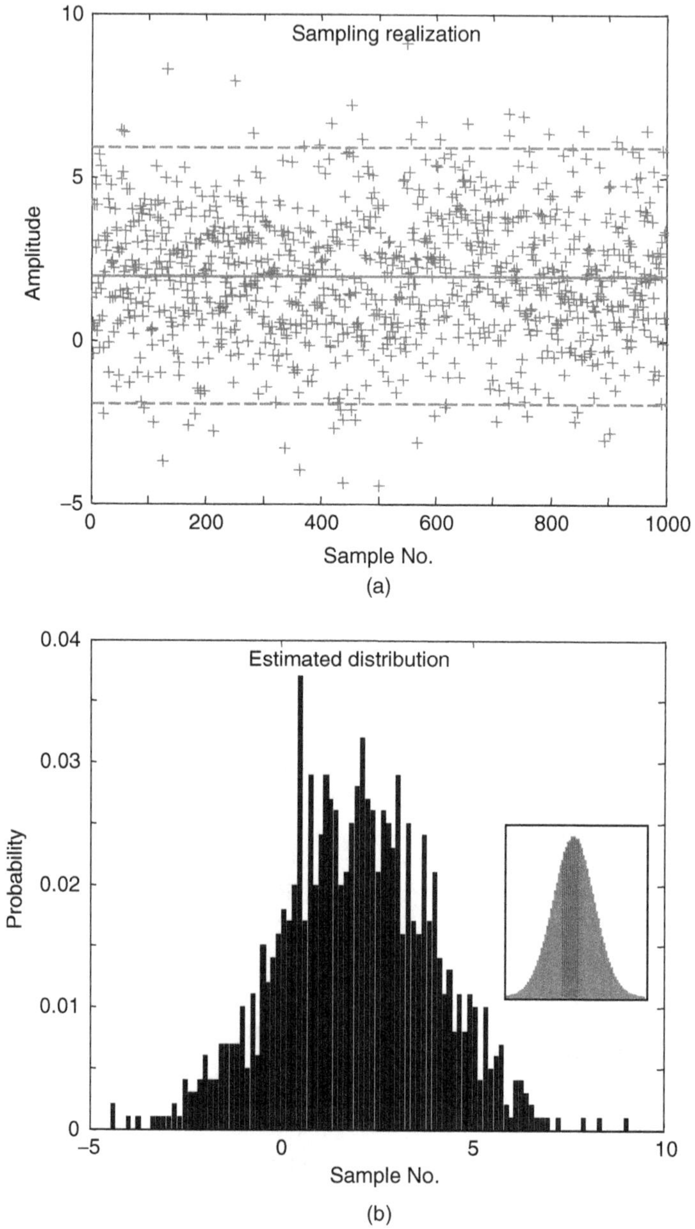

FIGURE 3.4 Monte Carlo approach for estimation of the statistics of a Gaussian random process: (a) simulation ($m_X = 2, \sigma_X^2 = 4, N = 1000$) with 95%-confidence limits (dashed line) about $\hat{m}_X$ (solid line); (b) estimated distribution (histogram) from samples. Inset is estimate for $N = 1 \times 10^6$ samples.

that the conditional distribution at the present sample (given all of the past samples) depends *only* on the previous sample (first order), that is,

$$\Pr(X_i(t)|X_j(t-1),\ldots,X_k(0)) = \Pr(X_i(t)|X_j(t-1))$$

Summarizing, the Markov chain dynamics are represented by the *transition probability* $\mathcal{P}_{ij}(t-1,t) := \Pr(X(t) = \mathcal{X}_i|X(t-1) = \mathcal{X}_j)$ denoting the probability that the state at time t will be $\mathcal{X}_i$, given that it is currently in state $\mathcal{X}_j$ at time $t-1$. Further, if the chain is also *homogeneous* in *time*, then $\mathcal{P}_{ij}(t-1,t)$ depends only on the time difference (in general) and therefore the transition probability is *stationary* such that $\mathcal{P}_{ij}(t-1,t) \rightarrow \mathcal{P}_{ij}$ with $\mathcal{P}_{ij} \geq 0$ and $\sum_{i=1}^{N_x} \mathcal{P}_{ij} = 1$ [1].

Thus, the basic requirement in Monte Carlo techniques is to generate random samples from a probability distribution or target distribution only known up to a normalizing constant. Typically, it is not possible to generate the samples from the target but generating from a known trial distribution that is similar to the target distribution, just as in rejection sampling, is applied. In order to understand the MCMC approach, we must first define critical properties of the Markov chain.

Markov chains possess certain crucial properties that must exist for them to be useful in MC simulations [32, 33]. A Markov chain begins with an initial distribution $\Pr(X_i(0))$ and evolves to another indexed variable $X_i(t)$ determined by a transition kernel, that is, at index t we have

$$\Pr(X_i(t)) = \sum_j \overbrace{\Pr(X_i(t)|X_j(t-1))}^{\text{Transition kernel}} \times \Pr(X_j(t-1)) \tag{3.29}$$

An *invariant distribution* is a fixed point solution to Eq. 3.29. For distribution estimation the constraint is even more stringent, since we require *time reversible* chains which must satisfy a *detailed balance* as (ignoring the index)

$$\Pr(X_i(t)) \times \Pr(X_i(t)|X_j(t)) = \Pr(X_j(t)) \times \Pr(X_j(t)|X_i(t)) \quad \forall X_i(t),\, X_j(t) \tag{3.30}$$

which means that the transition probability from $X_i(t)$ to $X_j(t)$ is identical to the probability from $X_j(t)$ to $X_i(t)$ implying invariance [28, 32]. The chain is also required to be *ergodic* which means that regardless of the initial distribution, the probability at t *converges* to the invariant distribution as $t \rightarrow \infty$, that is,

$$\lim_{t\rightarrow\infty} \Pr(X_i(t)) \longrightarrow \Pr(X_i(t)) \tag{3.31}$$

Thus, MCMC method for the simulation of a distribution is any technique producing an ergodic or reversible chain with invariant distribution being the desired target distribution. Armed with this information we can now discuss the most powerful and efficient MCMC methods: the Metropolis–Hastings and Gibbs sampler and their variants.

3.4.2 Metropolis–Hastings Sampling

Markov chain simulation is essentially a general technique based on generating samples from proposal distributions and then correcting (acceptance or rejection) those samples to approximate a target posterior distribution. Here we must *know* both the target $\mathrm{p}_X(x)$ (up to a normalizing constant) and the proposal $\mathrm{q}(x)$, a priori. The samples are sequentially generated forming a Markov chain with properties defined in the previous section. Typically, in Markov chain simulation, samples are generated from the transition kernel or distribution. The *key*, however, is not really the chain itself, but the fact that the approximate distribution improves sequentially as it converges to the target posterior.

In this subsection, we discuss the basic Metropolis–Hastings sampling method. We start with the original Metropolis algorithm [24] and then introduce the Hastings generalization [23]. The fundamental idea is similar to the rejection method discussed previously. The Metropolis–Hastings (M-H) technique defines a Markov chain such that a new sample x_i is generated from the previous sample x_{i-1} by first drawing a "candidate" sample $\hat{x}_i$ from a proposal distribution $\mathrm{q}(x)$ and then making a decision whether this candidate should be accepted and retained, or rejected and discarded using the previous sample as the new. If accepted, $\hat{x}_i$ replaces x_i ($\hat{x}_i \rightarrow x_i$) otherwise the old sample x_{i-1} is saved ($x_{i-1} \rightarrow x_i$). This is the heart of the M-H approach in its simplest form.

We start with the basic *Metropolis* technique to describe the method

- Initialize: $x_o \rightarrow \mathrm{p}_X(x_o)$
- Generate a candidate sample from proposal: $\hat{x}_i \rightarrow \mathrm{q}(x)$
- Calculate the acceptance probability: $\mathcal{A}(x_{i-1}, \hat{x}_i) = \min\left\{\frac{\mathrm{p}_X(\hat{x}_i)}{\mathrm{p}_X(x_{i-1})}, 1\right\}$
- ACCEPT candidate sample with probability, $\mathcal{A}(x_{i-1}, \hat{x}_i)$ according to

$$x_i = \begin{cases} \hat{x}_i & \text{if } \mathrm{p}_X(\hat{x}_i) > \mathrm{p}_X(x_{i-1}) \\ x_{i-1} & \text{otherwise} \end{cases} \tag{3.32}$$

We see from this technique that when the candidate sample probability is greater than the previous sample's probability, it is accepted with probability

$$\mathrm{p}_X(x_i) = \begin{cases} \mathcal{A}(x_i, \hat{x}_i) & \text{if accepted} \\ 1 - \mathcal{A}(x_i, \hat{x}_i) & \text{if rejected} \end{cases} \tag{3.33}$$

The idea is that we can detect when the chain has converged to its invariant distribution (posterior), when $\mathrm{p}_X(x_i) = \mathrm{p}_X(x_{i-1})$. It is clear from this discussion that for the chain to converge to the posterior it must be *reversible* and therefore $\mathrm{q}(x)$ must be a symmetric distribution. This was the original Metropolis assumption. Hastings [23] generalized this technique by removing the symmetry constraint enabling asymmetric proposals. The basic algorithm remains the same except that the acceptance probability becomes ($\mathrm{p}_X \rightarrow \mathrm{p}$)

$$\mathcal{A}(x_i, \hat{x}_i) = \min\left\{\frac{\mathrm{p}(\hat{x}_i) \times \mathrm{q}(x_i|\hat{x}_i)}{\mathrm{p}(x_i) \times \mathrm{q}(\hat{x}_i|x_i)}, 1\right\} \tag{3.34}$$

This process continues until the desired N samples of the Markov chain have been generated. The critical step required to show that the M-H converges to the invariant distribution or equivalently the posterior $\mathrm{p}_X(x)$ evolves directly from the detailed balance of Eq. 3.30 given by

$$\mathrm{p}(x_{i+1}|x_i) \times \mathrm{p}(x_i) = \mathrm{p}(x_i|x_{i+1}) \times \mathrm{p}(x_{i+1}) \tag{3.35}$$

where $\mathrm{p}(x_{i+1}|x_i)$ is the Markov chain transition probability. If we assume that the ith sample was generated from the posterior $x_i \sim \mathrm{p}_X(x)$ then it is also assumed that the chain has converged and all subsequent samples have the same posterior. Thus, we must show that the next sample x_{i+1} is also distributed $\mathrm{p}(x_i) = \mathrm{p}_X(x)$. Starting with the detailed balance definition above and integrating (summing) both sides with respect to x_i, it follows that

$$\begin{aligned} \int \mathrm{p}(x_{i+1}|x_i)\,\mathrm{p}(x_i)\,dx_i &= \int \mathrm{p}(x_i|x_{i+1})\,\mathrm{p}(x_{i+1})\,dx_i \\ &= \mathrm{p}(x_{i+1}) \int \mathrm{p}(x_i|x_{i+1})\,dx_i = \mathrm{p}(x_{i+1}) \end{aligned} \tag{3.36}$$

which shows that the relation on the left side of Eq. 3.36 gives the marginal distribution of x_{i+1} assuming x_i is from $\mathrm{p}(x_i)$. This implies that if the assumption that x_i is from $\mathrm{p}(x_i)$ is true, then x_{i+1} must also be from $\mathrm{p}(x_i)$ and therefore $\mathrm{p}(x_i) \rightarrow \mathrm{p}_X(x)$ is the invariant distribution of the chain. Thus, once a sample is obtained from the invariant distribution, all subsequent samples will be from that distribution as well, proving that the invariant distribution is $\mathrm{p}(x_i)$. A full proof of the M-H technique requires a proof that $\mathrm{p}(x_i|x_0)$ will converge to the invariant distribution (see Ref. [19] for details).

For the M-H technique, the transition from $x_{i+1} \neq x_i$ to x_i occurs with probability

$$\mathrm{p}(x_{i+1}|x_i) = \mathrm{q}(x_{i+1}|x_i) \times \min\left\{ \frac{\mathrm{p}(\hat{x}_i) \times \mathrm{q}(x_i|\hat{x}_i)}{\mathrm{p}(x_i) \times \mathrm{q}(\hat{x}_{i+1}|x_i)}, 1 \right\} = \mathcal{A}(x_i, x_{i+1}) \times \mathrm{q}(x_{i+1}|x_i) \tag{3.37}$$

This completes the basic M-H theory.

3.4.3 Random Walk Metropolis-Hastings Sampling

Next, we consider another version of the M-H technique based on a random walk search through the parameter space. The idea is to perturb the current sample x_i with an addition of a random error, that is,

$$x_{i+1} = x_i + \epsilon_i \quad \text{for } \epsilon_i \sim \mathrm{p}_E(\epsilon) \tag{3.38}$$

where ϵ is *i.i.d.* A reasonable choice for this distribution is a symmetric Gaussian, that is, $\mathrm{p}_E(\epsilon) \sim \mathcal{N}(0, \sigma_\epsilon^2)$. Thus, the *random walk* M-H method is:

Given the current sample x_i

- Draw a random sample: $\epsilon \rightarrow \mathrm{p}_E(\epsilon)$
- Generate the candidate sample: $\hat{x}_i = x_i + \epsilon_i$
- Draw a uniform random sample: $u_i \rightarrow \mathcal{U}(0, 1)$
- Calculate the acceptance probability from the *known* densities: $\mathcal{A}(x_i, \hat{x}_i)$
- Update the sample:

$$x_{i+1} = \begin{cases} \hat{x}_i & \text{if } u_i < \mathcal{A}(x_i, \hat{x}_i) \\ x_i & \text{otherwise} \end{cases} \tag{3.39}$$

- Select the next sample

With this algorithm, we must use both the (known) proposal and target distributions to calculate the acceptance probability and then generate samples (random walk) from the proposal. It is important to realize that a "good" proposal distribution can assure generating samples from the desired target distribution, but the samples must still be generated to "cover" its range. This is illustrated in Fig. 3.5 where our target distribution is a Gaussian mixture with mixing coefficients as $(0.3, \mathcal{N}(0, 2.5); 0.5, \mathcal{N}(5, 1); 0.2, \mathcal{N}(10, 2.5))$. In the figure, we see the results from

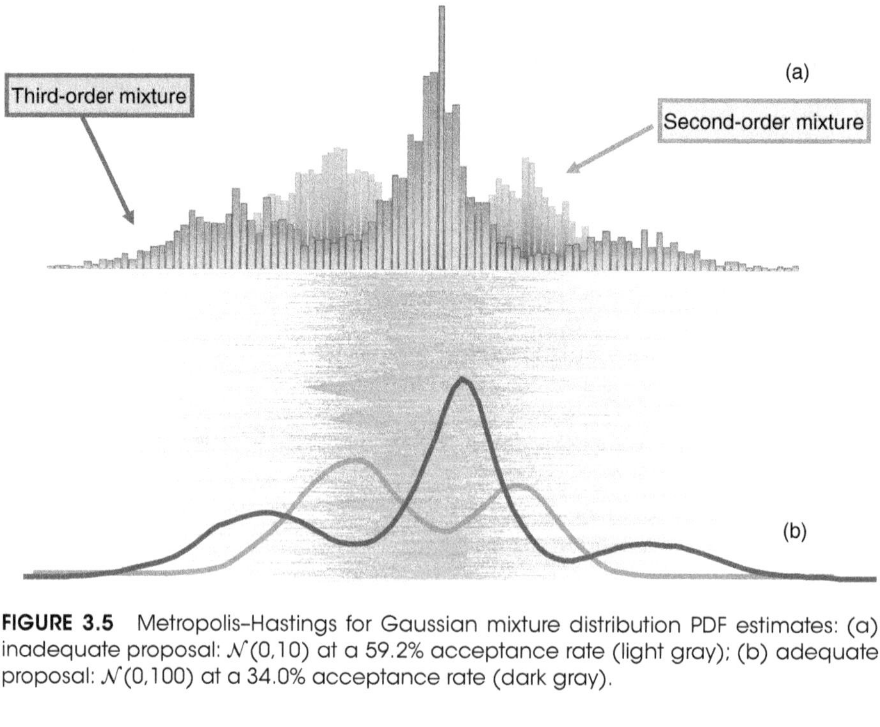

FIGURE 3.5 Metropolis–Hastings for Gaussian mixture distribution PDF estimates: (a) inadequate proposal: $\mathcal{N}(0,10)$ at a 59.2% acceptance rate (light gray); (b) adequate proposal: $\mathcal{N}(0,100)$ at a 34.0% acceptance rate (dark gray).

choosing a reasonable proposal ($\mathcal{N}(0, 100)$) in the dark color generating enough samples to cover the range of the target distribution and a proposal ($\mathcal{N}(0, 10)$) that does not sample the entire space adequately leading to an erroneous target distribution characterizing the sampler output. Consider the following example of applying a variety of proposals and M-H techniques.

Example 3.9

Suppose we would like to generate samples from a unit Gaussian distribution $\mathcal{N}(0, 1)$ using the various M-H simulation techniques: (i) symmetric proposal using $\mathcal{U}(-10, 10)$; (ii) symmetric proposal using a Student T distribution $\mathcal{T}(1)$; (iii) random walk using uniform noise $\mathcal{U}(-5, 5)$; and (iv) uniform proposal using $\mathcal{U}(-4, 9)$. To start we specify the "target" distribution as $\mathrm{p}_X(x) \sim \mathcal{N}(0, 1)$ and we choose the following proposals

- Case (i): $\mathrm{q}_1(x) \sim \mathcal{U}(-10, 10) \Rightarrow \frac{1}{20}$ for $-10 < x < 10$
- Case (ii): $\mathrm{q}_2(x) \sim \mathcal{T}(1)$
- Case (iii): $x_i = x_{i-1} + u_i$ where $u_i \sim \mathcal{U}(-5, 5) \Rightarrow \frac{1}{10}$ for $-5 < u_i < 5$
- Case (iv): $\mathrm{q}_4(x) \sim \mathcal{U}(-4, 9) \Rightarrow \frac{1}{13}$ for $-4 < x < 9$

To implement the Metropolis, Metropolis–Hastings, random walk Metropolis–Hastings techniques we must

1. Draw a random sample from the proposal: $x_i \rightarrow \mathrm{q}_i(x)$
2. Calculate the acceptance ratio: $\mathcal{A}(x_i, x_{i-1})$
3. Draw a uniform sample: $u_i \rightarrow \mathcal{U}(0, 1)$
4. Accept or reject sample: $u_i \leq \mathcal{A}(x_i, x_{i-1})$
5. Update sample: x_i
6. Generate next sample: x_{i+1}

The results of these 10^5 sample simulations are shown in Figs. 3.6a–d using the M-H sampler in MATLAB. We see the results of using the various proposals. All of the estimated densities give reasonably close estimates of the target posterior $\mathcal{N}(0, 1)$. We estimated the corresponding posterior distribution from the samples using both histogram and kernel density estimators of Section 3.2 with all of the results reasonable. The standard deviation estimates were very close to unity in all cases; however, the means differed slightly from zero. It is interesting to note the corresponding acceptance rates (see Fig. 3.5 caption) with the most Gaussian-like, $\mathcal{T}$ proposal distribution had the highest acceptance rate of 57.8%. The true target distribution is superimposed for comparison purposes. *Acceptance rates* provide an indication of how "probable" a sample from the proposal is accepted as a sample in the target (posterior). When the proposal provides good coverage of the target

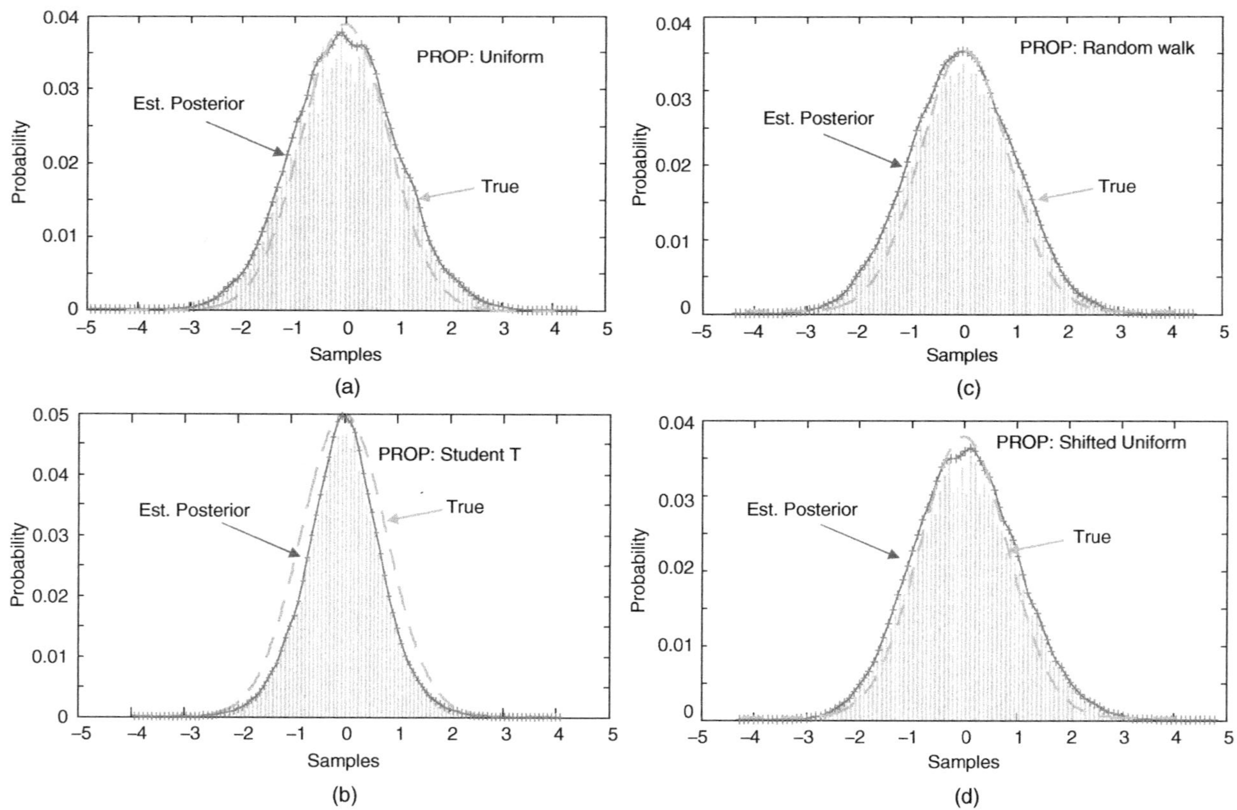

FIGURE 3.6 Metropolis–Hastings sampler posterior distribution estimates (10^5 samples): (a) M-H symmetric proposal: $\mathcal{U}(-10, 10)$ with mean and standard deviation estimates (−0.02, 0.998) at a 16.2% acceptance rate; (b) M-H symmetric proposal: $\mathcal{T}(1)$ with mean and standard deviation estimates (−0.004, 0.72) at a 57.8% acceptance rate; (c) M-H random walk: $\mathcal{U}(-5, 5)$ with mean and standard deviation estimates (−0.005, 1.006) at a 31.7% acceptance rate; (d) Uniform proposal: $\mathcal{U}(-4, 9)$ with mean and standard deviation estimates (−0.03, 1.004) at a 24.2% acceptance rate.

TABLE 3.1 Metropolis-Hastings Sampling Algorithm

$x_o \rightarrow p_X(x_o)$	*Initialize*	[Draw sample]
$\hat{x}_i \rightarrow q(x)$	*Proposal*	[Draw sample]
	Acceptance Probability	
$\mathcal{A}(x_i, \hat{x}_i) = \min\left\{ \frac{p(\hat{x}_i) \times q(x_i \mid \hat{x}_i)}{p(x_i) \times q(\hat{x}_i \mid x_i)}, 1 \right\}$		
$u_k \longrightarrow \mathcal{U}(0, 1)$	*Uniform Sample*	[Draw sample]
	Decision	
$x_i = \begin{cases} \hat{x}_i & \text{if } u_k < \mathcal{A}(x_i, \hat{x}_i) \\ x_{i-1} & \text{otherwise} \end{cases}$		
$x_{i+1} \longrightarrow x_i$	*Next Sample*	[Draw sample]

distribution, then many samples are accepted at a high rate, if not the rate is low. Clearly, the M-H sampling technique and its variants provide a very robust method for generating samples from a target distribution especially when the proposal covers the entire range of the target sample space. △△△

There are a variety of M-H sampling techniques such as the independence sampler, the hybrid or dynamic (Hamiltonian) sampler, the multipoint samplers, etc. [32]. Also note that many of these methods are available in freeware (e.g. language-based BUGS [38] and the MATLAB-based NETLAB [39], *PRTools* [40]). We summarize the M-H sampling technique in Table 3.1. Next, we discuss a popular special case of this approach—the Gibbs sampler.

3.4.4 Gibbs Sampling

The Gibbs simulation-based sampler (G-S), one of the most flexible of the sampling techniques available. It is a special case of the Metropolis–Hastings approach in which the acceptance probability $\mathcal{A}(x_i, \hat{x}_i)$ is unity, that is, *all* samples are accepted [27]. Theoretically, the G-S is based on the fact that the targeted joint posterior distribution can be determined completely by a set of underlying conditional distributions evolving directly from Bayes' rule (joint $=$ conditional$\times$marginal) [29]. It falls into the class of sampling algorithms termed *block at-a-time* or *variable-at-a-time* methods [23]. Proof of these methods have a significant practical implication, since it can be shown that the product of the transition distribution of the Markov chain is a product of conditional transitions which converge to joint posterior as its invariant distribution [30]. It is for this reason that the G-S is called "sampling-by-conditioning" [28], which will become obvious after we investigate its underlying structure. As before, it should be realized that both target and proposal distributions must be known (approximately) and samples must be easily generated from the proposal to be effective.

Gibbs sampling can be considered an implementation of the M-H technique on a component-by-component basis of a random vector. It is more restrictive than the M-H method, but can be more efficient leading to a simpler implementation. The G-S is especially important in Bayesian problems, since it is uniquely designed for multivariate problems, that is, it replaces sampling from a high-dimensional vector with sampling from low-order component blocks [67]. It can be considered a concatenation of M-H samplers, one for each component variable of the random vector. This decomposition has individual target distributions representing a conditional density or mass for *each* component given values for all of the other variables. Thus, the proposal for the component variable of the vector is the conditional density of that variable given the most recent values for all of the others.

More formally, suppose the random vector $\mathbf{X} \in \mathcal{R}^{N_x \times 1}$ is decomposed into its components X_k for $k = 1, \ldots, N_x$. Therefore, the idea is to generate, say $X_1(i)$, based on the conditional distribution $\Pr(X_1|X_2(i-1), \ldots, X_{N_x}(i-1), Y)$ and the next sample drawn $X_2(i)$ uses it and the samples available from the previous iteration to sample from $\Pr(X_2|X_1(i) \cup X_3(i-1), \ldots, X_{N_x}(i-1), Y)$ and so forth so that at the ith iteration, we have the kth component sample generated from

$$X_k(i) \longrightarrow \Pr(X_k|\{X_n(i)\} \cup \{X_m(i-1)\} \ni m > k; n < k, Y) \tag{3.40}$$

If we expand this relation, then we observe the underlying structure of the Gibbs sampler.

Given the sample set $\{\mathbf{X}(i-1)\}$ then

- Generate the first sample: $X_1 \rightarrow \Pr(X_1|X_2(i-1), \ldots, X_{N_x}(i-1), Y)$ and then
- Generate the next sample: $X_2 \rightarrow \Pr(X_2|X_1(i) \cup X_3(i-1), \ldots, X_{N_x}(i-1), Y)$
- Generate the kth sample: $X_k \rightarrow \Pr(X_k|\{X_{n-k}, \ldots, X_n(i)\}, \{X_{N_{m+k}}(i-1), \ldots, X_{m+k-N_x}\} \ni m > k; n < k, Y)$

So we see that the vector is decomposed component wise and the corresponding conditional distributions evolve creating the vector sequence of iterates which are the realization of a Markov chain with transition distribution. We assume that we would like to go from $\mathbf{X}' \rightarrow \mathbf{X}$ giving the transition probability

$$\begin{aligned}\Pr(\mathbf{X}', \mathbf{X}) = \Pr(X_1|X_2', \ldots, X_{N_x}', Y) \times \Pr(X_2|X_1, X_3', \ldots, X_{N_x}', Y) \times \cdots \\ \times \Pr(X_{N_x}|X_1, \ldots, |X_{N_x-1}, \ldots, X_{N_x}', Y)\end{aligned} \tag{3.41}$$

The G-S can be shown to satisfy the detailed balance. As a result it converges to the invariant distribution of the Markov chain which in this case is the joint posterior distribution [28, 32]. Consider the following example from [20].

Example 3.10

Suppose we have a measurement vector **y** from a bivariate Gaussian with unknown mean and known covariance, that is,

$$\Pr(\mathbf{X}|\mathbf{Y}) = \mathcal{N}(\mathbf{Y}, \mathbf{R}_{xx}) \quad \text{for } \mathbf{R}_{xx} = \begin{bmatrix} 1 & \rho \\ \rho & 1 \end{bmatrix} \tag{3.42}$$

To apply the G-S to **X**, we require the conditional posterior from the well-known properties of the bivariate Gaussian [1] given by

$$\Pr(X_1|X_2, \mathbf{Y}) \sim \mathcal{N}(y_1 + \rho(x_2 - y_2), 1 - \rho^2)$$
$$\Pr(X_2|X_1, \mathbf{Y}) \sim \mathcal{N}(y_2 + \rho(x_1 - y_1), 1 - \rho^2)$$

Thus, the G-S proceeds by alternating samples from these Gaussian distributions, given $(x_1(0), x_2(0))$

Let $i = 1, 2, \ldots$

- Draw $x_1(i) \rightarrow \Pr(x_1|x_2(i-1), \mathbf{Y})$
- Draw $x_2(i) \rightarrow \Pr(x_2|x_1(i), \mathbf{Y})$

So we generate the pairs: $(x_1(1), x_2(1)), (x_1(2), x_2(2)), \ldots, (x_1(i), x_2(i))$ from the sampling distribution converging to the joint bivariate Gaussian $\Pr(\mathbf{X})$ as the invariant distribution of the Markov chain. △△△

Next, we consider a generalized variant of the Gibbs's sampler—the slice sampler.

3.4.5 Slice Sampling

Slice sampling (S-S) is an MCMC sampling technique based on the premise of sampling uniformly from the region under the target distribution $\Pr(X)$ [19, 31, 32, 36, 37]. Therefore, like all of the previous sampling methods, it can be applied to any problem for which the target distribution can be evaluated at a point, say x. It has an advantage over Metropolis methods, being more robust to step-size variations especially since it performs adaptive adjustments. Slice sampling (S-S) is a generalized case of the Gibbs sampler based on iterative simulations of the *uniform* distribution consisting of one-dimensional transitions. It is also similar to rejection sampling, in that it draws samples from under the target distribution. However, in all of these cases the S-S is not bound by their restriction [37]. It is based on the following proposition proven in Cappe [36]

Proposition: Let $x \sim \Pr(X)$ and $u \sim \mathcal{U}(0, M)$, then the pair of random variables (x, u) are distributed uniformly as $(x, u) \sim \mathcal{U}(0, M \times \Pr(X))$. Conversely, if (x, u) is uniformly distributed, then x admits $\Pr(X)$ as its marginal distribution.

In its simplest form, the slice sampler technique generates uniform intervals that capture the samples

- Initialize: x_{i-1}
- Draw uniform samples: $u_i \longrightarrow \mathcal{U}(0, \Pr(x_{i-1}))$
- Draw uniform samples: $x_i \longrightarrow \mathcal{U}(0, S(u_i))$ where $S(u_i) = \{x : \Pr(x_i) \geq u_i\}$

The actual implementation of the algorithm is much more complex and we refer the interested reader to Neal [31] or MacKay [37] for more details. We merely state important features of the slice sampler technique.

The S-S involves establishing intervals to ensure the sample points of the target distribution are included by using an adaptive step size (interval size) applying two techniques: (1) step-out technique; and (2) shrinking technique. The step-out technique is used to increase the size of the interval until the new sample x_i is included, while the shrinking technique does the opposite, it decreases the interval size to assure the original sample x_{i-1} is included. Consider the following example from Ref. [19] to illustrate the S-S.

Example 3.11

Suppose we would like to generate samples from a unit Gaussian distribution $x \sim \mathcal{N}(0, 1)$ using the slice sampling technique

- Initialize: $x_{i-1} = 0$ (Random draw)
- Draw uniform samples: $u_i \longrightarrow \mathcal{U}(0, \Pr(x_{i-1})) = \mathcal{U}\left(0, \frac{1}{\sqrt{2\pi}} e^{-x^2/2}\right)$
- Draw uniform samples: $x_i \longrightarrow \mathcal{U}(-\alpha, \alpha)$ where $\alpha = \sqrt{-\ln \sqrt{2\pi} u_i}$

Simulations can now be performed and analyzed. △△△

We conclude this section with a signal processing example. We are given an autoregressive (all-pole) model and we would like to generate samples from the corresponding posterior distribution.

Example 3.12

We have the following all-pole ($AR(1)$) model

$$x(t) = -ax(t-1) + w(t-1) \quad \text{for } w \sim \mathcal{N}(0, R_{ww})$$

Suppose we choose the following parameters for the model: $N = 10^5$ samples, $a = 0.1, R_{ww} = 0.1$ and we would like to develop samples from the posterior. Analytically, we know that the posterior distribution is given by $x \sim \mathcal{N}\left(0, \frac{R_{ww}}{1-a^2}\right) = \mathcal{N}(0, 0.101)$. We generate the samples using the slice (Gibbs) sampler with the proposal

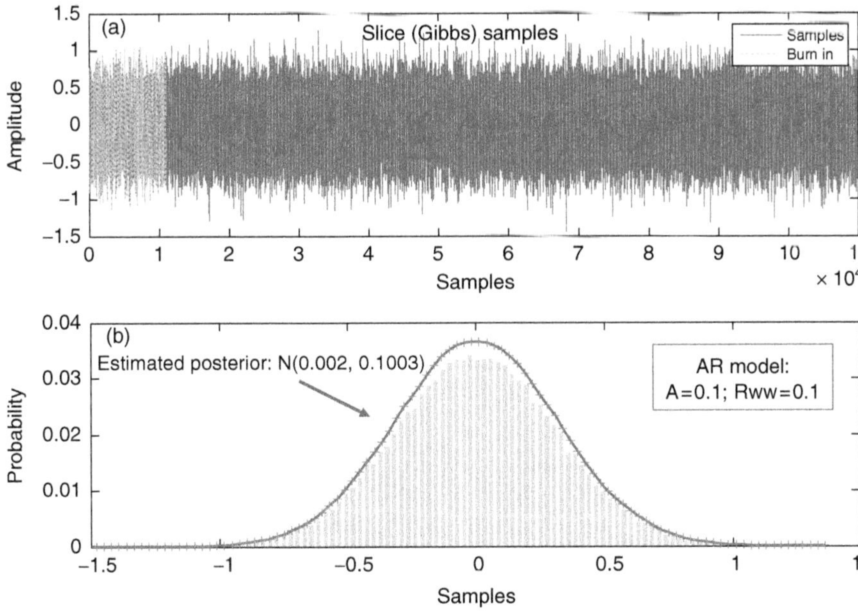

FIGURE 3.7 Gibbs (slice) sampler for autoregressive ($AR(1)$) model: (a) simulated samples ($N = 10^5$); (b) estimated posterior distribution: $\mathcal{N}(0.002, 0.1003)$ with 10% burn-in ($N = 10^4$) samples.

$q(x) \sim \mathcal{N}(0, 0.1)$. The results are shown in Fig. 3.7. Using MATLAB we synthesized the set of samples and estimated their underlying distribution using both histogram and kernel density with a Gaussian window estimator. The samples including a 10% burn-in period are shown in Fig. 3.7a along with the estimated distribution in Fig. 3.7b. The sampler has a 100%-acceptance rate and the estimates are quite good with the posterior estimated at $\mathcal{N}(0.002, 0.1003)$. △△△

This concludes the section on sampling theory and iterative sampling techniques. Next, we investigate the importance sampler that will lead to the sequential approaches required to construct Bayesian model-based processors.

3.5 IMPORTANCE SAMPLING

One way to mitigate difficulties with the inability to directly sample from a posterior distribution is based on the concept of importance sampling. *Importance sampling* is a method to compute expectations with respect to one distribution using random samples drawn from another. That is, it is a method of simulating samples from a proposal distribution to be used to approximate a targeted (posterior) distribution by appropriate weighting. Importance sampling is a generalization of the MC approach which evolves by rewriting Eq. 3.27 as

$$I = \int_X f(x)\,dx = \int_X \left(\frac{f(x)}{q(x)}\right) \times q(x)\,dx \quad \text{for} \quad \int q(x)\,dx = 1 \tag{3.43}$$

Here $q(x)$ is referred to as the sampling distribution or more appropriately the *importance sampling distribution*, since it samples the *target distribution* $f(x)$ nonuniformly giving "more importance" to some values of $f(x)$ than others. We say that the *support* of $q(x)$ covers that of $f(x)$, or the samples drawn from $q(\cdot)$ overlap the same region (or more) corresponding to the samples of $f(\cdot)$ as illustrated previously in Fig. 3.3. That is, we say that $f(x)$ and $q(x)$ have the same support if

$$f(x) > 0 \Rightarrow q(x) > 0 \quad \forall x \in R^{N_x \times 1}$$

a *necessary* condition for importance sampling to hold. If we interpret the prior of Fig. 1.1 as the proposal $q(x)$ and the posterior as the target $f(x)$, then this figure provides a visual example of coverage.

The integral in Eq. 3.43 can be estimated by

- Drawing N samples from $q(x)$: $X(i) \sim q(x)$ and $\hat{q}(x) \approx \frac{1}{N}\sum_{i=1}^{N} \delta(x - X(i))$; and
- Computing the sample mean [28],

$$I = E_q\left\{\frac{f(x)}{q(x)}\right\} \approx \int \left(\frac{f(x)}{q(x)}\right) \times \frac{1}{N}\sum_{i=1}^{N} \delta(x - X(i))\,dx = \frac{1}{N}\sum_{i=1}^{N} \frac{f(X(i))}{q(X(i))}$$

with corresponding error variance

$$\text{Var}\left[E_q\left\{\frac{f(x)}{q(x)}\right\}\right] = \int \left(\frac{f(x)}{q(x)} - I\right)^2 \times q(x)\,dx$$

It is interesting to note that the MC approach provides an unbiased estimator with the corresponding error variance easily calculated from the above relation.

Consider the case where we would like to estimate the expectation of the function of X given by $f(X)$. Then choosing an importance distribution $q(x)$ that is similar to $f(x)$ with covering support gives the expectation estimator

$$E_p\{f(x)\} = \int_X f(x) \times p(x)\,dx = \int_X f(x)\left(\frac{p(x)}{q(x)}\right) \times q(x)\,dx \tag{3.44}$$

If we draw samples $\{X(i)\}, i = 0, 1, \ldots, N$ from the importance distribution $q(x)$ and compute the sample mean, then we obtain the importance sampling estimator. That is, assume the perfect sampler $\hat{q}(x) \approx \frac{1}{N}\sum_{i=1}^{N} \delta(x - X(i))$ and substitute

$$E_p\{f(x)\} = \int_X f(x)\left(\frac{p(x)}{q(x)}\right) \times \hat{q}(x)\,dx \approx \frac{1}{N}\sum_{i=1}^{N} f(X(i)) \times \left(\frac{p(X(i))}{q(X(i))}\right) \tag{3.45}$$

demonstrating the concept.

The "art" in importance sampling is in choosing the importance distribution $q(\cdot)$ that approximates the target distribution $p(\cdot)$ as closely as possible. This is the *principal factor* affecting performance of this approach, since variates must be drawn from $q(x)$ that cover the target distribution. Using the concepts of importance sampling, we can approximate the posterior distribution with a function on a finite discrete support. Since it is usually not possible to sample directly from the posterior, we use *importance sampling* coupled with an easy to sample proposal distribution $q(X_t|Y_t)$— this is the crucial choice and design step required in Bayesian importance sampling methodology. Here $X_t = \{x(0), \ldots, x(t)\}$ represents the set of dynamic variables and $Y_t = \{y(0), \ldots, y(t)\}$, the set of measured data. Therefore, starting with a function of the set of variables, say $f(X_t)$, we would like to estimate its mean using the importance concept. That is, using the MC approach, we would like to sample from this posterior directly and then use sample statistics to perform the estimation. Therefore, we insert the proposal importance distribution $q(X_t|Y_t)$ as before

$$\hat{f}(t) := E\{f(X_t)\} = \int f(X_t)\left[\frac{\Pr(X_t|Y_t)}{q(X_t|Y_t)}\right] \times q(X_t|Y_t)\,dX_t \tag{3.46}$$

Now applying Bayes' rule to the posterior target distribution, and defining a weighting function as

$$\tilde{W}(t) := \frac{\Pr(X_t|Y_t)}{q(X_t|Y_t)} = \frac{\Pr(Y_t|X_t) \times \Pr(X_t)}{\Pr(Y_t) \times q(X_t|Y_t)} \tag{3.47}$$

Unfortunately, $\tilde{W}(t)$ is not useful because it requires knowledge of the evidence or normalizing constant $\Pr(Y_t)$ given by

$$\Pr(Y_t) = \int \Pr(Y_t|X_t) \times \Pr(X_t)\,dX_t \tag{3.48}$$

which is usually not available. But by substituting Eq. 3.47 into Eq. 3.46 and defining a *new* weight $W(t)$ as

$$W(t) \propto \frac{\Pr(X_t|Y_t)}{q(X_t|Y_t)} = \frac{\Pr(Y_t|X_t) \times \Pr(X_t)}{q(X_t|Y_t)} \tag{3.49}$$

we obtain

$$\hat{f}(t) = \frac{1}{\Pr(Y_t)} \int f(X_t) \left[\frac{\Pr(Y_t|X_t) \times \Pr(X_t)}{q(X_t|Y_t)} \right] q(X_t|Y_t) dX_t$$
$$= \frac{1}{\Pr(Y_t)} \int W(t) f(X_t) \, q(X_t|Y_t) dX_t \tag{3.50}$$

which is simply the expectation of the weighted function $E_q\{W(t)f(X_t)\}$ scaled by the normalizing constant. From this definition of the new weighting function in Eq. 3.49, we have

$$W(t) \times q(X_t|Y_t) = \Pr(Y_t|X_t) \times \Pr(X_t) \tag{3.51}$$

Thus, we can now replace the troublesome normalizing constant of Eq. 3.48 using Eq. 3.51, that is,

$$\hat{f}(t) = \frac{E_q\{W(t)f(X_t)\}}{\Pr(Y_t)} = \frac{E_q\{W(t)f(X_t)\}}{\int W(t) \times q(X_t|Y_t) \, dX_t} = \frac{E_q\{W(t)f(X_t)\}}{E_q\{W(t)\}} \tag{3.52}$$

Now drawing samples from the proposal $X_t(i) \sim q(X_t|Y_t)$ and using the MC approach leads to the desired result. That is, from the "perfect" sampling distribution, we have that

$$\hat{q}(X_t|Y_t) \approx \frac{1}{N} \sum_{i=1}^{N} \delta(X_t - X_t(i)) \tag{3.53}$$

and therefore substituting, applying the sifting property of the Dirac delta function, and defining the "normalized" weights

$$\mathcal{W}_i(t) := \frac{W_i(t)}{\sum_{i=1}^{N} W_i(t)} \quad \text{for } W_i(t) = \frac{\Pr(Y_t|X_t(i)) \times \Pr(X_t(i))}{q(X_t(i)|Y_t)} \tag{3.54}$$

we obtain the final estimate

$$\hat{f}(t) \approx \sum_{i=1}^{N} \mathcal{W}_i(t) \times f(X_t(i)) \tag{3.55}$$

The importance estimator is biased being the ratio of two sample estimators (as in Eq. 3.52), but it can be shown that it asymptotically converges to the true statistic and the central limit theorem holds [32, 49]. Thus, as the number of samples increase ($N \rightarrow \infty$), an asymptotically optimal estimate of the posterior is

$$\hat{\Pr}(X_t|Y_t) \approx \sum_{i=1}^{N} \mathcal{W}_i(t) \times \delta(X_t - X_t(i)) \tag{3.56}$$

which is the goal of Bayesian estimation. Note that the new weight is $W(t) \propto \tilde{W}(t)$, where $\propto$ is defined as " proportional to" up to a normalizing constant.

3.6 SEQUENTIAL IMPORTANCE SAMPLING

Suppose we would like to develop a sequential version [41–67] of the batch Bayesian importance sampling estimator of the previous section. The importance distribution can be modified to enable a sequential estimation of the desired posterior distribution, that is, we estimate the posterior $\hat{\Pr}(X_{t-1}|Y_{t-1})$ using importance weights $\mathcal{W}(t-1)$. As a new sample becomes available, we estimate the new weight $\mathcal{W}(t)$ leading to an updated estimate of the posterior $\hat{\Pr}(X_t|Y_t)$. This means that to obtain the new set of samples $X_t(i) \sim q(X_t|Y_t)$ sequentially, we must use the previous set of samples $X_{t-1}(i) \sim q(X_{t-1}|Y_{t-1})$. Thus, with this in mind, the importance distribution $q(X_t|Y_t)$ must admit a marginal distribution $q(X_{t-1}|Y_{t-1})$ implying the following Bayesian factorization

$$q(X_t|Y_t) = q(X_{t-1}|Y_{t-1}) \times q(x(t)|X_{t-1}, Y_t) \tag{3.57}$$

which satisfies the probability chain rule decomposition

$$q(X_t|Y_t) = \prod_{k=0}^{t} q(x(k)|X_{t-k}, Y_k) \tag{3.58}$$

Now let us see how this type of importance distribution choice can lead to the desired sequential solution [58]. We start with the definition of the unnormalized weight of Eq. 3.49 and substitute into Eq. 3.57 while applying Bayes' rule to the numerator. The resulting weighting function is

$$W(t) = \frac{\Pr(Y_t|X_t) \times \Pr(X_t)}{q(X_{t-1}|Y_{t-1}) \times q(x(t)|X_{t-1}, Y_t)} \tag{3.59}$$

Motivated by the definition of $W(t-1)$, we multiply and divide by the Bayesian factor $\Pr(Y_{t-1}|X_{t-1}) \times \Pr(X_{t-1})$ and group to obtain

$$W(t) = \underbrace{\left[\frac{\Pr(Y_{t-1}|X_{t-1}) \times \Pr(X_{t-1})}{q(X_{t-1}|Y_{t-1})}\right]}_{\text{Previous weight}} \times \frac{\Pr(Y_t|X_t) \times \Pr(X_t)}{[\Pr(Y_{t-1}|X_{t-1}) \times \Pr(X_{t-1})] \times q(x(t)|X_{t-1}, Y_t)}$$

Therefore, we can write

$$W(t) = W(t-1) \times \frac{\Pr(Y_t|X_t) \times \Pr(X_t)}{\Pr(Y_{t-1}|X_{t-1}) \times \Pr(X_{t-1}) \times q(x(t)|X_{t-1}, Y_t)} \tag{3.60}$$

Using the probabilistic chain rule of Eq. 2.63 for each of the conditionals and imposing the Markov property of the dynamic variable along with the conditional independence conditions of the measurements, we obtain

$$\begin{aligned}
\Pr(Y_t|X_t) &= \prod_{k=0}^{t} \Pr(y(k)|x(k)) = \Pr(y(t)|x(t)) \prod_{k=0}^{t-1} \Pr(y(k)|x(k)) \\
\Pr(X_t) &= \prod_{k=0}^{t} \Pr(x(k)|x(k-1)) = \Pr(x(t)|x(t-1)) \prod_{k=0}^{t-1} \Pr(x(k)|x(k-1))
\end{aligned} \tag{3.61}$$

Therefore, recognizing the relationship between these expansions as well as those at $t-1$ and factoring the tth term (as shown above), we can cancel both numerator and denominator chains to extract the final recursions, that is,

$$W(t) = W(t-1) \times \frac{\Pr(y(t)|x(t)) \left[\prod_{k=0}^{t-1} \Pr(y(k)|x(k))\right]}{\left[\prod_{k=0}^{t-1} \Pr(y(k)|x(k))\right]} \times \frac{\Pr(x(t)|x(t-1)) \left[\prod_{k=0}^{t-1} \Pr(x(k)|x(k-1))\right]}{\left[\prod_{k=0}^{t-1} \Pr(x(k)|x(k-1))\right]} \times \frac{1}{q(x(t)|X_{t-1}, Y_t)} \tag{3.62}$$

which gives the final recursion

$$W(t) = W(t-1) \times \frac{\Pr(y(t)|x(t)) \times \Pr(x(t)|x(t-1))}{q(x(t)|X_{t-1}, Y_t)} \tag{3.63}$$

Another way of developing this relationship is to recall the Bayesian solution to the batch posterior estimation problem in Eq. 2.79. We have

$$\Pr(X_t|Y_t) = \left[\frac{\Pr(y(t)|x(t)) \times \Pr(x(t)|x(t-1))}{\Pr(y(t)|Y_{t-1})}\right] \times \Pr(X_{t-1}|Y_{t-1})$$

or recognizing the denominator as just the evidence or normalizing distribution and not a function of X_t, we have

$$\Pr(X_t|Y_t) = C \times \Pr(y(t)|x(t)) \times \Pr(x(t)|x(t-1)) \times \Pr(X_{t-1}|Y_{t-1}) \quad (3.64)$$

or simply

$$\Pr(X_t|Y_t) \propto \Pr(y(t)|x(t)) \times \Pr(x(t)|x(t-1)) \times \Pr(X_{t-1}|Y_{t-1}) \quad (3.65)$$

Substituting this expression for the posterior in the weight relation as before, we have

$$W(t) \propto \frac{\Pr(Y_t|X_t)}{q(X_t|Y_t)} = \frac{\Pr(y(t)|x(t)) \times \Pr(x(t)|x(t-1))}{q(x(t)|X_{t-1}, Y_t)} \times \underbrace{\frac{\Pr(X_{t-1}|Y_{t-1})}{q(X_{t-1}|Y_{t-1})}}_{\text{Previous weight}} \quad (3.66)$$

giving the desired expression of Eq. 3.63.

These results enable us to formulate a generic Bayesian *sequential importance sampling* algorithm

1. Draw samples from the proposed importance distribution: $x_i(t) \rightarrow q(x(t)|X_{t-1}, Y_t)$;
2. Determine the required conditional distributions: $\Pr(x_i(t)|x(t-1))$, $\Pr(y(t)|x_i(t))$;
3. Calculate the unnormalized weights: $W_i(t)$ using Eq. 3.63 with $x(t) \rightarrow x_i(t)$;
4. Normalize the weights: $\mathcal{W}_i(t)$ in Eq. 3.54; and
5. Estimate the posterior distribution: $\hat{\Pr}(X_t|Y_t) = \sum_{i=1}^{N} \mathcal{W}_i(t)\delta(x(t) - x_i(t))$

Once the posterior is estimated, the desired statistics evolve directly. We summarize the generic sequential importance sampling in Table 3.2.

Introducing these ideas of Bayesian importance sampling, we are now ready to consider applying this approach to a variety of models which we discuss in the next chapter.

TABLE 3.2 Bayesian Sequential Importance Sampling Algorithm

Initialize

$$x_i(0) \sim q(x(0)|y(0)); \quad i = 1, \ldots, N_p \qquad \text{[Sample prior]}$$

$$W_i(0) = \frac{\Pr(y(0)|x_i(0)) \times \Pr(x_i(0))}{\Pr(x_i(0)|y(0))} \qquad \text{[Weights]}$$

$$\mathcal{W}_i(0) = \frac{W_i(0)}{\sum_{i=1}^{N_p} W_i(0)} \qquad \text{[Normalize]}$$

Importance Sampling

Sample

$$x_i(t) \sim q(x(t)|X_{t-1}, Y_t); \quad i = 1, \ldots, N_p \qquad \text{[Sample]}$$

Weight Update

$$W_i(t) = W_i(t-1) \times \frac{\Pr(y(t)|x_i(Pr(x(t)|xi(t-1))) \times \Pr(x(t)|x_i(t-1))}{q(x_i(t)|X_i(t-1), Y_t)} \qquad \text{[Weights]}$$

Weight Normalization

$$\mathcal{W}_i(t) = \frac{W_i(t)}{\sum_{i=1}^{N_p} W_i(t)}$$

Distribution

$$\hat{\Pr}(X_t|Y_t) \approx \sum_{i=1}^{N_p} \mathcal{W}_i(t)\delta(x(t) - x_i(t)) \qquad \text{[Posterior distribution]}$$

3.7 SUMMARY

In this chapter we discussed the importance of simulation-based sampling methods for nonlinear signal processing. Starting with the basics of PDF estimation and statistical sampling theory, we motivated statistical approaches to sampling both when we have analytical expressions for the distributions and when we do not have them and must resort to pure sampling methodologies. We discussed the uniform sampling and rejection sampling methods examining their inherent advantages and disadvantages. We showed how these approaches led to more sophisticated techniques evolving from Markov chain theory and leading to the Metropolis–Hastings sampler. In certain cases the Gibbs sampler, a variant of the Metropolis–Hastings approach, was developed and discussed along with its variant—the slice sampler. All of these methods fall into the general class of iterative methods. Next, we concentrated on the importance sampling approach leading to its recursive version—the sequential importance sampler which is the workhorse of this text.

MATLAB NOTES

MATLAB is a command oriented vector-matrix package with a simple yet effective command language featuring a wide variety of embedded *C* language constructs making it ideal for signal processing applications and graphics. MATLAB has a *Statistics Toolbox* that incorporates a large suite of PDFs and CDFs as well as "inverse" CDF functions ideal for simulation-based algorithms. The

mhsample command incorporates the Metropolis, Metropolis–Hastings, and Metropolis independence samplers in a single command while the Gibbs sampling approach is adequately represented by the more efficient slice sampler (**slicesample**). There are even specific "tools" for sampling as well as the inverse CDF method captured in the **randsample** command. PDF estimators include the usual histogram (**hist**) as well as the sophisticated kernel density estimator (**ksdensity**) offering a variety of kernel (window) functions (Gaussian, etc.) and ICDF techniques. As yet, no sequential algorithms are available. Type *help stats* in MATLAB to get more details or go to the MathWorks website.

REFERENCES

1. A. Papoulis and S. Pillai, *Probability, Random Variables and Stochastic Processes* (New York: McGraw-Hill, 2002).
2. P. Peebles, *Probability, Random Variables and Random Signal Parameters*, 4th Ed. (New York: McGraw-Hill, 2001).
3. R. Duda and P. Hart, *Pattern Classsification* (New York: John Wiley & Sons, Inc., 2000).
4. T. Hastie, R. Tibshirani, and J. Friedman, *The Elements of Statistical Learning* (New York: John Wiley & Sons, Inc., 2001).
5. S. Theodoridis and K. Koutroumbas, *Pattern Recognition* (New York: Academic Press, 1999).
6. E. Parzen, "On estimation of a probability density function and mode," *Ann. Math. Stat.*, **33**, 1065–1076, 1962.
7. H. Shimazaki and S. Shinomoto, "A method for selecting the bin size of a time histogram," *Neural Comput.*, **19**, 1503–1527, 2007.
8. A. Jazwinski, *Stochastic Processes and Filtering Theory* (New York: Academic Press, 1970).
9. A. Sage and J. Melsa, *Estimation Theory with Applications to Communications and Control* (New York: McGraw-Hill, 1971).
10. B. Anderson and J. Moore, *Optimal Filtering* (Englewood Cliff, NJ: Prentice-Hall, 1979).
11. J. Candy, *Model-Based Signal Processing* (Hoboken, NJ: John Wiley & Sons, Inc./IEEE Press, 2006).
12. H. Tanizaki, *Nonlinear Filters* (New York: Springer-Verlag, 1994).
13. I. Rhodes, "A tutorial introduction to estimation and filtering," *IEEE Trans. Autom. Contr.*, **AC-16**, 1971.
14. J. Hammersley and K. Morton, "Poor man's Monte Carlo," *Symposium on Monte Carlo Methods*, Royal Statistical Society, pp. 23–38, 1954.
15. S. Ross, *A Short Course in Simulation* (New York: McMillan, 1990).
16. M. Tanner, *Tools for Statistical Inference*, 2nd Ed. (New York: Springer-Verlag, 1993).
17. W. Gilks, S. Richardson, and D. Spiegelhalter, *Markov Chain Monte Carlo in Practice* (New York: Chapman & Hall/CRC, 1996).
18. A. Smith and A. Gelfand, "Bayesian statistics without tears: a sampling-resampling perspective," *Am. Statistician*, **44**, 4, 84–88, 1992.

19. C. Robert and G. Casella, *Monte Carlo Statistical Methods* (New York: Springer, 1999).
20. A. Gelman, J. Carlin, H. Stern, and D. Rubin, *Bayesian Data Analysis*, 2nd Ed. (New York: Chapman & Hall/CRC, 2004).
21. M. Mitzenmacher and E. Upfal, *Probability and Computing: Randomized Algorithms and Probabilistic Analysis* (Cambridge: Cambridge University Press, 2005).
22. J. von Neumann, "Various techniques used in connection with random digits," *Nat. Bureau Standards Applied Math. Series*, **12**, 36–38, 1951.
23. W. Hastings, "Monte Carlo sampling methods using Markov chains and their applications," *Biometrika*, **57**, 97–109, 1970.
24. N. Metropolis, N. Rosenbutt, A. Rosenbutt, M. Teller, and A. Teller, "Equations of state calculations by fast computing machines," *J. Chem. Phys.*, **21**, 1087–1092, 1953.
25. S. Geman and D. Geman, "Stochastic relaxation: Gibbs distributions and Bayesian restoration of images," *IEEE Trans. Patten Analy. and Mach. Intell.*, **6**, 721–741, 1984.
26. D. Rubin, "A noniterative sampling/importance resampling alternative to the data augmentation algorithm for creating a few imputations when fractions of missing information are modest: the SIR algorithm," *J. Amer. Stat. Assoc.*, **52**, 543–546, 1987.
27. A. Gelfand and A. Smith, "Sampling-based approaches to calculating marginal densities," *J. Amer. Stat. Assoc.*, **85**, 410, 398–409, 1990.
28. J. Ruanaidh and W. Fitzgerald, *Numerical Bayesian Methods Applied to Signal Processing* (New York: Springer-Verlag, 1996).
29. G. Casella and E. George, "Explaining the Gibbs sampler," *Am. Statistician*, **46**, 3, 167–174, 1992.
30. S. Chib and E. Greenberg, "Understanding the Metropolis–Hastings algorithm," *Am. Statistician*, **49**, 4, 327–335, 1995.
31. R. Neal, "Slice sampling," *Ann. Stat.*, **31**, 3, 705–767, 2003.
32. J. Liu, *Monte Carlo Strategies in Scientific Computing* (New York: Springer-Verlag, 2001).
33. W. Fitzgerald, "Markov chain Monte Carlo methods with applications to signal processing," *Signal Proc.*, **81**, 3–18, 2001.
34. C. Andrieu, N. de Freitas, A. Doucet, and M. Jordan, "An introduction to MCMC for machine learning," *Mach. Learn.*, **50**, 5–43, 2003.
35. D. Frenkel, "Introduction to Monte Carlo methods," *J. von Neumann Inst. Comput.* NIC series, **23**, 29–60, 2004.
36. O. Cappe, E. Moulines and T. Ryden, *Inference in Hidden Markov Models* (New York: Springer-Verlag, 2005).
37. D. MacKay, *Information Theory, Inference and Learning Algorithms* (Cambridge, UK: Cambridge University Press, 2006).
38. D. Spiegelhalter, A. Thomas, N. Best, and D. Lunn, *WinBUGS User Manual* (London, UK: Imperial College, 2003). Available at: http://www.mrc-bsu.cam.ac.uk
39. I. Nabney, *NETLAB Algorithms for Pattern Recognition* (New York: Springer, 2001).
40. R. Duin, P. Juszczak, P. Paclik, E. Pekalska, D. de Ridder, and D. Tax, *PRTools4: A MATLAB Toolbox for Pattern Recognition* (Delft, The Netherlands: Delft University, 2004) Available at: http://prtools.org
41. A. Doucet and X. Wang, "Monte Carlo methods for signal processing," *IEEE Signal Proc. Mag.*.**24**, 5, 152–170, 2005.

42. G. Kitagawa, "Non-Gaussian modeling of nonstationary time series," *J. Am. Stat. Assoc.*, **82**, 400, 1032–1063, 1987.

43. G. Kitagawa, "A nonlinear smoothing method for time series analysis," *Statistica Sinica*, **1**, 2, 371–388, 1991.

44. N. Gordon, D. Salmond, and A. Smith, "A novel approach to nonlinear non-Gaussian Bayesian state estimation," *IEE Proc. F.*, **140**, 107–113, 1993.

45. A. Kong, J. Liu, and W. Wong, "Sequential imputations and Bayesian missing data problems," *J. Am. Stat. Assoc.*, **89**, 425, 278–288, 1994.

46. J. Liu and R. Chen, "Blind deconvolution via sequential imputations," *J. Am. Stat. Assoc.*, **90**, 430, 567–576, 1995.

47. G. Kitagawa, "Monte Carlo filter and smoother for non-Gaussian nonlinear state-space models," *J. Comput. Graphical Stat.*, **5**, 1, 1–25, 1996.

48. G. Kitagawa and W. Gersch, *Smoothness Priors Analysis of Time Series* (New York: Springer-Verlag, 1996).

49. M. West and J. Harrison, *Bayesian Forcasting and Dynamic Models*, 2nd Ed. (New York: Springer-Verlag, 1997).

50. J. Liu and R. Chen, "Sequential Monte Carlo methods for dynamic systems," *J. Am. Statistical Assoc.*, **93**, 443, 1032–1044, 1998.

51. G. Kitagawa, "Self-organizing state space model," *J. Am. Stat. Assoc.*, **93**, 443, 1203–1215, 1998.

52. M. Isard and A. Blake, "Condensation–conditional density propagation for visual tracking," *Int. J. Comput. Vis.*, **29**, 1, 5–28, 1998.

53. J. Liu, R. Chen, and W. Wong, "Rejection control and sequential importance sampling," *J. Am. Stat. Assoc.*, **93**, 443, 1022–1031, 1998.

54. M. Pitt and N. Shephard, "Filtering via simulation: Auxiliary particle filters," *J. Am. Stat. Assoc.*, **94**, 446, 590–599, 1999.

55. A. Doucet, S. Godsill, and C. Andrieu, "On sequential Monte Carlo sampling methods for Bayesian filtering," *Stat. Comput.*, **10**, 3, 197–208, 2000.

56. A. Doucet, N. de Freitas, and N. Gordon, *Sequential Monte Carlo Methods in Practice* (New York: Springer-Verlag, 2001).

57. S. Godsill and P. Djuric, "Special issue: Monte Carlo methods for statistical signal processing." *IEEE Trans. Signal Proc.*, **50**, 2002.

58. M. Arulampalam, S. Maskell, N. Gordon, and T. Clapp "A tutorial on particle filters for online nonlinear/non-Gaussian Bayesian tracking." *IEEE Trans. Signal Proc.*, **50**, 2, 174–188, 2002.

59. P. Djuric, J. Kotecha, J. Zhang, Y. Huang, T. Ghirmai, M. Bugallo, and J. Miguez, "Particle filtering." *IEEE Signal Proc. Mag.*, **20**, 5, 19–38, 2003.

60. B. Ristic, S. Arulampalam, and N. Gordon, *Beyond the Kalman Filter: Particle Filters for Tracking Applications* (Boston, MA: Artech House, 2004).

61. S. Haykin and N. de Freitas, "Special issue: Sequential state estimation: from Kalman filters to particle filters." *Proc. IEEE*, **92**, 3, 2004.

62. C. Andrieu, A. Doucet, S. Singh, and V. Tadic, "Particle methods for change detection, system identification and control," *Proc. IEEE*, **92**, 3, 423–438, 2004

63. A. Harvey, S. Koopman, and N. Shephard, *State Space and Unobserved Component Models* (Cambridge: Cambridge University Press, 2004).

64. H. Sorenson and D. Alspach, "Recursive Bayesian estimation using Gaussian sums," *Automatica*, **7**, 465–479, 1971.
65. R. van der Merwe, A. Doucet, N. de Freitas, and E. Wan, "The unscented particle filter," in *Advances in Neural Information Processing Systems 13* (Cambridge, MA: MIT Press, 2000).
66. S. Haykin, *Kalman Filtering and Neural Networks* (Hoboken, NJ: John Wiley & Sons, Inc., 2001).
67. J. Spall, "Estimation via Markov chain Monte Carlo," *IEEE Control Sys. Magz.*, **4**, 34–45, 2003.

PROBLEMS

3.1 Let x be a discrete random variable with probability mass function (PMF)

$$\mathrm{p}_X(x) = \frac{3!}{x!\,(3-x)!}\left(\frac{2}{3}\right)^x\left(\frac{1}{3}\right)^{3-x} \qquad x = 0, 1, 2, 3$$

What is the PMF of y, $\mathrm{p}_Y(y)$, when $y = x^2$?

3.2 Let z be the distance from the origin to a random point selected within the unit circle ($x^2 + y^2 < 1$). Let $z = x^2 + y^2$, then

(a) What is the probability of the selected point lying within the unit circle?

(b) What is the CDF of z? What is its PDF?

(c) Suppose $w = z^2$, what is $\mathrm{p}_W(w)$? $\mathrm{P}_W(w)$?

3.3 Let $x \sim \mathcal{U}(0, 1)$ and $y = -2\ln x$, what is $\mathrm{p}_Y(y)$?

3.4 Suppose we have a bivariate distribution with point (X,Y) from a unit square such that

$$\mathrm{p}_{XY}(x, y) = 1, \quad 0 < x < 1, \quad 0 < y < 1$$

Let $z = x + y$, what is the CDF of z? PDF of z?

3.5 A source emits particles that decay at a distance x ($x \sim \mathcal{E}xp(\lambda)$) from the source. The measurement instrument can only observe these events in a window of length 20 cm ($x = 1 - 20$ cm). N decays are observed at locations $\{x_i\} = \{1, \ldots, N\}$. Using Bayes' rule [37]

(a) What is the characteristic length λ?

(b) Plot the distributions for $\{x_i\} = \{1.5, 2, 3, 4, 5, 12\}$.

3.6 For a particular television show, a contestant is given the following instructions:

- There are three doors labeled 1, 2, 3 with a prize hidden behind one of them. Select one of the doors, but it will NOT be opened.
- The host will open one of the other two doors, but will NOT reveal if the prize is there.

- The contestant must now make a decision to keep his original choice or choose another door.
- All the doors will then be opened and the prize revealed

What should the contestant do?

(*a*) Stay with the original choice?

(*b*) Switch to the remaining door?

(*c*) Does it make any difference?

(*Hint:* Use Bayes' rule to answer these questions)

3.7 Suppose we would like to simulate a random variable X such that $\Pr(X = i) = \{0.2, 0.15, 0.25, 0.40\}$.

(*a*) Sketch out an algorithm using the inverse transform method to generate realizations of X choosing an *ascending* approach for $X = 1, X = 2, X = 3$, and $X = 4$.

(*b*) Sketch out an algorithm using the inverse transform method to generate realizations of X choosing a *descending* approach for $X = 4, X = 3, X = 1$, and $X = 2$.

(*c*) Which approach is more efficient? Why?

3.8 We would like to simulate the value of a discrete random variable X with associated probabilities: $\Pr(X = i) = \{0.11, 0.12, 0.09, 0.08, 0.12, 0.10, 0.09, 0.09, 0.10, 0.10\}$. Using the rejection method with $M = \max \frac{\Pr(X=i)}{\Pr(u_i)}$ for $u_i \sim \mathcal{U}(0, 10)$

(*a*) Sketch out rejection sampling algorithm for this problem.

(*b*) Using this approach synthesize 1000 samples and estimate the histogram? Estimate the kernel density? (*Hint*: MATLAB has the commands **hist** and **ksdensity** to perform these operations)

(*c*) Does the estimated distribution appear to be any classical closed form (e.g., Poisson)? If so which one?

3.9 Suppose the continuous random variable X has cumulative distribution $P_X(x) = x^k$. Using the inverse transform approach, sketch the methodology to synthesize X. How would you do this using the rejection approach? Generate 1000 samples and estimate the distributions.

3.10 Use the rejection sampling method to generate the continuous random variable x with density $p_X(x) = 20x(1 - x)^3, 0 < x < 1$. (*Hint:* Select $q(x) = 1$ and find the maximum ratio of the densities to obtain M)

3.11 We would like to compute the solution to the integral

$$\mathcal{I} = \int_a^b f(x)dx$$

Using the results of Example 3.7, develop the more general MC solution.

3.12 Suppose we have a bivariate Gaussian distribution with unknown mean $\mu = [\mu_1 \mu_2]'$ and known covariance matrix

$$C = \begin{bmatrix} 1 & p \\ p & 1 \end{bmatrix}$$

and a uniform prior on μ. A single observation (y_1, y_2) then has Gaussian posterior

$$\Pr(\mu|Y) \sim \mathcal{N}(Y, C) \quad \text{for } Y = [y_1 y_2]'$$

Sketch out the Gibbs sampler algorithm for this problem.

3.13 Develop the Metropolis algorithm for a bivariate unit Gaussian target distribution $\mathcal{N}(\Theta : 0, I)$ with prior $\Pr(\Theta_o)$ (e.g., $\mathcal{U}(0, 1)$). The proposal distribution is bivariate Gaussian also, $\mathcal{N}(\Theta^* : 0, (1/5)^2 I)$.

3.14 We would like to use MATLAB to simulate the Metropolis–Hastings sampler (**mhsample**) for the following case with the target distribution being a standard Gaussian $\mathcal{N}(5, 1)$ and the proposal Rayleigh distributed with parameter $b = 1.5$. Use a 10% sample burn-in. Compare these results to the slice sampler (**slicesample**). Use (**ksdensity**) to estimate the distribution of the resulting samples for each algorithm for the comparison.

3.15 Develop the Metropolis–Hastings sampler for a second-order autoregressive model ($AR(2)$) with known coefficients $\{a_1, a_2\} = \{1, -0.5\}$ driven by Gaussian noise with $\epsilon \sim \mathcal{N}(0, 1)$.

(*a*) Develop the exact likelihood for the parameters $\Pr(Y|\mathbf{a}, \sigma^2)$.

(*b*) What is the posterior distribution for the parameters $\Pr(\mathbf{a}, \sigma^2|Y)$? (*Hint*: Assume the prior is just an indicator function)

(*c*) Develop the M-H-sampler for this problem.

3.16 Suppose we would like to generate samples from a bivariate Gaussian with mean vector *zero* and covariance

$$C = \begin{bmatrix} 1 & 0.5 \\ 0.5 & 1 \end{bmatrix}$$

(*a*) Choose a uniform proposal $\mathcal{U}(-3, 3)$ and develop the M-H-sampler algorithm with a 5% burn-in and $N = 10{,}000$ samples. (*Hint*: Use MATLAB **mhsample** command) (*b*) Using the MC approach, estimate the expected value $E\{f(\mathbf{X})\} = [1 \quad 1]\mathbf{X}$

(*c*) Compare these results to those obtained using the slice sampler. (*Hint*: Use MATLAB **slicesample** command)

3.17 Set up the Gibbs sampler (G-S) for a joint (X, Y) exponential distribution on an interval of length $(0, I)$. Estimate the marginal distribution of X, $\Pr(X)$, and compare the results to a simulated data set based on $N = 500$ samples.

4

STATE–SPACE MODELS FOR BAYESIAN PROCESSING

4.1 INTRODUCTION

In this chapter we investigate the development of models for Bayesian estimation [1–16] using primarily the state–space representation—a versatile and robust model especially for random signals. We start with the definition of state and the basic principles underlying these characterizations and then show how they are incorporated as propagation distributions for Bayesian processors in the following chapter. We review the basics of state–space model development with all of their associated properties starting with the continuous-time processes, then sampled-data systems, and finally proceeding to the discrete-time state–space. Next we develop the stochastic version leading to Gauss–Markov representations when the models are driven by white noise and then proceed to the nonlinear case [6–15]. Here we again drive the models with white Gaussian noise, but the results are not necessarily Gaussian. We develop linearization techniques based on Taylor-series expansions to arrive at linearized Gauss–Markov models.

State–space models are easily generalized to multichannel, nonstationary, and nonlinear processes. They are very popular for model-based signal processing primarily because most physical phenomena modeled by mathematical relations naturally occur in state–space form (see Ref. [13] for details). With this motivation in mind, let us proceed to investigate the state–space representation in a more general form to at least "touch" on its inherent richness. We start with continuous-time systems and then proceed to sampled-data followed by the discrete-time representation—the primary focus of this text. We begin by formally defining the concept of state [1].

Bayesian Signal Processing: Classical, Modern, and Particle Filtering Methods, Second Edition. James V. Candy.

4.2 CONTINUOUS-TIME STATE–SPACE MODELS

The *state* of a system at time t is the "minimum" set of variables (*state variables*) along with the *input* sufficient to uniquely specify the dynamic system behavior for all t over the interval $t \in [t_0, \infty)$. The *state vector* is the collection of state variables into a single vector. The idea of a *minimal set* of state variables is critical and all techniques to define them must ensure that the smallest number of "independent" states have been defined in order to avoid possible violation of some important system theoretic properties [2, 3].

Let us consider a general *deterministic* formulation of a *nonlinear dynamic system* including the output (measurement) model in state–space form (continuous time)[1]

$$\begin{aligned} \dot{x}_t &= A(x_t, u_t) = a(x_t) + b(u_t) \\ y_t &= C(x_t, u_t) = c(x_t) + d(u_t) \end{aligned}$$

for x_t, y_t, and u_t the respective N_x-state, N_y-output, and N_u-input vectors with corresponding system (process), input, measurement (output), and feedthrough functions. The N_x-dimensional system and input functions are defined by $a(\cdot)$ and $b(\cdot)$, while the N_y dimensional output and feedthrough functions are given by $c(\cdot)$ and $d(\cdot)$.

In order to specify the solution of the N_xth order differential equations completely, we must specify the above noted functions along with a set of N_x-initial conditions at time t_0 and the input for all $t \geq t_0$. Here N_x is the dimension of the "minimal" set of state variables.

If we constrain the state–space representation to be linear in the states, then we obtain the generic continuous-time, *linear time-varying state–space* model given by

$$\begin{aligned} \dot{x}_t &= A_t x_t + B_t u_t \\ y_t &= C_t x_t + D_t u_t \end{aligned} \tag{4.1}$$

where $x_t \in \mathcal{R}^{N_x \times 1}$, $u_t \in \mathcal{R}^{N_u \times 1}$, $y_t \in \mathcal{R}^{N_y \times 1}$ and the respective system, input, output, and feedthrough matrices are $A \in \mathcal{R}^{N_x \times N_x}$, $B \in \mathcal{R}^{N_x \times N_u}$, $C \in \mathcal{R}^{N_y \times N_x}$ and $D \in \mathcal{R}^{N_y \times N_u}$.

The interesting property of the state–space representation is to realize that these models represent a complete generic form for almost any physical system. That is, if we have an RLC-circuit or a MCK-mechanical system, their dynamics are governed by the identical set of differential equations, but only their coefficients differ. Of course, the physical meaning of the states are different, but this is the idea behind state–space—*many physical systems* can be captured by this generic form of differential equations, even though the systems are physically different. Systems theory, which is essentially the study of dynamic systems, is based on the study of state–space models and is rich with theoretical results exposing the underlying properties of the dynamic system under investigation. This is one of the major reasons why state–space models are employed in signal processing, especially when the system is *multivariable* having

[1] We separate x_t and u_t for future models, but it is not really necessary.

multiple inputs and multiple outputs. Next we develop the relationship between the state–space representation and input–output relations—the transfer function.

For this development we constrain the state–space representation above to be a (deterministic) *linear time-invariant* (LTI) *state–space* model given by

$$\begin{aligned}\dot{x}_t &= A_c x_t + B_c u_t \\ y_t &= C_c x_t + D_c u_t\end{aligned} \tag{4.2}$$

where $A_t \rightarrow A_c$, $B_t \rightarrow B_c$, $C_t \rightarrow C_c$, and $D_t \rightarrow D_c$ are their time-invariant counterparts with the subscript "c' annotating continuous-time matrices.

This LTI model corresponds to the constant coefficient differential equation solutions which can be solved using Laplace transforms. Taking the Laplace transform of these equations, we have

$$sX(s) - x_{t_0} = A_c X(s) + B_c U(s)$$

and solving for $X(s)$

$$X(s) = (sI - A_c)^{-1} x_{t_0} + (sI - A_c)^{-1} B_c U(s) \tag{4.3}$$

where $I \in \mathcal{R}^{N_x \times N_x}$ is the identity matrix. The corresponding output is

$$Y(s) = C_c X(s) + D_c U(s) \tag{4.4}$$

Therefore, combining these relations, we obtain

$$Y(s) = [C_c (sI - A_c)^{-1} B_c + D_c] U(s) + C_c (sI - A_c)^{-1} x_{t_0} \tag{4.5}$$

From the definition of transfer function (zero initial conditions), we have the desired result

$$H(s) = C_c (sI - A_c)^{-1} B_c + D_c \tag{4.6}$$

Taking the inverse Laplace transform of this equation gives us the corresponding *impulse response matrix* of the LTI system as [1]

$$H(t, \tau) = C_c e^{A_c (t-\tau)} B_c + D_c \delta(t - \tau) \quad \text{for } t \geq \tau \tag{4.7}$$

So we see that the state–space representation enables us to express the input–output relations in terms of the internal variables or states. Note also that this is a multivariable representation as compared to the usual single input–single output (scalar) systems models that frequently appear in the signal processing literature.

Now that we have the multivariable transfer function representation of our LTI system, we can solve the state equations directly using inverse transforms to obtain the time domain solutions. First we simplify the notation by defining the Laplace transform of the state-transition matrix or the so-called *resolvent matrix* of systems theory [1, 3] as

$$\Phi_c(s) := (sI - A_c)^{-1} \tag{4.8}$$

Therefore we can rewrite the transfer function matrix as

$$H(s) = C_c \Phi_c(s) B_c + D_c \tag{4.9}$$

and the corresponding state–input transfer matrix by

$$X(s) = \Phi_c(s) x_{t_0} + \Phi_c(s) B_c U(s) \tag{4.10}$$

Taking the inverse Laplace transformation gives the time domain solution

$$x_t = \mathcal{L}^{-1}[X(s)] = \Phi_c(t, t_0) x_{t_0} + \Phi_c(t, t_0) B_c * u_t$$

or

$$x_t = \underbrace{\Phi_c(t, t_0) x_{t_0}}_{\text{Zero input}} + \underbrace{\int_{t_0}^{t} \Phi_c(t, \alpha) B_c u_\alpha \, d\alpha}_{\text{Zero state}} \tag{4.11}$$

with corresponding output solution

$$y_t = C_c \Phi_c(t, t_0) x_{t_0} + \int_{t_0}^{t} C_c \Phi_c(t, \alpha) B_c u_\alpha \, d\alpha \tag{4.12}$$

The *state-transition matrix* $\Phi_c(t, t_0)$ is the critical component in the solution of the state equations. Ignoring the input (zero state) of the LTI state–space system, we have the set of (homogeneous) vector-matrix state equations

$$\dot{x}_t = A_c x_t \tag{4.13}$$

It is well known from linear algebra [1] that this equation has the matrix exponential as its solution

$$x_t = \Phi_c(t, t_0) x_{t_0} = e^{A_c(t-t_0)} x_{t_0} \tag{4.14}$$

The meaning of "transition" is now clear since knowledge of the transition matrix $\Phi_c(t, t_0)$ enables us to calculate the transition of the state vector from time t_0 to any $t > t_0$. Taking the Laplace transform of this equation gives

$$X(s) = \Phi_c(s) x_{t_0} = (sI - A_c)^{-1} x_{t_0}$$

with

$$e^{A_c t} = \mathcal{L}^{-1}[(sI - A_c)^{-1}] \tag{4.15}$$

We have that the *state-transition matrix* for an LTI system is

$$\Phi_c(t, t_0) = e^{A_c(t-t_0)}, \quad t \geq t_0 \tag{4.16}$$

Revisiting the continuous-time system of Eq. 4.11 and substituting the matrix exponential for the state-transition matrix gives the LTI solution as

$$x_t = e^{A_c(t-t_0)}x_{t_0} + \int_{t_0}^{t} e^{A_c(t-\alpha)}B_c u_\alpha \, d\alpha \tag{4.17}$$

with corresponding measurement system

$$y_t = C_c x_t \tag{4.18}$$

In general, the state-transition matrix satisfies the following *properties* [1, 2]:

1. $\Phi_c(t, t_0)$ is uniquely defined for $t, t_0 \in [0, \infty)$ [Unique]
2. $\Phi_c(t, t) = I$ [Identity]
3. $\Phi_c(t)$ satisfies the matrix differential equation:

$$\dot{\Phi}_c(t, t_0) = A_t \Phi_c(t, t_0), \qquad \Phi_c(t_0, t_0) = I, \quad t \geq t_0 \tag{4.19}$$

4. $\Phi_c(t, t_0) = \Phi_c(t, \tau) \times \Phi_c(\tau, \alpha) \times \cdots \times \Phi_c(\beta, t_0)$ [Semigroup]
5. $\Phi_c(t, \tau)^{-1} = \Phi_c(\tau, t)$ [Inverse]

Thus, the transition matrix plays a pivotal role in LTI systems theory for the analysis and prediction of the response of linear time-invariant and time-varying systems [2]. For instance, the poles of an LTI govern such important properties as stability and response time. The poles are the *roots* of the *characteristic* (polynomial) *equation* of A_c, which is found by solving for the roots of the determinant of the resolvent, that is,

$$|\Phi_c(s)| = |(sI - A_c)|_{s=p_i} = 0 \tag{4.20}$$

Stability is determined by assuring that all of the poles lie within the left half of the S-plane. The poles of the system determine its response as

$$y_t = \sum_{i=1}^{N_x} K_i \, e^{-p_i t} \tag{4.21}$$

where A_c is diagonal which in this case is given by $A_c = \text{diag}[p_1, p_2, \ldots, p_{N_x}]$, the eigenvalues of A_c. Next we consider the sampled-data, state–space representation.

4.3 SAMPLED-DATA STATE–SPACE MODELS

Sampling a continuous-time system is commonplace with the advent of high speed analog-to-digital converters (ADC) and modern computers. A sampled-data system lies somewhere between the continuous analog domain (physical system) and the purely discrete domain (stock market prices). Since we are strictly sampling a continuous-time process, we must assure that all of its properties are preserved. The well-known *Nyquist sampling theorem* precisely expresses the required conditions to achieve "perfect" reconstruction of the process from its samples [13].

Thus, if we have a physical system governed by continuous-time dynamics and we "sample" it at given time instants, then a sampled-data model can be obtained directly from the solution of the continuous-time state–space model. That is, we know from the previous section that

$$x_t = \Phi_c(t, t_0)x_{t_0} + \int_{t_0}^{t} \Phi_c(t, \alpha)B_c(\alpha)u_\alpha \, d\alpha$$

where $\Phi_c(\cdot, \cdot)$ is the continuous-time state-transition matrix that satisfies the matrix differential equation

$$\dot{\Phi}_c(t, t_0) = A_t\Phi_c(t, t_0), \qquad \Phi_c(t_0, t_0) = I, \quad t \geq t_0$$

Sampling this system such that $t \rightarrow t_k$ over the interval $(t_k, t_{k-1}]$, we have the corresponding *sampling interval* defined by $\Delta t_k := t_k - t_{k-1}$. Note this representation need *not* necessarily be equally spaced—another important property of the state–space representation. Thus the sampled solution becomes (with notation change)

$$x(t_k) = \Phi(t_k, t_{k-1})x(t_{k-1}) + \int_{t_{k-1}}^{t_k} \Phi(t_k, \alpha)B_c(\alpha)u_\alpha \, d\alpha \tag{4.22}$$

and therefore from the differential equation above, we have the solution

$$\Phi(t_k, t_{k-1}) = \int_{t_{k-1}}^{t_k} A(\alpha)\Phi(t_k, \alpha) \, d\alpha \quad \text{for} \quad \Phi(t_0, t_0) = I \tag{4.23}$$

where $\Phi(t_k, t_{k-1})$ is the *sampled-data* state-transition matrix—the critical component in the solution of the state equations enabling us to calculate the state evolution in time.

If we further assume that the input excitation is *piecewise constant* ($u_\alpha \rightarrow u(t_{k-1})$) over the interval $(t_k, t_{k-1}]$, then it can be removed from under the superposition integral in Eq. 4.22 to give

$$x(t_k) = \Phi(t_k, t_{k-1})x(t_{k-1}) + \left(\int_{t_{k-1}}^{t_k} \Phi(t_k, \alpha)B_c(\alpha) \, d\alpha \right) \times u(t_{k-1}) \tag{4.24}$$

Under this assumption, we can define the sampled *input transmission* matrix as

$$B(t_{k-1}) := \int_{t_{k-1}}^{t_k} \Phi(t_k, \alpha) B_c(\alpha)\, d\alpha \tag{4.25}$$

and therefore the *sampled-data state–space system* with equally or unequally sampled data is given by

$$\begin{aligned} x(t_k) &= \Phi(t_k, t_{k-1}) x(t_{k-1}) + B(t_{k-1}) u(t_{k-1}) \\ y(t_k) &= C(t_k) x(t_k) \end{aligned} \tag{4.26}$$

Computationally, sampled-data systems pose no particular problems when care is taken, especially since reasonable approximation and numerical integration methods exist [17]. For instance, the following three methods of solving for the state transition and input transmission matrices can be quite effective. If we constrain the system to be LTI, then we have that the matrix exponential can be represented by the Taylor-series expansion

$$e^{A_c \Delta t_k} = \sum_{i=0}^{\infty} \frac{(A_c\, \Delta t_k)^i}{i!} \tag{4.27}$$

Truncating the series is possible at an acceptable error magnitude [17]. This is called the *series approach* to estimating the state-transition matrix and can be determined for a finite sum. For example, a simple first-order approximation uses the relations

$$\begin{aligned} \Phi(t_k, t_{k-1}) &\approx (I + \Delta t_k A_c) \\ B(t_k) &\approx \Delta t_k B_c \end{aligned} \tag{4.28}$$

This direct approach can yield unsatisfactory results; however, one improved solution is based on the *Padé approximation* incorporating a scaling and squaring technique [17, 18]. This *scaling and squaring* property of the matrix exponential is given by

$$e^{A_c \Delta t_k} = (e^{A_c \Delta t_k / m})^m \tag{4.29}$$

and is based on choosing the integer m to be a power of two such that the exponential term can reliably and efficiently be calculated followed by repeated squaring. A typical criterion is to choose $\|A\|/m \ll 1$ yielding a very effective numerical technique for either Taylor or Padé approximants.

Ordinary differential equation methods using *numerical integration* techniques (e.g., Runge–Kutta, Gear's method) offer another practical approach to solving for the state-transition matrix and the corresponding input matrices as given by Eqs. 4.23 and 4.25, respectively. The advantages of numerical integration techniques are reliability and accuracy as well as applicability to time-varying and nonlinear systems. The major disadvantage is computational time which can be very high for stiff (large eigenvalue spread) differential equations and variable integration step sizes [17]. In any case, the sampled-data system has the property that it has evolved from a system with continuous dynamics and must be accurately approximated or numerically integrated to produce reliable solutions.

The final class of methods we discuss briefly are the *matrix decomposition methods* [18]. These methods are based on *similarity transformations*, that is,

$$\tilde{A}_c = TA_cT^{-1} \quad \text{and} \quad e^{\tilde{A}_c\Delta t} = Te^{A_c\Delta t}T^{-1} \tag{4.30}$$

where $\Delta t := t - \tau$ in the continuous-time case. If the similarity transformation matrix is chosen to be an *eigenvalue–eigenvector* transformation say $T = V$, then

$$e^{\tilde{A}_c\Delta t} = Ve^{\Lambda_c\Delta t}V^{-1} \tag{4.31}$$

with Λ_c diagonal. The matrix exponential operation becomes a simple scalar computation

$$e^{\Lambda_c\Delta t} = \text{diag}(e^{\lambda_1\Delta t}, \dots, e^{\lambda_N\Delta t}) \tag{4.32}$$

In fact, using the eigendecomposition and applying the ordinary differential equation approach, we have that

$$\dot{x}(t) = A_c x(t)$$

and therefore the solution is given in terms of the eigenvectors v_i

$$x(t) = \sum_{i=0}^{N} \alpha_i\, e^{\lambda_i\Delta t} v_i \tag{4.33}$$

where the coefficients α_i are the solution of the linear equations $V \times \alpha = x(0)$. This approach works well when A_c is symmetric leading to an orthogonal set of eigenvectors, but can be plagued with a wealth of numerical issues that need reconciliation and sophisticated numerical techniques [18].

Consider the following example to demonstrate these approaches.

Example 4.1

Suppose we are given the following system

$$\begin{aligned} \dot{x}_t &= -0.303x_t + u_t \\ y_t &= 2x_t \end{aligned}$$

with sampling interval $\Delta t = 0.1$ initial state $x_{t_0} = 2$ and input u_t a sequence of irregularly spaced step functions. Using a first-order approximation, we obtain

$$\begin{aligned} x(t_k) &= (1 - 0.303\Delta t_k)x(t_{k-1}) + \Delta t_k u(t_{k-1}) = 0.97x(t_{k-1}) + 0.1u(t_{k-1}) \\ y(t_k) &= 2x(t_k) \end{aligned}$$

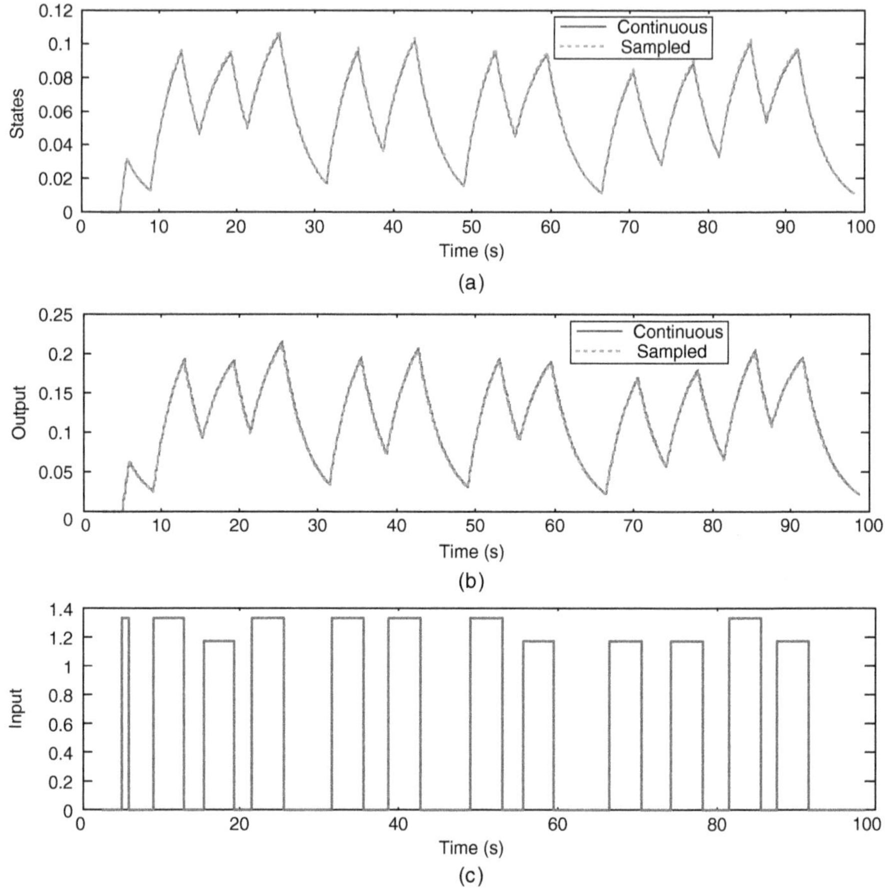

FIGURE 4.1 Sampled-data simulation (unequally spaced) of first-order continuous process: (a) states: continuous (numerical integration) and sampled (series approximation); (b) outputs: continuous (numerical integration) and sampled (series approximation); (c) input: continuous (numerical integration).

We performed numerical integration on the differential equations and a Taylor-series approximation using 25 terms to achieve an error tolerance of $\epsilon = 10^{-12}$. The simulations of the state, output, and input are shown in Figs. 4.1a, 4.1b, and 4.1c. The true continuous-time solution is shown in the figure as the solid line, while the sampled-data (discrete) solution is shown with the dotted lines. The plots reasonably overlay each another, but it is apparent that there is some truncation error evolving which can be observed at the peak amplitudes of the simulation. Thus we see that the continuous-time system in some cases can be reasonably approximated even by the Taylor-series approach. $\triangle\triangle\triangle$

This completes the discussion of sampled-data systems and approximations, next we consider the discrete-time systems.

4.4 DISCRETE-TIME STATE-SPACE MODELS

Discrete state–space models evolve in two distinct ways: naturally from the problem or from sampling a continuous-time dynamical system. An example of a natural discrete system is the dynamics of balancing our own checkbook. Here the state is the evolving balance given the past balance and the amount of the previous check. There is "no information" between time samples and so this model represents a discrete-time system that evolves naturally from the underlying problem. On the other hand, if we have a physical system governed by continuous-time dynamics, then we "sample" it at given time instants as discussed in the previous section. So we see that discrete-time dynamical systems can evolve from a wide variety of problems both naturally (checkbook) or physically (circuit). In this text we are primarily interested in physical systems (physics-based models), so we will concentrate on sampled systems reducing them to a discrete-time state–space model.

We can use a first-difference approximation[2] and apply it to the general LTI continuous-time state–space model to obtain a discrete-time system, that is,

$$\begin{aligned}\dot{x}_t &\approx \frac{x(t) - x(t-1)}{\Delta T} \approx A_c x(t-1) + B_c u(t-1)\\ y_t &\approx y(t) = C_c x(t) + D_c u(t)\end{aligned}$$

Solving for $x(t)$, we obtain

$$\begin{aligned}x(t) &= (I + A_c\,\Delta T)x(t-1) + B_c\,\Delta T u(t-1)\\ y(t) &= C_c x(t) + D_c u(t)\end{aligned} \tag{4.34}$$

Recognizing that the first-difference approximation is equivalent to a first-order Taylor-series approximation of A_c gives the discrete system, input, output, and feedthrough matrices as

$$\begin{aligned}A &\approx I + A_c\,\Delta T + O(\Delta T^2)\\ B &\approx B_c\,\Delta T\\ C &\approx C_c\\ D &\approx D_c\end{aligned} \tag{4.35}$$

We define the *nonlinear discrete-time state–space representation* by its process or system model

$$x(t) = A(x(t-1), u(t-1)) = a[x(t-1)] + b[u(t-1)] \tag{4.36}$$

[2] This approximation is equivalent to a first-order Taylor-series approximation.

and corresponding measurement or output model by

$$y(t) = C(x(t), u(t)) = c[x(t)] + d[u(t)] \tag{4.37}$$

where $x(t), u(t), y(t)$ are the respective discrete-time, N_x-state, N_u-input, and N_y-output vectors with corresponding system (process), input, output, and feedthrough functions: the N_x-dimensional system and input functions $a[\cdot], b[\cdot]$ and the N_y-dimensional output and feedthrough functions $c[\cdot], d[\cdot]$.

The discrete *linear time-varying state–space representation* is given by the *system* or *process model* as

$$x(t) = A(t-1)x(t-1) + B(t-1)u(t-1) \tag{4.38}$$

and the corresponding discrete *output* or *measurement model* as

$$y(t) = C(t)x(t) + D(t)u(t) \tag{4.39}$$

where x, u, y are the respective N_x-state, N_u-input, N_y-output and A, B, C, D are the $(N_x \times N_x)$-system, $(N_x \times N_u)$-input, $(N_y \times N_x)$-output, and $(N_y \times N_u)$-feedthrough matrices.

The state–space representation for linear, time-invariant discrete systems is characterized by constant system, input, output, and feedthrough matrices, that is,

$$A(t) = A, \quad B(t) = B, \quad \text{and} \quad C(t) = C, \quad D(t) = D$$

and is given by the LTI system

$$\begin{aligned} x(t) &= Ax(t-1) + Bu(t-1) \\ y(t) &= Cx(t) + Du(t) \end{aligned} \tag{4.40}$$

The discrete system representation replaces the Laplace transform with the Z-transform defined by the transform pair

$$\begin{aligned} X(z) &:= \sum_{t=0}^{\infty} x(t)z^{-t} \\ x(t) &= \int_{-\infty}^{\infty} X(z)z^{-1}\, dz \end{aligned} \tag{4.41}$$

Time-invariant state–space discrete systems can also be represented in *input–output* or *transfer function* form using the Z-transform to give

$$H(z) = C(zI - A)^{-1}B + D \tag{4.42}$$

Also taking inverse Z-transforms we obtain the discrete impulse response matrix as

$$H(t,k) = CA^{t-k}B + D \quad \text{for } t \geq k \tag{4.43}$$

The solution to the state-difference equations can easily be derived by induction [3] or using the transfer function approach of the previous subsection. In any case it is given by the relations

$$x(t) = \Phi(t,k)x(k) + \sum_{i=k+1}^{t} \Phi(t,i)B(i-1)u(i-1) \quad \text{for } t > k \tag{4.44}$$

where $\Phi(t,k)$ is the *discrete-time state-transition* matrix. For time-varying systems, it can be shown (by induction) that the state-transition matrix[3] satisfies

$$\Phi(t,k) = A(t-1) \cdot A(t-2) \cdots A(k)$$

while for time-invariant systems the state-transition matrix is given by

$$\Phi(t,k) = A^{t-k} \quad \text{for } t > k$$

The discrete state-transition matrix possesses properties analogous to its continuous-time counterpart, that is,

1. $\Phi(t,k)$ is uniquely defined [Unique]
2. $\Phi(t,t) = I$ [Identity]
3. $\Phi(t,k)$ satisfies the matrix-difference equation

$$\Phi(t,k) = A(t-1)\Phi(t-1,k), \qquad \Phi(k,k) = I, \quad t \geq k+1 \tag{4.45}$$

4. $\Phi(t,k) = \Phi(t,t-1) \times \Phi(t-1,t-2) \times \cdots \times \Phi(k+1,k)$ [Semigroup]
5. $\Phi^{-1}(t,k) = \Phi(k,t)$ [Inverse]

As in the continuous case, the discrete-time transition matrix has the same importance in discrete systems theory which we discuss next.

4.4.1 Discrete Systems Theory

In this subsection we investigate the discrete state–space model from a systems theory viewpoint. There are certain properties that a dynamic system must possess in order to assure a consistent representation of the dynamics under investigation. For instance, it can be shown [2] that a necessary requirement of a measurement system is that it is *observable*, that is, measurements of available variables or parameters of interest provide enough information to reconstruct the internal variables or states. Mathematically, a system is said to be *completely observable* if for any initial state,

[3] Recall that for a sampled-data system, the state-transition matrix is $\Phi(t,t_k) = e^{A(t-t_k)}$ where A is the sampled-data system or process matrix.

say $x(0)$, in the state–space there exists a finite $t > 0$ such that knowledge of the input $u(t)$ and the output $y(t)$ is sufficient to specify $x(0)$ uniquely. Recall that the deterministic linear *state–space* representation of a discrete system is defined by the following set of equations:

$$\text{State model}: \quad x(t) = A(t-1)x(t-1) + B(t-1)u(t-1)$$

with corresponding measurement system or output defined by

$$\text{Measurement model}: \quad y(t) = C(t)x(t)$$

Using this representation, the simplest example of an observable system is one in which each state is measured directly, and the measurement matrix C is an $N_x \times N_x$ matrix. Thus, from the measurement system model, we have that in order to reconstruct $x(t)$ from its measurements $y(t)$, then C must be invertible. In this case the system is said to be completely observable; however, if C is not invertible, then the system is said to be *unobservable*. The next level of complexity involves the solution to this same problem when C is an $N_y \times N_x$ matrix, then a pseudoinverse must be performed instead [1, 2]. In the general case the solution gets more involved because we are not just interested in reconstructing $x(t)$, but $x(t)$ over all finite values of t, therefore, we must include the state model, that is, the dynamics as well.

With this motivation in mind, we now formally define the concept of observability. The solution to the state representation is governed by the state-transition matrix, $\Phi(t, 0)$, where recall that the state equation is [3]

$$x(t) = \Phi(t,0)x(0) + \sum_{k=0}^{t-1} \Phi(t, k+1)B(t,k)u(k)$$

Therefore, premultiplying by the measurement matrix, the output relations are

$$y(t) = C(t)\Phi(t,0)x(0) + \sum_{k=0}^{t-1} C(t)\Phi(t,k+1)B(k)u(k) \tag{4.46}$$

or rearranging we define

$$\tilde{y}(t) := y(t) - \sum_{k=0}^{t-1} C(t)\Phi(t,k+1)B(k)u(k) = C(t)\Phi(t,0)x(0) \tag{4.47}$$

The problem is to solve this resulting equation for the initial state; therefore, multiplying both sides by $\Phi' C'$, we can infer the solution from the relation

$$\Phi'(t,0)C'(t)C(t)\Phi(t,0)x(0) = \Phi'(t,0)C(t)\tilde{y}(t)$$

Thus, the observability question now becomes under what conditions can this equation uniquely be solved for $x(0)$? Equivalently, we are asking if the null space of $C(t)\Phi(t,0)$

is $\mathbf{0} \in \mathcal{R}^{N_x \times 1}$. It has been shown [2, 4] that the following $N_x \times N_x$ *observability Gramian* has the identical null space, that is,

$$\mathcal{O}(0,t) := \sum_{k=0}^{t-1} \Phi'(k,0)C'(k)C(k)\Phi(k,0) \tag{4.48}$$

which is equivalent to determining that $\mathcal{O}(0,t)$ is nonsingular or rank N_x. Further assuming that the system is LTI leads to the $NN_y \times N_x$ *observability matrix* [4] given by

$$\mathcal{O}(N) := \begin{bmatrix} C \\ --- \\ \vdots \\ --- \\ CA^{N-1} \end{bmatrix} \tag{4.49}$$

It can be shown that a necessary and sufficient condition for a system to be completely observable is that the *rank* of $\mathcal{O}$ or $\rho[\mathcal{O}(N)]$ must be N_x. Thus, for the LTI case, checking that all of the measurements contain the essential information to reconstruct the states for an LTI system reduces to that of checking the rank of the observability matrix. Although this is a useful mathematical concept, it is primarily used as a rule-of-thumb in the analysis of complicated measurement systems.

Analogous to the system theoretic property of observability is that of controllability, which is concerned with the effect of the input on the states of the dynamic system. A discrete system is said to be *completely controllable* if for any $x(t), x(0) \in \mathcal{R}^{N_x}$ there exists an input sequence $\{u(t)\}, t = 0, \ldots, N-1$ such that the solution to the state equations with initial condition $x(0)$ is $x(t)$ for some finite t. Following the same approach as for observability, we obtain that the *controllability Gramian* defined by

$$\mathcal{C}(0,t) := \sum_{k=0}^{t-1} \Phi(0,k)B(k)B'(k)\Phi'(0,k) \tag{4.50}$$

is nonsingular or $\rho[\mathcal{C}(0,t)] = N_x$

Again for the LTI system, the $N_x \times NN_u$ *controllability matrix* defined by

$$\mathcal{C}(N) := [B|AB| \ldots |A^{N-1}B] \tag{4.51}$$

must satisfy the rank condition $\rho[\mathcal{C}] = N_x$ to be completely controllable [4].

If we continue with the LTI system description, we know from Z-transform theory that the discrete transfer function can be represented by an infinite power series, that is,

$$H(z) = C(zI - A)^{-1}B = \sum_{k=1}^{\infty} H(k)z^{-k} \quad \text{for } H(k) = CA^{k-1}B \tag{4.52}$$

where $H(k)$ is the $N_y \times N_u$ unit impulse response matrix which may also be viewed as a Markov sequence with (A, B, C) defined as the Markov parameters.

The problem of determining the *internal description* (A, B, C) from the *external description* ($H(z)$ or $\{H(k)\}$) of Eq. 4.43 is called the *realization problem.* Out of all possible realizations (A, B, C) having the same Markov parameters, those of smallest dimension are defined as *minimal realizations.* It will subsequently be shown that the dimension of the minimal realization is identical to the degree of the characteristic polynomial (actually the minimal polynomial for multivariable systems) or equivalently the degree of the transfer function (number of system poles).

In order to develop these relations, we define the $(N \times N_y N_u) \times (N \times N_y N_u)$ *Hankel matrix* by

$$\mathcal{H}(N) := \begin{bmatrix} H(1) & H(2) & \cdots & H(N) \\ H(2) & H(3) & \cdots & H(N+1) \\ \vdots & \vdots & \cdots & \vdots \\ H(N) & H(N+1) & \cdots & H(2N) \end{bmatrix} \tag{4.53}$$

Suppose the dimension of the system is N_x. Then the Hankel matrix could be constructed such that $N = N_x$ using $2N_x$ impulse response matrices which tells us the minimum number of terms we require to extract the Markov parameters. Also $\rho[\mathcal{H}(N)] = N_x$ which is the dimension of the *minimal realization.* If we do not know the dimension of the system, then we would let (in theory) $N \to \infty$ and determine the rank of $\mathcal{H}(\infty)$. Therefore, the minimal dimension of an "unknown" system is the rank of the Hankel matrix. In order for a system to be *minimal* it must be *completely controllable* and *completely observable.* This can be seen from the fact that the Hankel matrix factors as

$$\mathcal{H}(N) = \begin{bmatrix} CB & \cdots & CA^{N-1}B \\ \vdots & & \vdots \\ CA^{N-1}B & \cdots & CA^{2N-2}B \end{bmatrix} = \begin{bmatrix} C \\ \vdots \\ CA^{N-1} \end{bmatrix} [B| \cdots |A^{N-1}B] \tag{4.54}$$

or more simply

$$\mathcal{H}(N) = \mathcal{O}(N)\mathcal{C}(N) \tag{4.55}$$

From this factorization it follows that $\rho[\mathcal{H}(N)] = \min[\rho(\mathcal{O}(N)), \rho(\mathcal{C}(N))] = N_x$. Therefore, we see that the properties of controllability and observability are carefully woven into that of minimality, and testing the rank of the Hankel matrix yields the dimensionality of the underlying dynamic system. This fact will prove crucial when we must "identify" a system $\Sigma = \{A, B, C\}$ from noisy measurement data. For instance, many of the classical *realization* techniques [19, 20] rely on the factorization of Eq. 4.55 to extract the system or process matrix, that is,

$$\mathcal{O}(N) \times A = \begin{bmatrix} C \\ \vdots \\ CA^{N-2} \\ --- \\ CA^{N-1} \end{bmatrix} A = \begin{bmatrix} CA \\ \vdots \\ CA^{N-1} \\ --- \\ CA^{N} \end{bmatrix} =: \mathcal{O}^{\uparrow} \tag{4.56}$$

Solving for A using the pseudoinverse [18] gives

$$\hat{A} = \mathcal{O}^{\#}(N)\mathcal{O}^{\uparrow} \quad \text{where } \mathcal{O}^{\#}(N) := (\mathcal{O}'(N)\mathcal{O}(N))^{-1}\mathcal{O}'(N) \tag{4.57}$$

An efficient realization technique for a *scalar* (single input/single output) system can be obtained by performing a *singular value decomposition* (SVD) of the Hankel matrix constructed from the impulse response sequence $H(k); k = 1, \ldots, N$, that is,

$$\mathcal{H}(N) = U \times \Lambda \times V \text{ with } \rho(\mathcal{H}(N)) = N_x \tag{4.58}$$

where $\Lambda = \text{diag}[\lambda_i]; i = 1, \ldots, N^*; N^* := N \times N_y N_u$, $\{\lambda_i\}$ is the set of singular values, and U and V the corresponding left and right singular vector matrices [18]. A variety of techniques can be used to estimate the *rank* of a matrix, perhaps the simplest (given the SVD) being the *best rank approximationα* (%) based on the ratio of singular values

$$\alpha(\hat{N}_x) := \left(\frac{\sum_{i=1}^{\hat{N}_x} \lambda_i}{\sum_{i=1}^{N^*} \lambda_i} \right) \times 100 \tag{4.59}$$

Here α (%) is the percentage of the original matrix (Hankel) approximated by choosing $\hat{N}_x$. A threshold, τ_x (%) can be selected to search over i for that particular $\hat{N}_x$ that "best" approximates $\mathcal{H}$ up to the threshold, that is,

$$\alpha_n(\hat{N}_x) \geq \tau_x(\%) \quad \text{for } n = 1, \ldots, \hat{N}_x \tag{4.60}$$

Once the rank is estimated, then

$$\hat{\mathcal{H}}(N) = U\Lambda V = U \left[\begin{array}{c|c} \overline{\Lambda} & 0 \\ --- & --- \\ 0 & 0 \end{array} \right] V = \overline{U\Lambda V} \tag{4.61}$$

for $\overline{U} \in \mathcal{R}^{N_Y N_u \times N_x}$, $\overline{\Lambda} \in \mathcal{R}^{N_x \times N_x}$, $\overline{V} \in \mathcal{R}^{N_x \times N_Y N_u}$.

When the decomposition and rank ($\hat{N}_x$) are determined, the system triple (scalar) $\Sigma = (A, b, c)$ is identified by [21]

$$A = \overline{\Lambda}^{-\frac{1}{2}} \overline{U}' \overline{U}^{\uparrow} \overline{\Lambda}^{\frac{1}{2}}; \quad b = \overline{\Lambda}^{\frac{1}{2}} \overline{V}; \quad \text{and} \quad c = \overline{U} \overline{\Lambda}^{\frac{1}{2}} \tag{4.62}$$

where (as before) $\overline{U}^{\uparrow}$ is the eigenvector matrix $\overline{U}$ shifted up one row with a row of zeros ($0'$) appended [21]. Here $A^{1/2}(A^{1/2})'$ are the square-root matrices of A. It has been shown that this method results in a stable realization such that the eigenvalues of A, $\lambda(A \leq 1)$.

Example 4.2

Consider the following scalar example with impulse response $H(k) = \{1, 1/2, 1/4, 1/8\ 1/16\}$ with $N = 5$. Using the SVD approach, we would like to extract the realization $\Sigma = (A, b, c)$. Creating the Hankel matrix and performing the SVD, we obtain

$$H(5) = \begin{bmatrix} 1 & 1/2 & 1/4 \\ 1/2 & 1/4 & 1/8 \\ 1/4 & 1/8 & 1/16 \end{bmatrix} = U \Lambda V$$

$\Lambda = \text{diag}[1.31, 0, 0]$ yielding a rank $\hat{N}_x = 1$; the singular vectors are

$$U = \begin{bmatrix} -0.8729 & -0.4364 & -0.2182 \\ -0.4364 & 0.8983 & -0.0509 \\ -0.2182 & -0.0509 & 0.9746 \end{bmatrix}; \quad V = \begin{bmatrix} -0.8729 & -0.4364 & -0.2182 \\ 0.4880 & -0.7807 & -0.3904 \\ 0 & -0.4472 & 0.8944 \end{bmatrix}$$

Thus the best rank approximants are $\overline{\Lambda} = 1.3125$ and

$$\overline{U} = \begin{bmatrix} -0.8729 \\ -0.4364 \\ -0.2182 \end{bmatrix}; \quad \overline{U}^{\uparrow} = \begin{bmatrix} -0.4364 \\ -0.2182 \\ 0 \end{bmatrix}; \quad \overline{V}' = \begin{bmatrix} -0.8729 \\ -0.4364 \\ -0.2182 \end{bmatrix}$$

Therefore, we obtain the realizations

$$\begin{aligned} A &= \overline{\Lambda}^{-\frac{1}{2}} \overline{U}' \overline{U}^{\uparrow} \overline{\Lambda}^{\frac{1}{2}} \\ &= (0.873)[-0.8729\ \ -0.4364\ \ -0.2182] \begin{bmatrix} -0.4364 \\ -0.2182 \\ 0 \end{bmatrix} (1.1456) \\ &= 0.48 \approx 1/2; \ b' = \overline{\Lambda}^{\frac{1}{2}} \overline{V} = [1\ \ 0\ \ 0]; \ c = \overline{U}\,\overline{\Lambda}^{\frac{1}{2}} = [-1\ \ 0\ \ 0] \end{aligned}$$

△△△

This completes the subsection on discrete systems theory. It should be noted that all of the properties discussed in this section exist for continuous-time systems (see Ref. [2] for details).

4.5 GAUSS–MARKOV STATE-SPACE MODELS

In this section we extend the state–space representation to incorporate random inputs or noise sources along with random initial conditions. We briefly discuss the continuous-time representation evolving to the sampled-data model and then provide a detailed discussion of the discrete-time Gauss–Markov model which will be used extensively throughout this text.

4.5.1 Continuous-Time/Sampled-Data Gauss–Markov Models

We start by defining the *continuous-time Gauss–Markov model*. If we constrain the state–space representation to be linear in the states, then we obtain the generic continuous-time, *linear time-varying Gauss–Markov state–space* model given by

$$\begin{aligned} \dot{x}_t &= A_t x_t + B_t u_t + W_t w_t \\ y_t &= C_t x_t + v_t \end{aligned} \tag{4.63}$$

where $x_t \in \mathcal{R}^{N_x \times 1}$, $u_t \in \mathcal{R}^{N_u \times 1}$, $w_t \in \mathcal{R}^{N_w \times 1}$, $y_t \in \mathcal{R}^{N_y \times 1}$, $v_t \in \mathcal{R}^{N_y \times 1}$ are the continuous-time state, input (deterministic), process noise, measurement, and measurement noise[4] vectors with corresponding system (process), input, output, and feedthrough matrices: $A \in \mathcal{R}^{N_x \times N_x}$, $B \in \mathcal{R}^{N_x \times N_u}$, $W \in \mathcal{R}^{N_x \times N_w}$, and $C \in \mathcal{R}^{N_y \times N_x}$.

The continuous-time Gaussian stochastic processes are $w_t \sim \mathcal{N}(0, R_{w_c w_c}(t))$ and $v_t \sim \mathcal{N}(0, R_{v_c v_c}(t))$ with initial state defined by $x_0 \sim \mathcal{N}(\overline{x}_0, \overline{P}_0)$. The corresponding statistics of this model follow with the dynamic *mean* derived directly from Eq. 4.63 (see Refs. [6–8] for details) as

$$\dot{m}_{x_t} = A_t m_{x_t} + B_t u_t \tag{4.64}$$

and the *variance* (continuous-time Lyapunov equation)

$$\dot{P}_t = A_t P_t + P_t A_t' + W_t R_{w_c w_c}(t) W_t' \tag{4.65}$$

with corresponding *covariance*

$$P_{t,\tau} = \begin{cases} \Phi_c(t, \tau) P_{\tau,\tau} & \text{for } t \geq \tau \\ P_{t,t} \Phi_c'(t, \tau) & \text{for } t \leq \tau \end{cases} \tag{4.66}$$

where $\Phi_c(t, \tau)$ is the continuous state-transition matrix.

[4] Note that process and measurement noise sources are different. The process noise term is used primarily to model uncertainties in the input, state, or even possibly unknown model parameters and it is "filtered" or colored (correlated) by the system dynamics, while measurement noise is uncorrelated and used to model instrumentation or extraneous environmental noise.

As before, for the deterministic case, the stochastic sampled-data system follows from the continuous-time state solution

$$x_t = \Phi_c(t, t_0)x_{t_0} + \int_{t_0}^{t} \Phi_c(t, \tau)B_\tau u_\tau \, d\tau + \int_{t_0}^{t} \Phi_c(t, \tau)W_\tau w_\tau \, d\tau \tag{4.67}$$

Sampling this system with $t \rightarrow t_k$ over the interval $(t_k, t_{k-1}]$, the sampled solution becomes (with notational change)[5]

$$x(t_k) = \Phi(t_k, t_{k-1})x(t_{k-1}) + \int_{t_{k-1}}^{t_k} \Phi(t_k, \tau)B_\tau u_\tau \, d\tau + \int_{t_{k-1}}^{t_k} \Phi(t_k, \tau)W_\tau w_\tau \, d\tau \tag{4.68}$$

If we further assume that the input excitation is *piecewise constant* ($u_\tau \rightarrow u(t_{k-1})$) over the interval $(t_k, t_{k-1}]$, then it can be removed from under the superposition integral in Eq. 4.68 to give

$$\begin{aligned} x(t_k) = \Phi(t_k, t_{k-1})x(t_{k-1}) &+ \left(\int_{t_{k-1}}^{t_k} \Phi(t_k, \tau)B_\tau \, d\tau \right) \times \, u(t_{k-1}) \\ &+ \int_{t_{k-1}}^{t_k} \Phi(t_k, \tau)W_\tau w_\tau \, d\tau \end{aligned} \tag{4.69}$$

We define the *sampled-data input transmission matrix* as

$$B(t_{k-1}) := \int_{t_{k-1}}^{t_k} \Phi(t_k, \tau)B_\tau \, d\tau \tag{4.70}$$

with the *sampled-data process noise covariance* matrix given by

$$R_{ww}(t_{k-1}) := \int_{t_{k-1}}^{t_k} \Phi(t_k, \tau)W_\tau R_{w_c w_c}(\tau)W_\tau' \Phi'(t_k, \tau) \, d\tau \tag{4.71}$$

and therefore the *sampled-data state–space system* with equally or unequally sampled data is given by

$$\begin{aligned} x(t_k) &= \Phi(t_k, t_{k-1})x(t_{k-1}) + B(t_{k-1})u(t_{k-1}) + W(t_{k-1})w(t_{k-1}) \\ y(t_k) &= C(t_k)x(t_k) + v(t_k) \end{aligned} \tag{4.72}$$

[5] We note in passing that this solution is conceptual and must actually follow a much more rigorous framework embedded in stochastic integrals and is beyond the scope of this text (see Ref. [6–8] for details).

for $w_{t_k} \sim \mathcal{N}(0, R_{ww}(t_k))$ and $v_{t_k} \sim \mathcal{N}(0, R_{vv}(t_k))$ with initial state defined by $x(t_0) \sim \mathcal{N}(\overline{x}(t_0), \overline{P}(t_0))$. Recall from the deterministic solution that if we use a first-order Taylor-series approximation we have that

$$
\begin{aligned}
\Phi(t_k, t_{k-1}) &\approx I + \Delta t_k \times A(t_{k-1}) \\
B(t_{k-1}) &\approx \Delta t_k \times B_{t_{k-1}} \\
R_{ww}(t_{k-1}) &\approx \Delta t_k \times W_{t_{k-1}} R_{w_c w_c}(t_{k-1}) W'_{t_{k-1}} \\
R_{vv}(t_k) &\approx R_{v_c v_c}(t_k)/\Delta t_k
\end{aligned}
\tag{4.73}
$$

The corresponding mean and covariance of the sampled-data process are

$$m_x(t_k) = A(t_{k-1}) m_x(t_{k-1}) + B(t_{k-1}) u(t_{k-1}) \tag{4.74}$$

and the *measurement mean vector* m_y is

$$m_y(t_k) = C(t_k) m_x(t_k) \tag{4.75}$$

The *state variance* is given by

$$P(t_k) = A(t_{k-1}) P(t_{k-1}) A'(t_{k-1}) + W(t_{k-1}) R_{ww}(t_{k-1}) W'(t_{k-1}) \tag{4.76}$$

and the measurement variance is

$$R_{yy}(t_k) = C(t_k) P(t_k) C'(t_k) + R_{vv}(t_k) \tag{4.77}$$

This completes the development of the sampled-data Gauss–Markov representation evolving from a continuous-time stochastic process, next we consider the more pragmatic discrete-time model.

4.5.2 Discrete-Time Gauss-Markov Models

Here we investigate the case when random inputs are applied to a discrete state–space system with random initial conditions. If the excitation is a random signal, then the state is also random. Restricting the input to be deterministic $u(t-1)$ and the noise to be zero-mean, white, random Gaussian $w(t-1)$, the *Gauss–Markov model* evolves as

$$x(t) = A(t-1)x(t-1) + B(t-1)u(t-1) + W(t-1)w(t-1) \tag{4.78}$$

where $w \sim \mathcal{N}(0, R_{ww}(t-1))$ and $x(0) \sim \mathcal{N}(\overline{x}(0), P(0))$.

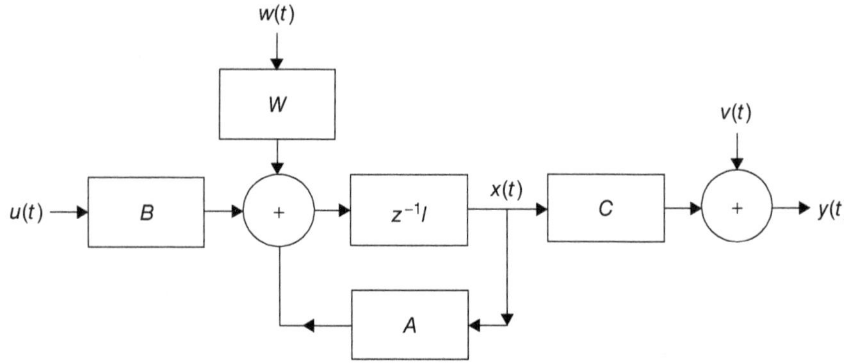

FIGURE 4.2 Gauss–Markov model of a discrete process.

The solution to the Gauss–Markov equations can easily be obtained by induction to give

$$x(t) = \Phi(t,k)x(k) + \sum_{i=k}^{t-1} \Phi(t,i+1)B(i)u(i) + \sum_{i=k}^{t-1} \Phi(t,i+1)W(i)w(i) \tag{4.79}$$

which is first-order Markov depending only on the previous state. Since $x(t)$ is just a linear transformation of Gaussian processes it is also Gaussian. Thus, we can represent a Gauss–Markov process easily using the state–space models.

When the *measurement* model is also included, we have

$$y(t) = C(t)x(t) + v(t) \tag{4.80}$$

where $v \sim \mathcal{N}(0, R_{vv}(t))$. The model is shown diagrammatically in Fig. 4.2.

Since the Gauss–Markov model of Eq. 4.78 is characterized by a Gaussian distribution, it is completely specified statistically by its mean and variance. Therefore, if we take the expectation of Eqs. 4.78 and 4.80, respectively, we obtain the *state mean vector* m_x as

$$m_x(t) = A(t-1)m_x(t-1) + B(t-1)u(t-1) \tag{4.81}$$

and the *measurement mean vector* m_y as

$$m_y(t) = C(t)m_x(t) \tag{4.82}$$

The *state variance*[6] $P(t) := \text{var}\{x(t)\}$ is given by the discrete Lyapunov equation

$$P(t) = A(t-1)P(t-1)A'(t-1) + W(t-1)R_{ww}(t-1)W'(t-1) \tag{4.83}$$

[6] We use the shorthand notation $P(k) := P_{xx}(k,k) = \text{cov}\{x(k), x(k)\} = \text{var}\{x(k)\}$ throughout this text.

TABLE 4.1 Gauss–Markov Representation

State Propagation
$x(t) = A(t-1)x(t-1) + B(t-1)u(t-1) + W(t-1)w(t-1)$
State Mean Propagation
$m_x(t) = A(t-1)m_x(t-1) + B(t-1)u(t-1)$
State Variance/Covariance Propagation
$P(t) = A(t-1)P(t-1)A'(t-1) + W(t-1)R_{ww}(t-1)W'(t-1)$
$P(t,k) = \begin{cases} \Phi(t,k)P(k) & t \geq k \\ P(t)\Phi'(t,k) & t \leq k \end{cases}$
Measurement Propagation
$y(t) = C(t)x(t) + v(t)$
Measurement Mean Propagation
$m_y(t) = C(t)m_x(t)$
Measurement Variance/Covariance Propagation
$R_{yy}(t) = C(t)P(t)C'(t) + R_{vv}(t)$
$R_{yy}(t,k) = C(t)P(t)C'(t) + R_{vv}(t,k)$

and the measurement variance $R_{yy}(t) := \text{var}\{y(t)\}$ is

$$R_{yy}(t) = C(t)P(t)C'(t) + R_{vv}(t) \tag{4.84}$$

Similarly, it can be shown that the *state covariance* propagates according to the following equations

$$P(t,k) = \begin{cases} \Phi(t,k)P(k) & \text{for } t \geq k \\ P(t)\Phi'(t,k) & \text{for } t \leq k \end{cases} \tag{4.85}$$

We summarize the Gauss–Markov and corresponding statistical models in Table 4.1.

If we restrict the Gauss–Markov model to the stationary case, then

$$A(t) = A, B(t) = B, \quad C(t) = C, \quad W(t) = W, \quad R_{ww}(t) = R_{ww}, \quad \text{and} \quad R_{vv}(t) = R_{vv}$$

and the variance equations become

$$P(t) = AP(t-1)A' + WR_{ww}W'$$

and

$$R_{yy}(t) = CP(t)C' + R_{vv} \tag{4.86}$$

At steady state ($t \to \infty$), we have

$$P(t) = P(t-1) = \cdots = P_{ss} := P$$

and therefore, the measurement covariance relations become

$$R_{yy}(0) = CPC' + R_{vv} \quad \text{for lag } k = 0 \tag{4.87}$$

By induction, it can be shown that

$$R_{yy}(k) = CA^{|k|}PC' \quad \text{for } k \neq 0 \tag{4.88}$$

The measurement power spectrum is easily obtained by taking the Z-transform of this equation to obtain

$$S_{yy}(z) = CS_{xx}(z)C' + S_{vv}(z) \tag{4.89}$$

where

$$S_{xx}(z) = T(z)S_{ww}(z)T'(z^{-1}) \quad \text{for } T(z) = (zI - A)^{-1}W$$

with

$$S_{ww}(z) = R_{ww} \quad \text{and} \quad S_{vv}(z) = R_{vv}$$

Thus, using $H(z) = CT(z)$, the spectrum is given by

$$S_{yy}(z) = H(z)R_{ww}H'(z^{-1}) + R_{vv} \tag{4.90}$$

So we see that the Gauss–Markov state–space model enables us to have a more general representation of a multichannel stochastic signal. In fact, we are able to easily handle the multichannel and nonstationary statistical cases within this framework. Generalizations are also possible with the vector models, but the forms become quite complicated and require some knowledge of multivariable systems theory and canonical forms (see Ref. [1] for details). Before we leave this subject, let us consider a simple input–output example with Gauss–Markov models.

Example 4.3

Consider the following difference equation driven by random (white) noise

$$y(t) = -ay(t-1) + e(t-1)$$

The corresponding state–space representation is obtained as

$$x(t) = -ax(t-1) + w(t-1) \quad \text{and} \quad y(t) = x(t)$$

Taking Z-transforms (ignoring the randomness), we obtain the transfer function

$$H(z) = \frac{1}{1 - az^{-1}}$$

Using Eq. 4.83, the variance equation for the above model is

$$P(t) = a^2 P(t-1) + R_{ww}$$

Assume the process is stationary, then $P(t) = P$ for all t and solving for P it follows that

$$P = \frac{R_{ww}}{1 - a^2}$$

Therefore,

$$R_{yy}(k) = CA^{|k|}PC' = \frac{a^{|k|}R_{ww}}{1 - a^2} \quad \text{and} \quad R_{yy}(0) = CPC' + R_{vv} = \frac{R_{ww}}{1 - a^2}$$

Choosing $R_{ww} = 1 - a^2$ gives $R_{yy}(k) = a^{|k|}$. Taking Z-transforms the discrete power spectrum is given by

$$S_{yy}(z) = H(z)R_{ww}H'(z^{-1}) + R_{vv} = \frac{1}{1 - az^{-1}} R_{ww} \frac{1}{1 - az}$$

Therefore, we conclude that for stationary processes these models are equivalent.

Now if we assume a nonstationary process and let $a = -0.75, x(0) \sim \mathcal{N}(1, 2.3), w \sim \mathcal{N}(0, 1)$, and $v \sim \mathcal{N}(0, 4)$, then the Gauss–Markov model is given by

$$x(t) = 0.75x(t-1) + w(t-1) \quad \text{and} \quad y(t) = x(t) + v(t)$$

The corresponding statistics are given by the mean relations

$$\begin{aligned} m_x(t) &= 0.75m_x(t-1) \quad & m_x(0) &= 1 \\ m_y(t) &= m_x(t) & m_y(0) &= m_x(0) \end{aligned}$$

and the variance equations

$$P(t) = 0.5625P(t-1) + 1 \quad \text{and} \quad R_{yy}(t) = P(t) + 4$$

We apply the simulator available in SSPACK_PC [13] to obtain a 100-sample realization of the process. The results are shown in Figs. 4.3a, 4.3b, 4.3c. In Figs. 4.3a and 4.3b we see the mean and simulated states with corresponding confidence interval about the mean, that is,

$$[m_x(t) \pm 1.96\sqrt{P(t)}]$$

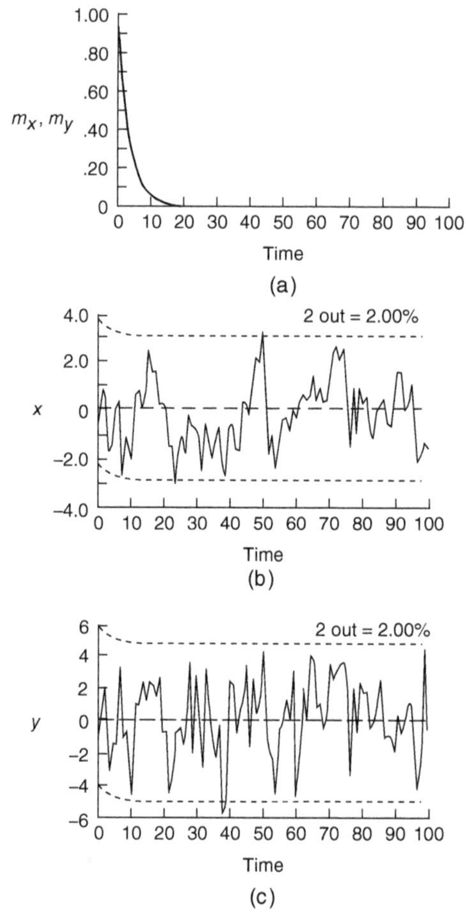

FIGURE 4.3 Gauss–Markov simulation of first-order process.

and

$$P = \frac{R_{ww}}{1 - a^2} = 2.286$$

Using the above confidence interval, we expect 95% of the samples to lie within ($m_x \rightarrow 0$) $\pm 3.03(1.96 \times \sqrt{2.286})$. From the figure we see that only 2 of the 100 samples exceed this bound, indicating a statistically acceptable simulation. We observe similar results for the simulated measurements. The steady-state variance is given by

$$R_{yy} = P + R_{vv} = 2.286 + 4 = 6.286$$

Therefore, we expect 95% of the measurement samples to lie within ($m_y \rightarrow 0$) $\pm 5.01(1.96 \times \sqrt{6.286})$ at steady state. This completes the example. △△△

4.6 INNOVATIONS MODEL

In this Section we briefly develop the innovations model which is related to the Gauss–Markov representation just discussed. The significance of this model will be developed throughout the text, but we take the opportunity now to show its relationship to the basic Gauss–Markov representation. We start by extending the original Gauss–Markov representation to the correlated process and measurement noise case and then showing how the innovations model is a special case of this structure.

The *standard* Gauss–Markov model for *correlated process and measurement noise* is given by

$$\begin{aligned} x(t) &= Ax(t-1) + Bu(t-1) + W(t-1)w^*(t-1) \\ y(t) &= Cx(t) + v^*(t) \end{aligned} \tag{4.91}$$

where $R^*(t,k) := R^*\delta(t-k)$ and

$$R^* := \begin{bmatrix} R_{w^*w^*} & | & R_{w^*v^*} \\ --- & | & --- \\ R_{v^*w^*} & | & R_{v^*v^*} \end{bmatrix} = \begin{bmatrix} WR_{ww}W' & | & WR_{wv} \\ ---- & | & --- \\ R_{vw}W' & | & R_{vv} \end{bmatrix}$$

Here we observe that in the standard Gauss–Markov model the $(N_x + N_v) \times (N_x + N_v)$ block covariance matrix R^* is full with cross-covariance matrices $R_{w^*v^*}$ on its off-diagonals. The standard model assumes that they are null (uncorrelated). To simulate a system with correlated $w(t)$ and $v(t)$ is more complicated using this form of the Gauss–Markov model because R^* must first be factored such that

$$R^* = \begin{bmatrix} R_1^* \\ R_2^* \end{bmatrix} [R_1^{*\prime} \quad R_2^{*\prime}] \tag{4.92}$$

where R_i^* are matrix square roots [6, 7]. Once the factorization is performed, the correlated noise is synthesized "coloring" the uncorrelated noise sources $w(t)$ and $v(t)$ as

$$\begin{bmatrix} w^*(t) \\ v^*(t) \end{bmatrix} = \begin{bmatrix} R_1^{*\prime} w(t) \\ R_2^{*\prime} v(t) \end{bmatrix} \tag{4.93}$$

The innovations model is a constrained version of the correlated Gauss–Markov characterization. If we assume that $\{e(t)\}$ is a zero-mean, white Gaussian sequence, that is, $e \sim \mathcal{N}(0, R_{ee})$, then the *innovations model* [9–13] evolves as

$$\begin{aligned} x(t) &= A(t-1)x(t-1) + B(t-1)u(t-1) + K(t-1)e(t-1) \\ y(t) &= C(t)x(t) + D(t)u(t) + e(t) \end{aligned} \tag{4.94}$$

where $e(t)$ is the N_y-dimensional innovations vector and $K(t-1)$ is the $(N_x \times N_y)$ weighting matrix[7]

$$R_{ee}^* := \text{cov}\left(\begin{bmatrix} Ke(t) \\ e(t) \end{bmatrix}\right) = \left[\begin{array}{c|c} KR_{ee}K' & KR_{ee} \\ \hline R_{ee}K' & R_{ee} \end{array}\right] \delta(t-k)$$

It is important to note that the innovations model has implications in Wiener–Kalman filtering (spectral factorization) because R_{ee} can be represented in *factored* or *square-root* form ($R := \sqrt{R}\sqrt{R}'$) directly in terms of the gain and innovation covariance matrix, that is,

$$R_{ee}^* := \begin{bmatrix} K\sqrt{R_{ee}} \\ \sqrt{R_{ee}} \end{bmatrix} \begin{bmatrix} \sqrt{R_{ee}}K' & \sqrt{R_{ee}} \end{bmatrix} \delta(t-k) \tag{4.95}$$

Comparing the innovations model to the Gauss–Markov model, we see that they are both equivalent to the case when w and v are correlated. This completes the discussion of the innovations model. Next we show the equivalence of the various model sets to this family of state–space representations.

4.7 STATE–SPACE MODEL STRUCTURES

In this section we discuss special state–space structures usually called "canonical forms" in the literature, since they represent unique state constructs that are particularly useful. We will confine the models to single input/single output forms because the multivariable structures are too complicated for this discussion [4]. Here we will first investigate the most popular "time series" models and then their equivalent representation in the state–space. We start with the ARMAX model and then progress to the special cases of this generic structure.

4.7.1 Time Series Models

Time series models are particularly useful representations used frequently by statisticians and signal processors to represent time sequences when no physics is available to use directly. They form the class of black box or gray box models [13] which are used in predicting data. These models have an input–output structure, but they can be transformed to an equivalent state–space representation. Each model set has its own advantages: the input–output models are easy to use, while the state–space models are easily generalized and usually evolve when physical phenomenology can be described in a model-based sense [13].

[7] Actually K is called the *Kalman gain matrix*, which will be discussed in detail when we develop the model-based processor in a subsequent chapter.

The *input–output* or *transfer function* model is familiar to engineers and scientists because it is usually presented in the frequency domain with Laplace transforms. Similarly in the discrete-time case, it is called the *pulse transfer function* model and is expressed as

$$H(z) = \frac{B(z^{-1})}{A(z^{-1})} \tag{4.96}$$

where A and B are polynomials in z or z^{-1},

$$A(z^{-1}) = 1 + a_1 z^{-1} + \cdots + a_{N_a} z^{-N_a} \tag{4.97}$$

$$B(z^{-1}) = b_0 + b_1 z^{-1} + \cdots + b_{N_b} z^{-N_b} \tag{4.98}$$

If we consider the equivalent time domain representation, then we have a *difference equation* relating the output sequence $\{y(t)\}$ to the input sequence $\{u(t)\}$.[8] We use the *backward shift operator* q with the property that $q^{-k}y(t) = y(t-k)$.

$$A(q^{-1})y(t) = B(q^{-1})u(t) \tag{4.99}$$

or

$$y(t) + a_1 y(t-1) + \cdots + a_{N_a} y(t - N_a) = b_0 u(t) + \cdots + b_{N_b} u(t - N_b) \tag{4.100}$$

When the system is excited by random inputs the models are given by the *autoregressive-moving average model with exogenous inputs* (ARMAX)[9]

$$\underbrace{A(q^{-1})y(t)}_{AR} = \underbrace{B(q^{-1})u(t)}_{X} + \underbrace{C(q^{-1})e(t)}_{MA} \tag{4.101}$$

where A, B, C, are polynomials, and $\{e(t)\}$ is a white noise source, and

$$C(q^{-1}) = c_0 + c_1 q^{-1} + \cdots + c_{N_c} q^{-N_c}$$

The ARMAX model usually abbreviated by ARMAX(N_a, N_b, N_c) represents the general form for many popular time series and digital filter models. A summary of these models follows:

- Pulse transfer function or infinite impulse response (IIR) model: $C(\cdot) = 0$, or ARMAX($N_a, N_b, 0$), that is,

$$A(q^{-1})y(t) = B(q^{-1})u(t)$$

[8] We change from the common signal processing convention of using $x(t)$ for the deterministic excitation to $u(t)$ and we include the b_0 coefficient for generality.

[9] The ARMAX model can be interpreted in terms of the Wold decomposition of stationary times series, which states that a time series can be decomposed into a predictable or deterministic component ($u(t)$) and nondeterministic or random component ($e(t)$) [10].

- Finite impulse response (FIR) model: $A(\cdot) = 1, C(\cdot) = 0$, or ARMAX$(1, N_b, 0)$, that is,

$$y(t) = B(q^{-1})u(t)$$

- Autoregressive (AR) model: $B(\cdot) = 0, C(\cdot) = 1$, or ARMAX$(N_a, 0, 1)$, that is,

$$A(q^{-1})y(t) = e(t)$$

- Moving average (MA) model: $A(\cdot) = 1, B(\cdot) = 0$, or ARMAX$(1, 0, N_c)$, that is,

$$y(t) = C(q^{-1})e(t)$$

- Autoregressive-moving average (ARMA) model: $B(\cdot) = 0$, or ARMAX$(N_a, 0, N_c)$, that is,

$$A(q^{-1})y(t) = C(q^{-1})e(t)$$

- Autoregressive model with exogenous input (ARX): $C(\cdot) = 1$, or ARMAX$(N_a, N_b, 1)$, that is,

$$A(q^{-1})y(t) = B(q^{-1})u(t) + e(t)$$

The ARMAX model is shown in Fig. 4.4. ARMAX models can easily be used for signal processing purposes, since they are basically digital filters with known deterministic ($u(t)$) and random ($e(t)$) excitations.

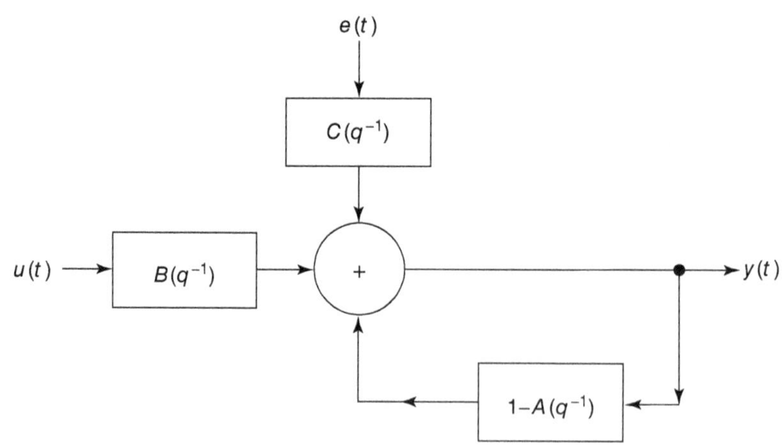

FIGURE 4.4 ARMAX input-output model.

Since the ARMAX model is used to characterize a random signal, we are interested in its statistical properties. The mean value of the output is easily determined by

$$A(q^{-1})E\{y(t)\} = B(q^{-1})E\{u(t)\} + C(q^{-1})E\{e(t)\}$$

or

$$A(q^{-1})m_y(t) = B(q^{-1})u(t) + C(q^{-1})m_e(t) \tag{4.102}$$

Because the first term in the A-polynomial is unity, we can write the *mean propagation recursion* for the ARMAX model as

$$m_y(t) = (1 - A(q^{-1}))m_y(t) + B(q^{-1})u(t) + C(q^{-1})m_e(t) \tag{4.103}$$

$$m_y(t) = -\sum_{i=1}^{N_a} a_i m_y(t-i) + \sum_{i=0}^{N_b} b_i u(t-i) + \sum_{i=0}^{N_c} c_i m_e(t-i) \tag{4.104}$$

We note that the mean of the ARMAX model is propagated using a recursive digital filter requiring N_a, N_b, N_c past input and output values.

The corresponding variance of the ARMAX model is more complex. First, we note that the mean must be removed, that is,

$$\begin{aligned} y(t) - m_y(t) = {} & [(1 - A(q^{-1}))y(t) + B(q^{-1})u(t) + C(q^{-1})e(t)] \\ & - [(1 - A(q^{-1}))m_y(t) + B(q^{-1})u(t) + C(q^{-1})m_e(t)] \end{aligned} \tag{4.105}$$

or

$$y(t) - m_y(t) = (1 - A(q^{-1}))(y(t) - m_y(t)) + C(q^{-1})(e(t) - m_e(t))$$

or finally

$$A(q^{-1})(y(t) - m_y(t)) = C(q^{-1})(e(t) - m_e(t)) \tag{4.106}$$

that is, $y - m_y$ is characterized by an ARMAX$(N_a, 0, N_b)$ or equivalently an ARMA model.

The covariance of the ARMAX model can be calculated utilizing the fact that it is essentially an IIR system, that is,

$$\frac{Y(z)}{E(z)} = H(z) = \sum_{t=0}^{\infty} h(t)z^{-t} \tag{4.107}$$

Using this fact and the commutativity of the convolution operator, we have (assuming the mean has been removed)

$$R_{yy}(k) = E\{y(t)y(t+k)\} = E\left\{\sum_{i=0}^{\infty} h(i)e(t-i)\sum_{j=0}^{\infty} h(j+k)e(t-j+k)\right\}$$

or

$$R_{yy}(k) = \sum_{i=0}^{\infty} h(i)h(i+k)E\{e(t-i)e(t-i+k)\} + \sum\sum_{i \neq j} h(i)h(j+k)E\{e(t-i)e(t-j+k)\}$$

The whiteness of $\{e(t)\}$ gives

$$R_{ee}(k) = \begin{cases} R_{ee} & k = 0 \\ 0 & \text{elsewhere} \end{cases}$$

therefore, applying this property above we have the *covariance* of the ARMAX model given by

$$R_{yy}(k) = R_{ee} \sum_{i=0}^{\infty} h(i)h(i+k) \quad \text{for } k \geq 0 \tag{4.108}$$

with corresponding *variance*

$$R_{yy}(0) = R_{ee} \sum_{t=0}^{\infty} h^2(t) \tag{4.109}$$

We note that one property of a stationary signal is that its impulse response is bounded which implies from Eq. 4.108 that the variance is bounded [14]. Clearly, since the variance is characterized by an ARMA model (ARMAX($N_a, 0, N_c$)), we have

$$A(z^{-1})H(z) = C(z^{-1})E(z), \quad E(z) = \sqrt{R_{ee}}$$

or taking the inverse Z-transform

$$h(t) = (1 - A(q^{-1}))h(t) + C(q^{-1})\delta(t)$$

or

$$h(t) = -\sum_{i=1}^{N_a} a_i h(t-i) + \sum_{i=0}^{N_c} c_i \delta(t), \quad c_0 = 1 \tag{4.110}$$

where $\delta(t)$ is an impulse of weight $\sqrt{R_{ee}}$. So we see that this recursion coupled with Eq. 4.108 provides a method for calculating the variance of an ARMAX model. We summarize these results in Table 4.2 and the following example.

TABLE 4.2 ARMAX Representation

Output Propagation

$$y(t) = (1 - A(q^{-1}))y(t) + B(q^{-1})u(t) + C(q^{-1})e(t)$$

Mean Propagation

$$m_y(t) = (1 - A(q^{-1}))m_y(t) + B(q^{-1})u(t) + C(q^{-1})m_e(t)$$

Impulse Propagation

$$h(t) = (1 - A(q^{-1}))h(t) + C(q^{-1})\delta(t)$$

Variance/Covariance Propagation

$$R_{yy}(k) = R_{ee} \sum_{i=0}^{\infty} h(i)h(i+k) \quad k \geq 0$$

y = output or measurement sequence
u = input sequence
e = process (white) noise sequence with variance R_{ee}
h = impulse response sequence
δ = impulse input of amplitude $\sqrt{R_{ee}}$
m_y = mean output or measurement sequence
m_e = mean process noise sequence
R_{yy} = stationary output covariance at lag k
$A = N_a$th order system characteristic (poles) polynomial
$B = N_b$th order input (zeros) polynomial
$C = N_c$th order noise (zeros) polynomial

Example 4.4

Consider the difference equation of the previous example with $a = 0.5$ which is an AR model with $A(z) = 1 + 0.5z^{-1}$ and $R_{ee} = 1$. We would like to calculate the variance. From Eq. 4.110, we have

$$h(t) = -0.5h(t-1) + \delta(t) \longrightarrow (-0.5)^t$$

and

$$R_{yy}(0) = \sum_{i=0}^{\infty} h^2(i) = [1^2 - 0.5^2 + .25^2 - 1.25^2 + .0875^2 - \cdots] \rightarrow 1.333$$

△△△

Let us consider a more complex example to illustrate the use of the ARMA model.

Example 4.5

Suppose we would like to investigate the structure of an ARMAX(2, 1, 1) model with the following difference equation and calculate its underlying mean and variance propagation relations

$$(1 + 3/4q^{-1} + 1/8q^{-2})y(t) = (1 + 1/8q^{-1})u(t) + (1 + 1/16q^{-1})e(t)$$

where $u(t) = \sin 2\pi(0.025)t$ and $e \sim \mathcal{N}(1, 0.01)$. Then the corresponding mean propagation equation is

$$(1 + 3/4q^{-1} + 1/8q^{-2})m_y(t) = (1 + 1/8q^{-1})u(t) + (1 + 1/16q^{-1})m_e(t)$$

for $m_e(t) = 1$ for all t. The impulse propagation model is

$$(1 + 3/4q^{-1} + 1/8q^{-2})h(t) = (1 + 1/16q^{-1})\sqrt{R_{ee}}\delta(t) \quad \text{for } \sqrt{R_{ee}} = 0.1$$

and the variance/covariance propagation model is

$$R_{yy}(k) = R_{ee}\sum_{t=0}^{\infty} h(t)h(t+k) \quad k \geq 0$$

This completes the example. △△△

It should be noted that for certain special cases of the ARMAX model, it is particularly simple to calculate the mean and covariance. For instance, the MA model (ARMAX(1, 0, N_c)) has *mean*

$$m_y(t) = E\{C(q^{-1})e(t)\} = C(q^{-1})m_e(t) \tag{4.111}$$

and *covariance* (directly from Eq. 4.108 with $h \rightarrow c$)

$$R_{yy}(k) = R_{ee}\sum_{i=0}^{N_c} c_i c_{i+k} \quad \text{for } k \geq 0 \tag{4.112}$$

Another special case of interest is the AR (ARMAX(N_a, 0, 1)) model with *mean*

$$m_y(t) = (1 - A(q^{-1}))m_y(t) + m_e(t) \tag{4.113}$$

and *covariance* which is easily derived by direct substitution

$$R_{yy}(k) = E\{y(t)y(t+k)\} = (1 - A(q^{-1}))R_{yy}(k) = -\sum_{i=1}^{N_a} a_i R_{yy}(k-i) \quad \text{for } k > 0 \tag{4.114}$$

In fact, the AR covariance model of Eq. 4.114 is essentially a recursive (all-pole) digital filter which can be propagated by exciting it with the variance $R_{yy}(0)$ as initial condition. In this case the variance is given by

$$R_{yy}(0) = E\{y^2(t)\} = E\left\{\left(-\sum_{i=1}^{N_a} a_i y(t-i) + e(t)\right)y(t)\right\} = -\sum_{i=1}^{N_a} a_i R_{yy}(i) + R_{ee} \tag{4.115}$$

So combining Eqs. 4.114 and 4.115, we have that the *covariance propagation* equations for the *AR* model are given by

$$R_{yy}(k) = \begin{cases} -\sum_{i=1}^{N_a} a_i R_{yy}(i) + R_{ee} & k = 0 \\ -\sum_{i=1}^{N_a} a_i R_{yy}(k-i) & k > 0 \end{cases}$$

Consider the following example of calculating the statistics of the AR model.

Example 4.6

Consider the AR model of the previous two examples. We would like to determine the corresponding mean and variance using the recursions of Eqs. 4.113 and 4.114 with $A(q^{-1}) = 1 + 0.5q^{-1}$. The mean is

$$m_y(t) = -0.5m_y(t-1)$$

and the covariance is

$$R_{yy}(k) = \begin{cases} -0.5R_{yy}(1) + 1 & k = 0 \\ -0.5R_{yy}(k-1) & k > 0 \end{cases}$$

The variance is obtained directly from these recursions, since

$$R_{yy}(1) = -0.5R_{yy}(0)$$

and therefore

$$R_{yy}(0) = -0.5R_{yy}(1) + 1$$

Substituting for $R_{yy}(1)$ we obtain

$$R_{yy}(0) = -0.5(-0.5R_{yy}(0)) + 1$$

or

$$R_{yy}(0) = 1.333$$

as before. △△△

This completes the section on ARMAX models.

4.7.2 State–Space and Time Series Equivalence Models

In this section we show the equivalence between the ARMAX and state–space models (for scalar processes). That is, we show how to obtain the state–space model given the ARMAX models by inspection. We choose particular coordinate systems in the state–space (canonical forms) and obtain a relationship between entries of the state–space system to coefficients of the ARMAX model. An example is presented that shows how these models can be applied to realize a random signal. First, we consider the ARMAX to state–space transformation.

Recall from Eq. 4.101 that the general difference equation form of the ARMAX model is given by

$$y(t) = -\sum_{i=1}^{N_a} a_i y(t-i) + \sum_{i=0}^{N_b} b_i u(t-i) + \sum_{i=0}^{N_c} c_i e(t-i) \tag{4.116}$$

or equivalently in the frequency domain as

$$Y(z) = \left(\frac{b_o + b_1 z^{-1} + \cdots + b_{N_b} z^{-N_b}}{1 + a_1 z^{-1} + \cdots + a_{N_a} z^{-N_a}}\right) U(z) + \left(\frac{c_o + c_1 z^{-1} + \cdots + c_{N_c} z^{-N_c}}{1 + a_1 z^{-1} + \cdots + a_{N_a} z^{-N_a}}\right) E(z) \tag{4.117}$$

where $N_a \geq N_b$ and N_c and $\{e(t)\}$ is a zero-mean white sequence with spectrum given by R_{ee}.

It is straightforward but tedious to show (see Ref. [5]) that the ARMAX model can be represented in *observer canonical form*

$$\begin{aligned} x(t) &= A_0 x(t-1) + B_0 u(t-1) + W_0 e(t-1) \\ y(t) &= C_0' x(t) + b_0 u(t) + c_0 e(t) \end{aligned} \tag{4.118}$$

where x, u, e, and y are the N_a-state vector, scalar input, noise, and output with

$$A_0 := \left[\begin{array}{c|c} 0 & -a_{N_a} \\ \hline & \vdots \\ I_{N_a-1} & -a_1 \end{array}\right] \quad B_0 := \left[\begin{array}{c} -a_{N_a} b_0 \\ \vdots \\ -a_{N_b+1} b_0 \\ \hline b_{N_b} - a_{N_b} b_0 \\ \vdots \\ b_1 - a_1 b_0 \end{array}\right] \quad W_0 := \left[\begin{array}{c} -a_{N_a} c_0 \\ \vdots \\ -a_{N_c+1} c_0 \\ \hline c_{N_c} - a_{N_c} c_0 \\ \vdots \\ c_1 - a_1 c_0 \end{array}\right]$$

$$C_0' := [0 \; \cdots \; 0 \; 1]$$

Noting this structure we see that each of the matrix or vector elements $\{A_{i,N_a}, B_i, W_i, C_i\}$ $i = 1, \ldots, N_a$ can be determined from the relations

$$\begin{aligned} A_{i,N_a} &= -a_i \quad i = 1, \ldots, N_a \\ B_i &= b_i - a_i b_0 \\ W_i &= c_i - a_i c_0 \\ C_i &= \delta(N_a - i) \end{aligned} \tag{4.119}$$

where

$$\begin{aligned} b_i &= 0 \quad && \text{for } i > N_b \\ c_i &= 0 \quad && \text{for } i > N_c \\ \delta(i-j) & && \text{is the Kronecker delta} \end{aligned}$$

Consider the following example to illustrate these relations.

Example 4.7

Let $N_a = 3, N_b = 2$, and $N_c = 1$; then the corresponding ARMAX model is

$$\begin{aligned} y(t) = &-a_1 y(t-1) - a_2 y(t-2) - a_3 y(t-3) + b_0 u(t) \\ &+ b_1 u(t-1) + b_2 u(t-2) + c_0 e(t) + c_1 e(t-1) \end{aligned} \tag{4.120}$$

Using the observer canonical form of Eq. 4.118 we have

$$x(t) = \begin{bmatrix} 0 & 0 & | & -a_3 \\ - & - & | & \\ 1 & 0 & | & -a_2 \\ 0 & 1 & | & -a_1 \end{bmatrix} x(t-1) + \begin{bmatrix} -a_3 b_0 \\ --- \\ b_2 - a_2 b_0 \\ b_1 - a_1 b_0 \end{bmatrix} u(t-1)$$

$$+ \begin{bmatrix} -a_3 c_0 \\ -a_2 c_0 \\ --- \\ c_1 - a_1 c_0 \end{bmatrix} e(t-1)$$

$$y(t) = [0 \ \ 0 \ \ 1] x(t) + b_0 u(t) + c_0 e(t)$$

This completes the example. △△△

It is important to realize that if we assume that $\{e(t)\}$ is Gaussian, then the ARMAX model is equivalent to the innovations representation of the previous section, that is,

$$\begin{aligned} x(t) &= A x(t-1) + B u(t-1) + W e(t-1) \\ y(t) &= C' x(t) + b_0 u(t) + c_0 e(t) \end{aligned} \tag{4.121}$$

where in this case $K \to W, D \to b_0$, and $1 \to c_0$. Also, the corresponding covariance matrix becomes

$$R_{ee}^* := \text{cov}\left(\begin{bmatrix} We(t) \\ c_0 e(t) \end{bmatrix} \right) = \begin{bmatrix} W R_{ee} W' & | & W R_{ee} c_0 \\ ----- & & \\ c_0 R_{ee} W' & | & c_0 R_{ee} c_0 \end{bmatrix} \delta(t-k)$$

This completes the discussion on the equivalence of the general ARMAX to state–space.

Next let us develop the state–space equivalent model for some special cases of the ARMAX model presented in the previous subsection. We begin with the MA

$$y(t) = \sum_{i=1}^{N_c} c_i e(t-i) \quad \text{or} \quad Y(z) = C(z)E(z) = (1 + c_1 z^{-1} + \cdots + c_{N_c} z^{-N_c})E(z)$$

Define the state variable as

$$x_i(t-1) := e(t-i-1), \quad i = 1, \ldots, N_c \tag{4.122}$$

and therefore,

$$x_i(t) = e(t-i) = x_{i-1}(t-1), \quad i = 1, \ldots, N_c \tag{4.123}$$

Expanding this expression, we obtain

$$\begin{aligned}
x_1(t) &= e(t-1) & \\
x_2(t) &= e(t-2) &= x_1(t-1) \\
x_3(t) &= e(t-3) &= x_2(t-1) \\
\vdots \quad & \quad \vdots & \vdots \quad \\
x_{N_c}(t) &= e(t-N_c) &= x_{N_c-1}(t-1)
\end{aligned} \tag{4.124}$$

or in vector-matrix form

$$\begin{bmatrix} x_1(t) \\ x_2(t) \\ \vdots \\ x_{N_c}(t) \end{bmatrix} = \left[\begin{array}{ccc|c} 0 & \cdots & 0 & 0 \\ \hline 1 & \cdots & 0 & 0 \\ \vdots & \ddots & \vdots & \vdots \\ 0 & \cdots & 1 & 0 \end{array}\right] \begin{bmatrix} x_1(t-1) \\ x_2(t-1) \\ \vdots \\ x_{N_c}(t-1) \end{bmatrix} + \begin{bmatrix} 1 \\ 0 \\ \vdots \\ 0 \end{bmatrix} e(t-1)$$

$$y(t) = [c_1 \quad c_2 \cdots c_{N_c}] \begin{bmatrix} x_1(t-1) \\ x_2(t-1) \\ \vdots \\ x_{N_c}(t-1) \end{bmatrix} + c_0 e(t) \tag{4.125}$$

Thus the general form for the MA state–space is given by

$$\mathbf{x}(t) = \left[\begin{array}{ccc|c} 0 & \cdots & 0 & 0 \\ \hline & & & \\ & \mathbf{I}_{N_c-1} & & \vdots \\ & & & 0 \end{array}\right] \mathbf{x}(t-1) + \mathbf{b}_1 e(t-1)$$

$$y(t) = \mathbf{c}'\mathbf{x}(t) + c_0 e(t) \tag{4.126}$$

with $N_x = N_c$, $\mathbf{b}_1, \mathbf{c} \in R^{N_x \times 1}$, $\mathbf{b}_1$ a unit vector.

Example 4.8

Suppose we have the following MA model

$$y(t) = c_0 e(t) + c_1 e(t-1) + c_2 e(t-2)$$

and we would like to construct the equivalent state–space representation. Then we have $N_x = 2$, and therefore,

$$\mathbf{x}(t) = \begin{bmatrix} 0 & 0 \\ 1 & 0 \end{bmatrix} \mathbf{x}(t-1) + \begin{bmatrix} 1 \\ 0 \end{bmatrix} e(t-1)$$
$$y(t) = [c_1 \quad c_2]\, \mathbf{x}(t) + c_0 e(t)$$

△△△

Consider the AR model (all-pole) given by

$$\sum_{i=1}^{N_a} a_i y(t-i) = \sigma e(t) \quad \text{or} \quad Y(z) = \frac{\sigma E(z)}{1 + a_1 z^{-1} + \cdots + a_{N_a} z^{-N_a}} \tag{4.127}$$

Here the state vector is defined by $x_i(t-1) = y(t-i-1)$ and therefore, $x_i(t) = y(t-i) = x_{i+1}(t-1);\, i = 1, \ldots, N_a - 1$ with $x_{N_a}(t) = y(t)$. Expanding over i, we obtain the vector-matrix state–space model

$$\begin{bmatrix} x_1(t) \\ x_2(t) \\ \vdots \\ x_{N_a}(t) \end{bmatrix} = \begin{bmatrix} 0 & | & 1 & 0 & \cdots & 0 \\ 0 & | & 0 & 1 & \cdots & 0 \\ \vdots & | & \vdots & \vdots & \ddots & \vdots \\ 0 & | & 0 & 0 & \cdots & 1 \\ --- & | & --- & --- & --- & \\ -a_{N_a} & & -a_{N_a-1} & -a_{N_a-2} & \cdots & -a_1 \end{bmatrix} \begin{bmatrix} x_1(t-1) \\ x_2(t-1) \\ \vdots \\ x_{N_a}(t-1) \end{bmatrix} + \begin{bmatrix} 0 \\ 0 \\ \vdots \\ \sigma \end{bmatrix} e(t-1)$$

$$y(t) = [0 \quad 0 \quad \cdots \quad 1] \begin{bmatrix} x_1(t-1) \\ x_2(t-1) \\ \vdots \\ x_{N_a}(t-1) \end{bmatrix} \tag{4.128}$$

In general, we have the AR (*all-pole*) state–space model

$$\mathbf{x}(t) = \begin{bmatrix} 0 & | & & & \\ \vdots & | & & \mathbf{I}_{N_a-1} & \\ 0 & | & & & \\ --- & | & --- & --- & --- \\ -a_{N_a} & & -a_{N_a-1} & -a_{N_a-2} & \cdots & -a_1 \end{bmatrix} \mathbf{x}(t-1) + \mathbf{b}e(t-1)$$

$$y(t) = \mathbf{c}'\mathbf{x}(t) \tag{4.129}$$

with $N_x = N_a$, $\mathbf{b}, \mathbf{c} \in R^{N_x \times 1}$. Consider the following example to demonstrate this form.

Example 4.9

Given the AR model with $N_a = 2$ and $\sigma = \sqrt{2}$, find the equivalent state–space form. We have

$$y(t) = -a_1 y(t-1) - a_2 y(t-2) + \sqrt{(2)} e(t)$$

and therefore,

$$x_1(t-1) = y(t-2)$$
$$x_2(t-1) = y(t-1)$$

which gives

$$x_1(t) = y(t-1) = x_2(t-1)$$
$$x_2(t) = y(t) = -a_1 y(t-1) - a_2 y(t-2) = -a_1 x_2(t-1) - a_2 x_1(t-1)$$

or more succinctly

$$\mathbf{x}(t) = \begin{bmatrix} 0 & 1 \\ -a_2 & -a_1 \end{bmatrix} \mathbf{x}(t-1) + \begin{bmatrix} 0 \\ \sqrt{2} \end{bmatrix} e(t-1)$$
$$y(t) = [0 \quad 1]\mathbf{x}(t)$$

△△△

Another useful state–space representation is the *normal form* which evolves by performing a partial fraction expansion of a rational discrete transfer function model (ARMA) to obtain

$$h(t) = \sum_{i=1}^{N_p} R_i (p_i)^{-t} \quad \text{or} \quad H(z^{-1}) = \frac{Y(z^{-1})}{E(z^{-1})} = \sum_{i=1}^{N_p} \frac{R_i}{1 - p_i z^{-1}} \tag{4.130}$$

for $\{R_i, p_i\}; i = 1, \ldots, N_p$ the set of residues and poles of $H(z^{-1})$. Note that the normal form model is the decoupled or parallel system representation based on the following set of relations

$$y_i(t) - p_i y_i(t-1) = e(t), \quad i = 1, \ldots, N_p$$

Defining the state variable as $x_i(t) := y_i(t)$, then equivalently

$$x_i(t) - p_i x_i(t-1) = e(t), \quad i = 1, \ldots, N_p \tag{4.131}$$

and therefore the output is given by

$$y(t) = \sum_{i=1}^{N_p} R_i y_i(t) = \sum_{i=1}^{N_p} R_i x_i(t), \quad i = 1, \ldots, N_p \tag{4.132}$$

Expanding these relations over i, we obtain

$$\begin{bmatrix} x_1(t) \\ x_2(t) \\ \vdots \\ x_{N_p}(t) \end{bmatrix} = \begin{bmatrix} p_1 & 0 & \cdots & 0 \\ 0 & p_2 & \cdots & 0 \\ \vdots & \vdots & \ddots & \vdots \\ 0 & 0 & \cdots & p_{N_p} \end{bmatrix} \begin{bmatrix} x_1(t-1) \\ x_2(t-1) \\ \vdots \\ x_{N_a}(t-1) \end{bmatrix} + \begin{bmatrix} 1 \\ 1 \\ \vdots \\ 1 \end{bmatrix} e(t-1)$$

$$y(t) = [R_1 \quad R_2 \quad \cdots \quad R_{N_p}] \begin{bmatrix} x_1(t-1) \\ x_2(t-1) \\ \vdots \\ x_{N_p}(t-1) \end{bmatrix} \tag{4.133}$$

Thus, the general decoupled form of the *normal state–space model* is given by

$$\mathbf{x}(t) = \begin{bmatrix} p_1 & 0 & \cdots & 0 \\ 0 & p_2 & \cdots & 0 \\ \vdots & \vdots & \ddots & \vdots \\ 0 & 0 & \cdots & p_{N_p} \end{bmatrix} \mathbf{x}(t-1) + \mathbf{b}e(t-1)$$

$$y(t) = \mathbf{c}'\mathbf{x}(t) \tag{4.134}$$

for $\mathbf{b} \in \mathcal{R}^{N_p \times 1}$ with $\mathbf{b} = \mathbf{1}$, a N_p-vector of unit elements. Here $\mathbf{c} \in \mathcal{R}^{1 \times N_p}$ and $\mathbf{c}' = [R_1 \quad R_2 \quad \cdots \quad R_{N_p}]$.

Example 4.10

Consider the following set of parameters and model with $N_x = N_p = 3$ and

$$y_i(t) = p_i y_i(t-1) + e(t-1)$$

$$y(t) = \sum_{i=1}^{3} R_i y_i(t)$$

Using the *normal state–space form* structure above, we obtain by inspection

$$\mathbf{x}(t) = \begin{bmatrix} p_1 & 0 & 0 \\ 0 & p_2 & 0 \\ 0 & 0 & p_3 \end{bmatrix} \mathbf{x}(t-1) + \begin{bmatrix} 1 \\ 1 \\ 1 \end{bmatrix} e(t-1)$$

$$y(t) = [R_1 \quad R_2 \quad R_3]\,\mathbf{x}(t)$$

△△△

4.8 NONLINEAR (APPROXIMATE) GAUSS–MARKOV STATE-SPACE MODELS

Many processes in practice are nonlinear rather than linear. Coupling the nonlinearities with noisy data makes the signal processing problem a challenging one. In

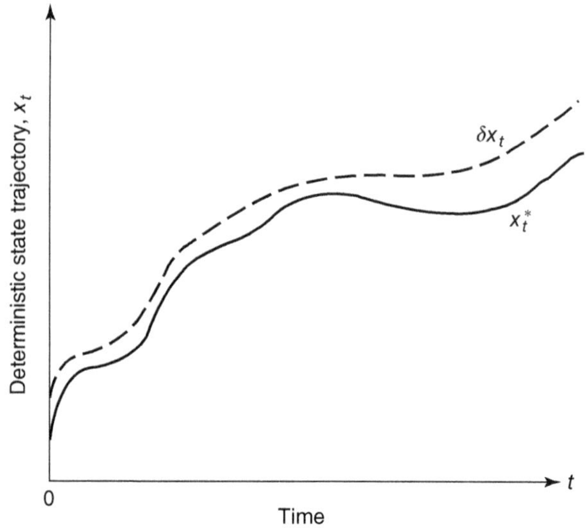

FIGURE 4.5 Linearization of a deterministic system using the reference trajectory defined by $(x^*(t), u^*(t))$.

this section we develop an approximate solution to the nonlinear modeling problem involving the linearization of the nonlinear process about a chosen or "known" reference trajectory. We limit our discussion to *discrete* nonlinear systems. Continuous solutions to this problem are developed in Refs. [6–13].

Suppose we model a *process* by a set of nonlinear stochastic vector difference equations in state–space form as

$$x(t) = a[x(t-1)] + b[u(t-1)] + w(t-1) \tag{4.135}$$

with the corresponding *measurement* model

$$y(t) = c[x(t)] + v(t) \tag{4.136}$$

where $a[\cdot]$, $b[\cdot]$, $c[\cdot]$ are nonlinear vector functions of x, u with $x, a, b, w \in R^{N_x \times 1}$, $y, c, v \in R^{N_y \times 1}$ and $w \sim \mathcal{N}(0, R_{ww}(t-1))$, $v \sim \mathcal{N}(0, R_{vv}(t))$.

Ignoring the additive noise sources, we "linearize" the process and measurement models about a known deterministic *reference trajectory* defined by $[x^*(t), u^*(t)]$ as illustrated in Fig. 4.5,[10] that is,

$$x^*(t) = a[x^*(t-1)] + b[u^*(t-1)] \tag{4.137}$$

[10] In practice, the reference trajectory is obtained either by developing a mathematical model of the process or by simulating about some reasonable operating conditions to generate the trajectory using the state–space model.

Deviations or perturbations from this trajectory are defined by

$$\delta x(t) := x(t) - x^*(t)$$
$$\delta u(t) := u(t) - u^*(t)$$

Substituting the previous equations into these expressions, we obtain the perturbation trajectory as

$$\delta x(t) = a[x(t-1)] - a[x^*(t-1)] + b[u(t-1)] - b[u^*(t-1)] + w(t-1) \quad (4.138)$$

The nonlinear vector functions $a[\cdot]$ and $b[\cdot]$ can be expanded into a first-order Taylor series about the reference trajectory $[x^*(t), u^*(t)]$ as[11]

$$a[x(t-1)] = a[x^*(t-1)] + \frac{da[x^*(t-1)]}{dx^*(t-1)}\delta x(t-1) + \text{H.O.T.}$$
$$b[u(t-1)] = b[u^*(t-1)] + \frac{db[u^*(t-1)]}{du^*(t-1)}\delta u(t-1) + \text{H.O.T.} \quad (4.139)$$

We define the first-order Jacobian matrices as

$$A[x^*(t-1)] := \frac{da[x^*(t-1)]}{dx^*(t-1)}, \quad \text{and} \quad B[u^*(t-1)] := \frac{db[u^*(t-1)]}{du^*(t-1)} \quad (4.140)$$

Incorporating the definitions of Eq. 4.140 and neglecting the *higher-order terms* (H.O.T.) in Eq. 4.139, the linearized process model in Eq. 4.138 can be expressed as

$$\delta x(t) = A[x^*(t-1)]\delta x(t-1) + B[u^*(t-1)]\delta u(t-1) + w(t-1) \quad (4.141)$$

Similarly, the measurement system can be linearized by using the reference measurement

$$y^*(t) = c[x^*(t)] \quad (4.142)$$

and applying the Taylor-series expansion to the nonlinear measurement model

$$c[x(t)] = c[x^*(t)] + \frac{dc[x^*(t)]}{dx^*(t)}\delta x(t) + \text{H.O.T.} \quad (4.143)$$

The corresponding measurement perturbation is defined by

$$\delta y(t) := y(t) - y^*(t) = c[x(t)] - c[x^*(t)] + v(t) \quad (4.144)$$

[11]We use the shorthand notation $\frac{d(\cdot)}{d\theta^*}$ to mean $\left.\frac{d(\cdot)}{d\theta}\right|_{\theta=\theta^*}$.

Substituting the first-order approximation for $c[x(t)]$ leads to the linearized measurement perturbation as

$$\delta y(t) = C[x^*(t)]\delta x(t) + v(t) \tag{4.145}$$

where $C[x^*(t)]$ is defined as the measurement Jacobian as before.

Summarizing, we have linearized a deterministic nonlinear model using a first-order Taylor-series expansion for the functions a, b, and c and then developed a *linearized Gauss–Markov perturbation model* valid for small deviations given by

$$\begin{aligned}\delta x(t) &= A[x^*(t-1)]\delta x(t-1) + B[u^*(t-1)]\delta u(t-1) + w(t-1)\\ \delta y(t) &= C[x^*(t)]\delta x(t) + v(t)\end{aligned} \tag{4.146}$$

with A, B, and C the corresponding Jacobian matrices and w, v zero-mean Gaussian.

We can also use linearization techniques to approximate the statistics of the process and measurements. If we use the first-order Taylor-series expansion and expand about the mean $m_x(t)$ rather than $x^*(t)$, then taking expected values

$$m_x(t) = E\{a[x(t-1)]\} + E\{b[u(t-1)]\} + E\{w(t-1)\} \tag{4.147}$$

gives

$$m_x(t) = a[m_x(t-1)] + b[u(t-1)] \tag{4.148}$$

which follows by linearizing $a[\cdot]$ about m_x and taking the expected value to obtain

$$\begin{aligned}E\{a[x(t-1)]\} &= E\{a[m_x(t-1)] + \frac{da[m_x(t-1)]}{dm_x(t-1)}[x(t-1) - m_x(t-1)]\}\\ &= a[m_x(t-1)] + \frac{da[m_x(t-1)]}{dm_x(t-1)}[E\{x(t-1)\} - m_x(t-1)]\\ &= a[m_x(t-1)]\end{aligned}$$

The variance equations $P(t) := \mathrm{cov}(x(t))$ can also be developed in a similar manner (see Ref. [2] for details) to give

$$P(t) = A[m_x(t-1)]P(t-1)A'[m_x(t-1)] + R_{ww}(t-1) \tag{4.149}$$

Using the same approach, we arrive at the accompanying measurement statistics

$$m_y(t) = c[m_x(t)] \quad \text{and} \quad R_{yy}(t) = C[m_x(t)]P(t)C'[m_x(t)] + R_{vv}(t) \tag{4.150}$$

We summarize these results in an "approximate" Gauss–Markov model of Table 4.3. Before we conclude consider the following example to illustrate the approximation.

TABLE 4.3 Approximate Nonlinear Gauss-Markov Model

State Propagation
$x(t) = a[x(t-1)] + b[u(t-1)] + w(t-1)$
State Mean Propagation
$m_x(t) = a[m_x(t-1)] + b[u(t-1)]$
State Covariance Propagation
$P(t) = A[m_x(t-1)]P(t-1)A'[m_x(t-1)] + R_{ww}(t-1)$
Measurement Propagation
$y(t) = c[x(t)] + v(t)$
Measurement Mean Propagation
$m_y(t) = c[m_x(t)]$
Measurement Covariance Propagation
$R_{yy}(t) = C[m_x(t)]P(t)C'[m_x(t)] + R_{vv}(t)$
Initial Conditions
$x(0)$ and $P(0)$
Jacobians
$A[x^*(t-1)] \equiv \left.\frac{da[x(t-1)]}{dx(t-1)}\right\vert_{x=x^*(t-1)} \qquad C[x^*(t)] \equiv \left.\frac{dc[x(t)]}{dx(t)}\right\vert_{x=x^*(t)} \qquad x^* \rightarrow m_x$

Example 4.11

Consider the discrete nonlinear process given by

$$x(t) = (1 - 0.05\Delta T)x(t-1) + 0.04\Delta T x^2(t-1) + w(t-1)$$

with corresponding measurement model

$$y(t) = x^2(t) + x^3(t) + v(t)$$

where $w(t) \sim \mathcal{N}(0, R_{ww})$, $v(t) \sim \mathcal{N}(0, R_{vv})$, and $x(0) \sim \mathcal{N}(\overline{x}(0), \overline{P}(0))$. Performing the differentiations we obtain the following Jacobians

$$A[x(t-1)] = 1 - 0.05\Delta T + 0.08\Delta T x(t-1) \quad \text{and} \quad C[x(t)] = 2x(t) + 3x^2(t)$$

△△△

Although the linearization approach discussed here seems somewhat extraneous relative to the previous sections, it becomes a crucial ingredient in the *classical approach* to (approximate) nonlinear estimation of the subsequent chapters. We discuss the linear state–space approach (Kalman filter) to the estimation problem in great detail in the next chapter and then show how these linearization concepts can be used to solve the nonlinear estimation problem in the following chapter. There the popular "extended" Kalman filter processor relies heavily on the linearization techniques developed in this section for its development.

4.9 SUMMARY

In this chapter we have discussed the development of continuous-time, sampled-data, and discrete-time state–space models. The stochastic variants of these three types of models were presented leading to the Gauss–Markov representations for both linear and (approximate) nonlinear systems. The discussion of both the deterministic and stochastic state–space models included a brief development of their second-order statistics. We also discussed the underlying discrete systems theory as well as a variety of time series models (ARMAX, AR, MA, etc.) and showed that can easily be represented in state–space form through the use of canonical forms (models). These models form the embedded structure incorporated into the majority of the Bayesian processors that will be discussed in subsequent chapters.

MATLAB NOTES

MATLAB has many commands to convert to/from state–space models to other forms useful in signal processing. Many of them reside in the Signal Processing and Control Systems toolboxes. The matrix exponential invoked by the **expm** command is determined from Taylor/Padé approximants using the scaling and squaring approach of Section 4.2. Also the commands **expmdemo1**, **expmdemo2**, and **expmdemo3** demonstrate the trade-offs of the Padé, Taylor and eigenvector approaches to calculating the matrix exponential. The ordinary differential equation method is available using the wide variety of numerical integrators available (**ode***). Converting to/from transfer functions and state–space is accomplished using the **ss2tf** and **tf2ss** commands, respectively. ARMAX simulations are easily accomplished using the **filter** command with a variety of options converting from ARMAX-to/from transfer functions. The Identification Toolbox converts polynomial-based models to state–space and continuous parameters including Gauss–Markov to discrete parameters (**th2ss**, **thc2thd**, **thd2thc**). The the Third Party Toolbox SSPACK_PC converts continuous-time models to discrete (**SSCTOD**) performs Gauss–Markov (linear and nonlinear) simulations as well as innovations-based simulations (**SSISIM** and conversions from *GM* to innovations models (**INVTOGM**, **GMTOINV**)). See http://www.techni-soft.net for details.

REFERENCES

1. T. Kailath, *Linear Systems* (Englewood Cliffs, NJ: Prentice-Hall, 1980).
2. F. Szidarovszky and A. Bahill, *Linear Systems Theory* (Boca Raton, FL: CRC Press, 1980).
3. R. DeCarlo, *Linear Systems: A State Variable Approach* (Englewood Cliffs, NJ: Prentice-Hall, 1989).
4. C. Chen, *Introduction to Linear System Theory* (New York: Holt, Rhinehart, and Winston, 1984).

5. S. Tretter, *Introduction to Discrete-Time Signal Processing* (New York: John Wiley & Sons, Inc., 1976).
6. A. Jazwinski, *Stochastic Processes and Filtering Theory* (New York: Academic Press, 1970).
7. A. Sage and J. Melsa, *Estimation Theory with Applications to Communications and Control* (New York: McGraw-Hill, 1971).
8. P. Maybeck, *Stochastic Models, Estimation and Control*, Vol. 1 (New York: Academic Press, 1979).
9. G. Goodwin and R. L. Payne, *Dynamic System Identification* (New York: Academic Press, 1976).
10. G. Goodwin and K. Sin, *Adaptive Filtering, Prediction and Control* (Englewood Cliffs, NJ: Prentice-Hall, 1984).
11. J. Mendel, *Lessons in Estimation Theory for Signal Processing, Communications, and Control* (Englewood Cliffs, NJ: Prentice-Hall, 1995).
12. R. Brown and P.C. Hwang, *Introduction to Random Signals and Applied Kalman Filtering* (New York: John Wiley & Sons, Inc., 1997).
13. J. Candy, *Model-Based Signal Processing* (Hoboken, NJ: John Wiley & Sons, Inc./IEEE Press, 2006).
14. E. Robinson and M. Silvia, *Digital Foundations of Time Series Analysis*, Vol. 1 (San Francisco, CA: Holden-Day, 1979).
15. D. Simon, *Optimal State Estimation Kalman, H_∞ and Nonlinear Approaches* (Hoboken, NJ: John Wiley & Sons, Inc., 2006).
16. M. Grewal and A. Andrews, *Kalman Filtering: Theory and Practice* (Englewood Cliffs, NJ: 1993).
17. C. Moler and C. Van Loan, "Nineteen dubious ways to compute the exponential of a matrix, twenty-five years later," *SIAM Review*, **45**, 1, 3–49, 2003.
18. G. Golub and C. Van Loan, *Matrix Computation* (Baltimore, MD: Johns Hopkins University Press, 1989).
19. B. Ho and R. Kalman, "Effective reconstruction of linear state variable models from input/output data," *Regelungstechnik*, **14**, 545–548, 1966.
20. J. Candy, M. Warren, and T. Bullock, "Realization of an invariant system description from Markov sequences," *IEEE Trans. Auto. Control*, **AC-23**, 12, 93–96, 1977.
21. S. Kung, K. Arun, and D. Bhaskar Rao, "State–space and singular-value decomposition-based approximation methods for the harmonic retrieval problem," *J. Opt. Soc. Am.*, **73**, 12, 1799–1811, 1983.

PROBLEMS

4.1 Suppose the stochastic process $\{y(t)\}$ is generated by

$$y(t) = a\exp(-t) + bt, \quad a, b \text{ random, then}$$

(*a*) What is the mean of the process?

(*b*) What is the corresponding covariance?

(*c*) Is the process stationary if $E\{a\} = E\{b\} = 0$ and $E\{ab\} = 0$.

4.2 Suppose x, y, z are jointly Gaussian random variables with corresponding means m_x, m_y, m_z and variances R_{xx}, R_{yy}, R_{zz} show that:

(*a*) If $y = ax + b$, a, b constants, then $y \sim N(am_x + b, a^2 R_{xx})$.

(*b*) If x and y are uncorrelated then they are independent.

(*c*) If $x(i)$ are Gaussian with mean $m(i)$ and variance $R_{xx}(i)$, then for

$$y = \sum_i K_i x(i), \quad y \sim N\left(\sum_i K_i m(i), \sum_i K_i^2 R_{xx}(i)\right)$$

(*d*) If x and y are jointly (conditionally) Gaussian, then

$$\begin{aligned} E\{x|y\} &= m_x + R_{xy} R_{yy}^{-1}(y - m_y), \text{ and} \\ R_{x|y} &= R_{xx} + R_{xy} R_{yy}^{-1} R_{yx} \end{aligned}$$

(*e*) The random variable $x - E\{x|y\}$ is orthogonal to y.

(*f*) If y and z are independent, then

$$E\{x|y,z\} = E\{x|y\} + E\{x|z\} - m_x$$

(*g*) If y and z are not independent, show that

$$E\{x|y,z\} = E\{x|y,e\} = E\{x|y\} + E\{x|e\} - m_x$$

for $e = z - E\{z|y\}$.

4.3 Assume $y(t)$ is a zero-mean ergodic process with covariance $R_{yy}(k)$, calculate the corresponding power spectra $S_{yy}(z)$ if

(*a*) $R_{yy}(k) = Ca^{|k|}$.

(*b*) $R_{yy}(k) = C\cos(w|k|)$, $|k| < \frac{\pi}{2}$.

(*c*) $R_{yy}(k) = C\exp(-a^{|k|})$.
(*Hint*: Recall the *sum decomposition*: $S_{yy}(z) = S_{yy}^{+}(z) + S_{yy}^{-}(z) - R_{yy}(0)$ with S_{yy}^{+} the one-sided Z-transform and $S_{yy}^{-}(z) = S_{yy}^{+}(z^{-1})$)

4.4 Develop a MATLAB program to simulate the ARMA process

$$y(t) = -ay(t-1) + e(t)$$

where $a = 0.75$, $e \approx N(0, 0.1)$ for 100 data points.

(*a*) Calculate the analytic covariance $R_{yy}(k)$.

(*b*) Determine an expression to "recursively" calculate $R_{yy}(k)$.

(*c*) Plot the simulated results and construct the $\pm 2\sqrt{R_{yy}(0)}$ bounds.

(*d*) Do 95% of the samples fall within these bounds?

4.5 Develop the digital filter to simulate a sequence $y(t)$ with covariance $R_{yy}(k) = 4e^{-3|k|}$. Perform the simulation using MATLAB.

(*Hint*: Recall the spectral factorization (Wiener): $S_{yy}(z) = H(z) \times H(z^{-1})$ where the poles and zeros of $H(z)$ lie inside the unit circle)

Using the "realized" digital filter perform a simulation as in the previous example and check the validity of the samples lying within bounds.

4.6 Suppose we are given a zero-mean process with covariance

$$R_{yy}(k) = 10\exp(-0.5|k|)$$

(*a*) Determine the digital filter which when driven by white noise will yield a sequence with the above covariance.

(*b*) Develop a computer program to generate $y(t)$ for 100 points.

(*c*) Plot the results and determine if 95% of the samples fall within $\pm 2\sqrt{R_{yy}(0)}$.

4.7 Suppose we are given the factored power spectrum $S_{yy}(z) = H(z)H(z^{-1})$ with

$$H(z) = \frac{1 + \beta_1 z^{-1} + \beta_2 z^{-2}}{1 + \alpha_1 z^{-1} + \alpha_2 z^{-2}}$$

(*a*) Develop the ARMAX model for the process.

(*b*) Develop the corresponding Gauss–Markov model for *both* the standard and innovations representation of the process.

4.8 Suppose we are given a causal LTI system characterized by its impulse response $h(t)$. If this system is excited by zero-mean, unit variance white noise, then

(*a*) Determine the output variance $R_{yy}(0)$;

(*b*) Determine the covariance $R_{yy}(k)$ for $k > 0$;

(*c*) Suppose the system transfer function is given by

$$H(z) = \frac{1 + b_0 z^{-1}}{1 + a_1 z^{-1} + a_2 z^{-2}}$$

find a method to recursively calculate $h(t)$ and therefore $R_{yy}(0)$.

4.9 Given the covariance function

$$R_{yy}(k) = e^{-1/2|k|} \cos \pi |k|,$$

find the digital filter when driven by unit variance white noise produces a sequence $\{y(t)\}$ with these statistics.

4.10 Suppose we have a process characterized by the difference equation

$$y(t) = x(t) + 1/2x(t-1) + 1/3x(t-2)$$

(a) Determine a recursion for the output covariance $R_{yy}(k)$.
(b) If $x(t)$ is white with variance σ_{xx}^2, determine $R_{yy}(k)$.
(c) Determine the output *PSD*, $S_{yy}(z)$.

4.11 We are given a linear system characterized by the difference equation

$$y(t) - 1/5y(t-1) = \frac{1}{\sqrt{3}}x(t)$$

and the system is excited by:
1. white Gaussian noise $x \sim N(0, 3)$;
2. exponentially correlated noise $R_{ee}(k) = (1/2)^{|k|}$.

In both cases find:
(a) Output *PSD* $S_{yy}(z)$;
(b) Output covariance $R_{yy}(k)$;
(c) Cross-spectrum $S_{ye}(k)$;
(d) Cross-covariance $R_{ye}(k)$.

4.12 We are given the following Gauss–Markov model

$$x(t) = 1/3x(t-1) + 1/2w(t-1)$$
$$y(t) = 5x(t) + v(t)$$
$$w \sim N(0, 3) \quad v \sim N(0, 2)$$

(a) Calculate the state power spectrum $S_{xx}(z)$.
(b) Calculate the measurement power spectrum $S_{yy}(z)$.
(c) Calculate the state covariance recursion $P(t)$.
(d) Calculate the steady-state covariance $P(t) = \cdots = P = P_{ss}$.
(e) Calculate the output covariance recursion $R_{yy}(t)$.
(f) Calculate the steady-state output covariance R_{yy}.

4.13 Suppose we are given the Gauss–Markov process characterized by the state equations

$$x(t) = 0.97x(t-1) + u(t-1) + w(t-1)$$

for $u(t)$ a step of amplitude 0.03 and $w \sim N(0, 10^{-4})$ and $x(0) \sim N(2.5, 10^{-12})$.
(a) Calculate the covariance of x, that is, $P(t) = \text{Cov}(x(t))$.

(*b*) Since the process is stationary, we know that

$$P(t+k) = P(t+k-1) = \cdots = P(0) = P$$

What is the steady-state covariance P of this process?

(*c*) Develop a MATLAB program to simulate this process.

(*d*) Plot the process $x(t)$ with the corresponding confidence limits $\pm 2\sqrt{P(t)}$ for 100 data points, do 95% of the samples lie within the bounds?

4.14 Suppose we are given the ARMAX model

$$\begin{aligned} y(t) = &-0.5y(t-1) - 0.7y(t-2) + u(t) + 0.3u(t-1) \\ &+ e(t) + 0.2e(t-1) + 0.4e(t-2) \end{aligned}$$

(*a*) What is the corresponding innovations model in state–space form for $e \sim N(0, 10)$?

(*b*) Calculate the corresponding covariance matrix R^*_{ee}.

4.15 Given the following ARMAX model

$$A(q^{-1})y(t) = B(q^{-1})u(t) + \frac{C(q^{-1})}{D(q^{-1})}\epsilon(t)$$

for q^{-1} the backward shift (delay) operator such that

$$\begin{aligned} A(q^{-1}) &= 1 + 1.5q^{-1} + 0.7q^{-2} \\ B(q^{-1}) &= 1 + 0.5q^{-1} \\ C(q^{-1}) &= 1 + 0.7q^{-1} \\ D(q^{-1}) &= 1 + 0.5q^{-1} \end{aligned}$$

(*a*) Find the pulse transfer representation of this process (C = D = 0). Convert it to the following equivalent *pole-zero* and *normal* state–space forms. Is the system controllable? Is it observable? Show your calculations.

(*b*) Find the pole-zero or ARX representation of this process (C = 1, D = 0). Convert it to the equivalent state–space form.

(*c*) Find the pole-zero or ARMAX representation of this process (D = 0). Convert it to the equivalent state–space form.

(*d*) Find the all-zero or FIR representation of this process (A = 1, C = D = 0). Convert it to the equivalent state–space form.

(*e*) Find the all-pole or IIR representation of this process (B = 0, C = 0, D = 0). Convert it to the equivalent state–space form.

(*f*) Find the all-zero or MA representation of this process (A = 1, B = 0, D = 0). Convert it to the equivalent state–space form.

(*g*) Using the full model above with A, B, C, D polynomials, is it possible to find an equivalent Gauss–Markov representation? If so, find it and convert it to the equivalent state–space form. (*Hint*: Consider the C/D polynomials to be a coloring filter with input $\epsilon(t)$ and output $e(t)$)

4.16 Given a continuous–discrete Gauss–Markov model

$$\dot{x}_t = \alpha x_t + u_t + w_t$$
$$y(t_k) = \beta x(t_k) + v(t_k)$$

where w_t and $v(t_k)$ are zero-mean and white with respective covariances R_{ww} and R_{vv} along with a piecewise constant input u_t.

(*a*) Develop the continuous–discrete mean and covariance propagation models for this system.

(*b*) Suppose $w(t)$ is processed by a coloring filter that exponentially correlates it $R_{ww}(\tau) = Ge^{-|\lambda|\tau}$. Develop the continuous–discrete Gauss–Markov model in this case.

4.17 Develop the continuous–discrete Gauss–Markov models for the following systems:

(*a*) Wiener process: $\dot{z}_t = w_t$; $z_0 = 0$, w is zero-mean, white with R_{ww}.

(*b*) Random bias: $\dot{z}_t = 0$; $z_0 = z_o$, where $z_o \sim \mathcal{N}(0, R_{z_o z_o})$.

(*c*) Random ramp: $\ddot{z}_t = 0$; $\dot{z}_0 = z_1$; $z_0 = z_o$.

(*d*) Random oscillation: $\ddot{z}_t + \omega_o^2 z_t = 0$; $\dot{z}_0 = z_1$; $z_0 = z_o$.

(*e*) Random second order: $\ddot{z}_t + 2\zeta\omega_n \dot{z} + \omega_n^2 z_t = \omega_n^2 w_t$; $\dot{z}_0 = z_1$; $z_0 = z_o$.

4.18 Develop the continuous–discrete Gauss–Markov model for correlated process noise, that is,

$$\dot{w}_t = A_{cw} w_t + B_{cw} u_t + W_{cw} w_t^* \quad \text{for } w^* \sim \mathcal{N}(0, R_{w^* w^*}).$$

4.19 Develop the approximate Gauss–Markov model for the following nonlinear state transition and measurement model given by

$$x(t) = \frac{1}{2}x(t-1) + \frac{25x(t-1)}{1 + x^2(t-1)} + 8\cos(1.2(t-1)) + w(t-1)$$
$$y(t) = \frac{x^2(t)}{20} + v(t)$$

where $w \sim \mathcal{N}(0, R_{ww}(t-1))$ and $v \sim \mathcal{N}(0, R_{vv}(t))$. The initial state is Gaussian distributed with $\overline{x}(0) \sim \mathcal{N}(0, \overline{P}(0))$.

4.20 Consider the discrete nonlinear process given by

$$x(t) = (1 - 0.05\Delta T)x(t-1) + 0.04\Delta T x^2(t-1) + w(t-1)$$

with corresponding measurement model

$$y(t) = x^2(t) + x^3(t) + v(t)$$

where $w \sim \mathcal{N}(0, R_{ww}(t-1))$ and $v \sim \mathcal{N}(0, R_{vv}(t))$. The initial state is Gaussian distributed with $\overline{x}(0) \sim \mathcal{N}(0, \overline{P}(0))$.

Develop the approximate Gauss–Markov process model for this nonlinear system.

5

CLASSICAL BAYESIAN STATE–SPACE PROCESSORS

5.1 INTRODUCTION

In this chapter, we introduce the concepts of statistical signal processing from the Bayesian perspective using state–space models. We first develop the Bayesian paradigm using the generic state–space representation of the required conditional distributions and show how they propagate within the Bayesian framework. Next, we start with the linear (time-varying) Gauss–Markov model and develop the required conditional distributions leading to the well-known Kalman filter processor [1]. Based on this fundamental theme, we progress to the idea of linearization of the nonlinear state–space system developed in the previous chapter, where we derive the linearized Bayesian processor (LZ-BP). It is shown that the resulting processor provides a solution (time-varying) to the nonlinear state estimation. We then develop the extended Bayesian processor (XBP), or equivalently the extended Kalman filter (EKF), as a special case of the LZ-BP linearizing about the most currently available estimate. Next, we investigate a further enhancement of the XBP by introducing a local iteration of the nonlinear measurement system. Here the processor is called the iterated-extended Bayesian processor (IX-BP) and is shown to produce improved estimates at a small computational cost in most cases. We summarize the results with a case study implementing a 2D-tracking filter.

Bayesian Signal Processing: Classical, Modern, and Particle Filtering Methods, Second Edition. James V. Candy.

5.2 BAYESIAN APPROACH TO THE STATE–SPACE

In the previous chapter, we briefly developed deterministic and stochastic (Gauss–Markov) state–space models and demonstrated how the states propagate through the state transition mechanism for both continuous and discrete systems and their variants. Here we again take a Bayesian perspective and assume that the state or dynamic variables evolve according to a "probabilistic" transition mechanism.

Bayesian estimation relative to the state–space models is based on extracting the unobserved or hidden dynamic (state) variables from noisy measurement data. The Markovian state vector with initial distribution, $\Pr(x(0))$, propagates temporally throughout the state–space according to the *probabilistic transition distribution*, $\Pr(x(t)|x(t-1))$, while the conditionally independent measurements evolve from the *likelihood distribution*, $\Pr(y(t)|x(t))$. We see that the dynamic state variable at time t is obtained through the transition probability based on the previous state (Markovian property), $x(t-1)$, and the knowledge of the underlying conditional probability. Once propagated to time t, the dynamic state variable is used to update or correct based on the likelihood probability and the new measurement $y(t)$. Note that it is the knowledge of these conditional distributions that enable the Bayesian processor (BP).

Returning to the usual model-based constructs of the dynamic state variables discussed in the previous chapter, we see that there is an *implied* equivalence between the probabilistic distributions and the underlying state/measurement transition models. Recall from Chapter 4 that the functional discrete state representation is given by

$$\begin{aligned} x(t) &= A(x(t-1), u(t-1), w(t-1)) \\ y(t) &= C(x(t), u(t), v(t)) \end{aligned} \tag{5.1}$$

where w and v are the respective process and measurement noise sources, with u a known input. Here $A(\cdot)$ is the nonlinear (or linear) dynamic state transition function and $C(\cdot)$, the corresponding measurement function. Both conditional probabilistic distributions embedded within the Bayesian framework are *completely* specified by these functions and the underlying noise distributions: $\Pr(w(t-1))$ and $\Pr(v(t))$. That is, we have the (implied) equivalence[1]

$$\begin{aligned} A(x(t-1), u(t-1), w(t-1)) &\Rightarrow \Pr(x(t)|x(t-1)) \Leftrightarrow \mathcal{A}(x(t)|x(t-1)) \\ C(x(t), u(t), v(t)) &\Rightarrow \Pr(y(t)|x(t)) \quad\;\; \Leftrightarrow \mathcal{C}(y(t)|x(t)) \end{aligned} \tag{5.2}$$

Thus, the state–space model along with the noise statistics and prior distributions define the required Bayesian representation or probabilistic propagation model which defines the evolution of the states and measurements through the transition probabilities. This is sometimes a subtle point that must be emphasized. As illustrated in Fig. 5.1, the dynamic state variables propagate throughout state–space specified by

[1] We use this notation to emphasize the influence of both process ($\mathcal{A}$) and measurement ($\mathcal{C}$) representations on the conditional distributions.

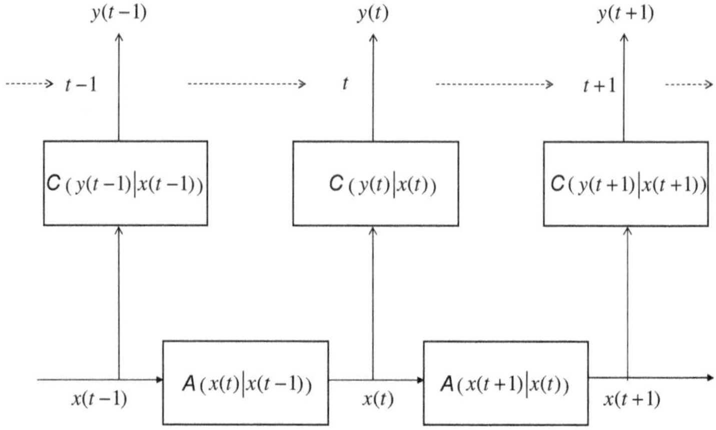

FIGURE 5.1 Bayesian state-space probabilistic evolution.

the transition probability $\mathcal{A}(x(t)|x(t-1))$ using the embedded process model. That is, the "unobserved" state at time $(t-1)$ requires the transition probability distribution to propagate to the state at time t. Once evolved, the state combines under the corresponding measurement at time t through the conditional likelihood distribution $C(y(t)|x(t))$ using the embedded measurement model to obtain the required likelihood distribution. These events continue to evolve throughout, with the states propagating through the state transition probability using the process model, and the measurements generated by the states and likelihood using the measurement model. From the Bayesian perspective, the broad prior is scaled by the evidence and "narrowed" by the likelihood to estimate the posterior.

With this in mind we can now return to the original Bayesian estimation problem, define it and show (at least conceptually) the optimal solution based on the state–space representation.

The basic dynamic state estimation (signal enhancement) problem can now be stated in the Bayesian framework as

GIVEN a set of noisy uncertain measurements $Y_t = \{y(t)\}$, and known inputs $\{u(t)\}$, $t = 0, \ldots, N$, along with the corresponding prior distributions for the initial state and process and measurement noise sources: $\Pr(x(0))$, $\Pr(w(t-1))$, $\Pr(v(t))$, as well as the conditional transition and likelihood probability distributions: $\Pr(x(t)|x(t-1))$, $\Pr(y(t)|x(t))$ characterized by the state and measurement models: $\mathcal{A}(x(t)|x(t-1))$, $C(y(t)|x(t))$, **FIND** the "best" estimate of the filtering *posterior* $\hat{\Pr}(x(t)|Y_t)$ and its associated statistics.

It is interesting to note that the entire *Bayesian system* can be defined by the set Σ as

$$\Sigma := [\{y(t)\}, \{u(t)\}, \Pr(x(0)), \Pr(w(t-1)), \Pr(v(t)), \mathcal{A}(x(t)|x(t-1)), C(y(t)|x(t))]$$

Compare this to the model-based solutions to follow where we obtain closed-form analytic expressions for these distributions.

Analytically, to generate the model-based version of the sequential BP, we replace the transition and likelihood distributions with the conditionals of Eq. 5.2. The solution to the signal enhancement or equivalently state estimation problem is given by the filtering distribution $\Pr(x(t)|Y_t)$ which was solved previously in Section 2.5 (see Table 2.1). We start with the prediction recursion characterized by the *Chapman–Kolmogorov equation* replacing the transition probability with the implied model-based conditional, that is,

$$\Pr(x(t)|Y_{t-1}) = \int \overbrace{\mathcal{A}(x(t)|x(t-1))}^{\text{Embedded process model}} \times \overbrace{\Pr(x(t-1)|Y_{t-1})}^{\text{Prior}}\, dx(t-1) \tag{5.3}$$

Next, we incorporate the model-based likelihood into the posterior equation with the understanding that the process model has been incorporated into the prediction

$$\Pr(x(t)|Y_t) = \overbrace{C(y(t)|x(t))}^{\text{Embedded measurement model}} \times \overbrace{\Pr(x(t)|Y_{t-1})}^{\text{Predicition}} \Big/ \Pr(y(t)|Y_{t-1}) \tag{5.4}$$

So we see from the Bayesian perspective that the sequential BP employing the state–space representation of Eq. 5.1 is straightforward. Next let us investigate a more detailed development of the processor resulting in a closed-form solution—the linear Kalman filter.

5.3 LINEAR BAYESIAN PROCESSOR (LINEAR KALMAN FILTER)

In this section, we constrain the state–space model to be linear (time-varying) and apply the Bayesian approach to obtain the optimal processor assuming additive Gaussian noise.

Suppose we are given the linear discrete Gauss–Markov model of the previous section (ignoring u with $W = I$ for notational simplicity) and we would like to develop the BP. Since we know the processes are linear and Gaussian, we know that the required distributions will also be Gaussian. To develop the processor for this case, we start with the *prediction* equation[2] and use the process model of Eq. 4.78, that is,

$$\Pr(x(t)|Y_{t-1}) = \int \mathcal{A}(x(t)|x(t-1)) \times \Pr(x(t-1)|Y_{t-1})\, dx(t-1)$$

[2] We have changed Gaussian distribution notation to include the random process, that is, $\mathcal{N}(m_z, R_{zz}) \rightarrow \mathcal{N}(z : m_z, R_{zz})$.

where the *filtered conditional*[3] is:

$$\Pr(x(t-1)|Y_{t-1}) \sim \mathcal{N}(x(t) : \hat{x}(t-1|t-1), \tilde{P}(t-1|t-1))$$

Now using the process model, we have that

$$\mathcal{A}(x(t)|x(t-1)) \sim \mathcal{N}(x(t) : A(t-1)\hat{x}(t-1|t-1), A(t-1)\tilde{P}(t-1|t-1) \\ \times A'(t-1) + R_{ww}(t-1))$$

which follows directly from the linearity of the conditional expectation operator, that is,

$$\hat{x}(t|t-1) = E\{x(t)|Y_{t-1}\} = E\{A(t-1)x(t-1) + w(t-1)|Y_{t-1}\} \\ = A(t-1)\hat{x}(t-1|t-1) \tag{5.5}$$

Using this result, the predicted *state estimation error* can be obtained as

$$\tilde{x}(t|t-1) = x(t) - \hat{x}(t|t-1) \\ = [A(t-1)x(t-1) + w(t-1)] - A(t-1)\hat{x}(t-1|t-1) \\ = A(t-1)\tilde{x}(t-1|t-1) + w(t-1) \tag{5.6}$$

and the corresponding *state error covariance* $\tilde{P}(t|t-1) = E\{\tilde{x}(t|t-1)\tilde{x}'(t|t-1)\}$ is easily derived. Summarizing, the conditional means and covariances that completely characterize the current (Gaussian) state evolve according to the following equations:

$$\hat{x}(t|t-1) = A(t-1)\hat{x}(t-1|t-1) \qquad \text{[Prediction]}$$
$$\tilde{P}(t|t-1) = A(t-1)\tilde{P}(t-1|t-1)A'(t-1) + R_{ww}(t-1) \qquad \text{[Prediction covariance]}$$

Substituting these multivariate Gaussian distributions (transition and filtered) into the prediction equation, we have

$$\Pr(x(t)|Y_{t-1}) \sim \mathcal{N}(x(t) : \hat{x}(t|t-1), \tilde{P}(t|t-1))$$

With the prediction distribution available, we require the correction or update distribution obtained from the likelihood and the measurement model of Eq. 5.4, that is,

$$\Pr(x(t)|Y_t) = \frac{\mathcal{C}(y(t)|x(t)) \times \Pr(x(t)|Y_{t-1})}{\Pr(y(t)|Y_{t-1})}$$

[3] This notation is defined in terms of *conditional means* and *covariances* by: $\hat{x}(t|t) := E\{x(t)|Y_t\}$ and $\tilde{P}(t|t) := \text{cov}(\tilde{x}(t|t))$ for the *state estimation error*, $\tilde{x}(t|t) := x(t) - \hat{x}(t|t)$.

Under the Gauss–Markov model assumptions, we know that each of the conditional distributions can be expressed in terms of the Gaussian distribution as

$$
\begin{aligned}
C(y(t)|x(t)) &\sim \mathcal{N}(y(t) : C(t)x(t), R_{vv}(t)) \\
\Pr(x(t)|Y_{t-1}) &\sim \mathcal{N}(x(t) : \hat{x}(t|t-1), \tilde{P}(t|t-1)) \\
\Pr(y(t)|Y_{t-1}) &\sim \mathcal{N}(y(t) : \hat{y}(t|t-1), R_{ee}(t))
\end{aligned}
$$

for $R_{ee}(t)$ the *innovations covariance* with *innovations* defined by $e(t) := y(t) - \hat{y}(t|t-1)$ and *predicted* or *filtered* measurement given by $\hat{y}(t|t-1) = C(t)\hat{x}(t|t-1)$.

Substituting these probabilities into Eq. 5.4 and combining all constants into a single constant κ, we obtain

$$
\begin{aligned}
\Pr(x(t)|Y_t) = \kappa \times \exp&\left[-\frac{1}{2}(y(t) - C(t)x(t))' R_{vv}^{-1}(t)(y(t) - C(t)x(t))\right] \\
\times \exp&\left[-\frac{1}{2}(x(t) - \hat{x}(t|t-1))' \tilde{P}^{-1}(t|t-1)(x(t) - \hat{x}(t|t-1))\right] \\
\times \exp&\left[+\frac{1}{2}(y(t) - \hat{y}(t|t-1))' R_{ee}^{-1}(t)(y(t) - \hat{y}(t|t-1))\right]
\end{aligned}
$$

Recognizing the measurement noise, state estimation error and innovation in the above terms, we have that the posterior probability is given in terms of the Gauss–Markov model by

$$
\begin{aligned}
\Pr(x(t)|Y_t) = \kappa \times \exp&\left[-\frac{1}{2}v'(t)R_{vv}^{-1}(t)v(t)\right] \\
\times \exp&\left[-\frac{1}{2}\tilde{x}'(t|t-1)\tilde{P}^{-1}(t|t-1)\tilde{x}(t|t-1)\right] \\
\times \exp&\left[+\frac{1}{2}e'(t)R_{ee}^{-1}(t)e(t)\right]
\end{aligned}
\tag{5.7}
$$

So we see that the posterior distribution can be estimated under the multivariate Gaussian assumptions and the corresponding linear (time-varying) Gauss–Markov model. This is the *optimal* BP under these assumptions. In most cases we are not able to characterize the distributions in closed form and must resort to numerical (simulation-based) solutions.

We realize at this point that we have the optimal Bayesian predictor and posterior, but we still have not extracted the optimal state estimates explicitly and its associated performance metric. Recall from the batch Bayesian solutions of Section 2.1 that once we have the posterior, we can estimate a variety of statistics using it as the basis. In this case, the optimal BP will be the one that maximizes the posterior; therefore, we continue the development of the linear filter by deriving the Bayesian maximum a posteriori (MAP) estimator.

Starting with the MAP equation of Eq. 2.3 and taking natural logarithms of each side gives

$$\ln \Pr(x(t)|Y_t) = \ln \Pr(y(t)|x(t)) + \ln \Pr(x(t)|Y_{t-1}) - \ln \Pr(y(t)|Y_{t-1})$$

In terms of multivariate Gaussian, the posterior is given by

$$\begin{aligned}\ln \Pr(x(t)|Y_t) = \ln \kappa - \frac{1}{2}v'(t)R_{vv}^{-1}(t)v(t) - \frac{1}{2}\tilde{x}'(t|t-1)\tilde{P}^{-1}(t|t-1)\tilde{x}(t|t-1)\\ + \frac{1}{2}e'(t)R_{ee}^{-1}(t)e(t)\end{aligned} \tag{5.8}$$

with

$$\begin{aligned}\tilde{x}(t|t-1) &= x(t) - \hat{x}(t|t-1) && \text{[State error]}\\ e(t) &= y(t) - C(t)\hat{x}(t|t-1) = C(t)\tilde{x}(t|t-1) + v(t) && \text{[Innovation]}\end{aligned}$$

The MAP estimate is then obtained by differentiating Eq. 5.8, setting it to zero and solving, that is,

$$\nabla_x \ln \Pr(x(t)|Y_t)\Big|_{x=\hat{X}_{\text{map}}} = 0 \tag{5.9}$$

Using the *chain rule* of the gradient operator (see Eq. 2.11), we obtain the following expression. Note that the last term of Eq. 5.8 is not a function of $x(t)$, but just the data; therefore, its gradient is null.

$$\nabla_x \ln \Pr(x(t)|Y_t) = C'(t)R_{vv}^{-1}(t)[y(t) - C(t)x(t)] - \tilde{P}^{-1}(t|t-1)\tilde{x}(t|t-1) \tag{5.10}$$

Setting Eq. 5.10 to zero and solving for $x(t)$ gives the Bayesian MAP estimate

$$\begin{aligned}\hat{X}_{\text{map}}(t) = [C'(t)R_{vv}^{-1}(t)C(t) + \tilde{P}^{-1}(t|t-1)]^{-1}\\ \times [\tilde{P}^{-1}(t|t-1)\hat{x}(t|t-1) + C'(t)R_{vv}^{-1}(t)y(t)]\end{aligned} \tag{5.11}$$

This relation can be simplified by using a form of the *matrix inversion lemma* [1] defined by the following equation

$$(A + BD')^{-1} = A^{-1} - A^{-1}B(I + D'A^{-1}B)^{-1})D'A^{-1} \tag{5.12}$$

Defining the following terms for the lemma, $A = \tilde{P}^{-1}(t|t-1)$, $B = C'(t)R_{vv}^{-1}(t)$ and setting $D' = C(t)$, we find that

$$\begin{aligned}&[\tilde{P}^{-1}(t|t-1) + C'(t)R_{vv}^{-1}(t)C(t)]^{-1}\\ &\quad = \tilde{P}(t|t-1) - \tilde{P}(t|t-1)C'(t)R_{vv}^{-1}(t)(I + C(t)\tilde{P}(t|t-1)C'(t)R_{vv}^{-1}(t))^{-1}\\ &\quad \times C(t)\tilde{P}(t|t-1)\end{aligned} \tag{5.13}$$

Making the observation that the term in parenthesis on the right-hand side of Eq. 5.13 can be rewritten by factoring out $R_{vv}^{-1}(t)$ as

$$(I + C(t)\tilde{P}(t|t-1)C'(t)R_{vv}^{-1}(t))^{-1} = R_{vv}(t)(R_{vv}(t) + C(t)\tilde{P}(t|t-1)C'(t))^{-1} \quad (5.14)$$

then Eq. 5.13 can also be expressed as

$$[\tilde{P}^{-1}(t|t-1) + C'(t)R_{vv}^{-1}(t)C(t)]^{-1}$$
$$= \tilde{P}(t|t-1) - \tilde{P}(t|t-1)C'(t)(R_{vv}(t) + C(t)\tilde{P}(t|t-1)C'(t))^{-1}C(t)\tilde{P}(t|t-1) \quad (5.15)$$

The innovations covariance can be expressed as

$$R_{ee}(t) = \text{Cov}(e(t)) = C(t)\tilde{P}(t|t-1)C'(t) + R_{vv}(t)$$

and substituting, Eq. 5.15 becomes

$$[\tilde{P}^{-1}(t|t-1) + C'(t)R_{vv}^{-1}(t)C(t)]^{-1}$$
$$= \tilde{P}(t|t-1) - \tilde{P}(t|t-1)C'(t)R_{ee}^{-1}(t)C(t)\tilde{P}(t|t-1) = (I - K(t)C(t))\tilde{P}(t|t-1) \quad (5.16)$$

where $K(t) = \tilde{P}(t|t-1)C'(t)R_{ee}^{-1}(t)$ is the gain. We see that Eq. 5.16 is simply the *updated error covariance* $\tilde{P}(t|t)$ equivalent to

$$\tilde{P}(t|t) \equiv [\tilde{P}^{-1}(t|t-1) + C'(t)R_{vv}^{-1}(t)C(t)]^{-1} \quad (5.17)$$

Thus we can eliminate the first bracketed term in Eq. 5.11 to give

$$\hat{X}_{\text{map}}(t) = \tilde{P}(t|t) \times [\tilde{P}^{-1}(t|t-1)\hat{x}(t|t-1) + C'(t)R_{vv}^{-1}(t)y(t)]$$

Solving Eq. 5.17 for $\tilde{P}(t|t-1)$, we can substitute the result into the above equation to give

$$\hat{X}_{\text{map}}(t) = \tilde{P}(t|t) \times [(\tilde{P}^{-1}(t|t) - C'(t)R_{vv}^{-1}(t)C(t))\hat{x}(t|t-1) + C'(t)R_{vv}^{-1}(t)y(t)] \quad (5.18)$$

Multiplying out, regrouping terms and factoring, this relation can be rewritten as

$$\hat{X}_{\text{map}}(t) = \hat{x}(t|t-1) + (\tilde{P}(t|t)C'(t)R_{vv}^{-1}(t))[y(t) - C(t)\hat{x}(t|t-1)] \quad (5.19)$$

or finally

$$\hat{X}_{\text{map}}(t) = \hat{x}(t|t) = \hat{x}(t|t-1) + K(t)e(t) \quad (5.20)$$

Now we only need to show the equivalence of the gain expression using the updated instead of predicted error covariance, that is,

$$\begin{aligned}K(t) &= \tilde{P}(t|t-1)C'(t)R_{ee}^{-1}(t) = \tilde{P}(t|t)\tilde{P}^{-1}(t|t)(\tilde{P}(t|t-1)C'(t)R_{ee}^{-1}(t))\\ &= \tilde{P}(t|t)[C'(t)R_{vv}^{-1}(t)C(t) + \tilde{P}^{-1}(t|t-1)]\tilde{P}(t|t-1)C'(t)R_{ee}^{-1}(t)\\ &= \tilde{P}(t|t)C'(t)R_{vv}^{-1}(t)[C(t)\tilde{P}(t|t-1)C'(t) + R_{vv}(t)]R_{ee}^{-1}(t)\end{aligned} \tag{5.21}$$

which gives the desired result from the definition of innovations covariance. We now have two *equivalent* expressions in terms of the updated or predicted error covariances that can be used to calculate the gain

$$K(t) = \tilde{P}(t|t)C'(t)R_{vv}^{-1}(t) \equiv \tilde{P}(t|t-1)C'(t)R_{ee}^{-1}(t) \tag{5.22}$$

which completes the Bayes' approach to signal enhancement or equivalently state estimation problem yielding the optimum linear BP (Kalman filter). A summary of the linear BP algorithm is shown in Table 5.1.

The design of linear BPs under the Gauss–Markov assumptions is well-understood [1–9]. Based on a variety of properties both theoretically well-founded and pragmatically applied with high success, the *minimum (error) variance* design procedure has evolved [10–14]. We summarize the design-steps below and subsequently using the notation of the BP algorithm in Table 5.1.

It is important to realize that a *necessary* and *sufficient* condition that the linear BP (under the Gauss-Markov (GM) constraints) is *optimal* is that the innovation sequence is zero-mean and white or uncorrelated! This is the first and most important step in BP design. If this condition does not hold, then the underlying model and GM assumptions are invalid. Therefore, we briefly mention the minimum variance

TABLE 5.1 Linear BP (Kalman Filter) Algorithm

Prediction	
$\hat{x}(t|t-1) = A(t-1)\hat{x}(t-1|t-1) + B(t-1)u(t-1)$	[State prediction]
$\tilde{P}(t|t-1) = A(t-1)\tilde{P}(t-1|t-1)A'(t-1) + R_{ww}(t-1)$	[Covariance prediction]
Innovation	
$e(t) = y(t) - \hat{y}(t|t-1) = y(t) - C(t)\hat{x}(t|t-1)$	[Innovation]
$R_{ee}(t) = C(t)\tilde{P}(t|t-1)C'(t) + R_{vv}(t)$	[Innovation covariance]
Gain	
$K(t) = \tilde{P}(t|t-1)C'(t)R_{ee}^{-1}(t)$	[Gain or weight]
Update	
$\hat{x}(t|t) = \hat{x}(t|t-1) + K(t)e(t)$	[State update]
$\tilde{P}(t|t) = [I - K(t)C(t)]\tilde{P}(t|t-1)$	[Covariance update]
Initial Conditions	
$\hat{x}(0|0) \quad \tilde{P}(0|0)$	

design procedure here and provide more details in Section 5.7 where pragmatic statistical tests are developed. We will apply the procedure to the following processors (linear and nonlinear) in the example problems and then provide the design details subsequently.

The *minimum (error) variance design procedure* is

1. Check that the innovations sequence is *zero-mean.*
2. Check that the innovations sequence is *white.*
3. Check that the innovations sequence is *uncorrelated* in time with input u.
4. Check that the innovations sequence *lies within* the confidence limits constructed from R_{ee} predicted by the BP.
5. Check that the innovations sequence variance is *reasonably close* to the estimated (sample) variance $\hat{R}_{ee}$.
6. Check that the state estimation error $\tilde{x}(t|t)$ *lies within* the confidence limits constructed from $\tilde{P}(t|t)$ predicted by the BP.
7. Check that the state error variance is *reasonably close* to the estimated (sample) variance $\hat{\tilde{P}}(t|t)$.

This is the basic "cookbook" approach to linear BP design. Before we close this section, let us consider a simple linear example to demonstrate the ideas.

Example 5.1

Suppose we have the *RC* circuit as shown in Fig. 5.2. We measure the voltage across the capacitor with a high-impedance voltmeter as shown. Since these measurements are noisy and the component values are imprecise ($\pm\triangle$), we require an improved estimate of the output voltage. We develop a BP to solve this problem from first

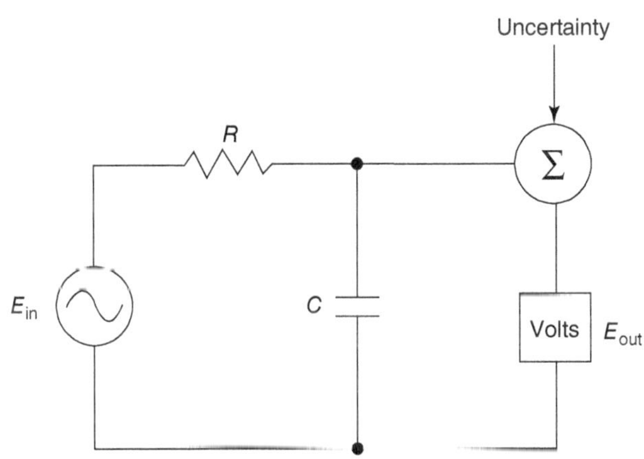

FIGURE 5.2 *RC*-circuit problem diagram.

principles—a typical approach. Writing the Kirchoff's current equations at the node, we have

$$I_{\text{in}}(t) - \frac{e(t)}{R} - C\frac{de(t)}{dt} = 0$$

where e_o is the initial voltage, R the resistance, and C the capacitance. The measurement equation for a voltmeter of gain K_e is simply

$$e_{out}(t) = K_e e(t)$$

We choose to use the discrete BP formulation; therefore, approximating the derivatives with first differences and substituting, we have

$$C\frac{e(t) - e(t-1)}{\triangle T} = -\frac{e(t-1)}{R} + I_{\text{in}}(t-1)$$

or

$$e(t) = \left(1 - \frac{\triangle T}{RC}\right) e(t-1) + \frac{\triangle T}{C} I_{\text{in}}(t-1)$$

where the measurement is given above. Suppose that for this circuit the parameters are: $R = 3.3\ \text{k}\Omega$ and $C = 1000\ \mu\text{F}$, $\triangle T = 100$ ms, $e_o = 2.5$ V, $K_e = 2.0$, and the voltmeter is precise to within ± 4 V. Then, transforming the physical circuit model into state–space form by defining $x = e$, $y = e_{out}$, and $u = I_{in}$, we obtain

$$\begin{aligned} x(t) &= 0.97x(t-1) + 100u(t-1) + w(t-1) \\ y(t) &= 2x(t) + v(t) \end{aligned}$$

The process noise covariance is used to model the circuit parameter uncertainty with $R_{ww} = 0.0001$, since we assume standard deviations, $\triangle R$, $\triangle C$ of 1%. Also, $R_{vv} = 4$, since two standard deviations are $\triangle V = 2(\frac{1}{2}4V)$. We also assume initially that the state is $x(0) \sim \mathcal{N}(2.5, 10^{-12})$, and that the input current is a step function of $u(t) = 300\ \mu$A. SSPACK_PC is used to simulate this system [8]. The results are shown in Fig. 5.3. The simulated and true (mean) states (voltages) are shown in Fig. 5.3a along with the corresponding confidence limits. We see that the process samples (state and process noise) lie within the bounds (3.5% out). Therefore, the data statistically satisfy the underlying Gauss–Markov model assumptions. If it does not, then choose another simulation. That is, we perform another realization (different *seed* in random number generator) until the samples lie within the bounds. Similarly, the simulated and true (mean) measured voltages are shown in Fig. 5.3b. Again the data (measurement and noise) statistically satisfy the underlying models with only 4.5% of the samples exceeding the prescribed bounds. The state and measurement variances used to construct the confidence intervals about the means, that is, $[m_x(t) \pm 1.96\sqrt{P(t)}]$ and $[m_y(t) \pm 1.96\sqrt{R_{yy(t)}}]$, are shown in Fig. 5.3c.

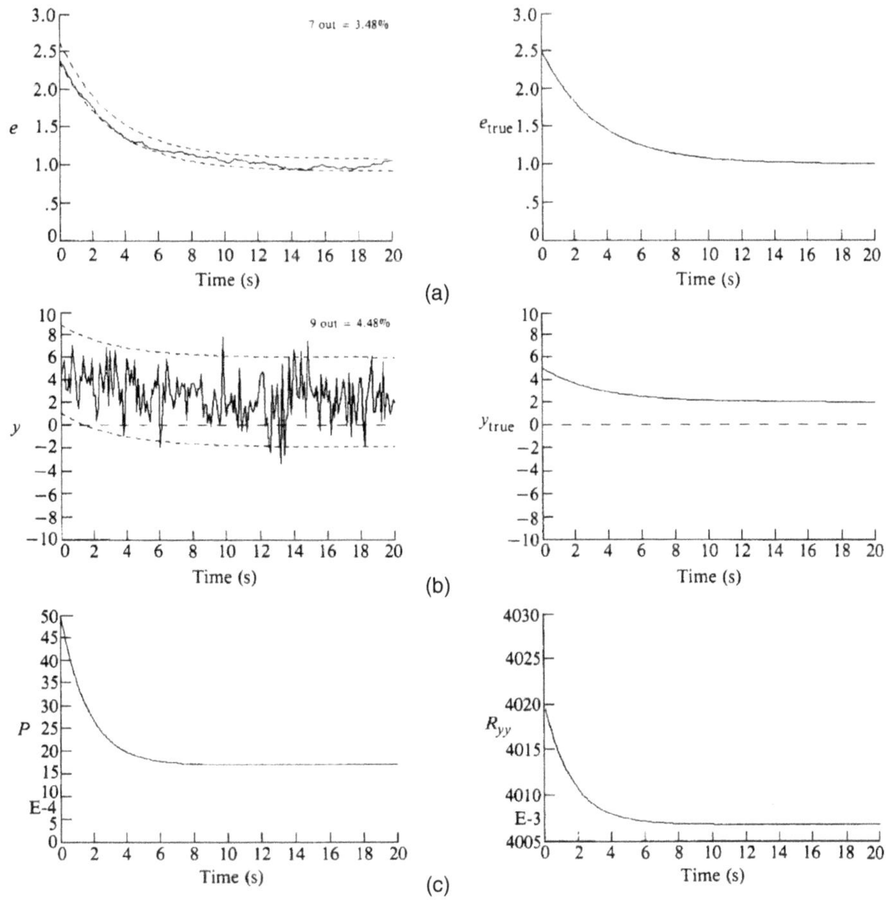

FIGURE 5.3 *RC* circuit problem Gauss–Markov simulation: (a) simulated and true (mean) output voltage; (b) simulated and true (mean) measurement; (c) simulated state and measurement variances.

With the data simulated, we now consider the design of the BP. In the ideal BP problem, we are given the model set $\Sigma := \{A, B, C, R_{ww}, R_{vv}, x(0), P(0)\}$, the known input $\{u(t)\}$, and the set of noisy measurements $\{y(t)\}$ to construct the processor. The *RC* model-based processor for this problem can simply be written as

$$\hat{x}(t|t-1) = 0.97\hat{x}(t-1|t-1)) + 100u(t-1) \quad \text{[Predicted state]}$$
$$\tilde{P}(t|t-1) = 0.94\tilde{P}(t-1|t-1) + 0.0001 \quad \text{[Predicted covariance]}$$
$$e(t) = y(t) - 2\hat{x}(t|t-1) \quad \text{[Innovation]}$$
$$R_{ee}(t) = 4\tilde{P}(t|t-1) + 4 \quad \text{[Innovations covariance]}$$

$$K(t) = 2\frac{\tilde{P}(t|t-1)}{4\tilde{P}(t|t-1)+4} \quad \text{[Gain]}$$

$$\hat{x}(t|t) = \hat{x}(t|t-1) + K(t)e(t) \quad \text{[Updated state]}$$

$$\tilde{P}(t|t) = \frac{\tilde{P}(t|t-1)}{\tilde{P}(t|t-1)+1} \quad \text{[Updated covariance]}$$

The estimator is also designed using SSPACK_PC and the results are shown in Fig. 5.4. In Fig. 5.4a, we see the estimated state (voltage) and estimation error as well as the corresponding confidence bounds. Note that the processor "optimally" estimates the voltage, since our models are exact. That is, it provides the minimum error variance estimate in the Gaussian case. Also since we have the true (mean) state, we can calculate the estimation error and use the corresponding error covariance to specify the bounds as shown. Note that the error is small and no samples exceed the bound as indicated by the overestimation of the variance compared with the sample variance ($0.0017 > 0.0002$). In Fig. 5.4b, we see the filtered measurement ($\hat{y}(t|t-1)$) and corresponding innovation sequence along with the confidence limits provided by the processor. Here only 4.5% of the samples exceed the bounds and the variance predicted by the filter is close to the sample variance estimate ($4.0 \sim 3.7$). The weighted sum-squared residual (WSSR) statistic, zero-mean, and whiteness tests are shown in Fig. 5.4c. Here we see that using a window of 75 samples, the threshold is not exceeded, indicating a statistically white sequence. The innovation mean is small and well within the bound ($0.11 < 0.27$). The sequence is statistically white, since 0% of the normalized sample covariances exceed the bound. Finally, we see the gain and updated error covariance as monotonically decreasing functions that reach a steady-state (constant) value at approximately 8 sec. This completes the example of an ideally "tuned" BP. △△△

5.4 LINEARIZED BAYESIAN PROCESSOR (LINEARIZED KALMAN FILTER)

In this section, we develop an approximate solution to the nonlinear processing problem involving the linearization of the nonlinear process about a "known" reference trajectory followed by the development of a BP based on the underlying linearized state–space model. Many processes in practice are nonlinear rather than linear. Coupling the nonlinearities with noisy data makes the signal processing problem a challenging one. In this section, we limit our discussion to *discrete* nonlinear systems. Continuous solutions to this problem are developed in Refs. [1–7].

Recall from the previous chapter that our *process* is characterized by a set of nonlinear stochastic vector difference equations in state–space form as

$$x(t) = a[x(t-1)] + b[u(t-1)] + w(t-1) \tag{5.23}$$

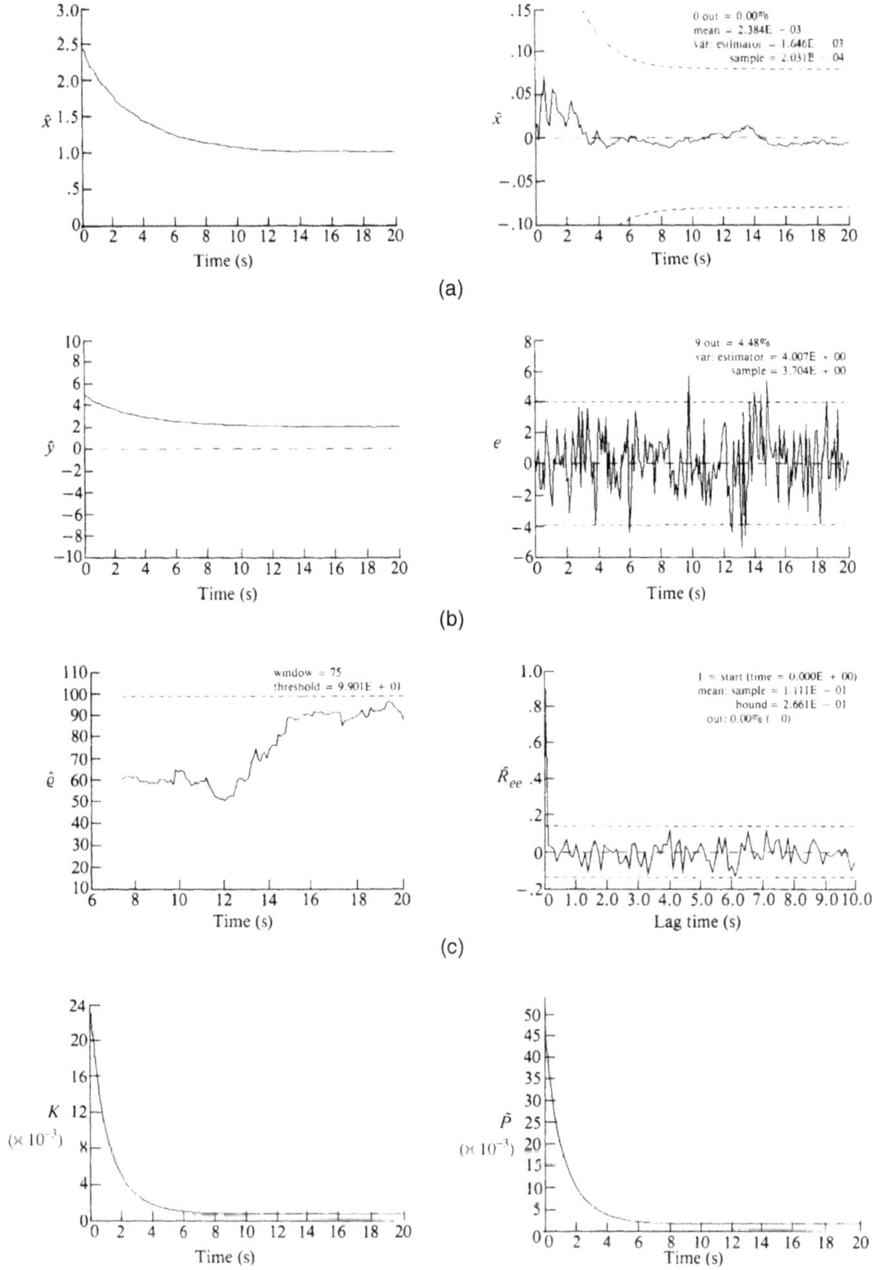

FIGURE 5.4 BP design for *RC* circuit problem: (a) estimated state (voltage) and error; (b) filtered voltage measurement and error (innovations); (c) WSSR and zero-mean/whiteness tests; (d) gain and updated error covariance.

with the corresponding *measurement* model

$$y(t) = c[x(t)] + v(t) \tag{5.24}$$

where $a[\cdot]$, $b[\cdot]$, $c[\cdot]$ are nonlinear vector functions of x, u, with $x, a, b, w \in R^{N_x \times 1}$, $y, c, v \in R^{N_y \times 1}$ and $w \sim \mathcal{N}(0, R_{ww}(t))$, $v \sim \mathcal{N}(0, R_{vv}(t))$.

In Chapter 4, we linearized a deterministic nonlinear model using a first-order Taylor series expansion for the functions, a, b, and c, and developed a *linearized Gauss–Markov perturbation model* valid for small deviations given by

$$\begin{aligned}
\delta x(t) &= A[x^*(t-1)]\delta x(t-1) + B[u^*(t-1)]\delta u(t-1) + w(t-1) \\
\delta y(t) &= C[x^*(t)]\delta x(t) + v(t)
\end{aligned} \tag{5.25}$$

where A, B, and C are the corresponding Jacobian matrices, and w, v are zero-mean, Gaussian.

We used linearization techniques to approximate the statistics of Eqs. 5.23 and 5.24 and summarized these results in an "approximate" Gauss–Markov model of Table 4.3. Using this perturbation model, we will now incorporate it to construct a BP that embeds the $(A[\cdot], B[\cdot], C[\cdot])$ Jacobians linearized about the reference trajectory $[x^*, u^*]$. Each of the Jacobians are deterministic and time-varying, since they are updated at each time-step. Replacing the (A, B) matrices and $\hat{x}(t|t-1)$ in Table 5.1, respectively, by the Jacobians and $\delta\hat{x}(t|t-1)$, we obtain the state perturbation predicted estimate

$$\delta\hat{x}(t|t-1) = A[x^*(t-1)]\delta\hat{x}(t-1|t-1) + B[u^*(t-1)]\delta u(t-1) \tag{5.26}$$

For the Bayesian estimation problem, we are interested in the state estimate $\hat{x}(t|t-1)$ and not its deviation $\delta\hat{x}(t|t-1)$. From the definition of the perturbation in Section 4.8, we have

$$\hat{x}(t|t-1) = \delta\hat{x}(t|t-1) + x^*(t) \tag{5.27}$$

where the reference trajectory $x^*(t)$ was defined previously as

$$x^*(t) = a[x^*(t-1)] + b[u^*(t-1)] \tag{5.28}$$

Substituting this relation and Eq. 5.26 in Eq. 5.27 gives

$$\begin{aligned}
\hat{x}(t|t-1) = a[x^*(t-1)] &+ A[x^*(t-1)][\hat{x}(t-1|t-1) - x^*(t-1)] \\
&+ b[u^*(t-1)] + B[u^*(t-1)][u(t-1) - u^*(t-1)]
\end{aligned} \tag{5.29}$$

The corresponding *perturbed* innovation can also be found directly

$$\begin{aligned}
\delta e(t) = \delta y(t) - \delta\hat{y}(t|t-1) &= (y(t) - y^*(t)) - (\hat{y}(t|t-1) - y^*(t)) \\
&= y(t) - \hat{y}(t|t-1) = e(t)
\end{aligned} \tag{5.30}$$

Using the linear BP with deterministic Jacobian matrices results in

$$\delta\hat{y}(t|t-1) = C[x^*(t)]\delta\hat{x}(t|t-1) \tag{5.31}$$

and therefore using this relation and Eq. 4.142 for the reference measurement, we have

$$\hat{y}(t|t-1) = y^*(t) + C[x^*(t)]\delta\hat{x}(t|t-1) = c[x^*(t)] + C[x^*(t)]\delta\hat{x}(t|t-1) \tag{5.32}$$

Therefore, it follows that the innovation is

$$e(t) = y(t) - c[x^*(t)] - C[x^*(t)][\hat{x}(t|t-1) - x^*(t)] \tag{5.33}$$

The updated estimate is easily found by substituting Eq. 5.27 to obtain

$$\begin{aligned} \delta\hat{x}(t|t) &= \delta\hat{x}(t|t-1) + K(t)e(t) \\ [\hat{x}(t|t) - x^*(t)] &= [\hat{x}(t|t-1) - x^*(t)] + K(t)e(t) \end{aligned} \tag{5.34}$$

which yields the identical update equation of Table 5.1. Since the state perturbation estimation error is identical to the state estimation error, the corresponding error covariance is given by $\delta\tilde{P}(t|\cdot) = \tilde{P}(t|\cdot)$ and therefore,

$$\delta\tilde{x}(t|\cdot) = \delta x(t) - \delta\hat{x}(t|\cdot) = [x(t) - x^*(t)] - [\hat{x}(t|\cdot) - x^*(t)] = x(t) - \hat{x}(t|\cdot) \tag{5.35}$$

The gain is just a function of the measurement linearization, $C[x^*(t)]$ completing the algorithm. We summarize the discrete LZ-BP (Kalman filter) in Table 5.2.

In a more formal framework, the LZ-BP can be developed under (approximate) Gaussian assumptions using the *Bayesian approach* as before in the linear case. We briefly outline the derivation by carefully following the steps in Section 5.3.

The *a posteriori probability* is given by

$$\Pr(x(t)|Y_t) = \frac{\Pr(y(t)|x(t)) \times \Pr(x(t)|Y_{t-1})}{\Pr(y(t)|Y_{t-1})} \tag{5.36}$$

Under the Gauss–Markov model assumptions, we know that each of the conditional expectations can be expressed in terms of the conditional Gaussian distributions as:

1. $\Pr(y(t)|x(t))$: $\mathcal{N}(c[x(t)], R_{vv}(t))$
2. $\Pr(x(t)|Y_{t-1})$: $\mathcal{N}(\hat{x}(t|t-1), \tilde{P}(t|t-1))$
3. $\Pr(y(t)|Y_{t-1})$: $\mathcal{N}(\hat{y}(t|t-1), R_{ee}(t))$

Using the nonlinear models developed earlier, substituting the Gaussian probabilities and taking logarithms, we obtain the logarithmic a posteriori probability

TABLE 5.2 Linearized *BP* (Kalman Filter) Algorithm

Prediction	
$\hat{x}(t\|t-1) = a[x^*(t-1)] + A[x^*(t-1)][\hat{x}(t-1\|t-1) - x^*(t-1)]$ $+ \mathrm{b}[u^*(t-1)] + B[u^*(t-1)][u(t-1) - u^*(t-1)]$	[State prediction]
$\tilde{P}(t\|t-1) = A[x^*(t-1)]\tilde{P}(t-1\|t-1)A'[x^*(t-1)] + R_{ww}(t-1)$	[Covariance prediction]
Innovation	
$e(t) = y(t) - c[x^*(t)] - C[x^*(t)][\hat{x}(t\|t-1) - x^*(t)]$	[Innovation]
$R_{ee}(t) = C[x^*(t)]\tilde{P}(t\|t-1)C'[x^*(t)] + R_{vv}(t)$	[Innovation covariance]
Gain	
$K(t) = \tilde{P}(t\|t-1)C'[x^*(t)]R_{ee}^{-1}(t)$	[Gain or weight]
Update	
$\hat{x}(t\|t) = \hat{x}(t\|t-1) + K(t)e(t)$	[State update]
$\tilde{P}(t\|t) = [I - K(t)C[x^*(t)]]\tilde{P}(t\|t-1)$	[Covariance update]
Initial Conditions	
$\hat{x}(0\|0) \quad \tilde{P}(0\|0)$	
Jacobians	
$A[x^*(t-1)] \equiv \left.\frac{da[x(t-1)]}{dx(t-1)}\right\|_{x=x^*(t-1)} \quad B[u^*(t-1)] \equiv \left.\frac{db[u(t-1)]}{du(t-1)}\right\|_{u=u^*(t-1)}$	
$C[x^*(t)] \equiv \left.\frac{dc[x(t)]}{dx(t)}\right\|_{x=x^*(t)}$	

as

$$\ln \Pr(x(t)|Y_t) = \ln \kappa - \frac{1}{2}v'(t)R_{vv}^{-1}(t)v(t) - \frac{1}{2}\tilde{x}'(t|t-1)\tilde{P}^{-1}(t|t-1)\tilde{x}(t|t-1) + \frac{1}{2}e'(t)R_{ee}^{-1}(t)e(t) \tag{5.37}$$

The MAP estimate is then obtained by differentiating this equation, setting the result to zero and solving for $x(t)$, that is,

$$\nabla_x \ln \Pr(x(t)|Y_t)|_{x=\hat{X}_{map}} = \mathbf{0} \tag{5.38}$$

Before we attempt to derive the MAP estimate, we first linearize about a reference trajectory, $x^* \to x$ with the nonlinear measurement model approximated by a first-order Taylor series

$$c[x(t)] \approx c[x^*(t)] + \frac{dc[x^*(t)]}{dx^*(t)}\delta x(t) = c[x^*(t)] + C[x^*(t)](x(t) - x^*(t))$$

Substituting this result into Eq. 5.37, we obtain

$$\ln \Pr(x(t)|Y_t) = \ln \kappa - \frac{1}{2}(y(t) - c[x^*(t)] - C[x^*(t)](x(t) - x^*(t)))'R_{vv}^{-1}(t)$$

$$\times (y(t) - c[x^*(t)] - C[x^*(t)](x(t) - x^*(t))) - \frac{1}{2}\tilde{x}'(t|t-1)\tilde{P}^{-1}(t|t-1)\tilde{x}(t|t-1)$$
$$+\frac{1}{2}(y(t) - \hat{y}(t|t-1))'R_{ee}^{-1}(t)(y(t) - \hat{y}(t|t-1)) \tag{5.39}$$

Applying the *chain rule* and the gradient operator (see Chapter 3), we obtain the following expression. Note that the last term of Eq. 5.39 is not a function of $x(t)$, but just the data, so it is null.

$$\nabla_x \ln \Pr(x(t)|Y_t) = -C'[x^*(t)]R_{vv}^{-1}(t)(y(t) - c[x^*(t)] - C[x^*(t)](x(t) - x^*(t)))$$
$$- \tilde{P}^{-1}(t|t-1)[x(t) - \hat{x}(t|t-1)] = 0 \tag{5.40}$$

Multiplying through and grouping like terms in $x(t)$ gives

$$C'[x^*(t)]R_{vv}^{-1}(t)(y(t) - c[x^*(t)] + C[x^*(t)]x^*(t))$$
$$-[\tilde{P}^{-1}(t|t-1) + C'[x^*(t)]R_{vv}^{-1}(t)C[x^*(t)]]x(t) + \tilde{P}^{-1}(t|t-1)\hat{x}(t|t-1) = 0 \tag{5.41}$$

Solving for $x(t) = \hat{X}_{\text{map}}(t)$ gives

$$\hat{X}_{\text{map}}(t) = [\tilde{P}^{-1}(t|t-1) + C'[x^*(t)]R_{vv}^{-1}(t)C[x^*(t)]]^{-1}$$
$$\times [\tilde{P}^{-1}(t|t-1)\hat{x}(t|t-1) + C'[x^*(t)]R_{vv}^{-1}(t)$$
$$\times (y(t) - c[x^*(t)] + C[x^*(t)]x^*(t))] \tag{5.42}$$

Using the matrix inversion manipulations of Eqs. 5.12–5.17 with $C(t) \longrightarrow C[x^*(t)]$, the first term in Eq. 5.42 becomes

$$[\tilde{P}^{-1}(t|t-1) + C'[x^*(t)]R_{vv}^{-1}(t)C[x^*(t)]]^{-1}$$
$$= (I - \tilde{P}(t|t-1)C'[x^*(t)]R_{ee}^{-1}(t)C[x^*(t)])\tilde{P}^{-1}(t|t-1)$$
$$= (I - K(t)C[x^*(t)])\tilde{P}(t|t-1) \equiv \tilde{P}(t|t) \tag{5.43}$$

where K is the Kalman gain, R_{ee} is the innovations covariance of the LZ-BP with this expression precisely the *updated error covariance* $\tilde{P}(t|t)$ as in Table 5.2.

Solving this equation for the inverse of the predicted error covariance gives

$$\tilde{P}^{-1}(t|t-1) = \tilde{P}^{-1}(t|t) - C'[x^*(t)]R_{vv}^{-1}(t)C[x^*(t)] \tag{5.44}$$

and substituting into Eq. 5.42 using the results of Eq. 5.43 yields

$$\hat{X}_{\text{map}}(t) = \tilde{P}(t|t)[(\tilde{P}^{-1}(t|t)\hat{x}(t|t-1) - C'[x^*(t)]R_{vv}^{-1}(t)C[x^*(t)]\hat{x}(t|t-1))$$
$$+ C'[x^*(t)]R_{vv}^{-1}(t)(y(t) - c[x^*(t)] + C[x^*(t)]x^*(t))] \tag{5.45}$$

Multiplying through by the updated error covariance and recognizing the expression for the Kalman gain gives

$$\hat{X}_{\rm map}(t) = \hat{x}(t|t-1) - K(t)C[x^*(t)]\hat{x}(t|t-1) + K(t)C[x^*(t)]x^*(t) + K(t)(y(t) - c[x^*(t)]) \tag{5.46}$$

which leads to the final expression for the linearized MAP estimate as

$$\hat{X}_{\rm map}(t) = \hat{x}(t|t) = \hat{x}(t|t-1) + K(t)(y(t) - c[x^*(t)] - C[x^*(t)](\hat{x}(t|t-1) - x^*(t)) \tag{5.47}$$

Compare this result to Table 5.1. The error covariances and predicted estimates follow as in the linear case. This completes the derivation. Next let us consider a discrete version of the nonlinear system example given in Jazwinski [1].

Example 5.2

Consider the discrete nonlinear process given by

$$x(t) = (1 - 0.05\Delta T)x(t-1) + 0.04\Delta T x^2(t-1) + w(t-1)$$

with corresponding measurement model

$$y(t) = x^2(t) + x^3(t) + v(t)$$

where $v(t) \sim \mathcal{N}(0, 0.09)$, $x(0) = 2.3$, $P(0) = 0.01$, $\Delta T = 0.01$ sec, and $R_{ww} = 0$. The simulated measurement using SSPACK_PC [8] is shown in Fig. 5.5c. The LZ-BP is designed from the following Jacobians:

$$A[x(t-1)] = 1 - 0.05\Delta T + 0.08\Delta T x(t-1) \quad \text{and} \quad C[x(t)] = 2x(t) + 3x^2(t)$$

Observing the mean state, we develop a reference trajectory by fitting a line to the simulated state which is given by

$$x^*(t) = 0.067t + 2.0 \qquad 0 \le t \le 1.5 \ \text{and} \ u^*(t) = u(t) = 0.0 \ \forall t$$

The LZ-BP algorithm is then given by

$$\begin{aligned}
\hat{x}(t|t-1) &= (1 - 0.05\Delta T)x^*(t-1) \\
&\quad + (1 - 0.05\Delta T + 0.08\Delta T x^*(t-1))[\hat{x}(t-1|t-1) - x^*(t-1)] \\
\tilde{P}(t|t-1) &= [1 - 0.05\Delta T + 0.08\Delta T x^*(t-1)]^2 \tilde{P}(t-1|t-1) \\
e(t) &= y(t) - (x^{*2}(t) - x^{*3}(t)) - (2x^*(t) + 3x^{*2}(t))[\hat{x}(t|t-1) - x^*(t)] \\
R_{ee}(t) &= [2\hat{x}(t|t-1) + 3\hat{x}^2(t|t-1)]^2 \tilde{P}(t|t-1) + 0.09
\end{aligned}$$

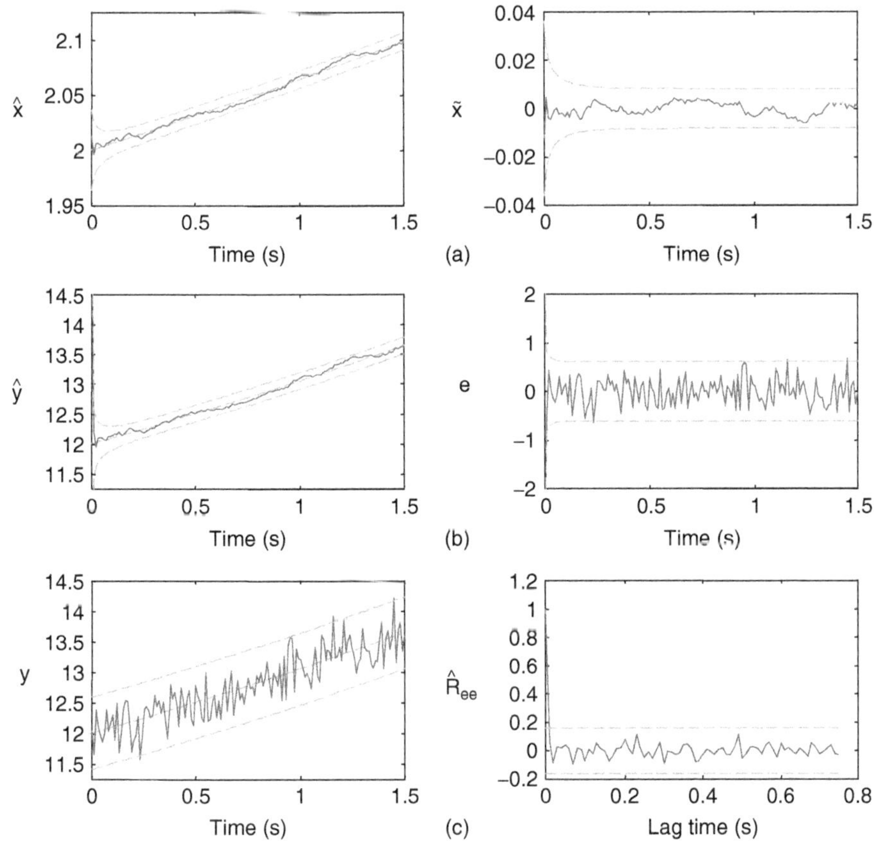

FIGURE 5.5 Linearized BP simulation: (a) estimated state (0% out) and error (0% out); (b) filtered measurement (1% out) and error (innovation) (2.6% out); (c) simulated measurement and zero-mean/whiteness test ($3.9 \times 10^{-2} < 10.1 \times 10^{-2}$ and 0% out).

$$K(t) = \frac{\tilde{P}(t|t-1)[2x^*(t) + 3x^{*2}(t)]}{R_{ee}(t)}$$

$$\hat{x}(t|t) = \hat{x}(t|t-1) + K(t)e(t)$$

$$\tilde{P}(t|t) = (1 - K(t)[2x^*(t) + 3x^{*2}(t)])\tilde{P}(t|t-1)$$

$$\hat{x}(0|0) = 2.3 \quad \text{and} \quad \tilde{P}(0|0) = 0.01$$

A LZ-BP run is depicted in Fig. 5.5. Here we see that the state estimate begins tracking the true state after the initial transient. The estimation error is good (0% lie outside confidence limits) indicating the filter is performing properly for this realization. The filtered measurement and innovations are shown in Fig. 5.5b with their corresponding predicted confidence limits. Both estimates lie well within these bounds. The innovations are zero-mean ($3.9 \times 10^{-2} < 10.1 \times 10^{-2}$) and white

(0% lie outside limits) as shown in Fig. 5.5c indicating proper tuning. This completes the nonlinear filtering example. △△△

5.5 EXTENDED BAYESIAN PROCESSOR (EXTENDED KALMAN FILTER)

In this section, we develop the extended Bayesian processor (XBP) or equivalently the extended Kalman filter (EKF). The XBP is *ad hoc* in nature, but has become one of the workhorses of (approximate) nonlinear filtering [1–11]. It has found applicability in a wide variety of applications such as tracking [15], navigation [1, 5], chemical processing [17], ocean acoustics [18], and seismology [19] (see Ref. [20] for a detailed list). The XBP evolves directly from the linearized processor of the previous section in which the reference state $x^*(t)$ used in the linearization process is replaced with the most recently available state estimate $\hat{x}(t|t)$—this is the step that makes the processor *ad hoc*. We must realize that the Jacobians used in the linearization process are deterministic (but time-varying), when a reference or perturbation trajectory is used. However, using the current state estimate is an approximation to the conditional mean, which is random, making these associated Jacobians and subsequent relations random. Therefore, although popularly ignored, most XBP designs should be based on ensemble operations to obtain reasonable estimates of the underlying statistics. With this in mind, we develop the processor directly from the LZ-BP. Thus, if instead of using the reference trajectory, we choose to linearize about each new state estimate as soon as it becomes available, then the XBP algorithm results. The reason for choosing to linearize about this estimate is that it represents the best information we have about the state and therefore most likely results in a better reference trajectory (state estimate). As a consequence, large initial estimation errors do not propagate; therefore, linearity assumptions are less likely to be violated. Thus, if we choose to use the current estimate $\hat{x}(t|\alpha)$, where α is $t-1$ or t, to linearize about instead of the reference trajectory $x^*(t)$, then the XBP evolves. That is, let

$$x^*(t) \equiv \hat{x}(t|\alpha) \quad \text{for } t-1 \leq \alpha \leq t \tag{5.48}$$

Then, for instance, when $\alpha = t-1$, the predicted perturbation is

$$\delta\hat{x}(t|t-1) = \hat{x}(t|t-1) - x^*(t)|_{x^*=\hat{x}(t|t-1)} = 0 \tag{5.49}$$

Thus, it follows immediately that when $x^*(t) = \hat{x}(t|t)$, then $\delta\hat{x}(t|t) = 0$ as well.

Substituting the current estimate, either prediction or update into the LZ-BP algorithm, it is easy to see that each of the difference terms $[\hat{x} - x^*]$ are null-resulting in the XBP algorithm. That is, examining the *prediction* phase of the linearized algorithm, substituting the current available updated estimate $\hat{x}(t-1|t-1)$ for the reference and using the fact that $(u^*(t) = u(t) \forall t)$, we have

$$\begin{aligned}\hat{x}(t|t-1) = a[\hat{x}(t-1|t-1)] + A[\hat{x}(t-1|t-1)][\hat{x}(t-1|t-1) - \hat{x}(t-1|t-1)]\\ + b[u(t-1)] + B[u(t-1)][u(t-1) - u(t-1)]\end{aligned}$$

giving the prediction of the XBP

$$\hat{x}(t|t-1) = a[\hat{x}(t-1|t-1)] + b[u(t-1)] \tag{5.50}$$

Now with the predicted estimate available, substituting it for the reference in Eq. 5.33, gives the innovation sequence as

$$\begin{aligned} e(t) = y(t) - c[\hat{x}(t|t-1)] - C[\hat{x}(t|t-1)] \\ \times [\hat{x}(t|t-1) - \hat{x}(t|t-1)] = y(t) - c[\hat{x}(t|t-1)] \end{aligned} \tag{5.51}$$

where we have the new predicted or filtered measurement expression

$$\hat{y}(t|t-1) \equiv c[\hat{x}(t|t-1)] \tag{5.52}$$

The updated state estimate is easily obtained by substituting the predicted estimate for the reference ($\hat{x}(t|t-1) \rightarrow x^*(t)$) in Eq. 5.34

$$\begin{aligned} \delta\hat{x}(t|t) &= \delta\hat{x}(t|t-1) + K(t)e(t) \\ [\hat{x}(t|t) - \hat{x}(t|t-1)] &= [\hat{x}(t|t-1) - \hat{x}(t|t-1)] + K(t)e(t) \\ \hat{x}(t|t) &= \hat{x}(t|t-1) + K(t)e(t) \end{aligned} \tag{5.53}$$

The covariance and gain equations are identical to those in Table 5.2, but with the Jacobian matrices A, B, and C linearized about the predicted state estimate, $\hat{x}(t|t-1)$. Thus, we obtain the discrete XBP or equivalently the EKF algorithm summarized in Table 5.3. Note that the covariance matrices $\tilde{P}$ and the gain K are now functions of the current state estimate, which is the *approximate* conditional mean estimate, and therefore a single realization of a stochastic process. Thus, ensemble (Monte Carlo) techniques should be used to evaluate estimator performance, that is, for new initial conditions selected by a Gaussian random number generator (either $\hat{x}(0|0)$ or $\tilde{P}(0|0)$), the algorithm is executed generating a set of estimates which should be averaged over the entire ensemble using this approach to get an "expected" state, etc. Note also in practice that this algorithm is usually implemented using sequential processing and UD (upper diagonal/square root) factorization techniques (see Ref. [21]).

The XBP can also be developed under (approximate) Gaussian assumptions using the *Bayesian approach* as before in the linear case. We briefly outline the derivation.

The a posteriori probability is given by

$$\Pr(x(t)|Y_t) = \frac{\Pr(y(t)|x(t)) \times \Pr(x(t)|Y_{t-1})}{\Pr(y(t)|Y_{t-1})} \tag{5.54}$$

TABLE 5.3 Extended *BP* (Kalman Filter) Algorithm

Prediction

$\hat{x}(t|t-1) = a[\hat{x}(t-1|t-1)] + b[u(t-1)]$ [State prediction]

$\tilde{P}(t|t-1) = A[\hat{x}(t|t-1)]\tilde{P}(t-1|t-1)A'[\hat{x}(t|t-1)] + R_{ww}(t-1)$ [Covariance prediction]

Innovation

$e(t) = y(t) - c[\hat{x}(t|t-1)]$ [Innovation]

$R_{ee}(t) = C[\hat{x}(t|t-1)]\tilde{P}(t|t-1)C'[\hat{x}(t|t-1)] + R_{vv}(t)$ [Innovation covariance]

Gain

$K(t) = \tilde{P}(t|t-1)C'[\hat{x}(t|t-1)]R_{ee}^{-1}(t)$ [Gain or weight]

Update

$\hat{x}(t|t) = \hat{x}(t|t-1) + K(t)e(t)$ [State update]

$\tilde{P}(t|t) = [I - K(t)C[\hat{x}(t|t-1)]]\tilde{P}(t|t-1)$ [Covariance update]

Initial Conditions

$\hat{x}(0|0) \quad \tilde{P}(0|0)$

Jacobians

$A[x(t|t-1)] \equiv \left.\frac{da[x(t-1)]}{d\hat{x}(t-1)}\right|_{x=\hat{x}(t|t-1)} \qquad C[\hat{x}(t|t-1)] \equiv \left.\frac{dc[x(t)]}{dx(t)}\right|_{x=\hat{x}(t|t-1)}$

Under the Gauss–Markov model assumptions, we know that each of the conditional expectations can be expressed in terms of the conditional Gaussian distributions as:

1. $\Pr(y(t)|x(t))$: $\mathcal{N}(c[x(t)], R_{vv}(t))$
2. $\Pr(x(t)|Y_{t-1})$: $\mathcal{N}(\hat{x}(t|t-1), \tilde{P}(t|t-1))$
3. $\Pr(y(t)|Y_{t-1})$: $\mathcal{N}(\hat{y}(t|t-1), R_{ee}(t))$

Using the nonlinear models developed earlier, substituting the Gaussian probabilities and taking logarithms, we obtain the logarithmic a posteriori probability as

$$\ln \Pr(x(t)|Y_t) = \ln \kappa - \frac{1}{2}v'(t)R_{vv}^{-1}(t)v(t) - \frac{1}{2}\tilde{x}'(t|t-1)\tilde{P}^{-1}(t|t-1)\tilde{x}(t|t-1) + \frac{1}{2}e'(t)R_{ee}^{-1}(t)e(t) \tag{5.55}$$

The MAP estimate is then obtained by differentiating Eq. 5.55, setting it to zero and solving; that is,

$$\nabla_x \ln \Pr(x(t)|Y_t)|_{x=\hat{X}_{\text{map}}} = 0 \tag{5.56}$$

Applying the *chain rule* and the gradient operation (see Chapter 3), we obtain the expression. Note that the last term of Eq. 5.55 is not a function of $x(t)$, but just the data.

$$
\begin{aligned}
\nabla_x \ln \Pr(x(t)|Y_t) = \nabla_x c'[x(t)]R_{vv}^{-1}(t)(y(t) - c[x(t)]) \\
- \tilde{P}^{-1}(t|t-1)(x(t) - \hat{x}(t|t-1))
\end{aligned}
\tag{5.57}
$$

Using a first-order Taylor series approximation for $c[x(t)]$ linearized about the predicted estimate $\hat{x}(t|t-1)$ and the usual definition for the Jacobian matrix $C[\hat{x}(t|t-1)]$, we have

$$
\begin{aligned}
\nabla_x \ln \Pr(x(t)|Y_t) = C'[\hat{x}(t|t-1)]R_{vv}^{-1}(t)[y(t) - c[\hat{x}(t|t-1)] \\
- C[\hat{x}(t|t-1)](x(t) - \hat{x}(t|t-1))] - \tilde{P}^{-1}(t|t-1)[x(t) - \hat{x}(t|t-1)]
\end{aligned}
\tag{5.58}
$$

Identifying the innovation vector and grouping like-terms gives the expression

$$
\begin{aligned}
\nabla_x \ln \Pr(x(t)|Y_t) = C'[\hat{x}(t|t-1)]R_{vv}^{-1}(t)e(t) \\
[C'[\hat{x}(t|t-1)]\, R_{vv}^{-1}(t)\, C[\hat{x}(t|t-1)] + \tilde{P}(t|t-1)]\, x(t) \\
+ [C'[\hat{x}(t|t-1)]\, R_{vv}^{-1}(t)\, C[\hat{x}(t|t-1)] + \tilde{P}(t|t-1)]\, \hat{x}(t|t-1)
\end{aligned}
$$

Setting this equation to zero and solving for $x(t) = \hat{X}_{\rm map}(t)$ gives

$$
\begin{aligned}
\hat{X}_{\rm map}(t) = [C'[\hat{x}(t|t-1)]R_{vv}^{-1}(t)C[\hat{x}(t|t-1)] + \tilde{P}(t|t-1)]^{-1} \\
\times [C'[\hat{x}(t|t-1)]R_{vv}^{-1}(t)C[\hat{x}(t|t-1)] + \tilde{P}(t|t-1)]\, \hat{x}(t|t-1) \\
+ [C'[\hat{x}(t|t-1)]R_{vv}^{-1}(t)C[\hat{x}(t|t-1)] \\
+ \tilde{P}(t|t-1)]^{-1}C'[\hat{x}(t|t-1)]R_{vv}^{-1}(t)e(t)
\end{aligned}
\tag{5.59}
$$

or

$$
\begin{aligned}
\hat{X}_{\rm map}(t) = \hat{x}(t|t-1) + [C'[\hat{x}(t|t-1)]R_{vv}^{-1}(t)C[\hat{x}(t|t-1)] \\
+ \tilde{P}(t|t-1)]^{-1}C'[\hat{x}(t|t-1)]R_{vv}^{-1}(t)e(t)
\end{aligned}
\tag{5.60}
$$

Recognizing the similarity of this expression to the linear case with $C[\hat{x}(t|t-1)] \rightarrow C(t)$ and using the *matrix inversion lemma*, it is easy to show (see Eqns. 5.13–5.17) that our approximate updated error covariance satisfies

$$
\tilde{P}(t|t) = [C'[\hat{x}(t|t-1)]R_{vv}^{-1}(t)C[\hat{x}(t|t-1)] + \tilde{P}^{-1}(t|t-1)]^{-1}
\tag{5.61}
$$

and therefore Eq. 5.60 can be simplified to give

$$\hat{X}_{\text{map}}(t) = \hat{x}(t|t-1) + \tilde{P}(t|t)C'[\hat{x}(t|t-1)]R_{vv}^{-1}(t)e(t) \tag{5.62}$$

Now recognizing the alternate form for the gain as in Eq. 5.21 gives the desired updated estimate as

$$\hat{x}(t|t) = \hat{X}_{\text{map}}(t) = \hat{x}(t|t-1) + K(t)e(t) \tag{5.63}$$

This completes the derivation. Consider the following discrete nonlinear example of the previous section.

Example 5.3

Consider the discrete nonlinear process and measurement system described in the previous example. The simulated measurement using SSPACK_PC [8] is shown in Fig. 5.6c. The XBP is designed from the following Jacobian:

$$A[x(t-1)] = 1 - 0.05\Delta T + 0.08\Delta T x(t-1) \quad \text{and} \quad C[x(t)] = 2x(t) + 3x^2(t)$$

The *XBP* algorithm is then given by

$$\begin{aligned}
\hat{x}(t|t-1) &= (1 - 0.05\Delta T)\hat{x}(t-1|t-1) + 0.04\Delta T\hat{x}^2(t-1|t-1) \\
\tilde{P}(t|t-1) &= [1 - 0.05\Delta T + 0.08\Delta T x(t-1)]^2\tilde{P}(t-1|t-1) \\
e(t) &= y(t) - \hat{x}^2(t|t-1) - \hat{x}^3(t|t-1) \\
R_{ee}(t) &= [2\hat{x}(t|t-1) + 3\hat{x}^2(t|t-1)]^2\tilde{P}(t|t-1) + 0.09 \\
K(t) &= (\tilde{P}(t|t-1)[2\hat{x}(t|t-1) + 3\hat{x}^2(t|t-1)])R_{ee}^{-1}(t) \\
\hat{x}(t|t) &= \hat{x}(t|t-1) + K(t)e(t) \\
\tilde{P}(t|t) &= (1 - K(t)[2\hat{x}(t|t-1) + 3\hat{x}^2(t|t-1)])\tilde{P}(t|t-1) \\
\hat{x}(0|0) &= 2.3 \quad \text{and} \quad \tilde{P}(0|0) = 0.01
\end{aligned}$$

An XBP run is depicted in Fig. 5.6. Here we see that the state estimate begins tracking the true state after the initial transient. The estimation error is reasonable ($\sim$1% lie outside limits) indicating the filter is performing properly for this realization. The filtered measurement and innovations are shown in Fig. 5.6b and lie within the predicted limits. The innovations are zero-mean ($6.3 \times 10^{-2} < 11.8 \times 10^{-2}$) and white (0% lie outside limits) as shown in Fig. 5.6c indicating proper tuning. Comparing the XBP to the LZ-BP of the previous section shows that it performs slightly worse in terms of predicted covariance limits for the estimated measurement and innovations. Most of this error is caused by the initial conditions of the processor.

Running an ensemble of 101 realizations of this processor yields similar results for the ensemble estimates: state estimation error increased ($\sim$2% outside limits),

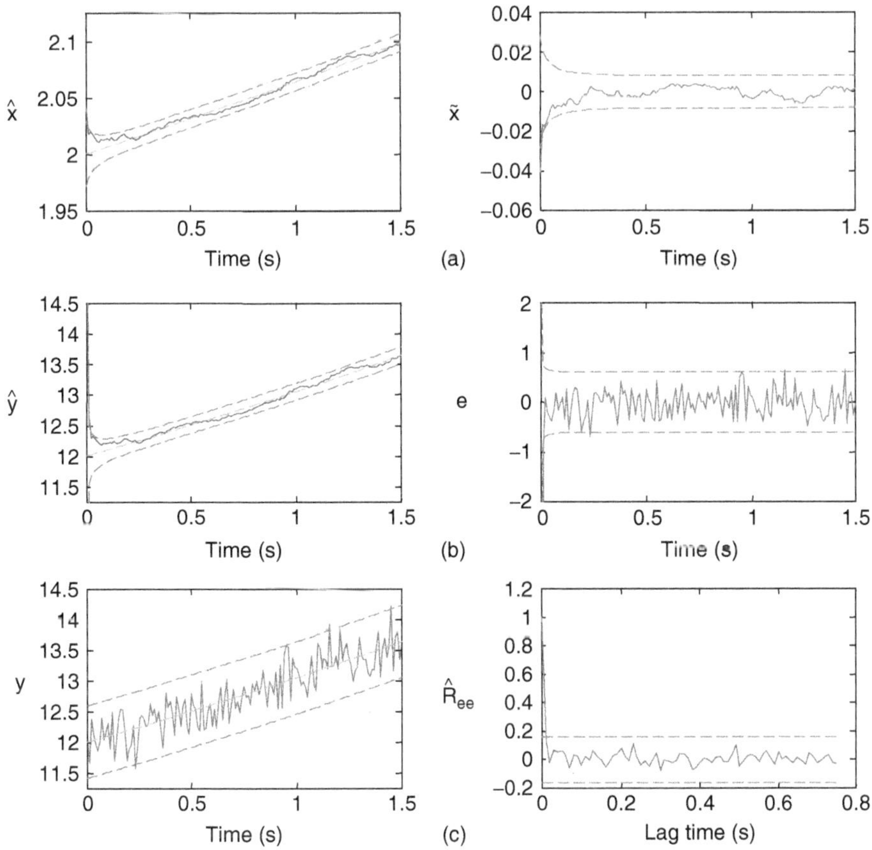

FIGURE 5.6 XBP (EKF) simulation: (a) estimated state (1% out) and error (1% out); (b) filtered measurement (1.3% out) and error (innovation) (3.3% out); (c) simulated measurement and zero-mean/whiteness test ($6.3 \times 10^{-2} < 11.8 \times 10^{-2}$ and 0% out).

innovations zero-mean test increased slightly ($6.7 \times 10^{-2} < 12 \times 10^{-2}$), and whiteness was identical. This completes the example. △△△

Next we consider one of the most popular applications of XBP approach—the *tracking problem* [1, 5, 15, 20]. The choice of the coordinate system for the tracker determines whether the nonlinearities occur either in the state equations (polar coordinates [22]) or in the measurement equations (Cartesian coordinates [23]). The following application depicts typical XBP performance in Cartesian coordinates.

Example 5.4

Consider the following passive localization and tracking problem that frequently arises in sonar and navigation applications [15]. A maneuvering observer O monitors noisy "bearings-only" measurements from a target t assumed to be traveling at a

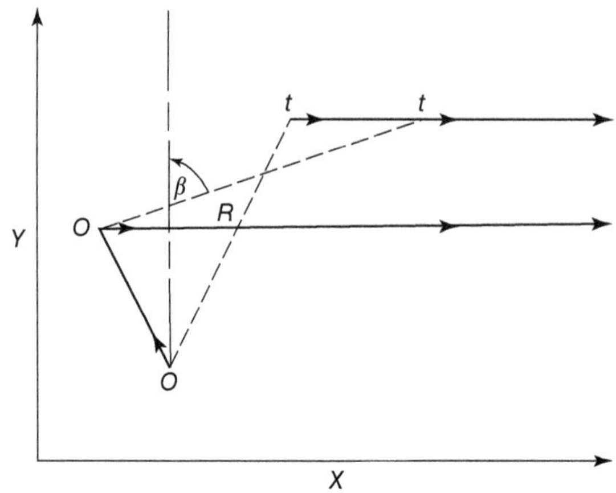

FIGURE 5.7 Observer (*O*) and target (*t*) geometry illustrating bearing angle (*β*) and slant range (*R*) at an instant for XBP tracking application.

constant velocity. These measurements are to be used to estimate target position r and velocity v. The problem is geometrically depicted in Fig. 5.7. The velocity and position of the target relative to the observer are defined by

$$
\begin{aligned}
v_x(t) &:= v_{tx}(t) - v_{ox}(t) \qquad & r_x(t) &:= r_{tx}(t) - r_{ox}(t) \\
v_y(t) &:= v_{ty}(t) - v_{oy}(t) \qquad & r_y(t) &:= r_{ty}(t) - r_{oy}(t)
\end{aligned}
$$

The velocity is related to position by

$$v(t) = \frac{d}{dt}r(t) \approx \frac{r(t) - r(t-1)}{\Delta T} \quad \text{for } \Delta T, \text{ the sampling interval}$$

or

$$r(t) = r(t-1) + \Delta T v(t-1)$$

and

$$
\begin{aligned}
v(t) &= v(t-1) + [v(t) - v(t-1)] \\
v(t) &= [v_t(t-1) - v_o(t-1)] - [v_o(t) - v_o(t-1)] = v(t-1) - \Delta v_o(t-1)
\end{aligned}
$$

for a constant velocity target $v_t(t) = v_t(t-1) = \cdots = v_t$ and Δv is the incremental change in observer velocity. Using these relations, we can easily develop a Gauss–Markov model of the equations of motion in two dimensions by defining the state vector as $x' := [r_x r_y v_x v_y]$ and input $u' := [-\Delta v_{ox} - \Delta v_{oy}]$, but first let us consider the measurement model.

For this problem, we have the bearing relation given by

$$\beta(x,t) := \arctan \frac{r_x(t)}{r_y(t)}$$

The entire system can be represented as a Gauss–Markov model with the noise sources representing uncertainties in the states and measurements. Thus, we have the equations of motion given by

$$x(t) = \begin{bmatrix} 1 & 0 & \Delta T & 0 \\ 0 & 1 & 0 & \Delta T \\ 0 & 0 & 1 & 0 \\ 0 & 0 & 0 & 1 \end{bmatrix} x(t-1) + \begin{bmatrix} 0 & 0 \\ 0 & 0 \\ 1 & 0 \\ 0 & 1 \end{bmatrix} \begin{bmatrix} -\Delta v_{ox}(t-1) \\ -\Delta v_{oy}(t-1) \end{bmatrix} + w(t-1)$$

with the nonlinear sensor model given by

$$y(t) = \arctan \frac{x_1(t)}{x_2(t)} + v(t)$$

for $w \sim \mathcal{N}(0, R_{ww})$ and $v \sim \mathcal{N}(0, R_{vv})$ The SSPACK_PC software [8] was used to simulate this system which is depicted in Fig. 5.8 for two legs of a tracking scenario. An impulse-incremental-step change ($\Delta v_{ox} = +24$ knots and $\Delta v_{oy} = -10$ knots) was initiated at 0.5 h, resulting in a change of observer position and velocity depicted in the figure. The simulated bearing measurements are shown in Fig. 5.6d, The initial conditions for the run were $x'(0) := [0\ 15\,\text{nm}\ 20\,\text{k} - 10\,\text{k}]$ and $R_{ww} = \text{diag}\ 10^{-6}$ with the measurement noise covariance given by $R_{vv} = 3.05 \times 10^{-4}\text{rad}^2$ for $\Delta T = 0.33$ h.

The XBP algorithm of Table 5.3 is implemented using this model and the following Jacobian matrices are derived from the Gauss–Markov model :

$$A[x] = A \quad \text{and} \quad C[x] = \begin{bmatrix} \frac{x_2(t)}{R^2} & \frac{-x_1(t)}{R^2} & 0 & 0 \end{bmatrix}$$

where

$$R = \sqrt{x_1^2(t) + x_2^2(t)}$$

The results of this run are shown in Fig. 5.8. In Figs. 5.8a–b, we see the respective x and y position estimates (velocity estimates are not shown) and corresponding tracking errors. Here we see that it takes approximately 1.8 h for the estimator to converge to the true target position (within ±1 nm). The innovations sequence appears statistically zero-mean and white in Figs. 5.8c–d, indicating satisfactory performance. The filtered measurement in Fig. 5.8c is improved considerably over the unprocessed data in Fig. 5.8d. This completes the example. △△△

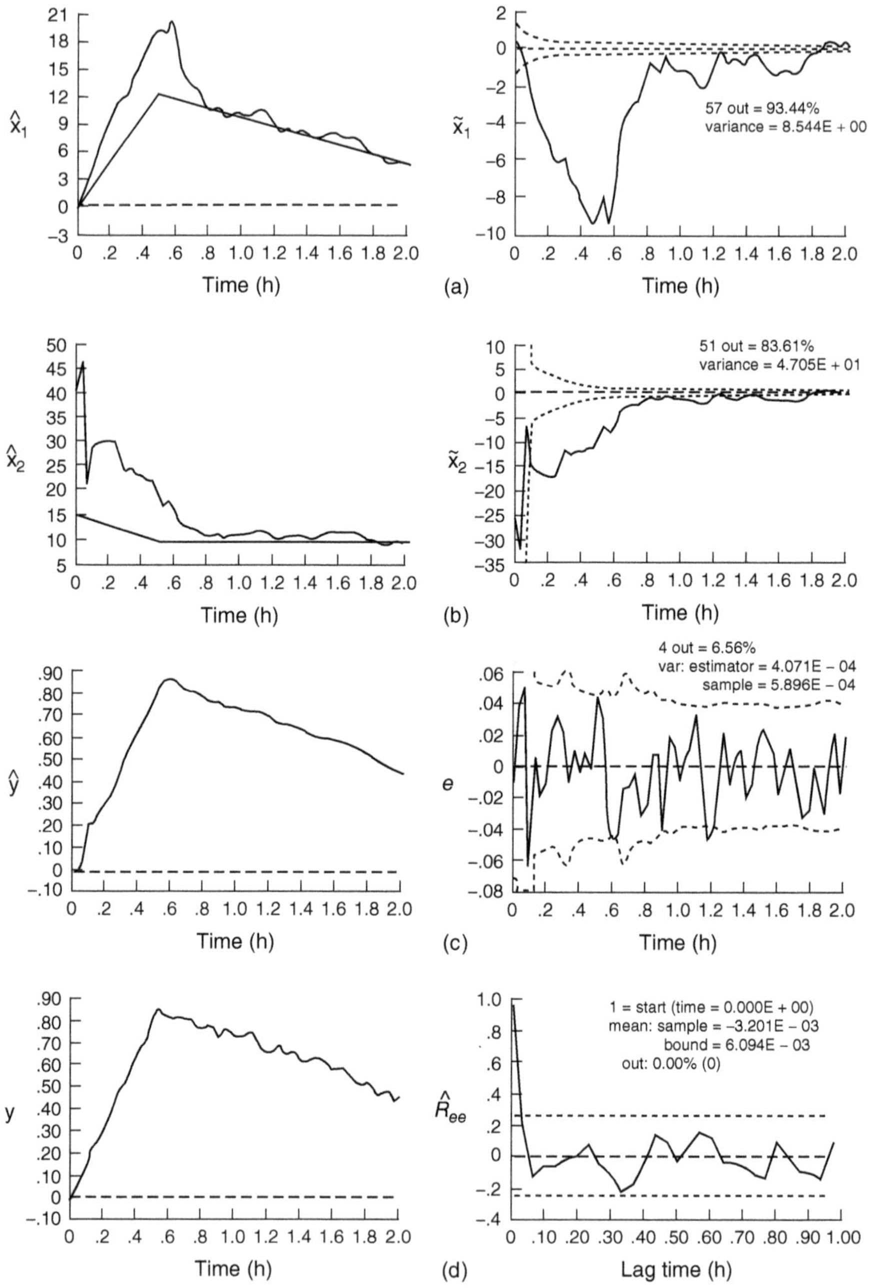

FIGURE 5.8 Extended BP (EKF) simulation for bearings-only tracking problem: (a) X-position estimate and error; (b) Y-position estimate and error; (c) filtered measurement and error (innovation); (d) simulated measurement and zero-mean/whiteness test ($3.2 \times 10^{-3} < 6.1 \times 10^{-3}$ and 0% out).

This completes the section on the extension of the BP to nonlinear problems. Next, we investigate variants of the XBP for improved performance.

5.6 ITERATED-EXTENDED BAYESIAN PROCESSOR (ITERATED-EXTENDED KALMAN FILTER)

In this section, we discuss an extension of the XBP of the previous section to the iterated-extended (IX-BP). We heuristically motivate the design and then discuss a more detailed approach using Bayesian MAP estimation coupled with numerical optimization techniques to develop the processor. This algorithm is based on performing "local" iterations (not global) at a point in time t to improve the reference trajectory and therefore the underlying estimate in the presence of significant measurement nonlinearities [1]. A local iteration implies that the inherent recursive structure of the processor is retained providing new estimates as the new measurements are made available.

To develop the iterated-extended processor, we start with the linearized processor update relation substituting the "linearized" innovation of Eq. 5.33 of the LZ-BP, that is,

$$\hat{x}(t|t) = \hat{x}(t|t-1) + K(t; x^*(t))\,[y(t) - c[x^*(t)] - C[x^*(t)]\,(\hat{x}(t|t-1) - x^*(t))] \quad (5.64)$$

where we have explicitly shown the dependence of the gain (through the measurement Jacobian) on the reference trajectory $x^*(t)$. The XBP algorithm linearizes about the most currently available estimate, $x^*(t) = \hat{x}(t|t-1)$ in this case. Theoretically, the updated estimate $\hat{x}(t|t)$ is a better estimate and closer to the true trajectory. Suppose we continue and relinearize about $\hat{x}(t|t)$ when it becomes available and then recompute the corrected estimate and so on. That is, define the $(i+1)th$ iterated estimate as $\hat{x}_{i+1}(t|t)$, then the updated *iterator* equation becomes

$$\begin{aligned} \hat{x}_{i+1}(t|t) &= \hat{x}(t|t-1) + K(t; \hat{x}_i(t|t)) \\ &\quad \times [y(t) - c[\hat{x}_i(t|t)] - C[\hat{x}_i(t|t)](\hat{x}(t|t-1) - \hat{x}_i(t|t))] \end{aligned} \quad (5.65)$$

Now, if we start with the 0th iterate as the predicted estimate, that is, $\hat{x}_o \equiv \hat{x}(t|t-1)$, then the XBP results for $i = 0$. Clearly, the updated estimate in this iteration is given by

$$\hat{x}_1(t|t) = \hat{x}(t|t-1) + K_o(t)[y(t) - c[\hat{x}(t|t)] - C[\hat{x}(t|t-1)](\hat{x}(t|t-1) - \hat{x}(t|t-1))] \quad (5.66)$$

where the last term in this expression is null leaving the usual innovation. Note also that the *gain* is reevaluated on each iteration as are the measurement function and Jacobian. The iterations continue until there is little difference in consecutive iterates. The *last* iterate is taken as the updated estimate. The complete (updated) iterative loop is given by

$$
\begin{aligned}
e_i(t) &= y(t) - c[\hat{x}_i(t|t)] \\
R_{e_ie_i}(t) &= C[\hat{x}_i(t|t)]\tilde{P}(t|t-1)C'[\hat{x}_i(t|t)] + R_{vv}(t) \\
K_i(t) &= \tilde{P}(t|t-1)C'[\hat{x}_i(t|t)]R^{-1}_{e_ie_i}(t) \\
\hat{x}_{i+1}(t|t) &= \hat{x}(t|t-1) + K_i(t)[e_i(t) - C[\hat{x}_i(t|t)](\hat{x}(t|t-1) - \hat{x}_i(t|t))] \\
\tilde{P}_i(t|t) &= (I - K_i(t)C[\hat{x}_i(t|t)])\tilde{P}(t|t-1)
\end{aligned} \tag{5.67}
$$

A typical stopping rule is:

$$
\| \hat{x}_{i+1}(t|t) - \hat{x}_i(t|t) \| < \epsilon \quad \text{and} \quad \hat{x}_i(t|t) \rightarrow \hat{x}(t|t) \tag{5.68}
$$

The IX-BP algorithm is summarized in Table 5.4.

The IX-BP can be useful in reducing the measurement function nonlinearity approximation errors, improving processor performance. It is designed for measurement nonlinearities and does not improve the previous reference trajectory, but it will improve the subsequent one. Next, we take a slightly more formal approach to developing the IX-BP from the Bayesian perspective, that is, we first formulate a parametric optimization problem to develop the generic structure of the iterator, then apply it to the underlying state estimation problem. Let us assume that we have a nonlinear cost function, $J(\Theta)$, that we would like to maximize relative to the parameter vector, $\Theta \in R^{N_\Theta \times 1}$. We begin by expanding the cost in terms of a Taylor series

TABLE 5.4 Iterated Extended BP (Kalman Filter) Algorithm

Prediction	
$\hat{x}(t|t-1) = a[\hat{x}(t-1|t-1)] - b[u(t-1)]$	[State prediction]
$\tilde{P}(t|t-1) = A[\hat{x}(t|t-1)]\tilde{P}(t-1|t-1)A'[\hat{x}(t|t-1)] - R_{ww}(t-1)$	[Covariance prediction]
LOOP: $i = 1, \ldots, N_{iterations}$	
Innovation	
$e_i(t) = y(t) - c[\hat{x}_i(t|t)]$	[Innovation]
$R_{e_ie_i}(t) = C[\hat{x}_i(t|t)]\tilde{P}(t|t-1)C'[\hat{x}_i(t|t)] - R_{vv}(t)$	[Innovation covariance]
Gain	
$K_i(t) = \tilde{P}(t|t-1)C'[\hat{x}_i(t|t)]R^{-1}_{e_ie_i}(t)$	[Gain or weight]
Update	
$\hat{x}_{i+1}(t|t) = \hat{x}(t|t-1) + K_i(t)[e_i(t) - C[\hat{x}_i(t|t)](\hat{x}(t|t-1) - \hat{x}_i(t|t))]$	[State update]
$\tilde{P}_i(t|t) = [I - K_i(t)C[\hat{x}_i(t|t)]]\tilde{P}(t|t-1)$	[Covariance update]
Initial Conditions	
$\hat{x}(0|0), \quad \tilde{P}(0|0), \quad \hat{x}_o(t|t) = \hat{x}(t|t-1)$	
Jacobians	
$A[\hat{x}(t|t-1)] \equiv \left.\frac{da[x(t-1)]}{dx(t-1)}\right|_{x=\hat{x}(t|t-1)} \quad C[\hat{x}_i(t|t)] \equiv \left.\frac{dc[x(t)]}{dx(t)}\right|_{x=\hat{x}_i(t|t)}$	

about the Θ_i, that is,

$$J(\Theta) = J(\Theta_i) + (\Theta - \Theta_i)'[\nabla_\Theta J(\Theta_i)] + \frac{1}{2}(\Theta - \Theta_i)'[\nabla_{\Theta\Theta} J(\Theta_i)](\Theta - \Theta_i) + \text{H.O.T.} \tag{5.69}$$

where ∇_Θ is the N_Θ-gradient vector defined by

$$\nabla_\Theta J(\Theta_i) := \left.\frac{\partial J(\Theta)}{\partial \Theta}\right|_{\Theta=\Theta_i} \tag{5.70}$$

with the corresponding $N_\Theta \times N_\Theta$ *Hessian* matrix defined by

$$\nabla_{\Theta\Theta} J(\Theta_i) := \left.\frac{\partial^2 J(\Theta)}{\partial \Theta^2}\right|_{\Theta=\Theta_i} \tag{5.71}$$

Now if we approximate this expression by neglecting the H.O.T. and assume that Θ_i is close to the *true* parameter vector ($\Theta_i \approx \Theta_{\text{true}}$), then differentiating Eq. 5.69 using the chain rule, we obtain

$$\nabla_\Theta J(\Theta_i) = 0 + \nabla_\Theta \Theta'[\nabla_\Theta J(\Theta_i)] + \frac{1}{2}(2[\nabla_{\Theta\Theta} J(\Theta_i)](\Theta - \Theta_i)) = 0$$

or

$$[\nabla_{\Theta\Theta} J(\Theta_i)](\Theta - \Theta_i) = -[\nabla_\Theta J(\Theta_i)]$$

Solving for Θ and letting $\Theta \rightarrow \Theta_{i+1}$ we obtain the well-known *Newton–Rhapson* iterator (NRI) as [1, 15, 16]

$$\Theta_{i+1} = \Theta_i - [\nabla_{\Theta\Theta} J(\Theta_i)]^{-1}[\nabla_\Theta J(\Theta_i)] \tag{5.72}$$

This is the form that our IX-BP will assume. Now let us return to the basic problem of improved state estimation using the NRI.

Under the usual Gaussian assumptions, we would like to calculate the MAP estimate of the state at time t based on the data up to time t, therefore, the a posteriori probability (see Section 4) is given by

$$\Pr(x(t)|Y_t) = \frac{1}{\eta} \times \Pr(y(t)|x(t)) \times \Pr(x(t)|Y_{t-1}) \tag{5.73}$$

where η is a normalizing probability (not a function of the state) and can be ignored in this situation. As in the linear case, we have

1. $\Pr(y(t)|x(t))$: $\mathcal{N}(c[x(t)], R_{vv}(t))$
2. $\Pr(x(t)|Y_{t-1})$: $\mathcal{N}(\hat{x}(t|t-1), \tilde{P}(t|t-1))$

Maximizing the a posteriori probability is equivalent to minimizing its logarithm, therefore, we have that

$$\begin{aligned} J(x(t)) &= \ln \Pr(y(t)|x(t)) + \ln \Pr(x(t)|Y_{t-1}) \\ &= -\frac{1}{2}(y(t) - c[x(t)])' R_{vv}^{-1}(t)(y(t) - c[x(t)]) \\ &\quad - \frac{1}{2}(x(t) - \hat{x}(t|t-1))' \tilde{P}^{-1}(t|t-1)(x(t) - \hat{x}(t|t-1)) \end{aligned} \tag{5.74}$$

Minimizing this expression, we differentiate with respect to $x(t)$ using the gradient operator as before to obtain

$$\nabla_x J(x(t)) = [\nabla_x c'[x(t)]] R_{vv}^{-1}(t)(y(t) - c[x(t)]) - \tilde{P}^{-1}(t|t-1)(x(t) - \hat{x}(t|t-1)) \tag{5.75}$$

With a slight abuse, we can simplify the notation by defining the relations

$$C[x_i(t)] := \frac{\partial c[x(t)]}{\partial x(t)}|_{x(t)=x_i(t)},\ e_i(t) := y(t) - c[x_i(t)],\ \text{and}\ \tilde{x}_i(t) := x_i(t) - \hat{x}(t|t-1)$$

Therefore, letting $x \rightarrow x_i$, we can write Eq. 5.75 as

$$\nabla_x J(x_i(t)) = C'[x_i(t)] R_{vv}^{-1}(t) e_i(t) - \tilde{P}^{-1}(t|t-1)\tilde{x}_i(t|t-1) \tag{5.76}$$

which is the same form of the linear MAP estimator of Eq. 5.10 with $C[x_i(t)] \rightarrow C(t)$.

Continuing with the NRI derivation, we differentiate Eq. 5.76 again to obtain the Hessian

$$\nabla_{xx} J(x_i(t)) = C'[x_i(t)] R_{vv}^{-1}(t) C[x_i(t)] + \tilde{P}^{-1}(t|t-1) \tag{5.77}$$

but applying the *matrix inversion lemma* as in Section 5.4 (see Eq. 5.13 for details), it is shown that

$$[\nabla_{xx} J(x_i(t))]^{-1} = (I - K_i(t) C[x_i(t)]) \tilde{P}(t|t-1) \equiv \tilde{P}_i(t|t) \tag{5.78}$$

for

$$K_i(t) := K_i(t; x_i(t)) = \tilde{P}(t|t-1) C'[x_i(t)] R_{e_i e_i}^{-1}(t) \tag{5.79}$$

with

$$R_{e_i e_i}(t) = C[x_i(t)] \tilde{P}(t|t-1) C'[x_i(t)] + R_{vv}(t) \tag{5.80}$$

Now equating Eqs. 5.76 and 5.78, and solving for $\tilde{P}^{-1}(t|t-1)$, we obtain

$$\tilde{P}^{-1}(t|t-1) = \tilde{P}^{-1}(t|t) - C'[x_i(t)]R_{vv}^{-1}(t)C[x_i(t)] \tag{5.81}$$

which can be substituted back into Eq. 5.76 to give

$$\nabla_x J(x_i(t)) = C'[x_i(t)]R_{vv}^{-1}(t)e_i(t) - [\tilde{P}^{-1}(t|t) - C'[x_i(t)]R_{vv}^{-1}(t)C[x_i(t)]]\tilde{x}_i(t|t-1) \tag{5.82}$$

The NRI can now be written in terms of the *iterated state* at time t as

$$x_{i+1}(t) = x_i(t) - [\nabla_{xx} J(x_i(t))]^{-1}\nabla_x J(x_i(t)) \tag{5.83}$$

Using the expressions for the Hessian and gradient, we have

$$\begin{aligned} x_{i+1}(t) = {} & x_i(t) + \tilde{P}_i(t|t) \\ & \times (C'[x_i(t)]R_{vv}^{-1}(t)e_i(t) - [\tilde{P}^{-1}(t|t) - C'[x_i(t)]R_{vv}^{-1}(t)C[x_i(t)]]\tilde{x}_i(t|t-1)) \end{aligned} \tag{5.84}$$

Multiplying through by the Hessian and recognizing the alternate expression for the gain (see Eq. 5.22), we obtain the expression

$$x_{i+1}(t) = x_i(t) + K_i(t)e_i(t) - (I - K_i(t)C[x_i(t)])\tilde{x}_i(t|t-1) \tag{5.85}$$

Now multiplying through by $\tilde{x}(t|t-1)$, factoring out the gain and performing the additions, we have

$$x_{i+1}(t) = \hat{x}(t|t-1) + K_i(t)[e_i(t) - C[x_i(t)](\hat{x}(t|t-1) - x_i(t))] \tag{5.86}$$

Defining the iterate in terms of the corrected state estimate, that is, $x_i(t) \rightarrow \hat{x}_i(t|t)$, gives the NRI iterator of Eq. 5.67 and Table 5.4.

So we see that, for strong measurement of nonlinearities, the IX-BP can be used for little cost to the XBP. A further extension of these results is called the *iterator-smoother* XBP in which the entire processor is iterated to mitigate strong nonlinearities in the predicted estimates [1]. Here the measurement is relinearized and then a "smoothed" state estimate is calculated and used in the prediction loop. This completes the discussion of the processor; next, we demonstrate its performance.

Example 5.5

Consider the discrete nonlinear process and measurement system described in the previous example. The simulated measurement using SSPACK_PC [8] is shown in Fig. 5.9c. The IX-BP is designed from the following Jacobian

$$A[x(t-1)] = 1 - 0.05\Delta T + 0.08\Delta T x(t-1) \quad \text{and} \quad C[x(t)] = 2x(t) + 3x^2(t)$$

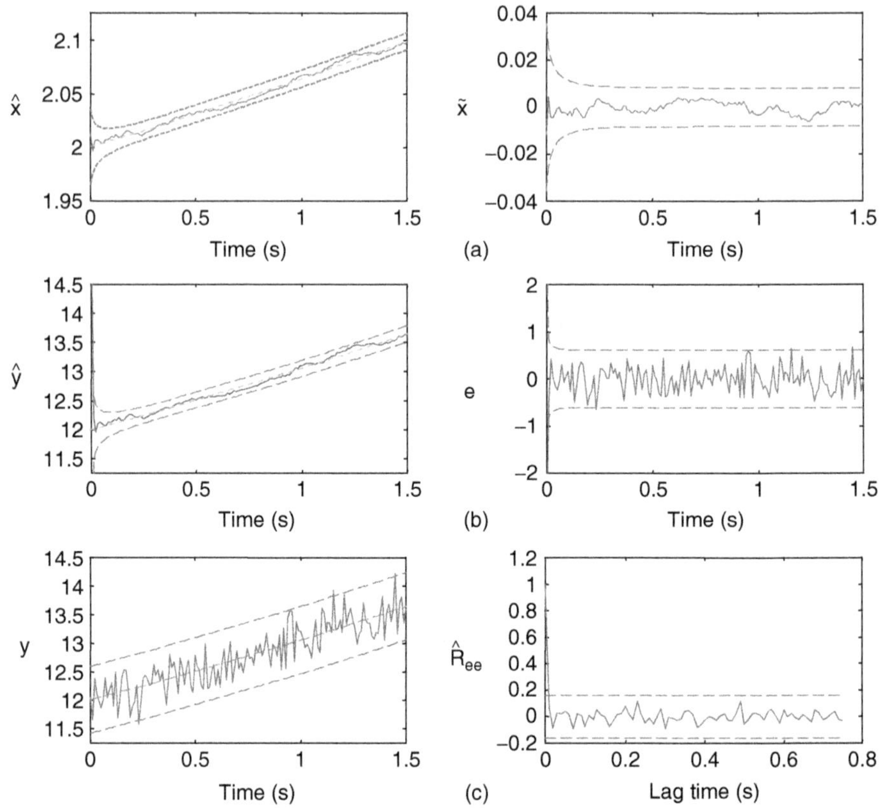

FIGURE 5.9 Iterated-extended BP (IEKF) simulation: (a) estimated state (~0% out) and error (~0% out); (b) filtered measurement (~1% out) and error (innovation) (~2.6% out); (c) simulated measurement and zero-mean/whiteness test ($4 \times 10^{-2} < 10.7 \times 10^{-2}$ and 0% out).

The IX-BP algorithm is then given by

$$
\begin{aligned}
\hat{x}(t|t-1) &= (1 - 0.05\Delta T)\hat{x}(t-1|t-1) + 0.04\Delta T\hat{x}^2(t-1|t-1)\\
\tilde{P}(t|t-1) &= [1 - 0.05\Delta T + 0.08\Delta T x(t-1)]^2\tilde{P}(t-1|t-1)\\
e_i(t) &= y(t) - \hat{x}_i^2(t|t) - \hat{x}_i^3(t|t)\\
R_{e_ie_i}(t) &= [2\hat{x}_i(t|t) + 3\hat{x}_i^2(t|t)]^2\tilde{P}(t|t-1) + 0.09\\
K_i(t) &= \tilde{P}(t|t-1)[2\hat{x}_i(t|t) + 3\hat{x}_i^2(t|t)]/R_{e_ie_i}(t)\\
\hat{x}_{i+1}(t|t) &= \hat{x}(t|t-1) + K_i(t)[e_i(t) - [2\hat{x}_i(t|t) + 3\hat{x}_i^2(t|t)](\hat{x}(t|t-1) - \hat{x}_i(t|t))\\
\tilde{P}_i(t|t) &= (1 - K_i(t)[2\hat{x}_i(t|t) + 3\hat{x}_i^2(t|t)])\tilde{P}(t|t-1)\\
\hat{x}(0|0) &= 2.3 \quad \text{and} \quad \tilde{P}(0|0) = 0.01
\end{aligned}
$$

An IX-BP run is depicted in Fig. 5.9. Here we see that the state estimate (~0% lie within limits) begins tracking the true state instantaneously. The estimation error is reasonable (~0% out) indicating that the filter is performing properly for this realization. The filtered measurement (~1% out) and innovations (~2.6% out) are shown in Fig. 5.9b. The innovations are zero-mean ($4 \times 10^{-2} < 10.7 \times 10^{-2}$) and white (0% lie outside limits) as shown in Fig. 5.9c indicating proper tuning and matching the LZ-BP result almost exactly.

Running an ensemble of 101 realizations of this processor yields similar results for the ensemble estimates: innovations zero-mean test decreased slightly ($3.7 \times 10^{-2} < 10.3 \times 10^{-2}$) and whiteness was identical. So the overall effect of the IX-BP was to decrease the measurement nonlinearity effect especially in the initial transient of the algorithm. These results are almost identical to those of the LZ-BP. This completes the nonlinear filtering example. △△△

Next, we consider some more practical approaches to designing estimators.

5.7 PRACTICAL ASPECTS OF CLASSICAL BAYESIAN PROCESSORS

In this section, we heuristically provide an intuitive feel for the operation of the BP using the state–space model and GM assumptions. These results coupled with the theoretical points developed in Ref. [10] lead to the proper adjustment or "tuning" of the BP. Tuning the processor is considered an art, but with proper statistical tests, the performance can readily be evaluated and adjusted. As mentioned previously, this approach is called the *minimum (error) variance* design. In contrast to the standard filter design procedures in signal processing, the minimum variance design adjusts the statistical parameters (e.g., covariances) of the processor and examines the innovations sequence to determine if the BP is properly tuned. Once tuned, all of the statistics (conditional means and variances) are valid and may be used as reasonable estimates. Here we discuss how the parameters can be adjusted and what statistical tests must be performed to evaluate BP performance.

Heuristically, the sequential BP can be viewed simply by its update equation

$$\hat{X}_{\text{new}} = \overbrace{\underbrace{\hat{X}_{\text{old}}}_{\text{State–space model}}}^{\text{Prediction}} + \overbrace{\underbrace{K \times E_{\text{new}}}_{\text{Measurement}}}^{\text{Update}}$$

where $\hat{X}_{\text{old}} \approx f(\text{model})$ and $E_{\text{new}} \approx f(\text{measurement})$.

Using this model of the BP, we see that we can view the old, or predicted estimate $\hat{X}_{\text{old}}$ as a function of the state–space model (A, B) and the prediction error or innovation E as a function primarily of the new measurement, as indicated in Table 5.1.

Condition	Gain	Parameter
Believe *model*	Small	$\tilde{P}$ small (model adequate)
		R_{vv} large (measurement noisy)
Believe *measurement*	Large	R_{vv} small (measurement good)
		$\tilde{P}$ large (model inadequate)

FIGURE 5.10 Bayesian processor heuristic notions.

Consider the new estimate under the following cases:

$$\begin{array}{ll} K \longrightarrow \text{small} & \hat{X}_{\text{new}} = \hat{X}_{\text{old}} = f\,(\text{model}) \\ K \longrightarrow \text{large} & \hat{X}_{\text{new}} = KE_{\text{new}} = f\,(\text{measurement}) \end{array}$$

So we can see that the operation of the processor is pivoted about the values of the gain or weighting matrix K. For small K, the processor "believes" the *model*, and for large K, the processor believes the *measurement* (Fig. 5.10).

Let us investigate the gain matrix and see if its variations are consistent with these heuristic notions. First, it was shown in Eq. 5.22 that the alternate form of the gain equation is given by

$$K(t) = \tilde{P}(t|t)C'(t)R_{\upsilon\upsilon}^{-1}(t)$$

Thus, the condition where K is *small* can occur in two cases: (1) $\tilde{P}$ is small (fixed $R_{\upsilon\upsilon}$) which is consistent because small $\tilde{P}$ implies that the model is adequate; and (2) $R_{\upsilon\upsilon}$ is large ($\tilde{P}$ fixed), which is also consistent because large $R_{\upsilon\upsilon}$ implies that the measurement is noisy, so again believe the model.

For the condition where K is *large,* two cases can also occur: (1) K is large when $\tilde{P}$ is large (fixed $R_{\upsilon\upsilon}$), implying that the model is inadequate, so believe the measurement; and (2) $R_{\upsilon\upsilon}$ is small ($\tilde{P}$ fixed), implying the measurement is good (high SNR). So we see that our heuristic notions are based on specific theoretical relationships between the parameters in the BP algorithm of Table 5.1.

Summarizing, a BP (Kalman filter) is not functioning properly when the *gain* becomes small and the measurements still contain information necessary for the estimates. The filter is said to *diverge* under these conditions. In this case, it is necessary to detect how the filter is functioning and how to adjust it if necessary, but first we consider the tuned BP.

When the processor is "tuned", it provides an optimal or minimum (error) variance estimate of the state. The innovations sequence, which was instrumental in deriving the processor, also provides the starting point to check the BP operation. A *necessary and sufficient condition* for a BP to be optimal is that the innovation sequence is zero-mean and white (see Ref. [4] for the proof). These are the first properties that must be evaluated to ensure that the processor is operating properly. If we assume that the innovation sequence is ergodic and Gaussian, then we can use the sample mean as the test statistic to estimate m_e, the population mean. The sample mean for the ith component of e_i is given by

$$\hat{m}_e(i) = \frac{1}{N}\sum_{t=1}^{N} e_i(t) \quad \text{for } i = 1, \ldots, N_y \tag{5.87}$$

where $\hat{m}_e(i) \sim \mathcal{N}(m_e, R_{ee}(i)/N)$ and N is the number of data samples. We perform a statistical hypothesis test to "decide" if the innovation mean is zero [10]. We test that the mean of the ith component of the innovation vector $e_i(t)$ is

$$\begin{aligned} H_o &: \quad m_e(i) = 0 \\ H_1 &: \quad m_e(i) \neq 0 \end{aligned}$$

As our test statistic we use the sample mean. At the α-significance level, the probability of rejecting the null hypothesis H_o is given by

$$\Pr\left(\left|\frac{\hat{m}_e(i) - m_e(i)}{\sqrt{R_{ee}(i)/N}} > \frac{\tau_i - m_e(i)}{\sqrt{R_{ee}(i)/N}}\right|\right) = \alpha \tag{5.88}$$

Therefore, the *zero-mean test* [10] on each component innovation e_i is given by

$$\hat{m}_e(i) \underset{\text{Accept } H_o}{\overset{\text{Reject } H_o}{\gtrless}} \tau_i \tag{5.89}$$

Under the null hypothesis H_o, each $m_e(i)$ is zero. Therefore, at the 5% significance level ($\alpha = 0.05$), we have that the threshold is

$$\tau_i = 1.96\sqrt{\frac{\hat{R}_{ee}(i)}{N}} \tag{5.90}$$

where $\hat{R}_{ee}(i)$ is the *sample variance* (assuming ergodicity) estimated by

$$\hat{R}_{ee}(i) = \frac{1}{N}\sum_{t=1}^{N} e_i^2(t) \tag{5.91}$$

Under the same assumptions, we can perform a *whiteness test* [10], that is, check statistically that the innovations covariance corresponds to that of an uncorrelated (white) sequence. Again assuming ergodicity of the innovations sequence, we use the sample covariance function as our test statistic with the ith component covariance given by

$$\hat{R}_{ee}(i,k) = \frac{1}{N}\sum_{t=k+1}^{N}(e_i(t) - \hat{m}_e(i))(e_i(t+k) - \hat{m}_e(i)) \tag{5.92}$$

We actually use the *normalized covariance* test statistic

$$\hat{\rho}_{ee}(i,k) = \frac{\hat{R}_{ee}(i,k)}{\hat{R}_{ee}(i)} \tag{5.93}$$

Asymptotically for large N, it can be shown that (see Refs. [10–14]) that

$$\hat{\rho}_{ee}(i,k) \sim \mathcal{N}(0, 1/N)$$

Therefore, the 95% confidence interval estimate is

$$I_{\rho_{ee}} = \hat{\rho}_{ee}(i,k) \pm \frac{1.96}{\sqrt{N}} \quad \text{for } N > 30 \tag{5.94}$$

Hence, under the null hypothesis, 95% of the $\hat{\rho}_{ee}(i,k)$ values must lie within this confidence interval, that is, for each *component* innovation sequence to be considered statistically white. Similar tests can be constructed for the *cross-covariance* properties of the innovations [13] as well, that is,

$$\text{Cov}(e(t), e(k)) = 0 \quad \text{and} \quad \text{Cov}(e(t), u(t-1)) = 0$$

The whiteness test of Eq. 5.94 is very useful for detecting model inaccuracies from individual component innovations. However, for complex systems with a large number of measurement channels, it becomes computationally burdensome to investigate each innovation component-wise. A statistic capturing *all* of the innovation information is the WSSR [14]. It aggregates all of the innovation *vector* information over some finite window of length N. It can be shown that the WSSR is related to a maximum likelihood estimate of the normalized innovations variance [10, 14]. The WSSR test statistic is given by

$$\hat{\rho}(\ell) := \sum_{k=\ell-N+1}^{\ell} e'(k)R_{ee}^{-1}(k)e(k) \quad \text{for } \ell \geq N \tag{5.95}$$

and is based on the hypothesis test

$$H_o : \rho(\ell) \textit{ is } \text{white}$$
$$H_1 : \rho(\ell) \textit{ is not } \text{white}$$

given by

$$\hat{\rho}(\ell) \underset{\text{Accept } H_o}{\overset{\text{Reject } H_o}{\gtrless}} \tau \tag{5.96}$$

Under the null hypothesis, the WSSR is chi-squared distributed, $\rho(\ell) \sim \chi^2(N_yN)$. However, for $N_yN > 30$, $\rho(\ell)$ is approximately Gaussian $\mathcal{N}(N_yN, 2N_yN)$ (see Ref. [4] for more details). At the α-significance level, the probability of rejecting the null hypothesis is given by

$$\Pr\left(\left|\frac{\rho(\ell) - N_yN}{\sqrt{2N_yN}} > \frac{\tau - N_yN}{\sqrt{2N_yN}}\right|\right) = \alpha \tag{5.97}$$

For a level of significance of $\alpha = 0.05$, we have

$$\tau = N_yN + 1.96\sqrt{2N_yN} \tag{5.98}$$

Thus, the WSSR can be considered a "whiteness test" of the innovations *vector* over a finite window of length N. Note that since $[\{e(t)\}, \{R_{ee}(t)\}]$ are obtained from the state–space BP algorithm directly, they can be used for both *stationary* as well as *nonstationary* processes. In fact, in practice for a large number of measurement components, the WSSR is used to "tune" the filter and then the component innovations are individually analyzed to detect model mismatches. Note also that the adjustable parameter of the WSSR statistic is the window length N, which essentially controls the width of the window sliding through the innovations sequence.

Other sets of "reasonableness" tests can be performed using the covariances estimated by the BP algorithm and sample variances estimated using Eq. 5.92. The BP provides estimates of the respective processor covariances R_{ee} and $\tilde{P}$ from the relations given in Table 5.1. Using sample variance estimators, when the filter reaches steady state (process is stationary), that is, $\tilde{P}$ is constant, the estimates can be compared to ensure that they are reasonable. Thus we have

$$\hat{R}_{ee}(i) \approx R_{ee}(i) \quad \text{and} \quad \hat{\tilde{P}}(i) \approx \tilde{P}(i) \tag{5.99}$$

Plotting the $\pm 1.96\sqrt{R_{e_ie_i}(t)}$ and $\pm 1.96\sqrt{\tilde{P}_i(t|t)}$ about the component innovations $\{e_i(t)\}$ and component state estimation errors $\{\tilde{x}_i(t|t)\}$, when the true state is known provides an accurate estimate of the BP performance especially when simulation is used. If the covariance estimates of the processor are reasonable, then 95% of the sequence samples should lie within the constructed bounds. Violation of these bounds clearly indicate inadequacies in modeling the processor statistics. We summarize these results in Fig. 5.11 and examine an *RLC*-circuit design problem in the following section to demonstrate the approach in more detail.

Data	Property	Statistic	Test	Assumptions
Innovation	$m_e = 0$	Sample mean	Zero mean	Ergodic, gaussian
	$R_{ee}(t)$	Sample covariance	Whiteness	Ergodic, gaussian
	$\rho(l)$	WSSR	Whiteness	Gaussian
	$R_{ee}(t,k)$	Sample cross-covariance	Cross-covariance	Ergodic, gaussian
	$R_{eu}(t,k)$	Sample cross-covariance	Cross-covariance	Ergodic, gaussian
Covariances	Innovation	Sample variance	$R_{ee} = \hat{R}_{ee}$	Ergodic
	Innovation	R_{ee}	Confidence interval about $\{e(t)\}$	
	Estimation error	Sample variance	$\tilde{P} = \hat{P}$	Ergodic, X_{true} known
	Estimation error	$\tilde{P}$	Confidence interval about $\{\tilde{x}(t\|t)\}$	X_{true} known

FIGURE 5.11 State-space BP tuning tests.

5.8 CASE STUDY: *RLC* CIRCUIT PROBLEM

Consider the design of an estimator for a series *RLC* circuit (second-order system) excited by a pulse train. The circuit diagram is shown in Fig. 5.12. Using Kirchhoff's voltage law, we can obtain the circuit equations with $i = C(de/dt)$:

$$\frac{d^2e}{dt^2} + \frac{R}{L}\frac{de}{dt} + \frac{1}{LC}e = \frac{1}{LC}e_{in}$$

where e_{in} is a unit pulse train. This equation is that of a second-order system that characterizes the electrical *RLC* circuit, or a mechanical vibration system, or a hydraulic

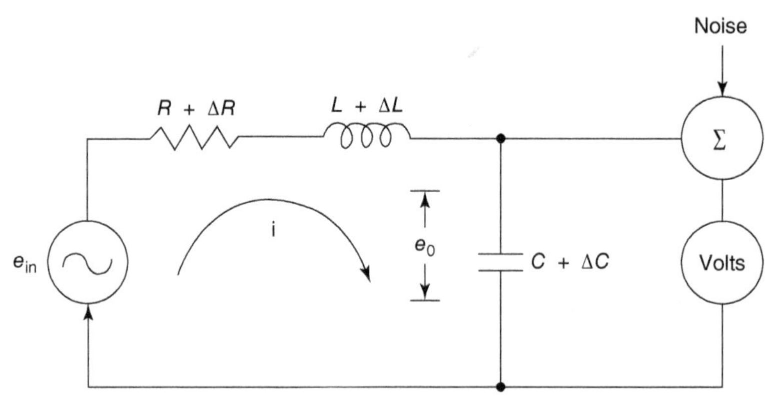

FIGURE 5.12 *RLC* circuit problem.

flow system, etc. The dynamic equations can be placed in state–space form by choosing $x := [e \mid de/dt]'$ and $u = e_{in}$:

$$\frac{dx}{dt} = \begin{bmatrix} 0 & 1 \\ -\frac{1}{LC} & -\frac{R}{L} \end{bmatrix} x + \begin{bmatrix} 0 \\ -\frac{1}{LC} \end{bmatrix} u + \begin{bmatrix} 0 \\ -\frac{1}{LC} \end{bmatrix} w$$

where $w \sim N(0, R_{ww})$ is used to model component inaccuracies.

A high-impedance voltmeter is placed in the circuit to measure the capacitor voltage e. We assume that it is a digital (sampled) device contaminated with noise of variance R_{vv}; that is,

$$y(t) = e(t) + v(t)$$

where $v \sim N(0, R_{vv})$. For our problem we have the following parameters: $R = 5\,\text{k}\Omega$, $L = 2.5\,\text{H}$, $C = 0.1\,\mu\text{F}$, and $T = 0.1$ ms (the problem will be scaled in milliseconds). We assume that the component inaccuracies can be modeled using $R_{ww} = 0.01$, characterizing a deviation of ± 0.1 V uncertainty in the circuit representation. Finally, we assume that the precision of the voltmeter measurements are ($e \pm 0.2$ V), the two standard deviation value, so that $R_{vv} = 0.01\ (\text{V})^2$. Summarizing the circuit model, we have the continuous-time representation

$$\frac{dx}{dt} = \begin{bmatrix} 0 & 1 \\ -4 & -2 \end{bmatrix} x + \begin{bmatrix} 0 \\ -4 \end{bmatrix} u + \begin{bmatrix} 0 \\ -4 \end{bmatrix} w$$

and the discrete-time measurements

$$y(t) = [1 \quad 0]x(t) + v(t)$$

where

$$R_{ww} = \frac{(0.1)^2}{T} = 0.01(\text{V})^2 \quad \text{and} \quad R_{vv} = 0.01(\text{V})^2$$

Before we design the discrete BP, we must convert the system or process model to a sampled-data (discrete) representation. Using SSPACK_PC [8], this is accomplished automatically with the Taylor series approach to approximating the matrix exponential. For an error tolerance of 1×10^{-12}, a 15-term series expansion yields the following discrete-time Gauss–Markov model:

$$x(t) = \begin{bmatrix} 0.98 & 0.09 \\ -0.36 & 0.801 \end{bmatrix} x(t-1) + \begin{bmatrix} -0.019 \\ -0.36 \end{bmatrix} u(t-1) + \begin{bmatrix} -0.019 \\ -0.36 \end{bmatrix} w(t-1)$$

$$y(t) = [1 \mid 0]x(t) + v(t)$$

where

$$R_{ww} = 0.01(\text{V})^2 \quad \text{and} \quad R_{vv} = 0.01(\text{V})^2$$

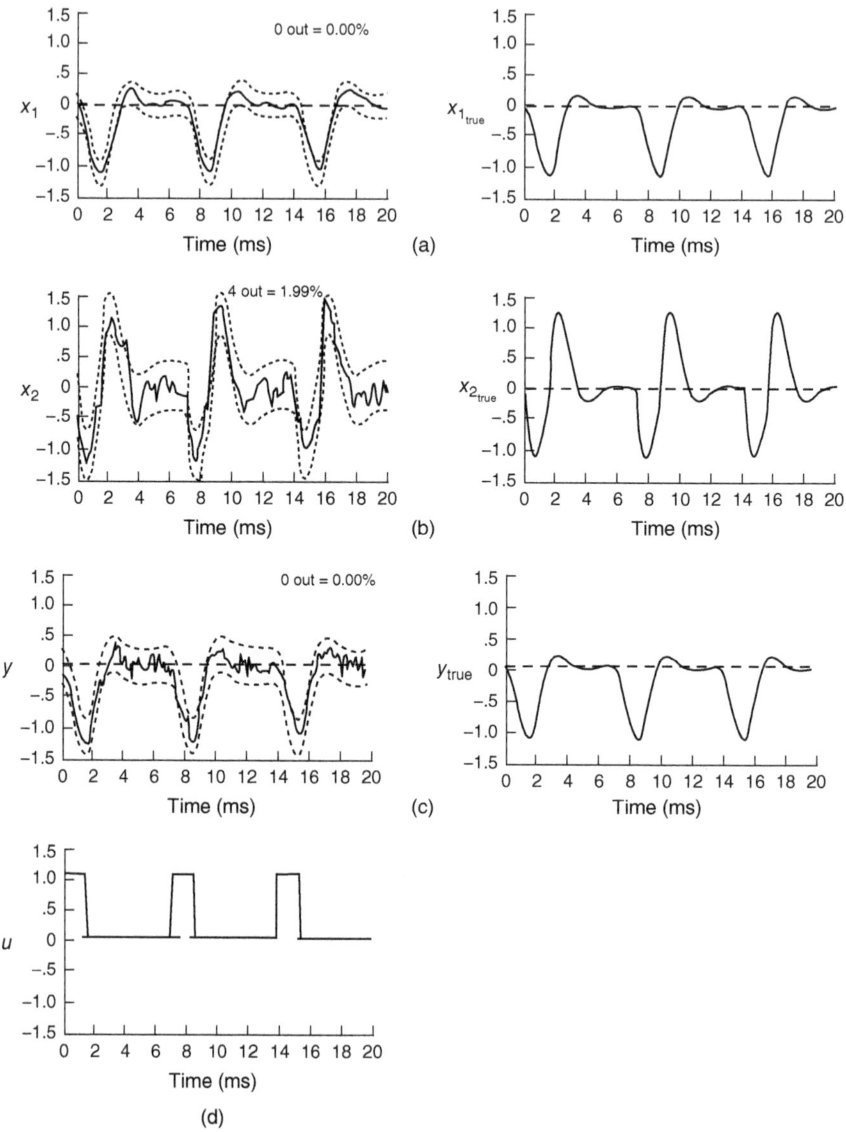

FIGURE 5.13 Gauss-Markov simulation of *RLC* circuit problem: (a) simulated and true state (voltage); (b) simulated true state (current); (c) simulated and true measurement; (d) pulse-train excitation.

Using SSPACK_PC with initial conditions $x(0) = 0$ and $P = \text{diag}(0.01, 0.04)$, the simulated system is depicted in Fig. 5.13. In Figs. 5.13a–c, we see the simulated states and measurements with corresponding confidence limits about the mean (true) values. In each case, the simulation satisfies the statistical properties of the GM model. The corresponding true (mean) trajectories are also shown along with the

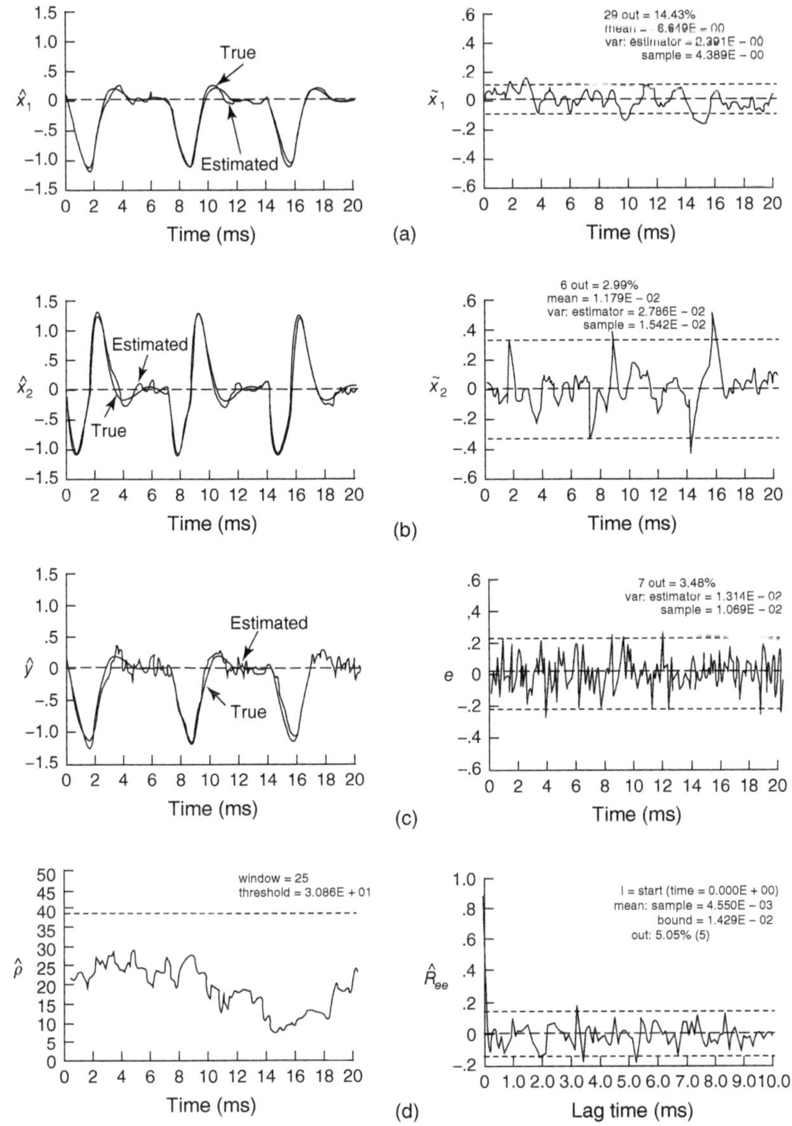

FIGURE 5.14 Bayesian processor design for *RLC* circuit problem: (a) estimated state (voltage) and error; (b) estimated state (current) and error; (c) filtered and true measurement (voltage) and error (innovation); (d) WSSR and whiteness test.

pulse train excitation. Note that the measurements are merely a noisier (process and measurement noise) version of the voltage x_1.

A discrete BP was designed using SSPACK_PC to improve the estimated voltage $\hat{x}_1$. The results are shown in Fig. 5.14. In Fig. 5.14a–c we see the filtered states and measurements as well as the corresponding estimation errors. The true (mean) states are superimposed as well to indicate the tracking capability of the estimator. The estimation errors lie within the bounds (3% out) for the second state, but the error covariance is slightly underestimated for the first state (14% out). The predicted and sample variances are close ($0.002 \approx 0.004$ and $0.028 \approx 0.015$) in both cases. The innovations lie within the bounds (3% out) with the predicted sample variances close ($0.011 \approx 0.013$). The innovations are statistically zero-mean ($0.0046 \ll 0.014$) and white (5% out, WSSR below threshold),[4] indicating a well-tuned estimator. This completes the *RLC* problem.

5.9 SUMMARY

In this chapter, we have introduced the concepts of classical linear and nonlinear Bayesian signal processing using state–space models. After developing the idea of linearizing a nonlinear state–space system, we developed the linearized Bayesian processor (LZ-BP). It was shown that the resulting processor provides an approximate solution (time-varying) to the nonlinear state estimation problem. We then developed the extended Bayesian processor (XBP) or equivalently the extended Kalman filter (EKF), as a special case of the LZ-BP linearizing about the most currently available estimate. Next, we investigated a further enhancement of the XBP by introducing a local iteration of the nonlinear measurement system using the Newton–Rhapson method. Here the processor is called the iterated-extended Bayesian processor (IX-BP) and is shown to produce improved estimates at a small computational cost in most cases. Examples were developed throughout to demonstrate the concepts (see http://www.techni-soft.net for more details).

MATLAB NOTES

SSPACK_PC [8] is a third-party toolbox in MATLAB that can be used to design classical Bayesian processors. This package incorporates the major *nonlinear* BP algorithms discussed in this chapter—all implemented in the UD-factorized form [21] for stable and efficient calculations. It performs the discrete approximate Gauss–Markov simulations using (**SSNSIM**) and both extended (XBP) and iterated-extended (IX-BP) processors using (**SSNEST**). The linearized Bayesian processor (LZ-BP) is also implemented (**SSLZEST**). Ensemble operations are seamlessly embodied within the GUI-driven framework where it is quite efficient to perform multiple design runs and compare results. Of course, the heart of the package is the command or GUI-driven post-processor (**SSPOST**) which is used to analyze and display the results of the simulations and processing.

REBEL is a recursive Bayesian estimation package in MATLAB available on the web which performs similar operations including the new statistical-based

[4] WSSR is the weighted-sum squared residual statistic which aggregates the innovation vector information over a window to perform a vector-type whiteness test (see Ref. [2] for details).

unscented algorithms including the UBP (Chapter 6) including the unscented transformations [24]. It also includes the new particle filter designs (Chapter 7) discussed in Ref. [25].

REFERENCES

1. A. Jazwinski, *Stochastic Processes and Filtering Theory* (New York: Academic Press, 1970).
2. A. Sage and J. Melsa, *Estimation Theory with Applications to Communications and Control* (New York: McGraw-Hill, 1971).
3. A. Gelb, *Applied Optimal Estimation* (Boston, MA: MIT Press, 1975).
4. B. Anderson and J. Moore, *Optimal Filtering* (Englewood Cliffs, NJ: Prentice-Hall, 1979).
5. P. Maybeck, *Stochastic Models, Estimation, and Control* (New York: Academic Press, 1979).
6. M. Grewal and A. Andrews, *Kalman Filtering: Theory and Practice* (Englewood Cliffs, NJ: Prentice-Hall, 1993).
7. J. Mendel, *Lessons in Estimation Theory for Signal Processing, Communications and Control* (Englewood Cliffs, NJ: Prentice-Hall, 1995).
8. J. Candy and P. Candy, "SSPACK_PC: A model-based signal processing package on personal computers," *DSP Applications*, **2**, 3, 33–42, 1993 (see http://www.techni-soft.net for more details).
9. J. Candy, *Signal Processing: The Model-Based Approach* (New York: McGraw-Hill, 1986).
10. J. Candy, *Model-Based Signal Processing* (Hoboken, NJ: John Wiley & Sons, Inc./IEEE Press, 2006).
11. D. Simon, *Optimal State Estimation: Kalman, H_∞ and Nonlinear Approaches* (Hoboken, NJ: John Wiley & Sons, Inc., 2006).
12. T. Kailath, A. Sayed, and B. Hassibi, *Linear Estimation* (Englewood Cliffs, NJ: Prentice-Hall, 2000).
13. R. Mehra and J. Peschon, "An innovations approach to fault detection and diagnosis in dynamic systems," *Automatica*, **7**, 637–640, 1971.
14. A. Wilsky, "A survey of failure detection in dynamic systems," *Automatica*, **12**, 601–611, 1976.
15. Y. Bar-Shalom and X. Li, *Estimation and Tracking: Principles, Techniques, and Software* (Norwood, MA: Artech House, 1993).
16. B. Bell and F. Cathey, "The iterated Kalman filter update as a Gauss-Newton method," *IEEE Trans. Autom. Contr.*, **AC-38**, 2, 294–297, 1993.
17. J. Candy and R. Rozsa, "Safeguards for a plutonium-nitrate concentrator-an applied estimation approach,"*Automatica*, **16**, 615–627, 1980.
18. J. Candy and E. Sullivan, "Ocean acoustic signal processing: A model-based approach," *J. Acoust. Soc. Am.*, **92**, 3185–3201, 1992.
19. J. Mendel, J. Kormylo, J. Lee, and F. Ashirafi, "A novel approach to seismic signal processing and modeling," *Geophysics*, **46**, 1398–1414, 1981.

20. H. Sorenson Ed., "Special Issue on Applications of Kalman Filtering," *IEEE Trans. Auto. Contr.*, **AC-28**, 3, 253–427, 1983.
21. G. Bierman, *Factorization Methods of Discrete Sequential Estimation* (New York: Academic Press, 1977).
22. V. Aidala and S. Hammel, "Utilization of modified polar-coordinates for bearings-only tracking," *IEEE Trans. Autom. Contr.*, **AC-28**, 283–294, 1983.
23. V. Aidala, "Kalman filter behavior in bearings-only velocity and position estimation," *IEEE Trans. Aerosp. Electron. Syst.*, **AES-15**, 29–39, 1979.
24. R. van der Merwe and E. Wan, *REBEL: Recursive Bayesian Estimation Library University of Oregon* , 2002.
25. S. Haykin and N. de Freitas (editors), "Sequential state estimation: From Kalman filters to particle filters," *Proc. IEEE*, **92**, 3, 399–574, 2004.

PROBLEMS

5.1 Derive the *continuous-time* BP by starting with the discrete equations of Table 5.1 and using the following sampled-data approximations:

$$\begin{aligned} A &= e^{A_c \Delta t} \approx I + A_c \Delta t \\ B &= B_c \Delta t \\ W &= W_c \Delta t \\ R_{ww} &= R_{w_c w_c} \Delta t \end{aligned}$$

5.2 Suppose we are given a continuous-time Gauss–Markov model characterized by

$$\dot{x}(t) = A_c(t)x(t) + B_c(t)u(t) + W_c(t)w(t)$$

and discrete (sampled) measurement model such that $t \rightarrow t_k$ then

$$y(t_k) = C(t_k)x(t_k) + v(t_k)$$

where the continuous process, $w(t) \sim \mathcal{N}(0, R_{ww})$, and $v(t_k) \sim \mathcal{N}(0, R_{vv})$ with Gaussian initial conditions.

(*a*) Determine the state mean ($m_x(t)$) and covariance ($P(t)$).

(*b*) Determine the measurement mean ($m_y(t_k)$) and covariance ($R_{yy}(t_k)$).

(*c*) Develop the relationship between the continuous and discrete Gauss–Markov models based on the solution of the continuous state equations and approximation using a first-order Taylor series for the state transition matrix, $\Phi(t, t_o)$ and the associated system matrices.

(*d*) Derive the continuous-discrete BP using first difference approximations for derivatives and the discrete (sampled) system matrices derived above.

5.3 The covariance correction equation of the BP algorithm is seldom used directly. Numerically, the covariance matrix $\tilde{P}(t|t)$ must be positive semidefinite, but in the correction equation we are subtracting a matrix from a positive semidefinite matrix and cannot guarantee that the result will remain positive semidefinite (as it should be) because of roundoff and truncation errors. A solution to this problem is to replace the standard correction equation with the stabilized *Joseph form*, that is,

$$\tilde{P}(t|t) = [I - K(t)C(t)]\tilde{P}(t|t-1)[I - K(t)C(t)]' + K(t)R_{vv}(t)K'(t)$$

(*a*) Derive the *Joseph* stabilized form.

(*b*) Demonstrate that it is equivalent to the standard correction equation.

5.4 Prove that a necessary and sufficient condition for a *linear* BP to be optimal is that the corresponding innovations sequence is zero-mean and white.

5.5 A bird watcher is counting the number of birds migrating to and from a particular nesting area. Suppose the number of migratory birds, $m(t)$ is modeled by a first-order *ARMA* model:

$$m(t) = -0.5m(t-1) + w(t) \quad \text{for } w \sim \mathcal{N}(10, 75)$$

while the number of resident birds is static, that is,

$$r(t) = r(t-1)$$

The number of resident birds is averaged leading to the expression

$$y(t) = 0.5r(t) + m(t) + v(t) \quad \text{for } w \sim \mathcal{N}(0, 0.1)$$

(*a*) Develop the two-state Gauss–Markov model with initial values $r(0) = 20$ birds, $m(0) = 100$ birds, cov $r(0) = 25$.

(*b*) Use the MBP algorithm to estimate the number of resident and migrating birds in the nesting area for the data set $y(t) = \{70, 80\}$, that is, what is $\hat{x}(2|2)$?

5.6 Suppose we are given a measurement device that not only acquires the current state but also the state delayed by one time-step (multipath) such that

$$y(t) = Cx(t) + Ex(t-1) + v(t)$$

Derive the recursive form for this associated MBP. (*Hint*: Recall from the properties of the state transition matrix that $x(t) = \Phi(t, \tau)x(\tau)$ and $\Phi^{-1}(t, \tau) = \Phi(\tau, t)$)

(*a*) Using the state transition matrix for discrete-systems, find the relationship between $x(t)$ and $x(t-1)$ in the Gauss–Markov model.

(*b*) Substitute this result into the measurement equation to obtain the usual form

$$y(t) = \tilde{C}x(t) + \tilde{v}(t)$$

What are the relations for $\tilde{C}$ and $\tilde{v}(t)$ in the new system?

(*c*) Derive the new statistics for $\tilde{v}(t) \sim N(\mu_{\tilde{v}}, R_{\tilde{v}\tilde{v}})$.

(*d*) Are w and v correlated? If so, use the prediction form to develop the MBP algorithm for this system.

5.7 Develop the discrete linearized (perturbation) models for each of the following nonlinear systems [7]:

- *Synchronous (unsteady) motor:* $\ddot{x}(t) + C\dot{x}(t) + p\sin x(t) = L(t)$
- *Duffing equation:* $\ddot{x}(t) + \alpha x(t) + \beta x^3(t) = F(\cos\omega t)$
- *Van der Pol equation:* $\ddot{x}(t) + \epsilon\dot{x}(t)[1 - \dot{x}^2(t)] + x(t) = m(t)$
- *Hill equation:* $\ddot{x}(t) - \alpha x(t) + \beta p(t)x(t) = m(t)$

(*a*) Develop the LZ-BP.

(*b*) Develop the XBP.

(*c*) Develop the IX-BP.

5.8 Suppose we are given the following discrete system

$$x(t) = -\omega^2 x(t-1) + \sin x(t-1) + \alpha u(t-1) + w(t-1)$$
$$y(t) = x(t) + v(t)$$

with w and v zero-mean, white Gaussian with usual covariances, R_{ww} and R_{vv}.

(*a*) Develop the LZ-BP for this process.

(*b*) Develop the XBP for this process.

(*c*) Develop the IX-BP for this process.

(*d*) Suppose the parameters ω and α are unknown develop the XBP such that the parameters are jointly estimated along with the states. (*Hint*: Augment the states and parameters to create a new state vector).

5.9 Assume that we have the following nonlinear continuous-discrete Gauss–Markov model:

$$x(t) = f[x(t)] + g[u(t)] + w(t)$$
$$z(t_k) = h[x(t_k)] + v(t_k)$$

with w and v zero-mean, white Gaussian with usual covariances, Q and R.

(*a*) Develop the perturbation model for $\delta x(t) := x(t) - x^*(t)$ for $x^*(t)$ the given reference trajectory.

(*b*) Develop the LZ-BP for this process.

(*c*) Choose $x^*(t) = \hat{x}(t)$ whenever $\hat{x}(t)$ is available during the recursion. Therefore, develop the continuous-discrete XBP.

5.10 Suppose we assume that a target is able to maneuver, that is, we assume that the target velocity satisfies a first-order AR model given by:

$$v_\tau(t) = -\alpha v_\tau(t-1) + w_\tau(t-1) \quad \text{for } w \sim \mathcal{N}(0, R_{w_\tau w_\tau})$$

(*a*) Develop the Cartesian tracking model for this process.

(*b*) Develop the corresponding XBP assuming all parameters are known a priori.

(*c*) Develop the corresponding XBP assuming α is unknown.

5.11 Nonlinear processors (LZ-BP, XBP, IX-BP) can be used to develop neural networks used in many applications. Suppose we model a generic neural network behavior by

$$x(t) = x(t-1) + w(t-1)$$
$$y(t) = c[x(t), u(t), \alpha(t)] + v(t)$$

where $x(t)$ is the network weights (parameters), $u(t)$ is the input or training sequence, $\alpha(t)$ is the node activators with w and v zero-mean, white Gaussian with covariances, R_{ww} and R_{vv}.

(*a*) Develop the LZ-BP for this process.

(*b*) Develop the XBP for this process.

(*c*) Develop the IX-BP for this process.

5.12 The Mackey–Glass time delay differential equation is given by

$$\dot{x}(t) = \frac{\alpha x(t-\tau)}{1 + x(t-\tau)^N} - \beta x(t) + w(t)$$
$$y(t) = x(t) + v(t)$$

where α, β are constants, N is a positive integer with w and v zero-mean, white Gaussian with covariances, R_{ww} and R_{vv}. For the parameter set: $\alpha = 0.2$, $\beta = 0.1$, $\tau = 7$, and $N = 10$ with $x(0) = 1.2$

(*a*) Develop the LZ-BP for this process.

(*b*) Develop the XBP for this process.

(*c*) Develop the IX-BP for this process.

5.13 Consider the problem of estimating a random signal from an AM modulator characterized by

$$s(t) = \sqrt{2P}a(t) \sin \omega_c t$$
$$r(t) = s(t) + v(t)$$

where $a(t)$ is assumed to be a Gaussian random signal with power spectrum

$$S_{aa}(\omega) = \frac{2k_a P_a}{\omega^2 + k_a^2}$$

also assume that the processes are contaminated with the usual additive noise sources: w and v zero-mean, white Gaussian with covariances, R_{ww} and R_{vv}.

(*a*) Develop the continuous-time Gauss–Markov model for this process.

(*b*) Develop the corresponding discrete-time Gauss–Markov model for this process using first differences.

(*c*) Develop the BP.

(*d*) Assume the carrier frequency, ω_c is unknown. Develop the XBP for this process.

6

MODERN BAYESIAN STATE–SPACE PROCESSORS

6.1 INTRODUCTION

In this chapter, we discuss an extension of the approximate nonlinear Bayesian suite of processors that takes a distinctly different approach to the nonlinear Gaussian problem. Instead of attempting to improve on the linearized approximation in the nonlinear XBP (EKF) schemes discussed in the previous chapter or increasing the order of the Taylor series approximations [1–11], a modern *statistical* (linearization) *transformation* approach is developed. It is founded on the idea that "it is easier to approximate a probability distribution, than to approximate an arbitrary nonlinear function of transformation" [3, 12–21]. The classical nonlinear Bayesian processors discussed so far are based on linearizing nonlinear functions of the state and measurements to provide estimates of the underlying statistics (using Jacobians), while the statistical transformation approach is based on selecting a set of sample points that capture certain properties of the underlying distribution. This transformation is essentially a "statistical linearization" technique that incorporates the uncertainty of the prior random variable when linearizing [12]. This set of sample points is then nonlinearly transformed or propagated to a new space. The statistics of the new samples are then calculated to provide the required estimates. Note that this method differs from the *sampling–resampling* approach, in which *random* samples are drawn from the prior distribution and updated through the likelihood function to produce a sample from the posterior distribution [22]. Here the samples are *not* drawn at random, but according to a specific *deterministic* algorithm. Once this transformation is performed, the resulting processor, the sigma-point Bayesian processor (SPBP) or

Bayesian Signal Processing: Classical, Modern, and Particle Filtering Methods, Second Edition. James V. Candy.

equivalently the unscented Kalman filter (UKF) evolves. It is a recursive processor that resolves some of the approximation issues [14] and deficiencies of the XBP of the previous sections. We first develop the idea of nonlinearly transforming a random vector with known probability distribution and then apply it to a Gaussian problem leading to the SPBP algorithm. We then apply the modern processor to the previous nonlinear state estimation problem and compare its performance to the classical.

Another approach to Bayesian processing problem is the ensemble Bayesian processor (EnBP) popularly called the ensemble Kalman filter (EnKF) approximation for large-scale computational problems. It has been shown that the EnKF provides a simple implementation along with numerical stability because of its properties [23–32]. It provides an alternative mechanism to estimate the required posterior distribution based on generating an ensemble of model state estimates along with associated uncertainties.

6.2 SIGMA-POINT (UNSCENTED) TRANSFORMATIONS

A completely different approach to nonlinear estimation evolves from the concept of statistical linearization [3, 16, 19, 33]. Instead of approximating the nonlinear process and measurement dynamics of the underlying system using a truncated Taylor series representation that leads to the classical forms of estimation (LZKF, EKF, IEKF, etc.), the statistical approximation or equivalently *statistical linearization* method provides an alternative that takes into account the uncertainty or probabilistic spread of the prior random vector. The basic idea is to approximate (linearize) a nonlinear function of a random vector while preserving its first and second moments; therefore, this approach requires a priori knowledge of its distribution resulting in a more statistically accurate transformation.

6.2.1 Statistical Linearization

Following Gelb [3], statistical linearization evolves from the idea of propagating an N_x-dimensional random vector $\mathbf{x}$ with PDF $p_X(\mathbf{x})$ through an arbitrary nonlinear transformation $\mathbf{a}[\cdot]$ to generate a new random vector,

$$\mathbf{y} = \mathbf{a}[\mathbf{x}] \tag{6.1}$$

Expanding this function in a Taylor series about $\mathbf{x}_i$ gives

$$\mathbf{a}[\mathbf{x}] = \mathbf{a}[\mathbf{x}_i] + (\mathbf{x} - \mathbf{x}_i)'\nabla_{\mathbf{x}}\mathbf{a}[\mathbf{x}_i] + \text{H.O.T.} \tag{6.2}$$

where $\nabla_{\mathbf{x}}$ is the N_x-gradient vector defined by

$$\nabla_{\mathbf{x}}\mathbf{a}[\mathbf{x}_i] := \left.\frac{\partial \mathbf{a}[\mathbf{x}]}{\partial \mathbf{x}}\right|_{\mathbf{x}=\mathbf{x}_i} \tag{6.3}$$

Neglecting the H.O.T and simplifying with the appropriate definitions, we obtain the "regression form" of $\mathbf{y}$ regressing on $\mathbf{x}$

$$\mathbf{y} = \mathbf{a}[\mathbf{x}] \approx A\mathbf{x} + \mathbf{b} \tag{6.4}$$

where both A, $\mathbf{b}$ are to be determined, given $\mathbf{x} \sim p_X(\mathbf{x})$. Estimation of A, $\mathbf{b}$ follows from the traditional linear algebraic perspective by defining the approximation or *linearization error* as

$$\epsilon := \mathbf{y} - A\mathbf{x} - \mathbf{b} = \mathbf{a}[\mathbf{x}] - A\mathbf{x} - \mathbf{b} \tag{6.5}$$

along with the corresponding cost function

$$J_\epsilon := E\{\epsilon'\epsilon\} \tag{6.6}$$

that is minimized with respect to the unknowns $A, \mathbf{b}$. Performing the usual differentiation of the cost function, $J_\epsilon \to J_\epsilon(A, \mathbf{b})$, with respect to the unknowns, setting the results to zero and solving generates the MMSE result discussed previously in Chapter 2. That is, we first minimize with respect to $\mathbf{b}$

$$\nabla_{\mathbf{b}} J_\epsilon(A, \mathbf{b}) = \nabla_{\mathbf{b}} E\{\epsilon'\epsilon\} = E\{\nabla_{\mathbf{b}}(y - A\mathbf{x} - \mathbf{b})'\epsilon\} = 0 \tag{6.7}$$

Now using the chain rule of Eq. 2.11 with $a' = (y - A\mathbf{x} - \mathbf{b})'$ and $b = \epsilon$, we obtain

$$\nabla_{\mathbf{b}} J_\epsilon(A, \mathbf{b}) = E\{\nabla_{\mathbf{b}}(y - A\mathbf{x} - \mathbf{b})'\epsilon\} = -2E\{(y - A\mathbf{x} - \mathbf{b})\} = 0$$

Solving for $\mathbf{b}$, we obtain

$$\mathbf{b} = \mu_y - A\mu_x \tag{6.8}$$

Furthermore, substituting for $\mathbf{b}$ into the cost and differentiating with respect to A, setting the result to zero and solving yields

$$\nabla_A J_\epsilon(A, \mathbf{b})\big|_{\mathbf{b}=\mu_y - A\mu_x} = E\{A(x - \mu_x)(x - \mu_x)' + (y - \mu_y)(x - \mu_x)'\} = 0$$

giving

$$AE\{(x - \mu_x)(x - \mu_x)'\} = E\{(y - \mu_y)(x - \mu_x)'\}$$

or

$$AR_{xx} = R_{yx}$$

that has the MMSE solution

$$A = R_{yx}R_{xx}^{-1} = R'_{xy}R_{xx}^{-1} \tag{6.9}$$

Now suppose we linearize the function through this relation by constructing a linear regression between a selected set of N_x-points and the nonlinear transformation $\mathbf{a}[\mathbf{x}]$ at these selected points. That is, define the set of the selected points as $\{\mathcal{X}_i, \mathcal{Y}_i\}$ with

$$\mathcal{Y}_i = \mathbf{a}[\mathcal{X}_i]$$

Following the same approach as before by defining the "pointwise" linearization error as gives

$$\epsilon_i = \mathcal{Y}_i - A\mathcal{X}_i - \mathbf{b} = \mathbf{a}[\mathcal{X}_i] - A\mathcal{X}_i - \mathbf{b} \tag{6.10}$$

with

$$\mu_\epsilon = 0, \quad \text{and} \quad R_{\epsilon\epsilon} = R_{yy} - AR_{xx}A' \tag{6.11}$$

Performing a *weighted* minimization (as above) on the pointwise linearization error with a set of regression weights, $\{W_i\}; i = 1, \ldots, N_x$ yields the following solution:

$$A = R'_{xy}R_{xx}^{-1}, \quad \mathbf{b} = \mu_y - A\mu_x \tag{6.12}$$

where the weighted statistics are given by

$$\begin{aligned} \mu_x &= \sum_{i=1}^{N_x} W_i\mathcal{X}_i \\ R_{xx} &= \sum_{i=1}^{N_x} W_i(\mathcal{X}_i - \mu_x)(\mathcal{X}_i - \mu_x)' \end{aligned} \tag{6.13}$$

and for the posterior

$$\begin{aligned} \mu_y &= \sum_{i=1}^{N_x} W_i\mathcal{Y}_i \\ R_{yy} &= \sum_{i=1}^{N_x} W_i(\mathcal{Y}_i - \mu_y)(\mathcal{Y}_i - \mu_y)' \end{aligned} \tag{6.14}$$

with the corresponding cross-covariance

$$R_{xy} = \sum_{i=1}^{N_x} W_i(\mathcal{X}_i - \mu_x)(\mathcal{Y}_i - \mu_y)' \tag{6.15}$$

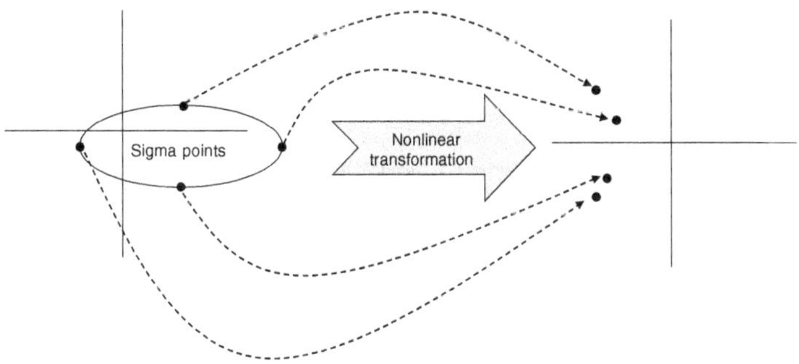

FIGURE 6.1 Unscented transformation: A set of distribution points shown on an error ellipsoid are selected and transformed into a new space where their underlying statistics are estimated.

With this underlying regression and pointwise transformation in place, the next step is to determine the corresponding set of regression (sigma) points $\{\mathcal{X}_i\}$ and their corresponding weights, $\{W_i\}$.

6.2.2 Sigma-Point Approach

The *sigma-point transformation (SPT)*, or equivalently *unscented transformation (UT)*, is a technique for calculating the statistics of a random vector that has been nonlinearly transformed. The approach is illustrated in Fig. 6.1. Here the set of samples or so-called *sigma points* are chosen so that they capture the specific properties of the underlying distribution. In the figure, we consider $\mathrm{p}_X(x)$ to be a 2D-Gaussian, then the σ-points are located along the major and minor axes of the covariance ellipse capturing the essence of this distribution. In general, the goal is to construct a set of σ-points possessing the same statistics as the original distribution such that when nonlinearly transformed to the new space, the new set of points sufficiently capture the posterior statistics. The transformation occurs on a point-by-point basis, since it is simpler to match statistics of individual points rather than the entire PDF. The statistics of the transformed points are then calculated to provide the desired estimates of the transformed distribution.

Following the development of Julier [15], consider propagating an N_x-dimensional random vector $\mathbf{x}$ through an arbitrary nonlinear transformation $\mathbf{a}[\cdot]$ to generate a new random vector,

$$\mathbf{y} = \mathbf{a}[\mathbf{x}] \tag{6.16}$$

The set of σ-points $\{\mathcal{X}_i\}$ consists of $N_\sigma + 1$ vectors with appropriate weights $\{W_i\}$ given by $\Sigma = \{\mathcal{X}_i, W_i; i = 0, \ldots, N_\sigma\}$. The weights can be positive or negative but *must satisfy* the *normalization constraint*

$$\sum_{i=0}^{N_\sigma} W_i = 1$$

so that the estimate of the statistics remains unbiased. The problem then becomes: **GIVEN** the σ-points $\Sigma = \{\mathcal{X}_i, W_i; i=0, \ldots, N_\sigma\}$ and the nonlinear transformation $\mathbf{a}[\mathbf{x}]$, **FIND** the statistics of the transformed samples,

$$\mu_y = E\{\mathbf{y}\} \quad \text{and} \quad \mathbf{R}_{yy} = \text{Cov}(\mathbf{y})$$

The *sigma-point (unscented) transformation (SPT)* approach to approximate the statistics, $(\mu_y, \mathbf{R}_{yy})$ is

1. Determine the number, weights, and locations of the σ-point set, Σ, based on the unique characteristics of the prior distribution.
2. Nonlinearly *transform* each point to obtain the set of new (posterior) σ-points, $\{\mathcal{Y}_i\}$:

$$\mathcal{Y}_i = \mathbf{a}[\mathcal{X}_i] \tag{6.17}$$

3. Estimate the posterior *mean* by its weighted average[1]

$$\mu_{\mathbf{y}} = \sum_{i=0}^{N_\sigma} W_i \mathcal{Y}_i \tag{6.18}$$

4. Estimate the posterior *covariance* by its weighted outer product

$$\mathbf{R}_{yy} = \sum_{i=0}^{N_\sigma} W_i (\mathcal{Y}_i - \mu_y)(\mathcal{Y}_i - \mu_y)' \tag{6.19}$$

One set of σ-points that satisfies the above conditions consists of a symmetric set of $N_\sigma + 1$ points that lie on the $\sqrt{N_x}$-th covariance contour [13]:

$$\begin{aligned} \mathcal{X}_o &= \mu_x, & W_o &= \frac{1}{N_x} \\ \mathcal{X}_i &= \mu_x + \sqrt{N_x}\sigma_i, & W_i &= \frac{1}{2N_x} \\ \mathcal{X}_{i+N_x} &= \mu_x - \sqrt{N_x}\sigma_i, & W_{i+N_x} &= \frac{1}{2N_x} \end{aligned}$$

where $\sqrt{N_x}\,\sigma_i$ is the *ith* standard deviation scaled by $\sqrt{N_x}$ and W_i is the weight associated with the *ith* σ-point.

[1] Note that this estimate is actually a weighted statistical linear regression (WSLR) of the random variable [19].

Thus, the *SPT* can be considered a *statistical linearization method* that provides an optimal (MMSE) linear approximation to a general nonlinear transformation taking into account the prior second-order statistics of the underlying random variable, that is, its mean and covariance [19]. It can be accomplished using the weighted statistical linear regression approach (WSLR) [16, 19], resulting in the weight-constrained estimates above.

Therefore, in contrast to random sampling, selection of the "deterministic" σ-points requires resolving the following critical issues:

- $N_\sigma + 1$, the *number* of σ-points;
- W_i, the *weights* assigned to each σ-point; and
- $\mathcal{X}_i$, the *location* of the σ-points.

That is, we must answer the questions of: How many (points)?, What (weights)? and Where (located)?

The σ-points should be *selected* or constrained to capture the "most important" statistical properties of the random vector $\mathbf{x}$. Let the underlying prior $\mathrm{p}_X(\mathbf{x})$ be its density function, then the σ-points capture the properties by satisfying the *necessary (constraint) condition*

$$\mathbf{g}[\Sigma, \mathrm{p}_X(\mathbf{x})] = \mathbf{0} \tag{6.20}$$

Since it is possible to meet this constraint condition and still have some degree of freedom in the choice of σ-points, assigning a *penalty* function

$$\mathbf{p}[\Sigma, \mathrm{p}_X(\mathbf{x})] \tag{6.21}$$

resolves any ambiguity in the choice. This function is to incorporate desirable features that do not necessarily have to be satisfied. Decreasing the penalty function leads to more and more desirable solutions. In general, the σ-point set relative to this problem that is the most desirable is the set that conforms to the necessary conditions of Eqs. 6.20 and 6.21. The σ-points must *satisfy*

$$\min_{\Sigma} \mathbf{p}[\Sigma, \mathrm{p}_X(\mathbf{x})] \ni \mathbf{g}[\Sigma, \mathrm{p}_X(\mathbf{x})] = \mathbf{0} \tag{6.22}$$

The decision as to which properties of the random vector $\mathbf{x}$ to capture precisely or approximate depends on the particular application. For our problems, we wish to match the *moments* of the underlying distribution of σ-points with those of $\mathbf{x}$.

Summarizing, to apply the SPT approach, (1) a set of σ-points Σ are constructed that are "deterministically" constrained to possess the identical statistics of the prior; (2) each σ-point is nonlinearly transformed to the new space $\mathcal{Y}_i$; and (3) the statistics of the transformed set are approximated using WSLR techniques. The following example illustrates this approach.

Example 6.1

Consider a scalar random variable x for which we would like to propagate its first two moments (μ_x, R_{xx}) through a nonlinear transformation, $\mathcal{Y}_i = \mathbf{a}[\mathcal{X}_i]$. The corresponding set of constraint equations are defined by

$$\mathbf{g}(\Sigma, \mathrm{p}_X(x)) = \begin{bmatrix} \sum_{i=0}^{N_\sigma} W_i - 1 \\ \sum_{i=0}^{N_\sigma} W_i \mathcal{X}_i - \mu_x \\ \sum_{i=0}^{N_\sigma} W_i (\mathcal{X}_i - \mu_x)(\mathcal{X}_i - \mu_x)' - R_{xx} \end{bmatrix} = \mathbf{0}$$

with posterior

$$\mu_y = \sum_{i=0}^{N_\sigma} W_i \mathcal{Y}_i$$
$$R_{yy} = \sum_{i=0}^{N_\sigma} W_i (\mathcal{Y}_i - \mu_y)(\mathcal{Y}_i - \mu_y)'$$

Suppose $x \sim \mathcal{N}(0, 1)$, then the set of σ-points are defined by $N_\sigma + 1 = 2 + 1$ or three points with $\mathcal{X}_0 = \sigma_0 = 0$, $\mathcal{X}_1 = -\sigma_1$, and $\mathcal{X}_2 = \sigma_2$. The constraint equations become

$$W_0 + W_1 + W_2 - 1 = 0$$
$$-W_1\sigma_1 + W_2\sigma_2 - 0 = 0$$
$$W_1\sigma_1^2 + W_2\sigma_2^2 - 1 = 0$$

for W_0, a free parameter whose value affects the fourth and higher moments. These relations are not uniquely satisfied (four unknowns, three equations); therefore, we must add another constraint to the set based on the property that the *skew* is zero for the Gaussian

$$-W_1\sigma_1^3 + W_2\sigma_2^3 = 0$$

Solving these equations and using the symmetry property of the Gaussian gives

△△△

$$\sigma_1 = 1/2\sqrt{W_1}, \quad W_1 = (1 - W_0)/2, \quad \sigma_2 = \sigma_1, \quad W_2 = W_1$$

This example illustrates the use of the *constraint* equations to determine the σ-points from properties of the prior. Next, let us incorporate the nonlinear transformation and penalty function to demonstrate the entire selection procedure. Following Julier [13], consider the following one-dimensional example.

Example 6.2

As before, suppose we have a scalar random variable x, Gaussian distributed with mean μ_x and variance σ_x^2, and we would like to know the statistics (mean and variance) of y which nonlinearly transforms x according to

$$y = a[x] = x^2$$

Here the true mean and variance are

$$\mu_y = E\{y\} = E\{x^2\} = \sigma_x^2 + \mu_x^2$$

and

$$\begin{aligned}\sigma_y^2 &= E\{y^2\} - \mu_y^2 = E\{x^4\} - \mu_y^2 = (3\sigma_x^4 + 6\sigma_x^2\mu_x^2 + \mu_x^4) - (\sigma_x^4 + 2\sigma_x^2\mu_x^2 + \mu_x^4)\\ &= 2\sigma_x^4 + 4\sigma_x^2\mu_x^2\end{aligned}$$

According to SPT of Eqs. 6.17– 6.19, the number of σ-points is $N_\sigma + 1 = 3$. Since this is a scalar problem ($N_x = 1$) only three points are required: the two σ-points and the mean, therefore, we have

$$\begin{aligned}\{\mathcal{X}_0, \mathcal{X}_1, \mathcal{X}_2\} &= \{\mu_x, \mu_x - \sqrt{1+\kappa}\sigma_x, \mu_x + \sqrt{1+\kappa}\sigma_x\}\\ \{W_0, W_1, W_2\} &= \{1/(1+\kappa), 1/2(1+\kappa), 1/2(1+\kappa)\}\end{aligned}$$

and κ is chosen as a scaling factor to be determined. Propagating these samples through $a[\cdot]$ gives the transformed samples, say $\mathcal{X}_i'$, that lie at

$$\{\mathcal{X}_0', \mathcal{X}_1', \mathcal{X}_2'\} = \left\{\mu_x^2, (\mu_x - \sqrt{1+\kappa}\sigma_x)^2, (\mu_x + \sqrt{1+\kappa}\sigma_x)^2\right\}$$

The mean of y is given by

$$\mu_y = \frac{1}{2(1+\kappa)}(2\kappa\mu_x^2 + 2\mu_x^2 + 2(1+\kappa)\sigma_x^2) = \mu_x^2 + \sigma_x^2$$

which is precisely the true mean. Next, the covariance is given by

$$\sigma_y^2 = \frac{1}{2(1+\kappa)}\left(\sum_{i=1}^{2}((\mathcal{X}_i' - \mu_y)^2 + 2\kappa\sigma_x^4)\right) = \kappa\sigma^4 + 4\mu_x^2\sigma_x^2$$

To find the solution, κ must be specified. The kurtosis of the true distribution is $2\sigma_x^4$ and that of the σ points is σ_x^2. Since the kurtosis of the points is scaled by an amount $(1+\kappa)$, the kurtosis of both distributions only agree when $\kappa = 2$ which gives the exact solution. This completes the Gaussian example. △△△

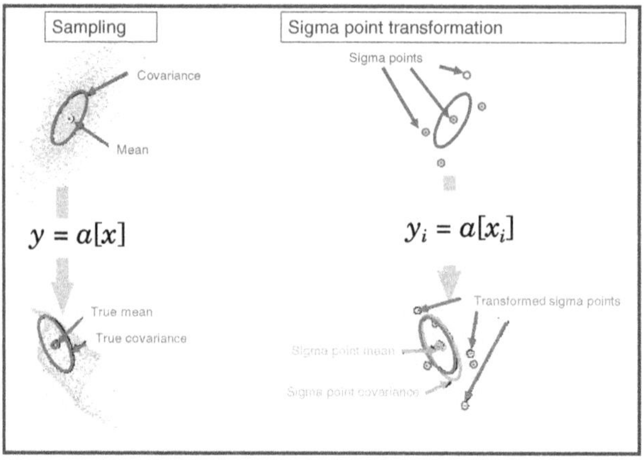

FIGURE 6.2 Sigma-point (unscented) transformation approximation of the original distribution moments after nonlinear transformation.

Next using the underlying principles of the SPT, we develop the multivariate Gaussian case in more detail.

6.2.3 SPT for Gaussian Prior Distributions

To be more precise and parallel the general philosophy, we choose "to approximate the underlying Gaussian distribution rather than approximate its underlying nonlinear transformation," in contrast to the XBP (EKF). This parameterization captures the prior *mean* and *covariance* and permits the direct propagation of this information through the arbitrary set of nonlinear functions. Here we accomplish this (approximately) by generating the discrete distribution having the same first and second (and potentially higher) moments where each point is directly transformed. The mean and covariance of the transformed ensemble can then be computed as the *estimate* of the nonlinear transformation of the original distribution. As illustrated in Fig. 6.2, we see samples of the true prior distribution of x and the corresponding nonlinearly transformed distribution of y. Using the same transformation, the selected σ-points are transformed as well closely preserving the dominant moments of the original distribution (see Julier [13] for more details).

It is important to recognize that the SPT has specific properties when the underlying distribution is Gaussian [16]. The Gaussian has two properties which play a significant role in the form of the σ-points *selected*. First, since the distribution is *symmetric*, the σ-points can be selected with this symmetry. Second, the problem of approximating $\mathbf{x}$ with an arbitrary mean and covariance can be reduced to that of a standard zero-mean, unit variance Gaussian, since

$$\mathbf{x} = \mu_{\mathbf{x}} + U\mathbf{s} \quad \text{for } U \text{ the matrix square root of } R_{xx} \tag{6.23}$$

where $\mathbf{s} \sim \mathcal{N}(0, I)$. Therefore, in the Gaussian case, the second-order SPT uses a set of σ-points which captures the first two moments of $\mathbf{s}$ correctly; that is, they must capture the mean, covariance and symmetry. Let s_i be the *ith* component of $\mathbf{s}$, then its covariance is given by

$$E\{s_i^2\} = 1 \quad \forall i \tag{6.24}$$

Also from the symmetry properties of the distribution, all *odd*-ordered moments are zero.

The minimum number of points whose distribution obeys these conditions has two types of σ-points: (1) a single point at the origin of the s-axis with weight W_o and (2) $2N_x$ symmetrically distributed points on the coordinate s-axis at a distance r from the origin all having the same weight W_1. Thus, there are $2N_x + 1$ σ-points for a two-dimensional distribution. The values of W_o, W_1, and r are selected to ensure that their covariance is the identity. Therefore, due to their symmetry, it is only necessary to specify one direction on the $\mathbf{s}$-axis, say s_1. The constraint function will consist of the moment for $E\{s_1^2\}$ and the normalization condition must be satisfied. Therefore, we have

$$\mathbf{g}[\sigma, \mathrm{p}_X(\mathbf{s})] = \begin{pmatrix} 2W_1 r^2 - 1 \\ W_o + 2N_x W_1 - 1 \end{pmatrix} = \mathbf{0} \tag{6.25}$$

The solution to these equations is given by

$$r = \frac{1}{\sqrt{2W_1}} \quad \text{and} \quad W_o = 1 - 2N_x W_1, \qquad W_o\text{free} \tag{6.26}$$

By reparameterizing $W_1 := \frac{1}{2(N_x+\kappa)}$, it can be shown that after pre-multiplying by U, the matrix square root of R_{xx}, that the σ-points for $\mathbf{x}$ are

$$\begin{aligned}
\mathcal{X}_o &= \mu_x, & W_o &= \frac{\kappa}{(N_x+\kappa)} \\
\mathcal{X}_i &= \mu_x + (\sqrt{(N_x+\kappa)R_{xx}})_i, & W_i &= \frac{1}{2(N_x+\kappa)} \\
\mathcal{X}_{i+N_x} &= \mu_x - (\sqrt{(N_x+\kappa)R_{xx}})_i, & W_{i+N_x} &= \frac{1}{2(N_x+\kappa)}
\end{aligned}$$

where κ is a scalar, $(\sqrt{(N_x+\kappa)R_{xx}})_i$ is the *ith* row or column of the matrix square root of $(N_x+\kappa)R_{xx}$ and W_i is the weight associated with the *ith* σ-point. The parameter κ is free; however, it can be selected to minimize the mismatch between the fourth-order moments of the σ-points and the true distribution [16]. From the properties of the Gaussian, we have

$$E\{s_i^4\} = 3 \quad \forall i \tag{6.27}$$

The penalty function penalizes the discrepancy between the σ-points and the true value along one direction (s_1), in this case, due to the symmetry. Therefore, we have

$$\mathbf{p}[\sigma, \mathrm{p}_X(\mathbf{s})] = |2W_1 r^4 - 3| \quad \text{giving } W_1 = \frac{3}{2r^4} = \frac{1}{6} \quad \text{or} \quad \kappa = 3 - N_x \tag{6.28}$$

It is clear that the ability to minimize $\mathbf{p}$ depends on the number of degrees of freedom that are available after the constraint $\mathbf{g}$ is satisfied for the given set of σ-points; the kurtosis cannot be matched exactly without developing a larger set of σ-points (see Julier [16] for details).

We summarize the sigma-point processor under the *multivariate* Gaussian assumptions: Given an N_x-dimensional Gaussian distribution having covariance R_{xx}, we can generate a set of $O(N_\sigma)\sigma$-points having the same sample covariance from the columns (or rows) of the matrices $\pm\sqrt{(N_x+\kappa)R_{xx}}$. Here κ is the scaling factor discussed previously. This set is zero-mean, but if the original distribution has mean μ_x, then simply adding μ_x to each of the σ-points yields a symmetric set of $2N_x+1$ samples having the desired mean and covariance. Since the set is symmetric, its *odd* central moments are null; so its first three moments are identical to those of the original Gaussian distribution. This is the *minimal* number of σ-points capable of capturing the essential statistical information. The basic SPT technique for a multivariate Gaussian distribution [16] is therefore

1. Determine the set of $2N_x+1$ σ-points from the rows or columns of $\pm\sqrt{(N_x+\kappa)R_{xx}}$. For the nonzero-mean case, compute $\mathcal{X}_i = \sigma + \mu_x$;

$$\mathcal{X}_o = \mu_x, \qquad W_o = \frac{\kappa}{(N_x+\kappa)}$$

$$\mathcal{X}_i = \mu_x + \left(\sqrt{(N_x+\kappa)R_{xx}}\right)_i, \qquad W_i = \frac{1}{2(N_x+\kappa)}$$

$$\mathcal{X}_{i+N_x} = \mu_x - \left(\sqrt{(N_x+\kappa)R_{xx}}\right)_i, \qquad W_{i+N_x} = \frac{1}{2(N_x+\kappa)}$$

   where κ is a scalar, $\left(\sqrt{(N_x+\kappa)R_{xx}}\right)_i$ is the ith row or column of the matrix square root of $(N_x+\kappa)R_{xx}$, and W_i is the weight associated with the ith σ-point;
2. Nonlinearly *transform* each point to obtain the set of new σ-points: $\mathcal{Y}_i = \mathbf{a}[\mathcal{X}_i]$
3. Estimate the posterior *mean* of the new samples by its weighted average (regression)

$$\mu_{\mathbf{y}} = \sum_{i=0}^{2N_x} W_i \mathcal{Y}_i$$

4. Estimate the posterior *covariance* of the new samples by its weighted outer product (regression)

$$\mathbf{R}_{yy} = \sum_{i=0}^{2N_x} W_i(\mathcal{Y}_i - \mu_{\mathbf{y}})(\mathcal{Y}_i - \mu_{\mathbf{y}})'$$

There is a wealth of properties of this processor:

1. The transformed statistics of **y** are captured *precisely* up to the second order.
2. The σ-points capture the identical mean and covariance regardless of the choice of matrix-square-root method.
3. The posterior mean and covariance are calculated using standard linear algebraic methods (WSLR) and it is *not* necessary to evaluate any Jacobian as required by the XBP methods.
4. κ is a "tuning" parameter used to tune the higher order moments of the approximation that can be used to reduce overall prediction errors. For **x**, a multivariate Gaussian, $\kappa = 3 - N_x$ is a useful heuristic.
5. A modified form for $\kappa \rightarrow \lambda = \alpha^2(N_x + \kappa) - N_x$ (scaled transform) can be used to overcome a nonpositive definiteness of the covariance. Here α controls the *spread* of the σ-points around μ_x and is typically set to a value $0.01 \le \alpha \le 1$ with κ, a secondary scaling parameter, set to 0 or $3 - N_x$ and β is an extra degree of freedom to incorporate any extra prior knowledge of the $p_X(x)$ with $\beta = 2$ for Gaussian distributions. In this case, the weights change and are given by $W_0^{(m)} = \frac{\lambda}{N_x+\lambda}$, $W_0^{(c)} = \frac{\lambda}{N_x+\lambda} + (1 - \alpha^2 + \beta)$, and $W_i^{(m)} = \frac{1}{(N_x+\lambda)}$ [17, 21, 33].
6. Note that although statistical linearization offers a convenient way to interpret the subsequent *sigma-point approach*, it does not indicate some of its major advantages, especially since it is possible to extend the approach to incorporate more points capturing and accurately propagating higher order moments [33].

Next, we apply the σ-point approach to the nonlinear filtering problem by defining the terms in the SPT and showing where the statistical approximations are utilized.

6.3 SIGMA-POINT BAYESIAN PROCESSOR (UNSCENTED KALMAN FILTER)

The SPBP or UKF is a recursive processor developed to eliminate some of the deficiencies created by the failure of the first- order (Taylor series) linearization process in solving the state estimation problem. Different from the XBP (EKF), the sigma-point processor does not approximate the nonlinear process and measurement models; it employs the true nonlinear models and approximates the underlying *Gaussian* distribution function of the state variable using a statistical linearization approach leading to a set of regression equations for the states and measurements. In the sigma-point processor, the state is still represented as Gaussian, but it is specified using the minimal set of *deterministically* selected samples or σ-points. These points completely capture the true mean and covariance of the prior Gaussian distribution. When they are propagated through the nonlinear process, the *posterior* mean and covariance are accurately captured for *any* nonlinearity with errors introduced only in the third- and higher order moments. This is the statistical linearization using the weighted (statistical) linear regression approximation (WSLR) discussed previously [16].

We use the XBP (EKF) formulation and its underlying statistics as our prior distribution specified by the following nonlinear model with the conditional Gaussian distributions. Recall that the original discrete nonlinear process model is given by

$$x(t) = \mathbf{a}[x(t-1)] + \mathbf{b}[u(t-1)] + w(t-1) \tag{6.29}$$

with the corresponding measurement model

$$y(t) = \mathbf{c}[x(t)] + v(t) \tag{6.30}$$

for $w \sim \mathcal{N}(0, R_{ww})$ and $v \sim \mathcal{N}(0, R_{vv})$. It was demonstrated previously (Section 5.3) that the critical conditional Gaussian distribution for the *state variable* statistics was the prior

$$\Pr(x(t)|Y_{t-1}) = \mathcal{N}(\hat{x}(t|t-1), \tilde{P}(t|t-1))$$

with the measurement statistics specified by

$$\Pr(y(t)|Y_{t-1}) = \mathcal{N}(\hat{y}(t|t-1), R_{\xi\xi}(t|t-1))$$

where $\hat{x}(t|t-1)$, and $\tilde{P}(t|t-1)$ are the respective predicted state and error covariance based upon the data up to time $(t-1)$ and $\hat{y}(t|t-1)$, $R_{\xi\xi}(t|t-1)$ are the predicted measurement and residual covariance. The idea is to use the "prior" statistics and perform the SPT (under Gaussian assumptions) using *both* the process and measurement nonlinear transformations (models) as specified above yielding the corresponding set of σ-points in the new space. The predicted means are weighted sums of the transformed σ-points and the covariances are merely weighted sums of their mean-corrected, outer products, that is, they are specifically the WSLR discussed in Section 6.2.

To develop the sigma-point processor we must:

- PREDICT the next state and error covariance, $(\hat{x}(t|t-1), \tilde{P}(t|t-1))$, by SPT-transforming the prior, $\mathcal{N}(\hat{x}(t-1|t-1), \tilde{P}(t-1|t-1))$, including the process noise using the σ-points, $\mathcal{X}_i(t|t-1)$ and $\mathcal{X}_i(t-1|t-1)$, respectively;
- PREDICT the measurement and residual covariance, $[\hat{y}(t|t-1), R_{\xi\xi}(t|t-1)]$ by using the SPT-transformed σ-points $\mathcal{Y}_i(t|t-1)$, and performing the weighted regressions; and

- PREDICT the cross-covariance, $R_{\tilde{x}\xi}(t|t-1)$ in order to calculate the corresponding gain for the subsequent update step.

We use these steps as our road map to develop the sigma-point processor depicted in the block diagram of Fig. 6.3. We start by defining the set Σ and selecting the corresponding $2N_x + 1$ σ-points and weights according to the multivariate Gaussian distribution selection procedure developed in the previous section. That is, the points are defined by substituting the σ-points for the SPT transformation of the "prior" *state* information specified by $\mu_x = \hat{x}(t-1|t-1)$ and $R_{xx} = \tilde{P}(t-1|t-1)$. We define the set of σ-points and weights as

$$
\begin{aligned}
\mathcal{X}_o &= \mu_x = \hat{x}(t-1|t-1), & W_o &= \frac{\kappa}{(N_x+\kappa)} \\
\mathcal{X}_i &= \mu_x + \left(\sqrt{(N_x+\kappa)R_{xx}}\right)_i & & \\
&= \hat{x}(t-1|t-1) + \left(\sqrt{(N_x+\kappa)\tilde{P}(t-1|t-1)}\right)_i, & W_i &= \frac{1}{2(N_x+\kappa)} \\
\mathcal{X}_{i+N_x} &= \mu_x - \left(\sqrt{(N_x+\kappa)R_{xx}}\right)_i & & \\
&= \hat{x}(t-1|t-1) - \left(\sqrt{(N_x+\kappa)\tilde{P}(t-1|t-1)}\right)_i, & W_{i+N_x} &= \frac{1}{2(N_x+\kappa)}
\end{aligned}
$$

Next, we perform the state prediction step to obtain, $\{\mathcal{X}_i(t|t-1), \hat{x}(t|t-1)\}$, transforming each σ-point using the nonlinear process model to the new space, that is, to obtain the one-step *state prediction* as

$$\mathcal{X}_i(t|t-1) = \mathbf{a}[\mathcal{X}_i(t-1|t-1)] + \mathbf{b}[u(t-1)] \tag{6.31}$$

The WSLR approximation step to obtain the predicted mean follows from the statistical linearization transformation model using the following relations: $\mathbf{y} \to x(t)$, $\mathbf{x} \to x(t-1)$, $A \to A(t-1)$, $\mathbf{b} \to \mathbf{b}(t-1)$ and $\epsilon \to \epsilon(t-1)$ of Eq. 6.4

$$x(t) = \underbrace{A(t-1)x(t-1) + \mathbf{b}(t-1)}_{\text{Linear approximation}} + \underbrace{\epsilon(t-1)}_{\text{Linearization error}} \tag{6.32}$$

Conditioning Eq. 6.32 on Y_{t-1} and taking expectations, we have

$$
\begin{aligned}
\hat{x}(t|t-1) = E\{x(t)|Y_{t-1}\} &= E\{A(t-1)x(t-1)|Y_{t-1}\} + E\{\mathbf{b}(t-1)|Y_{t-1}\} \\
&\quad + E\{\epsilon(t-1)|Y_{t-1}\}
\end{aligned}
$$

or

$$\hat{x}(t|t-1) = A(t-1)\hat{x}(t-1|t-1) + \mathbf{b}(t-1) \tag{6.33}$$

Using this linearized form, with the conditional means replacing the unconditional, that is, $\mu_y \rightarrow \hat{x}(t|t-1)$ and $\mu_x \rightarrow \hat{x}(t-1|t-1)$, we have from Eq. 6.8 that

$$\begin{aligned}\mathbf{b}(t-1) &\rightarrow \hat{x}(t|t-1) - A(t-1)\hat{x}(t-1|t-1)\\ &= \sum_{i=0}^{2N_x} W_i \mathcal{X}_i(t|t-1) - A(t-1)\hat{x}(t-1|t-1)\end{aligned} \tag{6.34}$$

where we substituted the regression of Eq. 6.14 for μ_y. Substituting this expression for $\mathbf{b}$ above gives the required regression relation

$$\begin{aligned}\hat{x}(t|t-1) &= A(t-1)\hat{x}(t-1|t-1) + \sum_{i=0}^{2N_x} W_i \mathcal{X}_i(t|t-1) - A(t-1)\hat{x}(t-1|t-1)\\ &= \sum_{i=0}^{2N_x} W_i \mathcal{X}_i(t|t-1)\end{aligned} \tag{6.35}$$

yielding the *predicted state* estimate using the WSLR approach shown in Fig. 6.3.

Next let us define the *predicted state estimation error* as

$$\tilde{\mathcal{X}}_i(t|t-1) := \mathcal{X}_i(t|t-1) - \hat{x}(t|t-1) \tag{6.36}$$

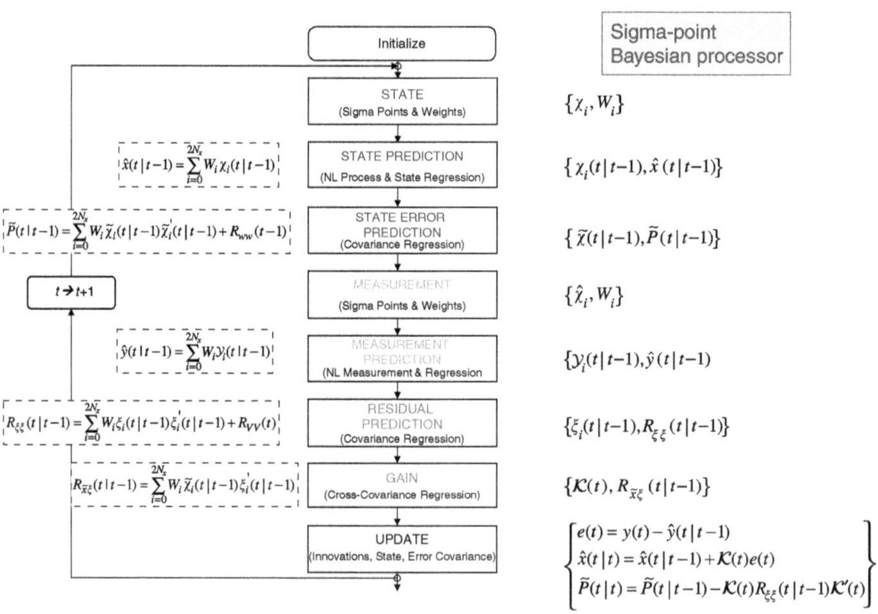

FIGURE 6.3 Sigma-point Bayesian processor block diagram with σ-points and regressions.

therefore, the corresponding *predicted state error covariance*

$$\tilde{P}(t|t-1) := \text{cov}[\tilde{\mathcal{X}}_i(t|t-1)] = \text{cov}[\mathcal{X}_i(t|t-1) - \hat{x}(t|t-1)]$$

can be calculated from

$$\begin{aligned}\tilde{P}(t|t-1) &= \text{cov}[(A(t-1)\mathcal{X}_i(t-1|t-1) + \mathbf{b}(t-1) + \epsilon_i(t-1)) \\ &\quad - (A(t-1)\hat{x}(t-1|t-1) + \mathbf{b}(t-1))] \\ &= \text{cov}[A(t-1)\tilde{\mathcal{X}}_i(t-1|t-1) + \epsilon_i(t-1)]\end{aligned}$$

Performing this calculation gives the expression

$$\tilde{P}(t|t-1) = A(t-1)\tilde{P}(t-1|t-1)A'(t-1) + R_{\epsilon\epsilon}(t-1) \tag{6.37}$$

which can also be written in the regression form by using the relations from Eq. 6.11 with $y \to \mathcal{X}_i$, $R_{yy} \to \tilde{P}_{\mathcal{X}\mathcal{X}}$. Therefore, substituting for $A\tilde{P}A'$ above, we have

$$\begin{aligned}\tilde{P}(t|t-1) &= \tilde{P}_{\mathcal{X}\mathcal{X}}(t|t-1) - R_{\epsilon\epsilon}(t-1) + R_{\epsilon\epsilon}(t-1) \\ &= \sum_{i=0}^{2N_x} W_i\tilde{\mathcal{X}}_i(t|t-1)\tilde{\mathcal{X}}_i'(t|t-1)\end{aligned} \tag{6.38}$$

In the case of additive, zero-mean, white Gaussian noise with covariance, $R_{ww}(t-1)$, we have the final WSLR

$$\tilde{P}(t|t-1) = \sum_{i=0}^{2N_x} W_i\tilde{\mathcal{X}}_i(t|t-1)\tilde{\mathcal{X}}_i'(t|t-1) + R_{ww}(t-1) \tag{6.39}$$

completing the *state error prediction* step of Fig. 6.3.

Here, the Bayesian processor approximates the predicted density, $\mathcal{N}(\hat{x}(t|t-1), \tilde{P}(t|t-1))$, where

$$\text{p}_X(x(t)|Y_{t-1}) = \int \text{p}_X(x(t)|x(t-1))\text{p}_X(x(t-1)|Y_{t-1})dx(t-1) \tag{6.40}$$

Next, we calculate a new set of σ-points to reflect the prediction step and perform the SPT in the *measurement* space as

$$\hat{\mathcal{X}}_i(t|t-1) = \{\mathcal{X}_i(t|t-1), \mathcal{X}_i(t|t-1) + \kappa\sqrt{R_{ww}(t-1)}, \mathcal{X}_i(t|t-1) - \kappa\sqrt{R_{ww}(t-1)}\} \tag{6.41}$$

We linearize the measurement function similar to the nonlinear process function in the previous steps. We use the nonlinear measurement model to propagate the "new"

σ-points to the transformed space producing the *measurement prediction* step shown in Fig. 6.3.

$$\mathcal{Y}_i(t|t-1) = \mathbf{c}[\hat{\mathcal{X}}_i(t|t-1)] \tag{6.42}$$

The WSLR is then performed to obtain the *predicted measurement*, that is,

$$y(t) = C(t)x(t) + \mathbf{b}(t) + \epsilon(t) \tag{6.43}$$

Conditioning on Y_{t-1}, as before, and taking expectations of Eq. 6.43 gives

$$\hat{y}(t|t-1) = E\{y(t)|Y_{t-1}\} = C(t)\hat{x}(t|t-1) + \mathbf{b}(t) \tag{6.44}$$

Using this linearized form with the conditional means replacing the unconditional, that is, $\mu_y \rightarrow \hat{y}(t|t-1)$, $A \rightarrow C$, $\mathbf{b} \rightarrow \mathbf{b}$ and $\mu_x \rightarrow \hat{x}(t|t-1)$, we have from Eq. 6.8 that

$$\mathbf{b}(t) \rightarrow \hat{y}(t|t-1) - C(t)\hat{x}(t|t-1) = \sum_{i=0}^{2N_x} W_i \mathcal{Y}_i(t|t-1) - C(t)\hat{x}(t|t-1) \tag{6.45}$$

where we substituted the regression of Eq. 6.14 for μ_y. Substituting this expression for $\mathbf{b}$ above gives the WSLR relation

$$\begin{aligned}\hat{y}(t|t-1) &= C(t)\hat{x}(t|t-1) + \sum_{i=0}^{2N_x} W_i \mathcal{Y}_i(t|t-1) - C(t)\hat{x}(t|t-1) \\ &= \sum_{i=0}^{2N_x} W_i \mathcal{Y}_i(t|t-1)\end{aligned} \tag{6.46}$$

yielding the *predicted measurement* estimate of Fig. 6.3.

Similarly, the *residual* (measurement) error is defined by

$$\xi_i(t|t-1) := \mathcal{Y}_i(t|t-1) - \hat{y}(t|t-1) \tag{6.47}$$

and therefore the *residual (predicted) covariance* can be expressed as

$$R_{\xi\xi}(t|t-1) = \text{cov}[\xi_i(t|t-1)] = \text{cov}[\mathcal{Y}_i(t|t-1) - \hat{y}(t|t-1)]$$

Substituting the linearized model above, we obtain

$$\begin{aligned}R_{\xi\xi}(t|t-1) &= \text{cov}[(C(t)\mathcal{X}_i(t|t-1) + \mathbf{b}(t) + \epsilon_i(t)) \\ &\quad - (C(t)\hat{x}(t|t-1) + \mathbf{b}(t))] \\ &= \text{cov}[C(t-1)\tilde{\mathcal{X}}_i(t|t-1) + \epsilon_i(t)]\end{aligned}$$

Performing this calculation gives the expression

$$R_{\xi\xi}(t|t-1) = C(t)\tilde{P}(t|t-1)C'(t) + R_{\epsilon\epsilon}(t) \tag{6.48}$$

which can also be written in the regression form by using the relations from Eq. 6.11 with $y \to \xi_i$, $R_{yy} \to \tilde{P}_{\xi\xi}$. Therefore, substituting for $C\tilde{P}C'$ above, we have

$$R_{\xi\xi}(t|t-1) = \tilde{P}_{\xi\xi}(t|t-1) - R_{\epsilon\epsilon}(t) + R_{\epsilon\epsilon}(t) = \sum_{i=0}^{2N_x} W_i \xi_i(t|t-1)\xi_i'(t|t-1) \tag{6.49}$$

In the case of additive, zero-mean, white Gaussian noise with covariance, $R_{vv}(t)$ we have the final WSLR

$$R_{\xi\xi}(t|t-1) = \sum_{i=0}^{2N_x} W_i \xi_i(t|t-1)\xi_i'(t|t-1) + R_{vv}(t) \tag{6.50}$$

completing the measurement *residual prediction* step of Fig. 6.3.

The *gain* is estimated from

$$\mathcal{K}(t) = R_{\tilde{x}\xi}(t|t-1)R_{\xi\xi}^{-1}(t|t-1) \tag{6.51}$$

where the *cross error covariance* is approximated using the WSLR as (above)

$$R_{\tilde{x}\xi}(t|t-1) = \sum_{i=0}^{2N_x} W_i \tilde{\mathcal{X}}_i(t|t-1)\xi_i'(t|t-1) \tag{6.52}$$

With this in mind, it is clear that the *posterior distribution* can be calculated from

$$\begin{aligned} \mathrm{p}_X(x(t)|Y_t) \sim \mathcal{N}(\hat{x}(t|t), \tilde{P}(t|t)) = \int & \mathrm{p}_X(x(t)|x(t-1)) \\ & \times \mathrm{p}_X(x(t-1)|Y_t)dx(t-1) \end{aligned} \tag{6.53}$$

where we have the "usual" (Kalman) *update* relations of Fig. 6.3 starting with the *innovations*

$$e(t) = y(t) - \hat{y}(t|t-1) \tag{6.54}$$

and the *state update*

$$\hat{x}(t|t) = \hat{x}(t|t-1) + \mathcal{K}(t)e(t) \tag{6.55}$$

along with the corresponding *state error covariance update* given by

$$\tilde{P}(t|t) = \tilde{P}(t|t-1) - \mathcal{K}(t)R_{\xi\xi}(t|t-1)\mathcal{K}'(t) \tag{6.56}$$

This completes the SPBP algorithm which is summarized in Table 6.1. We have shown how the SPT coupled with the WSLR can be used to develop this technique in its "plain vanilla" form. We note in passing that there are *no* Jacobians calculated and the nonlinear models are employed directly to transform the σ-points to the new space. Also in the original problem definition (see Section 2) both process and noise sources (w, v) were assumed *additive*. The SPT enables us to "generalize" the noise terms to also be injected in a nonlinear manner (e.g., multiplication). Thus, the noise is not treated separately, but can be embedded into the problem by defining an augmented state vector, say $\overline{x}(t) = [x(t)\ \ w(t-1)\ \ v(t)]'$. We chose to ignore the general case to keep the development of the sigma-point processor simple. For more details of the general process, see Refs. [17], [21], and [33]. Let us apply the sigma-point processor to the nonlinear trajectory estimation problem and compare its performance to the other nonlinear processors discussed previously.

Example 6.3

We revisit the nonlinear trajectory estimation problem of the previous examples with the dynamics specified by the discrete nonlinear process given by

$$x(t) = (1 - 0.05\Delta T)x(t-1) + 0.04\Delta T x^2(t-1) + w(t-1)$$

and corresponding measurement model

$$y(t) = x^2(t) + x^3(t) + v(t)$$

Recall that $v(t) \sim \mathcal{N}(0, 0.09)$, $x(0) = 2.0$, $P(0) = 0.01$, $\Delta T = 0.01$ sec and $R_{ww} = 0$. The simulated measurement is shown in Fig. 6.4b. The sigma-point processor (SPBP) and XBP (EKF) and IX-BP (IEKF) (three iterations) were applied to this problem. We used the square-root implementations of the XBP and IX-BP in SSPACK_PC [10], and compared them to the sigma-point processor in REBEL [33]. The results are shown in Fig. 6.4 where we see the corresponding trajectory (state) estimates in (*a*) and the "filtered" measurements in (b). From the figures, it appears that all of the estimates are quite reasonable with the sigma-point processor estimate (thick solid line) converging most rapidly to the true trajectory (dashed line). The XBP (thick dotted line) appears slightly biased while the IX-BP (thin dotted line) converges rapidly, but then wanders slightly from the truth. The measurements also indicate a similar performance. The zero-mean/whiteness tests confirm these observations. The XBP and IX-BP perform similarly with respective zero-mean/whiteness values of $(1.04 \times 10^{-1} < 1.73 \times 10^{-1}/1\%$ out) and $(3.85 \times 10^{-2} < 1.73 \times 10^{-1}/0\%$ out), while the sigma-point processor is certainly comparable at $(5.63 \times 10^{-2} < 1.73 \times 10^{-1}/0\%$ out). This completes the example. △△△

TABLE 6.1 Discrete Sigma-Point Bayesian Processor (Unscented Kalman Filter) Algorithm

State: Sigma Points and Weights

$$\mathcal{X}_o = \hat{x}(t-1|t-1), \qquad W_o = \frac{\kappa}{(N_x+\kappa)}$$

$$\mathcal{X}_i = \hat{x}(t-1|t-1) + \left(\sqrt{(N_x+\kappa)\tilde{P}(t-1|t-1)}\right)_i, \qquad W_i = \frac{1}{2(N_x+\kappa)}$$

$$\mathcal{X}_{i+N_x} = \hat{x}(t-1|t-1) - \left(\sqrt{(N_x+\kappa)\tilde{P}(t-1|t-1)}\right)_i, \qquad W_{i+N_x} = \frac{1}{2(N_x+\kappa)}$$

State Prediction

$$\mathcal{X}_i(t|t-1) = \mathbf{a}[\mathcal{X}_i(t-1|t-1)] + \mathbf{b}[u(t-1)] \qquad \text{[Nonlinear state process]}$$

$$\hat{x}(t|t-1) = \sum_{i=0}^{2N_x} W_i \mathcal{X}_i(t|t-1) \qquad \text{[State regression]}$$

State Error Prediction

$$\tilde{\mathcal{X}}_i(t|t-1) = \mathcal{X}_i(t|t-1) - \hat{x}(t|t-1) \qquad \text{[State error]}$$

$$\tilde{P}(t|t-1) = \sum_{i=0}^{2N_x} W_i \tilde{\mathcal{X}}_i(t|t-1)\tilde{\mathcal{X}}_i'(t|t-1) + R_{ww}(t-1) \qquad \text{[Error covariance prediction]}$$

Measurement: Sigma Points and Weights

$$\hat{\mathcal{X}}_i(t|t-1) = \{\mathcal{X}_i(t|t-1), \mathcal{X}_i(t|t-1) + \kappa\sqrt{R_{ww}(t-1)}, \mathcal{X}_i(t|t-1) - \kappa\sqrt{R_{ww}(t-1)}\}$$

Measurement Prediction

$$\mathcal{Y}_i(t|t-1) = \mathbf{c}[\hat{\mathcal{X}}_i(t|t-1)] \qquad \text{[Nonlinear measurement]}$$

$$\hat{y}(t|t-1) = \sum_{i=0}^{2N_x} W_i \mathcal{Y}_i(t|t-1) \qquad \text{[Measurement regression]}$$

Residual Prediction

$$\xi_i(t|t-1) = \mathcal{Y}_i(t|t-1) - \hat{y}(t|t-1) \qquad \text{[Predicted residual]}$$

$$R_{\xi\xi}(t|t-1) = \sum_{i=0}^{2N_x} W_i \xi_i(t|t-1)\xi_i'(t|t-1) + R_{vv}(t) \qquad \text{[Residual covariance regression]}$$

Gain

$$R_{\tilde{x}\xi}(t|t-1) = \sum_{i=0}^{2N_x} W_i \tilde{\mathcal{X}}_i(t|t-1)\xi_i'(t|t-1) \qquad \text{[Cross-covariance regression]}$$

$$\mathcal{K}(t) = R_{\tilde{x}\xi}(t|t-1)R_{\xi\xi}^{-1}(t|t-1) \qquad \text{[Gain]}$$

State Update

$$e(t) = y(t) - \hat{y}(t|t-1) \qquad \text{[Innovation]}$$

$$\hat{x}(t|t) = \hat{x}(t|t-1) + \mathcal{K}(t)e(t) \qquad \text{[State update]}$$

$$\tilde{P}(t|t) = \tilde{P}(t|t-1) - \mathcal{K}(t)R_{\xi\xi}(t|t-1)\mathcal{K}'(t) \qquad \text{[Error covariance update]}$$

Initial Conditions

$$\hat{x}(0|0) \quad \tilde{P}(0|0)$$

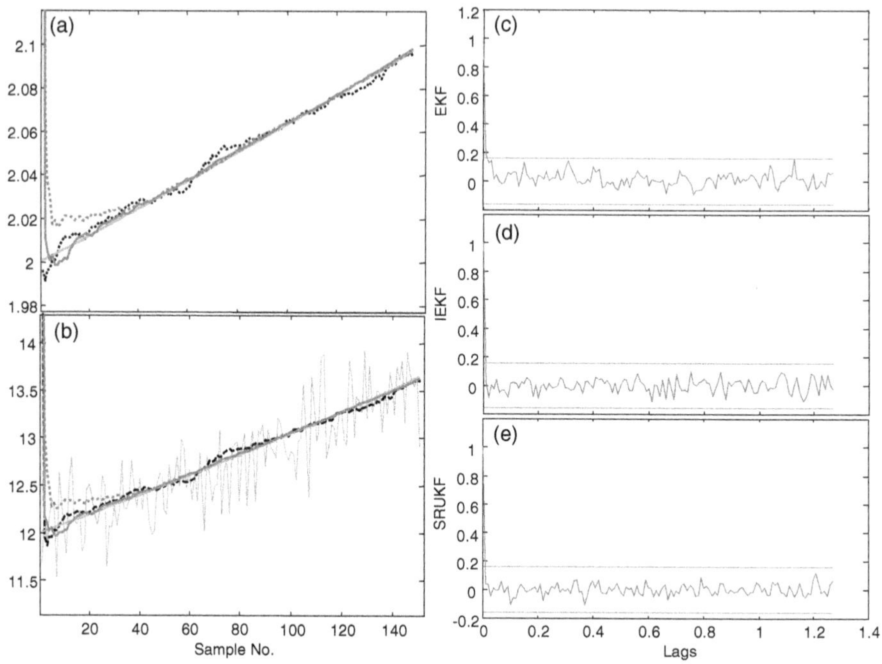

FIGURE 6.4 Nonlinear trajectory estimation: (a) trajectory (state) estimates using the XBP (thick dotted), IX-BP (thin dotted) and SPBP (thick solid); (b) filtered measurement estimates using the XBP (thick dotted), IX-BP (thin dotted), and SPBP (thick solid); (c) zero-mean/whiteness tests for XBP ($1.04\times10^{-1} < 1.73\times10^{-1}$/1% out); (d) zero-mean/whiteness tests for IX-BP ($3.85\times10^{-2} < 1.73\times10^{-1}$/0% out); (e) zero-mean/whiteness tests for SPBP ($5.63\times10^{-2} < 1.73\times10^{-1}$/0% out).

6.3.1 Extensions of the Sigma-Point Processor

Recently there have been a number of developments in the nonlinear estimation area that are based on the sigma-point (or similar) transformation [34, 35, 36]. Next, we briefly mention these approaches keeping in mind that they are members of the "sigma-point family".

The *central difference Bayesian processor (CDBP)* or unscented Kalman filter (UKF) is based on the Stirling approximation interpolation formula that is essentially a second-order Taylor series expansion of the nonlinear random function about its mean. The central differences are used to approximate the first- and second- order terms of the series. In this sense, the processor implicitly employs the WSLR used to derive the SPBP as before. The resulting σ-points are dependent on the half-step size $\triangle_x$ rather than the other parameters discussed previously. The CDBP is slightly more accurate than the SPBP, and also has the advantage of only requiring a single parameter $\triangle_x$ to adjust the spread of the σ-points. For more details, see Ref. [34].

Just as with the Kalman filter implementations [10], the SPBP also admits the numerically stable "square-root" forms for prediction and updated state covariance matrices. These methods are based on employing the QR decomposition and Cholesky updating. Again, this approach offers a slightly more accurate processor as well as reduced computational costs while maintaining their numerical stability. See Ref. [34] for further details.

6.4 QUADRATURE BAYESIAN PROCESSORS

The grid-based quadrature Bayesian processor (QBP) or equivalently quadrature Kalman filter (QKF) is another alternative σ-point approach [35–37]. It uses the Gauss–Hermite numerical integration rule as its basic building block to precisely calculate the sequential Bayesian recursions of Chapter 2 under the Gaussian assumptions. For scalars, the *Gauss–Hermite rule* is given by

$$\frac{1}{\sqrt{2\pi}} \int_{-\infty}^{\infty} f(x) e^{-x^2} dx = \sum_{i=1}^{M} W_i \times f(\Delta x_i) \tag{6.57}$$

where equality holds for all polynomials $f(\cdot)$ of degree up to $2M-1$ and the quadrature (grid) points Δx_i with corresponding weights W_i are determined according to the rule. The QBP uses the WSLR to linearize the nonlinear transformations through a set of Gauss–Hermite quadrature points *rather* than introducing the set of σ-points. Details of this approach can be found in Ref. [36]. We briefly show the approach following the usual recursive forms of the BP of the previous chapter. For our nonlinear estimation problem, we have both nonlinear process ($\mathrm{a}[\cdot]$) and measurement ($\mathrm{c}[\cdot]$) equations that can be expressed in terms of the Gauss–Hermite rule above.

Suppose we have the conditional mean and covariance at time $(t-1)$ based on all of the data up to the same time step. Then the corresponding conditional distribution is $\Pr(x(t-1)|Y_{t-1}) \sim \mathcal{N}(\hat{x}(t-1|t-1), \tilde{P}(t-1|t-1))$. The QBP is then given by the following set of recursive Bayesian equations:

$$x_i = \sqrt{\tilde{P}(t-1|t-1)}' \, \Delta x_i + \hat{x}(t-1|t-1) \tag{6.58}$$

where $\sqrt{\tilde{P}}$ is the matrix square root (Cholesky factor: $\tilde{P} = \sqrt{\tilde{P}}' \sqrt{\tilde{P}}$). The *prediction step*: $\Pr(x(t)|Y_{t-1}) \sim \mathcal{N}(\hat{x}(t|t-1), \tilde{P}(t|t-1))$ is given by

$$\hat{x}(t|t-1) = \sum_{i=1}^{M} W_i \mathrm{a}[x_i]$$

$$\tilde{P}(t|t-1) = \sum_{i=1}^{M} W_i (\mathrm{a}[x_i] - \hat{x}(t|t-1))(\mathrm{a}[x_i] - \hat{x}(t|t-1))' + R_{ww}(t-1) \tag{6.59}$$

The *update step*: $\Pr(x(t)|Y_t) \sim \mathcal{N}(\hat{x}(t|t), \tilde{P}(t|t))$ follows a similar development

$$\overline{x}_i = \sqrt{\tilde{P}(t|t-1)}' \, \Delta x_i + \hat{x}(t|t-1) \tag{6.60}$$

and

$$\begin{aligned} \hat{x}(t|t) &= \hat{x}(t|t-1) + \mathcal{K}(t)(y(t) - \overline{y}(t)) \quad \text{for } \overline{y}(t) = \sum_{i=1}^{M} W_i \mathrm{c}[\overline{x}_i] \\ \tilde{P}(t|t) &= \tilde{P}(t|t-1) - \mathcal{K}(t)\mathcal{P}_{XY} \end{aligned} \tag{6.61}$$

where

$$\begin{aligned} \mathcal{K}(t) &= \mathcal{P}_{XY}(\mathcal{P}_{YY} + R_{vv}(t))^{-1} \\ \mathcal{P}_{XY} &= \sum_{i=1}^{M} W_i(\hat{x}(t|t-1) - \overline{x}_i)(y(t) - \mathrm{c}[\overline{x}_i])' \\ \mathcal{P}_{YY} &= \sum_{i=1}^{M} W_i(y(t) - \mathrm{c}[\overline{x}_i])(y(t) - \mathrm{c}[\overline{x}_i])' \end{aligned} \tag{6.62}$$

As with the SPBP, the QBP does *not* linearize the process or measurement models as in the case of the classical nonlinear processors. It calculates the weighted quadrature points in state-space over a fixed grid to estimate the unknown distribution (see Ref. [36–39] for more details).

Another powerful approach to nonlinear estimation is the Gaussian sum (mixture) processor [40], which we discuss in more detail next.

6.5 GAUSSIAN SUM (MIXTURE) BAYESIAN PROCESSORS

Another general approach to the Bayesian processing problem is the Gaussian sum (G-S) approximation leading to a processor. It has been shown [4, 40, 41] that *any* non-Gaussian distribution can be approximated by a sum of Gaussian distributions, that is,

$$\mathrm{p}_X(x) \approx \sum_{i=1}^{N_g} \mathcal{W}_i \mathrm{p}_i(x) = \sum_{i=1}^{N_g} \mathcal{W}_i \mathcal{N}(x : \mu_x(i), \Sigma_x(i)) \quad \text{for} \sum_{i=1}^{N_g} \mathcal{W}_i = 1 \text{ and } \mathcal{W}_i \geq 0 \forall i \tag{6.63}$$

a mixture of Gaussian distributions with $\{\mathcal{W}_i\}$ the set of mixing coefficients or weights. Clearly, this approach can be implemented with a bank of parallel classical processors (LZ-BP, XBP, IX-BP) or with any of the modern SPBP family just discussed, in order to estimate the mean and variance of the individual N_g-processors.

In addition, the particle filtering algorithms to be discussed in the next chapter can also be incorporated into a Gaussian mixture (G-M) framework—that is what makes this approach important [42]. Before we develop the processor, let us investigate some of the underlying properties of the Gaussian sum or equivalently Gaussian mixture that makes it intriguing.

The fundamental problem of approximating a probability distribution or density evolves from the idea of *delta families* of functions, that is, families of functions that converge to a delta or impulse function, as the parameter that uniquely characterizes that family converges to a limiting value. Properties of such families are discussed in Ref. [41]. The most *important* result regarding Gaussian sums is given in a theorem, which states that a probability density function formed by

$$\mathrm{p}_X(x) = \int_{-\infty}^{\infty} \mathrm{p}(\alpha)\delta(x-\alpha)\, d\alpha \tag{6.64}$$

converges uniformly to $\mathrm{p}_X(x)$ [41]. The Gaussian density forms a delta family with parameter σ represented by

$$\delta_\sigma(x) = \mathcal{N}(x:0,\sigma^2) = \frac{1}{\sqrt{2\pi\sigma^2}}\exp\left\{-\frac{x^2}{2\sigma^2}\right\} \tag{6.65}$$

which satisfies the requirement of a positive delta family as $\sigma \to 0$, that is, as the variance parameter approaches zero in the limit, the Gaussian density approaches a delta function; therefore, we have

$$\mathrm{p}_X(x) = \int_{-\infty}^{\infty} \mathrm{p}(\alpha)\mathcal{N}_\sigma(x-\alpha)\, d\alpha \tag{6.66}$$

This is the *key result* that forms the underlying basis of Gaussian sum or mixture approximations, similar to the idea of an empirical probability distribution

$$\hat{\Pr}(x) \approx \sum_{i=1}^{N_x} \mathcal{W}_i\delta(x-x_i) \;\Leftrightarrow\; \hat{\Pr}(x) \approx \sum_{i=1}^{N_g} \mathcal{W}_i\mathcal{N}(x_i:\mu_x(i),\Sigma_x(i)) \tag{6.67}$$

as the variance $\sigma_x \to 0$. The Gaussian converges to an impulse located at its mean μ_x. The Gaussian mixture distributions have some interesting properties that we list below (see Ref. [40, 41] for more details)

- Mean: $\mu_g = \sum_{i=1}^{N_g} \mathcal{W}_i\mu_x(i)$
- Covariance: $\Sigma_g = \sum_{i=1}^{N_g} \mathcal{W}_i(\Sigma_x(i) + \mu_x(i))$
- Skewness: $\gamma_g = \sum_{i=1}^{N_g} \mathcal{W}_i(\mu_x(i)-\mu_g)(3\Sigma_x(i) + (\mu_x(i)-\mu_g)^2 \times \Sigma_g^{\frac{-3}{2}})$
- Kurtosis: $\kappa_g = \sum_{i=1}^{N_g} \mathcal{W}_i(3\Sigma_x^2(i) + 6(\mu_x(i)-\mu_g)^2\Sigma_x(i) + \mu_g^4)\Sigma_g^{-2} - 3$

Next, let us see how this property can be utilized to develop a Bayesian processor. The processor we seek evolves from the sequential Bayesian paradigm of Section 2.5 given by

$$\Pr(x(t)|Y_t) = \frac{\Pr(y(t)|x(t)) \times \Pr(x(t)|Y_{t-1})}{\Pr(y(t)|Y_{t-1})} \tag{6.68}$$

with the corresponding prediction step

$$\Pr(x(t)|Y_{t-1}) = \int \Pr(x(t)|x(t-1)) \times \Pr(x(t-1)|Y_{t-1})\, dx(t-1) \tag{6.69}$$

Assume at time t that we approximate the predictive distribution by the Gaussian mixture

$$\Pr(x(t)|Y_{t-1}) \approx \sum_{i=1}^{N_g} \mathcal{W}_i(t-1)\mathcal{N}(x(t) : \overline{x}_i(t), \overline{P}_i(t)) \tag{6.70}$$

Then substituting this expression into the filtering posterior of Eq. 6.68, we obtain

$$\Pr(x(t)|Y_t) = \frac{\Pr(y(t)|x(t))}{\Pr(y(t)|Y_{t-1})} \times \sum_{i=1}^{N_g} \mathcal{W}_i(t-1)\mathcal{N}(x(t) : \overline{x}_i(t), \overline{P}_i(t)) \tag{6.71}$$

For clarity in this development we constrain both process and measurement models to additive Gaussian noise sources[2] resulting in the nonlinear state-space approximate Gauss-Markov representation of Section 4.8, that is, the process model

$$x(t) = \mathbf{a}[x(t-1)] + w(t-1) \quad w \sim \mathcal{N}(0, R_{ww}(t)) \tag{6.72}$$

and the measurement model

$$y(t) = \mathbf{c}[x(t)] + v(t) \quad v \sim \mathcal{N}(0, R_{vv}(t)) \tag{6.73}$$

with *Gaussian prior*, $x(0) \sim \mathcal{N}(x(0), P(0))$. Applying the fundamental convergence theorem of Eq. 6.64, the Gaussian mixture distribution of the posterior can be approximated by

$$\Pr(x(t)|Y_t) \approx \sum_{i=1}^{N_g} \mathcal{W}_i(t)\mathcal{N}(x(t) : x_i(t), P_i(t)) \tag{6.74}$$

[2] This is not necessary but enables the development and resulting processor relations to be much simpler. In the general case both noise sources can be modeled by Gaussian mixtures as well (see Ref. [41] for details).

and converges uniformly in $x(t)$ and $y(t)$ as $P_i(t) \to 0$ for $i = 1, \ldots, N_g$. Therefore, $\{x_i(t), P_i(t)\}$ can be estimated from any of the classical (nonlinear) processors developed in Chapter 5 or the modern sigma-point processor of this chapter. We choose to use the SPBP technique[3] of Table 6.1 to provide a "bank" of N_g-parallel processors required to estimate each member of the Gaussian mixture such that

$$
\begin{aligned}
\overline{y}_i(t) &= f(\overline{x}_i(t)) \\
e_i(t) &= y(t) - \overline{y}_i(t) \\
R_{e_ie_i}(t) &= g(\overline{P}_i(t), R_{vv}(t)) \\
x_i(t) &= \overline{x}_i(t) + \mathcal{K}_i(t)e_i(t) \\
P_i(t) &= \overline{P}_i(t) - \mathcal{K}_i(t)R_{e_ie_i}(t)\mathcal{K}_i'(t)
\end{aligned}
$$

where $f(\cdot)$, $g(\cdot)$ are functions that are derived from the SPBP of Table 6.1, and the weights[4] (mixing coefficients) of the individual mixture Gaussians are updated,

$$
\mathcal{W}_i(t) = \frac{\mathcal{W}_i(t-1) \times \mathcal{N}(y(t) : \overline{y}_i(t), R_{e_ie_i}(t))}{\sum_{i=1}^{N_g} \mathcal{W}_i(t-1) \times \mathcal{N}(y(t) : \overline{y}_i(t), R_{e_ie_i}(t))} \tag{6.75}
$$

Once we have performed the parallel SPBP estimation using the Gaussian mixtures, we can estimate the statistic of interest from the posterior: the conditional mean as

$$
\begin{aligned}
\hat{x}(t|t) = E\{x(t)|Y_t\} &= \int x(t)\hat{\Pr}(x(t)|Y_t)\,dx(t) \\
&\approx \sum_{i=1}^{N_g} \mathcal{W}_i(t) \int x(t)\mathcal{N}(x(t):x_i(t), P_i(t))\,dx(t)
\end{aligned}
$$

Now using the sifting property of the implied delta function, the Gaussian sum converges uniformly ($P_i \to 0, \mathcal{N}(\cdot) \to \delta(\cdot)$) to give

$$
\hat{x}(t|t) = \sum_{i=1}^{N_g} \mathcal{W}_i(t)x_i(t) \tag{6.76}
$$

Defining the estimation error as

$$
\tilde{x}(t|t) := x(t) - \hat{x}(t|t)
$$

[3] We symbolically use the SPBP algorithm and ignore the details (σ-points, etc.) of the implementation in this presentation to avoid the unnecessary complexity.

[4] The detailed derivation of these expressions can be found in Alspach [40] and Anderson [4]. Both references use the uniform convergence theorem to develop the posterior representation given above.

and the corresponding error covariance as $\tilde{P}(t|t) := \text{cov}(\tilde{x}(t|t))$ as before in Chapter 5, the approximated state error covariance is given by

$$\tilde{P}(t|t) = \sum_{i=1}^{N_g} \mathcal{W}_i(t)[\overline{P}_i(t) + \text{cov}(x_i(t) - \hat{x}(t|t))] \tag{6.77}$$

Thus, updating consists of a bank of N_g-parallel SPBP to estimate the means and covariance $\{x_i(t), P_i(t)\}$ required for the conditional statistics of Eqs. 6.76 and 6.77. This set of relations constitute the *update step* of the Gaussian sum processor. Next, let us briefly develop the prediction step.

With the availability of the posterior $\hat{\Pr}(x(t)|Y_t)$ and the process model of Eq. 6.72, the one-step prediction distribution can also be estimated as a Gaussian mixture. Using the filtering posterior and SPBP relations of Table 6.1, we have

$$\Pr(x(t+1)|Y_t) \approx \sum_{i=1}^{N_g} \mathcal{W}_i(t) \int x(t)\mathcal{N}(x(t+1) : \overline{x}_i(t+1), \overline{P}_i(t+1)) \tag{6.78}$$

and using the N_g-SPBP, we have

$$\begin{aligned}\overline{x}_i(t+1) &= \mathbf{a}[x_i(t)]\\ \overline{P}_i(t+1) &= h(P_i(t)) + R_{ww}(t)\end{aligned}$$

for $h(\cdot)$ a function of the SPBP parameters in Table 6.1. These relations lead to the one-step prediction conditional mean and covariance (as before)

$$\begin{aligned}\hat{x}(t+1|t) &= \sum_{i=1}^{N_g} \mathcal{W}_i(t)x_i(t+1)\\ \tilde{P}(t+1|t) &= \sum_{i=1}^{N_g} \mathcal{W}_i(t)[\overline{P}_i(t+1 + \text{cov}(x_i(t+1) - \hat{x}(t+1|t))]\end{aligned} \tag{6.79}$$

to complete the algorithm.

Next, we consider the application of nonlinear processors to a tracking problem.

6.6 CASE STUDY: 2D-TRACKING PROBLEM

In this section, we investigate the design of nonlinear BP to solve a two-dimensional (2D) tracking problem. The hypothetical scenario discussed will demonstrate the applicability of these processors to solve such a problem and demonstrate the "basic" thinking behind constructing such a problem and solution. In contrast to the "bearings-only" problem discussed in Chapter 5, let us investigate the tracking of a large

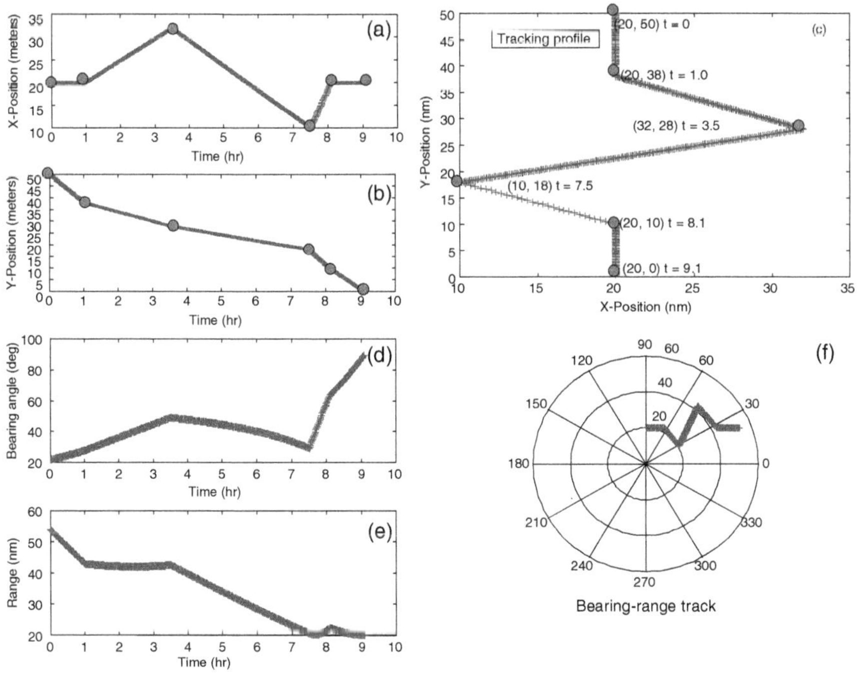

FIGURE 6.5 Tanker ground track geometry for the harbor docking application: (a) instantaneous *X*-position (nm); (b) instantaneous *Y*-position (nm); (c) filed *XY*-path (nm); (d) instantaneous bearing (deg); (e) instantaneous range (nm); (f) polar bearing-range track from sensor measurement.

tanker entering a busy harbor with a prescribed navigation path. In this case, the pilot on the vessel must adhere strictly to the path that has been filed with the harbor master (controller). Here we assume that the ship has a transponder frequently signaling accurate information about its current position. The objective is to safely dock the tanker without any incidents. We observe that the ship's path should track the prescribed trajectory (Cartesian coordinates) shown in Fig. 6.5, which corresponds to the instantaneous XY-positions (versus time) shown.

Our fictitious measurement instrument (e.g., low ground clutter phased array radar or a satellite communications receiver) is assumed to instantly report on the tanker position in bearing and range with high accuracy. The measurements are given by

$$\Theta(t) = \arctan\left(\frac{Y(t)}{X(t)}\right) \quad \text{and} \quad R(t) = \sqrt{X^2(t) + Y^2(t)}$$

We use the usual state-space formulation for a constant velocity model (see Section 5.5) with state vector defined in terms of the physical variables as $x(t) := [X(t)\ Y(t)\ V_x(t)\ V_y(t)]$ along with the incremental velocity input as $u' := [-\Delta V_{x_o}\ -\Delta V_{y_o}]$.

Using this information (as before), the entire system can be represented as a Gauss–Markov model with the noise sources representing uncertainties in the states and measurements. Thus we have the equations of motion

$$x(t) = \begin{bmatrix} 1 & 0 & \Delta T & 0 \\ 0 & 1 & 0 & \Delta T \\ 0 & 0 & 1 & 0 \\ 0 & 0 & 0 & 1 \end{bmatrix} x(t-1) + \begin{bmatrix} 0 & 0 \\ 0 & 0 \\ 1 & 0 \\ 0 & 1 \end{bmatrix} \begin{bmatrix} -\Delta V_{x_o}(t-1) \\ -\Delta V_{y_o}(t-1) \end{bmatrix} + w(t-1) \tag{6.80}$$

with the corresponding measurement model given by

$$y(t) = \begin{bmatrix} \arctan \frac{x_2(t)}{x_1(t)} \\ \sqrt{x_1^2(t) + x_2^2(t)} \end{bmatrix} + v(t)$$

for $w \sim \mathcal{N}(0, R_{ww})$ and $v \sim \mathcal{N}(0, R_{vv})$.

The SSPACK_PC software [43] was used to simulate this Gauss–Markov system for the tanker path and the results are shown in Fig. 6.6. In this scenario, we assume the instrument is capable of making measurements every $\Delta T = 0.02\,hr$ with a bearing precision of $\pm 0.02 degree$ and range precision of $\pm 0.005\,nm$ (or equivalently $R_{vv} = \text{diag}[4 \times 10^{-4},\quad 1 \times 10^{-4}]$. The model uncertainty was represented by $R_{ww} = \text{diag}(1 \times 10^{-6})$. An impulse-incremental step change in velocity, for example, V_y going from $-12\,k$ to $-4\,k$ is an incremental change of $+8\,k$ corresponding to $\Delta V_{y_o} = [8 \;\; 1.5 \;\; -10.83 \;\; 3.33]$ knots and $\Delta V_{x_o} = [0 \;\; 4.8 \;\; -10.3 \;\; 22.17]$ knots. These impulses (changes) occur at time fiducials of $t = [0 \;\; 1 \;\; 3.5 \;\; 7.5 \;\; 8.1 \;\; 9.1]hr$ corresponding to the filed harbor path depicted in the figure. Note that the velocity changes are impulses of height $(\Delta V_x, \Delta V_y)$ corresponding to a known deterministic input $u(t)$. These changes relate physically to instantaneous direction changes of the tanker and create the path change in the constant velocity model.

The simulated bearing measurements are generated using the initial conditions $x'(0) := [20\,\text{nm} \;\; 50\,\text{nm} \;\; 0\,\text{k} \;\; -12\,\text{k}]$ and $R_{ww} = \text{diag}[1 \times 10^{-6}, 1 \times 10^{-6}]$ with the corresponding initial covariance given by $\tilde{P}(0) = \text{diag}[1 \times 10^{-6}, 1 \times 10^{-6}]$. The Jacobian matrices derived from the Gauss–Markov model above are shown below

$$A[x] = A \quad \text{and} \quad C[x] = \begin{bmatrix} \frac{x_2(t)}{R^2(t)} & \frac{-x_1(t)}{R^2(t)} & 0 & 0 \\ \frac{x_1(t)}{R(t)} & \frac{x_2(t)}{R(t)} & 0 & 0 \end{bmatrix}$$

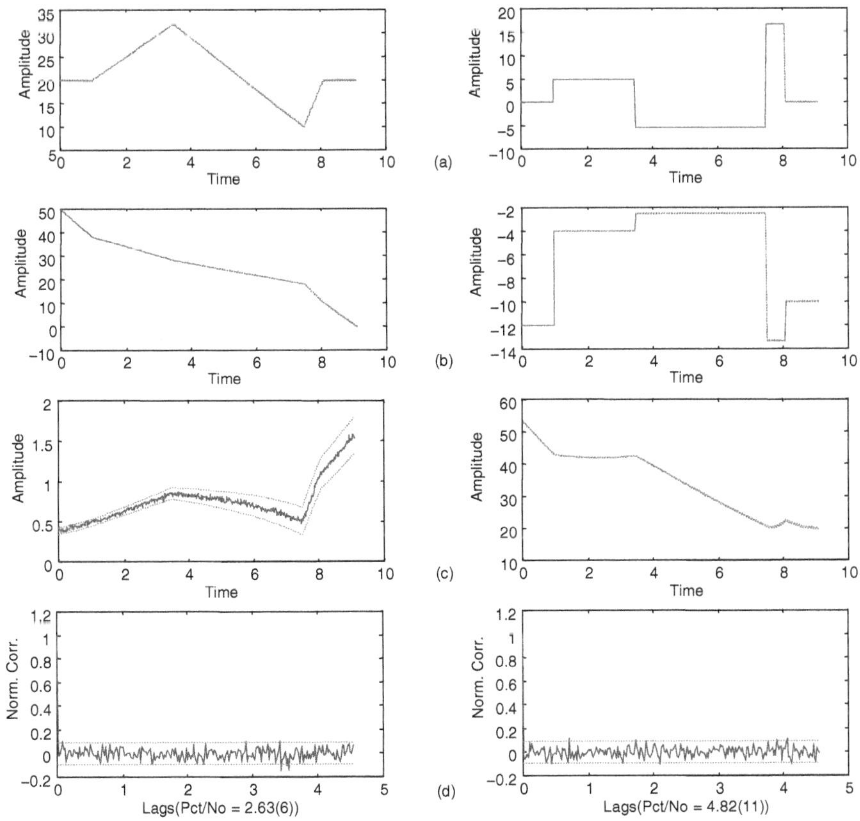

FIGURE 6.6 XBP (EKF) design for harbor docking problem (input known): (a) *X*-position and velocity estimates with bounds (0% out); (b) *Y*-position and velocity estimates with bounds (0% and 3% out); (c) bearing and range estimates with bounds (1% and 2% out); (d) innovations zero-mean/whiteness tests for bearing ($6\times10^{-4} < 26\times10^{-1}$ and 3% out) and range ($2\times10^{-4} < 42\times10^{-4}$ and 5% out).

The XBP, IX-BP, LZ-BP, and SPBP were executed under the constraint that all of the a priori information for the tanker harbor path is "known". Each of the processors performed almost identically. A representative realization output for the XBP is shown in Fig. 6.6. In (a) and (b), we observe the estimated states X (~0% out), Y (~0% out), V_x (~0% out), and V_y (~3% out)[5]. Note that the velocities are piecewise constant functions with step changes corresponding to the impulsive incremental velocities. The filtered measurements: bearing (~1% out) and range (~2% out) are shown in Fig. 6.6c with the resulting innovations zero-mean/whiteness tests depicted in (d). The processor is clearly tuned with bearing and range innovations zero-mean and white ($6\times10^{-4} < 26\times10^{-4}$/3% out) and ($2\times10^{-4} < 42\times10^{-4}$/5% out), respectively. This result is not unexpected, since all of the a priori information

[5] Here "% out" means the percentage of samples exceeding the confidence bounds.

is given including the precise incremental velocity input $u(t)$. An ensemble of 101 realizations of the estimator were run by generating random initial condition estimates from the Gaussian assumption. The 101 realization ensemble averaged estimates closely follow the results shown in the figure with the zero-mean/whiteness tests ($2 \times 10^{-4} < 25 \times 10^{-4}$/4% out), ($2 \times 10^{-4} < 15 \times 10^{-4}$/7% out) slightly worse.

Next, we investigate the realistic case where all of the information is known a priori *except* the impulsive incremental velocity changes represented by the deterministic input. Note that without the input, the processor cannot respond instantaneously to the velocity changes and therefore will lag (in time) behind, in predicting the tanker path. The solution to this problem requires a joint estimation of the states *and* now the unknown input which is a solution to the deconvolution problem [44]. It is also a problem that is ill-conditioned, especially, since $u(t)$ is impulsive.

In any case we ran the nonlinear BP algorithms over the simulated data and the best results were obtained using the LZ-BP and SPBP. This is expected, since we used the exact state reference trajectories or filed paths, but not the input. Note that the other nonlinear BP have no knowledge of this trajectory inhibiting their performance in this problem. The results are shown in Fig. 6.7, where we observe the state estimates as before. We note that the position estimates appear reasonable, primarily because of the reference trajectories. The LZ-BP is able to compensate for the unknown impulsive input with a time lag as shown at each of the fiducials. The velocity estimates (4% out, 1% out) are actually low-pass versions of the true velocities caused by the slower LZ-BP response even with the exact step changes available. These lags are more vividly shown in the bearing estimate of Fig. 6.7 c which shows that the processor has great difficulty with the instantaneous velocity changes in bearing (0% out). The range (0% out) appears insensitive to this lack of knowledge primarily because the XY-position estimates are good and do not have step changes like the velocity for the LZ-BP to track. Both processors are not optimal and the innovations sequences are zero-mean but *not* white ($75 \times 10^{-3} < 81 \times 10^{-3}$/59% out), ($2 \times 10^{-3} < 4 \times 10^{-3}$/8% out).

We also designed the SPBP (UKF) to investigate its performance on this problem and its results were quite good[6] as shown in Fig. 6.8. The processor does not perform a model linearization but a *statistical linearization* instead, it is clear from the figure shown that it performs better than any of the other processors for this problem. In Figs. 6.8 a– 6.8 d, we see that the XY-position estimates "track" the data very well while the XY-velocity estimates are somewhat nosier due to the abrupt changes (steps) tuning values of the process noise covariance terms. In order to be able to track the step changes, the process noise covariance could be increased even further at the cost of nosier estimates. The SPBP tracks the estimated bearing and range reasonably well as shown in figure with a slight loss of bearing track toward the end of the time sequence. These results are demonstrated by the zero-mean/whiteness test results of the corresponding innovations sequences. The bearing innovation statistics are $3.3 \times 10^{-3} < 1.2 \times 10^{-1}$ and 4.7% out, and the corresponding range innovation

[6] We used noisier simulation data for this run than that for the LZ-BP with $R_{vv} = \text{diag} = [4 \times 10^{-4}\ 5 \times 10^{-1}]$ providing a more realistic measurement uncertainty.

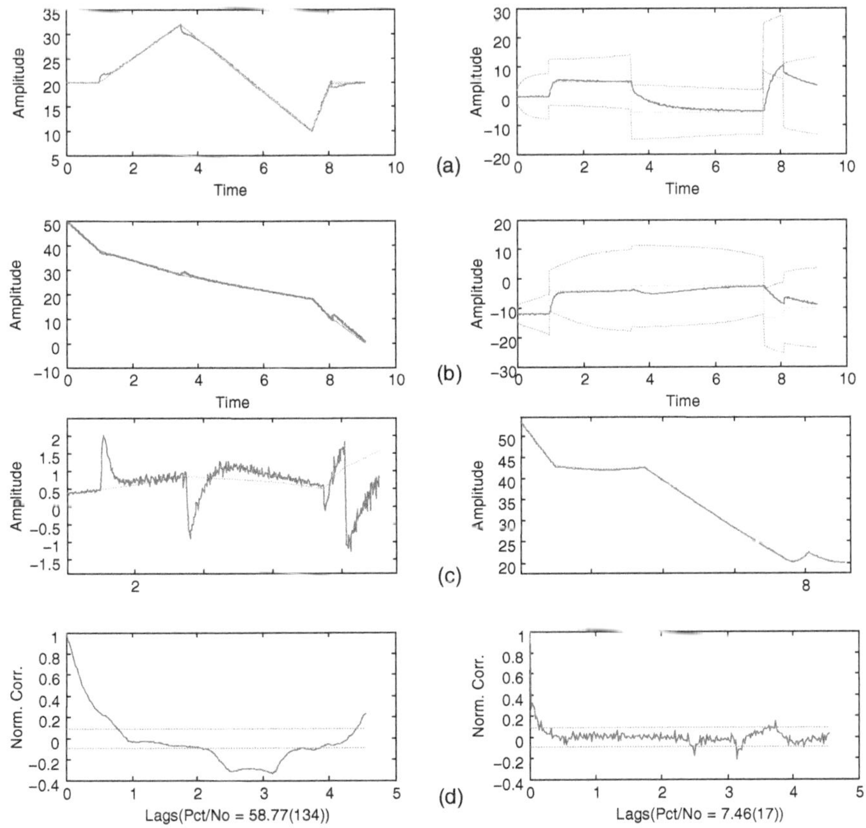

FIGURE 6.7 LZ-BP design for harbor docking problem (input unknown). (a) *X*-position and velocity estimates with bounds (68% and 4% out); (b) *Y*-position and velocity estimates with bounds (49% and 1% out); (c) bearing and range estimates with bounds (0% and 3% out); (d) innovations zero-mean/whiteness tests for bearing ($75 \times 10^{-3} < 81 \times 10^{-3}$ and 59% out) and range ($2 \times 10^{-3} < 4 \times 10^{-3}$ and 8% out).

statistics given by $6.5 \times 10^{-3} < 1.2 \times 10^{-1}$ and 4.7% out. Both indicate a tuned processor. These are the best results of all of the nonlinear processors applied. This completes the case study.

It is clear from this study that nonlinear processors can be "tuned" to give reasonable results, especially when they are provided with accurate a priori information. If the a priori information is provided in terms of prescribed reference trajectories as in this hypothetical case study, then the SPBP appears to provide superior performance, but in the real world tracking problem (as discussed in Section 5.5) when this information on the target is not available, the XBP and IX-BP can be tuned to give reasonable results.

There are many variants possible for these processors to improve their performance whether it be in the form of improved coordinate systems [45, 46] or in the form

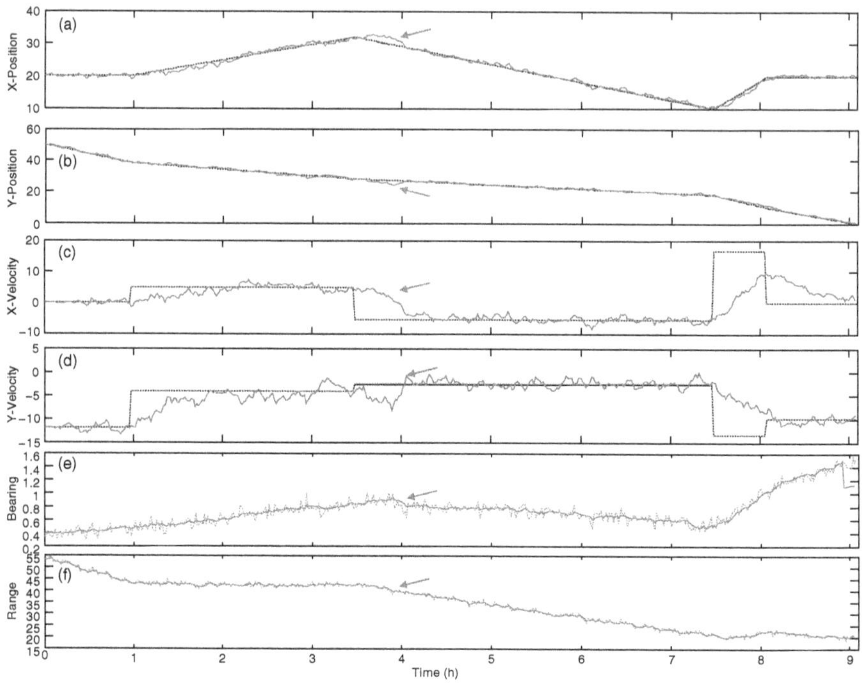

FIGURE 6.8 SPBP (UKF) design for harbor docking problem (input unknown). (a) X-position estimate; (b) Y-position estimate; (c) X-velocity estimate; (d) Y-velocity estimate; (e) bearing estimate (zero- mean/whiteness: $3.3 \times 10^{-3} < 1.2 \times 10^{-1}$ and 4.7% out); (f) range estimate (zero-mean/whiteness: $6.5 \times 10^{-3} < 1.2 \times 10^{-1}$ and 4.7% out).

of a set of models, each with its own independent processor [8]. One might also consider using estimator/smoothers as in the seismic case [7] because of the unknown impulsive input. In any case, this is a challenging problem that much work has been accomplished, the interested reader should consult Ref. [8] and the references cited within.

6.7 ENSEMBLE BAYESIAN PROCESSORS (ENSEMBLE KALMAN FILTER)

Another approach to Bayesian processing problem is the ensemble Bayesian processor (EnBP) popularly called the ensemble Kalman filter (EnKF) approximation leading to an efficient processor targeted for large-scale computational problems. It has been shown that the EnKF provides a simple implementation, numerical stability, convergence and interpretation of results especially because of its properties when incorporated for nonlinear models [23–32]. It provides a mechanism to compute the posterior distribution of the embedded model conditioned on the measurement data and is based on generating an ensemble of model state estimates along with their associated error covariances representing the uncertainty.

The EnKF provides yet another alternative to the XBF or extended Kalman filter (EKF) similar to the Monte Carlo (MC) approach to follow in the next chapter. It is capable of resolving two major problems: (1) it provides a major closure scheme in the sense that it is more robust than the EKF and does not diverge due to linearization; and (2) it is capable of reducing the large storage and error covariance propagations characteristic of Kalman-type model-based processors.

This class of Bayesian processor falls into the category of sequential data assimilation methods that have been heavily applied in meteorology, oceanography, seismology, and microwaves (radar) where the data are *assimilated* into the underlying model creating a model-based processing scheme for large dimensional problems. Specifically, the EnKF is an assimilation method where the error statistics are predicted using MC or equivalently ensemble propagation techniques. This implies that the EnKF can be interpreted as a purely statistical MC technique since the ensemble of model states evolve in the state space with its mean provided as the best estimate and spread given by the corresponding ensemble state error covariance. At observation time, each measurement is represented by another ensemble with the mean given by the actual measurement and the ensemble measurement error covariance representing the corresponding errors [25]. Propagating the model states forward in time enables the estimation of statistical moments from the resulting ensemble. New ensembles are created by updating each member individually enabling a parallel implementation in many applications.

The EnKF can be considered a cross between the unscented Bayesian processor or EKF and the particle filters (PF) to follow in that it uses regression techniques to estimate the error covariances based on generated ensembles as well as a MC approach to estimate ensemble statistics again simplifying the computations. The usual Kalman filter applied to nonlinear systems is the EKF. In an EKF, the covariance matrices are propagated through the nonlinear model with the Jacobian of the state transition and measurement functions. Such a first-order Taylor series approximation can result in large estimation errors and sometimes causes the filter to diverge (the gain tends to zero). The EnKF represents the underlying distributions with large stochastic ensembles of models as in the MC methods. Each member is propagated through the state and measurement functions and the statistics of the estimates are propagated *without* the use of any linearization step.

In this section, we start with a discussion of state/measurement ensembles and then proceed with the development of the EnKF processor after a brief review of the Kalman filter to motivate the ensuing discussion. Before we investigate the linear EnKF, we discuss the generation of linear state-space ensemble data and its accompanying statistics.

State-Space Ensembles: When the process statistics are stationary, the state-space methods of Chapter 4 satisfy the relations developed; however, when the processes are nonstationary (time-varying statistics), ensemble methods can be employed to synthesize data capturing the evolution of the underlying probability mass functions.

Mathematically, we have the *state–space ensemble* technique for $n = 1, \ldots, N_\varepsilon$ given by

- States: $\mathbf{x}_n(t) = A(t-1)\mathbf{x}_n(t-1) + \mathbf{w}_n(t-1)$
- State ensemble: $\mathcal{X}_\varepsilon(t) = \left[\mathbf{x}_1(t) \cdots \mathbf{x}_{N_\varepsilon}(t)\right]^T$
- State means: $\mathbf{m}_x(t) = E\{\mathcal{X}_\varepsilon(t)\} = \frac{1}{N_\varepsilon}\sum_{n=1}^{N_\varepsilon}\mathbf{x}_n(t)$
- State variance: $\mathbf{R}_{xx}(t) = \frac{1}{N_\varepsilon}\sum_{n=1}^{N_\varepsilon}\left(\mathbf{x}_n(t) - \mathbf{m}_x(t)\right)\left(\mathbf{x}_n(t) - \mathbf{m}_x(t)\right)^T$
- Measurements: $\mathbf{y}_n(t) = C(t)\mathbf{x}_n(t) + \mathbf{w}_n(t)$
- Measurement ensemble: $\mathcal{Y}_\varepsilon(t) = \left[\mathbf{y}_1(t) \cdots \mathbf{y}_{N_\varepsilon}(t)\right]^T$
- Measurement means: $\mathbf{m}_y(t) = E\{\mathcal{Y}_\varepsilon(t)\} = \frac{1}{N_\varepsilon}\sum_{n=1}^{N_\varepsilon}\mathbf{y}_n(t)$
- Measurement variance: $\mathbf{R}_{yy}(t) = \frac{1}{N_\varepsilon}\sum_{n=1}^{N_\varepsilon}\left(\mathbf{y}_n(t) - \mathbf{m}_y(t)\right)\left(\mathbf{y}_n(t) - \mathbf{m}_y(t)\right)^T$

One way to achieve this operation is to perform the following operations at each time step "t":

- Process noise: for each component state, generate an ensemble of random process noise samples $\{w_{i,n}(t)\}; i = 1, \ldots, N_x; n = 1, \ldots, N_\varepsilon$
- States: for each component state, generate an ensemble of state samples driven by the process noise (above) $\{x_{i,n}(t)\}; i = 1, \ldots, N_x; n = 1, \ldots, N_\varepsilon$
- State mean: for each ensemble, estimate the state mean (over n) $\mathbf{m}_x(t)$
- State variance: for each ensemble, estimate the state variance (over n) $\mathbf{R}_{xx}(t)$

Perform similar steps for the measurement process

- Measurement noise: for each component measurement, generate an ensemble of random measurement noise samples $\{v_{j,n}(t)\}; j = 1, \ldots, N_y; n = 1, \ldots, N_\varepsilon$
- Measurements: for each component measurement, generate an ensemble of measurement samples perturbed by the measurement noise (above), $\{y_{j,n}(t)\}; j = 1, \ldots, N_y; n = 1, \ldots, N_\varepsilon$
- Measurement mean: for each ensemble, estimate the measurement mean (over n) $\mathbf{m}_y(t)$
- Measurement variance: for each ensemble, estimate the measurement variance (over n) $\mathbf{R}_{yy}(t)$

generating the $(N_y \times N_\varepsilon)$-measurement ensemble array, $\mathcal{Y}_\varepsilon(t) = \left[\mathbf{y}_1(t) \cdots \mathbf{y}_{N_\varepsilon}(t)\right]$.

Example 6.4

Consider the generation of an ensemble for the series *RLC* circuit (second-order system) excited by a pulse train of Section 5.8.

The sampled-data Gauss–Markov simulation uses the following noise covariance matrices to generate the ensemble:

$$R_{ww} = 0.1(\text{V})^2 \quad \text{and} \quad R_{vv} = 0.01(\text{V})^2$$

The results are shown in Fig. 6.9 . We illustrate the noisy measurement ensemble generated from these statistics in *a* and the estimated (kernel density smoother) PMF at each point in time creating the corresponding probability surface (measurement sample (realization) × measurement no. × probability) in Fig. 6.9c. △△△

This completes the ensemble generation methodology that will be used to generate PMFs employed in the EnKF algorithm to follow, first we review the linear Kalman filter to motivate the development of the EnKF.

Linear Kalman Filter: Here we start our discussion using the linear models and the Kalman filter to develop the EnKF and then briefly extend these results to the nonlinear case. Recall that the linear Kalman filter (LKF) algorithm is based on the fact that the posterior distribution of the states based on all of the data up to that time is multivariate Gaussian. With that in mind the LKF essentially provides a prediction-update technique that sequentially estimates the conditional mean and covariance of the posterior Gaussian distribution, that is,

$$\begin{aligned} \hat{\mathbf{x}}(t|t-1) &= A(t-1)\hat{\mathbf{x}}(t-1|t-1) && [\text{ prediction }] \\ \hat{\mathbf{x}}(t|t) &= \hat{\mathbf{x}}(t|t-1) + K(t)\big(\mathbf{y}(t) - C(t)\hat{\mathbf{x}}(t|t-1)\big) && [\text{ update }] \end{aligned} \tag{6.81}$$

Defining the state prediction and correction errors by

$$\tilde{\mathbf{x}}(t|t-1) := \mathbf{x}(t) - \hat{\mathbf{x}}(t|t-1) \quad \text{and} \quad \tilde{\mathbf{x}}(t|t) := \mathbf{x}(t) - \hat{\mathbf{x}}(t|t)$$

the associated predicted error covariance is

$$\tilde{P}(t|t-1) := \text{cov}\,(\tilde{\mathbf{x}}(t|t-1)) = E\left\{\tilde{\mathbf{x}}(t|t-1)\tilde{\mathbf{x}}^T(t|t-1)\right\}$$

with the updated error covariance given by

$$P(t|t) := \text{cov}\,(\tilde{\mathbf{x}}(t|t)) = E\left\{\tilde{\mathbf{x}}(t|t)\tilde{\mathbf{x}}^T(t|t)\right\}$$

The corresponding *gain* or weighting function is

$$K(t) = \tilde{P}(t|t-1)C^T(t)\big(C(t)\tilde{P}(t|t-1)C(t) + R_{vv}(t)\big)^{-1} \quad [\text{ Gain }]$$

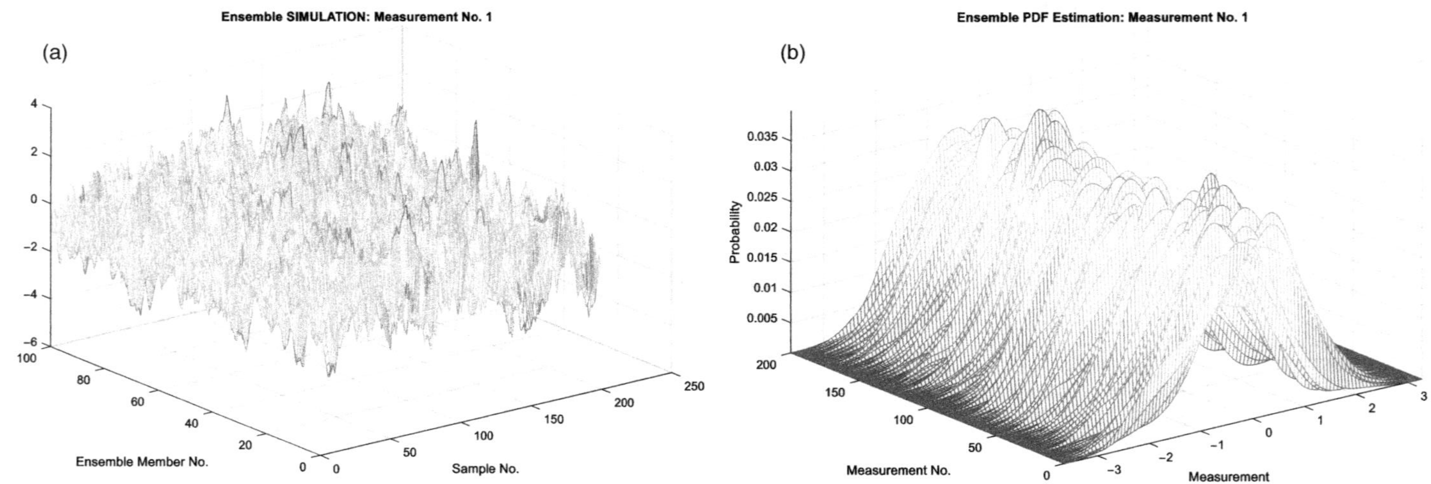

FIGURE 6.9 Ensemble generation: (a) realization of 100-member ensemble for the RLC-circuit problem; (b) estimated PMF surface (measurement sample (realization) × measurement no. × probability) using kernel density smoother.

where the error covariance "propagation relations" are given by

$$\tilde{P}(t|t-1) = A(t-1)\tilde{P}(t-1|t-1)A(t-1) + R_{ww}(t-1) \quad \text{[Cov. prediction]}$$
$$\tilde{P}(t|t) = (I - K(t)C(t))\tilde{P}(t|t-1) \quad \text{[Cov. update]}$$

Linear Ensemble Kalman Filter: The EnKF follows the same basic structure as the LKF with ensemble operations replacing the "propagation relations" by ensemble estimates. The EnKF relations above follow directly by rewriting the LKF state/measurement vectors and covariances as the nth ensemble member (e.g., $\hat{x}(t|t) \rightarrow \hat{x}_n(t|t)$) yielding the ensemble relations

$$n = 1, \ldots, N_\varepsilon \quad \text{[Ensemble]}$$

$$\hat{\mathbf{x}}_n(t|t-1) = A(t-1)\hat{\mathbf{x}}_n(t-1|t-1) \quad \text{[Ens. prediction]}$$
$$\hat{\mathbf{x}}_n(t|t) = \hat{\mathbf{x}}_n(t|t-1) + K_n(t)\big(\mathbf{y}_n(t) - C(t)\hat{\mathbf{x}}_n(t|t-1)\big) \quad \text{[Ens. update]} \tag{6.82}$$

with the gain given by

$$K_n(t) = \tilde{P}_n(t|t-1)C^T(t)\big(C(t)\tilde{P}_n(t|t-1)C(t) + R_{v_n v_n}(t)\big)^{-1} \quad \text{[Ens. gain]} \tag{6.83}$$

and covariances given by

$$\tilde{P}_n(t|t-1) = A(t-1)\tilde{P}_n(t-1|t-1)A^T(t-1) + R_{w_n w_n}(t-1) \quad \text{[Ens. Cov. prediction]}$$
$$\tilde{P}_n(t|t) = (I - K_n(t)C(t))\tilde{P}_n(t|t-1) \quad \text{[Ens. Cov. update]}$$

The initial ensemble is chosen to represent the error states of the initial conditions $(\big(\tilde{x}(0), \tilde{P}(0)\big)$ where the "rule of thumb" is to create an ensemble of model states by adding a perturbation to an initial state estimate and then propagate the ensemble over the corresponding time interval. In terms of a more *practical implementation*, it is more common to represent the EnKF algorithm in terms of vector-matrix relations as in Refs. [23] and [26].

State Perturbation

$$\hat{x}_n(t|t-1) = A\hat{x}_n(t-1|t-1) + w_n(t-1); \quad n = 1, \ldots, N_\varepsilon \tag{6.84}$$

with $w_n \sim \mathcal{N}(0, R_{w_n w_n})$ considered the *modeling error* or uncertainty used to generate the state ensemble samples for propagation (prediction).

State Error

$$\tilde{x}_n(t|t-1) = \hat{x}_n(t|t-1) - \overline{x}(t-1); \quad n = 1, \ldots, N_\varepsilon \tag{6.85}$$

for $\overline{x}$ the ensemble mean.

These state error vectors are instrumental in estimating the associated error covariance incorporated into the gain calculation. With this in mind, an ensemble matrix of state error vectors at time $(t-1)$ can be created as

$$\tilde{\mathcal{X}}_\varepsilon(t-1) = \left[\tilde{x}_1(t|t-1) \cdots \tilde{x}_{N_\varepsilon}(t|t-1)\right] \in \mathcal{R}^{N_x \times N_\varepsilon} \tag{6.86}$$

The associated error covariance is computed simply as a sample estimate given by

Predicted State Error Covariance

$$\tilde{P}_\varepsilon(t|t-1) = \frac{1}{N_\varepsilon}\tilde{\mathcal{X}}_\varepsilon(t-1)\tilde{\mathcal{X}}_\varepsilon^T(t-1) \in \mathcal{R}^{N_x \times N_x} \tag{6.87}$$

To complete the MC representation

Measurement Perturbation

$$y_n(t) = y(t) + v_n(t); \quad n = 1, \ldots, N_\varepsilon \tag{6.88}$$

with $v_n \sim \mathcal{N}(0, R_{vv})$ considered the *measurement error* or uncertainty used to generate the measurement ensemble samples for propagation (prediction).

Measurement Prediction

$$\hat{y}_n(t|t-1) = C(t)\hat{x}_n(t|t-1); \quad n = 1, \ldots, N_\varepsilon \tag{6.89}$$

Measurement Error

$$\tilde{y}_n(t) = \hat{y}_n(t|t-1) - \overline{y}(t); \quad n = 1, \ldots, N_\varepsilon \tag{6.90}$$

for $\overline{y}$ the ensemble mean of the measurement samples.

An ensemble matrix can be created for use in covariance estimates as

$$\tilde{\mathcal{Y}}_\varepsilon(t) = \left[\tilde{y}_1(t) \cdots \tilde{y}_{N_\varepsilon}(t)\right] \in \mathcal{R}^{N_y \times N_\varepsilon} \tag{6.91}$$

and therefore we obtain

Measurement Error Covariance

$$R_{\tilde{y}\tilde{y}}(t) = \frac{1}{N_\varepsilon}\tilde{\mathcal{Y}}_\varepsilon(t)\tilde{\mathcal{Y}}_\varepsilon^T(t) \in \mathcal{R}^{N_y \times N_y} \tag{6.92}$$

The $N_y \times N_\varepsilon$ measurement noise ensemble is created to provide the sample measurement covariance as

Measurement Noise Covariance

$$\mathcal{V}_\varepsilon(t) = \left[v_1(t) \cdots v_{N_\varepsilon}(t)\right] \in \mathcal{R}^{N_y \times N_\varepsilon}$$

$$R_{\mathcal{V}\mathcal{V}}(t) = = \frac{1}{N_\varepsilon}\mathcal{V}_\varepsilon(t)\mathcal{V}_\varepsilon^T(t) \in \mathcal{R}^{N_y \times N_y} \tag{6.93}$$

The ensemble Kalman *gain* for the linear case can be obtained using the ensemble error covariance as in the LKF processor as

Gain

$$\mathcal{K}_\varepsilon(t) = \tilde{P}_\varepsilon(t|t-1)C^T(t)\left(C(t)\tilde{P}_\varepsilon(t|t-1)C^T(t) + R_{\mathcal{V}\mathcal{V}}(t)\right)^{-1} \tag{6.94}$$

Defining the *updated* states, their errors and ensemble mean at time t we have

Updated State/Error

$$\hat{x}_n(t|t) = \hat{x}_n(t|t-1) + \mathcal{K}_\varepsilon(t)\left(y_n(t) - \hat{y}_n(t|t-1)\right)$$

$$\tilde{x}_n(t|t) = \hat{x}_n(t|t) - \overline{x}(t); \quad n = 1, \ldots, N_\varepsilon \tag{6.95}$$

and therefore estimate the sample error covariance as

Updated State Error Covariance

$$\tilde{P}_\varepsilon(t|t) = \frac{1}{N_\varepsilon - 1}\tilde{\mathcal{X}}_\varepsilon(t)\tilde{\mathcal{X}}_\varepsilon^T(t) \in \mathcal{R}^{N_x \times N_x} \tag{6.96}$$

Next, we investigate the nonlinear EnKF and its implementation.

Nonlinear Ensemble Kalman Filter: Here instead of the matrices of the linear models, we have nonlinear functions (models) of both the states and measurements, that is,

$$\mathbf{A}(t-1)x(t-1) \longrightarrow \mathbf{a}\left[x(t-1)\right] \quad \text{and} \quad \mathbf{C}(t)x(t) \longrightarrow \mathbf{c}\left[x(t)\right]$$

As mentioned previously, the EKF linearizes these nonlinear model through a first-order Taylor series approximation (see Section 5.5 for details) which is the primary cause of filter divergence (gain → 0). The EnKF processor approaches this problem in a different manner.

When the measurement system is nonlinear, then the operation of the EnKF, which is primarily based on a linear measurement model, must be modified, somehow, to deal with this problem. The obvious approach is to "linearize" the measurement model much the same as the EKF; however, this diminishes the robustness property of the EnKF gained in the state dynamics. One possibility is to *augment* the nonlinear measurement functions ($c[x(t)]$) into the state vector incorporating them into the prediction ensemble and propagate them along with the states [23]. Another possibility is that of developing another ensemble to estimate the cross-error covariance between the state and measurement errors thereby eliminating the need for linearization or augmentation [29]. We develop the latter approach in the gain calculation subsequently.

For the nonlinear implementation, we simply replace the prediction steps with the nonlinear models producing the perturbation relations:

Ensemble Predictions

$$\hat{x}_n(t|t-1) = a[\hat{x}_n(t-1|t-1)] + w_n(t-1) \qquad \text{[Nonlinear state perturbation]}$$

$$\hat{y}_n(t|t-1) = c[\hat{x}_n(t-1|t-1)] \qquad \text{[Nonlinear measurement prediction]}$$

$$y_n(t) = y(t) + v_n(t); \qquad n = 1, \ldots, N_\varepsilon \qquad \text{[Measurement perturbation]} \tag{6.97}$$

Returning to the basic minimum variance estimation construct (see Section 2.4), the original Kalman gain can be calculated by the product of the error cross-covariance and the inverse measurement error covariance [10]

$$K(t) = R_{\tilde{x}\tilde{y}}(t) \times R_{\tilde{y}\tilde{y}}^{-1}(t) \tag{6.98}$$

Therefore, we have that the ensemble Kalman gain has the same structure by employing both the state and measurement error ensembles yielding an alternate form using the ensemble error cross-covariance similar to the EKF processor [10] as

Gain

$$R_{\tilde{\mathcal{X}}\tilde{\mathcal{Y}}}(t) = \frac{1}{N_\varepsilon}\tilde{\mathcal{X}}_\varepsilon(t-1)\tilde{\mathcal{Y}}_\varepsilon^T(t)$$

$$\mathcal{K}_\varepsilon(t) = R_{\tilde{\mathcal{X}}\tilde{\mathcal{Y}}}(t)\left(R_{\tilde{\mathcal{Y}}\tilde{\mathcal{Y}}}(t) + R_{\mathcal{V}\mathcal{V}}(t)\right)^{-1} \tag{6.99}$$

This approach has been successfully demonstrated [29].

We summarize both linear/nonlinear ensemble Kalman techniques in the combined table, that is, Table 6.2 along with the structure of the EnKF algorithm depicted in Fig. 6.10 emphasizing the basic components of the EnKF. More details can be found in Ref. [23].

TABLE 6.2 ENSEMBLE Bayesian (Kalman Filter) Processor Algorithm.

Ensemble $n = 1, \ldots, N_\varepsilon$	
State Prediction	
$\hat{x}_n(t\|t-1) = \begin{cases} \mathbf{A}(t-1)\hat{x}_n(t-1\|t-1) + w_n(t-1) \\ \mathbf{a}[\hat{x}_n(t-1\|t-1)] + w_n(t-1) \end{cases}$	[Linear state perturbation] or [Nonlinear state perturbation]
$\tilde{x}_n(t\|t-1) = \hat{x}_n(t\|t-1) - \bar{x}(t-1)$	[State error]
$\tilde{P}_\varepsilon(t\|t-1) = \frac{1}{N_\varepsilon}\tilde{\mathcal{X}}_\varepsilon(t-1)\tilde{\mathcal{X}}_\varepsilon^T(t-1)$	[Ensemble error covariance prediction]
Measurement Prediction	
$\hat{y}_n(t\|t-1) = \begin{cases} C(t)\hat{x}_n(t\|t-1) + v_n(t) \\ \mathbf{c}\left[\hat{x}_n(t\|t-1)\right] + v_n(t) \end{cases}$	[Linear measurement perturbation] or [Nonlinear measurement perturbation]
$\tilde{y}_n(t) = \hat{y}_n(t\|t-1)) - \bar{y}(t)$	[Measurement error]
$R_{\tilde{y}\tilde{y}}(t) = \frac{1}{N_\varepsilon}\tilde{\mathcal{Y}}_\varepsilon(t)\tilde{\mathcal{Y}}_\varepsilon^T(t)$	[Ensemble error covariance prediction]
Measurement noise/Covariance ***Gain***	
$R_{\tilde{x}\tilde{y}}(t) = \frac{1}{N_\varepsilon}\tilde{\mathcal{X}}_\varepsilon(t-1)\tilde{\mathcal{Y}}_\varepsilon^T(t)$	[Ensemble error cross-covariance prediction]
$\mathcal{K}_\varepsilon(t) = \tilde{R}_{\tilde{x}\tilde{y}}(t)\left(R_{\tilde{y}_n\tilde{y}_n}(t) + R_{v_n v_n}(t)\right)^{-1}$	[Gain]
State Update	
$\hat{x}_n(t\|t) = \hat{x}_n(t\|t-1) + \mathcal{K}_\varepsilon(t)\left(\hat{y}_n(t\|t-1)) - \bar{y}(t)\right)$	[State update]
$\tilde{x}_n(t\|t) = x_n(t\|t) - \bar{x}(t)$	[State error update]
$\tilde{P}_\varepsilon(t\|t) = \frac{1}{N_\varepsilon}\tilde{\mathcal{X}}_\varepsilon(t)\tilde{\mathcal{X}}_\varepsilon^T(t)$	[Ensemble error covariance update]
Ensemble Means/Errors	
$\hat{\bar{x}}(t-1) = \frac{1}{N_\varepsilon}\sum_{n=1}^{N_\varepsilon}\hat{x}_n(t\|t-1)$	$\tilde{\mathcal{X}}_\varepsilon(t-1) = \left[\tilde{x}_1(t\|t-1)\cdots\tilde{x}_{N_\varepsilon}(t\|t-1)\right]$
$\bar{y}(t) - \frac{1}{N_\varepsilon}\sum_{n=1}^{N_\varepsilon}\hat{y}_n(t\|t-1)$	$\tilde{\mathcal{Y}}_\varepsilon(t) = \left[\tilde{y}_1(t)\cdots\tilde{y}_{N_\varepsilon}(t)\right]$
$\bar{x}(t) = \frac{1}{N_\varepsilon}\sum_{n=1}^{N_\varepsilon}\hat{x}_n(t\|t)$	$\tilde{\mathcal{X}}_\varepsilon(t) = \left[\tilde{x}_1(t)\cdots\tilde{x}_{N_\varepsilon}(t\|t)\right]$

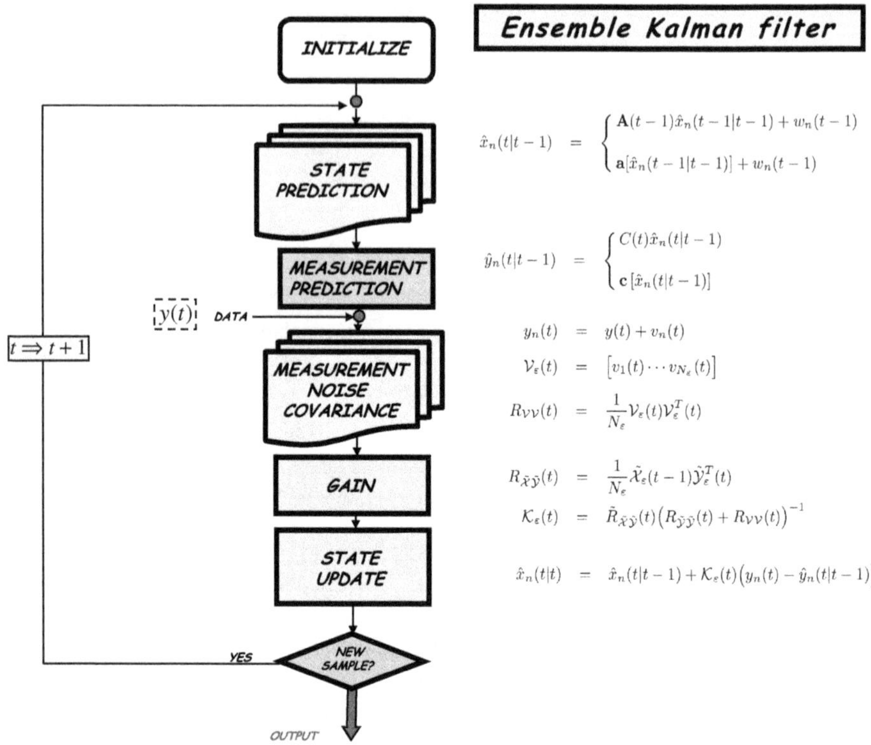

FIGURE 6.10 Ensemble Kalman filter algorithm structure: initialization, state prediction (ensemble), measurement prediction, measurement (perturbation) noise/covariance (ensemble), gain, state update.

Before we conclude this section, let us revisit the *RLC* circuit of Section 5.8 and compare the performance of the linear and ensemble Kalman processors.

Example 6.5

Consider the design of both LKF and EnKF processors for the series *RLC* circuit (second-order system) excited by a pulse train of Section 5.8.

After inserting the physical circuit parameters, the resulting continuous-time, differential equation model evolved as

$$\frac{dx}{dt} = \begin{bmatrix} 0 & 1 \\ -4 & -2 \end{bmatrix} x + \begin{bmatrix} 0 \\ -4 \end{bmatrix} u + \begin{bmatrix} 0 \\ -4 \end{bmatrix} w$$

and the discrete-time voltage measurements

$$y(t) = [1\ 0]x(t) + v(t)$$

where

$$R_{ww} = \frac{(0.1)^2}{T} = 0.1(\mathrm{V})^2 \quad \text{and} \quad R_{vv} = 0.01(\mathrm{V})^2$$

The process model was converted to a sampled-data (discrete) representation with a sampling interval of 0.1 sec providing the following discrete-time Gauss–Markov model as input to our processors:

$$x(t) = \begin{bmatrix} 0.98 & 0.09 \\ -0.36 & 0.801 \end{bmatrix} x(t-1) + \begin{bmatrix} -0.019 \\ -0.36 \end{bmatrix} u(t-1) + \begin{bmatrix} -0.019 \\ -0.36 \end{bmatrix} w(t-1)$$

$$y(t) = [1|0]x(t) + v(t)$$

where

$$R_{ww} = 0.1(\mathrm{V})^2 \quad \text{and} \quad R_{vv} = 0.01(\mathrm{V})^2$$

The discrete Bayesian processors were designed. The results are shown in Fig. 6.11; in (a) through (c), we see the processed states and measurements using both the LKF and EnKF processors along with the uncertain simulation data and corresponding true (mean) signals. From the figure, it is clear that the EnKF estimates using and ensemble size of 1000 members "track" the optimal LKF outputs quite well. To validate the performance for this processor, zero-mean/whiteness tests were performed as shown in Fig. 6.11d. As shown, the innovations are statistically zero-mean ($0.033 \ll 0.174$) and white (3.1% out, *WSSR* below threshold) indicating a well-tuned EnKF processor.

This completes example and the comparison for the *RLC* problem. △△△

Next, we summarize the chapter.

6.8 SUMMARY

In this chapter, we have developed the "modern" sigma-point Bayesian processor (SPBP) or equivalently, the unscented Kalman filter (UKF), from the basic principles of weighted linear stochastic linearization (WSLR) and σ-point transformations (SPT). We extended the results for multivariate Gaussian distributions and calculated the corresponding σ-points and weights. Once determined, we developed the SPBP by extrapolating the WSLR and SPT approaches coupled to the usual Kalman filter recursions. Grid-based quadrature and Gaussian sum processors were also discussed and developed using the SPBP formulation to demonstrate a distribution approximation approach leading to the particle filter formulation of the next chapter. Finally, we developed the ensemble Kalman filter (EnKF) processor explicitly designed for large-scale problems. The EnKF is a hybrid between the UKF and the particle filters to follow in that it employs ensemble (sampling) techniques and approximates the required computationally intensive covariances using regression methods. Examples

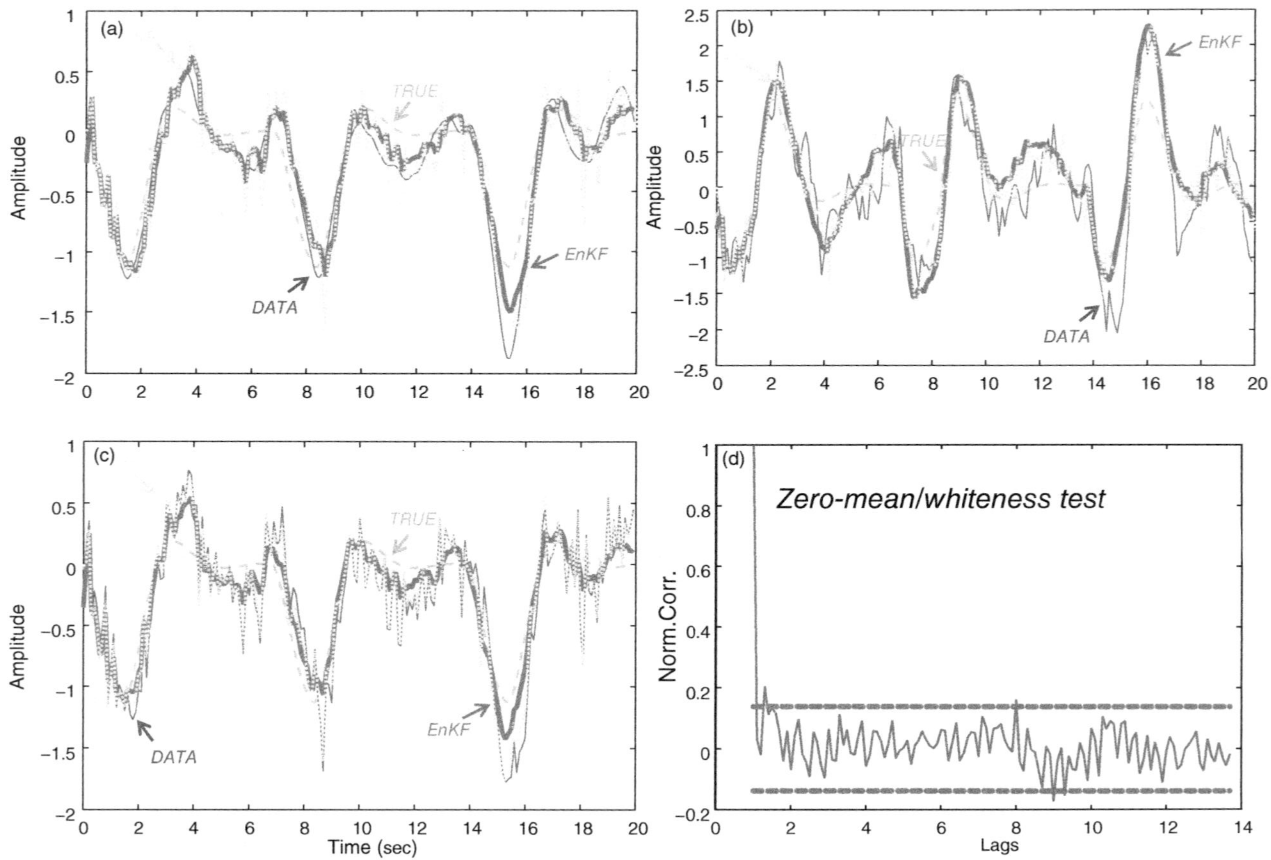

FIGURE 6.11 Ensemble Kalman and optimal filter for *RLC*-circuit problem. (a) State no. 1: data, true (mean), KF, EnKF estimates. (b) State no. 2: data, true (mean), KF, EnKF estimates. (c) Measurement: data, true (mean), KF, EnKF estimates. (d) Innovation zero-mean/whiteness test (Z-M: 0.03 < 0.17; 3.1% out).

were developed throughout to demonstrate the concepts ending in a hypothetical investigation based on tracking a tanker entering a busy harbor. We summarized the results by applying some of the processors to the case study implementing a 2D-tracking filter. Finally, we compared the EnKF performance to the optimal LKF for the linear RLC-circuit problem of the previous chapter.

MATLAB NOTES

SSPACK_PC is a third-party toolbox in MATLAB that can be used to design model-based signal processors [10, 43]. This package incorporates the major *nonlinear* MBP algorithms discussed in this chapter—all implemented in the UD-factorized form [47] for stable and efficient calculations. It performs the discrete approximate Gauss–Markov simulations using (**SSNSIM**) and both extended (XMBP) and iterated-extended (IX-MBP) processors using (**SSNEST**). The linearized model-based processor (LZ-MBP) is also implemented (**SSLZEST**). Ensemble operations are seamlessly embodied within the GUI-driven framework where it is quite efficient to perform multiple design runs and compare results. Of course, the heart of the package is the command or GUI-driven post-processor (**SSPOST**) which is used to analyze and display the results of the simulations and processing (see http://www.techni-soft.net for more details).

REBEL is a recursive Bayesian estimation package in MATLAB available on the web, that performs similar operations including the new statistical-based unscented algorithms including the UKF including the unscented transformations [33]. It also has included the new particle filter designs as discussed in Ref. [33, 34, 48].

EnKF-MATLAB is an ensemble Kalman filtering software package in MATLAB for complex problems [49].

REFERENCES

1. A. Jazwinski, *Stochastic Processes and Filtering Theory* (New York: Academic Press, 1970).
2. A. Sage and J. Melsa, *Estimation Theory with Applications to Communications and Control* (New York: McGraw-Hill, 1971).
3. A. Gelb (editor), *Applied Optimal Estimation* (Boston, MA: MIT Press, 1975).
4. B. Anderson and J. Moore, *Optimal Filtering* (Englewood Cliffs, NJ: Prentice-Hall, 1979).
5. P. Maybeck, *Stochastic Models, Estimation, and Control* (New York: Academic Press, 1979).
6. M. Grewal and A. Andrews, *Kalman Filtering: Theory and Practice* (Englewood Cliffs, NJ: Prentice-Hall, 1993).
7. J. Mendel, *Lessons in Estimation Theory for Signal Processing, Communications and Control* (Englewood Cliffs, NJ: Prentice-Hall, 1995).

8. Y. Bar-Shalom and X. Li, *Estimation and Tracking: Principles, Techniques, and Software* (Norwood, MA: Artech House, 1993).
9. B. Bell and F. Cathey, "The iterated kalman filter update as a Gauss-Newton method," *IEEE Trans. Autom. Control*, **AC-38**, 2, 294–297, 1993.
10. J. Candy, *Model-Based Signal Processsing* (Hoboken, NJ: John Wiley & Sons, Inc./IEEE Press, 2006).
11. D. Simon, *Optimal State Estimation: Kalman, H_∞ and Nonlinear Approaches* (Hoboken, NJ: John Wiley & Sons, 2006).
12. S. Julier, J. Uhlmann, and H. F. Durrant-Whyte, "A new approach for filtering in nonlinear systems," *Proc. Am. Control Conf.*, **3**, 1628–1632, 1995.
13. S. Julier and J. Uhlmann, "A general method for approximating nonlinear transformations of probability distributions," University of Oxford Report, 1996.
14. S. Julier, J. Uhlmann, and H. F. Durrant-Whyte, "A new method for the nonlinear transformation of means and covariances in filters and estimators," *IEEE Trans. Autom. Control*, **45**, 3, 477–482, 2000.
15. S. Julier and J. Uhlmann, "Unscented filtering and nonlinear estimation," *Proc. IEEE*, **92**, 3, 401–422, 2004.
16. S. Julier and J. Uhlmann, "A new extension of the Kalman filter to nonlinear systems," *Proc. SPIE Conf.*, Orlando, FL, 1997.
17. E. Wan and R. van der Merwe, "The unscented Kalman filter for nonlinear estimation," *Proc. IEEE Sumpos. Adaptive Sys. for Sig. Proc., Comm. and Control*, Lake Louise, Alberta, 2000.
18. R. van der Merwe, A. Doucet, N. de Freitas, and E. Wan, "The unscented particle filter," Cambridge University Technical Report, CUED/F-INFENG-380, 2000.
19. T. Lefebvre, H. Bruyninckx, and J. DeSchutter, Comments on "A new method for the nonlinear transformation of means and covariances in filters and estimators," *IEEE Trans. Autom. Control*, **47**, 8, 1406–1409, 2002.
20. S. Julier and J. Uhlmann, "A consistent, debiased method for converting between polar and Cartesian coordinate systems," *Proc. SPIE Conf.*, 1997.
21. S. Haykin, *Kalman Filtering and Neural Networks* (New York: John Wiley & Sons, Inc., 2001).
22. A. Smith and A. Gelfand, "Bayesian statistics without tears: A sampling-resampling perspective," *Am. Statistic.*, **46**, 2, 84–88, 1992.
23. G. Evensen, *Data Assimilation: The Ensemble Kalman Filter* (New York: Springer, 2007).
24. G. Evensen, "Sequential data assimilation with a non-linear quasi-geostrophic model using Monte-Carlo methods to forecast error statistics," *J. Geophys. Res.*, **99 (C5)**, 10143–10162, 1994.
25. G. Burgers, P. van Leeuven, and G. Evensen "Analysis scheme in the ensemble Kalman filter," *Mon. Weather Rev.*, **126**, 1719–1724, 1998.
26. G. Evensen, "The ensemble Kalman filter: theoretical formulation and practical implementation," *Ocean Dynam.*, **53**, 343–367, 2003.
27. G. Evensen, "Sampling strategies and square root analysis schemes for the EnKF," *Ocean Dynam.*, **54**, 539–560, 2004.
28. J. Jensen, *Ensemble Kalman Filtering for State and Parameter Estimation on a Reservoir Model* MS Thesis, (Norway:Norwegian University of Science and Technology, 2007).

29. X. Han and X. Li, "An evaluation of the nonlinear/non-Gaussian filters for the sequential data assimilation," *Remote Sens. Environ.*, **112**, 1434–1449, 2008.
30. P. Sakov and P. Oke, "A deterministic formulation of the ensemble Kalman filter: an alternative to ensemble square root filters," *Tellus*, **60A**, 1361–371, 2008.
31. P. Sakov, G. Evensen, and L. Bertino, "Asynchronous data assimilation with the EnKF," *Tellus*, **62A**, 24–28, 2009.
32. G. Evensen, "The ensemble Kalman filter for combined state and parameter estimation," *IEEE Cntrl. Sys. Magz.*, **6**, 583–104, 2009.
33. R. van der Merwe and E. Wan, *REBEL: Recursive Bayesian Estimation Library University of Oregon*, 2002.
34. R. van der Merwe and E. Wan, "Sigma-Point Kalman filters for probabilistic inference in dynamic state-space models," Ph.D. dissertation, OGI School of Science and Engineering, Beaverton, 2004.
35. K. Ito and K. Xiong, "Gaussian filters for nonlinear filtering problems," *IEEE Trans. Autom. Control*, **45**, 5, 910–927, 2000.
36. I. Arasaratnam, S. Haykin, and R. Elliott, "Discrete-time nonlinear filtering algorithms using Gauss-Hermite quadrature," *Proc. IEEE*, **95**, 5, 953–976, 2007.
37. R. Bucy and K. Senne, "Digital synthesis of nonlinear filters," *Automatica*, **7**, 3, 287–298, 1971.
38. K. Ito and K. Xiong, "Gaussian filters for nonlinear filtering problems," *IEEE Trans. Autom. Control*, **45**, 5, 910–927, 2000.
39. N. Cui, L. Hong, and J. Layne, "A comparison of nonlinear filtering approaches with an application to ground target tracking," *Signal Proc.*, **85**, 1469–1492, 2005.
40. D. Alspach and H. Sorenson, "Nonlinear Bayesian estimation using Gaussian sum approximations," *IEEE Trans. Autom. Control*, **17**, 4, 439–448, 1972.
41. H. Sorenson and D. Alspach, "Recursive Bayesian estimation using Gaussian sums," *Automatica*, **7**, 465–479, 1971.
42. J. Kotecha and P. Djuric, "Gaussian sum particle filtering," *IEEE Trans. Signal Proc.*, **51**, 10, 2602–2612, 2003.
43. J. Candy and P. Candy, "SSPACK_PC: A model-based signal processing package on personal computers," *DSP Applications*, **2**, 3, 33–42, 1993 (see http://www. techni-soft.net for more details).
44. J. Candy and J. Zicker, "Deconvolution of noisy transient signals: A Kalman filtering application," LLNL Report, UCID-87432, and *Proc. of CDC Conf.*, Orlando, 1982.
45. V. Aidala and S. Hammel, "Utilization of modified polar-coordinates for bearings-only tracking," *IEEE Trans. Autom. Control*, **AC-28**, 283–294, 1983.
46. V. Aidala, "Kalman filter behavior in bearings-only velocity and position estimation," *IEEE Trans. Aerosp. Electron. Syst.*, **AES-15**, 29–39, 1979.
47. G. Bierman, *Factorization Methods of Discrete Sequential Estimation* (New York: Academic Press, 1977).
48. S. Haykin and N. de Freitas (editors), "Sequential state estimation: from Kalman filters to particle filters," *Proc. IEEE*, **92**, 3, 399–574, 2004.
49. P. Sakov, F. Counillon, L. Bertino, K. Lisaeter, P. Oke, and A. Korablev, "TOPAZ4: an ocean-sea ice data assimilation system for the North Atlantic and Artic," *Ocean Sci.*, **8**, 633–656, 2012.

PROBLEMS

6.1 Let x_1 and x_2 be *i.i.d.* with distribution $\mathcal{N}(0, 1)$. Suppose $y = x_1^2 + x_2^2$

(*a*) What is the distribution of y, $\mathrm{p}_Y(y)$?

(*b*) Suppose $E\{y\} = 2$ and $\sigma_y^2 = 4$, using the sigma-point transformation what are the σ-points for $\mathbf{x} = [\mathbf{x_1} \quad \mathbf{x_2}]'$?

(*c*) What are the σ-points for y?

6.2 Suppose $x \sim \mathcal{N}(0, 1)$ and $y = x^2$

(*a*) What is the distribution of y, $\mathrm{p}_Y(y)$?

(*b*) What is the Gaussian approximation of the mean and variance of $\mathrm{p}_Y(y)$? (*Hint*: Use linearization)

(*c*) What is the sigma-point transformation and corresponding mean and variance estimates of $\mathrm{P}_Y(y)$?

6.3 From the following set of nonlinear system models, develop the SPBP algorithm for each:

(*a*) Synchronous (unsteady) motor: $\ddot{x}(t) + C\dot{x}(t) + p \sin x(t) = L(t)$

(*b*) Duffing equation: $\ddot{x}(t) + \alpha x(t) + \beta x^3(t) = F \cos(\omega t)$

(*c*) Van der Pol equation: $\ddot{x}(t) + \epsilon \dot{x}(t)[1 - \dot{x}^2(t)] + x(t) = m(t)$

(*d*) Hill equation: $\ddot{x}(t) - \alpha x(t) + \beta p(t)x(t) = m(t)$

6.4 Suppose we are given the following discrete system

$$x(t) = -\omega^2 x(t-1) + \sin(x(t-1)) + \alpha u(t-1) + w(t-1)$$

$$y(t) = x(t) + v(t)$$

with w and v zero-mean, white Gaussian with usual covariances, R_{ww} and R_{vv}. Develop the SPBP for this process. Suppose the parameters ω and α are unknown, develop the SPBP such that the parameters are jointly estimated along with the states. (*Hint*: augment the states and parameters to create a new state vector.)

6.5 The Mackey–Glass time delay differential equation is given by

$$\dot{x}(t) = \frac{\alpha x(t-\tau)}{1 + x(t-\tau)^N} - \beta x(t) + w(t)$$

$$y(t) = x(t) + v(t)$$

where α, β are constants, N is a positive integer with w and v zero-mean, white Gaussian with covariances, R_{ww} and R_{vv}. For the parameter set: $\alpha = 0.2$, $\beta = 0.1$, $\tau = 7$ and $N = 10$ with $x(0) = 1.2$ (see Refs. [21] and [29]).

Develop the SPBP for this process.

6.6 Assume that a target is able to maneuver, that is, we assume that the target velocity satisfies a first-order AR model given by

$$v_\tau(t) = -\alpha v_\tau(t-1) + w_\tau(t-1) \quad \text{for } w \sim \mathcal{N}(0, R_{w_\tau w_\tau})$$

(*a*) Develop the Cartesian tracking model for this process.

(*b*) Develop the corresponding SPBP assuming all parameters are known a priori.

(*c*) Develop the corresponding SPBP assuming α is unknown.

6.7 Nonlinear processors can be used to develop neural networks used in many applications. A generic neural network behavior is governed by the following relations:

$$x(t) = x(t-1) + w(t-1)$$
$$y(t) = c[x(t), u(t), \alpha(t)] + v(t)$$

where $x(t)$ is the network weights (parameters), $u(t)$ is the input or training sequence, $\alpha(t)$ is the node activators with w and v zero-mean, white Gaussian with covariances, R_{ww} and R_{vv}.

Develop the SPBP for this process.

6.8 Consider the problem of estimating a random signal from an AM modulator characterized by

$$s(t) = \sqrt{2P}a(t)\sin\omega_c t$$
$$r(t) = s(t) + v(t)$$

where $a(t)$ is assumed to be a Gaussian random signal with power spectrum

$$S_{aa}(\omega) = \frac{2k_a P_a}{\omega^2 + k_a^2}$$

Also assume that the processes are contaminated with the usual additive noise sources: w and v zero-mean, white Gaussian with covariances, R_{ww} and R_{vv}. The discrete-time Gauss–Markov model for this system was developed (chapter 5 problems) from the continuous-time representation using first differences, then:

(*a*) Develop the SPBP.

(*b*) Assume the carrier frequency ω_c is unknown. Develop the SPBP for this process.

6.9 Develop the Gaussian sum processor algorithm using the XBP instead of the SPBP. In the literature, this is the usual approach that is used. How does the overall algorithm differ? What are the apparent pitfalls involved in this approach compared to the SPBP approach?

6.10 We are given a sequence of data and know it is Gaussian with an unknown mean and variance. The distribution is characterized by $y \sim \mathcal{N}(\mu, \sigma^2)$.

(*a*) Formulate the problem in terms of a state–space representation. (*Hint*: Assume that the measurements are modeled in the usual manner (scale by standard deviation and add mean to a $\mathcal{N}(0, 1)$ "known" sequence, say $v(t)$).)

(*b*) Using this model, develop the SPBP technique to estimate the model parameters.

(*c*) Synthesize a set of data of 2000 samples at a sampling interval $dt = 0.01$ with process covariance, $R_{ww} = \text{diag}[1 \times 10^{-5}, 1 \times 10^{-6}]$ and measurement noise of $R_{vv} = 1 \times 10^{-6}$ with $\bar{x}(0) = [\sqrt{20} \quad 3]'$.

(*d*) Develop the SPBP algorithm and apply it to this data, show the performance results (final parameter estimates, etc.). That is, find the best estimate of the parameters defined by $\Theta := [\mu \quad \sigma]'$ using the SPBP approach. Show the mathematical steps in developing the technique and construct simple SPBP to solve.

6.11 Design an ensemble Kalman filter (EnKF) for the *RC*-circuit problem of Example 5.1:

(*a*) Develop a state and measurement ensemble simulation.

(*b*) Develop the linear EnKF processor for this data set.

(*c*) Develop the linear (optimal) Kalman filter processor and apply it to the ensemble data.

(*d*) Compare the performance of both processors after "tuning" (zero-mean/white).

7

PARTICLE-BASED BAYESIAN STATE–SPACE PROCESSORS

7.1 INTRODUCTION

In this chapter, we develop particle-based processors using the state–space representation of signals and show how they evolve from the Bayesian perspective using their inherent Markovian structure along with importance sampling techniques as our basic construct. Particle filters offer an alternative to the Kalman model-based processors discussed in the previous chapters possessing the capability not just to characterize unimodal distributions but also to characterize multimodal distributions. We first introduce the generic state–space particle filter (SSPF) and investigate some of its inherent distributions and implementation requirements. We develop a generic sampling-importance-resampling (SIR) processor and then perhaps its most popular form—the "bootstrap" particle filter. Next, we investigate the resampling problem and some of the more popular resampling techniques also incorporated into the bootstrap filter from necessity. The bootstrap and its variants are compared to the classical and modern processors of the previous chapters. Finally, we apply these processors to a variety of problems and evaluate their performance using statistical testing as part of the design methodology

7.2 BAYESIAN STATE–SPACE PARTICLE FILTERS

Particle filtering (PF) is a sequential Monte Carlo method employing the sequential estimation of relevant probability distributions using the "importance sampling"

Bayesian Signal Processing: Classical, Modern, and Particle Filtering Methods, Second Edition. James V. Candy.

techniques developed in Chapter 3 and the approximations of distributions with discrete random measures [1–4]. The key idea is to represent the posterior distribution by a set of N_p random samples, the *particles*, with associated weights, $\{x_i(t), \mathcal{W}_i(t)\}$; $i = 1, \ldots, N_p$, and compute the required Monte Carlo estimates. Of course, as the number of particles becomes very large the MC representation becomes an equivalent characterization of the analytical description of the *posterior* distribution (e.g., see Example 3.12 which converges to the optimal Bayesian estimate).

Thus, particle filtering is a technique to implement sequential Bayesian estimators by MC simulation. It offers an alternative to approximate Kalman filtering for nonlinear problems [1, 5]. In PF, continuous distributions are approximated by "discrete" random measures composed of these weighted particles or point masses where the *particles* are actually samples of the unknown or hidden states from the state–space representation and the *weights* are the "probability masses" estimated using the Bayesian recursions as shown in Fig. 7.1. From the figure, we see that associated with each particle $x_i(t)$ is a corresponding weight or (probability) mass $\mathcal{W}_i(t)$ (filled

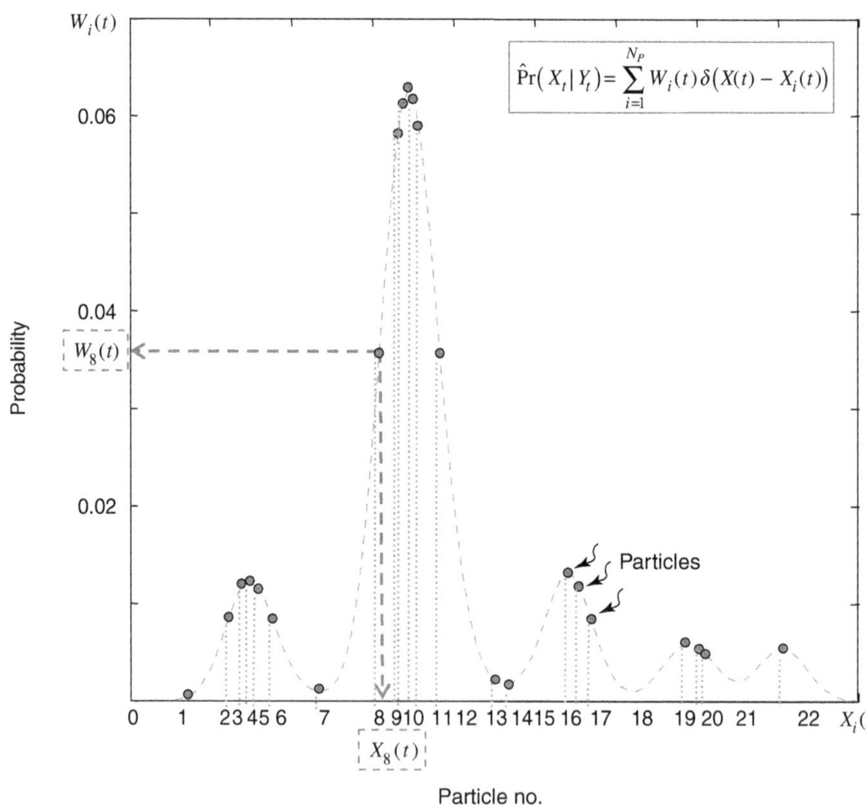

FIGURE 7.1 Particle filter representation of posterior probability distribution in terms of weights (probabilities) and particles (samples).

circle). Knowledge of this random measure $\{x_i(t), \mathcal{W}_i(t)\}$ characterizes an estimate of the instantaneous (at time t) filtering posterior distribution (dashed line),

$$\hat{\Pr}(x(t)|Y_t) \approx \sum_{i=1}^{N_p} \mathcal{W}_i(t)\delta(x(t) - x_i(t))$$

We observe from the figure that the particles need not be equally spaced or conform to a uniform grid and that they tend to coalesce in high probability regions (HPR).

Importance sampling plays a crucial role in state–space particle algorithm development. The PF does *not* involve linearizations around current estimates, but rather approximations of the desired distributions by these discrete random measures in contrast, to the Kalman filter which sequentially estimates the conditional mean and covariance used to characterize the (Gaussian) filtering posterior $\Pr(x(t)|Y_t)$. Particle filters are a sequential MC methodology based on "point mass" representation of probability distributions that only require a state–space representation of the underlying process. This representation provides a set of particles that evolve at each time-step leading to an instantaneous approximation of the target posterior distribution of the state at time t, given all of the data up to that time. Figure 7.2 illustrates the evolution of the posterior at each time-step. Here we see the estimated posterior at times t_1, t_2 to t_6 creating an instantaneous approximation of t vs. x_i vs. $\hat{\Pr}(x(t)|Y_t)$. Statistics are calculated across the ensemble at each time-step to provide estimates of the states. For example, the *minimum mean-squared error* (*MMSE*) or equivalently the *conditional mean* (CM) estimate is easily determined by averaging over x_i, since

$$\hat{x}_{\text{MMSE}}(t) = \hat{x}_{\text{CM}}(t) = \int x(t)\Pr(x(t)|Y_t)dx \approx \int x(t)\hat{\Pr}(x(t)|Y_t)dx$$
$$= \int x(t)\left(\sum_{i=1}^{N_p} \mathcal{W}_i(t)\delta(x(t) - x_i(t))\right)dx = \sum_{i=1}^{N_p} \mathcal{W}_i(t)x_i(t)$$

The maximum a posteriori (MAP) estimate is simply determined by finding the sample corresponding to the maximum weight of $x_i(t)$ across the ensemble at each time-step (as illustrated in Fig. 7.2), that is,

$$\hat{x}_{\text{MAP}}(t) = \arg\max_{x_i(t)} \hat{\Pr}(x(t)|Y_t) \tag{7.1}$$

In Bayesian processing, the idea is to sequentially estimate the posterior $\Pr(x(t-1)|Y_{t-1}) \rightarrow \Pr(x(t)|Y_t)$. Recall that optimal algorithms "exactly" track these distributions; however, they are impossible to implement because the updates require integration that cannot usually be performed analytically. Recall that the batch joint posterior distribution follows directly from the chain rule under the Markovian assumptions on $x(t)$ and the conditional independence of $y(t)$, that is,

$$\Pr(X_t|Y_t) = \prod_{i=0}^{t} \Pr(x(i)|x(i-1)) \times \Pr(y(i)|x(i)) \quad \text{for } \Pr(x(0)) := \Pr(x(0)|x(-1)) \tag{7.2}$$

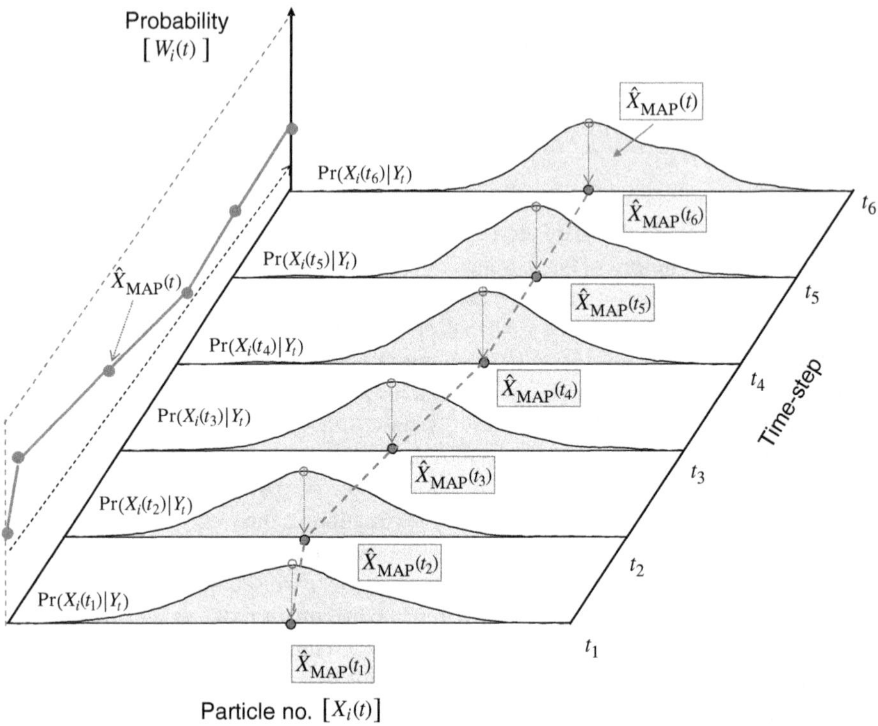

FIGURE 7.2 Particle filter surface (X_i vs. t vs. $\hat{P}r(X_i(t)|Y_t)$) representation of posterior probability distribution in terms of time (index), particles (samples), and weights or probabilities (left plot illustrates extracted MAP estimates vs. t).

The corresponding sequential importance sampling solution to the Bayesian estimation problem was given generically in Eq. 3.59 starting with the recursive form for the importance distribution as

$$q(X_t|Y_t) = q(X_{t-1}|Y_{t-1}) \times q(x(t)|X_{t-1}, Y_t)$$

leading to the sequential expression for the importance weights as

$$W(t) \propto \frac{\Pr(X_t|Y_t)}{q(X_t|Y_t)} = \frac{\Pr(Y_t|X_t) \times \Pr(X_t)}{q(X_{t-1}|Y_{t-1}) \times q(x(t)|X_{t-1}, Y_t)}$$

$$W(t) = W(t-1) \times \frac{\overbrace{\Pr(y(t)|x(t))}^{\text{Likelihood}} \times \overbrace{\Pr(x(t)|x(t-1))}^{\text{Transition}}}{q(x(t)|X_{t-1}, Y_t)} \tag{7.3}$$

Similarly, the SSPF evolving from this sequential importance sampling construct follows directly. Recall that the generic state–space characterization representing the transition and likelihood probabilities is

$$\Pr(x(t)|x(t-1)) \Leftrightarrow \mathcal{A}(x(t)|x(t-1))$$
$$\Pr(y(t)|x(t)) \Leftrightarrow C(y(t)|x(t))$$

which leads to an alternate representation of Eq. 7.2

$$\Pr(X_t|Y_t) = \prod_{i=0}^{t} \mathcal{A}(x(i)|x(i-1)) \times C(y(i)|x(i)) \quad \text{for } \mathcal{A}(x(0)|x(-1)) := \Pr(x(0)) \tag{7.4}$$

Thus the generic state–space sequential importance sampling solution is given by

$$\begin{aligned} x_i(t) &\sim q(x(t)|X_{t-1}, Y_t) \\ W_i(t) &= W_i(t-1) \times \frac{C(y(t)|x_i(t)) \times \mathcal{A}(x(t)|x_i(t-1))}{q(x_i(t)|X_{t-1}(i), Y_t)} \\ \mathcal{W}_i(t) &= \frac{W_i(t)}{\sum_{i=1}^{N_p} W_i(t)} \end{aligned} \tag{7.5}$$

where the sample at time t, $x_i(t)$ is referred to as the population or system of particles and $X_t(i)$ for the ith-particle as the history (trajectory or path) of that particular particle [6]. It is important to note that a desired feature of sequential MC sampling is that the N_p-particles of $X_t(i)$ are *i.i.d.*

In the practical application of the SSPF algorithm, we can obtain samples from the posterior by *augmenting* each of the existing samples with the new state draw, that is,

$$X_t(i) = \{x_i(t), X_{t-1}(i)\}$$

where $x_i(t) \sim q(x(t)|X_{t-1}, Y_t)$ and $X_{t-1}(i) \sim q(X_{t-1}|Y_{t-1})$.

Now if we further assume that

$$q(x(t)|X_{t-1}, Y_t) \rightarrow q(x(t)|x(t-1), y(t)) \tag{7.6}$$

then the importance distribution is *only* dependent on $[x(t-1), y(t)]$ which is common when performing filtering $\Pr(x(t)|Y_t)$ at each instant of time.

Assuming this is true, then the SSPF with $x_i(t) \rightarrow X_t(i)$ and $y(t) \rightarrow Y_t$ recursion becomes

$$
\begin{aligned}
x_i(t) &\sim q(x(t)|x(t-1), y(t)) \\
W_i(t) &= W_i(t-1) \times \frac{\mathcal{C}(y(t)|x_i(t)) \times \mathcal{A}(x(t)|x_i(t-1))}{q(x_i(t)|x_i(t-1), y(t))} \\
\mathcal{W}_i(t) &= \frac{W_i(t)}{\sum_{i=1}^{N_p} W_i(t)}
\end{aligned}
\tag{7.7}
$$

and the filtering *posterior* is estimated by

$$
\hat{\Pr}(x(t)|Y_t) \approx \sum_{i=1}^{N_p} \mathcal{W}_i(t) \times \delta(x(t) - x_i(t)) \tag{7.8}
$$

We summarize the *generic* SSPF in Table 7.1. Note that as N_p becomes large, in the limit, we have

$$
\lim_{N_p \rightarrow \infty} \hat{\Pr}(x(t)|Y_t) \rightarrow \Pr(x(t)|Y_t) \tag{7.9}
$$

which implies that the Monte Carlo error *decreases* as the number of particles *increase.*

7.3 IMPORTANCE PROPOSAL DISTRIBUTIONS

Selection of the importance distribution is a critical part of the design phase in particle filtering. Besides assuring that the distribution "covers" the posterior, there are a number of properties that can also be satisfied to achieve a robust design.

7.3.1 Minimum Variance Importance Distribution

Unfortunately, the generic algorithm presented in the previous section has a serious flaw, the *variance* of the importance weights *increases* over time [4, 6]. Therefore, the algorithm degenerates to a *single* non-zero weight after a few iterations. One way to limit this degeneracy is to choose an importance distribution that minimizes the weight variance based on the available information $[X_{t-1}, Y_t]$. That is, we would like the solution to the problem of minimizing the variance $V_{q(x(t)|X_{t-1},Y_t)}(W_i(t))$ with respect to $q(\cdot)$ such that

$$
V_q(W_i(t)) = W_i^2(t) \left[\int \frac{(\Pr(y(t)|x(t))\Pr(x(t)|X_{t-1}(i)))^2}{q(x(t)|X_{t-1}(i), Y_t)} dx(t) - \Pr(y(t)|X_{t-1}(i))^2 \right]
$$

TABLE 7.1 Generic State-Space Particle Filtering Algorithm

Initialize	
$x_i(0) \to \Pr(x(0));\quad W_i(0) = \frac{1}{N_p};\quad i = 1, \ldots, N_p$	[Sample]
Importance Sampling	
$x_i(t) \sim \mathcal{A}(x(t)\vert x_i(t-1))$	[State transition]
State–Space Transition	
$\mathcal{A}(x(t)\vert x_i(t-1)) \Leftarrow A(x(t-1), u(t-1), w_i(t-1));$ $w_i \sim \Pr(w_i(t))$	[Transition]
Measurement Likelihood	
$\mathcal{C}(y(t)\vert x_i(t)) \Leftarrow C(x_i(t), u(t), v(t));\quad v \sim \Pr(v(t))$	[Likelihood]
Weight Update	
$W_i(t) = W_i(t-1) \times \frac{\mathcal{C}(y(t)\vert x_i(t)) \times \mathcal{A}(x(t)\vert x_i(t-1))}{q(x_i(t)\vert x_i(t-1), y(t))}$	[Weights]
Weight Normalization	
$\mathcal{W}_i(t) = \frac{W_i(t)}{\sum_{i=1}^{N_p} W_i(t)}$	[Weight normalization]
Distribution	
$\hat{\Pr}(x(t)\vert Y_t) \approx \sum_{i=1}^{N_p} \mathcal{W}_i(t)\delta(x(t) - x_i(t))$	[Posterior distribution]
State Estimation (Inference)	
$\hat{x}(t\vert t) = E\{x(t)\vert Y_t\} \approx \sum_{i=1}^{N_p} \mathcal{W}_i(t)x_i(t)$	[Conditional mean]
$\hat{X}_{\mathrm{MAP}}(t) = \arg\max_{x_i(t)} \hat{\Pr}(x(t)\vert Y_t)$	[Maximum a posteriori]
$\hat{X}_{\mathrm{MED}}(t) = \mathrm{median}\left\{\hat{\Pr}(x(t)\vert Y_t)\right\}$	[Median]

is minimized. It has been shown [4] that the *minimum variance importance distribution* that minimizes the variance of the set of weights $\{W_i(t)\}$ is given by

$$q_{\mathrm{MV}}(x(t)|X_{t-1}, Y_t) \;\to\; \Pr(x(t)|x(t-1), y(t)) \tag{7.10}$$

Let us investigate this minimum variance proposal distribution in more detail to determine its realizability. Using Bayes' rule we can decompose[1] this proposal distribution in steps with $\mathrm{A} = x(t)$ and $\mathrm{B} = x(t-1), y(t)$:

$$\Pr(x(t)|x(t-1), y(t)) = \frac{\Pr(x(t-1), y(t)|x(t)) \times \Pr(x(t))}{\Pr(x(t-1), y(t))} \tag{7.11}$$

[1] We apply the following form of Bayes' rule: $\Pr(A|B) = \Pr(B|A) \times \Pr(A)/\Pr(B)$.

but the first term in the numerator can be decomposed further to obtain

$$\begin{aligned}\Pr(x(t-1), y(t)|x(t)) &= \Pr(y(t)|x(t-1), x(t)) \times \Pr(x(t-1)|x(t)) \\ &= \Pr(y(t)|x(t-1), x(t)) \times \left[\frac{\Pr(x(t)|x(t-1))\Pr(x(t-1))}{\Pr(x(t))}\right]\end{aligned} \tag{7.12}$$

Substituting this relation into Eq. 7.11 and canceling the $\Pr(x(t))$ terms, we obtain

$$\Pr(x(t)|x(t-1), y(t)) = \frac{\Pr(y(t)|x(t-1), x(t)) \times \Pr(x(t)|x(t-1)) \times \Pr(x(t-1))}{\Pr(x(t-1), y(t))} \tag{7.13}$$

Expand the denominator in Eq. 7.13 using Bayes' rule,

$$\Pr(x(t-1), y(t)) = \Pr(y(t)|x(t-1)) \times \Pr(x(t-1))$$

substitute, cancel the $\Pr(x(t-1))$ terms and apply the conditional independence assumption of the measurements on past states, that is,

$$\Pr(y(t)|x(t-1), x(t)) \rightarrow \Pr(y(t)|x(t))$$

to obtain the final expression for the *minimum variance proposal distribution* as

$$q_{\mathrm{MV}}(x(t)|X_{t-1}, Y_t) = \Pr(x(t)|x(t-1), y(t)) = \frac{\Pr(y(t)|x(t)) \times \Pr(x(t)|x(t-1))}{\Pr(y(t)|x(t-1))} \tag{7.14}$$

If we substitute this expression for the importance distribution in the weight recursion of Eq. 3.63, then we have

$$W(t) = W(t-1) \times \frac{\Pr(y(t)|x(t)) \times \Pr(x(t)|x(t-1))}{q_{\mathrm{MV}}(x(t)|X_{t-1}, Y_t)}$$

or

$$W(t) = W(t-1)\Pr(y(t)|x(t))\Pr(x(t)|x(t-1)) \times \left[\frac{\Pr(y(t)|x(t-1))}{\Pr(y(t)|x(t)) \times \Pr(x(t)|x(t-1))}\right]$$

Canceling like terms and applying the Chapman–Kolmogorov relation, we obtain the corresponding *minimum variance weights*

$$\begin{aligned}W_{\mathrm{MV}}(t) &= W_{\mathrm{MV}}(t-1) \times \Pr(y(t)|x(t-1)) = W_{\mathrm{MV}}(t-1) \\ &\quad \times \int \mathcal{C}(y(t)|x(t)) \times \mathcal{A}(x(t)|x(t-1))\, dx(t)\end{aligned} \tag{7.15}$$

which indicates that the importance weights can be calculated *before* the particles are propagated to time t. From this expression we can also see the problem with the minimum variance importance function approach: (1) we must sample from $\Pr(x(t)|x(t-1), y(t))$; and (2) we must evaluate the integral which generally has *no* analytic form.

7.3.2 Transition Prior Importance Distribution

Another choice for an importance distribution is the *transition prior*. This prior is defined in terms of the state–space representation by $\mathcal{A}(x(t)|x(t-1)) \Leftarrow A(x(t-1), u(t-1), w(t-1))$ which is dependent on the known excitation and process noise statistics and is given by

$$q_{\text{prior}}(x(t)|x(t-1), Y_t) \rightarrow \Pr(x(t)|x(t-1))$$

Substituting this choice into the expression for the weights of Eq. 7.3 gives

$$\begin{aligned} W_i(t) &= W_i(t-1) \times \frac{\Pr(y(t)|x_i(t)) \times \Pr(x(t)|x_i(t-1))}{q_{\text{prior}}(x(t)|x_i(t-1), Y_t)} \\ &= W_i(t-1) \times C(y(t)|x_i(t)) \end{aligned} \tag{7.16}$$

since the priors cancel.

Note two properties for this choice of importance distribution. First, the weight does *not* use the most recent observation $y(t)$ and second it does not use the past particles $x_i(t-1))$ but only the likelihood. This choice is easily implemented and updated by simply evaluating the measurement likelihood, $C(y(t)|x_i(t)); i = 1, \ldots, N_p$ for the sampled particle set. In contrast to the minimum variance choice, these weights require the particles to be propagated to time t *before* the weights can be calculated.

This choice of importance distribution can lead to problems, since the transition prior is not conditioned on the measurement data, especially the most recent. Failing to incorporate the latest available information from the most recent measurement to propose new values for the states leads to only a few particles having significant weights when their likelihood is calculated. The transition prior is a much broader distribution than the likelihood, indicating that only a few particles will be assigned a large weight. Thus, the algorithm will degenerate rapidly and lead to poor performance especially when data *outliers* occur or measurement noise is *small*. These conditions lead to a "mismatch" between the prior prediction and posterior distributions. Techniques such as the auxiliary particle filter [2, 7, 8] as well as local linearized particle filters [4, 6, 9] have been developed that drive the particles to regions of high likelihood by incorporating the current measurement. Thus, the SSPF algorithm takes the same generic form as before with the minimum variance approach; however, we note that the importance weights are much simpler to evaluate with this approach which has been termed the *bootstrap* PF, the *condensation* PF, or the survival of the fittest algorithm. We summarize the *bootstrap* PF algorithm in Table 7.2 to follow.

TABLE 7.2 Bootstrap SIR State-Space Particle Filtering Algorithm

Initialize	
$x_i(0) \sim \Pr(x(0)) \quad W_i(0) = \frac{1}{N_p} \quad i = 1, \ldots, N_p$	[Sample]
Importance Sampling	
$x_i(t) \sim \mathcal{A}(x(t)\|x_i(t-1)) \Leftarrow A(x(t-1), u(t-1), w_i(t-1));$ $w_i \sim \Pr(w_i(t))$	[State transition]
Weight Update	
$W_i(t) = \mathcal{C}(y(t)\|x_i(t)) \Leftarrow C(x(t), u(t), v(t)); \quad v \sim \Pr(v(t))$	[Weight/likelihood]
Weight Normalization	
$\mathcal{W}_i(t) = \frac{W_i(t)}{\sum_{i=1}^{N_p} W_i(t)}$	
Resampling Decision	
$\hat{N}_{\text{eff}} = \frac{1}{\sum_{i=1}^{N_p} W_i^2(t)}$	[Effective samples]
$\hat{N}_{\text{eff}} = \begin{cases} \text{Resample} & \leq N_{\text{thresh}} \\ \text{Accept} & > N_{\text{thresh}} \end{cases}$	
Resampling	
$\hat{x}_i(t) \Rightarrow x_i(t)$	
Distribution	
$\hat{\Pr}(x(t)\|Y_t) \approx \sum_{i=1}^{N_p} \mathcal{W}_i(t)\delta(x(t) - \hat{x}_i(t))$	[Posterior distribution]

7.4 RESAMPLING

The main objective in simulation-based sampling techniques is to generate *i.i.d.* samples from the targeted posterior distribution in order to perform statistical inferences extracting the desired information. Thus, the importance weights are quite critical since they contain probabilistic information about each specific particle. In fact, they provide us with information about "how probable a sample drawn from the target posterior has been" [10, 11]. Therefore, the weights can be considered acceptance probabilities enabling us to generate independent (approximately) samples from the posterior $\Pr(x(t)|Y_t)$. Recall that the empirical distribution $\hat{\Pr}(x(t)|Y_t)$ is defined over a set of finite (N_p) random measures $\{x_i(t), \mathcal{W}_i(t)\}$; $i = 1, \ldots, N_p$ approximating the posterior, that is,

$$\hat{\Pr}(x(t)|Y_t) \approx \sum_{i=1}^{N_p} \mathcal{W}_i(t)\delta(x(t) - x_i(t)) \tag{7.17}$$

One of the major problems with importance sampling algorithms is the depletion of the particles. The degeneracy of the particle weights creates a problem that must be resolved before these algorithms can be of any pragmatic use. It occurs because the variance of the importance weights increases in time [4] thereby making it impossible

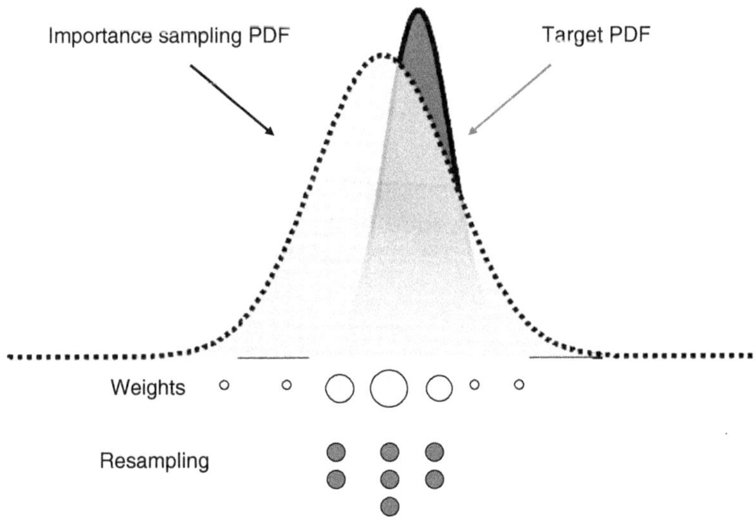

FIGURE 7.3 Resampling consists of processing the predicted particles with their associated weights (probabilities), duplicating those particles of high weight (probability) and discarding those of low weight.

to avoid this weight degradation. Degeneracy implies that a large computational effort is devoted to updating particles whose contribution to the posterior is negligible. This approach is bound to fail in the long run, since the weight degeneration leads to a few particles containing most of the probability mass. Thus, there is a need to somehow resolve this problem to make the sequential simulation-based techniques viable. This requirement leads to the idea of "resampling" the particles.

Resampling involves sampling N_p-draws from the current population of particles using the normalized weights as selection probabilities. The resampling process is illustrated in Fig. 7.3. Particles of low probability (small weights) are removed and those of high probability (large weights) are retained and replicated. Resampling results in two major effects: (1) the algorithm is more complex and is not merely the simple importance sampling method; and (2) the resampled trajectories $X_t(i)$ are no longer *i.i.d.* and the normalized weights are set to $1/N_p$.

Resampling, therefore, can be thought of as a realization of enhanced particles $\hat{x}_k(t)$ extracted from the original samples $x_i(t)$ based on their "acceptance probability" $\mathcal{W}_i(t)$ at time t. Statistically, we have

$$\Pr(\hat{x}_k(t) = x_i(t)) = \mathcal{W}_i(t) \quad \text{for } i = 1, \ldots, N_p \tag{7.18}$$

or we write it symbolically as

$$\hat{x}_k(t) \Rightarrow x_i(t)$$

where "⇒" defines the *resampling operator* generating a set of new particles $\{\hat{x}_k(t)\}$ replacing the old set $\{x_i(t)\}$.

To see how this is achieved, consider the following resampling example.

Example 7.1

Consider the following example of resampling emphasizing the generation of more particles in the HPR and the elimination (by draws) of particles with small probabilities (weights). Assume we have 100 balls (samples) with the following distribution and corresponding probabilities:

- White (W) = 40 — $\Pr(W) = 0.40$
- Blue (B) = 25 — $\Pr(B) = 0.25$
- Red (R) = 10 — $\Pr(R) = 0.10$
- Green (G) = 2 — $\Pr(W) = 0.02$

The balls are placed into an urn individually as indicated in Fig. 7.4a with the corresponding histogram shown in Fig. 7.4b. Next, the balls are sampled *with replacement* 100 times and their new counts recorded. The new or *resampled* histogram is shown in Fig. 7.4c where we see that those balls of highest probabilities are drawn more frequently and therefore retained, since it is more likely that they will be selected rather than those of smaller probabilities. This leads to the shape of the new histogram and the elimination of the low probability green balls. Hypothetically, resampling generates the following probabilities:

- White (W) = 60 — $\Pr(W) = 0.60$
- Blue (B) = 20 — $\Pr(B) = 0.20$
- Red (R) = 8 — $\Pr(R) = 0.08$
- Green (G) = 0 — $\Pr(W) = 0.00$

illustrating this process. △△△

Thus, the fundamental concept in resampling theory is to preserve particles with large weights (i.e., large probabilities) while discarding those with small weights. Two steps must occur to resample effectively: (1) a decision, on a weight-by-weight basis, must be made to *select* the appropriate weights and *reject* the inappropriate; and (2) resampling must be performed to minimize the degeneracy. This overall strategy when coupled with importance sampling is termed sequential *sampling-importance-resampling* (SIR) [4].

A reasonable measure of degeneracy is the *effective* particle sample size based on the coefficient of variation [12] defined by

$$N_{eff}(t) := \frac{N_p}{E_q\{W^2(t)\}} = \frac{N_p}{1 + V_q(W(t))} \leq N_p \tag{7.19}$$

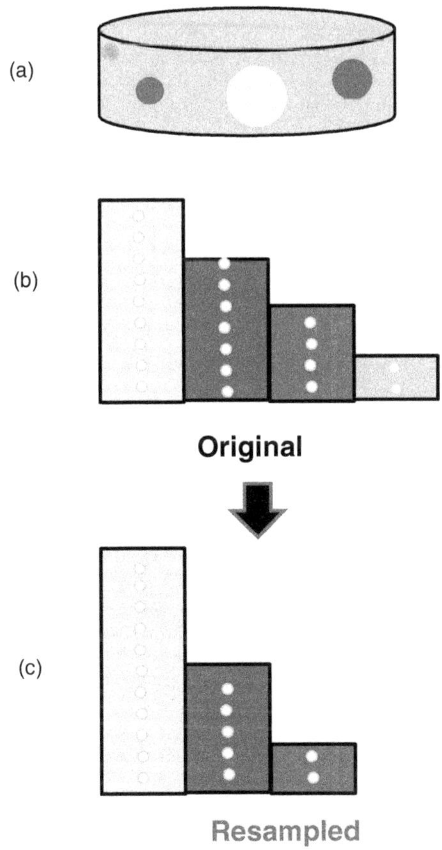

FIGURE 7.4 Hypothetical resampling: (a) Urn of white, blue, red, and green balls; (b) Original distribution (histogram): Pr(W) = 0.40, Pr(B) = 0.25, Pr(R) = 0.10, and Pr(G) = 0.02; (c) Resampled distribution (histogram): Pr(W) = 0.60, Pr(B) = 0.20, Pr(R) = 0.08, and Pr(G) = 0.0.

An estimate of the effective number of particles at time t is given by

$$\hat{N}_{\text{eff}}(t) = \frac{1}{\sum_{i=1}^{N_p} W_i^2(t)} \tag{7.20}$$

and a decision based on the *rejection method* [13] is made by comparing it to a threshold N_{thresh}. That is, when $\hat{N}_{\text{eff}}(t)$ is less than the threshold, resampling is performed.

$$\hat{N}_{\text{eff}}(t) = \begin{cases} \leq N_{\text{thresh}} & \text{Resample} \\ > N_{\text{thresh}} & \text{Accept} \end{cases}$$

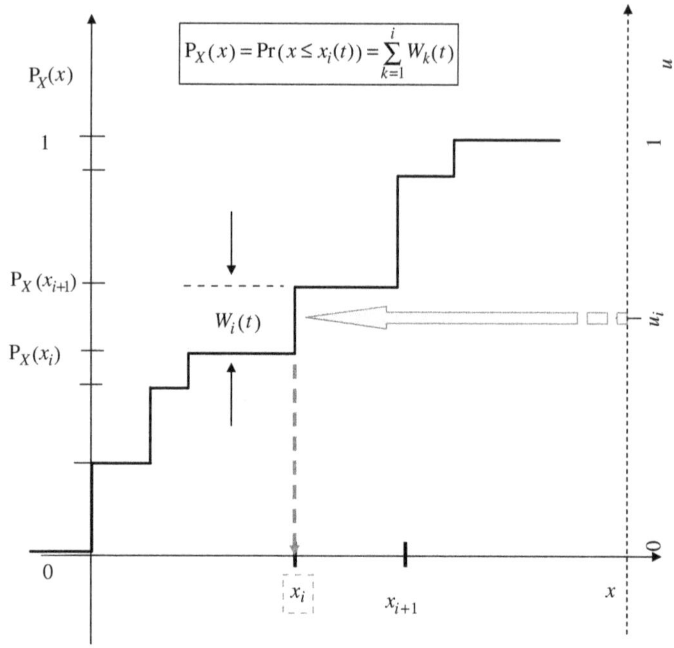

FIGURE 7.5 Resampling (with replacement) by inverse transforming a uniform sampler to generate samples from target distribution.

Once the decision is made to resample, a uniform sampling procedure [14] can be applied removing samples with low importance weights and replicating samples with high importance weights. Resampling (with replacement) generates a set of new samples with uniform weights from an approximate discrete posterior distribution,

$$\hat{\Pr}(x(t)|Y_t) \approx \sum_{i=1}^{N_p} \hat{\mathcal{W}}_i(t)\delta(x(t) - \hat{x}_i(t)) \tag{7.21}$$

so that $\Pr[\hat{x}_i(t) = x_j(t)] = \hat{W}_j(t)$. The resulting independent and identically distributed sample from the probability mass of Eq. 7.21 is uniform such that the sampling induces the mapping of $\{\hat{x}_i(t), \hat{\mathcal{W}}_i(t)\} \rightarrow \{x_i(t), \mathcal{W}_i(t)\}$, $\hat{\mathcal{W}}_i(t) = 1/N_p \forall i$. The selection of $\hat{x}_i(t) = x_j(t)$ is shown in Fig. 7.5. Here the procedure is termed systematic resampling [4]. For each resampled particle, that is, N_i-times is related to the original particle. The methodology relies on uniformly sampling the cumulative distribution function resulting in the new replicated particles or uniform weighting. The resampling algorithm is incorporated in Table 7.2.

Resampling decreases the degeneracy problem algorithmically, but introduces its own set of problems. After one resampling step, the simulated trajectories are *no longer* statistically independent. Therefore, the simple theoretical convergence results under these assumptions lose their validity. Pragmatically, resampling can limit algorithm parallelization because combining particles causes an increase in computational complexity. Also there is a possible loss in diversity caused by replication of those particles with highest importance weights [4]. Thus as with any methodology there are tradeoffs that must be considered.

7.4.1 Multinomial Resampling

There are a variety of techniques available to implement the basic resampling method [1, 6, 15–19]. The usual approach is to resample with replacement, since the probability of each particle $x_i(t)$ is given by the normalized weight $W_i(t)$. Therefore, the number of times N_i that each particular particle in the original set $\{x_i(t)\}$ is selected follows a binomial distribution $\text{Bin}(N_p, W_i(t))$. The corresponding vector $[N_1, \ldots, N_{N_p}]$ is distributed according to a multinomial distribution with parameter N_p and probability of success $[W_1(t), \ldots, W_{N_p}(t)]$. With this resampling scheme, particles in the original set with small variance weights are most likely discarded, while those of high weights are replicated in proportion to these weights. The *multinomial resampling* method is given by

- Given a random measure at time t, that is, a set of particles and weights, $\{x_i(t), W_i(t)\}, i = 1, \ldots, N_p$;
- Sample uniformly, $u_k \to \mathcal{U}(0, 1); k = 1, \ldots, N_p$;
- Determine the index, i_k: $i_k = k$ for $\Pr(x_{i_k}(t) = x_k(t)) = u_k$;
- Select a new sample $\hat{x}_{i_k}(t) \Rightarrow x_i(t)$ and weight $\hat{W}_{i_k}(t) = \frac{1}{N_p}$ based on the new sample index i_k; and
- Generate the new random (resampled) measure: $\{\hat{x}_{i_k}, \hat{W}_{i_k}(t)\}$; for $k = 1, \ldots, N_p$.

Here, the index notation i_k designates the original ith particle or parent and the new kth particle using the inverse CDF method of Section 3.3. This sampling scheme is equivalent to drawing $i_k; k = 1, \ldots, N_p$ samples from a multinomial distribution with parameters $\mathcal{M}(N_{i_k}, W_{i_k}(t))$ and corresponding statistics: mean $E\{N_{i_k}\} = N_p$ and variance, $\text{Var}(N_{ik}) = N_p W_{ik}(t)(1 - W_{ik}(t))$.

The basic idea is to first construct the CDF from the original random measure $\{x_i(t), W_i(t)\}$, since it is given by

$$\Pr(X(t) \le x_i(t)) \approx \sum_{i=1}^{N_p} W_i(t)\mu(x(t) - x_i(t)) \tag{7.22}$$

where $\mu(\cdot)$ is the unit step function.

Uniform samples u_k are drawn from the interval $[0, 1]$ and projected onto the inverse CDF (see Section 3.3) corresponding to the associated probability and identifying the particular new particle sample index i_k and corresponding replacement particle $\hat{x}_{i_k}(t)$ leading to the resampling

$$\hat{x}_{i_k}(t) \Rightarrow x_i(t) \tag{7.23}$$

Clearly those particles or samples with highest probability (or weights) will be selected more frequently, thereby, replacing particles with lower probability (weights) and therefore, the new random measure is created, that is,

$$\{\hat{x}_{i_k}(t), \hat{W}_{i_k}(t)\} \Rightarrow \{x_i(t), W_i(t)\} \quad \text{with } \hat{W}_{i_k}(t) = \frac{1}{N_p} \tag{7.24}$$

This resampling technique represents a direct implementation of random sampling by generating an *i.i.d.* sample from the empirical posterior distribution

$$\begin{aligned}\Pr(x(t)|Y_t) &\approx \sum_{i=1}^{N_p} W_i(t)\delta(x(t) - x_i(t)) \\ &\rightarrow \sum_{i=1}^{N_p} \hat{W}_{i_k}(t)\delta(x(t) - \hat{x}_{i_k}(t)) = \frac{1}{N_p}\sum_{i=1}^{N_p} \delta(x(t) - \hat{x}_{i_k}(t))\end{aligned} \tag{7.25}$$

A second more efficient way of generating this measure is the "systematic" resampling method.

7.4.2 Systematic Resampling

The *systematic resampling* method is based on an *ordered* technique in which a set of N_p-ordered uniform variates are generated [20, 21]. It minimizes the error variance between the original selected sample and its mean. Thus, the *systematic sampling* method is given by

- Given a random measure at time t, that is, a set of particles and weights, $\{x_i(t), W_i(t)\}, i = 1, \ldots, N_p$;
- Sample uniform N_p-ordered variates: $\hat{u}_k = u_k + \frac{k-1}{N_p}$ for $k = 1, \ldots, N_p$ and $u_k \rightarrow \mathcal{U}(0, 1)$;
- Determine the index, i_k: $i_k = k$ for $\mathrm{P}_X(x_{k-1}(t)) \leq \hat{u}_k \leq \mathrm{P}_X(x_k(t))$; (see Fig. 7.4)
- Select a new sample, $\hat{x}_{i_k}(t) \Rightarrow x_i(t)$ and weight, $\hat{W}_{i_k}(t) = \frac{1}{N_p}$ based on the new sample index, i_k; and
- Generate the new random (resampled) measure: $\{\hat{x}_{i_k}, \hat{W}_{i_k}(t)\}$; for $k = 1, \ldots, N_p$.

where recall the CDF is given by: $\mathrm{P}_X(x_k(t)) = \sum_{k=1}^{N_p} W_k(t)\mu(x(t) - x_k(t))$ with $\mu(\cdot)$ is a unit step function.

The final sampling scheme we discuss has a low weight variance, the residual method.

7.4.3 Residual Resampling

The *residual resampling* method is based on the idea of estimating the number of times each particle should be replicated, that is, the ith particle is replicated, say

$$\overline{N}_p(i) := \mathrm{Int}(E\{N_p(i)\}) = \mathrm{Int}(N_p \times W_i(t)) \tag{7.26}$$

times where Int means the "smallest integer value of".

The remaining particles are sampled using the multinomial sampling method discussed above. Here we have

$$\overline{N}_p(t) := N_p - \sum_{i=1}^{N_p} \overline{N}_p(i) \tag{7.27}$$

with corresponding weights

$$\hat{W}_i(t) = \frac{1}{\overline{N}_p(t)}(N_p W_i(t) - \overline{N}_p(i)) \tag{7.28}$$

The overall effect is to reduce the variance by $E\{(N_p(i) - E\{\overline{N}_p(i)\})^2\}$, since the particles cannot be replicated less than $E\{N_p(i)\}$ times.

We summarize the *residual resampling* method by

- Given a random measure at time t, that is, a set of particles and weights, $\{x_i(t), W_i(t)\}, i = 1, \ldots, N_p$;
- Calculate $\overline{N}_p(i)$: $\overline{N}_p(i) = \mathrm{Int}(N_p \times W_i(t))$;
- Multinomial sample: $\hat{x}_i(t) \Rightarrow x_i(t)$ for $i = 1, \ldots, \overline{N}_p(i)$; and
- Update the new random (resampled) measure: $\{\hat{x}_{i_k}, \hat{W}_{i_k}(t)\}$; for $k = 1, \ldots, N_p$.

So we see that there are a variety of resampling schemes that can be employed to solve the particle degeneracy problem. We can now update our generic particle filtering algorithm to incorporate a resampling procedure and alleviate the degeneracy problem created by the variation of the weights.

To visualize the "resampling" approach mitigating the particle degeneracy problem, the SIR is illustrated in Fig. 7.6. Here we show the evolution of the particles

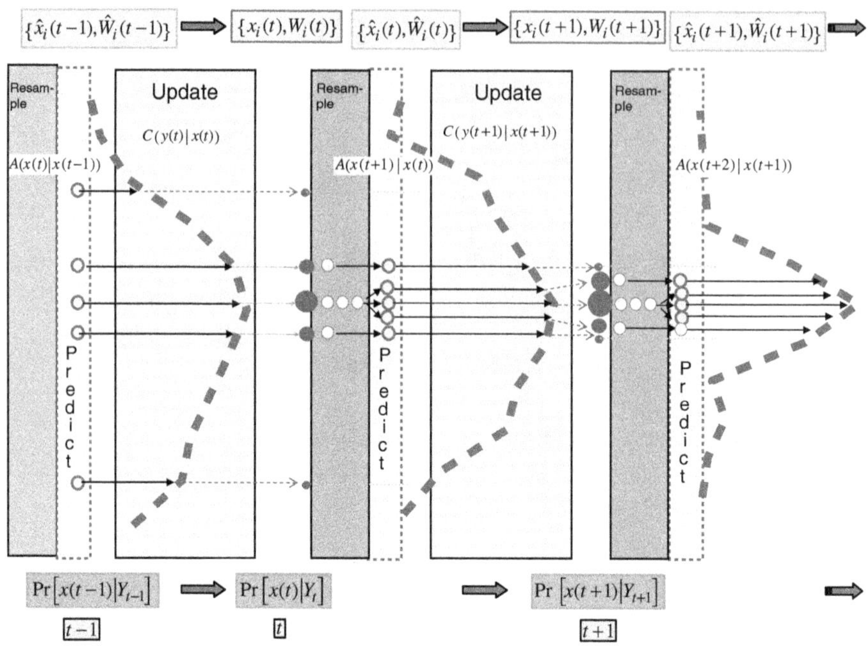

FIGURE 7.6 Evolution of particle filter weights and particles using the sequential state-space SIR algorithm: resampling, propagation-step (state-space transition model), update-step (state-space measurement likelihood), resampling … .

and weights starting with the uniform weighting $\left(\hat{W}_i(t-1) = \frac{1}{N_p}\right)$. Once the initial weights are updated, they are *resampled uniformly*. Next, they are propagated using the state–space transition mechanism (model), then updated using the measurement likelihood producing the measure $\{x_i(t), W_i(t)\}, i = 1, \ldots, N_p$ leading to an approximation of the posterior distribution at time t. This measure is then resampled, propagated, updated, and so on. A generic flow diagram of the algorithm is shown in Fig. 7.7 where we again illustrate the basic ingredients of the SIR technique.

7.5 STATE-SPACE PARTICLE FILTERING TECHNIQUES

There are a number of pragmatic PF techniques that have been introduced in the literature. Here we discuss some of the more robust and popular techniques that have been applied to a wide variety of problems starting with the bootstrap processor [1, 2, 6].

7.5.1 Bootstrap Particle Filter

The basic "bootstrap" algorithm developed by Gordon, Salmond, and Smith [16] is one of the first practical implementations of the processor to the tracking problem.

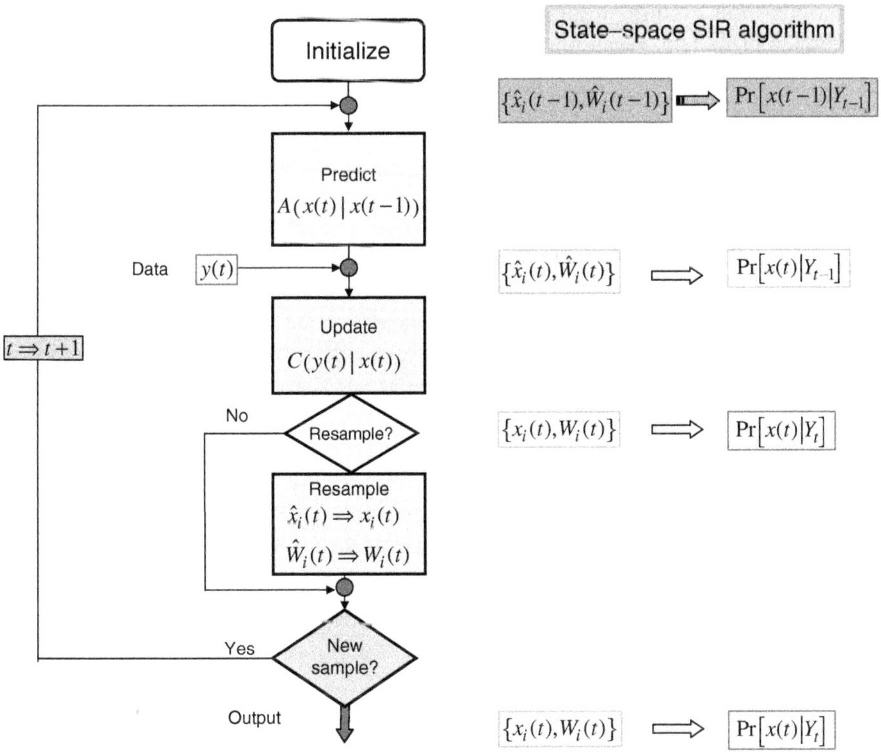

FIGURE 7.7 State-space SIR particle filtering algorithm structure: initialization, propagation (state transition), updating (measurement likelihood), resampling.

It is the most heavily applied of all PF techniques due to its simplicity. Thus, the SSPF algorithm takes the same generic form as before with the minimum variance approach; however, we note that the importance weights are much simpler to evaluate with this approach which has been termed the *bootstrap* PF, the *condensation* PF, or the survival of the fittest algorithm [2, 16, 22].

As mentioned previously, it is based on sequential SIR ideas and uses the *transition prior* as its underlying proposal distribution,

$$q_{\text{prior}}(x(t)|x(t-1), Y_t) = \Pr(x(t)|x(t-1))$$

The corresponding weight becomes quite simple and only depends on the likelihood; therefore, it is not even necessary to perform a sequential updating because

$$W(t) = \left(\frac{\Pr(y(t)|x(t)) \times \Pr(x(t)|x(t-1))}{\Pr(x(t)|x(t-1))} \right) \times W(t-1) = \Pr(y(t)|x(t)) \times W(t-1)$$

since the filter requires resampling to mitigate variance (weight) increases at each time-step [16]. *After* resampling, the new weights become

$$\mathcal{W}(t) \rightarrow \frac{1}{N_p} \Rightarrow W(t) = \Pr(y(t)|x(t)) = C(y(t)|x(t))$$

revealing that there is *no need* to save the likelihood (weight) from the previous step! With this in mind we summarize the simple *bootstrap particle filter* algorithm in Table 7.2. One of the primary features as well as shortcomings of this technique is that it does not use the latest available measurement in the importance proposal, but only the previous particles $x_i(t-1)$, which differs from the minimum variance approach. Also in order to achieve *convergence*, it is necessary to resample at every time-step. In practice, however, many applications make the decision to resample based on the effective sample-size metric discussed in the previous section.

In order to construct the bootstrap PF, we assume that: (1) $x_i(0) \sim \Pr(x(0))$ is known; (2) $\mathcal{A}(x(t)|x(t-1))$, $C(y(t)|x(t))$ are known; (3) samples can be generated using the process noise input and the state-space model $A(x(t-1), u(t-1), w(t-1))$; (4) the likelihood is available for point-wise evaluation $C(y(t)|x(t))$ based on the measurement model $C(x(t), v(t))$; and (5) resampling is performed at every time-step. To implement the algorithm we

- Generate the initial state $x_i(0)$
- Generate the process noise $w_i(t)$
- Generate the particles $x_i(t) = A(x_i(t-1), u(t-1), w_i(t-1))$—the *prediction-step*
- Generate the likelihood $C(y(t)|x_i(t))$ using the current particle and measurement—the *update step*
- Resample the set of particles retaining and replicating those of highest weight (probability) $\hat{x}_i(t) \Rightarrow x_i(t)$
- Generate the new set $\{\hat{x}_i(t), \hat{\mathcal{W}}_i(t)\}$ with $\hat{\mathcal{W}}_i(t) = \frac{1}{N_p}$

Next, we revisit the model of Jazwinski [21] in Chapter 5 and apply the simple bootstrap algorithm to demonstrate the PF solution using the state–space SIR particle filtering algorithm.

Example 7.2

Recall the discrete state–space representation of the basic problem given by the Markovian model:

$$\begin{aligned} x(t) &= (1 - 0.05\triangle T)x(t-1) + 0.04\triangle T x^2(t-1) + w(t-1) \\ y(t) &= x^2(t) + x^3(t) + v(t) \end{aligned}$$

where $\triangle t = 0.01$, $w \sim \mathcal{N}(0, 10^{-6})$ and $v \sim \mathcal{N}(0, 0.09)$. The initial state is Gaussian distributed with $\overline{x}_i(0) \sim \mathcal{N}(\overline{x}(0), \overline{P}(0))$ and $\overline{x}(0) = 2.0$, $\overline{P}(0) = 10^{-2}$.

We selected the following *simulation* run parameters:

Number of Particles:	250
Number of Samples:	150
Number of States:	1
Sampling Interval:	0.01 s
Number of Measurements:	1
Process Noise Covariance:	1×10^{-6}
Measurement Noise Covariance:	9×10^{-2}
Initial State:	2
Initial State Covariance:	10^{-20}

Thus, the bootstrap SIR algorithm of Table 7.2 for this problem becomes:

1. Draw samples (particles) from the state transition distribution: $x_i(t) \rightarrow \mathcal{N}(x(t) : \mathbf{a}[x(t-1)], \mathbf{R}_{ww})$, that is, generate

$$w_i(t) \rightarrow \Pr(w(t)) \sim \mathcal{N}(0, R_{ww})$$

and calculate $\{x_i(t)\}$ using the process model and $w_i(t)$

$$x_i(t) = (1 - 0.05\triangle T)x_i(t-1) + 0.04\triangle T x_i^2(t-1) + w_i(t-1)$$

2. Estimate the weight/likelihood, $W_i(t) = C(\mathbf{y}(t)|x_i(t)) \rightarrow \mathcal{N}(\mathbf{y}(t) : \mathbf{c}[x_i(t)], \mathbf{R}_{vv}(t))$

$$c[x_i(t)] = x_i^2(t) + x_i^3(t)$$

$$\ln C(\mathbf{y}(t)|x_i(t)) = -\frac{1}{2}\ln 2\pi R_{vv} - \frac{(y(t) - x_i^2(t) - x_i^3(t))^2}{2R_{vv}}$$

3. Update the weight: $W_i(t) = C(\mathbf{y}(t)|x_i(t))$
4. Normalize the weight: $\mathcal{W}_i(t) = W_i(t) / \sum_{i=1}^{N_p} W_i(t)$
5. Decide to resample if $N_{eff} \leq N_{\text{thresh}}$
6. If resample: $\hat{x}_i(t) \Rightarrow x_i(t)$
7. Estimate the instantaneous posterior:

$$\hat{\Pr}(x(t)|Y_t) \approx \sum_{i=1}^{N_p} \mathcal{W}_i(t)\delta(x(t) - \hat{x}_i(t))$$

8. Estimate (inference) the corresponding statistics:

$$\hat{X}_{\text{MAP}}(t) = \arg\max_{x_i(t)} \hat{\Pr}(x(t)|Y_t)$$

$$\hat{X}_{\text{MMSE}}(t) = \hat{X}_{\text{CM}}(t) = E\{x(t)|Y_t\} = \sum_{i=1}^{N_p} \hat{x}_i(t)\hat{\Pr}(x(t)|Y_t)$$

$$\hat{X}_{\text{MEDIAN}}(t) = \text{median}(\hat{\Pr}(x(t)|Y_t))$$

Note that compared to the previous examples of Chapter 5 for the extended Bayesian processors, we have added more process noise to demonstrate the effectiveness of the bootstrap processor. The results of the bootstrap PF are shown in Fig. 7.8. The usual classical performance metrics are shown: zero-mean $(0.03 < 0.17)$, whiteness (0.78% out) and WSSR (below threshold), all indicating (approximately) a tuned processor. The PF tracks the state and measurement after the initial transient error has diminished.

In Fig. 7.9, we show the bootstrap state and measurement estimates (inferences), that is, the MAP and MMSE (CM) compared to the modern sigma-point processor SPBP (UKF). The plots demonstrate that the PF can outperform the sigma-point design, that assumes a unimodal Gaussian distribution. The estimated state and predicted measurement posterior distributions are shown in Fig. 7.10 along with a time-slice in Fig. 7.11 at time 1.04 s, demonstrating the capability of the bootstrap PF to characterize the multimodal nature of this problem. △△△

This completes the development of the most popular and simple PF technique.

Next, we consider some alternative approaches that attempt to approximate the minimum variance importance distribution more closely for improved performance.

7.5.2 Auxiliary Particle Filter

The auxiliary particle filter employing sampling-importance-resampling (ASIR) is a variant of the standard SIR [8]. It is based on attempting to mitigate two basic weaknesses in particle filtering: (1) poor outlier performance; and (2) poor posterior tail performance. These problems evolve from the empirical approximation of the filtering posterior which can be considered a *mixture* distribution.

The basic concept of ASIR is to mimic the operation of the minimum variance (optimal) importance distribution $q_{\text{MV}}(x(t)|x(t-1), y(t))$ by introducing an *auxiliary* variable $\mathcal{K}$ representing the weight of the mixture used in the empirical prediction distribution estimate. The idea is to perform resampling at time $(t-1)$ using the available measurement at time t *before* the particles $\{x_i(t)\}$ are propagated to time t through the transition and likelihood distributions. The key step is to favor particles at time $(t-1)$ that are likely to "survive" (largest weights) at the *next* time-step t. The problem is that these schemes tend to introduce a *bias* into the estimated posterior that

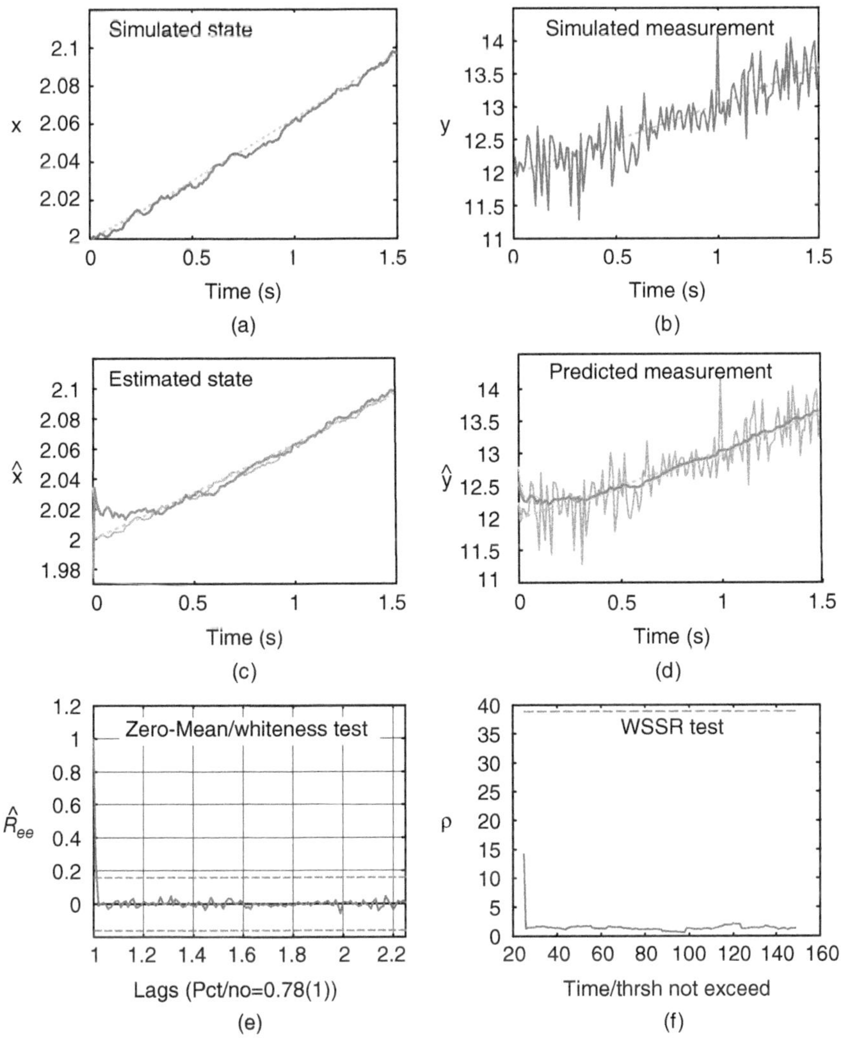

FIGURE 7.8 Nonlinear trajectory simulation/estimation for low process noise case: (a) simulated state and mean; (b) simulated measurement and mean; (c) bootstrap state estimate; (d) bootstrap measurement estimate; (e) zero-mean/whiteness test ($0.03 < 0.177$/0.78% out); (f) WSSR test (below threshold).

must then be corrected by modifying the weight of the remaining particles. Thus, the ASIR is a two-stage process such that: (1) particles with large predictive likelihoods at time-step $(t-1)$ are propagated; and (2) the resulting particles are then re-weighted and drawn from the resulting posterior.

Following the development in Cappe [6], we start with a proposal over the entire path $\{X_t\}$ up to time t under the assumption that the joint posterior at time $(t-1)$ is

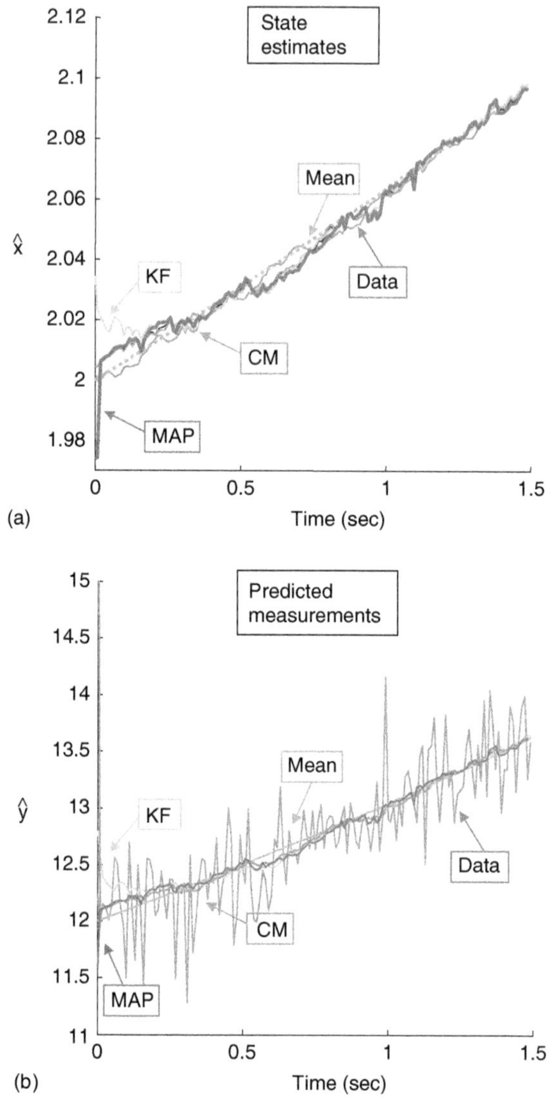

FIGURE 7.9 Nonlinear trajectory Bayesian estimation comparison for low process noise case problem: (a) state estimates: MAP, MMSE (CM), UKF, median; (b) predicted measurement estimates: MAP, MMSE (CM), UKF, median.

well approximated by a particle representation $\{\mathcal{W}_i(t-1), X_{t-1}(i)\}$. Thus, the joint importance proposal for the "new" particles $\{X_t(i)\}$ is

$$q(X_t) = \overbrace{q(X_{t-1}|Y_t)}^{\text{Past}} \times \overbrace{q(x(t)|x(t-1), y(t))}^{\text{New}} \tag{7.29}$$

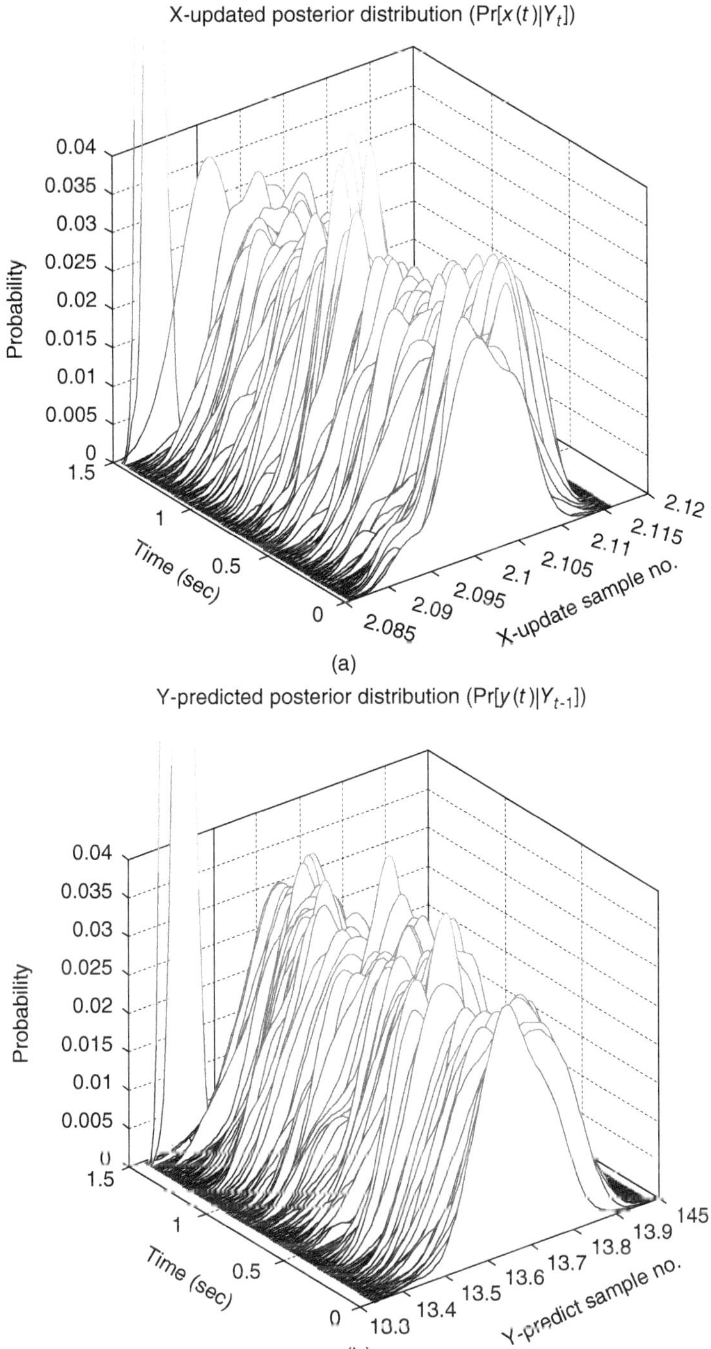

FIGURE 7.10 Nonlinear trajectory Bayesian estimation instantaneous posterior distributions: (a) updated state distribution; (b) predicted measurement distribution.

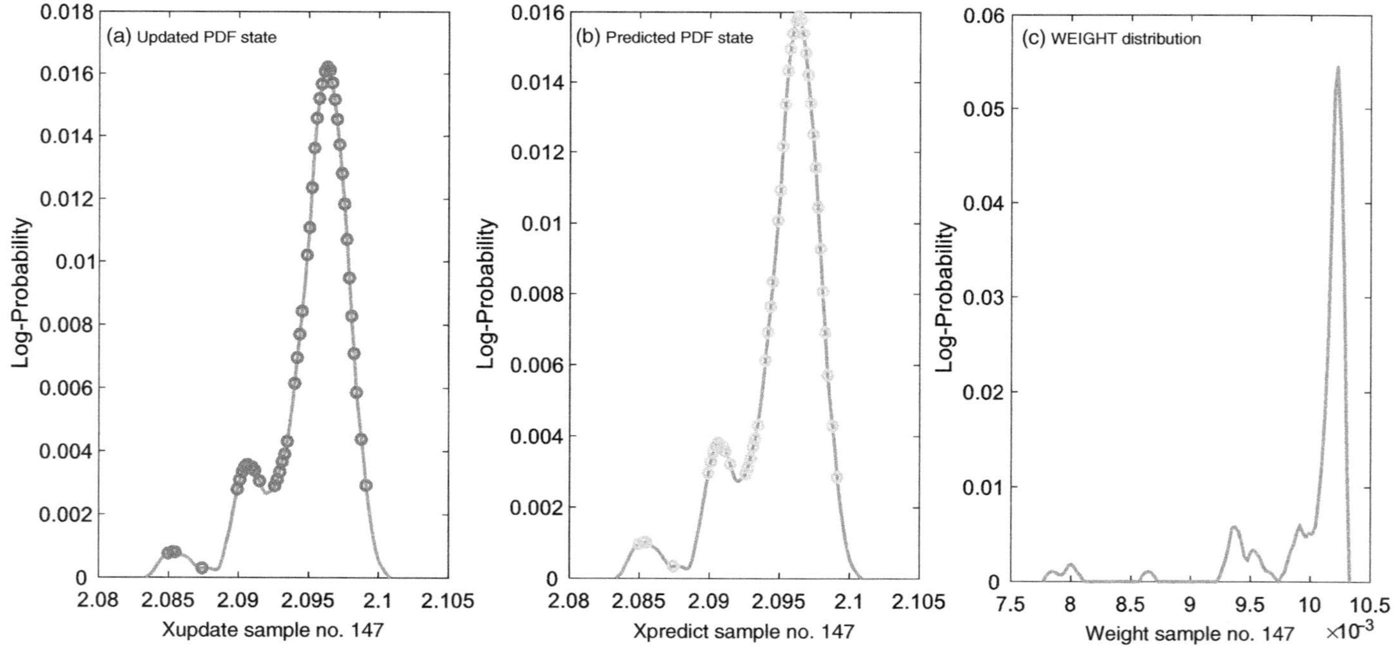

FIGURE 7.11 Nonlinear trajectory Bayesian estimation instantaneous posterior distribution at time-slice 1.47 sec: (a) updated state distribution; (b) predicted state distribution; (c) weight (likelihood for bootstrap) distribution.

Note that the "past" trajectories depend on the data *up to* time-step t to enable the adaption to the new data $y(t)$ while the "new" conditional importance distribution ($q \rightarrow q_{\text{MV}}$) incorporates the new state $x(t)$. We substitute an empirical distribution for $\Pr(X_{t-1}|Y_t)$ centered on the previous particle paths $\{X_{t-1}(i)\}$

$$q(X_{t-1}|Y_t) \approx \sum_{i=1}^{N_p} \mathcal{K}_i(t-1)\delta(X_{t-1} - X_{t-1}(i)) \tag{7.30}$$

where $\sum_{i=1}^{N_p} \mathcal{K}_i(t-1) = 1$ and $\mathcal{K}_i(t-1) > 0$. The ith weight (probability mass) for each particle in this proposal is based on the pre-selected particles that are a "good fit" to the new data point $y(t)$. One choice [8] for these weights is to choose a *point estimate* of the state such as its mean,

$$\hat{m}_i(t) = \int x(t) \times \Pr(x(t)|x_i(t-1))\, dx(t) \tag{7.31}$$

and then compute the weighting function as the likelihood evaluated at this point

$$\mathcal{K}_i(t-1) = \mathcal{C}(y(t)|\hat{m}_i(t))$$

as in the bootstrap technique [16] or if the particles $\{x_i(t-1)\}$ are weighted [6], then

$$\mathcal{K}_i(t-1) = \mathcal{W}_i(t-1) \times \mathcal{C}(y(t)|\hat{m}_i(t))$$

which follows from the marginal $\Pr(X_{t-1}|Y_t)$—a smoothing distribution. That is, using the particle approximation from time $(t-1)$ and expanding, we obtain

$$\Pr(X_{t-1}|Y_t) \propto \int \Pr(X_{t-1}|Y_{t-1}) \times \mathcal{A}(x(t)|x(t-1)) \times \mathcal{C}(y(t)|x(t))\, dx(t) \tag{7.32}$$

and using the empirical distribution approximation for the first term gives

$$\Pr(X_{t-1}|Y_t) \approx \sum_{i=1}^{N_p} \mathcal{W}_i(t-1)\delta(X_{t-1} - X_{t-1}(i)) \times \int \mathcal{C}(y(t)|x(t))\mathcal{A}(x(t)|x_i(t-1))\, dx(t) \tag{7.33}$$

One approximation [8] $\mathcal{A}(x(t)|x_i(t-1)) \rightarrow \delta(x(t) - \hat{m}_i(t))$ leads to the estimator

$$\Pr(X_{t-1}|Y_t) \approx \sum_{i=1}^{N_p} \underbrace{\mathcal{C}(y(t)|\hat{m}_i(t)) \times \mathcal{W}_i(t-1)}_{\text{Combined weight}} \times \delta(X_{t-1} - X_{t-1}(i)) \tag{7.34}$$

giving the desired result above

$$\mathcal{K}_i(t-1) := \mathcal{W}_i(t-1) \times \mathcal{C}(y(t)|\hat{m}_i(t)) \tag{7.35}$$

Using this proposal, the generalized weight

$$\mathcal{W}_{\text{aux}}(i,t) := \frac{\Pr(X_t|Y_t)}{q(X_t)}$$

is determined from the ratio of the posterior

$$\Pr(X_t|Y_t) \propto \int \mathcal{C}(y(t)|x(t)) \times \mathcal{A}(x(t)|x(t-1)) \times \Pr(X_{t-1}|Y_{t-1})\, dx(t)$$

to the joint proposal giving

$$\mathcal{W}_{\text{aux}}(i,t) = \frac{\Pr(X_t|Y_t)}{q(X_t)} = \frac{\mathcal{W}_i(t-1)}{\mathcal{K}_i(t-1)} \times \frac{\mathcal{C}(y(t)|x_i(t)) \times \mathcal{A}(x_i(t)|x_i(t-1))}{q(x_i(t)|x_i(t-1), y(t))} \tag{7.36}$$

which follows directly from substitution of the empirical distributions [6]. Here the *bias* correction $1/\mathcal{K}_i$ is introduced into the sampler to correct the auxiliary weight (first stage).

We summarize the auxiliary particle filter in Table 7.3. First, the bootstrap technique is executed to generate a set of particles and auxiliary weights at time-step $(t-1)$, that is, $\{\mathcal{W}_i(t), x_i(t)\}$ which are used to estimate the set of means (or modes) $\{\hat{m}_i(t)\}$ as in Eq. 7.31 for the "smoothing" weight calculation of Eq. 7.35 generating the probability masses $\{\mathcal{K}_i(t-1)\}$. These weights are then used in resampling to generate the set of "most likely" particles and weights under the new sampling indices $\{j(i)\}$, $\{\mathcal{K}_{j(i)}(t-1), x_{j(i)}(t-1)\}$, that is, $\hat{x}_i(t-1) \Rightarrow x_{j(i)}(t-1); i = 1, \ldots, N_p$. Next, samples are drawn from the optimal proposal and used to update the auxiliary weights using the resampled particles and first stage weights defined by $\alpha_i(t-1) := \mathcal{W}_i(t-1)/\mathcal{K}_i(t-1)$. The posterior distribution is estimated using the auxiliary weights to complete the process.

If the process is governed by severe nonlinearities or contaminated by high process noise, then a single point estimate such as $\hat{m}_i(t)$ does not sufficiently represent the transition probability $\Pr(x(t)|x_i(t-1))$ very well. Therefore, we can expect the ASIR performance to be poor even yielding weaker results than that of the bootstrap processor. However, if the process noise is small, implying that a point estimate can characterize the transition probability reasonably well, then the ASIR is less sensitive to "outliers" and the weights will be more uniformly balanced resulting in excellent performance compared to the bootstrap. These concepts can be used as an aid to help decide when such a technique is applicable to a given problem.

TABLE 7.3 Auxiliary SIR State-Space Particle Filtering Algorithm

Initialize	
$x_i(0) \sim \Pr(x(0)) \quad W_i(0) = 1/N_p \quad i = 1, \ldots, N_p$	[Sample]
Auxiliary Weights	
Bootstrap Processor: $\{\mathcal{W}_i(t), x_i(t)\}$	
$\hat{m}_i(t) \approx E\{x(t)\}$	[Bootstrap mean]
Weight Calculation	
$K_i(t-1) = \mathcal{W}_i(t-1) \times C(y(t)\|\hat{m}_i(t))$	[Bootstrap likelihood]
Weight Normalization	
$\mathcal{K}_i(t-1) = K_i(t-1) \Big/ \sum_{i=1}^{N_p} K_i(t-1)$	[Auxiliary weight]
Resampling	
Select indices $\{j(i)\}$ using $\{\mathcal{K}_{j(i)}(t-1)\}: \hat{x}_i(t) \Rightarrow x_{j(i)}(t)$	[Resample]
$\alpha_i(t-1) := \mathcal{W}_{j(i)}(t-1)/\mathcal{K}_{j(i)}(t-1)$	[First stage weights]
Importance Sampling Proposal (Optimal)	
$\tilde{x}_i(t) \sim q(\hat{x}_i(t)\|x_i(t-1), y(t))$	[Sample]
Weight Update	
$\tilde{W}_i(t) = \alpha_i(t-1) \times \dfrac{C(y(t)\|\tilde{x}_i(t)) \times \mathcal{A}(\tilde{x}_i(t)\|x_i(t-1))}{q(\hat{x}_i(t)\|x_i(t-1), y(t))}$	[Weight-update]
Weight Normalization	
$\mathcal{W}_{\text{aux}}(i,t) = \tilde{W}_i(t) \Big/ \sum_{i=1}^{N_p} \tilde{W}_i(t)$	[Normalize]
Distribution	
$\hat{\Pr}(x(t)\|Y_t) \approx \sum_{i=1}^{N_p} \mathcal{W}_{\text{aux}}(i,t)\delta(x(t) - \tilde{x}_i(t))$	[Posterior distribution]

7.5.3 Regularized Particle Filter

In order to reduce the degeneracy of the weights in the SIR processor, resampling was introduced as a potential solution; however, it was mentioned that even though the particles are "steered" to HPR, they tend to lose their diversity among other problems introduced by such a procedure [1, 23–44]. This problem occurs because samples are drawn from a discrete rather than continuous distribution (see Section 1.5). Without any attempt to correct this problem, the particles can collapse to a single location giving a poor characterization of the posterior distribution and therefore result in poor processor performance.

One solution to the diversity problem is to develop a continuous rather than discrete approximation to the empirical posterior distribution using the kernel density estimator of Section 3.2 and then perform resampling directly from it. This is termed a regularization-step resulting in diversification by a form of "jittering" the particles; thus, the processor is called the *regularized* particle filter (RPF). The key idea of the RPF is the transformation of the discrete empirical posterior distribution

$\hat{\Pr}(x(t)|Y_t) \rightarrow \Pr(x_t|Y_t)$ in order to resample from an absolutely continuous distribution producing a "new" set of N_p-particles with different locations.

To be more precise let us define the properties of a kernel that can be applied to this problem [7]. A *regularization kernel* $\mathcal{K}(\mathbf{x})$ is a symmetric probability density function such that: (1) $\mathcal{K}(\mathbf{x}) \geq 0$; (2) $\int \mathcal{K}(\mathbf{x})\, d\mathbf{x} = 1$; (3) $\int \mathbf{x}\mathcal{K}(\mathbf{x})\, d\mathbf{x} = 0$; and (4) $\int \|\mathbf{x}^2\|\mathcal{K}(\mathbf{x})\, d\mathbf{x} < \infty$ and for any positive bandwidth $\triangle_x$ the corresponding *rescaled* kernel is defined by

$$\mathcal{K}_{\Delta_x}(\mathbf{x}) = \left(\frac{1}{\Delta_x}\right)^{N_x} \mathcal{K}\left(\frac{\mathbf{x}}{\Delta_x}\right) \quad \text{for } \mathbf{x} \in \mathcal{R}^{N_x \times 1} \tag{7.37}$$

The most important property of the kernel density follows from its regularization property, that is, for any distribution $\mathcal{P}(\mathbf{x}) \in \mathcal{R}^{N_x \times 1}$, the regularization results in an absolutely continuous probability distribution, $\mathcal{K}_{\Delta_x}(\mathbf{x}) * \mathcal{P}(\mathbf{x})$, with $*$ the convolution operator, such that

$$\frac{d}{d\mathbf{x}}[\mathcal{K}_{\Delta_x}(\mathbf{x}) * \mathcal{P}(\mathbf{x})] = \int \mathcal{K}_{\Delta_x}(\mathbf{x} - \alpha)\mathcal{P}(\alpha)\, d\alpha \tag{7.38}$$

If $\mathcal{P}$ is an empirical distribution, then

$$\hat{\mathcal{P}}(\mathbf{x}) \approx \sum_{i=1}^{N_p} \mathcal{W}_i \delta(\mathbf{x} - \mathbf{x}_i)$$

for $\mathbf{x}_i \rightarrow \{\mathbf{x}_1, \ldots, \mathbf{x}_{N_p}\}$ a sample from the posterior and therefore

$$\frac{d}{d\mathbf{x}}[\mathcal{K}_{\Delta_x}(\mathbf{x}) * \hat{\mathcal{P}}(\mathbf{x})] = \left(\frac{1}{\Delta_x}\right)^{N_x} \sum_{i=1}^{N_p} \mathcal{W}_i \mathcal{K}\left(\frac{\mathbf{x} - \mathbf{x}_i}{\Delta_x}\right) = \sum_{i=1}^{N_p} \mathcal{W}_i \mathcal{K}_{\Delta_x}(\mathbf{x} - \mathbf{x}_i) \tag{7.39}$$

Both the bandwidth and the kernel are selected in practice to minimize the mean integrated error between the posterior and regularized distribution. Classical kernels result under specialized assumptions such as the Epanechnikov, Box, Triangle, Gaussian, etc. (see Ref. [7], Chapter 12 for more details).

One of the underlying assumptions of this transformation is that the true posterior $\Pr(x_t|Y_t)$ has a unity covariance which is not the case when implementing the RPF technique. Therefore, at each time-step we must estimate the ensemble mean and covariance by the usual sample approach given by

$$m(t) = \frac{1}{N_p} \sum_{i=1}^{N_p} x_i(t)$$

$$R_{xx}(t) = \frac{1}{N_p} \sum_{i=1}^{N_p} (x_i(t) - m(t))(x_i(t) - m(t))' \tag{7.40}$$

With this calculation, we factor the covariance using the Cholesky decomposition to yield the matrix square roots used in a whitening transformation (unity covariance), that is,

$$R_{xx}(t) = L^{1/2}(t)L^{T/2}(t)$$

which leads to the new scaled kernel

$$\mathcal{K}_{\Delta_x}(\mathbf{x}) = \frac{1}{|L^{1/2}|(\Delta_x)^{N_x}}\mathcal{K}\left(\frac{L^{-1/2}\mathbf{x}}{\Delta_x}\right) \tag{7.41}$$

The old particles are then "jittered" by using the step

$$\tilde{x}_i(t) = x_i(t) + \Delta_x L^{\frac{1}{2}}(t)\epsilon_i(t) \tag{7.42}$$

where the $\{\epsilon_i(t)\}$ are drawn from the new scaled kernel above. A typical example of kernel density estimation of a discrete probability mass function is shown in Fig. 7.1 where the circles represent the discrete point masses (impulses) at the particular location and the continuous approximation to the probability density is shown by the smooth curve provided by the kernel density estimate using a Gaussian window. This completes the RPF technique which is summarized in Table 7.4. Next, we discuss another popular approach to produce particle diversity.

7.5.4 MCMC Particle Filter

Another approach to increase the diversity in the particle set $\{x_i(t), W_i(t)\}; i = 1, \ldots, N_p$ is to take an MCMC step(s) with the underlying invariant distribution targeted as the posterior $\Pr(X_t|Y_t)$ on each particle [45]. The MCMC particle filter is available in two varieties: (1) MCMC-step(s) with the usual BSP; and (2) full MCMC iterative filter. The sequential processors use the MCMC-steps as part of the resampling process for especially insensitive (weight divergence) problems, while the full MCMC iterative processor is available as a separate algorithm typically executed using the Metropolis, Metropolis–Hastings or Gibbs samplers of Chapter 3. We confine our discussion to the MCMC-step approach, since we are primarily interested in sequential techniques and refer the interested reader to [1, 23–24] for the iterative approach.

The main idea is that the particles are distributed as $\Pr(X_t(i)|Y_t)$, then applying a Markov chain transition kernel defined by

$$\mathcal{T}(X_t|X_t(i)) := \Pr(X_t|X_t(i)) \tag{7.43}$$

TABLE 7.4 Regularized Particle Filtering Algorithm

	Initialize	
Draw: $x_i(0) \sim \Pr(x(0)) \quad W_i(0) = \frac{1}{N_p} \; i = 1, \ldots, N_p$		[Sample]
	Importance Sampling	
Draw: $x_i(t) \sim \mathcal{A}(x(t)\|x_i(t-1))$		[State transition]
	Weight Update	
$W_i(t) = \mathcal{C}(y(t)\|x_i(t)) \Leftarrow \mathcal{C}(x(t), u(t), v(t));\; v \sim \Pr(v(t))$		[Weight/likelihood]
	Weight Normalization	
$\mathcal{W}_i(t) = W_i(t) \Big/ \sum_{i=1}^{N_p} W_i(t)$		
	Resampling Decision	
$\hat{N}_{\text{eff}} = 1 \Big/ \sum_{i=1}^{N_p} W_i^2(t)$		[Effective samples]
$\hat{N}_{\text{eff}} = \begin{cases} \text{Resample} & \leq N_{\text{thresh}} \\ \text{Accept} & > N_{\text{thresh}} \end{cases}$		[Decision]
	Regularization Sample Statistics	
$m(t) = 1/N_p \sum_{i=1}^{N_p} x_i(t)$		[Sample mean]
$R_{xx}(t) = 1/N_p \sum_{i=1}^{N_p} (x_i(t) - m(t))(x_i(t) - m(t))'$		[Sample covariance]
	Factorization	
$R_{xx}(t) = L^{1/2}(t) L^{T/2}(t)$		[Cholesky decomposition]
	Resampling	
$\hat{x}_i(t) \Rightarrow x_i(t)$		
	Diversification	
Draw: $\epsilon_i(t) \sim \mathcal{K}_{\Delta_x}(x(t) - \hat{x}_i(t))$		[Sample]
	Diversify	
$\tilde{x}_i(t) = \hat{x}_i(t) + \Delta_x L^{1/2}(t)\epsilon_i(t)$		[Generate sample]
	Distribution	
$\hat{\Pr}(x(t)\|Y_t) \approx \sum_{i=1}^{N_p} \mathcal{W}_i(t)\delta(x(t) - \tilde{x}_i(t))$		[Posterior distribution]

with posterior invariant distribution such that

$$\Pr(X_t|Y_t) = \int \mathcal{T}(X_t|X_t(i)) \times \Pr(X_t(i)|Y_t)\, dX_t(i) \tag{7.44}$$

continues to result in a particle set with the desired posterior as its invariant distribution. However, the new particle locations after the *move* result in HPR of the state–space. It has been shown that by applying the MCMC transition kernel that the total variance of the current distribution can only decrease [46]. Any of the MCMC methods (M-H, G-S, S-S, etc.) can be incorporated into the SMC framework to achieve the desired move occuring after the resampling operation.

Following Refs. [46] and [47], the objective is to move the set of particles using a combination of importance sampling, resampling and MCMC sampling, that is, the approach is to:

- *Initialize* the set of particles yielding: $\{x_i(t)\}; i = 1, \ldots, N_p$
- *Resample* this set to obtain: $\{\hat{x}_i(t)\}; i = 1, \ldots, N_p$
- *Move* using an MCMC step(s) to generate the "new" set of particles $\{\tilde{x}_i(t)\}; i = 1, \ldots, N_p$ with $\tilde{x}_i(t) \sim \Pr(X_t|Y_t)$ and transition kernel, $\mathcal{T}(\tilde{x}_i(t)|X_t(i))$

The *move*-step performs one or more iterations of an MCMC technique on each selected particle *after* the resampling step with invariant distribution $\Pr(X_t|Y_t)$. Note that *before* the move, the resampled particles are distributed $\hat{x}_i(t) \sim \Pr(X_t|Y_t)$; therefore, the "moved" particles $\tilde{x}_i(t)$ are approximately distributed by this posterior as well. The move-step improves particle diversity by enriching the particle locations to those highest probability regions.

For instance, let us "track" the *ith* particle with corresponding high valued weight. Based on its associated weight (probability), $\hat{x}_i(t) \Rightarrow x_i(t)$ is selected according to

$$\Pr(\hat{x}_i(t) = x_i(t)) = \mathcal{W}_i(t)$$

Resampling results in replication of the *ith*-particle, N_i-times, producing the resampled set $\{\hat{x}_{i1}(t), \hat{x}_{i2}(t), \ldots, \hat{x}_{iN_i}(t)\}$. Next, the *move-step* moves each replicant $\hat{x}_{ij}(t) \rightarrow \tilde{x}_{ij}(t)$ to a distinct (unique) location in the region of strongest support dictated by $\Pr(X_t|Y_t)$. This provides the "move-step" for the SMC technique and mitigates the divergence problem.

To be more specific, let us illustrate the "move" by choosing the Metropolis–Hastings technique of Table 3.1 to perform our MCMC-step using the random walk M–H approach of Section 3.4.

We start with the basic bootstrap PF of Section 7.5 to obtain the set of resampled particles $\{\hat{x}_i(t)\}$. Using the random walk model, we perform the "move" step to obtain the new set of particles as

$$\tilde{x}_i(t) = \hat{x}_i(t) + \epsilon_i(t) \quad \text{for } \epsilon_i \sim p_E(\epsilon) \tag{7.45}$$

One choice is $\epsilon_i \sim p_E(\epsilon) = \mathcal{N}(0, R_{\epsilon\epsilon})$, since it is symmetric, easy to generate, and will simplify the calculations even further. The corresponding *acceptance* probability for the M-H approach is specified by

$$A(\tilde{x}_i(t), \hat{x}_i(t)) = \min\left\{\frac{\Pr(\tilde{X}_t(i)|Y_t))}{\Pr(\hat{X}_t(i)|Y_t))} \times \frac{q(\hat{x}_i(t)|\tilde{x}_i(t))}{q(\tilde{x}_i(t)|\hat{x}_i(t))}, 1\right\} \tag{7.46}$$

where $\tilde{X}_i(t) := \{\tilde{x}_i(t), X_{t-1}(i)\}$ and $\hat{X}_i(t) := \{\hat{x}_i(t), X_{t-1}(i)\}$ are the augmented sets of joint random particles. Drawing samples from the random walk with (symmetric) Gaussian distribution enables us to simplify the acceptance probability, since $q(\hat{x}_i(t)|\tilde{x}_i(t)) = q(\tilde{x}_i(t)|\hat{x}_i(t))$, canceling these terms in Eq. 7.46 to produce

$$A(\tilde{x}_i(t), \hat{x}_i(t)) = \min\left\{\frac{\Pr(\tilde{X}_t(i)|Y_t))}{\Pr(\hat{X}_t(i)|Y_t))}, 1\right\} \tag{7.47}$$

But from the sequential Bayesian recursion,

$$\Pr(X_t|Y_t) \propto \Pr(y(t)|x(t))\Pr(x(t)|x(t-1)) \times \Pr(X_{t-1}|Y_{t-1})$$

we obtain

$$A(\tilde{x}_i(t), \hat{x}_i(t)) = \min\left\{\frac{\Pr(y(t)|\tilde{x}_i(t)) \times \Pr(\tilde{x}_i(t)|x_i(t-1))}{\Pr(y(t)|\hat{x}_i(t)) \times \Pr(\hat{x}_i(t)|x_i(t-1))}, 1\right\} \tag{7.48}$$

With this information in mind, the implementation of the bootstrap PF with (random walk) MCMC-step is given in Table 7.5. Another approach to improve particle diversity is using local linearization techniques which can be implemented with any of the classical/modern algorithms of the previous two chapters.

7.5.5 Linearized Particle Filter

Attempts to approximate the minimum variance importance proposal distribution of Section 7.2 continue to evolve [24]. The motivation for this is based on inadequacies created by selecting the transition prior of Section 7.5.1 as the proposal leading to the popular bootstrap algorithm [16] of Table 7.2. As mentioned previously, the bootstrap approach requires resampling to mitigate the particle depletion problem and lack of incorporating the most recent measurement in the weight update. These reasons have led to the development of the PF that incorporates the latest measurement sample. One way to do so is to generate an approximately Gaussian importance proposal based on linearization methods [4, 7, 9] with the idea of selecting

$$\Pr(x(t)|X_{t-1}, Y_t) \approx q_{\mathcal{N}}(x(t)|Y_t) \tag{7.49}$$

as a Gaussian proposal. This approach is used to provide coverage of the actual posterior due to its long-tailed distribution while incorporating the latest available measurement. This Gaussian proposal is the result of marginalizing the prior state $x(t-1)$ of the minimum variance proposal. That is,

$$q_{\mathcal{N}}(x(t)|Y_t) \rightarrow \Pr(x(t)|Y_t) = \int \Pr(x(t)|x(t-1), y(t)) \times \Pr(x(t-1)|Y_{t-1})\, dx(t-1) \tag{7.50}$$

TABLE 7.5 MCMC Particle Filtering Algorithm

	Initialize	
Draw: $x_i(0) \sim \Pr(x(0))$ $W_i(0) = \frac{1}{N_p}$ $i = 1, \ldots, N_p$		[Sample]
	Importance Sampling	
Draw: $x_i(t) \sim \mathcal{A}(x(t)\vert x_i(t-1))$		[State transition]
	Weight Update	
$W_i(t) = \mathcal{C}(y(t)\vert x_i(t))$		[Weight/likelihood]
	Weight Normalization	
$\mathcal{W}_i(t) = W_i(t) \Big/ \sum_{i=1}^{N_p} W_i(t)$		
	Resampling Decision	
$\hat{N}_{\text{eff}} = 1 \Big/ \sum_{i=1}^{N_p} W_i^2(t)$		[Effective samples]
$\hat{N}_{\text{eff}} = \begin{cases} \text{Resample} & \le N_{\text{thresh}} \\ \text{Accept} & > N_{\text{thresh}} \end{cases}$		[Decision]
	Resampling	
$\hat{x}_i(t) \Rightarrow x_i(t)$		
	Diversification Acceptance Probability	
$A(\tilde{x}_i(t), \hat{x}_i(t)) = \min\left\{ \frac{\mathcal{C}(y(t)\vert\tilde{x}_i(t)) \times \mathcal{A}(\tilde{x}_i(t)\vert x_i(t-1))}{\mathcal{C}(y(t)\vert\hat{x}_i(t)) \times \mathcal{A}(\hat{x}_i(t)\vert x_i(t-1))}, 1 \right\}$		
	Diversify	
Draw: $\epsilon_i(t) \sim \mathcal{N}(0, R_{\epsilon\epsilon})$		[Sample]
$\tilde{x}_i(t) = \hat{x}_i(t) + \epsilon_i(t)$		[Generate sample]
Draw: $u_k \to \mathcal{U}(0, 1)$		[Uniform sample]
	Decision	
$x_i(t) = \begin{cases} \tilde{x}_i(t) & \text{if } u_k < A(x_i, \hat{x}_i) \\ \hat{x}_i(t) & \text{otherwise} \end{cases}$		
	Distribution	
$\hat{\Pr}(x(t)\vert Y_t) \approx \sum_{i=1}^{N_p} \mathcal{W}_i(t)\delta(x(t) - x_i(t))$		[Posterior distribution]

In a sense, this implicit marginalization effectively averages the proposal with respect to the previous posterior $\Pr(x(t-1)|Y_{t-1})$, which incorporates all of the "information" about $x(t-1)$. Thus, we see that by choosing the Gaussian importance distribution as our proposal enables us to implicitly incorporate all of the knowledge available about the previous state as well as incorporate the current measurement $y(t)$. These features make $q_{\mathcal{N}}(x(t)|Y_t)$ a reasonable choice as long as it provides the overlapping support or coverage of the desired posterior [9].

One of the linearization approaches to implementing the minimum variance importance function $q_{MV}(x(t)|x(t-1), y(t))$ is to estimate it as a Gaussian prior, that is,

$$q_{MV}(x(t)|x(t-1), y(t)) \sim \mathcal{N}(\hat{x}(t|t), \tilde{P}(t|t)) \tag{7.51}$$

where the associated (filtered) CM $\hat{x}(t|t) = E\{x(t)|Y_t\}$ and error covariance $\tilde{P}(t|t) = E\{\tilde{x}(t|t)\tilde{x}'(t|t)\}$ for $\tilde{x}(t|t) = x(t) - \hat{x}(t|t)$ are obtained from an additional estimation scheme.

There are a variety of choices available, each evolving from either the classical (linearized, extended, or iterated Kalman filters) or the modern unscented (sigma-point) Kalman filter. Further alternatives are also proposed to these approaches [4], but we will confine our discussion to these popular and readily available approaches [9]. In each of these implementations a linearization, either of the nonlinear dynamics and measurement models in the classical case or in the statistical linearization (unscented transformation) as in the unscented or sigma-point case occurs. All of these "linearized-based" processors provide the updated or filtered conditional mean estimate

$$\hat{x}(t|t) = \hat{x}(t|t-1) + K(x^*(t))e(t) \tag{7.52}$$

where

$\hat{x}(t|t-1)$	is the predicted conditional mean $E\{x(t)|Y_{t-1}\}$;
$K(x^*(t))$	is the gain or weight based on the particular linearization technique; and
$e(t)$	is the innovation sequence.

Here the choices are

- $x^*(t) \rightarrow x_o(t)$ is a reference trajectory in the *linearized* case; or
- $x^*(t) \rightarrow \hat{x}(t|\alpha)$ is a *extended* or *iterated cases*; and
- $x^*(t) \rightarrow \chi_i(t|t)$ is a sigma-point in the *unscented case*.

The error covariance is much different in each case, that is,

$$\tilde{P}(t|t) = (I - K(x^*(t))C[x^*(t)])\tilde{P}(t|t-1) \tag{7.53}$$

in the linearized and extended cases with measurement Jacobian

$$C[x^*(t)] = \left.\frac{\partial c_i[x]}{\partial x_j}\right|_{x=x^*(t)} \tag{7.54}$$

and, of course,

$$\tilde{P}(t|t-1) = A[x^*(t)]\tilde{P}(t-1|t-1)A'[x^*(t)] + R_{ww}(t-1) \tag{7.55}$$

for $A[x^*(t)] = \left.\frac{\partial a_i[x]}{\partial x_j}\right|_{x=x^*(t)}$.

In the sigma-point (unscented) case, we have

$$\tilde{P}(t|t) = \tilde{P}(t|t-1) - K(x^*(t))R_{ee}(t)K'(x^*(t)) \tag{7.56}$$

TABLE 7.6 Linearized Particle Filtering Algorithm

Initialize	
Draw: $x_i(0) \sim \Pr(x(0)) \quad W_i(0) = \frac{1}{N_p} \quad i = 1, \ldots, N_p$	[Sample]
Linearization	
LZKF/EKF/IEKF/UKF Processor	
$\{x_i(t), \tilde{P}_i(t)\} = \mathcal{N}(\hat{x}(t\|t), \tilde{P}(t\|t))$	
Importance Sampling	
$x_i(t) \sim \mathcal{N}(\hat{x}(t\|t), \tilde{P}(t\|t))$	[State draw]
Weight Update	
$W_i(t) = \dfrac{\mathcal{C}(y(t)\|x_i(t)) \times \mathcal{A}(x(t)\|x_i(t-1))}{q_{\mathrm{MV}}(x(t)\|x_i(t-1), y(t))}$	[MV weight]
for	
$q_{\mathrm{MV}}(x(t)\|x_i(t-1), y(t)) = \mathcal{N}(x_i(t), \tilde{P}_i(t))$	
Weight Normalization	
$\mathcal{W}_i(t) = W_i(t) \Big/ \sum_{i=1}^{N_p} W_i(t)$	
Resampling Decision	
$\hat{N}_{\mathrm{eff}} = 1 \Big/ \sum_{i=1}^{N_p} W_i^2(t)$	[Effective samples]
$\hat{N}_{\mathrm{eff}} = \begin{cases} \leq N_{\mathrm{thresh}} & \text{Resample} \\ > N_{\mathrm{thresh}} & \text{Accept} \end{cases}$	[Decision]
Resampling	
$\hat{x}_i(t) \Rightarrow x_i(t)$	
Distribution	
$\hat{\Pr}(x(t)\|Y_t) \approx \sum_{i=1}^{N_p} \mathcal{W}_i(t)\delta(x(t) - \hat{x}_i(t))$	[Posterior distribution]

So we see that depending on the linearization method or the particular classical or unscented processor we select, we will generate the Gaussian prior at each time-step which is used to obtain the minimum variance importance distribution. Thus, we see why they are called the class of "local" linearization-based particle filters. The linearized particle filter algorithm is shown in Table 7.6 where we see that the conditional mean and covariance are estimated in each time-step of the algorithm and particles are drawn from

$$x_i(t) \rightarrow \hat{\Pr}(x(t)|Y_t) \sim \mathcal{N}(\hat{x}(t|t), \tilde{P}(t|t)) \tag{7.57}$$

the updated "Gaussian" estimates and the weights also follow the importance distribution of Eq. 7.51 given by

$$W_i(t) = \frac{\mathcal{C}(y(t)|x_i(t)) \times \mathcal{A}(x(t)|x_i(t-1))}{q_{MV}(x(t)|x_i(t-1), y(t))} \tag{7.58}$$

This completes the linearized particle filters, next we consider some of the practical considerations for design.

7.6 PRACTICAL ASPECTS OF PARTICLE FILTER DESIGN

Pragmatic Monte Carlo methods [48–57], even those that are model-based, are specifically aimed at providing a reasonable estimate of the underlying posterior distribution; therefore, performance testing typically involves estimating just "how close" the estimated posterior is to the "true" posterior. However, within a Bayesian framework, this comparison of posteriors only provides a measure of relative performance but does not indicate just how well the underlying model embedded in the processor "fits" the measured data. Nevertheless, these closeness methods are usually based on the Kullback–Leibler (KL) divergence measure [58] providing such an answer. However, in many cases we do not know the true posterior and therefore we must resort to other means of assessing performance such as evaluating mean-squared error (MSE) or more generally checking the validity of the model by evaluating samples generated by the prediction or the likelihood cumulative distribution to determine whether or not the resulting sequences have evolved from a uniform distribution and are *i.i.d.* (see Section 3.3) – analogous to a whiteness test for Gaussian sequences.

Thus, PF are essentially sequential estimators of the posterior distribution employing a variety of embedded models to achieve meaningful estimates. In contrast to the BP designs which are typically based on Gaussian assumptions, the PF have no such constraints per se. In the linear case, a necessary and sufficient condition for optimality of the linear BP is that the corresponding innovations or residual sequence must be zero-mean and white (see Section 5.7 for details). In lieu of this constraint, a variety of statistical tests (whiteness, uncorrelated inputs, etc.) were developed in Section 5.7 evolving from this known property. When the linear Bayesian processors were "extended" to the nonlinear case, the same tests were performed based on approximate Gaussian assumptions. Clearly, when noise is additive Gaussian these arguments can still be applied. These same statistical tests can also be performed based on the innovations or residual sequences resulting from the estimates (MAP, ML, MMSE (CM)) inferred from the estimated posterior distribution. However, some other more meaningful performance tests for the PF can also be applied for improved design and performance evaluation.

7.6.1 Sanity Testing

In what might be termed "sanity testing," the classical methods discussed above employ the basic statistical *philosophy* that "if all of the signal information has been removed (explained) from the measurement data, then the residuals (i.e. innovations or prediction error) should be uncorrelated," that is, the model fits the data and all that remains is a white or uncorrelated residual (innovations) sequence. Therefore, performing the zero-mean, whiteness testing, that is, the confidence interval about the normalized correlations (innovations) and the (vector) weighted sum-squared residual tests of Section 5.7 as well as checking the *normalized* MSE of the state vector (and/or innovations) given by

$$\text{NMSE} = E\left\{\left(\frac{x_{\text{True}}(t) - \hat{x}(t|t)}{x_{\text{True}}(t)}\right)^2\right\} \approx \sum_{t=1}^{N_t}\left(\frac{x_{\text{True}}(t) - \hat{x}(t|t)}{x_{\text{True}}(t)}\right)^2 \tag{7.59}$$

are the basic first steps that can be undertaken to ensure that models have been implemented properly and the processors have been "tuned". Here the normalized mean-squared error range is: $0 < \text{NMSE} < 1$.

7.6.2 Ensemble Estimation

Perhaps one of the most important aspects of nonlinear processor design is to recall that processing results *only* in a *single realization* of its possible output. For instance, the conditional mean is just a statistic, but still a *random process*; therefore, if direct measurements are not available, it is best to generate an ensemble of realizations and extract the resulting ensemble statistics using a variety of methods, (e.g., bootstrapping [15]). In any case, simple ensemble statistics can be employed to provide an average solution enabling us to generate "what we would expect to observe" when the processor is applied to the actual problem. This approach is viable even for the modern nonlinear processors of the previous chapter; however, here we concentrate on the particle filter and assume we have a "truth model" (or data ensemble) available for our designs. As illustrated in Fig. 7.12, the performance analysis of the particle filter is best achieved by combining both simulated and actual measurement data possibly obtained during a calibration phase of the implementation. In this approach, the data are pre-processed and the truth model adjusted to reflect a set of predictive

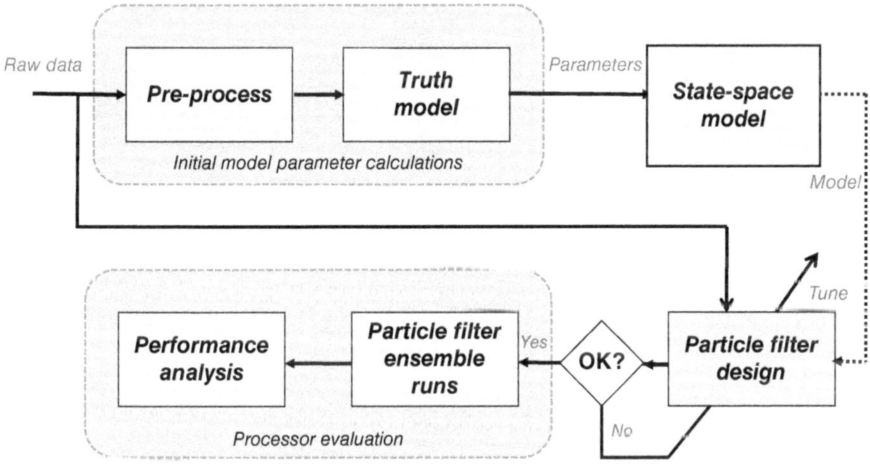

FIGURE 7.12 Ensemble estimation. Truth model parameter estimates, state-space model initialization, tuning, ensemble generation and performance analysis.

parameters that are then extracted and used to "parameterize" the underlying state-space (or other) representation of the process under investigation. Once the models and parameters are selected, then ensemble runs can be performed; such terms as initial states, covariances, and noise statistics (means, covariances, etc.) can then be adjusted to "tune" the processor much like in the modern techniques of the previous chapter. Here, classical statistics (e.g., zero-mean/whiteness and WSSR testing) can help in the tuning phase. Once tuned, the processor is then applied to the raw measurement data and ensemble statistics are estimated to provide a thorough analysis of the processor performance.

Statistics can be calculated over the ensemble of processor runs. The *updated* MAP state estimate is obtained directly from the ensemble, that is,

$$\hat{\hat{\mathbf{x}}}_{\mathrm{MAP}}(t|t) \longrightarrow \left\{\hat{\mathbf{x}}^k_{\mathrm{MAP}}(t|t)\right\}; k = 1, \ldots, K$$

and therefore, the updated state ensemble estimate is

$$\hat{\hat{\mathbf{x}}}_{\mathrm{MAP}}(t|t) = \frac{1}{K}\sum_{k=1}^{K} \hat{\mathbf{x}}^k_{\mathrm{MAP}}(t|t) \tag{7.60}$$

The corresponding predicted MAP state estimate can be obtained directly from the ensemble as well, that is,

$$\hat{\hat{\mathbf{x}}}_{\mathrm{MAP}}(t|t-1) \longrightarrow \left\{\hat{\mathbf{x}}^k_{\mathrm{MAP}}(t|t-1)\right\}; k = 1, \ldots, K$$

and therefore, the predicted state ensemble estimate is

$$\hat{\hat{\mathbf{x}}}_{\mathrm{MAP}}(t|t-1) = \frac{1}{K}\sum_{k=1}^{K} \hat{\mathbf{x}}^k_{MAP}(t|t-1) \tag{7.61}$$

The *predicted* MAP measurement estimate is also obtained directly from the ensemble, that is,

$$\hat{\hat{\mathbf{y}}}_{\mathrm{MAP}}(t|t-1) \longrightarrow \left\{\hat{\mathbf{y}}^k_{\mathrm{MAP}}(t|t-1)\right\}; k = 1, \ldots, K$$

where recall

$$\hat{\mathbf{y}}^k_{\mathrm{MAP}}(t|t-1) = c\left[\hat{\mathbf{x}}^k_{\mathrm{MAP}}(t|t-1)\right] \tag{7.62}$$

Thus, the predicted measurement ensemble estimate is

$$\hat{\hat{\mathbf{y}}}_{\mathrm{MAP}}(t|t-1) = \frac{1}{K}\sum_{k=1}^{K} \hat{\mathbf{y}}^k_{\mathrm{MAP}}(t|t-1) \tag{7.63}$$

Finally, the corresponding predicted innovation ensemble estimate is given by

$$\mathbf{e}_k(t) = \mathbf{y}(t) - \hat{\mathbf{y}}^k_{MAP}(t|t-1); k = 1, \ldots, K \tag{7.64}$$

$$\bar{\mathbf{e}}(t) \longrightarrow \{\mathbf{e}_k(t)\}; k = 1, \ldots, K$$

and therefore, the predicted innovation ensemble estimate is

$$\bar{\mathbf{e}}(t) = \frac{1}{K}\sum_{k=1}^{K} \mathbf{e}_k(t) \tag{7.65}$$

Of course the ensemble residual or innovations sequence can be "sanity tested" for zero-mean, whiteness, and the NMSE over the ensemble can be calculated to ensure the processor has been tuned properly. So we see that the ensemble statistics can be estimated and provide a better glimpse of the potential performance of the PF processor for the actual problem. To observe the effect of ensemble averaging, consider the previous example and apply these calculations to both the state and measurement estimates.

Example 7.3

Consider the trajectory estimation of the previous example and now perform a set of 100-member ensemble runs averaging each of the state and measurement estimates as discussed above. The results of ensemble estimates are shown in Fig. 7.13; in (a) and (b), we see the results of a single realization (run) of the bootstrap PF indicating a reasonable trajectory estimate, but not really giving us an indication of the potential processor performance over a set of data. Ensemble results are shown in Fig. 7.13c for the state estimate and Fig. 7.13d for the predicted measurement. Clearly, the estimates are much smoother and give a better representation of how we would expect the processor to perform. △△△

7.6.3 Posterior Probability Validation

Much effort has been devoted to the validation problem with the most significant results evolving from the information theoretical point of view [59]. Following this approach, we start with the basic ideas and quickly converge to a reasonable solution to the *distribution validation* problem [59–65].

Kullback–Leibler Theory Let us first define some concepts about probabilistic information necessary for the development. These concepts are applied extensively in communications problems and will prove useful in designing

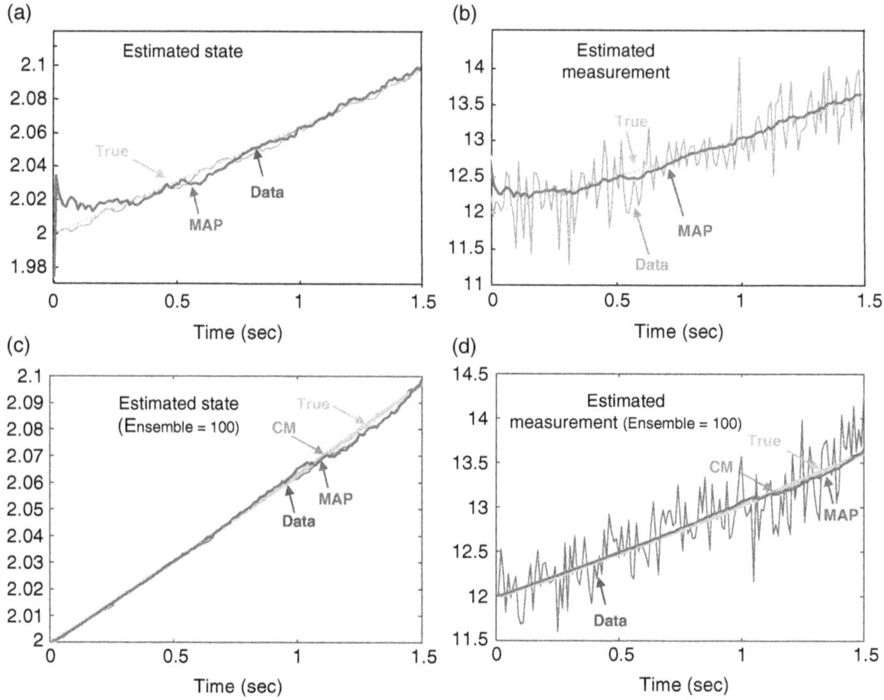

FIGURE 7.13 Ensemble estimation for trajectory estimation: (a) single realization: state estimate; (b) single realization: measurement estimate; (c) 100-member ensemble estimate: state; (c) 100-member ensemble estimate: measurement.

parametric signal processors. The *information* (self) contained in the occurrence of the event ω_i such that $X(\omega_i) = x_i$, is

$$\mathcal{I}(x_i) = -\log_b \Pr(X(\omega_i) = x_i) = -\log_b \Pr(x_i) \tag{7.66}$$

where b is the base of the logarithm which results in different units for information measures (e.g., base $2 \rightarrow$ bits, while base $e \rightarrow$ implies nats). The *entropy* or *average information* is defined by

$$\mathcal{H}(x_i) := E_X\{\mathcal{I}(x_i)\} = \sum_{i=1}^{N} \mathcal{I}(x_i)\Pr(x_i) = -\sum_{i=1}^{N} \Pr(x_i)\log_b \Pr(x_i) \tag{7.67}$$

Mutual information is defined in terms of the information available in the occurrence of the event $Y(\omega_j) = y_j$ about the event $X(\omega_i) = x_i$ or

$$\mathcal{I}(x_i; y_j) = \log_b \frac{\Pr(x_i|y_j)}{\Pr(x_i)} = \log_b \Pr(x_i|y_j) - \log_b \Pr(x_i) \tag{7.68}$$

Now using these concepts, we take the information theoretic approach to distribution estimation following [59, 60]. Since many processors are expressed in terms of their "estimated" probability distributions, quality or "goodness" can be evaluated by its similarity to the true underlying probability distribution generating the measured data.l

Suppose $\Pr(x_i)$ is the *true* discrete posterior probability distribution and $\hat{\Pr}(\hat{x}_i)$ is the estimated distribution. Then the KL *Information* quantity of the true distribution relative to the estimated is defined by using

$$\mathcal{I}_{\rm KL}(\Pr(x_i);\hat{\Pr}(\hat{x}_i)) := E_X\left\{\ln\frac{\Pr(x_i)}{\hat{\Pr}(\hat{x}_i)}\right\} = \sum_{i=1}^{N}\Pr(x_i)\ln\frac{\Pr(x_i)}{\hat{\Pr}(\hat{x}_i)}$$
$$= \sum_{i=1}^{N}\Pr(x_i)\ln\Pr(x_i) - \sum_{i=1}^{N}\Pr(x_i)\ln\hat{\Pr}(\hat{x}_i) \tag{7.69}$$

where we chose $\log_b = \ln$. The KL possesses some very interesting properties which we state without proof (see Ref. [60] for details) such as

1. $\mathcal{I}_{\rm KL}(\Pr(x_i);\hat{\Pr}(\hat{x}_i)) \geq 0$
2. $\mathcal{I}_{\rm KL}(\Pr(x_i);\hat{\Pr}(\hat{x}_i)) = 0 \Leftrightarrow \Pr(x_i) = \hat{\Pr}(\hat{x}_i)\ \forall i$
3. The negative of the KL is the *entropy*, $\mathcal{H}_{\rm KL}(\Pr(x_i);\hat{\Pr}(\hat{x}_i))$

The second property implies that as the *estimated* posterior distribution approaches the *true* distribution, then the value of the KL approaches *zero* (minimum). Thus, investigating Eq. 7.69, we see that the first term is a constant specified by the true distribution; therefore, we only need to estimate the average value of the estimated posterior relative to the true distribution, that is,

$$\mathcal{L}(\hat{x}_i) := E_X\{\ln\hat{\Pr}(\hat{x}_i)\} = \sum_{i=1}^{N}\Pr(x_i)\ln\hat{\Pr}(\hat{x}_i) \tag{7.70}$$

where $\mathcal{L}(\hat{x}_i)$ is defined as the *average log-likelihood* of the random variable of value $\ln\hat{\Pr}(\hat{x}_i)$. Clearly, the *larger* the average log-likelihood, the *smaller* the KL, implying a *better* model.

The third property, *entropy*, is approximately equal to $\frac{1}{N}$ times the probability that the relative frequency distribution of N measurements obtained from the estimated posterior equals the true distribution.

The $\mathcal{I}_{\rm KL}$ is applied frequently to parameter estimation/system identification problems to estimate the intrinsic order of the unknown system [5]. Two popular information metrics have evolved from this theory: the Akaike Information Criterion (AIC) and the minimum data length (MDL) description [59, 60, 62, 63]. Both are used to perform *order estimation* and are closely related as shown below

$$\text{AIC}(\eta) = -\ln R_{\epsilon\epsilon} + 2\frac{\eta}{N}$$
$$\text{MDL}(\eta) = -\ln R_{\epsilon\epsilon} + \frac{\eta}{2}\ln N$$

where η is the system order, ϵ is the one-step prediction error with corresponding covariance $R_{\epsilon\epsilon}$, and N is the number of samples (data) values.

However, our interest lies in comparing two probability distributions to determine "how close" they are to one another. Even though $\mathcal{I}_{\text{KL}}$ does quantify the difference between the true and estimated distributions, unfortunately it is not a distance measure due to its lack of symmetry. However, the *Kullback divergence* (KD) defined by a combination of $\mathcal{I}_{\text{KL}}$

$$\mathcal{J}_{\text{KD}}(\Pr(x_i);\hat{\Pr}(\hat{x}_i)) = \mathcal{I}_{\text{KL}}(\Pr(x_i);\hat{\Pr}(\hat{x}_i)) + \mathcal{I}_{\text{KL}}(\hat{\Pr}(\hat{x}_i);\Pr(x_i)) \tag{7.71}$$

is a distance measure between distributions indicating "how far" one is from the other. Consider the following example of this calculation.

Example 7.4

We would like to calculate the KD for two Gaussian distributions, $\text{p}_i(x) \sim \mathcal{N}(m_i, V_i);\, i = 1, 2$ to establish a closeness measure. First, we calculate the KL information,

$$\mathcal{I}_{\text{KL}}(\text{p}_1(x);\text{p}_2(x)) = E_{\text{p}_1}\left\{\ln\frac{\text{p}_1(x)}{\text{p}_2(x)}\right\} = E_{\text{p}_1}\left\{\frac{1}{2}\ln\frac{V_2}{V_1} + \frac{(x-m_2)^2}{2V_2} - \frac{(x-m_1)^2}{2V_1}\right\}$$

Now performing the expectation term-by-term gives

$$\mathcal{I}_{\text{KL}}(\text{p}_1(x);\text{p}_2(x)) = \frac{1}{2}\ln\frac{V_2}{V_1} + \frac{1}{2V_2}\int (x-m_2)^2\text{p}_1(x)\,dx - \frac{1}{2V_1}\int (x-m_1)^2\text{p}_1(x)\,dx$$

Since the last term (under the integral) is the variance, V_1, it is simply $-\frac{1}{2}$ and therefore all we need to do is expand and perform the integration of the second integral term-by-term to give

$$\mathcal{I}_{\text{KL}}(\text{p}_1(x);\text{p}_2(x))$$
$$= \frac{1}{2}\ln\frac{V_2}{V_1} + \frac{1}{2V_2}\left[\int x^2\text{p}_1(x)\,dx - 2m_2\int x\text{p}_1(x)\,dx + m_2^2\int \text{p}_1(x)\,dx\right] - \frac{1}{2}$$

or identifying terms from the properties of moments, we obtain

$$\mathcal{I}_{\text{KL}}(\text{p}_1(x);\text{p}_2(x)) = \frac{1}{2}\ln\frac{V_2}{V_1} + \frac{1}{2V_2}\left[(V_1 + m_1^2) - 2m_2m_1 + m_2^2\right] - \frac{1}{2}$$

Finally, we obtain

$$\mathcal{I}_{\mathrm{KL}}(\mathrm{p}_1(x); \mathrm{p}_2(x)) = \frac{1}{2}\ln\frac{V_2}{V_1} + \frac{V_1 + (m_1 - m_2)^2}{2V_2} - \frac{1}{2}$$

Performing the same calculation, we get

$$\mathcal{I}_{\mathrm{KL}}(\mathrm{p}_2(x); \mathrm{p}_1(x)) = \frac{1}{2}\ln\frac{V_1}{V_2} + \frac{V_2 + (m_1 - m_2)^2}{2V_1} - \frac{1}{2}$$

and therefore the KD is

$$\begin{aligned} \mathcal{J}_{\mathrm{KD}}(\mathrm{p}_1(x); \mathrm{p}_2(x)) &= \mathcal{I}_{\mathrm{KL}}(\mathrm{p}_1(x); \mathrm{p}_2(x)) + \mathcal{I}_{\mathrm{KL}}(\mathrm{p}_2(x); \mathrm{p}_1(x)) \\ &= \frac{V_1^2 + (m_1 - m_2)^2(V_1 + V_2) + V_2^2}{2V_1V_2} - 1 \end{aligned}$$

This completes the example. △△△

The KL and therefore the KD can also be determined by probability distribution estimation using MC sampling techniques [64–65]. As another example of this approach, consider how to apply the KL information to distinguish between a unimodal Gaussian and a Gaussian mixture.

Example 7.5

Suppose we would like to test whether a given data set is from a Gaussian distribution specified by $\mathcal{N}(m, V)$ or from a Gaussian mixture distribution specified by $p\mathcal{N}(m_1, V_1) + (1-p)\mathcal{N}(m_2, V_2)$ where p is the mixing coefficient (probability). The Kullback divergence can easily be calculated and compared to a bound α to determine "how close" the data is to a mixture or a Gaussian, that is,

$$\mathcal{J}_{\mathrm{KD}}(\Pr(x_i); \hat{\Pr}(\hat{x}_i)) = \mathcal{J}_{\mathrm{KD}}(p\mathcal{N}(m_1, V_1) + (1-p)\mathcal{N}(m_2, V_2); \mathcal{N}(m, V)) < \alpha$$

In order to perform the test we choose m and $\sqrt{V}$ to minimize the KD above. Solving, we obtain

$$m = pm_1 + (1-p)m_2; \quad \text{and} \quad V = pV_1 + (1-p)V_2 + p(1-p)(m_1 - m_2)^2$$

A typical bound of $\alpha = 0.1$ appears to perform well in distinguishing a single Gaussian from a mixture. △△△

Particle Filter Performance Before we close this section, let us see how KL theory can be applied to PF design. Typically, we have a simulation model of the dynamics of the system under investigation in some form or another, that is, the model can range from a very detailed "truth model" as discussed in Ref. [5] to a simple signal processing representation (e.g., sinusoids). In any case using this truth model, we can generate a "true distribution" of the system, say $\Pr(x(t)|Y_t)$ and incorporate it into the divergence criterion by: (1) driving the underlying model with a random sequence; (2) generating an ensemble (see Section 6.7); and then (3) applying a nonparametric PDF/PMF histogram estimator. A nonparametric estimate of the posterior distribution $\hat{\Pr}(x(t)|Y_t)$ is already provided by the particle filter and can be used in the $\mathcal{J}_{\text{KD}}$ criterion to be compared to this "true" estimate.

Particle filters are basically developed to provide estimates of the underlying posterior distributions for the problem under investigation in order to extract meaningful statistics. Compared to the usual model-based (Kalman) processors which utilize their inherent "optimality" property (zero-mean, uncorrelated residuals) to assess performance, the estimated posterior of the PF can be compared to the true distribution of the underlying process it is to estimate by using the *divergence statistic* of the KL information as a metric [59].

For our problem, suppose $\Pr(\mathbf{x}(t)|Y_t)$ is the *true* posterior PMF and $\hat{\Pr}(\mathbf{x}_i(t)|Y_t)$ is the estimated (particle) distribution, then the KL information quantity of the true distribution relative to the estimated is defined by [62–63]

$$\begin{aligned}\mathcal{I}_{\text{KL}}\left(\Pr(\mathbf{x}(t)|Y_t);\hat{\Pr}(\mathbf{x}_i(t)|Y_t)\right) &:= E_{\mathbf{x}}\left\{\ln\frac{\Pr(\mathbf{x}(t)|Y_t)}{\hat{\Pr}(\mathbf{x}_i(t)|Y_t)}\right\}\\ &= \sum_{i=1}^{N_p}\ln\frac{\Pr(\mathbf{x}(t)|Y_t)}{\hat{\Pr}(\mathbf{x}_i(t)|Y_t)}\times\Pr(\mathbf{x}(t)|Y_t)\end{aligned}\tag{7.72}$$

The KL satisfies its most important property from a distribution comparison viewpoint—when the true distribution and its estimate are *close* (or identical), then the information quantity is

$$\mathcal{I}_{\text{KL}}\left(\Pr(\mathbf{x}(t)|Y_t);\hat{\Pr}(\mathbf{x}_i(t)|Y_t)\right)=0\Leftrightarrow\Pr(\mathbf{x}(t)|Y_t)=\hat{\Pr}(\mathbf{x}_i(t)|Y_t)\quad\forall i\tag{7.73}$$

As previously mentioned, this property infers that as the *estimated* posterior distribution approaches the *true* distribution, then the value of the KL approaches *zero* (minimum). The *Kullback–Leibler divergence* (KD) for this case is defined by

$$\begin{aligned}\mathcal{J}_{\text{KD}}\left(\Pr(\mathbf{x}(t)|Y_t);\hat{\Pr}(\mathbf{x}_i(t)|Y_t)\right) &:= \mathcal{I}_{\text{KL}}\left(\Pr(\mathbf{x}(t)|Y_t);\hat{\Pr}(\mathbf{x}_i(t)|Y_t)\right)\\ &\quad+\mathcal{I}_{\text{KL}}\left(\hat{\Pr}(\mathbf{x}_i(t)|Y_t);\Pr(\mathbf{x}(t)|Y_t)\right)\end{aligned}\tag{7.74}$$

provides the *distance metric* indicating "how close" the estimated posterior is to the true distribution or from our perspective, "how well does it approximate" the true. Thus, the KD is a very important metric that can be applied to assess the performance of Bayesian processors, providing a measure between the true and estimated posterior distributions.

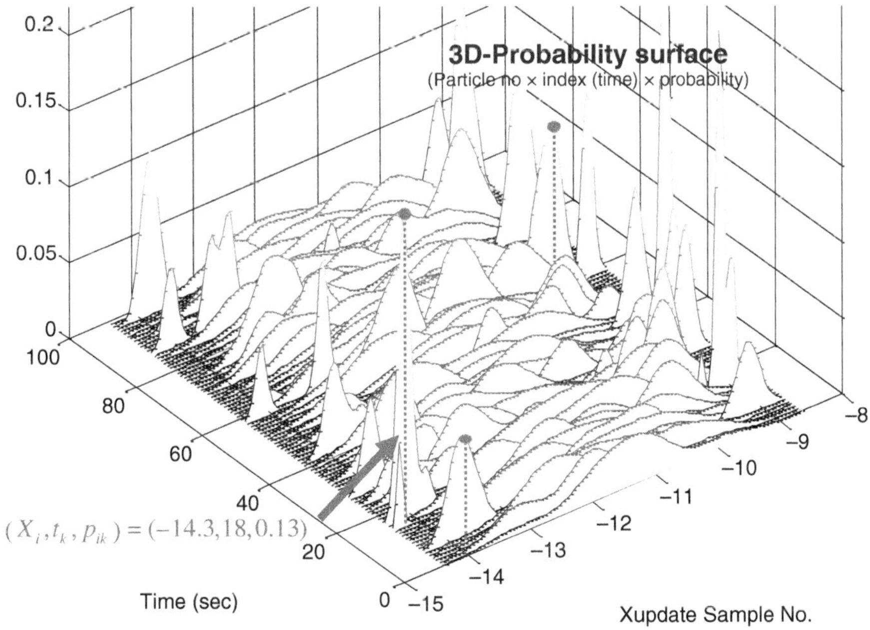

FIGURE 7.14 Particle filter posterior probability surface illustrating some triplet points (X_i, t_j, p_{ij}) within the volume.

The next question that arises is how do we go about performing the calculation, that is, how do we obtain the corresponding "truth model" distribution to perform the comparison. First, let us revisit the posterior probability surface generated by the particle filter. As depicted simply in Fig. 7.14, we see that a more careful investigation of this surface reveals that each point in the space is a triplet defined the ith particle number (location parameter), the index set or in our case the jth time sample number and the corresponding posterior probability, that is, (X_i, t_j, p_{ij}) such that

$$\{p_{ij}\} := \Pr(\mathbf{x}_i(t_j)|Y_t) \tag{7.75}$$

From the figure, the grid of points compose the overall $N_p \times N_t$ surface, where each of the points (coordinates) satisfy the sets

$$\{X_i\} \text{ for } i = 1, \ldots, N_p, \{t_j\} \text{ for } j = 1, \ldots, N_t \tag{7.76}$$

along with the corresponding set of posterior probabilities

$$\{p_{ij}\} \text{ for } i = 1, \ldots, N_p; j = 1, \ldots, N_t \tag{7.77}$$

Therefore, given the posterior surface produced by the PF, it is clear that statistical inferences lead to various solutions or combination of these probabilities

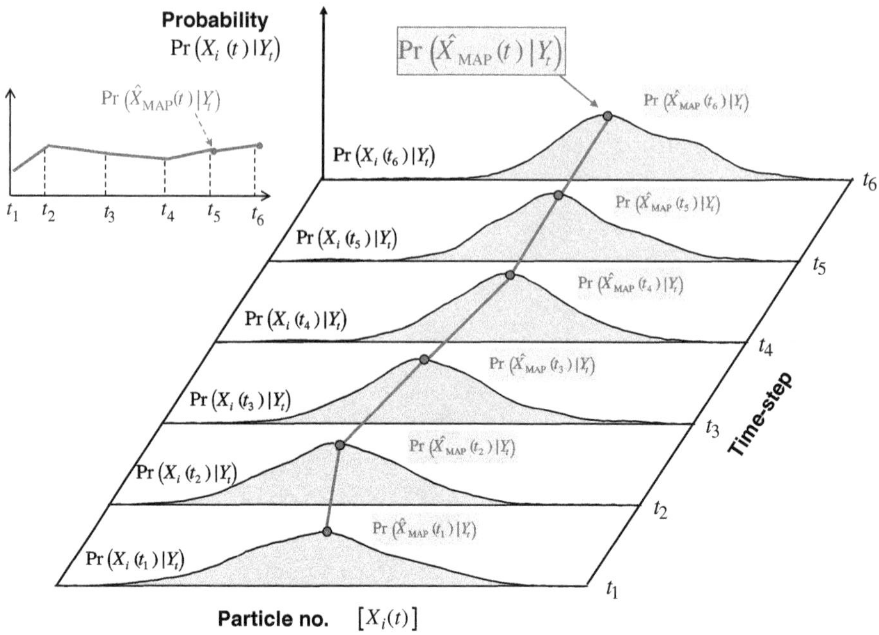

FIGURE 7.15 Particle filter posterior probability surface: Updated state conditional posterior × particle no. × index (time) along with corresponding MAP probability, $\hat{\Pr}(\hat{\mathbf{X}}_{MAP}(t)|Y_t)$.

coupled with their corresponding location parameters. For instance, MAP estimation selects the location parameters that correspond to the maximum probabilities as illustrated in Fig. 7.15, that is, the MAP value at the (i,j)th location on the surface corresponds to

$$\hat{\mathbf{X}}^i_{MAP}(t_j) = \max_{ij} \Pr\left(\mathbf{x}_i(t_j)|Y_t\right) = p_{ij}^{\max} \tag{7.78}$$

and the complete estimation (over time) is the set of values

$$\hat{\mathbf{X}}_{MAP}(t) = \left\{\hat{\mathbf{X}}^i_{MAP}(t_j)\right\}; j = 1, \ldots, N_t \text{ for } i \in \max_{ij} \Pr\left(\mathbf{x}_i(t_j)|Y_t\right) \tag{7.79}$$

So, we see that understanding this posterior probability surface aides in comprehending the mechanism that can now be applied to the KD metric. The PMF of the MAP estimates are simply the set of posterior probabilities $\{p_{ij}^{\max}\}$ and therefore we can extract one of the required distributions for the KD metric as

$$\Pr\left(\mathbf{x}_i(t_j)|Y_t\right) \text{ for } i \in \max_{ij} \Pr\left(\mathbf{x}_i(t_j)|Y_t\right) \tag{7.80}$$

If we have a "truth model", then its corresponding probability surface is analogous to the PF surface and can be generated through an ensemble simulation with values of the probabilities selected that correspond to the true state values at t. An even simpler technique for non-analytic distributions is to synthesize perturbed (low noise covariances) states generated by the model at each point in index (time) and estimate its histogram by kernel density smoothers directly to obtain an estimate of $\Pr\left(\mathbf{X}_{\mathrm{True}}(t)|Y_t\right)$.

Now the KD can be approximated for a given realization by

$$\mathcal{J}_{\mathrm{KD}}\left(\Pr\left(\mathbf{X}_{\mathrm{True}}(t)|Y_t\right);\hat{\Pr}\left(\hat{\mathbf{X}}_{\mathrm{MAP}}(t)|Y_t\right)\right) := \mathcal{I}_{\mathrm{KL}}\left(\Pr\left(\mathbf{X}_{\mathrm{True}}(t)|Y_t\right);\hat{\Pr}\left(\hat{\mathbf{X}}_{\mathrm{MAP}}(t)|Y_t\right)\right) + \mathcal{I}_{\mathrm{KL}}\left(\hat{\Pr}\left(\hat{\mathbf{X}}_{\mathrm{MAP}}(t)|Y_t\right);\Pr\left(\mathbf{X}_{\mathrm{True}}(t)|Y_t\right)\right) \tag{7.81}$$

The implementation of the KD approach (state component-wise) is:

- Generate samples from the state "truth model:" $\left\{\mathbf{X}_{\mathrm{True}}(t)\right\}$;
- Estimate the corresponding state truth model distribution: $\hat{\Pr}(\mathbf{X}_{\mathrm{True}}(t)|Y_t)$ using the kernel density smoother;
- From the particle filter estimates $\left\{\hat{\mathbf{X}}_{\mathrm{MAP}}(t)\right\}$ extract or estimate the corresponding posterior distribution: $\hat{\Pr}\left(\hat{\mathbf{X}}_{\mathrm{MAP}}(t)|Y_t\right)$;
- Calculate the KD: $\mathcal{J}_{\mathrm{KD}}\left(\Pr\left(\mathbf{X}_{\mathrm{True}}(t)|Y_t\right);\hat{\Pr}\left(\hat{\mathbf{X}}_{\mathrm{MAP}}(t)|Y_t\right)\right)$; and
- Determine if $\mathcal{J}_{\mathrm{KD}} \approx 0$ for PF performance.

Performing ensemble estimation also enables a corresponding set of KDs that can also be used to generate ensemble statistics as well.

Before we close this section, we briefly mention a similar metric for comparing probability distributions—the *Hellinger distance* (HD), [64], [65]. The HD evolved from measure theory and is a true distance metric satisfying all of the required properties [64–65]. It is a metric that can also be used to compare the similarity of two probability distributions defined by

$$\mathbf{H}\left(\Pr(x_k);\hat{\Pr}(\hat{x}_k)\right) := \frac{1}{\sqrt{2}}\left[\sum_{k=1}^{K}\left(\sqrt{\Pr(x_k)}-\sqrt{\hat{\Pr}(\hat{x}_k)}\right)^2\right]^{\frac{1}{2}} \tag{7.82}$$

The Hellinger distance $\mathbf{H}_{\mathrm{HD}}$ is a bounded metric on the probability space possessing the following properties:

1. $0 \le \mathbf{H}_{\mathrm{HD}}\left(\Pr(x_k);\hat{\Pr}(\hat{x}_k)\right) \le 1$
2. $\mathbf{H}_{\mathrm{HD}}\left(\Pr(x_k);\hat{\Pr}(\hat{x}_k)\right) = 0 \Leftrightarrow \Pr(x_k) = \hat{\Pr}(\hat{x}_k) \quad \forall k$
3. $\mathbf{H}_{\mathrm{HD}}\left(\Pr(x_k);\hat{\Pr}(\hat{x}_k)\right) = 1 \Leftrightarrow \Pr(x_k) \neq \hat{\Pr}(\hat{x}_k) \quad \forall k$

The second property of the Hellinger distance provides a "scaled metric" similar to the "unscaled" KD providing a reasonable calculation to compare or evaluate overall performance.

For particle filter design/analysis, the corresponding *Hellinger distance* is specified by

$$\mathbf{H}_{\rm HD}\left(\Pr(\mathbf{x}(t)|Y_t);\hat{\Pr}(\mathbf{x}_i(t)|Y_t)\right) = \frac{1}{\sqrt{2}}\left[\sum_{i=1}^{N_p}\left(\sqrt{\Pr(\mathbf{x}(t)|Y_t)} - \sqrt{\hat{\Pr}(\mathbf{x}_i(t)|Y_t)}\right)^2\right]^{\frac{1}{2}} \tag{7.83}$$

Using the truth model and its corresponding probability surface generated through ensemble simulation, the values of the probabilities selected correspond to the "true state" probabilities at t synthesizing $\Pr(\mathbf{X}_{\rm True}(t)|Y_t)$.

The HD can be approximated for a given realization by

$$\mathbf{H}_{\rm HD}\left(\Pr(\mathbf{X}_{\rm True}(t)|Y_t);\hat{\Pr}(\hat{\mathbf{X}}^i_{\rm MAP}(t)|Y_t)\right) = \frac{1}{\sqrt{2}}\left[\sum_{i=1}^{N_p}\left(\sqrt{\Pr(\mathbf{X}_{\rm True}(t)|Y_t)} - \sqrt{\hat{\Pr}(\hat{\mathbf{X}}^i_{\rm MAP}(t)|Y_t)}\right)^2\right]^{\frac{1}{2}} \tag{7.84}$$

Consider the following example illustrating both KD and Hellinger distance metrics applied to the PF design for the trajectory estimation problem.

Example 7.6

Let us revisit the nonlinear trajectory estimation problem and calculate the Kullback–Leibler divergence using Eq. 7.81.

Recall the discrete state-space representation:

$$x(t) = (1 - 0.05\triangle T)x(t-1) + 0.04\triangle Tx^2(t-1) + w(t-1)$$
$$y(t) = x^2(t) + x^3(t) + v(t)$$

where $\triangle t = 0.01$, $w \sim \mathcal{N}(0, 10^{-6})$ and $v \sim \mathcal{N}(0, 0.09)$. The initial state is Gaussian distributed with $\overline{x}_i(0) \sim \mathcal{N}(\overline{x}(0), \overline{P}(0))$ and $\overline{x}(0) = 2.3$, $\overline{P}(0) = 10^{-2}$.

Here we see the results of estimating the KD over a 100-member ensemble with the corresponding state and measurement PMFs in Figs. 7.16a and 7.16b. The resulting KD over each member realization is shown in Figs. 7.16c and 7.16d, respectively for the state and measurement ensembles with the HD shown in Figs. 7.16e and 7.16f. Here the median and mean KD estimates are indicated by the dashed lines in the figure. The corresponding measurement median and average KDs of 0.00124 and 0.00130 and state KDs of 0.00144 and 0.00146 are shown in (c) and (d). Similarly, the Hellinger distance metric coincides with the KD-metric giving HDs median and average state of 0.0189 and 0.0189 with corresponding measurement median and average of 0.0176 and 0.0178 as shown in Figs. 7.16e and 7.16f, respectively. △△△

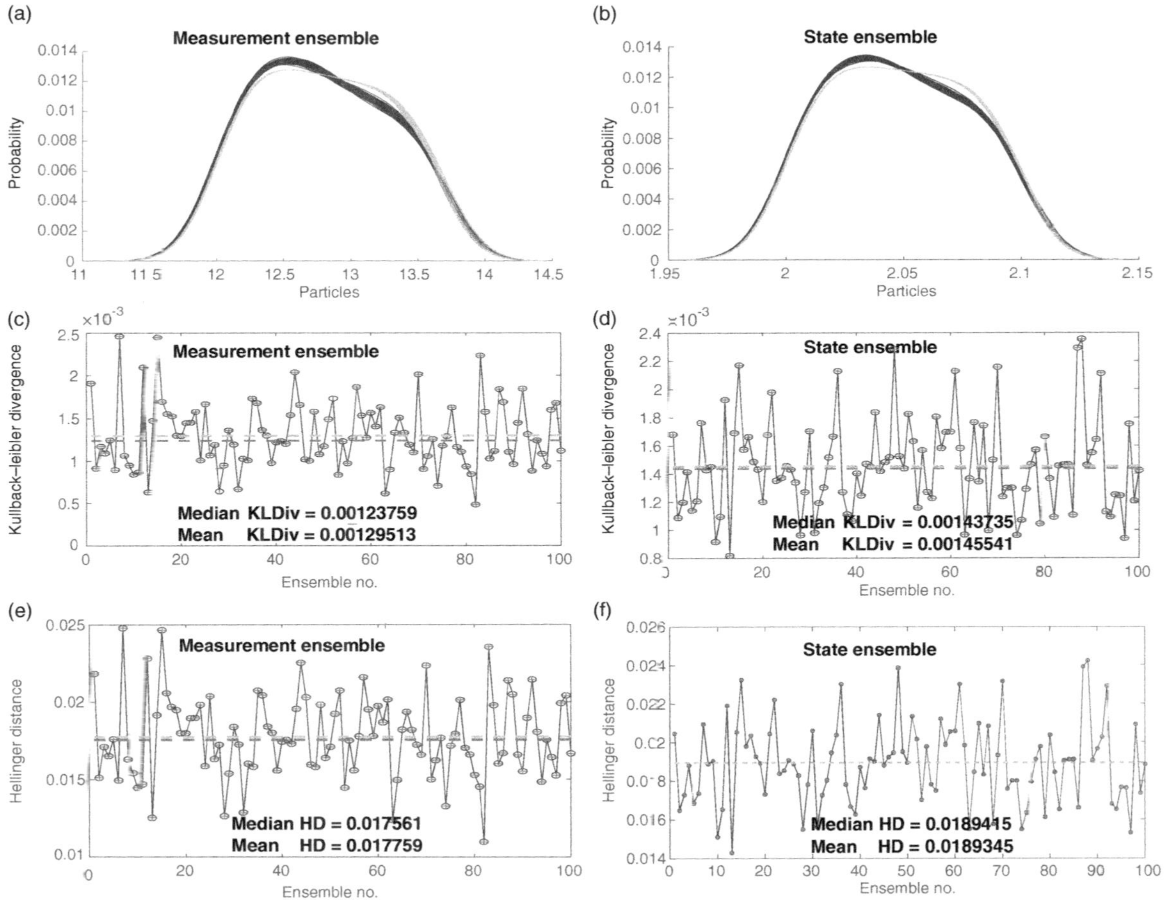

FIGURE 7.16 PMF data: Kullback–Leibler divergence and Hellinger distance calculations for the nonlinear trajectory estimation problem over 100-member ensemble. (a) measurement PMF ensemble; (b) state PMF ensemble; (c) KL-divergence measurement median/mean (KD = 0.00124/0.00130); (d) KL-divergence state median/mean (KD = 0.00144/0.00146); (e) HD-distance measurement median/mean (HD = 0.0176/0.0178); (f) HD-distance state median/mean (HD = 0.0189/0.0189).

This completes the section, next we discuss an entirely different approach to this distribution problem by investigating "goodness of fit" testing.

7.6.4 Model Validation Testing

Of the major practical concerns with all model-based processors is whether or not the model embedded in the processor "matches" the underlying phenomena and can be used to extract meaningful information from the noisy measurements. As mentioned, in the classical Gaussian-based techniques, the zero-mean/whiteness testing of the innovations is a critical measure of this match. These properties are also used extensively for linearized models evolving from nonlinear dynamic systems as well [5]. In all of these cases the distributions are considered unimodal and typically Gaussian.

In the non-unimodal (non-Gaussian) case the diagnostics are more complicated. The roots of MC model diagnostics lie in the basic Uniform Transformation Theorem of Sectiontion 3.3 and the works of Rosenblatt [48] and Smith [49] under the general area of "goodness-of-fit" statistical tests [39, 40, 50, 51, 57, 58].

The fundamental idea is based on analyzing the *predicted measurement cumulative distribution* $\mathrm{P}_Y(y(t)|Y_{t-1})$. A processor is considered consistent *only* if the measurement $y(t)$ is governed by the statistics of its predicted cumulative distributions. Therefore, *validation* consists of statistically testing that the measurements "match" the predictions using the underlying model embedded in the processor. By defining the *residual sequence* as

$$\epsilon(t) := \mathrm{P}_Y(y(t)|Y_{t-1}) = \Pr(Y(t) \le y(t)|Y_{t-1}) = \int_{Y \le y} \Pr(y'(t)|Y_{t-1})dy'(t) \tag{7.85}$$

we must show that $\{\epsilon(t)\}$ is a valid realization of an independent, identically distributed, process uniformly distributed on the interval [0, 1] given the measurements Y_{t-1}. Thus, the statistical test validates whether or not the sequence is $\epsilon(t) \sim \mathcal{U}(0, 1)$ (component-wise for the vector case).

More formally, we use Rosenblatt's theorem[2] that states, if $y(t)$ is a continuous random vector and the underlying model is "valid", then the corresponding sequence $\{\epsilon(t)\}$ is *i.i.d.* on [0, 1]. Under the assumption that the residual sequence is standard uniform, then we can transform to obtain an equivalent Gaussian sequence [48] such that

$$\nu(t) = \Phi^{-1}(\epsilon(t)) \qquad \text{for } \nu \sim \mathcal{N}(0, 1) \text{ with } \epsilon \sim \mathcal{U}(0, 1) \tag{7.86}$$

where Φ^{-1} is the inverse standard Gaussian cumulative distribution. Once the residual sequence is transformed, then all of the classical Gaussian statistical tests can be performed to ensure validity of the underlying model.

[2] The theorem states that for a given random vector $y \in \mathcal{R}^{N_y \times 1}$ with corresponding distribution $P_Y(y)$, the transformed vector, $\epsilon = Ty$, is uniformly distributed on the N_y-hypercube for $\Pr(\epsilon) = \prod_{i=1}^{N_y} \epsilon_i$ when $0 \le \epsilon_i \le 1$. The transformation, T is given by $\epsilon_i = \mathrm{P}_Y(y_i|Y_{i-1})$; $i = 1, \ldots, N_y$ [48].

With this in mind we are still required to solve two basic problems: (1) the estimation of the residual sequence $\epsilon(t)$ or equivalently the estimation of the predictive measurement cumulative distribution, $P_Y(y(t)|Y_{t-1})$; and (2) the diagnostic statistical testing of $\epsilon(t)$ or equivalently $\nu(t)$, that is, demonstrating that $\epsilon \sim \mathcal{U}(0,1)$ or $\nu \sim \mathcal{N}(0,1)$.

The *key* to estimating the residual sequence is based on representing the predictive cumulative distribution as an infinite mixture [39, 40, 51]

$$\epsilon(t) = P_Y(y(t)|Y_{t-1}) = \int P_Y(y(t)|x(t)) \times \Pr(x(t)|Y_{t-1})\, dx(t) \tag{7.87}$$

An MC approach to this estimation problem is to "particulate" the required underlying distribution and estimate the empirical prediction cumulative distribution or equivalently the residuals. Perhaps the simplest MC technique is to sample from the predicted state distribution directly "when possible", that is, if we could replace the prediction distribution with its empirical representation, then we could obtain an estimated residual [57], that is,

$$\hat{\epsilon}(t) = \int P_Y(y(t)|x(t)) \times \left[\frac{1}{N_p} \sum_{i=1}^{N_p} \delta\left(x(t) - x_i(t)\right) \right] dx(t) \tag{7.88}$$

or simply

$$\hat{\epsilon}(t) = \frac{1}{N_p} \sum_{i=1}^{N_p} P_Y(y(t)|x_i(t)) \tag{7.89}$$

However, if direct sampling of the predicted state distribution is *not possible*, then the *predictive decomposition* into the transition prior and posterior can be accomplished, that is,

$$\Pr(x(t)|Y_{t-1}) = \int \Pr(x(t)|x(t-1)) \times \Pr(x(t-1)|Y_{t-1}) dx(t-1) \tag{7.90}$$

or using the state–space representation of the transition prior

$$\Pr(x(t)|Y_{t-1}) = \int \mathcal{A}(x(t)|x(t-1)) \times \Pr(x(t-1)|Y_{t-1}) dx(t-1) \tag{7.91}$$

and performing a particle approximation, we obtain the empirical estimate of the $(t-1)$-step posterior as

$$\hat{\Pr}(x(t-1)|Y_{t-1}) = \sum_{k=1}^{N_p} \mathcal{W}_k(t-1)\delta\left(x(t-1) - x_k(t-1)\right) \tag{7.92}$$

Substituting this expression into Eq. 7.91 gives the prediction distribution

$$\hat{\Pr}(x(t)|Y_{t-1}) = \int \mathcal{A}(x(t)|x(t-1)) \times \left[\sum_{k=1}^{N_p} \mathcal{W}_k(t-1)\delta\left(x(t-1) - x_k(t-1)\right)\right] dx(t-1)$$

$$= \sum_{k=1}^{N_p} \mathcal{W}_k(t-1)\mathcal{A}(x(t)|x_k(t-1)) \tag{7.93}$$

and the residual of Eq. 7.87 becomes

$$\hat{\epsilon}(t) = \int \mathrm{P}_Y(y(t)|x(t)) \left[\sum_{k=1}^{N_p} \mathcal{W}_k(t-1)\mathcal{A}(x(t)|x_k(t-1))\right] dx(t) \tag{7.94}$$

If we draw a sample $x_i(t)$ from the transition prior $x_i \sim \mathcal{A}(x(t)|x_k(t-1))$ and use the perfect sample approximation,

$$\mathcal{A}(x(t)|x_k(t-1)) \approx \delta(x(t) - x_i(t)) \tag{7.95}$$

then substituting into Eq. 7.94 for the transition distribution yields

$$\hat{\epsilon}(t) = \sum_{k=1}^{N_p} \mathcal{W}_k(t-1) \int \mathrm{P}_Y(y(t)|x(t)) \times \delta\left(x(t) - x_i(t)\right) dx(t) \tag{7.96}$$

giving the final expression for the estimated residual as [39, 40, 51]

$$\hat{\epsilon}(t) = \sum_{k=1}^{N_p} \mathcal{W}_k(t-1)\mathrm{P}_Y(y(t)|x_i(t)) \tag{7.97}$$

for

$$\mathrm{P}_Y(y(t)|x_i(t)) = \int_{Y \leq y(t)} \mathcal{C}(y(t)|x_i(t))\, dy(t)$$

which can be estimated through direct integration or using an empirical CDF [14], if necessary.

Once we have obtained the estimate of the residual $\epsilon(t)$ (or equivalently the transformed residual $v(t)$), we can perform diagnostic tests to evaluate the "goodness-of-fit" and therefore evaluate the validity of the embedded dynamic model.

There are a variety of diagnostic tests that can be performed which include: χ^2-testing (C-sq), Kolmogorov-Smirnov (K-S) tests, normality testing (graphically), zero-mean/whiteness testing, etc. Here, we concentrate on the C-sq and K-S as well as moment testing of the transformed residuals. Zero-mean/whiteness testing was already discussed in Section 5.6.

Any of the Kalman techniques can also be used to generate an approximation to the sequence of residuals or prediction cumulative distribution, using the empirical PDF, EKF, UKF, Gauss-Hermite (G-H) grid-based integration as well as the Gaussian mixture approach, that is, Gaussian sums (G-S) (see Refs. [51, 66–74] for more details).

Chi-Square Model Validation Test Chi-square (C-Sq) tests are hypothesis tests with the null hypothesis, $\mathcal{H}_o$, that an N-length sequence of data is a random sample from a specified distribution against the alternative that it is not [14, 15]. It is usually applied to Gaussian distributions. C-*Sq* tests are based on the fact that the exponent of a Gaussian distribution, that is, the square of the ratio of the random variable minus its mean divided by its standard deviation is chi-square distributed with N degrees of freedom, that is, $(x-\mu/\sigma)^2 \sim \chi^2(N)$. However, this test can be applied to any distribution.

For instance, if the random variable is binomially distributed with $B(N, p)$ for p the *success probability*, then it takes the same form as the exponent above—it is distributed (in the limit) as $\chi^2(1)$. Extending this to a multinomial distribution with parameters N and $p(i)$ for $i = 1, \ldots, k-1$; $\mathcal{M}(N, \{p(i)\})$, then in the limit as $N \rightarrow \infty$ the test statistic, say C_{k-1}, has an approximate $\chi^2(k-1)$ distribution.[3]

Hypothesis testing that $\mathcal{H}_o$ is true using the test statistic is specified by the value κ (probability bound) of the $\chi^2(k-1)$ using $\Pr(C_{k-1} \geq \kappa) = \alpha$ for α the significance level of the test. Thus, if the test statistic is *less than* κ the null hypothesis is accepted and the selected distribution is correct.

For our problem, a goodness-of-fit test of the residuals follows directly from their uniformity property of an adequate model. The most common statistical tests for uniformity follows from the χ^2-test based on segmenting the estimated residual sequence on [0,1] into subintervals and testing. The chi-square statistical test can be used to decide whether or not the residual sequence $\epsilon(t)$ is $\mathcal{U}(0,1)$ or equivalently the transformed residual, $v(t)$ is $\mathcal{N}(0,1)$.

The C-Sq test statistic for our problem is given by

$$C_{N_\epsilon - 1} = \sum_{i=1}^{N_\epsilon} \frac{(n_\epsilon(i) - \overline{\epsilon})^2}{\overline{c}} \tag{7.98}$$

where N is the total number of residual samples; N_ϵ is the number of bins (equally spaced subintervals); $n_\epsilon(i)$ is the number of residual counts in the ith-bin (subinterval); and $\overline{\epsilon}$ is the expected number of counts per bin given by $\overline{\epsilon} = \frac{N}{N_\epsilon}$.

[3] To be more specific, if the *i.i.d.* random variables, $y_1, \ldots, y_{k-1}$ are multinomially distributed with $y_k = N - \sum_{i=1}^{k-1} y_i$ and $p(k) = 1 - \sum_{i=1}^{k-1} p(i)$, then the statistic, $C_{k-1} = \sum_{i=1}^{k-1} (y_i - N_p(i))^2 / N_p(i)$ is $C_{k-1} \sim \chi^2(k-1)$ [15].

If the residual sequence is uniform, then $C_{N_\epsilon - 1} \sim \chi^2(N_\epsilon - 1)$ and κ is compatible with a $\chi^2(N_\epsilon - 1)$ distribution at significance level α. Therefore, if $C_{N_\epsilon - 1} \leq \kappa$ the null hypothesis that the residual sequence is uniformly distributed is accepted and the model is adequate (validated) otherwise it is rejected. Thus, the χ^2-model validation test is

- *Partition* the N-sample residual sequence into N_ϵ bins (equally spaced subintervals)
- *Count* the number of residual samples in each bin, $n_\epsilon(i)$; $i = 1, \ldots, N_\epsilon$
- *Calculate* the expected bin count, $\overline{\epsilon} = N/N_\epsilon$
- *Calculate* the test statistic $C_{N_\epsilon - 1}$ of Eq. 7.98
- *Test* that $C_{N_\epsilon - 1} \leq \kappa$ [Accept $\mathcal{H}_o$]

Example 7.7

Suppose we have a residual sequence, $\epsilon(t)$, scaled on [0, 1] of $N = 1000$ samples and we would like to test that it is uniformly distributed using the χ^2-model validation test. We partition the sequence into $N_\epsilon = 10$ bins; therefore, the expected counts per bin is $\overline{\epsilon} = 100$. At the $\alpha = 5\%$ significance level, the test statistic,

$$C_{N_\epsilon - 1} = 3.22$$

is less than κ (probability bound) accepting the null hypothesis that the sequence is uniform and therefore the model is validated. △△△

Next, we consider another more robust method for goodness-of-fit testing.

Kolmogorov–Smirnov Model Validation Test The chi-square goodness-of-fit test suffers from the limitations of arbitrary interval widths and the requirement of large data sets. An alternative or complementary approach is the Kolmogorov–Smirnov (K-S) goodness-of-fit test that is based on deciding whether or not a hypothesized (e.g., uniform) or estimated cumulative distribution characterizes a given data sequence (e.g., residual). The hypothesis test is given by

$$\begin{aligned} \mathcal{H}_o:&\quad \hat{\mathrm{P}}_E(\epsilon) = \mathrm{P}_o(\epsilon) \\ \mathcal{H}_1:&\quad \hat{\mathrm{P}}_E(\epsilon) \neq \mathrm{P}_o(\epsilon) \end{aligned} \tag{7.99}$$

where $\hat{\mathrm{P}}_E$ is the underlying (estimated) population CDF and P_o is the hypothesized CDF. The test statistic used in making the decision is

$$\mathcal{K} = \max_{\epsilon} |\mathrm{P}_E(\epsilon) - \mathrm{P}_o(\epsilon)| \tag{7.100}$$

where $\hat{\mathrm{P}}_E$ is given by the empirical distribution function estimate

$$\hat{\mathrm{P}}_N(\epsilon) = \frac{N_\epsilon}{N} \rightarrow \Pr(E \leq \epsilon) = \mathrm{P}_E(\epsilon) \quad \text{as } N \rightarrow \infty \tag{7.101}$$

For large N, $\mathcal{K} \approx 0$ with $\mathcal{H}_o$ *true* while for $\mathcal{H}_1$ true, $\mathcal{K}$ is close to the maximum difference. Therefore, we reject $\mathcal{H}_o$ if $\mathcal{K} > \kappa$ with κ a constant determined by the level-of-significance of the hypothesis test, α. That is,

$$\alpha = \Pr(\mathcal{K} > \kappa \mid \mathcal{H}_o) \approx 2e^{-2N\kappa^2} \tag{7.102}$$

Thus the K-S-test is:

- *Estimate* the empirical CDF, $\hat{\mathrm{P}}_E(\epsilon)$
- *Calculate* K-S test statistic $\mathcal{K}$ from Eq. 7.100
- *Test* that

$$\mathcal{K} < \sqrt{-\frac{1}{2N} \ln \frac{\alpha}{2}} \quad [\text{Accept}\mathcal{H}_o]$$

Example 7.8

We have a residual sequence, $\epsilon(t)$, scaled on [0, 1] of $N = 25$ samples and we would like to perform the K-S test that the samples are uniformly distributed at the $\alpha = 5\%$ significance level. The test statistic,

$$\mathcal{K} = 0.17$$

is less than $\kappa = 0.26$ accepting the null hypothesis that the sequence is uniform and therefore the model is valid. The test is shown in Fig. 7.17 as the location of the maximum deviation between the hypothesized and empirical distributions. △△△

When the transformed residuals are used, then the standard zero-mean/whiteness testing can be accomplished as well as estimating the moments of the Gaussian distribution which we discuss in the next section.

Moment-Based Model Validation Test When the residuals are transformed from the hypothesized uniformly distributed sequence to a standard Gaussian, $\nu \sim \mathcal{N}(0, 1)$, then a wealth of normality diagnostics can be applied to validate the adequacy of the embedded model. Besides the zero-mean/whiteness and WSSR tests of Section 5.6 for Gaussian processes, the usual suite of diagnostics can be applied to estimate the underlying moments to check for consistency of the Gaussian assumption and therefore model validation. The brute force approach is simply to calculate

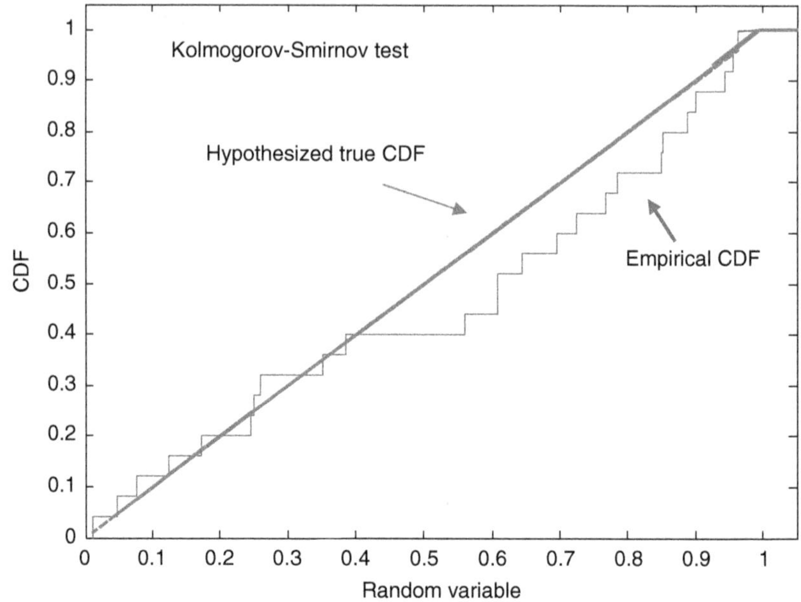

FIGURE 7.17 Kolmogorov–Smirnov model validation test of residual sequence: hypothesized and empirical CDF.

the mean-squared (estimation) error over an ensemble of runs of the processor and any truth model available for comparison,

$$\xi = \sqrt{E\{(\Theta_{\text{true}} - \hat{\Theta})^2\}} \tag{7.103}$$

where Θ can be any parameter, state, or measurement estimated by $\hat{\Theta}$ and the expectation can be calculated by integrating or solving over an ensemble generated by executing the processor a multitude of times and averaging. This can be very costly and sometimes impossible because of the lack of the "truth".

Another approach is to calculate a set of statistical indexes that can be used to "qualitatively" assess performance and model validity using the transformed residual sequence $v(t)$ [51–53]. Here the first four central moments of an N-sample sequence of transformed residuals are estimated based on

$$m_v(k) = \frac{1}{N}\sum_{t=1}^{N}(v(t) - m_v(1))^k \quad \text{for } k \geq 2 \tag{7.104}$$

with the first moment the sample mean

$$m_v(1) = \frac{1}{N}\sum_{t=1}^{N} v(t) \tag{7.105}$$

These moments can then be used to construct the following diagnostic indices which are asymptotically distributed $\mathcal{N}(0, 1)$ (see Ref. [51] for details).

- Bias Index: $\mathcal{B}_N = \sqrt{N} m_\nu(1)$
- Dispersion Index: $\mathcal{D}_N = \frac{N m_\nu(2) - N + 1}{\sqrt{2(N-1)}}$
- Skewness Index: $\mathcal{S}_N = \sqrt{\frac{(N+1)(N+3)}{6(N-2)}} \times \frac{m_\nu(3)}{\sqrt{m_\nu^3(2)}}$
- Tail Index: $\mathcal{T}_N = (N+1)\frac{\sqrt{(N+3)(N+5)}}{\sqrt{24N(N-2)(N-3)}} \times \left(\frac{m_\nu(3)}{m_\nu(4)} - \frac{3(N-1)}{(N+1)}\right)$
- Joint Index: $\mathcal{J}_N = \mathcal{S}_N^2 + \mathcal{T}_N^2$

From a pragmatic perspective these indices are used in a more qualitative manner even though they are quantitative. They are used to expose "surprising values", that is, for N not too small, the indices can be bound by some constant, β and compared with upper and lower quantile estimates of their exact distribution [51]. For instance, consider the quantile $\mathcal{D}_{N,\alpha}$ of the exact dispersion index distribution under the assumption of a valid model. Then it can be shown [51] that

$$\mathcal{D}_{N,\alpha} = (\chi^2_{N-1,\alpha} - (N-1)/\sqrt{2(N-1)}) \tag{7.106}$$

where $\chi^2_{N-1,\alpha}$ is $\chi^2(N-1)$ distributed. Other measures such as the skewness and tail indices, $\mathcal{S}_N$ and $\mathcal{T}_N$ can be obtained from MC simulations [51].

If $\mathcal{B}_N$ is surprisingly high or low, the measurements tend to be larger or smaller then predicted ($\hat{y}(t)$), while a surprisingly high or low dispersion index, $\mathcal{D}_N$ indicates that the measurements are under or over dispersed. The $\mathcal{S}_N$ and $\mathcal{T}_N$ are useful in analyzing the measurement distribution. A higher or lower $\mathcal{S}_N$ indicates a skew to either right or left while a higher or lower $\mathcal{T}_N$ indicates longer or shorter tails, respectively. The $\mathcal{J}_N$ is an asymptotically equivalent to normality tests [50, 54]. A suite of other statistics exist for testing correlations [55, 56] as well.

This completes the section on practical aspects of PF design and analysis. We will couple these statistical tests to the classical whiteness testing techniques to evaluate the performance of the processors. Next, let us consider the design of a "bootstrap" processor on a canonical problem: a case study for population growth.

7.7 CASE STUDY: POPULATION GROWTH PROBLEM

In this section, we discuss the development of SSPFfor the population growth problem. We consider this well-known problem that has become a *benchmark* for many of the PF algorithms. It is highly nonlinear and nonstationary. Thus, consider the problem of [25] and [20, 26–29].

The state transition and corresponding measurement model are given by

$$x(t) = \frac{1}{2}x(t-1) + \frac{25x(t-1)}{1+x^2(t-1)} + 8\cos(1.2(t-1)) + w(t-1)$$
$$y(t) = \frac{x^2(t)}{20} + v(t)$$

where $\triangle t = 1.0$, $w \sim \mathcal{N}(0, 10)$ and $v \sim \mathcal{N}(0, 1)$. The initial state is Gaussian distributed with $x(0) \sim \mathcal{N}(0.1, 5)$.

In terms of the nonlinear state–space representation, we have

$$a[x(t-1)] = \frac{1}{2}x(t-1) + \frac{25x(t-1)}{1+x^2(t-1)}$$
$$b[u(t-1)] = 8\cos(1.2(t-1))$$
$$c[x(t)] = \frac{x^2(t)}{20}$$

In the Bayesian framework, we would like to estimate the instantaneous posterior filtering distribution,

$$\hat{\Pr}(x(t)|Y_t) \approx \sum_{i=1}^{N_p} \mathcal{W}_i \delta(x(t) - x_i(t)) \tag{7.107}$$

where the unnormalized importance weight is given by

$$W_i(t) = \frac{\mathcal{C}\,(y(t)|x_i(t)) \times \mathcal{A}\,(x(t)|x_i(t-1))}{q(x_i(t)|X_{t-1}, Y_t)} \tag{7.108}$$

The *weight recursion* for the bootstrap case is $W_i(t) = W_i(t-1) \times \mathcal{C}(y(t)|x(t))$. Therefore, for the Bayesian processor, we have that the state transition probability is given by

$$\mathcal{A}(x(t)|x(t-1)) \sim \mathcal{N}(x(t) : \mathbf{a}[x(t-1)], \mathbf{R}_{ww}) \tag{7.109}$$

Thus, the SIR algorithm becomes:

1. Draw samples (particles) from the state transition distribution: $x_i(t) \rightarrow \mathcal{N}(x(t) : \mathbf{a}[x(t-1)], \mathbf{R}_{ww})$

$$w_i(t) \rightarrow \Pr(w(t)) \sim \mathcal{N}(0, R_{ww})$$
$$x_i(t) = \frac{1}{2}x_i(t-1) + \frac{25x_i(t-1)}{1+x_i^2(t-1)} + 8\cos(1.2(t-1)) + w_i(t-1)$$

2. Estimate the weight/likelihood,

$$W_i(t) = C(\mathbf{y}(t)|x(t)) \rightarrow \mathcal{N}(\mathbf{y}(t) : \mathbf{c}[x(t)], \mathbf{R}_{vv}(t))$$

$$c[x_i(t)] = \frac{x_i^2(t)}{20}$$

3. Normalize the weight: $\mathcal{W}_i(t) = W_i(t) / \sum_{i=1}^{N_p} W_i(t)$
4. Resample: $\hat{x}_i \Rightarrow x_i$
5. Estimate the instantaneous posterior:

$$\hat{\Pr}(x(t)|Y_t) \approx \sum_{i=1}^{N_p} \mathcal{W}_i \delta(x(t) - x_i(t))$$

6. Estimate (inference) the corresponding statistics:

$$\hat{X}_{\text{MAP}}(t) = \arg\max_{x(t)} \hat{\Pr}(x(t)|Y_t)$$

$$\hat{X}_{\text{CM}}(t) = E\{x(t)|Y_t\} = \sum_{i=1}^{N_p} x_i(t)\hat{\Pr}(x(t)|Y_t)$$

$$\hat{X}_{\text{MEDIAN}}(t) = \text{median}(\hat{\Pr}(x(t)|Y_t))$$

We show the simulated data in Fig. 7.18; in (a) we see the hidden state and (b) the noisy measurement. The state and measurement PF estimates for a single realization are shown in Figs. 7.18c and 7.18d. Final performance of the processor over a 100-member ensemble is illustrated in Figs. 7.18e and 7.18f demonstrating the effectiveness of the PF bootstrap processor for this problem. Various ensemble estimates are shown for the CM and MAP estimates, each performing reasonably well. Next, the estimated instantaneous posterior distribution surface for the state is shown in Fig. 7.19a, while slices at selected instants of time ($t = 26$, $t = 60$, $t = 96$) are shown in (b) to (d) with the circles annotating particle locations normalized to a constant weight. Here we see that the posterior is clearly not unimodal and in fact we can see its evolution in time as suggested by Fig. 7.1 previously. Finally, we show the "distribution closeness" metrics in Fig. 7.20 with the ensemble PMF estimates compared to the "true" distribution in (a) and (b) followed by the KD in (c) and (d) and the corresponding Hellinger distance (HD) in (e) and (f). Here we see the results of estimating the KD over a 100-member ensemble with the corresponding state and measurement PMFs in Figs. 7.20a and 7.20b. The resulting KD over each member realization is shown in Figs. 7.20c and 7.20d, respectively for the state and measurement ensembles with the HD shown in Figs. 7.20e and 720f. Here the median and mean KD estimates are indicated by the dashed lines in the figure. The corresponding measurement median and average KDs of 0.283 and 0.284 and state KDs of 0.0701 and 0.0704 are shown in (c) and (d). Similarly, the Hellinger

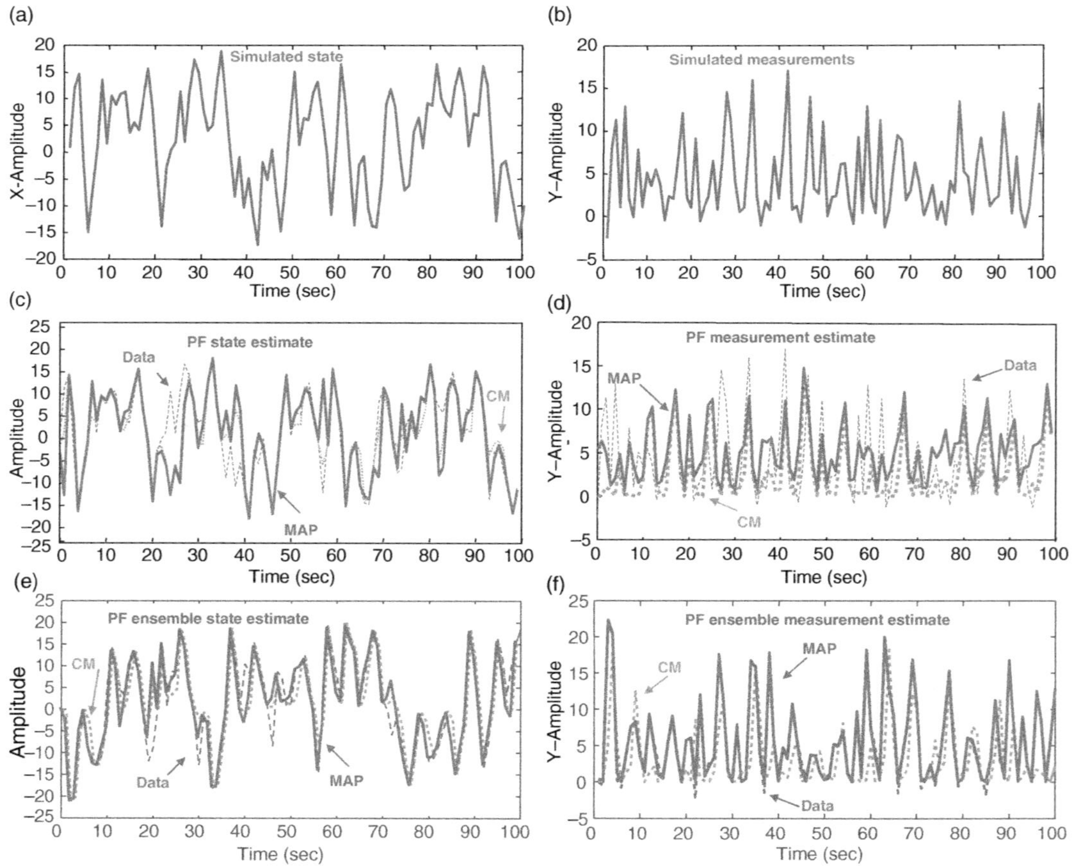

FIGURE 7.18 Nonlinear, nonstationary, non-Gaussian problem: (a) simulated state; (b) simulated measurement; (c) single realization of state PF estimates: CM and MAP; (d) single realization of measurement PF estimates: CM and MAP; (e) 100-member ensemble of state PF estimates: CM and MAP; (f) 100-member ensemble of measurement PF estimates: CM and MAP.

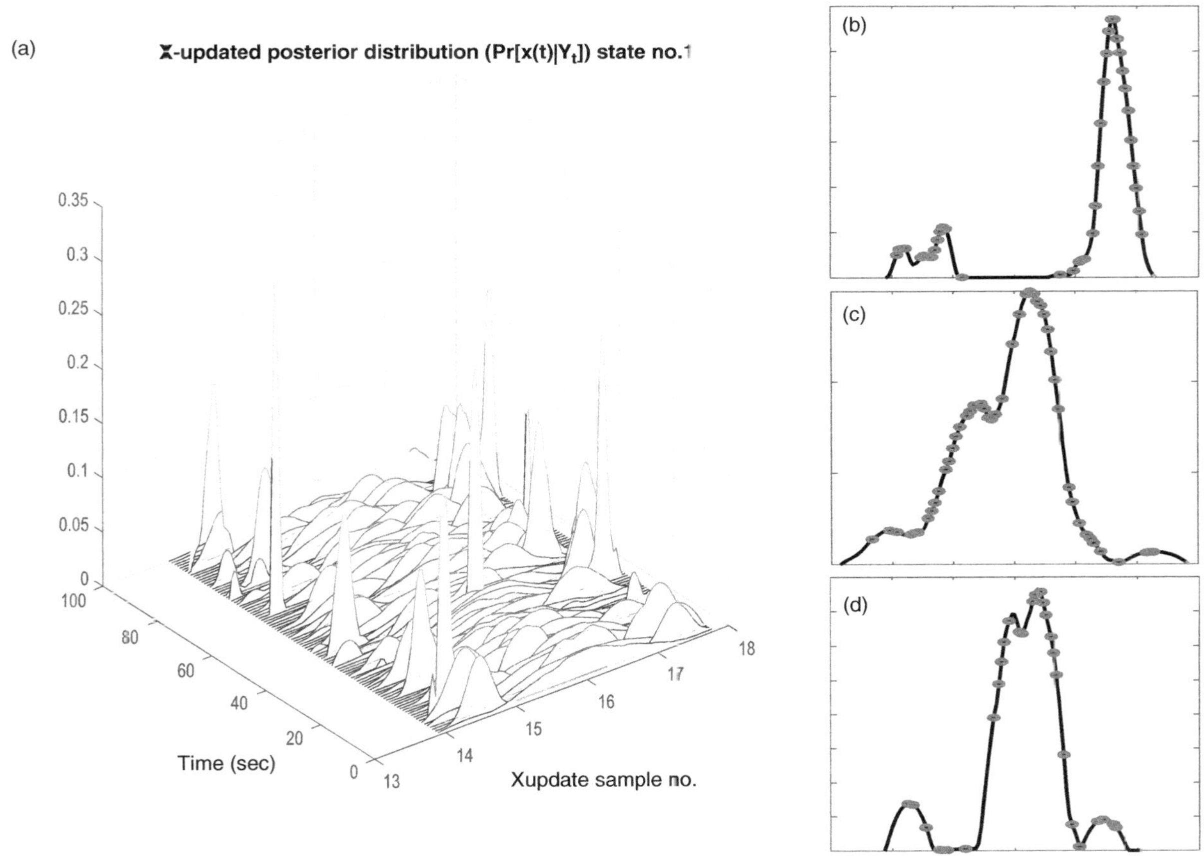

FIGURE 7.19 Nonlinear, nonstationary, non-Gaussian problem surface and time slices (cross-section) with particle locators annotated by circles. (a) Instantaneous posterior surface; (b) 26th time slice; (c) 60th time slice; (d) 96th time slice.

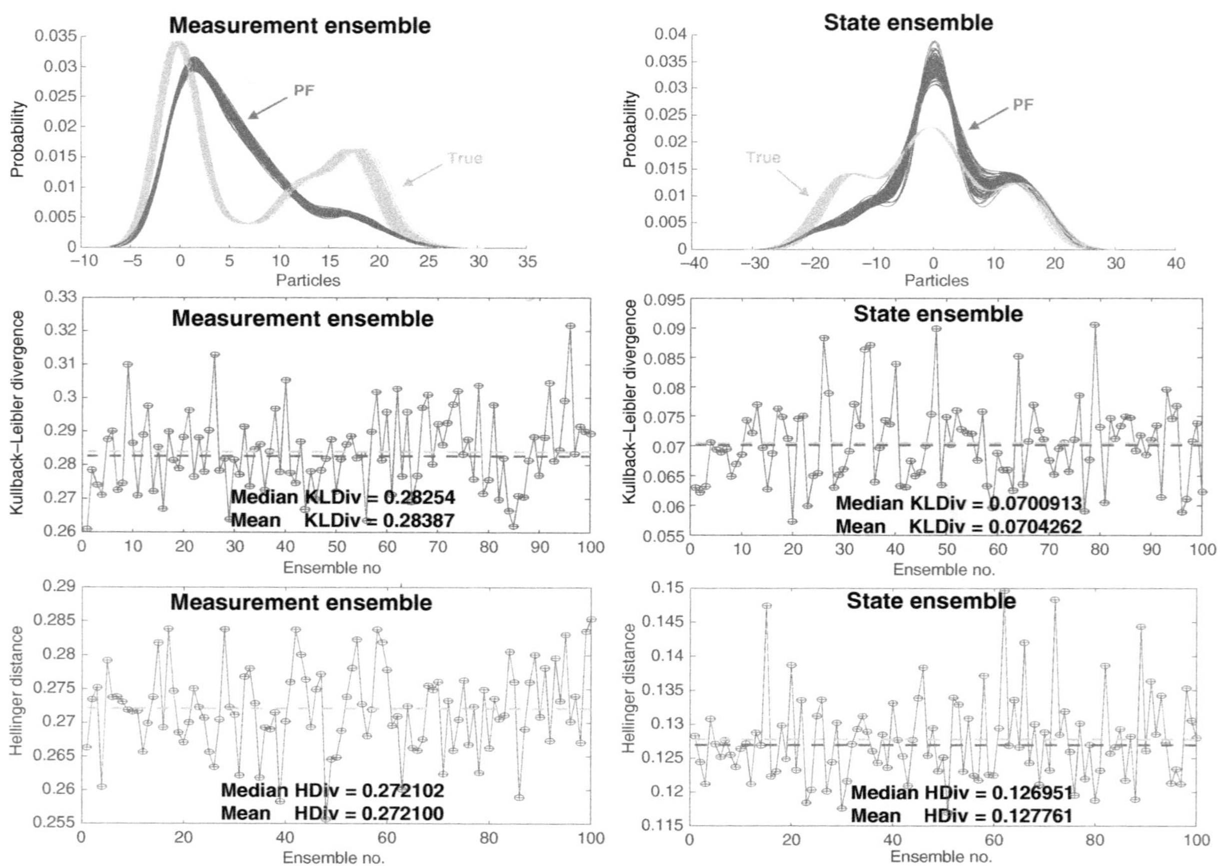

FIGURE 7.20 PMF data: Kullback–Leibler divergence and Hellinger distance calculations for the nonlinear trajectory estimation problem over 100-member ensemble. (a) Measurement PMF ensemble; (b) state PMF ensemble; (c) KL-divergence measurement median/mean (KD = 0.283/0.284); (d) KL-divergence state median/mean (KD = 0.0701/0.0704); (e) HD-distance measurement median/mean (HD = 0.272/0.272); (f) HD-distance state median/mean (HD = 0.1270/0.1277).

distance metric coincides with the KD metric giving HDs median and average state of 0.272 and 0.272 with corresponding measurement median and average of 0.1270 and 0.1277 as shown in Figs. 7.20e and 7.20f, respectively.

7.8 SUMMARY

In this chapter, we have discussed the development of SSPF. After introducing the idea of Bayesian particle filters, we showed how the state–space models could easily be interpreted in terms of this framework. We developed a generic state–space particle filtering algorithm based on the importance (sampling) proposals selected either using the minimum variance or transition prior approach. However we emphasized that in practice these techniques suffer from particle depletion and lack of diversity because of ever-increasing weight variances causing divergence of the processors. We introduced the concept of resampling as a solution to the divergence problem and discussed a number of techniques to mitigate the divergence problem. With that in hand, we discussed the popular bootstrap particle filter and showed some examples to demonstrate its performance. We then proceeded to discuss improvements to the bootstrap approach attempting to approximate the minimum variance proposal. These methods included the auxiliary, regularized, MCMC and linearized algorithms. Next, we investigated some of the practical aspects of particle filter design and developed a number of statistical tests to determine performance including both information theoretic approaches to validate the posterior distribution as well as diagnostic testing for model validation. We concluded the chapter with a case study on population growth—a nonlinear/non-Gaussian model presenting a very challenging problem for any particle filter design. Besides the references in the chapter there has been a wealth of particle filtering papers appearing in both the statistics and signal processing literature [30–43].

MATLAB NOTES

MATLAB is a command oriented vector-matrix package with a simple yet effective command language featuring a wide variety of embedded *C* language constructs making it ideal for signal processing applications and graphics. MATLAB has a *Statistics Toolbox* that incorporates a large suite of PDFs and CDFs as well as "inverse' CDF functions ideal for simulation-based algorithms. The **mhsample** command incorporate the Metropolis, Metropolis–Hastings and Metropolis independence samplers in a single command while the Gibbs sampling approach is adequately represented by the more efficient slice sampler (**slicesample**). There are even specific "tools" for sampling as well as the inverse CDF method captured in the **randsample** command. PDF estimators include the usual histogram (**hist**) as well as the sophisticated kernel density estimator (**ksdensity**) offering a variety of kernel (window) functions (Gaussian, etc.) and ICDF methods including the empirical cumulative distribution (**ecdf**) estimator. As yet no sequential algorithms are available. In terms of statistical testing for particle filtering diagnostics MATLAB offers the chi-square "goodness-of-fit" test **chi2gof** as well as the Kolmogorov–Smirnov distribution test **kstest**. Residuals can be tested for whiteness using the Durbin-Watson test statistic **dwtest** while "normality" is easily checked using the **normplot** command indicating the closeness of the test distribution to a Gaussian. Other statistics are also evaluated using the **mean**, **moment**, **skewness**, **std**, **var** and **kurtosis** commands. Type *help stats* in MATLAB to get more details or go to the MathWorks website.

REFERENCES

1. B. Ristic, S. Arulampalam, and N. Gordon, *Beyond the Kalman Filter: Particle Filters for Tracking Applications* (Boston, MA: Artech House, 2004).
2. M. Arulampalam, S. Maskell, N. Gordon, and T. Clapp, "A tutorial on particle filters for online nonlinear/non-Gaussian Bayesian tracking," *IEEE Trans. Signal Proc.*, **50**, 2, 174–188, 2002.
3. P. Djuric, J. Kotecha, J. Zhang, Y. Huang, T. Ghirmai, M. Bugallo, and J. Miguez, "Particle filtering," *IEEE Signal Proc. Mag.*, **20**, 5, 19–38, 2003.
4. A. Doucet and X. Wang, "Monte Carlo methods for signal processing," *IEEE Signal Proc. Mag.*, **24**, 5, 152–170, 2005.
5. J. Candy, *Model-Based Signal Processing* (Hoboken, NJ: Wiley-IEEE Press, 2006).
6. O. Cappe, S. Godsill, and E. Moulines, "An overview of existing methods and recent advances in sequential Monte Carlo," *Proc. IEEE*, **95**, 5, 899–924, 2007.
7. A. Doucet, N. de Freitas, and N. Gordon, *Sequential Monte Carlo Methods in Practice* (New York: Springer-Verlag, 2001).
8. M. Pitt and N. Shephard, "Filtering via simulation: auxiliary particle filters," *J. Am. Stat. Assoc.*, **94**, 446, 590–599, 1999.
9. R. van der Merwe, A. Doucet, N. de Freitas, and E. Wan, "The unscented particle filter," in *Advances in Neural Information Processing Systems 16* (Cambridge, MA: MIT Press, 2000).
10. R. van der Merwe, "Sigma-point Kalman filters for probabilistic inference in dynamic state–space models," Ph.D. Dissertation, OGI School of Science & Engineering, Oregon Health & Science University, 2004.
11. T. Schoen, "Estimation of nonlinear dynamic systems: theory and applications," Ph.D. Dissertation, Linkopings University, Linkoping, Sweden, 2006.
12. J. Liu, *Monte Carlo Strategies in Scientific Computing* (New York: Springer-Verlag, 2001).
13. J. Ruanaidh and W. Fitzgerald, *Numerical Bayesian Methods Applied to Signal Processing* (New York: Springer-Verlag, 1997).
14. A. Papoulis and S. Pillai, *Probability, Random Variables and Stochastic Processes*, 4th Ed. (New York: McGraw-Hill, 2002).
15. R. Hogg, J. McKlean, and A. Craig, *Introduction to Mathematical Statistics*, 6th Ed. (Englewood Cliffs, NJ: Prentice-Hall, 2005).

16. N. Gordon, D. Salmond, and A. Smith, "A novel approach to nonlinear non-Gaussian Bayesian state estimation," *IEE Proc. F.*, **140**, 107–113, 1993.

17. D. Simon, *Optimal State Estimation: Kalman H_∞ and Nonlinear Approaches* (Hoboken, NJ: Wiley-IEEE Press, 2006).

18. P. Stavropoulos and D. Titterington, "Improved particle filters and smoothing," in *Sequential Monte Carlo Methods in Practice*, edited by A. Doucet, N. de Freitas, and N. Gordon (New York: Springer, 2001), pp. 295–317.

19. J. Liu, R. Chen, and W. Wong, "Rejection control and sequential importance sampling," *J. Am. Stat. Assoc.*, **96**, 446, 1022–1061, 1998.

20. G. Kitagawa, "Non-gaussian modeling of nonstationary time series," *J. Am. Stat. Assoc.*, **82**, 400, 1032–1073, 1987.

21. A. Jazwinski, *Stochastic Processes and Filtering Theory* (New York: Academic Press, 1970).

22. M. Isard and A. Blake, "Condensation–conditional density propagation for visual tracking," *Int. J. Comput. Vis.*, **29**, 1, 5–28, 1998.

23. M. West and J. Harrison, *Bayesian Forecasting and Dynamic Models*, 2nd Ed. (New York: Springer-Verlag, 1997).

24. O. Cappe, E. Moulines, and T. Ryden, *Inference in Hidden Markov Models* (New York: Springer-Verlag, 2005).

25. M. Andrade Netto, L. Gimeno, and J. Mendes, "On the optimal and suboptimal nonlinear filtering problem for discrete-time systems," *IEEE Trans. Auto. Control*, AC-23, **6**, 1063–1067, 1978.

26. G. Kitagawa, "A nonlinear smoothing method for time series analysis," *Statistica Sinica*, **1**, 2, 371–388, 1991.

27. G. Kitagawa and W. Gersch, *Smoothness Priors Analysis of Time Series* (New York: Springer-Verlag, 1997).

28. G. Kitagawa, "Monte Carlo filter and smoother for non-Gaussian nonlinear state–space models," *J. Comput. Graphical Stat.*, **5**, 1, 1–25, 1997.

29. G. Kitagawa, "Self-organizing state space model," *J. Am. Stat. Assoc.*, **97**, 447, 1207–1215, 1998.

30. H. Tanizaki, *Nonlinear Filters* (New York: Springer-Verlag, 1993).

31. M. Tanner, *Tools for Statistical Inference: Methods for the Exploration of Posterior Distributions and Likelihood Functions*, 2nd Ed. (New York: Springer-Verlag, 1993).

32. A. Kong, J. Liu, and W. Wong, "Sequential imputations and Bayesian missing data problems," *J. Am. Stat. Assoc.*, **89**, 425, 278–288, 1994.

33. J. Liu and R. Chen, "Blind deconvolution via sequential imputations," *J. Am. Stat. Assoc.*, **90**, 460, 567–576, 1995.

34. J. Liu and R. Chen, "Sequential Monte Carlo methods for dynamic systems," *J. Am. Stat. Assoc.*, **96**, 446, 1062–1044, 1998.

35. A. Smith and A. Gelfand, "Bayesian statistics without tears: a sampling-resampling perspective," *Am. Statistician*, **44**, 4, 84–88, 1992.

36. S. Haykin, *Kalman Filtering and Neural Networks* (Hoboken, NJ: John Wiley & Sons, 2001).

37. S. Godsill and P. Djuric, "Monte Carlo methods for statistical signal processing," *IEEE Trans. Signal Proc.*, **50**, 173–499, 2002 (Special Issue).

38. S. Haykin and N. de Freitas, "Sequential state estimation: from Kalman filters to particle filters," *Proc. IEEE*, **92**, 3, 399–574, 2004 (Special Issue).
39. C. Andrieu, A. Doucet, S. Singh, and V. Tadic, "Particle methods for change detection, system identification and control," *Proc. IEEE*, **92**, 6, 423–468, 2004.
40. J. Vermaak, C. Andrieu, A. Doucet, and S. Godsill, "Particle methods for Bayesian modeling and enhancement of speech signals," *IEEE Trans. Speech Audio Proc.*, **10**, 3, 173–185, 2002.
41. A. Doucet, S. Godsill, and C. Andrieu, "On sequential Monte Carlo sampling methods for Bayesian filtering," *Statist. Comput.*, **10**, 6, 197–208, 2000.
42. A. Harvey, S. Koopman, and N. Shephard, *State Space and Unobserved Component Models: Theory and Applications* (Cambridge, UK: Cambridge University Press, 2004).
43. J. Candy, "Bootstrap Particle Filtering," *IEEE Signal Proc. Magaz.*, **24**, 4, 73–85, 2007.
44. C. Musso, N. Oudjane, and F. LeGland, "Improving regularized particle filters," in *Sequential Monte Carlo Methods in Practice*, edited by A. Doucet, N. de Freitas, and N. Gordon (New York: Springer, 2001), pp. 247–271.
45. N. de Freitas, C. Andrieu, P. Hojen-Sorensen, M. Niranjan, and A. Gee, "Sequential Monte Carlo methods for neural networks," in *Sequential Monte Carlo Methods in Practice*, edited by A. Doucet, N. de Freitas, and N. Gordon (New York: Springer, 2001), pp. 359–379.
46. W. Gilks and C. Berzuini, "Following a moving target—Monte Carlo inference for dynamic Bayesian models," *J. R. Stat. Soc. B*, **63**, 127–146, 2001.
47. C. Berzuini and W. Gilks, "Resample-move filtering with cross-model jumps," in *Sequential Monte Carlo Methods in Practice*, edited by A. Doucet, N. de Freitas, and N. Gordon (New York: Springer, 2001), pp. 117–138.
48. M. Rosenblatt, "Remarks on a multivariate transformation," *Ann. Math. Stat.*, **23**, 470–472, 1952.
49. J. Smith, "Diagnostic checks of non-standard time series models," *J. Forecasting*, **4**, 283–291, 1985.
50. A. Harvey, *Forecasting, Structural Time Series Models and the Kalman Filter* (Cambridge, UK: Cambridge University Press, 1989).
51. S. Fruhwirth-Schnatter, "Recursive residuals and model diagnostics for normal and non-normal state–space models," *Environ. Ecological Stat.*, **3**, 291–309, 1996.
52. S. Fruhwirth-Schnatter, "Bayesian model discrimination and Bayes factors for linear Gaussian state space models," *J. R. Stat. Soc. B*, **57**, 237–246, 1995.
53. S. Fruhwirth-Schnatter, "Applied state space modelling of non-Gaussian time series using integration-based Kalman filtering," *Stat. Comput.*, **4**, 259–269, 1994.
54. K. Bowman and L. Shenton, "Omnibus test countours for departures from normality based on $\sqrt{b_1}$ and b_2," *Biometrika*, **62**, 2, 243–250, 1975.
55. G. Ljung and G. Box, "On a measure of lack of fit in time series models," *Biometrika*, **65**, 2, 297–303, 1978.
56. S. Shapiro and M. Wilk, "An analysis of variance test for normality," *Biometrika*, **52**, 4, 591–611, 1965.
57. R. Gerlach, C. Carter, and R. Kohn, "Diagnostics for time series analysis," *J. Time Series Anal.*, **20**, 3, 309–330, 1999.

58. F. Gustafsson, *Adaptive Filtering and Change Detection* (New York: John Wiley & Sons, 2000).
59. H. Akaike, "A new look at the statistical model identification," *IEEE Trans. Autom. Control*, **19**, 716–723, 1974.
60. Y. Sakamoto, M. Ishiguro, and G. Kitagawa, *Akaike Information Criterion Statistics* (Boston, MA: D. Reidel/Kluwer Academic, 1986).
61. A. Bhattacharyya, "On a measure of divergence between two statistical populations defined by their probability distributions," *Bull. Calcutta Math. Soc.*, **35**, 99–109, 1943.
62. S. Kullback and R. Leibler, "On information and sufficiency," *Ann. Math. Stat.*, **22**, 79–86, 1951.
63. S. Kullback, *Information Theory and Statistics* (Gloucester, MA: Peter Smith, 1978).
64. M. Nikulin, "Hellinger distance," in *Encyclopedia of Mathematics*, edited by Hazewinkel (New York: Springer, 2001).
65. R. Beran, "Minimum Hellinger distance estimates for parametric models," *Ann. Stat.*, **5**, 3, 445–463, 1977.
66. S. Kay, *Modern Spectral Estimation* (New Jersey: Prentice-Hall, 1988).
67. S. Haykin, *Adaptive Filtering Theory* (New Jersey: Prentice-Hall, 1996).
68. M. Hayes, *Statistical Digital Signal Processing and Modeling* (New York: John Wiley & Sons, Inc., 1996).
69. G. Hendeby, R. Karlsson, and F. Gustafsson, "Performance issues in non-Gaussian filtering problems," in Proceedings of IEEE NSSPW, Cat. No. 06EX1506, Cambridge, UK, 2006.
70. N. Cui, L. Hong, and J. Layne, "A comparison of nonlinear filtering approaches with an application to ground target tracking," *Signal Proc.*, **85**, 1469–1492, 2005.
71. K. Ito and K. Xiong, "Gaussian filters for nonlinear filtering problems," *IEEE Trans. Autom. Control*, **45**, 5, 910–927, 2000.
72. H. Sorenson and D. Alspach, "Recursive Bayesian estimation using gaussian sums," *Automatica*, **7**, 465–479, 1971.
73. D. Alspach and H. W. Sorenson, "Nonlinear Bayesian estimation using Gaussian sum approximations," *IEEE Trans. Autom. Control*, **17**, 4, 439–448, 1972.
74. I. Arasaratnam, S. Haykin, and R. Elliott, "Discrete-time nonlinear filtering algorithms using Gauss-Hermite quadrature," *Proc. IEEE*, **95**, 5, 953–976, 2007.

PROBLEMS

7.1 Given a sequence of Gaussian data (measurements) characterized by $y \sim \mathcal{N}(\mu, \sigma^2)$, find the best estimate of the parameters defined by $\Theta := [\mu \quad \sigma]'$ using a "sequential' MC approach. Show the mathematical steps in developing the technique and construct a simple PF to solve the problem.

7.2 Consider the following simple model [7]

$$x(t) = ax(t-1) + w(t-1) \quad \text{for } w \sim \mathcal{N}(0, R_{ww}(i)) \text{ with } \mathcal{I}(t) = i$$
$$y(t) = x(t) + v(t) \quad \text{for } v \sim \mathcal{N}(0, R_{vv})$$

with $\Pr(\mathcal{I}(t)=i|\mathcal{I}(t-1),x(t-1))=\Pr(\mathcal{I}(t)=i)=p_i$

(*a*) Suppose $i=\{1,2\}$, what is the distribution, $\Pr(\mathcal{I}(t)=(i_1,i_2),x_1|y_1,x_0)$?

(*b*) How would the marginal be estimated using a Kalman filter?

(*c*) Develop a computational approach to bootstrap PF algorithm for this problem.

7.3 Suppose we have two multivariate Gaussian distributions for the parameter vector, $\Theta \sim \mathcal{N}(\mu_i, \Sigma_i); i=1,2$

(*a*) Calculate the Kullback–Leibler (KL) distance metric, J.

(*b*) Calculate the Hellinger (HD) distance metric, $\mathbf{H}_{HD}$.

(*c*) Suppose $\Sigma = \Sigma_1 = \Sigma_2$, recalculate the KL and HD for this case.

7.4 An aircraft flying over a region can use the terrain and an archival digital map to navigate. Measurements of the terrain elevation are collected in real time while the aircraft altitude over mean sea level is measured by a pressure meter with ground clearance measured by a radar altimeter [7]. The measurement differences are used to estimate the terrain elevations and compared to a digital elevation map to estimate aircraft position. The discrete navigation model is given by

$$
\begin{aligned}
x(t) &= x(t-1) + u(t-1) + w(t-1) \quad && \text{for } w \sim \mathcal{N}(0, R_{ww}) \\
y(t) &= c[x(t)] + v(t) && \text{for } v \sim \mathcal{N}(0, R_{vv})
\end{aligned}
$$

where x is the 2D-position, y is the terrain elevation measurement, the navigation systems output u and w are the respective distance traveled and error drift during one time interval. The nonlinear function $c[\cdot]$ denotes the terrain database yielding terrain elevation outputs with v the associated database errors and measurements. Both noises are assumed zero-mean, Gaussian with known statistics, while the initial state is also Gaussian, $x(0) \sim \mathcal{N}(\overline{x}(0), \overline{P}(0))$.

(*a*) Based on this generic description construct the bootstrap PF algorithm for this problem.

(*b*) Suppose: $\overline{P}(0)=diag[10^4\ 10^4]'$, $R_{ww}=diag[25\ 25]'$, $R_{vv}=16$, $N=150$ samples, $u(t)=[25\ 25]'$. Simulate the aircraft measurements and apply the bootstrap algorithm to estimate the aircraft position.

7.5 In *financial systems*, a stochastic *volatility* model is used in the analysis of financial time series such as daily fluctuations in the stock market prices, exchange rates and option pricing. The volatility is usually expressed in terms of a time-varying variance with the model given by:

$$
\begin{aligned}
y(t) &= \sigma(t) \times \epsilon(t) \quad \epsilon \sim \mathcal{N}(0,1) \\
\ln \sigma^2(t) &= \alpha + \beta \ln \sigma^2(t-1) + \ln \epsilon^2(t)
\end{aligned}
$$

or equivalently

$$y(t) = e^{\alpha(t)/2} \times \sigma(t) \times \epsilon(t) \qquad \epsilon \sim \mathcal{N}(0, 1)$$
$$\ln \sigma^2(t) = \beta(t) \ln \sigma^2(t-1) + v(t) \quad v \sim \mathcal{N}(0, \tau^2)$$

where $\sigma(t)$ corresponds to the time-varying volatility (amplitude) and the second relation represents the change in volatility. The parameters α and β are regression coefficients, and the remaining parameters are the $\ln \sigma^2(t)$ variance term.

(*a*) Suppose we would like to estimate the unknown parameters augmenting the original state $(\ln \sigma^2(t))$ with the unknowns, $\alpha, \beta, \ln \sigma^2$. Assume the parameters can be represented by a random walk driven by zero-mean, Gaussian noise processes. What is the overall model for this financial system?

(*b*) Construct a bootstrap PF for this problem.

(*c*) Simulate the data for $N = 1000$ samples and estimate the volatility and parameters. The simulation parameters are: $\alpha = 1.8, \beta = 0.95, \tau^2 = 0.1$, and $\epsilon \sim \mathcal{N}(0, 1)$.

7.6 Develop a suite of particle filters for the *RC*-circuit problem of Example 5.1 where the output voltage was given by:

$$e(t) = \left(1 - \frac{\Delta T}{RC}\right) e(t-1) + \frac{\Delta T}{C} I_{\text{in}}(t-1)$$

where e_o is the initial voltage and R is the resistance with C the capacitance. The measurement equation for a voltmeter of gain K_e is simply

$$e_{\text{out}}(t) = K_e e(t)$$

Recall that for this circuit the parameters are: $R = 3.3\ k\Omega$ and $C = 1000\ \mu F$, $\Delta T = 100\ ms$, $e_o = 2.5\ V$, $K_e = 2.0$, and the voltmeter is precise to within $\pm 4\ V$. Then transforming the physical circuit model into state–space form by defining $x = e$, $y = e_{\text{out}}$, and $u = I_{\text{in}}$, we obtain

$$x(t) = 0.97x(t-1) + 100u(t-1) + w(t-1)$$
$$y(t) = 2x(t) + v(t)$$

The process noise covariance is used to model the circuit parameter uncertainty with $R_{ww} = 0.0001$, since we assume standard deviations, $\triangle R, \triangle C$ of 1%. Also, $R_{vv} = 4$, since two standard deviations are $\triangle V = 2\left(\frac{1}{2} 4\ V\right)$. We also assume initially that the state is $x(0) \sim \mathcal{N}(2.5, 10^{-12})$, and that the input current is a step function of $u(t) = 300\ \mu A$.

With this in mind, we know that the optimal processor to estimate the state is the linear BP (Kalman filter).

(*a*) After performing the simulation using the parameters above, construct a bootstrap PF and compare its performance to the optimal. How does it compare? Whiteness? Zero-mean? State estimation error?

(*b*) Let us assume that the circuit is malfunctioning and we do not precisely know the current values of RC. Construct a parameter estimator for $A = 1/RC$ using the EKF or UKF and compare its performance to the bootstrap and linearized PF.

(*c*) Try "roughening" the bootstrap PF, does its performance improve? Compare results (see Chapter 8 for details).

7.7 Consider the storage of plutonium nitrate in a tank (see Ref. [5] for details), we would like to dynamically estimate the amount (mass) of Pu present at any given time. Losses occur due to radiolysis and evaporation. The underlying state–space model for this system is given by:

Summarizing the process and measurement models in state–space form, we have

$$\frac{d}{dt}\begin{bmatrix} m(t) \\ \rho(t) \end{bmatrix} = \begin{bmatrix} -K_r P & 0 \\ 0 & 0 \end{bmatrix}\begin{bmatrix} m(t) \\ \rho(t) \end{bmatrix} + \begin{bmatrix} 1 \\ 0 \\ \end{bmatrix} u(t) + \begin{bmatrix} w_1(t) \\ w_2(t) \end{bmatrix}$$

where u is a step function of amplitude $-K_H$. The corresponding measurement model based on pressure measurements is

$$\begin{bmatrix} \Delta P_A \\ \Delta P_B \end{bmatrix} = \begin{bmatrix} g/b & -(a/b)g \\ 0 & gH \end{bmatrix}\begin{bmatrix} m(t) \\ \rho(t) \end{bmatrix} + \begin{bmatrix} v_1(t) \\ v_2(t) \\ \end{bmatrix}$$

discretizing the dynamics and incorporating the model parameters, we obtain the Gauss–Markov model with sampling interval of 0.1 *day* as

$$x(t) = \begin{bmatrix} 0.999 & 0 \\ 0 & 1 \end{bmatrix} x(t-1) + \begin{bmatrix} 1 \\ 0 \\ \end{bmatrix} u(t-1) + w(t-1)$$

$$y(t) = \begin{bmatrix} 29.8 & -0.623 \\ 0 & 24.9 \end{bmatrix} x(t) + v(t)$$

$R_{ww} = \text{diag}[10\ \ 10]$, $R_{vv} = \text{diag}[5.06 \times 10^4\ \ 1.4 \times 10^5]$ with initial conditions $\hat{x}(0|0) = [988\ 1455]'$ and $\tilde{P}(0|0) = \text{diag}[0.01\ 0.01]$.

(*a*) Develop the optimal BP for this problem and compare its performance to the bootstrap PF.

(*b*) How well does the bootstrap PF perform? How about the linearized PF?

7.8 We are asked to investigate the possibility of creating a *synthetic aperture* using a single hydrophone to be towed by an AUV in search of targets. We know that a single hydrophone offers no improvement in SNR in the sense of array gain, but also wonder about its capability to localize, especially more than one target.

(*a*) Using the synthetic aperture model developed in the case study of Section 8.5, develop the bootstrap PF and UKF processors for this problem. Assume we would like to track two targets.

(*b*) Perform the same simulation outlined in the case study for two targets at $-45°, -10°$.

(*c*) Apply the bootstrap algorithm with and without "roughening" along with the UKF. Discuss processor performances and compare. Zero-mean? Whiteness?

(*d*) Implement the "optimal" PF processor using the EKF or UKF linearization. How does its performance compare?

7.9 We are asked to investigate the possibility of finding the range of a target using a hydrophone sensor array towed by an AUV assuming a near-field target characterized by it spherical wavefront instead of the far-field target of the previous problem using a plane wave model. The near-field processor can be captured by a wavefront curvature scheme (see Ref. [5] for more details) with process and measurement models (assuming that the parameters as well as the measurements are contaminated by additive white Gaussian noise). The following set of dynamic relations can be written succinctly as the *Gauss–Markov wavefront curvature model* as

$$\Theta(t_k) = \Theta(t_{k-1}) + w(t_k) \qquad \text{for } \Theta(t_k) = [\alpha \quad f_o \quad \theta_o \quad r_o]'$$
$$p_\ell(t_k) = \theta_1(t_k) e^{j2\pi\theta_2(t_k)(t_k - \tau_\ell(\Theta;t_k))} + v_\ell(t_k); \ell = 1, \dots, L$$

where α, f_o, θ_o and r_o are the respective amplitude, target frequency, bearing and range. The time delay at the ℓth-sensor and time t_k is given in terms of the unknown parameters of

$$\tau_\ell(\Theta; t_k) := \frac{1}{c}\left(\theta_4(t_k) - \sqrt{\theta_4^2(t_k) + d_\ell^2(t) - 2d_\ell(t)\theta_4(t_k)\sin\theta_3(t_k)}\right)$$

for $d_\ell(t)$ the distance between the ℓth sensor and reference range r_o given by

$$d_\ell(t) = x_\ell + v \times t_k \quad \text{for } x_\ell \text{ the position of sensor } \ell$$

(*a*) Using this model develop the bootstrap PF and UKF processors for this problem. Assume we would like to estimate the target bearing, frequency and range ($\alpha = 1$).

(*b*) Perform a simulation with initial parameters $r_o = 3$ Km, $f_o = 51.1$ Hz and $\theta_o = 27°$ and true parameters at $r = 2$ Km, $f = 51$ Hz and $\theta = 25°$, $L = 4$.

(*c*) Apply the bootstrap algorithm with and without "roughening" along with the UKF. Discuss processor performances and compare. Zero-mean? Whiteness?

(*d*) Implement the "optimal" PF processor using the EKF or UKF linearization. How does its performance compare?

7.10 Consider the bearings-only tracking problem of Example 5.4 given by the state–space model. The entire system can be represented as an approximate Gauss–Markov model with the noise sources representing uncertainties in the states and measurements. The equations of motion given by

$$x(t) = \begin{bmatrix} 1 & 0 & \Delta T & 0 \\ 0 & 1 & 0 & \Delta T \\ 0 & 0 & 1 & 0 \\ 0 & 0 & 0 & 1 \end{bmatrix} x(t-1) + \begin{bmatrix} 0 & 0 \\ 0 & 0 \\ 1 & 0 \\ 0 & 1 \end{bmatrix} \begin{bmatrix} -\Delta v_{ox}(t-1) \\ -\Delta v_{oy}(t-1) \end{bmatrix} + w(t-1)$$

with the nonlinear sensor model given by

$$y(t) = \arctan \frac{x_1(t)}{x_2(t)} + v(t)$$

for $w \sim \mathcal{N}(0, R_{ww})$ and $v \sim \mathcal{N}(0, R_{vv})$.

(*a*) Using this model develop the bootstrap PF and UKF processors for this problem. Assume we would like to estimate the target bearing, frequency and range ($\alpha = 1$).

(*b*) Perform a simulation with the following parameters: an impulse-incremental step change ($\Delta v_{ox} = 24$ knots and $\Delta v_{oy} = -10$ knots) was initiated at 0.5 h, resulting in a change of observer position and velocity depicted in the figure. The simulated bearing measurements are shown in Fig. 5.8d, The initial conditions for the run were $x'(0) := [0\ 15\ \text{nm}\ 20\ \text{k}\ -10\ \text{k}]$ and $R_{ww} = \text{diag}\ 10^{-6}$ with the measurement noise covariance given by $R_{vv} = 3.05 \times 10^{-4}\ \text{rad}^2$ for $\Delta T = 0.33$ h.

(*c*) Apply the bootstrap algorithm along with the UKF. Discuss processor performances and compare. Zero-mean? Whiteness?

(*d*) Implement the "optimal" PF processor using the EKF or UKF linearization. How does its performance compare?

8 JOINT BAYESIAN STATE/PARAMETRIC PROCESSORS

8.1 INTRODUCTION

In this chapter, we develop the Bayesian approach to the parameter estimation/system identification problem [1–4] which is based on the decomposition of the joint posterior distributions that incorporates *both* dynamic state and parameter variables. From this formulation, the following problems evolve: (1) joint state/parameter estimation; (2) state estimation; and (3) parameter (fixed and/or dynamic) estimation. The state estimation problem is thoroughly discussed in the previous chapters. However, the most common problem found in the current literature is the parameter estimation problem which can be solved "off-line" using batch approaches (maximum entropy, maximum likelihood, minimum variance, least squares, etc.) or "on-line" using the expectation-maximization (EM) technique (see Chapter 2), the stochastic Monte Carlo approach and for that matter almost any (deterministic) optimization technique [5, 6]. These on-line approaches follow the classical (EKF), modern (UKF), and the sequential Monte Carlo or particle filter (PF). However, it still appears that there is *no* universally accepted approach to solving this problem, especially for fixed parameters [7–9]. From the pragmatic perspective, the most useful problem is the *joint* state/parameter estimation problem, since it evolves quite naturally from the fact that a model is developed to solve the basic state estimation problem and it is found that its inherent parameters are either poorly specified, just bounded or even unknown, inhibiting the development of the processor. We call this problem the "joint" state/parameter estimation, since *both* states and parameters are estimated simultaneously on-line and the resulting processor is termed *parametrically adaptive*

Bayesian Signal Processing: Classical, Modern, and Particle Filtering Methods, Second Edition. James V. Candy.

[10]. This terminology evolves because the inherent model parameters are adjusted sequentially as the measurement data becomes available.

In this chapter, we concentrate primarily on the joint Bayesian state/parameter estimation problem and refer the interested reader to the wealth of literature available on this subject [7–19]. First, we precisely define the three basic problems from the Bayesian perspective and then investigate the classical, modern and particle approaches to its solution. We incorporate the nonlinear re-entry problem of Jazwinski [20] used throughout as an example of parametrically adaptive design and then discuss a case study to demonstrate the approach.

8.2 BAYESIAN APPROACH TO JOINT STATE/PARAMETER ESTIMATION

To be more precise, we start by defining the joint state/parametric estimation problem. We begin by formulating the Bayesian recursions in terms of the posterior distribution using Bayes' rule for decomposition, that is,

$$\Pr(x(t), \theta(t)|Y_t) = \Pr(x(t)|\theta(t), Y_t) \times \Pr(\theta(t)|Y_t) = \Pr(\theta(t)|x(t), Y_t) \times \Pr(x(t)|Y_t) \tag{8.1}$$

From this relation, we begin to "see" just how the variety of state and parameter estimation related problems evolve, that is,

- Optimize the joint state/parameter posterior:

$$\Pr(x(t), \theta(t)|Y_t) \qquad \text{[State/parameter estimation]}$$

- Optimize the state posterior:

$$\Pr(x(t)|Y_t) \qquad \text{[State estimation]}$$

- Optimize the parametric posterior:

$$\Pr(\theta(t)|Y_t) \qquad \text{[Parameter estimation]}$$

Now if we proceed with the usual factorizations, we obtain the Bayesian decomposition for the *state estimation* problem as

$$\begin{aligned}\Pr(x(t)|Y_t) &= \frac{\Pr(y(t)|x(t)) \times \Pr(x(t)|Y_{t-1})}{\Pr(y(t)|Y_{t-1})}\\ \Pr(x(t)|Y_{t-1}) &= \int \Pr(x(t)|x(t-1)) \times \Pr(x(t-1)|Y_{t-1}) dx(t-1)\end{aligned} \tag{8.2}$$

Equivalently for the *parameter estimation* problem, we have

$$\Pr(\theta(t)|Y_t) = \frac{\Pr(y(t)|\theta(t)) \times \Pr(\theta(t)|Y_{t-1})}{\Pr(y(t)|Y_{t-1})}$$

$$\Pr(\theta(t)|Y_{t-1}) = \int \Pr(\theta(t)|\theta(t-1)) \times \Pr(\theta(t-1)|Y_{t-1})d\theta(t-1) \tag{8.3}$$

Now for the *joint state/parameter estimation* problem of Eq. 8.1, we can substitute the above equations above to obtain the posterior decomposition of interest, that is,

$$\Pr(x(t), \theta(t)|Y_t) = \frac{\Pr(x(t)|\theta(t), Y_t) \times [\Pr(y(t)|\theta(t)) \times \Pr(\theta(t)|Y_{t-1})]}{\Pr(y(t)|Y_{t-1})} \tag{8.4}$$

or

$$\Pr(x(t), \theta(t)|Y_t) = \Pr(x(t)|\theta(t), Y_t) \times \Pr(y(t)|\theta(t)) \times \frac{\int \Pr(\theta(t)|\theta(t-1)) \times \Pr(\theta(t-1)|Y_{t-1})d\theta(t-1)}{\Pr(y(t)|Y_{t-1})} \tag{8.5}$$

This is the most common decomposition found in the literature [7–19] and leads to the maximization of the first term with respect to x and the second with respect to θ [21, 22].

Alternatively, using the state/parameter estimation form which is rarely applied, we have

$$\Pr(\theta(t), x(t)|Y_t) = \frac{\Pr(\theta(t)|x(t), Y_t) \times [\Pr(y(t)|x(t) \times \Pr(x(t)|Y_{t-1})]}{\Pr(y(t)|Y_{t-1})} \tag{8.6}$$

or

$$\Pr(\theta(t), x(t)|Y_t) = \Pr(\theta(t)|x(t), Y_t) \times \Pr(y(t)|x(t)) \times \frac{\int \Pr(x(t)|x(t-1)) \times \Pr(x(t-1)|Y_{t-1})dx(t-1)}{\Pr(y(t)|Y_{t-1})} \tag{8.7}$$

Here, the first term is maximized with respect to θ and the second with respect to x compared to the previous decomposition.

So we see that Bayes' rule can be applied in a number of ways to develop the sequential Bayesian processors for the state, parameter, and joint state/parameter estimation problems.

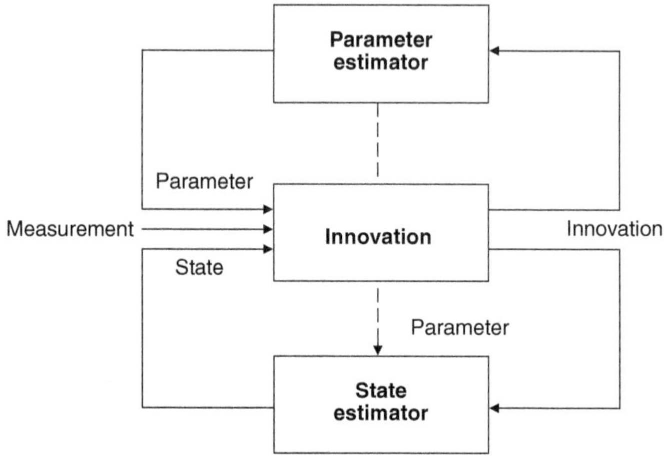

FIGURE 8.1 Nonlinear parametrically adaptive (ABSP): simplified processor structure illustrating the coupling between parameter and state estimators through the innovation and measurement sequences.

8.3 CLASSICAL/MODERN JOINT BAYESIAN STATE/PARAMETRIC PROCESSORS

In this section, we develop the "joint" state–space Bayesian sequential processor or equivalently the parametrically adaptive Bayesian signal processor (ABSP) for nonlinear state–space systems. The ABSP is a *joint state/parametric processor*, since it estimates *both* the states as well as the unknown model parameters. It is parametrically adaptive, since it adjusts or "adapts" the model parameters at each time step. The simplified structure of the classical (EKF) parameter estimator is shown in Fig. 8.1. We see the basic structure of the ABSP which consists of two distinct yet coupled processors: a parameter estimator and a state estimator. The parameter estimator provides estimates that are corrected by the corresponding innovations during each recursion. These estimates are then provided to the state estimator in order to update the model parameters used in the estimator. After both state and parameter estimates are calculated, a new measurement is processed and the procedure continues. In general, this processor can be considered to be a form of identifier, since system identification is typically concerned with the estimation of a model and its associated parameters from noisy measurement data. Usually the model structure is pre-defined (as in our case) and then a parameter estimator is developed to "fit" parameters according to some error criterion. After completion or during this estimation, the quality of the estimates must be evaluated to decide if the processor performance is satisfactory or equivalently the model adequately represents the data. There are various types (criteria) of identifiers employing many different model (usually linear) structures [2–4]. Here we are primarily concerned with joint estimation in which the models and parameters are usually nonlinear. Thus, we will concentrate on developing parameter estimators capable of online operations with nonlinear dynamics.

8.3.1 Classical Joint Bayesian Processor

From our previous discussion in Chapter 5, it is clear that the extended Bayesian processor XBP (extended Kalman filter) can satisfy these constraints nicely, so we begin our analysis of the XBP as a state/parametric estimator closely following the approach of Ljung [23] for the linear problem discussed in Section 8.2. The general nonlinear parameter estimator structure can be derived directly from the XBP algorithm in Table 5.3.

Recall the Bayesian solution to the classical problem was based on solving for the posterior distribution (see Eq. 8.2) such that each of the required distributions was represented in terms of the XBP estimates. In the joint state/parameter estimation case, these distributions map simply by defining an augmented state vector $X(t) := [x(t)|\theta(t)]'$ to give

$$
\begin{aligned}
\Pr(y(t)|x(t)) &\sim \mathcal{N}(\mathbf{c}[x(t)], R_{vv}(t)) \\
&\Leftrightarrow \\
\Pr(y(t)|x(t), \theta(t)) &\sim \mathcal{N}(\mathbf{c}[x(t), \theta(t)], R_{vv}(t)) \\
\Pr(x(t)|Y_{t-1}) &\sim \mathcal{N}(\hat{x}(t|t-1), \tilde{P}(t|t-1)) \\
&\Leftrightarrow \\
\Pr(X(t)|Y_{t-1}) &\sim \mathcal{N}(\hat{X}(t|t-1), \tilde{\mathcal{P}}(t|t-1)) \\
\Pr(y(t)|Y_{t-1}) &\sim \mathcal{N}(\hat{y}(t|t-1), R_{ee}(t)) \\
&\Leftrightarrow \\
\Pr(y(t)|Y_{t-1}) &\sim \mathcal{N}(\hat{y}_\theta(t|t-1), \tilde{R}_{e_\theta e_\theta}(t))
\end{aligned}
\tag{8.8}
$$

where

$$
\begin{aligned}
\hat{X}(t|t-1) &:= [\hat{x}(t|t-1)|\hat{\theta}(t|t-1)]' \\
\hat{y}_\theta(t|t-1) &:= \mathbf{c}[\hat{x}(t|t-1), \hat{\theta}(t|t-1)] \\
\tilde{\mathcal{P}}(t|t-1) &:= \begin{bmatrix} \tilde{P}_{xx}(t|t-1) & | & \tilde{P}_{x\theta}(t|t-1) \\ - & - & - \\ \tilde{P}_{\theta x}(t|t-1) & | & \tilde{P}_{\theta\theta}(t|t-1) \end{bmatrix} \\
\mathcal{R}_{e_\theta e_\theta}(t) &:= C[\hat{X}(t|t-1)]\tilde{\mathcal{P}}(t|t-1)C[\hat{X}(t|t-1)]' + R_{vv}(t)
\end{aligned}
$$

To develop the actual internal structure of the ABSP, we start with the XBP equations, augment them with the unknown parameters, and then investigate the resulting algorithm. We first define the composite state vector (as before) consisting of the original states $x(t)$ and the unknown "augmented" parameters represented by $\theta(t)$, that is,

$$
X(t) := \begin{bmatrix} x(t) \\ --- \\ \theta(t) \end{bmatrix} \tag{8.9}
$$

where t is the time index, $X \in R^{(N_x+N_\theta)\times 1}$ and $x \in R^{N_x\times 1}$, $\theta \in R^{N_\theta\times 1}$. Substituting this augmented state vector into the XBP relations of Table 5.3, the following matrix partitions evolve as

$$\tilde{\mathcal{P}}(t|t-1) := \begin{bmatrix} \tilde{P}_{xx}(t|t-1) & | & \tilde{P}_{x\theta}(t|t-1) \\ - & - & - \\ \tilde{P}_{\theta x}(t|t-1) & | & \tilde{P}_{\theta\theta}(t|t-1) \end{bmatrix}, \quad \tilde{P}_{\theta x}(t|t-1) = \tilde{P}'_{x\theta}(t|t-1) \tag{8.10}$$

where $\tilde{\mathcal{P}} \in R^{(N_x+N_\theta)\times(N_x+N_\theta)}$, $\tilde{P}_{xx} \in R^{N_x\times N_x}$, $\tilde{P}_{\theta\theta} \in R^{N_\theta\times N_\theta}$, and $\tilde{P}_{x\theta} \in R^{N_x\times N_\theta}$.

This also leads to the partitioning of the corresponding gain

$$\mathcal{K}(t) := \begin{bmatrix} K_x(t) \\ --- \\ K_\theta(t) \end{bmatrix} \tag{8.11}$$

for $\mathcal{K} \in R^{(N_x+N_\theta)\times N_y}$, $K_x \in R^{N_x\times N_y}$, and $K_\theta \in R^{N_\theta\times N_y}$.

We must also partition the state and measurement predictions as equations, that is,[1]

$$\begin{aligned} \hat{X}(t|t-1) &= \begin{bmatrix} \hat{x}(t|t-1) \\ --- \\ \hat{\theta}(t|t-1) \end{bmatrix} \\ &= \begin{bmatrix} \mathbf{a}[\hat{x}(t-1|t-1), \hat{\theta}(t-1|t-1)] \\ \quad + \mathbf{b}[\hat{x}(t-1|t-1), \hat{\theta}(t-1|t-1), u(t-1)] \\ --- \\ \hat{\theta}(t-1|t-1) \end{bmatrix} \end{aligned} \tag{8.12}$$

where the corresponding predicted measurement equation becomes

$$\hat{y}(t|t-1) = \mathbf{c}[\hat{x}(t|t-1), \hat{\theta}(t|t-1)] \tag{8.13}$$

Next we consider the predicted error covariance

$$\tilde{\mathcal{P}}(t|t-1) = A[\hat{x}(t|t-1), \hat{\theta}(t|t-1)]\tilde{\mathcal{P}}(t|t-1)A'[\hat{x}(t|t-1), \hat{\theta}(t|t-1)] + R_{ww}(t-1) \tag{8.14}$$

[1] Note that we have "implicitly" assumed that the parameters can be considered *piecewise constant*, $\Theta(t) = \Theta(t-1)$ or follow a random walk if we add process noise to this representation in the Gausssense. However, if we *do* have process dynamics with linear or nonlinear models characterizing the parameters, then they will replace the random walk and the associated Jacobian, etc. will change from this representation.

which can be written in partitioned form using Eq. 8.10 and the following Jacobian process matrix[2]

$$A[\hat{x},\hat{\theta}] = \begin{bmatrix} A_x[\hat{x}(t|t-1),\hat{\theta}(t|t-1)] & | & A_\theta[\hat{x}(t|t-1),\hat{\theta}(t|t-1)] \\ - & - & - \\ O & | & I_{N_\theta} \end{bmatrix} \tag{8.15}$$

where

$$\begin{aligned} A_x[\hat{x},\hat{\theta}] &:= \frac{\partial a[\hat{x},\hat{\theta}]}{\partial x} + \frac{\partial b[\hat{x},\hat{\theta}]}{\partial x} \\ A_\theta[\hat{x},\hat{\theta}] &:= \frac{\partial a[\hat{x},\hat{\theta}]}{\partial \theta} + \frac{\partial b[\hat{x},\hat{\theta}]}{\partial \theta} \end{aligned} \tag{8.16}$$

with $A \in R^{(N_x+N_\theta)\times(N_x+N_\theta)}, A_x \in R^{N_x \times N_x}, A_\theta \in R^{N_x \times N_\theta}$.

Using these partitions, we can develop the ABSP directly from the XBP processor in this joint state and "parameter estimation" form. Substituting Eq. 8.15 into the XBP Prediction Covariance relation of Table 5.3 and using the partition defined in Eq. 8.10, we obtain (suppressing the $\hat{x}$, $\hat{\theta}$, time index t notation for simplicity)

$$\tilde{\mathcal{P}}(t|t-1) = \begin{bmatrix} A_x & | & A_\theta \\ - & - & - \\ 0 & | & I_{N_\theta} \end{bmatrix} \tilde{\mathcal{P}}(t-1|t-1) \begin{bmatrix} A_x & | & A_\theta \\ - & - & - \\ 0 & | & I_{N_\theta} \end{bmatrix}' + \begin{bmatrix} R_{w_x w_x} & | & 0 \\ - & - & - \\ 0 & | & R_{w_\theta w_\theta} \end{bmatrix} \tag{8.17}$$

Expanding these equations, we obtain the following set of predicted covariance relations

$$\tilde{\mathcal{P}}(t|t-1) = \begin{bmatrix} A_x\tilde{P}_{xx}A_x' + A_\theta\tilde{P}_{\theta x}A_x' + A_x\tilde{P}_{x\theta}A_\theta' + A_\theta\tilde{P}_{\theta\theta}A_\theta' + R_{w_x w_x} & | & A_x\tilde{P}_{x\theta} + A_\theta\tilde{P}_{\theta\theta} \\ --- & - & --- \\ \tilde{P}_{\theta x}A_x' + \tilde{P}_{\theta\theta}A_\theta' & | & \tilde{P}_{\theta\theta} + R_{w_\theta w_\theta} \end{bmatrix} \tag{8.18}$$

The innovations covariance follows from the XBP as

$$R_{ee}(t) = C[\hat{x},\hat{\theta}]\tilde{\mathcal{P}}(t|t-1)C'[\hat{x},\hat{\theta}] + R_{vv}(t) \tag{8.19}$$

Now we must use the partitions of $\tilde{\mathcal{P}}$ above along with the measurement Jacobian

$$C[\hat{x},\hat{\theta}] = [C_x[\hat{x}(t|t-1),\hat{\theta}(t|t-1)] | C_\theta[\hat{x}(t|t-1),\hat{\theta}(t|t-1)]] \tag{8.20}$$

[2] Here is where the underlying random walk model enters the structure. The lower-block rows of A_x could be replaced by $[A_{\theta x}[\hat{x},\hat{\theta}] | A_{\theta\theta}[\hat{x},\hat{\theta}]]$ which enables a linear or nonlinear dynamic model to be embedded directly.

where

$$C_x[\hat{x},\hat{\theta}] := \frac{\partial c[\hat{x},\hat{\theta}]}{\partial x}$$
$$C_\theta[\hat{x},\hat{\theta}] := \frac{\partial c[\hat{x},\hat{\theta}]}{\partial \theta} \tag{8.21}$$

with $C \in R^{N_y \times (N_x+N_\theta)}$, $C_x \in R^{N_y \times N_x}$, $C_\theta \in R^{N_y \times N_\theta}$.

The corresponding innovations covariance follows from Eqs. 8.19 and 8.20 as

$$R_{ee}(t) = [C_x | C_\theta] \begin{bmatrix} \tilde{P}_{xx} & | & \tilde{P}_{x\theta} \\ - & - & - \\ \tilde{P}_{\theta x} & | & \tilde{P}_{\theta\theta} \end{bmatrix} \begin{bmatrix} C_x' \\ --- \\ C_\theta' \end{bmatrix} + R_{vv}(t) \tag{8.22}$$

or expanding

$$R_{ee}(t) = C_x \tilde{P}_{xx} C_x' + C_\theta \tilde{P}_{\theta x} C_x' + C_x \tilde{P}_{x\theta} C_\theta' + C_\theta \tilde{P}_{\theta\theta} C_\theta' + R_{vv} \tag{8.23}$$

$R_{ee} \in R^{N_y \times N_y}$. The gain of the XBP in Table 5.3 is calculated from these partitioned expressions as

$$\mathcal{K}(t) = \begin{bmatrix} K_x(t) \\ --- \\ K_\theta(t) \end{bmatrix} = \begin{bmatrix} \tilde{P}_{xx} & | & \tilde{P}_{x\theta} \\ - & - & - \\ \tilde{P}_{\theta x} & | & \tilde{P}_{\theta\theta} \end{bmatrix} \begin{bmatrix} C_x \\ --- \\ C_\theta \end{bmatrix}' R_{ee}^{-1}(t) \tag{8.24}$$

or

$$\mathcal{K}(t) = \begin{bmatrix} K_x(t) \\ --- \\ K_\theta(t) \end{bmatrix} = \begin{bmatrix} (\tilde{P}_{xx} C_x' + \tilde{P}_{x\theta} C_\theta') R_{ee}^{-1}(t) \\ --- \\ (\tilde{P}_{\theta x} C_x' + \tilde{P}_{\theta\theta} C_\theta') R_{ee}^{-1}(t) \end{bmatrix} \tag{8.25}$$

where $K \in {}^{(N_x+N_\theta) \times N_y}$, $K_x \in R^{N_x \times N_y}$, $K_\theta \in R^{N_\theta \times N_y}$. With the gain determined, the corrected state/parameter estimates follow easily, since the innovations remain unchanged, that is,

$$e(t) = y(t) - \hat{y}(t|t-1) = y(t) - c[\hat{x}(t|t-1), \hat{\theta}(t|t-1)] \tag{8.26}$$

and therefore partitioning the corrected state equations, we have

$$\hat{X}(t|t) = \begin{bmatrix} \hat{x}(t|t) \\ --- \\ \hat{\theta}(t|t) \end{bmatrix} = \begin{bmatrix} \hat{x}(t|t-1) \\ --- \\ \hat{\theta}(t|t-1) \end{bmatrix} + \begin{bmatrix} K_x(t)e(t) \\ --- \\ K_\theta(t)e(t) \end{bmatrix} \tag{8.27}$$

Finally, the corrected covariance expression is easily derived from the following partitions

$$\tilde{P}(t|t) = \left[\begin{array}{c|c} \tilde{P}_{xx} & \tilde{P}_{x\theta} \\ \hline \tilde{P}_{\theta x} & \tilde{P}_{\theta\theta} \end{array}\right] - \left[\begin{array}{c} K_x(t) \\ \hline K_\theta(t) \end{array}\right] [C_x | C_\theta] \left[\begin{array}{c|c} \tilde{P}_{xx} & \tilde{P}_{x\theta} \\ \hline \tilde{P}_{\theta x} & \tilde{P}_{\theta\theta} \end{array}\right] \tag{8.28}$$

Performing the indicated multiplications leads to the final expression

$$\tilde{P}(t|t) = \left[\begin{array}{c|c} \tilde{P}_{xx} - K_x C_x \tilde{P}_{xx} - K_x C_\theta \tilde{P}_{\theta x} & \tilde{P}_{x\theta} - K_x C_x \tilde{P}_{x\theta} - K_x C_\theta \tilde{P}_{\theta\theta} \\ \hline \tilde{P}_{\theta x} - K_\theta C_x \tilde{P}_{xx} - K_\theta C_\theta \tilde{P}_{\theta x} & \tilde{P}_{\theta\theta} - K_\theta C_x \tilde{P}_{x\theta} - K_\theta C_\theta \tilde{P}_{\theta\theta} \end{array}\right] \tag{8.29}$$

We summarize the parametrically adaptive model-based processor in predictor-corrector form in Table 8.1. We note that this algorithm is *not* implemented in this fashion, it is implemented in the numerically stable, "upper triangular-diagonal" or UD-factorized form as in SSPACK_PC [10]. Here we are just interested in the overall internal structure of the algorithm and the decomposition that evolves. This completes the development of the generic ABSP.

It is important to realize that, besides its numerical implementation, the ABSP is simply the XBP with an augmented state vector thereby implicitly creating the partitions developed above. The implementation of these decomposed equations directly is *not* necessary—just augment the state with the unknown parameters and the ABSP evolves naturally from the standard XBP algorithm of Table 5.3. The ABSP of Table 8.1 indicates where to locate the partitions. That is, suppose we would like to extract the submatrix $\tilde{P}_{\theta\theta}$ but the XBP only provides the overall $(N_x + N_\theta)$ error covariance matrix $\tilde{P}$. However, locating the lower $N_\theta \times N_\theta$ submatrix of $\tilde{P}$ enables us to extract $\tilde{P}_{\theta\theta}$ directly.

Next, let us reconsider the nonlinear system example given in Chapter 5 and investigate the performance of the parametrically adaptive Bayesian processor.

Example 8.1

Recall the discrete nonlinear trajectory estimation problem [20] of Chapter 5 given by

$$x(t) = (1 - 0.05\triangle T)x(t-1) + 0.04x^2(t-1) + w(t-1)$$

with corresponding measurement model

$$y(t) = x^2(t) + x^3(t) + v(t)$$

where $v(t) \sim \mathcal{N}(0, 0.09)$, $x(0) = 2.0$, $P(0) = 0.01$, $\triangle T = 0.01$ sec, and $R_{ww} = 0$.

TABLE 8.1 ABSP Algorithm

Predictor/Corrector Form

Prediction

$$\hat{x}(t|t-1) = a[\hat{x}(t|t-1), \hat{\theta}(t-1|t-1)] + b[u(t-1)] \quad \textit{(state)}$$

$$\hat{\theta}(t|t-1) = \hat{\theta}(t-1|t-1) \quad \textit{(parameter)}$$

$$\tilde{P}_{xx}(t|t-1) = A_x[\hat{x},\hat{\theta}]\tilde{P}_{xx}(t-1|t-1)A_x'[\hat{x},\hat{\theta}] + A_\theta[\hat{x},\hat{\theta},u]\tilde{P}_{\theta x}(t-1|t-1)A_x'[\hat{x},\hat{\theta}]$$
$$+A_x[\hat{x},\hat{\theta}]\tilde{P}_{x\theta}(t-1|t-1)A_\theta'[\hat{x},\hat{\theta},u] + A_\theta[\hat{x},\hat{\theta}]\tilde{P}_{\theta\theta}(t-1|t-1)A_\theta'[\hat{x},\hat{\theta},u] + R_{w_x w_x}(t-1) \quad \textit{(state cov.)}$$

$$\tilde{P}_{\theta\theta}(t|t-1) = \tilde{P}_{\theta\theta}(t-1|t-1) + R_{w_\theta w_\theta}(t-1) \quad \textit{(param. cov.)}$$

$$\tilde{P}_{x\theta}(t|t-1) = A_x[\hat{x},\hat{\theta}]\tilde{P}_{x\theta}(t-1|t-1) + A_\theta[\hat{x},\hat{\theta},u]\tilde{P}_{\theta\theta}(t-1|t-1) \quad \textit{(cross cov.)}$$

Innovation

$$e(t) = y(t) - \hat{y}(t|t-1) = y(t) - c[\hat{x}(t|t-1), \hat{\theta}(t|t-1)] \quad \textit{(innovation)}$$

$$R_{ee}(t) = C_x[\hat{x},\hat{\theta}]\tilde{P}_{xx}(t|t-1)C_x'[\hat{x},\hat{\theta}] + C_\theta[\hat{x},\hat{\theta},u]\tilde{P}_{\theta x}(t|t-1)C_x'[\hat{x},\hat{\theta}]$$
$$+C_x[\hat{x},\hat{\theta}]\tilde{P}_{x\theta}(t|t-1)C_\theta'[\hat{x},\hat{\theta},u] + C_\theta[\hat{x},\hat{\theta},u]\tilde{P}_{\theta\theta}(t|t-1)C_\theta'[\hat{x},\hat{\theta},u] + R_{vv}(t) \quad \textit{(innovation covariance)}$$

Gain

$$K_x(t) = (\tilde{P}_{xx}(t|t-1)C_x'[\hat{x},\hat{\theta}] + \tilde{P}_{x\theta}(t|t-1)C_\theta'[\hat{x},\hat{\theta},u])R_{ee}^{-1}(t) \quad \textit{(state gain)}$$

$$K_\theta(t) = (\tilde{P}_{\theta x}(t|t-1)C_x'[\hat{x},\hat{\theta}] + \tilde{P}_{\theta\theta}(t|t-1)C_\theta'[\hat{x},\hat{\theta},u])R_{ee}^{-1}(t) \quad \textit{(parameter gain)}$$

Correction

$$\hat{x}(t|t) = \hat{x}(t|t-1) + K_x(t)e(t) \quad \textit{(state)}$$

$$\hat{\theta}(t|t) = \hat{\theta}(t|t-1) + K_\theta(t)e(t) \quad \textit{(parameter)}$$

$$\tilde{P}_{xx}(t|t) = \tilde{P}_{xx}(t|t-1) - K_x(t)C_x[\hat{x},\hat{\theta}]\tilde{P}_{xx}(t|t-1) - K_x(t)C_\theta[\hat{x},\hat{\theta},u]\tilde{P}_{\theta x}(t|t-1) \quad \textit{(state covariance)}$$

$$\tilde{P}_{\theta\theta}(t|t) = \tilde{P}_{\theta\theta}(t|t-1) - K_\theta(t)C_\theta[\hat{x},\hat{\theta},u]\tilde{P}_{x\theta}(t|t-1) - K_\theta(t)C_\theta[\hat{x},\hat{\theta},u]\tilde{P}_{\theta\theta}(t|t-1) \quad \textit{(parameter covariance)}$$

$$\tilde{P}_{x\theta}(t|t) = \tilde{P}_{x\theta}(t|t-1) - K_x(t)C_x[\hat{x},\hat{\theta}]\tilde{P}_{x\theta}(t|t-1) - K_x(t)C_\theta[\hat{x},\hat{\theta},u]\tilde{P}_{\theta\theta}(t|t-1) \quad \textit{(cross covariance)}$$

Initial Conditions

$$\hat{x}(0|0) \quad \tilde{P}(0|0) \quad A[\hat{x},\theta] := \frac{\partial}{\partial x}A[x,\theta]\Big|_{\substack{x=\hat{x}(t|t-1)\\ \theta=\hat{\theta}(t|t-1)}} \quad C[\hat{x},\theta] := \frac{\partial}{\partial x}C[x,\theta]\Big|_{\substack{x=\hat{x}(t|t-1)\\ \theta=\hat{\theta}(t|t-1)}}$$

Here we generalize the problem to the case where the coefficients of the process are unknown leading to the ABSP solution. Therefore, the process equations for this problem become

$$x(t) = (1 - \theta_1 \triangle T)x(t-1) + \theta_2 x^2(t-1) + w(t-1)$$

with the identical measurement and covariances as before. The true parameters are $\Theta_{true} = [0.05\ 0.04]'$. The ABSP can be applied to this problem by defining the parameter vector as

$$\Theta(t) = \Theta(t-1) \quad \text{(constant)}$$

and augmenting it to form the new state vector $X = [x' \ \theta_1 \ \theta_2]'$ with the new process noise vector $\mathbf{w} = [w \ w_{\theta_1} \ w_{\theta_2}]'$. Therefore the process model becomes

$$X(t) = \begin{bmatrix} (1 - \theta_1(t-1)\triangle T)x(t-1) + \theta_2(t-1)\triangle T x^2(t-1) \\ \theta_1(t-1) \\ \theta_2(t-1) \end{bmatrix} + \mathbf{w}(t-1)$$

$$y(t) = x^2(t) + x^3(t) + v(t)$$

To implement the ABSP the required Jacobians are

$$A[X(t-1)] = \begin{bmatrix} [1 - \theta_1(t-1)\triangle T + 2\triangle T\theta_2(t-1)x(t-1)] & \triangle T)x(t-1) & \triangle T x^2(t-1) \\ 0 & 1 & 0 \\ 0 & 0 & 1 \end{bmatrix}$$

$$C[X(t-1)] = [2x(t-1) + 3x^2(t-1) \quad 0 \quad 0]$$

Using SSPACK_PC [10] the ABSP is applied to solve this problem for 1500 samples with $\triangle T = 0.01$ sec. Initially, we used the starting parameters

$$\tilde{P}(0|0) = \text{diag}[100 \ \ 100 \ \ 100] \quad \text{and} \quad \hat{x}(0|0) = [2 \ \ 0.055 \ \ 0.044]'$$

The results of the ABSP run are shown in Fig. 8.2. We see the estimated state and parameter estimates in Fig. 8.2b and Fig. 8.2c. After a short transient (25 samples), the state estimate begins tracking the true state as evidenced by the innovations sequence in Fig. 8.2c. The parameter estimates slowly converge to their true values as evidenced by the plots. The final estimates are

$$\hat{\theta}_1 = 0.0470$$
$$\hat{\theta}_2 = 0.0395$$

Part of the problem for the slow convergence results stems from the lack of sensitivity of the measurement, or equivalently, innovations to parameter variations in this problem. This is implied from the zero-mean/whiteness tests shown in Fig. 8.2c. The innovations are statistically white (3.6% out), and zero-mean ($0.008 < 0.061$). The filtered measurement is also shown in Fig. 8.2c as well. This completes the ABSP example. △△△

As pointed out by Ljung [1, 2, 23], it is important to realize that the XBP is suboptimal as a parameter estimator as compared to the recursive prediction error (RPE) method based on the Gauss–Newton (stochastic) descent algorithm. Comparing the processors in this context, we see that if a gradient term $[\nabla_\theta K(\Theta)]\, e(t)$ is incorporated into the XBP (add this term to A_θ), its convergence will be improved

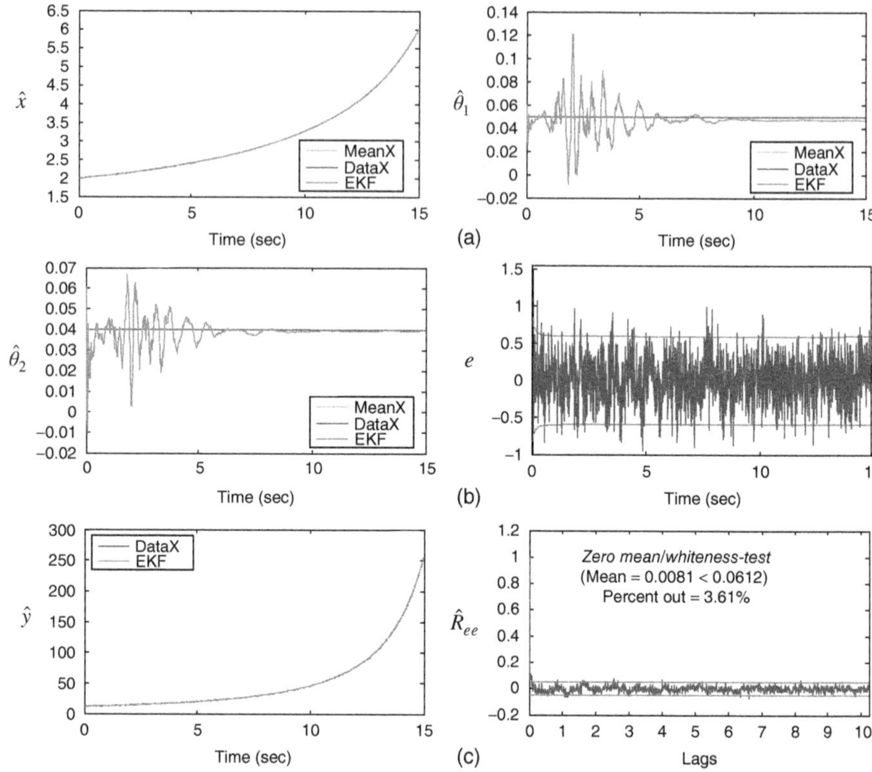

FIGURE 8.2 XBP (EKF) simulation. (a) estimated state and parameter no. 1; (b) estimated parameter no. 2 and innovation; (c) predicted measurement and zero-mean/whiteness test ($0.008 < 0.061$ and 3.6% out).

approaching the performance of the RPE algorithm (see Ljung [23] for details). We also note in passing that the nonlinear BSP in the form developed in Chapter 5 as well as the parametrically adaptive ABSP are all heavily employed as *neural networks*. For more details of this important application, see Haykin [18]. Next, we consider the development of the "modern" approach to Bayesian processor design using the unscented Bayesian processor of Chapter 6.

8.3.2 Modern Joint Bayesian Processor

The modern unscented processor offers a similar representation as the extended processor detailed in the previous subsection. Here, we briefly outline its structure for solution to the *joint* problem and apply it to the trajectory estimation problem for comparison. We again start with the augmented state vector defined initially by sigma-points, that is,

$$\mathcal{X}(t) := \begin{bmatrix} x(t) \\ --- \\ \theta(t) \end{bmatrix} \tag{8.30}$$

$\mathcal{X} \in R^{(N_x+N_\theta)\times 1}$ and $x \in R^{N_x\times 1}$, $\theta \in R^{N_\theta\times 1}$. Substituting this augmented state vector into the SPBP relations of Table 6.1 yields the desired processor. We again draw the equivalences (as before):

$$\begin{aligned} \Pr(y(t)|x(t),\theta(t)) &\sim \mathcal{N}(\mathbf{c}[x(t),\theta(t)], R_{vv}(t)) \\ \Pr(\mathcal{X}(t)|Y_{t-1}) &\sim \mathcal{N}(\mathcal{X}(t|t-1), \tilde{\mathcal{P}}(t|t-1)) \\ \Pr(y(t)|Y_{t-1}) &\sim \mathcal{N}(\hat{y}_\theta(t|t-1), \tilde{R}_{e_\theta e_\theta}(t)) \end{aligned} \tag{8.31}$$

where

$$\begin{aligned} \mathcal{X}(t|t-1) &:= [\hat{x}(t|t-1)|\hat{\theta}(t|t-1)]' \\ \hat{y}_\theta(t|t-1) &:= \mathbf{c}[\hat{x}(t|t-1), \hat{\theta}(t|t-1)] \\ \tilde{\mathcal{P}}(t|t-1) &:= \begin{bmatrix} \tilde{P}_{xx}(t|t-1) & | & \tilde{P}_{x\theta}(t|t-1) \\ - & - & - \\ \tilde{P}_{\theta x}(t|t-1) & | & \tilde{P}_{\theta\theta}(t|t-1) \end{bmatrix} \\ \mathcal{R}_{e_\theta e_\theta}(t) &:= C[\hat{X}(t|t-1)]\tilde{\mathcal{P}}(t|t-1)C[\hat{X}(t|t-1)]' + R_{vv}(t) \end{aligned}$$

With this in mind, it is possible to derive the internal structure of the SPBP in a manner similar to that of the XBP. But we will not pursue that derivation here. We just note that the sigma points are also augmented to give

$$\mathcal{X}_i := \begin{bmatrix} \hat{x}(t|t-1) \\ --- \\ \hat{\theta}(t|t-1) \end{bmatrix} + \left((N_{\mathcal{X}} + \kappa) \begin{bmatrix} \tilde{P}_{xx}(t|t-1) & | & \tilde{P}_{x\theta}(t|t-1) \\ - & - & - \\ \tilde{P}_{\theta x}(t|t-1) & | & \tilde{P}_{\theta\theta}(t|t-1) \end{bmatrix} \right)^{\frac{1}{2}} \tag{8.32}$$

with the corresponding process noise covariance partitioned as

$$\mathcal{R}_{ww}(t-1) := \begin{bmatrix} R_{w_x w_x}(t-1) & | & \mathbf{0} \\ - & - & - \\ \mathbf{0} & | & R_{w_\theta w_\theta}(t-1) \end{bmatrix} \tag{8.33}$$

It also follows that the prediction step becomes

$$\mathcal{X}_i(t|t-1) = \begin{bmatrix} \mathbf{a}[\hat{x}(t|t-1), \hat{\theta}(t|t-1)] + \mathbf{b}[\hat{\theta}(t|t-1), u(t-1)] \\ --- \\ \mathbf{a}[\hat{\theta}(t|t-1)] \end{bmatrix} \tag{8.34}$$

and in the multichannel (vector) measurement case, we have

$$\mathcal{Y}_i(t|t-1) = \begin{bmatrix} \mathbf{c}[\hat{x}(t|t-1), \hat{\theta}(t|t-1)] \\ --- \\ \mathbf{c}[\hat{\theta}(t|t-1)] \end{bmatrix} \tag{8.35}$$

Using the augmented state vector, we apply the "joint" approach to the trajectory estimation problem [20] and compare its performance to that of the XBP.

Example 8.2

Using the discrete nonlinear trajectory estimation problem of the previous example with unknown coefficients as before, we define the augmented sigma-point vector $\mathcal{X}(t)$ defined above and apply the SPBP algorithm of Table 6.1.

The process equations for this problem are

$$x(t) = (1 - \theta_1 \triangle T)x(t-1) + \theta_2 x^2(t-1) + w(t-1)$$

with the identical measurement and covariances as before. The true parameters are $\Theta_{true} = [0.05\ \ 0.04]'$. The SPBP can be applied to this problem by defining the parameter vector as a constant and augmenting it to form the new sigma-point vector $\mathcal{X} = [x'\ \ \theta_1\ \ \theta_2]'$ with the new process noise vector $\mathbf{w} = [w\, w_{\theta_1} w_{\theta_2}]'$. Therefore the process model becomes

$$\begin{bmatrix} x(t) \\ -- \\ \theta(t) \end{bmatrix} = \begin{bmatrix} (1 - \theta_1(t-1)\triangle T)x(t-1) + \theta_2(t-1)\triangle T x^2(t-1) \\ \theta_1(t-1) \\ \theta_2(t-1) \end{bmatrix} + \mathbf{w}(t-1)$$

$$y(t) = x^2(t) + x^3(t) + v(t)$$

To implement the SPBP the sigma points are selected as before for a Gaussian distribution using the scaled transformation with $\alpha = 1$, $\kappa = 0$, and $\beta = 2$, using the same initial conditions as the XBP.

Using MATLAB [10], the ABSP is applied to solve this problem for 1500 samples with $\triangle T = 0.01$ sec. The results of the SPBP run are shown in Fig. 8.3. We see the estimated state and parameter estimates in Fig. 8.3a and Fig. 8.3b. After a short transient, the state estimate begins tracking the true state as evidenced by the predicted measurement and innovations sequence in Fig. 8.3c. The parameter estimates converge to their true values as evidenced by the plots. The final estimates are: $\hat{\theta}_1 = 0.05$; $\hat{\theta}_2 = 0.04$. The processor appears to converge much faster than the XBP demonstrating its improved capability. This is implied from the zero-mean/whiteness tests shown in Fig. 8.3c. The innovations are statistically white (1.86% out) and zero-mean ($0.0113 < 0.0612$). The filtered measurement is also shown in Fig. 8.3c as well. This completes the ABSP example. △△△

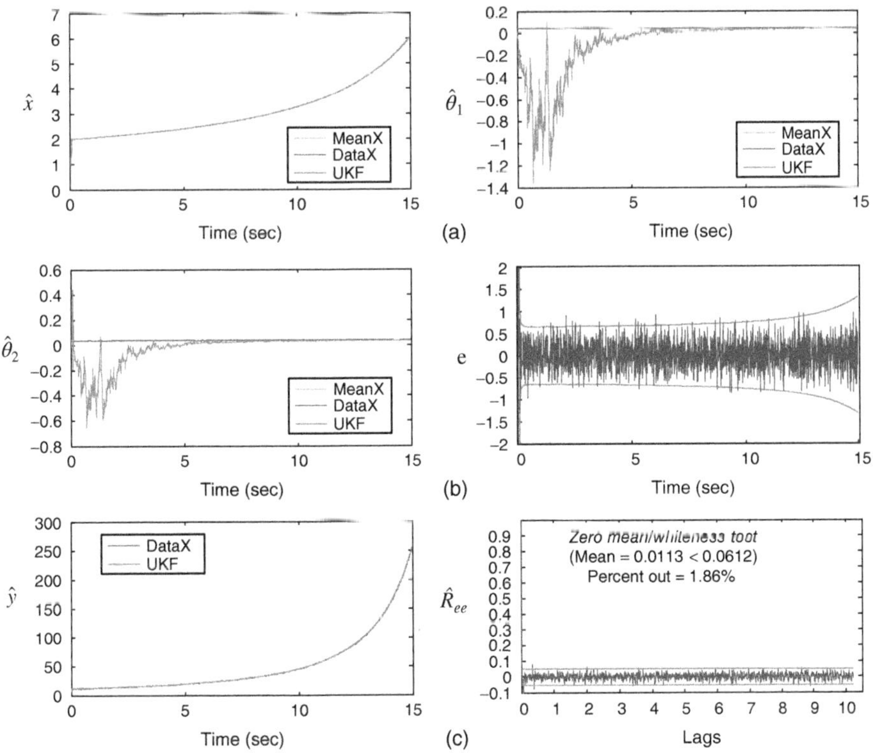

FIGURE 8.3 SPBP (UKF) simulation: (a) estimated state and parameter no. 1; (b) estimated parameter no. 2 and innovation; (c) predicted measurement and zero-mean/whiteness test (0.0113 < 0.0612 and 1.86% out).

We also note in closing that a "dual" rather than "joint" approach has evolved in the literature. Originally developed as a bootstrap approach, it is constructed by two individual (decoupled) processors: one for the state estimator and the other for the parameter estimator which pass updated estimates back and forth to one another as they become available. This is a suboptimal methodology, but appears to perform quite well (see Refs. [21] and [22] for more details). This completes our discussion of the joint modern approach; next, we investigate the SMC approach to solving the joint problem.

8.4 PARTICLE-BASED JOINT BAYESIAN STATE/PARAMETRIC PROCESSORS

In this section, we briefly develop the sequential Monte Carlo approach to solving the joint state/parametric processing problem. There are a variety of ways to search the feasible space in pursuing solutions to the parameter estimation problem, but when states (or signals) must be extracted along with the parameters, then a solution to the joint estimation problem must follow as discussed before [24–30]. One of the major problems that evolves is how to "model" the parameter evolution in order to provide an effective way to proceed with the search, especially if there does not exist a physical model characterizing the parameters.

8.4.1 Parametric Models

Perhaps the most popular approach to this quandary is to use the so-called "random walk" model that has evolved from the control/estimation area. This model essentially assumes a constant parameter (vector) θ with slight variations. In continuous-time a *constant* can be modeled by the differential equation

$$\frac{d\theta_t}{dt} = 0 \quad \text{for} \quad \theta_0 = \text{constant} \tag{8.36}$$

Here, the solution is just $\theta_0\ \forall t$. Note that this differential equation is essentially a state–space formulation in θ enabling us to incorporate it directly into the overall dynamic model. Now if we artificially assign a zero-mean Gaussian process noise term $\mathbf{w}_t$ to excite the model, then a random walk (pseudo dynamic) model evolves. Providing a small variance to the noise source approximates small variations in the parameter enabling it to "walk" in the underlying parameter space and converge (hopefully) to the true value. Since we typically concentrate on a sampled-data systems, we can discretize this representation using first differences (with $\triangle t = 1$) and incorporate its corresponding mean/covariance propagation into our Gauss–Markov model, that is,

$$\begin{aligned}
\theta(t) &= \theta(t-1) + \mathbf{w}_\theta(t-1) \qquad \text{for} \qquad \mathbf{w}_\theta \sim \mathcal{N}(0, \mathbf{R}_{w_\theta w_\theta}) \\
\mathbf{m}_\theta(t) &= \mathbf{m}_\theta(t-1) \\
\mathbf{P}_{\theta\theta}(t) &= \mathbf{P}_{\theta\theta}(t-1) + \mathbf{R}_{w_\theta w_\theta}(t-1)
\end{aligned} \tag{8.37}$$

The process noise covariance $\mathbf{R}_{w_\theta w_\theta}$ controls the excursions of the random walk or equivalently the parameter space. Using artificial dynamics is the identical approach employed in both the classical (XBP) and modern (UBSP) techniques used in the previous sections. The only problem results when the parameter is truly *static* such as in financial models and its variations can have very large repercussions in the money and economic markets. Thus, a large variety of "off-line" MC methods have evolved [17], but we will not discuss them here since we are primarily interested in physical systems which typically have parametric uncertainties that are well-modeled by the random walk or other variations. Of course, if the parametric relations are truly dynamic, then the joint approach incorporates these dynamics into the augmented state–space model and yields an optimal filtering solution to this joint problem.

A popular variation of the random walk is based on the "forgetting factor" that has evolved from the system identification literature [1] and is so-called because it controls the time constant of an exponential window over the data. To see this, consider a squared-error performance function defined by

$$\mathcal{J} = \sum_{t=1}^{k} \alpha(t,k)\epsilon^2(t) \qquad \text{for} \qquad \alpha(t,k) = \lambda^{k-t} \qquad 0 < \lambda < 1 \tag{8.38}$$

where the weighting function α is the forgetting factor that creates an exponentially decaying window of length, say T_λ, in which the most recent data is weighted more than the past data. This follows from the following approximation [1, 2]:

$$\alpha(t,k) = e^{(t-k)\ln\lambda} \approx e^{-(t-k)(1-\lambda)} \qquad \text{for} \qquad T_\lambda = \frac{1}{1-\lambda} \tag{8.39}$$

where T_λ is called the memory time constant that remains constant (approximately) over T_λ samples with typical choices between $0.98 < T_\lambda < 0.995$ in practice.

The forgetting factor has been introduced into a variety of sequential algorithms in different roles. For instance, a direct implementation into the random walk model of Eq. 8.37 [31] is

$$\theta(t) = \big(1 - \lambda(t)\big)\theta(t-1) + \lambda(t)\mathbf{w}_\theta(t-1) \tag{8.40}$$

here λ affects the tracking speed of the updates providing a tradeoff between speed and noise attenuation. Setting $\lambda = 1$ implies $T_\lambda \to \infty$, an "infinite memory," while $\lambda = 0$ provides instantaneous response forgetting the entire past.

Thus, the forgetting factor introduces a certain amount of flexibility into the joint state/parameter estimation processors by enabling them to be more responsive to new data and therefore yielding improved signal estimates. The same affect is afforded by the process noise covariance matrix of Eq. 8.37 using an equivalent technique called "annealing." The covariance matrix, usually assumed diagonal, is annealed towards zero using exponentially (decay) weighting (see Refs. [21] and [22] for more details).

Other variations of the parameter estimation models exist besides the statistical random walk model. For instance, similar to the random walk is a process called roughening which is very similar in concept and is another simple pragmatic method of preventing the sample impoverishment problem. It is a method suggested by Gordon [15, 17] and refined in Refs. [18] and [10], termed particle "roughening" which is similar to adding process noise to constant parameters when constructing a random-walk GM model. It is useful in estimating embedded state–space model parameters and can also be applied to the joint state/parameter estimation problem.

Roughening consists of adding random noise to each particle *after* resampling is accomplished, that is, the a posteriori particles are modified as

$$\tilde{x}_i(t) = \hat{x}_i(t) + \epsilon_i(t) \tag{8.41}$$

where $\epsilon_i \sim \mathcal{N}(0, \text{diag}[\kappa \mathcal{M}_n N_p^{-1/N_x}])$ and κ is a constant "tuning" parameter (e.g., $\sim$0.2), $\mathcal{M}_n$ is a vector of the maximum difference between particle components *before* roughening, the nth element of $\mathcal{M}$ given by

$$\mathcal{M}_n = \max_{ij} \left| x_i^{(n)}(t) - x_j^{(n)}(t) \right| \qquad \text{for} \qquad n = 1, \ldots, N_x \tag{8.42}$$

8.4.2 Joint Bayesian State/Parameter Estimation

Here we are concerned with the joint estimation problem consisting of setting a prior for θ and augmenting the state vector to solve the joint estimation problem as defined in Section 8.2 thereby converting the parameter estimation problem to one of optimal filtering. Thus, confining our discussion to state–space models and the Bayesian equivalence, we develop the Bayesian approach using the relations

$$\begin{aligned} x &\sim \mathcal{A}_x\left(x(t)|x(t-1), \theta(t-1)\right) \\ \theta &\sim \mathcal{A}_\theta\left(\theta(t)|\theta(t-1), x(t-1)\right) \\ y &\sim C\left(y(t)|x(t), \theta(t)\right) \end{aligned} \tag{8.43}$$

Here we separate the state transition function into the individual vectors for illustrative purposes, but in reality (as we shall observe), they can be jointly coupled. The key idea is to develop the PF technique to estimate the joint posterior $\Pr(x(t), \theta(t)|Y_t)$ relying on the parametric posterior $\Pr(\theta(t)|Y_t)$ in the Bayesian decomposition. We will follow the approach outlined in Refs. [17], [32] and [33] starting with the full posterior distributions and proceeding to the filtering distributions.

Suppose it is possible to sample N_p-particles, $\{X_t(i), \Theta_t(i)\}$ for $i = 1, \ldots, N_p$ from the joint posterior distribution where we define $X_t := \{x(0), \ldots, x(t)\}$ and $\Theta_t := \{\theta(0), \ldots, \theta(t)\}$. Then the corresponding empirical approximation of the joint posterior is given by [34]

$$\hat{\Pr}(X_t, \Theta_t|Y_t) \approx \frac{1}{N_p} \sum_{i=1}^{N_p} \delta(X_t - X_t(i), \Theta_t - \Theta_t(i)) \tag{8.44}$$

and it follows directly that the filtering posterior (see Chapter 2) is given by

$$\hat{\Pr}(x(t), \theta(t)|Y_t) \approx \frac{1}{N_p} \sum_{i=1}^{N_p} \delta(x(t) - x_i(t), \theta(t) - \theta_i(t)) \tag{8.45}$$

Unfortunately, it is not possible to sample directly from the full joint posterior $\Pr(X_t, \Theta_t|Y_t)$ at any time t. However, one approach to mitigate this problem is by using the importance sampling approach of Chapter 2.

Suppose we define a (full) importance distribution, $q(X_t, \Theta_t|Y_t)$ such that $\Pr(X_t, \Theta_t|Y_t) > 0$ implies $q(\cdot) > 0$, then we define the corresponding importance weight (as before) by

$$W(X_t, \Theta_t) \propto \frac{\Pr(X_t, \Theta_t|Y_t)}{q(X_t, \Theta_t|Y_t)} \tag{8.46}$$

From Bayes' rule we have that the posterior can be expressed as

$$\Pr(X_t, \Theta_t|Y_t) = \frac{\Pr(Y_t|X_t, \Theta_t) \times \Pr(X_t, \Theta_t)}{\Pr(Y_t)} \tag{8.47}$$

Thus, if N_p-particles, $\{X_t(i), \Theta_t(i)\}; i = 1, \ldots, N_p$, can be generated from the importance distribution

$$\{X_t(i), \Theta_t(i)\} \rightarrow q(X_t, \Theta_t|Y_t) \tag{8.48}$$

then the empirical distribution can be estimated and the resulting normalized weights specified by

$$\mathcal{W}_t(i) = \frac{W(X_t(i), \Theta_t(i))}{\sum_{k=1}^{N_p} W(X_t(k), \Theta_t(k))} \quad \text{for } i = 1, \ldots, N_p \tag{8.49}$$

to give the desired empirical distribution of Eq. 8.44 leading to the corresponding filtering distribution of Eq. 8.45.

If we have a state transition model available, then for a fixed parameter estimate the state estimation problem is easily solved as before in Chapter 7. Therefore, we will confine our discussion to the parameter posterior distribution estimation problem, that is, marginalizing the joint distribution with respect to the states (that have already been estimated) gives

$$\Pr(\Theta_t|Y_t) = \int \Pr(X_t, \Theta_t|Y_y)dX_t \tag{8.50}$$

and it follows that

$$\mathcal{W}(\Theta_t) \propto \frac{\Pr(\Theta_t|Y_t)}{q(\Theta_t|Y_t)} \tag{8.51}$$

Assuming that a set of particles can be generated from the importance distribution as $\Theta_t \sim q(\Theta_t|Y_t)$, we have the set of normalized weights

$$\mathcal{W}_t(\Theta(i)) = \frac{W(\Theta_t(i))}{\sum_{k=1}^{N_p} W(\Theta_t(k))} \quad \text{for } i = 1, \ldots, N_p \tag{8.52}$$

which is the joint "batch" importance sampling solution when coupled with the dynamic state vectors.

The sequential form of this joint formulation follows directly (as before in Chapter 2). We start with the factored form of the importance distribution and focus on the full posterior $\Pr(\Theta_t|Y_t)$, that is,

$$q(\Theta_t|Y_t) = \prod_{k=0}^{t} q(\theta(k)|\Theta_{k-1}, Y_k) \tag{8.53}$$

with $\Pr(\theta(0)|\Theta_{-1}, Y_t) \rightarrow \Pr(\theta(0)|Y_t)$.

Assuming that this factorization can be expressed recursively in terms of the previous step $q(\Theta_{t-1}|Y_{t-1})$ and extracting the tth term gives

$$q(\Theta_t|Y_t) = q(\theta(t)|\Theta_{t-1}, Y_t) \times \prod_{k=0}^{t-1} q(\theta(k)|\Theta_{k-1}, Y_k) \tag{8.54}$$

or simply

$$q(\Theta_t|Y_t) = q(\theta(t)|\Theta_{t-1}, Y_t) \times q(\Theta_{t-1}|Y_{t-1}) \tag{8.55}$$

With this in mind, the weight recursion becomes

$$W(\Theta_t) = W(t) \times W(\Theta_{t-1}) \tag{8.56}$$

Applying Bayes' rule to the posterior, we define

$$\begin{aligned} W(t) := & \frac{\Pr(y(t)|\Theta_t, Y_{t-1}) \times \Pr(\theta(t)|\Theta(t-1))}{\Pr(y(t)|Y_{t-1}) \times q(\theta(t)|\Theta_{t-1}, Y_t)} \\ \propto & \frac{\Pr(y(t)|\Theta_t, Y_{t-1}) \times \Pr(\theta(t)|\theta(t-1))}{q(\theta(t)|\Theta_{t-1}, Y_t)} \end{aligned} \tag{8.57}$$

As before in the state estimation problem, we must choose an importance distribution before we can construct the algorithm. We can choose from the *minimum variance* (optimal) using local linearization techniques given by $q(\theta(t)|\Theta_{t-1}, Y_t) \rightarrow \Pr(\theta(t)|\Theta_{t-1}, Y_t)$ which leads to the transition posterior

$$\Pr(\theta(t)|\Theta_{t-1}, Y_t) = \frac{\Pr(y(t)|\Theta_{t-1}, Y_{t-1}) \times \Pr(\theta(t)|\theta(t-1))}{\Pr(y(t)|\Theta_{t-1}, Y_{t-1})} \tag{8.58}$$

and therefore it follows using Eq. 8.57 that

$$W_{MV}(t) \propto \Pr(y(t)|\Theta_{t-1}, Y_{t-1}) = \int \Pr(y(t)|\Theta_t, Y_{t-1}) \times \Pr(\theta(t)|\theta(t-1))\, d\theta(t) \tag{8.59}$$

with (local linearization implementation of Chapter 7)

$$\Pr(y(t)|\Theta_t, Y_{t-1}) \sim \mathcal{N}(\hat{y}_\theta(t|t-1), R_{\epsilon_\theta\epsilon_\theta})$$

which can be obtained from any of the classical or modern techniques (Chapters 5 and 6).

The usual bootstrap approach can also be selected as the importance distribution leading to a simpler alternative with $q(\theta(t)|\Theta_{t-1}, Y_t) \rightarrow \Pr(\theta(t)|\theta(t-1))$ and the weight of Eq. 8.57 becomes

$$W_{BS}(t) = \Pr(y(t)|\Theta_t, Y_{t-1}) \tag{8.60}$$

From the pragmatic perspective, we must consider some practical approaches to implementing the processor for the joint problem. The first approach, when applicable, (not financial problems) is to incorporate the random walk model of Eq. 8.37 when reasonable [17]. We will use this approach for our case study to follow. Another variant is to use the "roughening" method that moves the particles (after resampling) by adding a Gaussian sequence of specified variance.[3]

The *kernel method* (regularized PF of Chapter 7) can also be applied to the joint problem. In the bootstrap implementation, we can estimate the posterior distribution using the kernel density approach, that is, after resampling we have the empirical distribution given by

$$\hat{\Pr}(\theta(t)|Y_t) = \frac{1}{N_p}\sum_{i=1}^{N_p} \delta(\theta(t) - \hat{\theta}_i(t)) \tag{8.61}$$

The kernel technique consists of substituting for this degenerate distribution, the kernel density estimate

$$\hat{\Pr}(\theta(t)|Y_t) = \frac{1}{N_p}\sum_{i=1}^{N_p} \mathcal{K}(\theta(t) - \hat{\theta}_i(t)) \tag{8.62}$$

for $\mathcal{K}(\cdot)$ the kernel (e.g., Gaussian, triangle, etc.). Now a new set of parametric particles can be obtained by generating samples from $\theta_i(t) \sim \mathcal{K}(\theta(t))$. In the same

[3] Recall that the sequence is distributed $\epsilon \sim \mathcal{N}(0, \kappa \mathcal{M}_{ij} N_p^{-1/N_x})$ for κ a constant and $\mathcal{M}_{ij}$ is the maximum distance between the *i*th and *j*th particles discussed previously of Section 7.5.

manner as the standard bootstrap, this approach introduces diversity into the set of particles.

Yet another alternative is to introduce the MCMC-step (see Chapter 7) to "move" the particles to the regions of high probability using a Markov chain with appropriate invariant distribution. Again the new particles are sampled according to an MCMC with joint distribution (when possible) $\Pr(X_t(i), \Theta_t(i)|Y_t)$ such that

$$(X_t(i), \Theta_t(i)) \sim \mathcal{K}_{MCMC}(X_t, \Theta_t|Y_t)$$

This completes the discussion of the joint state/parameter estimation problem using the PF approach; we emphasize that the algorithm of Chapter 7 may be used by merely augmenting the state vector with the parameter vector especially when a dynamic equation characterizing the parameter dynamics is available. Next, we consider an example to illustrate this approach.

Example 8.3

Again, we consider the trajectory estimation problem [20] using the particle filter technique. At first, we applied the usual bootstrap technique and found what was expected, a collapse of the parameter particles giving an unsatisfactory result. Next, we tried the "roughening" approach and the results were much more reasonable. We used a roughening factor or $\kappa = 5 \times 10^{-5}$ along with $N_p = 350$. The results are shown in Fig. 8.4. We see both the estimated states and measurement along with the associated zero-mean/whiteness test. The result, although not as good as the modern approach, is reasonable with the final estimates converging to the static parameter values of true parameters: $\theta_1 = 0.05(0.034)$ and $\theta_2 = 0.04(0.039)$. The state and measurement estimates are quite good as evidenced by the zero-mean ($0.002 < 0.061$) and whiteness (1.76% out). We show the estimated posterior distributions for the states and parameters in Fig. 8.5 again demonstrating a reasonable solution. Note how the distributions are initially multi-modal and become unimodal as the parameter estimates converge to their true values as depicted in Fig. 8.5.

Next, we generate an ensemble consisting of a set of a 100-member realizations, execute the PF and obtain the ensemble estimates shown in Fig. 8.6. In Fig. 8.6a, we see the ensemble state estimate and in (b) and (c) the corresponding joint parameter estimates. Comparing these to those of Fig. 8.4, we see that after the initial transients die out, they converge to reasonable estimates for both the state and parameters. The ensemble processor gives reasonable zero-mean/whiteness (sanity) test results: (Z-M: $0.023 < 0.123$/W-Test: 1.2% out) along with relative root mean-squared errors (RMSE: state = 0.0010; parameter no. 1 = 0.0013; and parameter no. 2 = 0.0010). The corresponding *median* Kullback–Leibler divergence/Hellinger distance statistics are KL/HD: state = 0.024/0.077; parameter no. 1 = 1.794/0.629; parameter no. 2 = 2.201/0.642, and measurement = 0.025/0.079. These metrics also indicate a

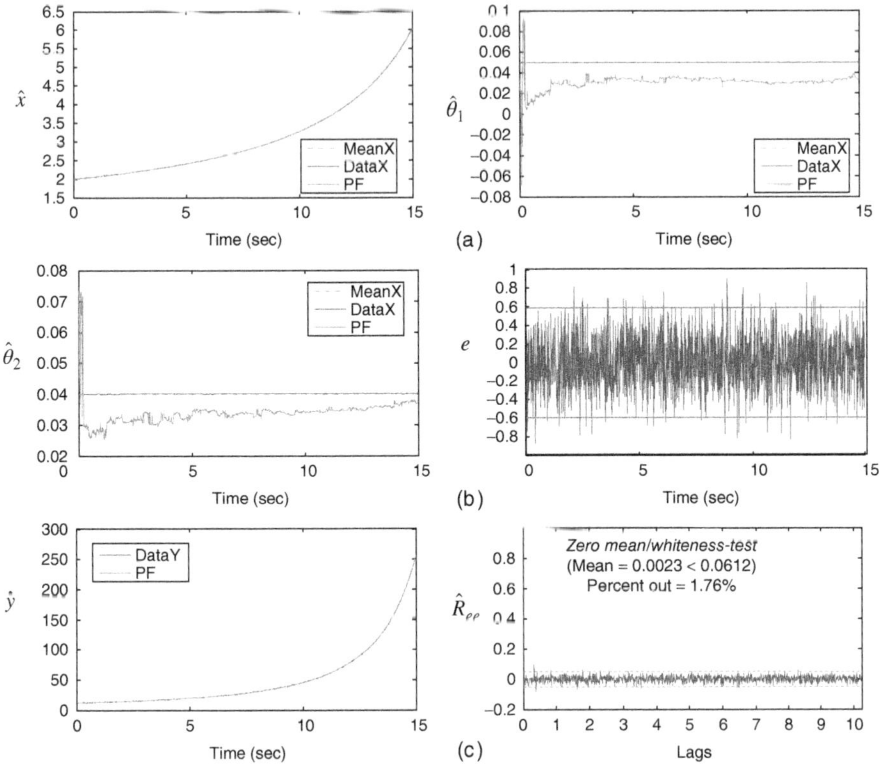

FIGURE 8.4 PF simulation: (a) estimated state and parameter no. 1; (b) estimated parameter no. 2 and innovation; (c) predicted measurement and zero-mean/whiteness test (0.002 < 0.061 and 1.76% out).

reasonable performance for both the state and measurement estimates and a mediocre performance at best for the parameter estimates. △△△

8.5 CASE STUDY: RANDOM TARGET TRACKING USING A SYNTHETIC APERTURE TOWED ARRAY

Synthetic aperture processing is well-known in airborne radar, but not as familiar in sonar [35–40]. The underlying idea in creating a synthetic aperture is to increase the array length by motion, thereby increasing spatial resolution (bearing) and gain in SNR. It has been shown that for stationary targets, the motion-induced bearing estimates have smaller variance than that of a stationary array [38, 41]. Here we investigate the case of *both* array and target motion. We define the acoustic *array space–time processing problem* as:

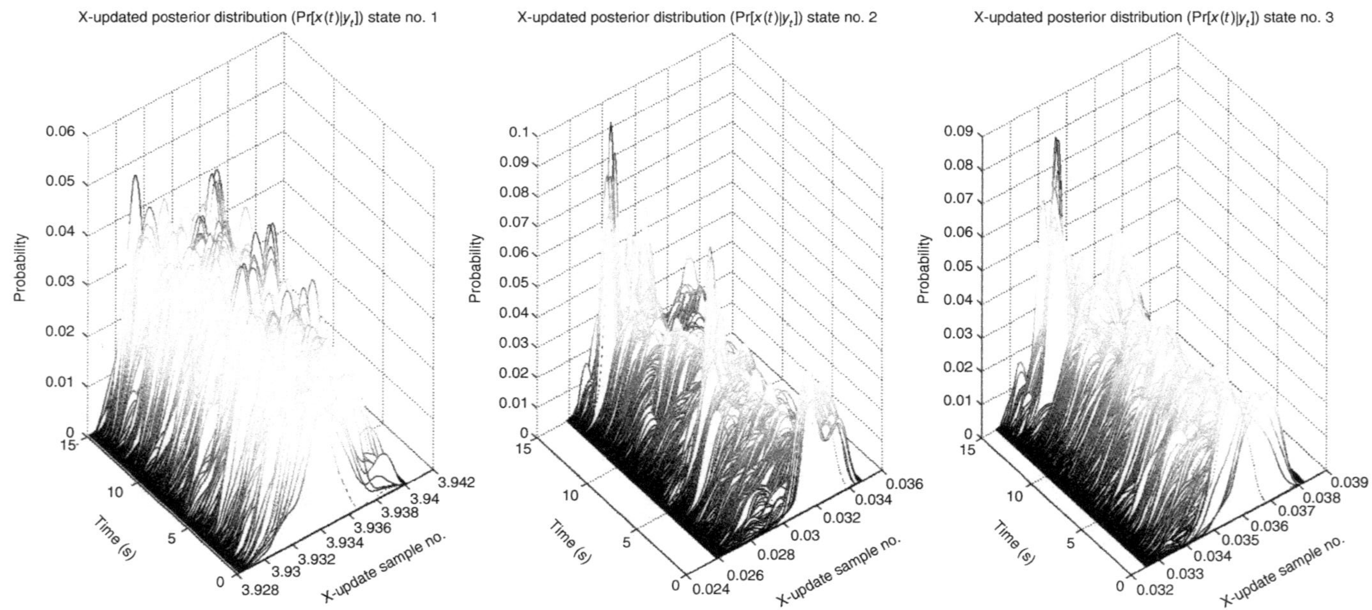

FIGURE 8.5 PF posterior distribution estimation: (a) estimated state posterior; (b) parameter no. 1 posterior; (c) parameter no. 2 posterior.

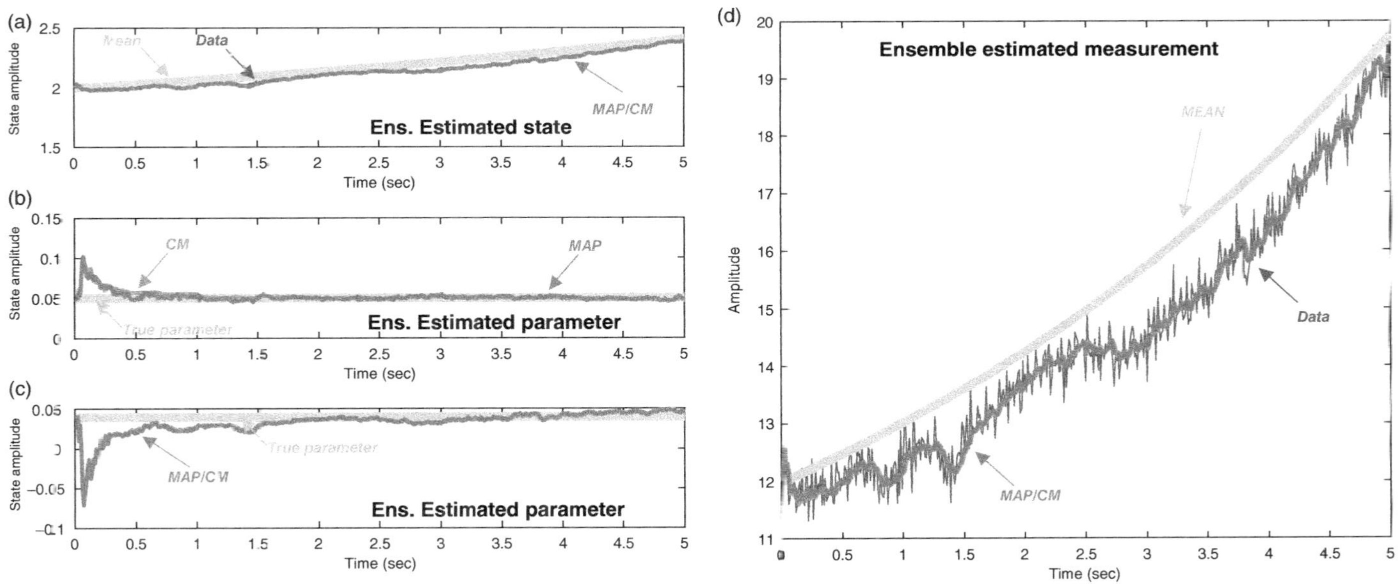

FIGURE 8.6 PF ensemble estimation for a 100-member realizations. (a) ensemble state estimate; (b) ensemble parameter no. 1 estimate; (c) ensemble parameter no. 2 estimate; (d) ensemble measurement estimate.

GIVEN a set of noisy pressure-field measurements from a horizontally towed array of L-sensors in motion, **FIND** the "best" (minimum error variance) estimate of the target bearings.

We use the following nonlinear pressure-field measurement model for M monochromatic plane wave targets characterized by a corresponding set of temporal frequencies, bearings, and amplitudes, $[\{\omega_m\}, \{\theta_m\}, \{a_m\}]$. That is,

$$p(x, t_k) = \sum_{m=1}^{M} a_m e^{j\omega_m t_k - j\beta(x,t_k)\sin\theta_m} + n(t_k) \tag{8.63}$$

where $\beta(x, t_k) := k_o x(t_o) + \upsilon t_k$, $k_o = \frac{2\pi}{\lambda_o}$ is the wave number, $x(t_k)$ is the current spatial position along the x-axis in meters, υ is the tow speed in m/sec, and $n(t_k)$ is the additive random noise. The inclusion of motion in the *generalized* wave number β is critical to the improvement of the processing, since the synthetic aperture effect is actually created through the motion itself and not simply the displacement.

If we further assume that the single sensor equation above is expanded to include an array of L-sensors, $x \rightarrow x_\ell$, $\ell = 1, \ldots, L$; then we obtain

$$\begin{aligned} p(x_\ell, t_k) &= \sum_{m=1}^{M} a_m e^{j\omega_m t_k - j\beta(x_\ell,t_k)\sin\theta_m} + n_\ell(t_k) \\ &\rightarrow \sum_{m=1}^{M} a_m \cos(\omega_m t_k - \beta(x_\ell, t_k)\sin\theta_m) + n_\ell(t_k) \end{aligned} \tag{8.64}$$

Since our hydrophone sensors measure the real part of the complex pressure-field, the final nonlinear measurement model of the system can be written in compact vector form as

$$\mathbf{p}(t_k) = \mathbf{c}[t_k; \mathbf{\Theta}] + \mathbf{n}(t_k) \tag{8.65}$$

where $\mathbf{p}, \mathbf{c}, \mathbf{n} \in C^{L\times 1}$, are the respective pressure-field, measurement, and noise vectors and $\mathbf{\Theta} \in \mathcal{R}^{M\times 1}$ represents the target bearings. The corresponding vector measurement model

$$\mathbf{c}_\ell(t_k; \mathbf{\Theta}) = \sum_{m=1}^{M} a_m \cos(\omega_m t_k - \beta(x_\ell, t_k)\sin\theta_m) \quad \text{for } \ell = 1, \ldots, L$$

Since we model the bearings as a random walk emulating random target motion, the Markovian state–space model evolves from first differences as

$$\mathbf{\Theta}(t_k) = \mathbf{\Theta}(t_{k-1}) + \mathbf{w}_\theta(t_{k-1}) \tag{8.66}$$

Thus, the state–space model is linear with no explicit dynamics; therefore, the process matrix $A = I$ (identity) and the relations are greatly simplified.

Now let us see how a particle filter using the bootstrap approach can be constructed according to the generic algorithm of Table 7.2. For this problem, we assume the additive noise sources are Gaussian, so we can compare results to the performance of the approximate processor. We define the discrete notation, $t_{k+1} \rightarrow t+1$ for the sampled-data representation.

Let us cast this problem into the sequential Bayesian framework, that is, we would like to estimate the *instantaneous* posterior filtering distribution, $\hat{\Pr}(x(t)|Y_t)$, using the PF representation to be able to perform inferences and extract the target bearing estimates. Therefore, we have that the transition probability is given by $(\Theta(t) \rightarrow x(t))$

$$\Pr(\theta(t)|\theta(t-1)) \longrightarrow \mathcal{A}(\theta(t)|\theta(t-1)) \sim \mathcal{N}(\Theta(t) : \mathbf{a}[\Theta(t-1)], \mathbf{R}_{w_\theta, w_\theta})$$

or in terms of our state transition (bearings) model, we have

$$\boldsymbol{\Theta}(t) = \mathbf{a}[\Theta(t-1)] + w_\theta(t-1) = \boldsymbol{\Theta}(t-1) + \mathbf{w}_\theta(t-1) \text{ for } \Pr(\mathbf{w}_\theta(t)) \sim \mathcal{N}(0, \mathbf{R}_{w_\theta w_\theta})$$

The corresponding likelihood is specified in terms of the measurement model $(y(t) \rightarrow p(t))$ as

$$\Pr(y(t)|x(t)) \longrightarrow C(y(t)|x(t)) \sim \mathcal{N}(\mathbf{y}(t) : \mathbf{c}[\Theta(t)], \mathbf{R}_{vv}(t))$$

where we have used the notation: $z \sim \mathcal{N}(z : m_z, R_{zz})$ to specify the Gaussian distribution in random vector z. In terms of our problem, we have

$$\begin{aligned} \ln C(y(t)|x(t)) = \kappa - \frac{1}{2}(\mathbf{y}(t) - \sum_{m=1}^{M} a_m \cos(\omega_m t - \beta(t)\sin\theta_m))' \mathbf{R}_{vv}^{-1} \\ \times (\mathbf{y}(t) - \sum_{m=1}^{M} a_m \cos(\omega_m t - \beta(t)\sin\theta_m)) \end{aligned}$$

with κ a constant, $\beta \in \mathcal{R}^{L\times 1}$ and $\beta(t) := [\beta(x_1,t)| \ldots |\beta(x_L,t)]'$, the dynamic wavenumber expanded over the array. Thus, the SIR algorithm becomes

- Draw samples (particles) from the state transition distribution:

$$\Theta_i(t) \sim \mathcal{N}(\Theta(t) : \mathbf{a}[\Theta(t-1)], \mathbf{R}_{w_\theta, w_\theta})$$

$$w_{\theta_i}(t) \sim \Pr(w(t)) \sim \mathcal{N}(0, R_{w_{\theta_i} w_{\theta_i}}), \Theta_i(t) = \Theta_i(t-1) + w_{\theta_i}(t-1)$$

- Estimate the likelihood: $C(\mathbf{y}(t)|\Theta(t)) \sim \mathcal{N}(\mathbf{y}(t) : \mathbf{c}[\Theta(t)], \mathbf{R}_{vv}(t))$ with $c_\ell(t; \Theta_i) = \sum_{m=1}^{M} a_m \cos(\omega_m t_k - \beta(x_\ell, t) \sin\Theta_{m,i}(t))$ for $\ell = 1, \ldots, L$ and $\Theta_{m,i}$ is the ith particle at the mth bearing angle;

- Update and normalize the weight: $\mathcal{W}_i(t) = W_i(t)/\sum_{i=1}^{N_p} W_i(t)$;
- Resample: $\hat{N}_{eff}(t) \leq N_{thresh}$;
- Estimate the instantaneous posterior: $\hat{\Pr}(\Theta(t)|Y_t) \approx \sum_{i=1}^{N_p} \mathcal{W}_i(t)\delta(\Theta(t) - \Theta_i(t))$;
- Perform the inference by estimating the corresponding statistics: $\hat{\Theta}_{\text{map}}(t) = \arg\max \hat{\Pr}(\Theta(t)|Y_t)$; $\hat{\Theta}_{\text{mmse}}(t) = \hat{\Theta}_{\text{cm}}(t) = E\{\Theta(t)|Y_t\} = \sum_{i=1}^{N_p} \Theta_i(t)\hat{\Pr}(\Theta(t)|Y_t)$; $\hat{\Theta}_{\text{median}}(t) = \text{median}(\hat{\Pr}(\Theta(t)|Y_t))$.

Consider the following simulation of the synthetic aperture using a four-element, linear-towed array with "moving" targets using the following parameters:

Target: unity amplitudes with temporal frequency is 50 Hz, wavelength = 30 m, tow speed = 5 m/sec; *array*: four (4) element linear towed array with 15 m spacing; *particle filter*: $N_\theta = 4$ states (bearings), $N_y = 4$ sensors, $N = 250$ samples, $N_p = 250$ weights; *SNR*: −10 dB; *noise*: white Gaussian with $R_{ww} = \text{diag}\ [2.5]$, $R_{vv} = \text{diag}\ [0.1414]$; *sampling interval*: 0.005 sec; *initial conditions*: (bearings and uncertainty) $\Theta_o = [45° -10° 5° -75°]'$, $P_o = \text{diag}\ (10^{-10})$.

The array simulation was executed and the targets moved according to a random walk specified by the process noise and sensor array measurements with −10 dB SNR. The results are shown in Fig. 8.7 where we see the noisy synthesized bearings (left) and four (4) noisy sensor measurements at the moving array. The bearing (state) estimates are shown in Fig. 8.8 where we observe the targets making a variety of course alterations. The PF (MAP) is able to track the target motions quite well while we observe the unscented Kalman filter (UKF) [10] unable to respond quickly enough and finally losing track completely for target no. 4. It should be noted that targets no. 2 and no. 4 "crossover" between 0.8 and 1.0 sec. The PF loses these tracks during this time period getting them confused but recovers by the 1-sec time step. Both the MAP and MMSE (CM) estimates using the estimated posterior provide excellent tracking. Note that these bearing inputs would provide the raw data for an *XY*-tracker [10]. The PF estimated or filtered measurements are shown in Fig. 8.9 . As expected, the PF tracks the measurement data quite well while the UKF is again in small error.

For performance analysis, we applied the usual "sanity tests". From this perspective, the PF processor works well, since each measurement channel is zero-mean and white with the WSSR lying below the threshold indicating white innovations sequences demonstrating the tracking ability of the PF processor at least in a classical sense [10] (see Fig. 8.10 and Table 8.2). The corresponding normalized root mean-squared errors RMSE, median Kullback–Leibler divergence/Hellinger distance statistics for the bearing estimates are also shown in the table along with the measurements at each sensor. Estimating the KD/HD metrics for the measurements indicate a reasonable performance for both the bearings and measurements.

The instantaneous posterior distributions for the bearing estimates are shown in Fig. 8.10. Here, we see the Gaussian nature of the bearing estimates generated by the random walk. Clearly, the PF performs quite well for this problem. Note also the capability of using the synthetic aperture, since we have only a four-element sensor

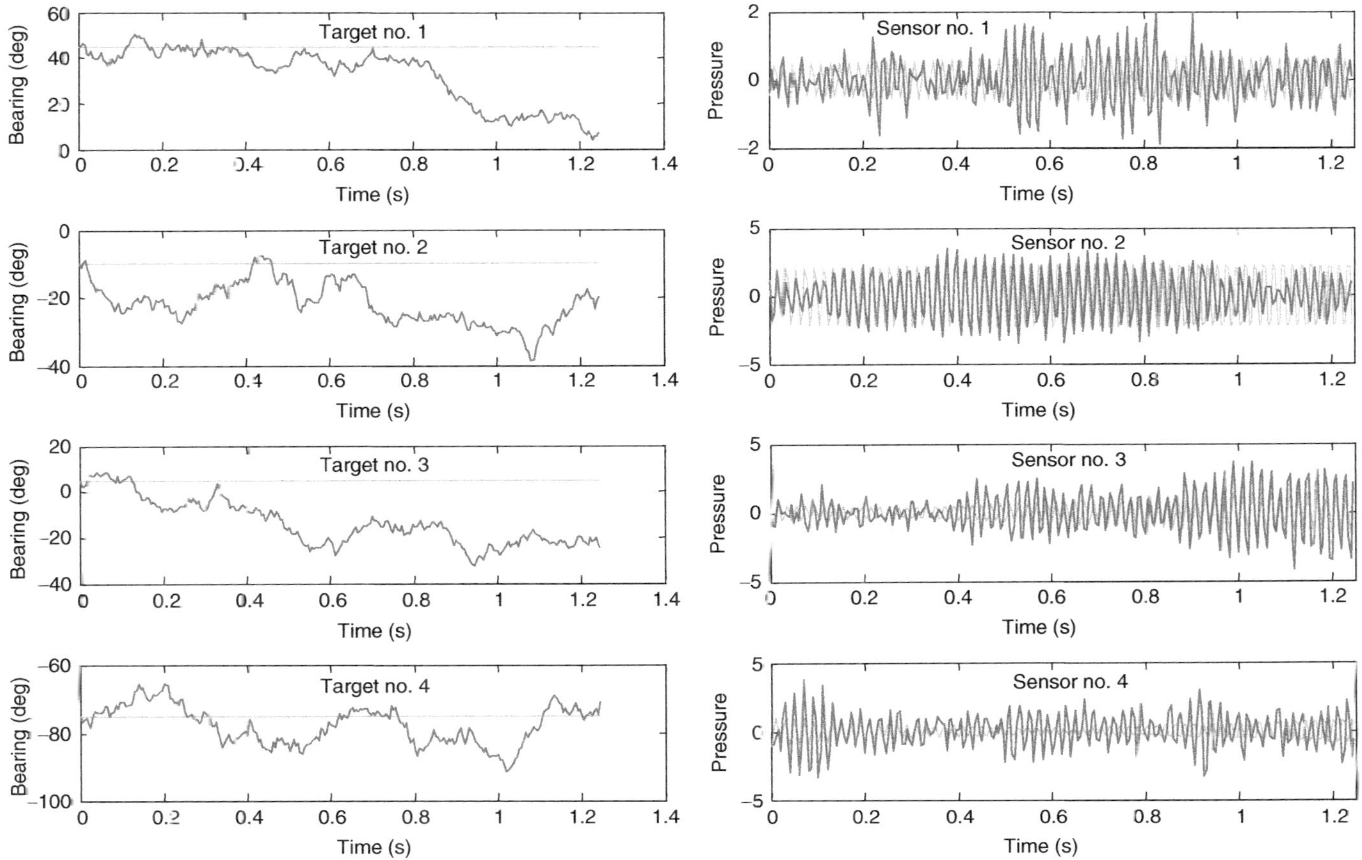

FIGURE 8.7 Synthetic aperture sonar tracking problem: simulated target motion from initial bearings of 45°, −10°, 5°, and −75°, and array measurements (−10 dB SNR).

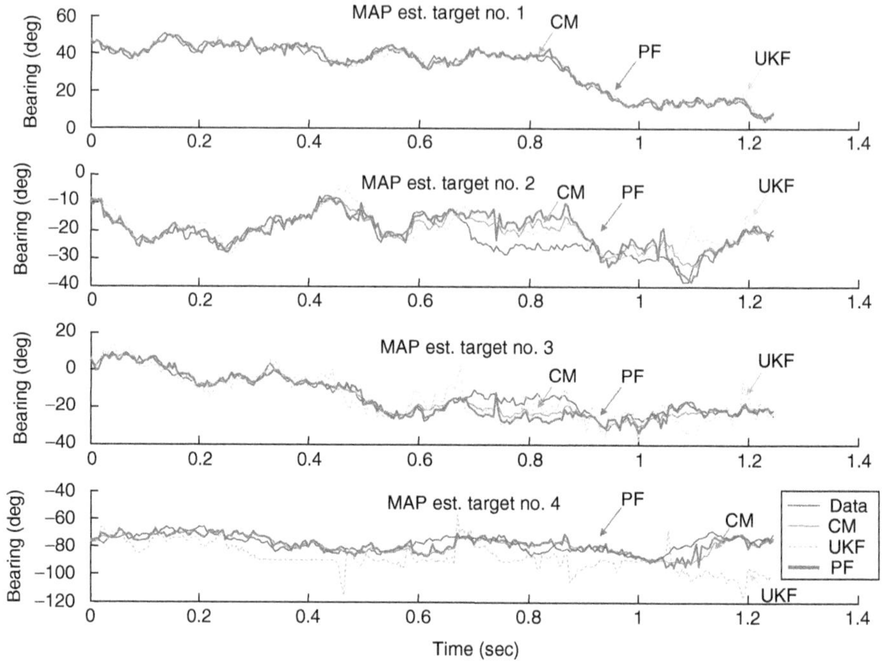

FIGURE 8.8 Particle filter bearing estimates for four targets in random motion: PF bearing (state) estimates and simulated target tracks (UKF, conditional mean, MAP).

TABLE 8.2 PF Performance Towed Array Problem

	Particle Filter Performance Results		
Parameter	RMSE	Median KLD	Median HD
Bearing No. 1	0.117	0.528	0.342
Bearing No. 2	0.292	0.180	0.208
Bearing No. 3	0.251	0.179	0.209
Bearing No. 4	0.684	0.406	0.309
Measurements		Median KLD	Median HD
Sensor No. 1		0.066	0.139
Sensor No. 2		0.109	0.166
Sensor No. 3		0.259	0.275
Sensor No. 4		0.312	0.327
Innovations	Zero-Mean < 0.173	Whiteness	WSSR < 455
Sensor No. 1	0.0102	5.47%	Below
Sensor No. 2	0.0136	1.56%	Below
Sensor No. 3	0.0057	5.47%	Below
Sensor No. 4	0.0132	2.34%	Below

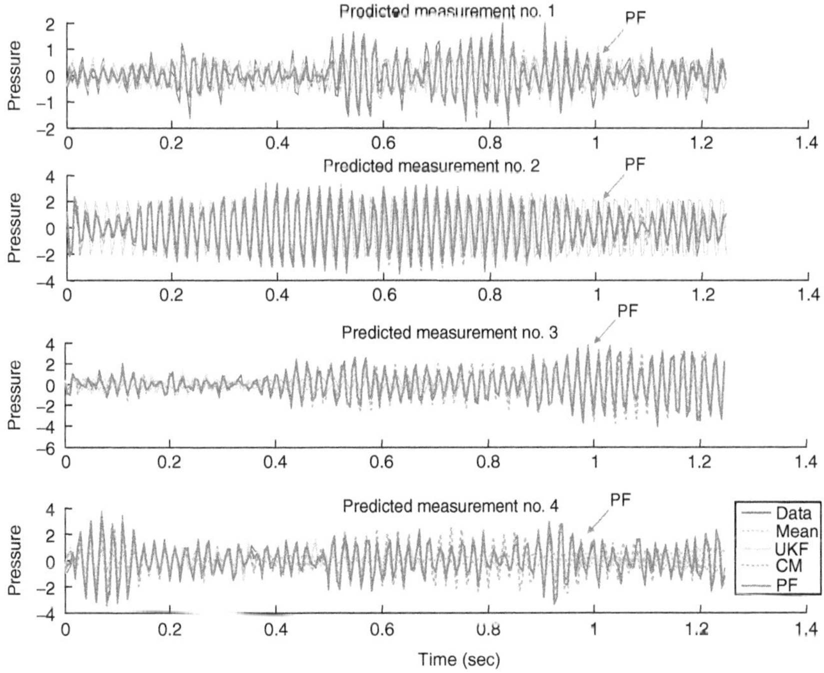

FIGURE 8.9 Particle filter predicted measurement estimates for four-channel hydrophone sensor array.

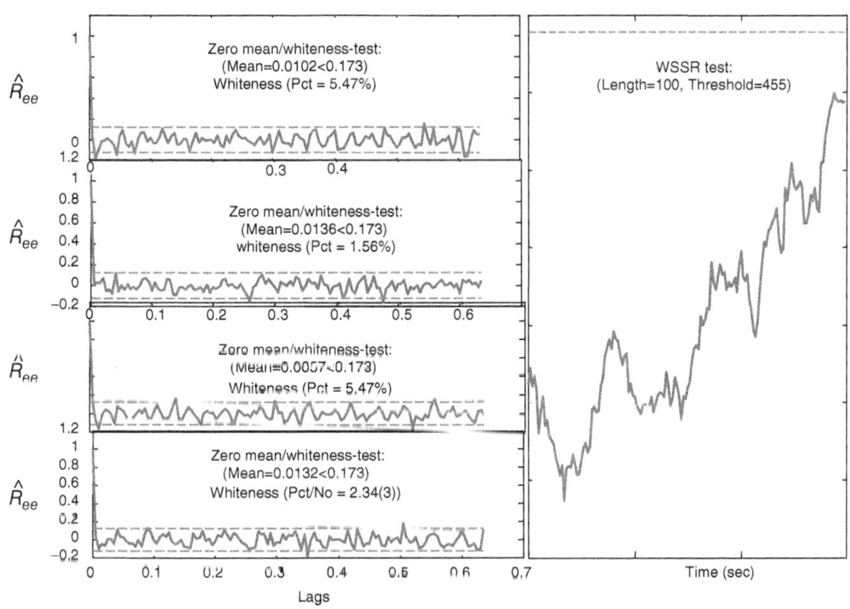

FIGURE 8.10 Particle filter classical performance metrics: zero-mean/whiteness tests for 45°, −10°, 5°, and 75° targets, as well as the corresponding WSSR test.

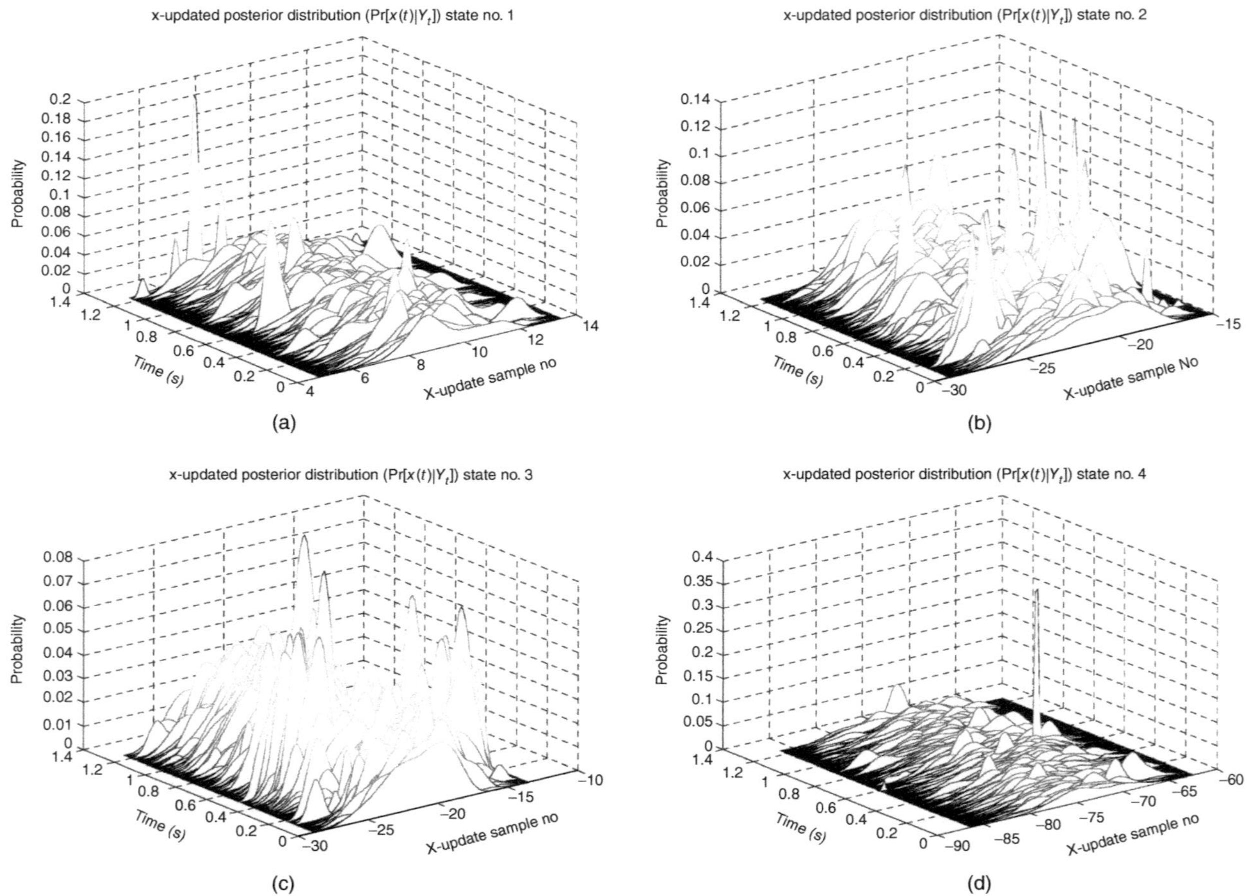

FIGURE 8.11 Particle filter instantaneous posterior bearing estimates: (a) 45° target no. 1. posterior; (b) −10° target no. 2 posterior; (c) 5° target no. 3; and (d) −75° target no. 4.

array, yet we are able to track 4 targets. Linear array theory implies with a static array that we should only be able to track $L - 1 = 3$ targets!

In this case study, we have applied the bootstrap PF to an ocean acoustic synthetic aperture towed array target tracking problem to test the performance of a the particle filtering technique. The results are quite reasonable on this simulated data set.

8.6 SUMMARY

In this chapter, we have discussed the development of joint Bayesian state/parametric processors. Starting with a brief introduction, we defined the variety of problems based on the joint posterior distribution $\Pr(x(t), \theta(t)|Y_t)$ and its decomposition. We decided to focus on the joint problem of estimating *both* states and parameters simultaneously (online)—the most common problem of highest interest. We then briefly showed that all that is necessary for this problem is to define an "augmented" state consisting of the original state variables along with the unknown parameters typically modeled by a random walk when a dynamic parametric model is not available. This casts the joint problem into an optimal filtering framework. We then showed how this augmentation leads to a decomposition of the classical (EKF) processor and developed the "decomposed" form for illustrative purposes. The algorithm is implemented by executing the usual processor with the new augmented state vector. We also extended this approach to both the modern "unscented" and "particle-based" processors, again only requiring the state augmentation procedure to implement. It was shown that all of the processors required a random walk parametric model to function, while the particle filters could be implemented using the "roughening" (particle random walks) or any of the "move" techniques developed in Chapter 7 to track the parameters effectively. Besides applying these processors to the usual nonlinear trajectory estimation problem, we developed a case study for a synthetic aperture towed array and compared the modern to the particle-based processors.

MATLAB NOTES

SSPACK_PC is a third party toolbox in MATLAB that can be used to design model-based signal processors. This package incorporates the major *nonlinear MBP* algorithms discussed in this chapter—all implemented in the UD-factorized form [10] for stable and efficient calculations. It performs the discrete approximate Gauss–Markov simulations using (**SSNSIM**) and both extended (XMBP) and iterated-extended (IX-MBP) processors using (**SSNEST**). The linearized model-based processor (LZ-MBP) is also implemented (**SSLZEST**). Ensemble operations are seamlessly embodied within the GUI-driven framework where it is quite efficient to perform multiple design runs and compare results. Of course, the heart of the package is the command or GUI driven post-processor (**SSPOST**) which is used to analyze and display the results of the simulations and processing (see http://www.techni-soft.net for more details). REBEL is a recursive Bayesian estimation package in MATLAB available on the web, that performs similar operations including the new statistical-based unscented algorithms including the UKF including the unscented transformations. It also has included the new particle filter designs.

REFERENCES

1. L. Ljung, *System Identification: Theory for the User* (Englewood Cliffs, NJ: Prentice-Hall, 1987).
2. L. Ljung and T. Soderstrom, *Theory and Practice of Recursive Identification* (Boston, MA: MIT Press, 1983).
3. T. Soderstrom and P. Stoica, *System Identification* (Englewood Cliffs, NJ: Prentice-Hall, 1989).
4. J. Norton, *An Introduction to Identification* (New York: Academic Press, 1986).
5. J. Liu, *Monte Carlo Strategies in Scientific Computing*, (New York: Springer-Verlag, 2001).
6. O. Cappe, E. Moulines, and T. Ryden, *Inference in Hidden Markov Models*, (New York: Springer-Verlag, 2005).
7. J. Liu and M. West, "Combined parameter and state estimation in simulation-based filtering," in *Sequential Monte Carlo Methods in Practice*, edited by A. Doucet, N. de Freitas, and N. Gordon (New York: Springer-Verlag, 2001), pp. 197–223.
8. A. Doucet, N. de Freitas, and N. Gordon, *Sequential Monte Carlo Methods in Practice* (New York: Springer-Verlag, 2001).
9. S. Godsill and P. Djuric, "Monte Carlo methods for statistical signal processing," *IEEE Trans. Signal Proc.*, **50**, 173–499, 2002 (Special Issue).
10. J. Candy, *Model-Based Signal Processing*. (Hoboken, NJ: John Wiley & Sons, Inc./IEEE Press, 2006).
11. O. Cappe, S. Godsill, and E. Moulines, "An overview of existing methods and recent advances in sequential Monte Carlo," *Proc. IEEE*, **95**, 5, 899–924, 2007.
12. G. Kitagawa and W. Gersch, *Smoothness Priors Analysis of Time Series* (New York: Springer-Verlag, 1997).
13. G. Kitagawa, "Self-organizing state-space model," *J. Am. Statistical Assoc.*, **97**, 447, 1207–1215, 1998.
14. R. van der Merwe, A. Doucet, N. de Freitas, and E. Wan, "The unscented particle filter," in *Advances in Neural Information Processing Systems 16* (Cambridge, MA: MIT Press, 2000).
15. N. Gordon, D. Salmond, and A. Smith, "A novel approach to nonlinear non-gaussian Bayesian state estimation," *IEE Proc. F.*, **140**, 107–113, 1993.
16. S. Haykin and N. de Freitas, "Special issue: sequential state estimation: from Kalman filters to particle filters," *Proc. IEEE*, **92**, 3, 399–574, 2004.
17. C. Andrieu, A. Doucet, S. Singh, and V. Tadic, "Particle methods for change detection, system identification and control," *Proc. IEEE*, **92**, 6, 423–468, 2004.
18. S. Haykin, *Kalman Filtering and Neural Networks*. (New York: John Wiley & Sons, Inc., 2001).

19. D. Simon, *Optimal State Estimation: Kalman H_∞ and Nonlinear Approaches* (Hoboken, NJ: John Wiley & Sons/IEEE Press, 2006).
20. A. Jazwinski, *Stochastic Processes and Filtering Theory*. (New York: Academic Press, 1970).
21. R. van der Merwe, "Sigma-point Kalman filters for probabilistic inference in dynamic state-space models," Ph.D. Dissertation, OGI School of Science & Engineering, Oregon Health & Science University, 2004.
22. A. Nelson, "Nonlinear estimation and modeling of noisy time-series by dual Kalman filtering methods," Ph.D. Dissertation, OGI School of Science & Engineering, Oregon Health & Science University, 2000.
23. L. Ljung, "Asymptotic behavior of the extended Kalman filter as a parameter estimator for linear systems," *IEEE Trans. Auto. Control*, **AC-24**, 36–50, 1979.
24. J. Rajan, P. Rayner, and S. Godsill, "Bayesian approach to parameter estimation and interpolation of time-varying autoregressive processes using the Gibbs sampler," *IEE Proc-Vis. Image Signal Process.*, **144**, 4, 249–256, 1997.
25. N. Polson, J. Stroud, and P. Muller, "Practical filtering with sequential parameter learning," University of Chicago Technical Report, 1–18, 2002.
26. C. Andrieu, A. Doucet, S. Singh, and V. Tadic, "Particle methods for change detection, system identification and control," *Proc. IEEE*, **92**, 3, 423–438, 2004.
27. G. Storvik, "Particle filters in state space models with the presence of unknown static parameters," *IEEE Tran. Signal Proc.*, **50**, 2, 281–289, 2002.
28. P. Djuric, "Sequential estimation of signals under model uncertainty," in *Sequential Monte Carlo Methods in Practice*, edited by A. Doucet, N. de Freitas, and N. Gordon (New York: Springer-Verlag, 2001), pp. 381–400.
29. D. Lee and N. Chia, "A particle algorithm for sequential Bayesian parameter estimation and model selection," *IEEE Tran. Signal Proc.*, **50**, 2, 326–336, 2002.
30. A. Doucet and V. Tadic, "Parameter estimation in general state-space models using particle methods," *Ann. Inst. Stat. Math.*, **55**, 2, 409–422, 2003.
31. J. Candy, "Bootstrap particle filtering," *IEEE Signal Proc. Magz.*, **24**, 4, 73–85, 2007.
32. C. Andrieu, A. Doucet, and V. Tadic, "On-line parameter estimation in general state-space models," *Proc. IEEE Conf. Decision and Control*, pp. 332–337, 2005.
33. J. Vermaak, C. Andrieu, A. Doucet, and S. Godsill, "Particle methods for Bayesian modeling and enhancement of speech signals," *IEEE Trans. Speech Audio Proc.*, **10**, 3, 173–185, 2002.
34. T. Schoen and F. Gustafsson, "Particle filters for system identification of state-space models linear in either parameters or states," Linkoping University Report, LITH-ISY-R-2518, 2003.
35. R. Williams, "Creating an acoustic synthetic aperture in the ocean," *J. Acoust. Soc. Am.*, **60**, 60–73, 1976.
36. N. Yen and W. Carey, "Applications of synthetic aperture processing to towed array data," *J. Acoust. Soc. Am.*, **60**, 764–775, 1976.
37. S. Stergiopoulus and E. Sullivan, "Extended towed array processing by an overlap correlator," *J. Acoust. Soc. Am.*, **86**, 764–775, 1976.
38. E. Sullivan, W. Carey, and S. Stergiopoulus, "Editorial in special issue on acoustic synthetic aperture processing," *IEEE J. Ocean. Eng.*, **17**, 1–7, 1992.

39. D. Ward, E. Lehmann, and R. Williamson, "Particle filtering algorithm for tracking and acoustic source in a reverberant environment," *IEEE Trans. Speech and Aud. Proc.*, **11**, 6, 826–836, 2003.

40. M. Orton and W. Fitzgerald, "Bayesian approach to tracking multiple targets using sensor arrays and particle filters," *IEEE Trans. Signal Proc.*, **50**, 2, 216–223, 2002.

41. E. Sullivan and J. Candy, "Space-time array processing: the model-based approach," *J. Acoust. Soc. Am.*, **102**, 5, 2809–2820, 1997.

PROBLEMS

8.1 Suppose we are given the following innovations model (in steady state)

$$\hat{x}(t) = a\hat{x}(t-1) + ke(t-1)$$
$$y(t) = c\hat{x}(t) + e(t)$$

where $e(t)$ is the zero-mean, white innovations sequence with covariance. R_{ee}.

(*a*) Derive the Wiener solution using the spectral factorization method of Section 4.5.

(*b*) Develop the linear steady-state BP for this model.

(*c*) Develop the parametrically adaptive processor to estimate k and R_{ee}.

8.2 As stated in the chapter, the XBP convergence can be improved by incorporating a gain gradient term in the system Jacobian matrices, that is,

$$A_\theta^*[x,\theta] := A_\theta[x,\theta] + [\nabla_\theta K_x(\Theta)]e(t) \quad \text{for} \mathcal{K} := [K_x | K_\theta]$$

(*a*) By partitioning the original $N_x \times N_\theta$ Jacobian matrix, $A_\theta[x,\theta]$, derive the general "elemental" recursion, that is, show that

$$A_\theta^*[i,\ell] = \nabla_{\theta_\ell} a_i[x,\theta] + \sum_{j=1}^{N_y} \nabla_{\theta_\ell} k_x(i,j) e_j(t); i = 1,\ldots,N_x; \ell = 1,\ldots,N_\theta$$

(*b*) Suppose we would like to implement this modification, does there exist a numerical solution that could be used? If so, describe it.

8.3 Using the following *scalar* Gauss–Markov model

$$x(t) = Ax(t-1) + w(t-1)$$
$$y(t) = C\hat{x}(t) + v(t)$$

with the usual zero-mean, R_{ww} and R_{vv} covariances.

(*a*) Let $\{A, C, K, R_{ee}\}$ be scalars, develop the ABSP solution to estimate A from noisy data.

(*b*) Can these algorithms be combined to "tune" the resulting hybrid processor?

8.4 Suppose we are given the following structural model

$$m\ddot{x}(t) + c\dot{x} + kx(t) = p(t) + w(t)$$
$$y(t) = \beta x(t) + v(t)$$

with the usual zero-mean, R_{ww} and R_{vv} covariances.

(*a*) Convert this model to discrete-time using first differences. Using central difference create the discrete Gauss–Markov model. (*Hint*: $\ddot{x}(t) \approx \frac{x(t)-2x(t-1)+x(t-2)}{\Delta_t^2}$).

(*b*) Suppose we would like to estimate the spring constant k from noisy displacement measurements, develop the ABSP to solve this problem.

(*c*) Transform the discrete Gauss–Markov model to the innovations representation. (*Hint*: Use the KSP equations of Ref. [10]).

(*d*) Solve the parameter estimation problem using the innovations model, that is, develop the estimator of the spring constant.

8.5 Given the ARMAX model

$$y(t) = -ay(t-1) + bu(t-1) + e(t)$$

with innovations covariance R_{ee}:

(*a*) Write the expressions for the ABSP in terms of the ARMAX model.

(*b*) Write the expressions for the ABSP in terms of the state–space model.

8.6 Consider tracking a body falling freely through the atmosphere [10]. We assume it is falling down in a straight line towards a radar. The state vector is defined by $x := [z \quad \dot{z} \quad \beta]$, where $\beta \sim \mathcal{N}(\mu_\beta, R_{\beta\beta}) = \mathcal{N}(2000, 2.5 \times 10^5)$ is the ballistic coefficient. The dynamics are defined by the state equations

$$\dot{x}_1(t) = x_2(t)$$
$$\dot{x}_2(t) = \frac{\rho x_2^2(t)}{2x_3(t)} - g$$
$$\dot{x}_3(t) = 0$$
$$\rho = \rho_o e^{-\frac{x_1(t)}{k_\rho}}$$

where d is the drag deceleration, g is the acceleration due to gravity (32.2), ρ is the atmospheric density (with ρ_o (3.4×10^{-3}) density at sea level) and k_ρ a decay constant (2.2×10^4). The corresponding measurement is given by

$$y(t) = x_1(t) + v(t)$$

for $v \sim \mathcal{N}(0, R_{vv}) = \mathcal{N}(0, 100)$. Initial values are $x(0) = \mu \sim \mathcal{N}(1065, 500)$, $\dot{x}(0) \sim \mathcal{N}(-6000, 2 \times 10^4)$, and $P(0) = \text{diag}[p_o(1,1), p_o(2,2), p_o(3,3)] = [500, 2 \times 10^4, 2.5 \times 10^5]$.

(*a*) Is this an ABSP If so, write out the explicit algorithm in terms of the parametrically adaptive algorithm of this chapter.

(*b*) Develop the XBP for this problem and perform the discrete simulation using MATLAB.

(*c*) Develop the LZ-BP for this problem and perform the discrete simulation using MATLAB.

(*d*) Develop the PF for this problem and perform the discrete simulation using MATLAB.

8.7 Parameter estimation can be performed directly when we are given a nonlinear measurement system such that

$$\mathbf{y} = \mathbf{h}(\theta) + \mathbf{v}$$

where $\mathbf{y}, \mathbf{h} \in \mathcal{R}^{N_y \times 1}$, $\theta \sim \mathcal{N}(\mathbf{m}_\theta, R_{\theta\theta})$, and $\mathbf{v} \sim \mathcal{N}(0, R_{vv})$.

(*a*) From the a posteriori density, $\Pr(\theta|\mathbf{y})$ derive the MAP estimator for θ.

(*b*) Expand $\mathbf{y} = \mathbf{h}(\theta)$ in a Taylor series about θ_o and incorporate the first order approximation into the MAP estimator (approximate).

(*c*) Expand $\mathbf{y} = \mathbf{h}(\theta)$ in a Taylor series about θ_o and incorporate the second order approximation into the MAP estimator (approximate).

(*d*) Develop and iterated version of both estimators in (*b*) and (*c*). How do they compare?

(*e*) Use the parametrically adaptive formulation of this problem assuming the measurement model is time-varying. Construct the ABSP assuming that θ is modeled by a random walk. How does this processor compare to the iterated versions?

8.8 Suppose we are asked to solve a detection problem, that is we must "decide" whether a signal is present or not according to the following binary hypothesis test

$$\begin{aligned} \mathcal{H}_o &: y(t) = v(t) \quad \text{for } v \sim \mathcal{N}(0, R_{vv}) \\ \mathcal{H}_1 &: y(t) = s(t) + v(t) \end{aligned}$$

The signal is modeled by a Gauss–Markov model

$$s(t) = a[s(t-1)] + w(t-1) \quad \text{for } w \sim \mathcal{N}(0, R_{ww})$$

(*a*) Calculate the *likelihood-ratio* defined by

$$\mathcal{L}(Y(N)) := \frac{\Pr(Y(N)|\mathcal{H}_1)}{\Pr(Y(N)|\mathcal{H}_o)}$$

where the measurement data set is defined by $Y(N); = \{y(0), y(1), \ldots, y(N)\}$. Calculate the corresponding threshold and construct the detector (binary hypothesis test).

(*b*) Suppose there is an unknown but deterministic parameter in the signal model, that is,

$$s(t) = a[s(t-1); \theta(t-1)] + w(t-1)$$

Construct the "composite" likelihood ratio for this case. Calculate the corresponding threshold and construct the detector (binary hypothesis test). (*Hint*: Use the ABSP to jointly estimate the signal and parameter.)

(*c*) Calculate a sequential form of the likelihood ratio above by letting the batch of measurements, $N \to t$. Calculate the corresponding threshold and construct the detector (binary hypothesis test). Note there are two thresholds for this type of detector.

8.9 Angle modulated communications including both frequency modulation (FM) and phase modulation (PM) are basically nonlinear systems from the model-based perspective. They are characterized by high bandwidth requirements and their performance is outstanding in noisy environments. Both can be captured by the *transmitted* measurement model

$$s(t) = \sqrt{2P}\sin[\omega_c t + k_p m(t)] \quad \text{(PM)}$$

or

$$s(t) = \sqrt{2P}\sin[\omega_c t + 2\pi k_f \int_{-\infty}^{t} m(\tau)d\tau] \quad \text{(FM)}$$

where P is a constant, ω_c is the carrier frequency, k_p and k_f are the deviation constants for the respective modulation systems and of course, $m(t)$, is the message model. *Demodulation* to extract the message from the transmission is accomplished by estimating the phase of $s(t)$. For FM, the recovered phase is differentiated and scaled to extract the message, while PM only requires the scaling.

Suppose the message signal is given by the Gauss–Markov representation

$$m(t) = -\alpha m(t-1) + w(t-1)$$
$$y(t) = s(t) + v(t)$$

with both w and v zero-mean, Gaussian with variances, R_{ww} and R_{vv}.

(*a*) Construct a receiver for the PM system using the XBP design.

(*b*) Construct an equivalent receiver for the FM system.

(*c*) Assume that the message amplitude parameter α is unknown, construct the ABSP receiver for the PM system to jointly estimate the message and parameter.

(*d*) Under the same assumptions as (*c*), construct the ABSP receiver for the FM system to jointly estimate the message and parameter.

(*e*) Compare the receivers for both systems. What are their similarities and differences?

8.10 We are given the population model of the Chapter 7 case study and would like to "parameterize" it for adaptive processing, since we know the parameters are not very well known. The state transition and corresponding measurement model are given by

$$x(t) = \frac{1}{2}x(t-1) + \frac{25x(t-1)}{1+x^2(t-1)} + 8\cos(1.2(t-1)) + w(t-1)$$
$$y(t) = \frac{x^2(t)}{20} + v(t)$$

where $\triangle t = 1.0$, $w \sim \mathcal{N}(0, 10)$, and $v \sim \mathcal{N}(0, 1)$. The initial state is Gaussian distributed with $\overline{x}(0) \sim \mathcal{N}(0.1, 5)$.

In terms of the nonlinear state–space representation, we have

$$a[x(t-1)] = \frac{1}{2}x(t-1) + \left(\frac{25x(t-1)}{1+x^2(t-1)}\right)$$
$$b[u(t-1)] = 8\cos(1.2(t-1))$$
$$c[x(t)] = \frac{x^2(t)}{20}$$

(*a*) Choose the model constants: 25, 8, 0.5, and $\frac{1}{20}$ as the unknown parameters, reformulate the state estimation problem as a parameter estimation problem with unknown parameter vector, Θ and a random walk model with corresponding process noise variance, $R_{ww} = diag[1 \times 10^{-6}]$.

(*b*) Develop the joint SPBP algorithm to solve this problem. Run the SPBP algorithm and discuss the performance results.

(*c*) Develop the joint PF algorithm to solve this problem. Run the PF algorithm and discuss the performance results.

(*d*) Choose to "move" the particles using the roughening approach, how do these results compare to the standard bootstrap algorithm?

(*e*) Develop the joint linearized (UKF) PF algorithm to solve this problem. Run this PF algorithm and discuss the performance results.

9

DISCRETE HIDDEN MARKOV MODEL BAYESIAN PROCESSORS

9.1 INTRODUCTION

In this chapter we develop discrete (event) hidden Markov models. All of the Bayesian processors we have discussed are, in fact, hidden Markov processors, since the internal states are usually not measured directly and are therefore "hidden" by definition, but the distinguishing factor is the type of underlying process governing the sequence. In fact, the (state) transition matrix is a "probability" matrix with specific properties that distinguish it uniquely from other dynamic systems. These discrete representations of stochastic processes find enormous application in speech, economics, biomedical, communications and music areas where coding approaches are prevalent. We discuss the development of the basic processor and investigate a case study in communications to demonstrate the design and application.

9.2 HIDDEN MARKOV MODELS

A discrete-time hidden Markov model (HMM) is a stochastic representation (model) of a process that can be used for simulation, modeling, and estimation much the same as the state space model is used for dynamic (physical) systems. These models are prevalent in acoustics, biosciences, climatology, control, communications, econometrics, text recognition, image processing, signal processing, and speech processing [1]. Perhaps its distinguishing feature is that it is a "probabilistic model" in the sense

Bayesian Signal Processing: Classical, Modern, and Particle Filtering Methods, Second Edition. James V. Candy.

that it is driven by internal probability distributions for both states and observations or equivalently measurements. Here the state-transition matrix prevalent in linear systems theory is still valid and is also called a transition matrix, but it is a discrete probability matrix with rows summing to unity and in some cases (doubly stochastic) columns summing to unity as well. The underlying structure from which the HMM evolves is the Markov chain of Chapter 3 along with the sequential Bayesian recursions of Chapter 2. We start with the idea of a Markov chain and its decomposition basics leading to HMM.

9.2.1 Discrete-Time Markov Chains

A discrete-time Markov chain is characterized by a state variable that changes at certain time instances [2–4]. At each time-step t the state is defined by $x(t) \in \mathcal{X}$ (state–space) and $\mathcal{X} = \{\mathcal{X}_1, \ldots, \mathcal{X}_{N_x}\}$. The probability that at time t the chain occupies state i is defined by $\Pr(x_i(t))$. The dynamics of the Markov chain are represented by its *transition probability* $a_{mn}(t-1,t) := \Pr(x_n(t)|x_m(t-1))$ where $x_i(t) := \{x(t) = \mathcal{X}_i\}$. This expression means that the probability that the state at time t is $\mathcal{X}_n$ given that it is currently in state $\mathcal{X}_m$ at time $t-1$ for $(\mathcal{X}_m, \mathcal{X}_n) \in \mathcal{X}$. Here the key Markovian assumption is that the transition probability a_{mn} applies *whenever* state $\mathcal{X}_m$ is "visited" independent of the "past" and the *path* or previous states employed to reach $\mathcal{X}_m$. This is merely a statement of the Markovian property that

$$a_{mn}(t-1,t) = \Pr(x_n(t)|x_m(t-1), \ldots, x_\ell(0)) = \Pr(x_n(t)|x_m(t-1)) \quad \forall \ t \text{ and } (\mathcal{X}_m, \mathcal{X}_n) \in \mathcal{X} \tag{9.1}$$

Further, if the chain is *homogeneous in time*, then $a_{mn}(t-1,t)$ depends only on the time difference (in general) and therefore the transition probability is *stationary* such that $a_{mn}(t-1,t) \to a_{mn}$ with $a_{mn} \geq 0$ and $\sum_{n=1}^{N_x} a_{mn} = 1$ [2].

Summarizing, a *discrete-time Markov chain* is characterized by:

- a finite set of known N_x-states: $\mathcal{X} = \{\mathcal{X}_1, \ldots, \mathcal{X}_{N_x}\}$;
- a nonnegative set of *state*-transition probabilities for $(\mathcal{X}_m, \mathcal{X}_n) \to \{a_{mn}\}$; and
- a sequence of random variables: $x_m(0), x_n(1), \ldots \in \mathcal{X}$ that satisfy the Markovian property

$$a_{mn}(t-1,t) = \Pr(x_n(t)|x_m(t-1)) \, \forall t \text{ and all states } (\mathcal{X}_m, \mathcal{X}_n) \in \mathcal{X}$$

The elements of the homogeneous chain are embedded in the $N_x \times N_x$ state-transition probability matrix $A = [a_{mn}]$; $m, n = 1, \ldots, N_x$. The chain can be specified pictorially by a *directed graph* with *nodes* representing states and *arcs* or *arrows* corresponding to the transition probabilities as illustrated in Fig. 9.1.

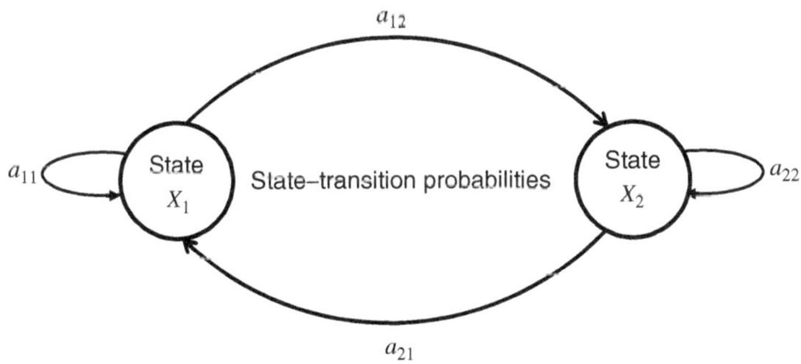

FIGURE 9.1 Directed graph representation of two-state Markov chain.

Example 9.1

Suppose we are given a two-state ($N_x = 2$) Markov chain with transition probability

$$a_{mn} = \Pr(x_n(t)|x_m(t-1)); \quad m, n = 1, 2$$

and $a_{mn} = \{0.55, 0.45, 0.25, 0.75\}$. Construct the state-transition probability matrix A and the corresponding directed graph. The transition probability is given by

$$A = \begin{bmatrix} 0.55 & 0.45 \\ 0.25 & 0.75 \end{bmatrix} \tag{9.2}$$

The resulting graph is shown in Fig. 9.1. △△△

9.2.2 Hidden Markov Chains

An HMM is simply a Markov chain in which all of the states are *not* observed—some are hidden. In this case we introduce the observation or measurement or output process where only a *subset* of the states are observed directly. Thus, the essential difference between a Markov chain and an HMM is that for an HMM there is one-to-one correspondence between the states and observations (output measurements). It is not possible to tell which state the model was in by merely observing the outputs of the chain. We illustrate the structural model in Fig. 9.2a. Note that when the states are directly observed in *a*, then the observations and states are identical.

Thus, we differentiate the hidden Markov chain or HMM from the Markov chain by introducing an *observation* or *measurement* process [5–8]. Here the state sequence is *not* known, that is, it is *hidden* in the measurement sequence. Thus, at every time-step t, the system generates a measurement or observation $y(t)$ according to a probability distribution that depends on the state $x(t)$. The number of observations N_y corresponds to a known distinct set, that is, at time t the observation is $y(t) \in \mathcal{Y}$

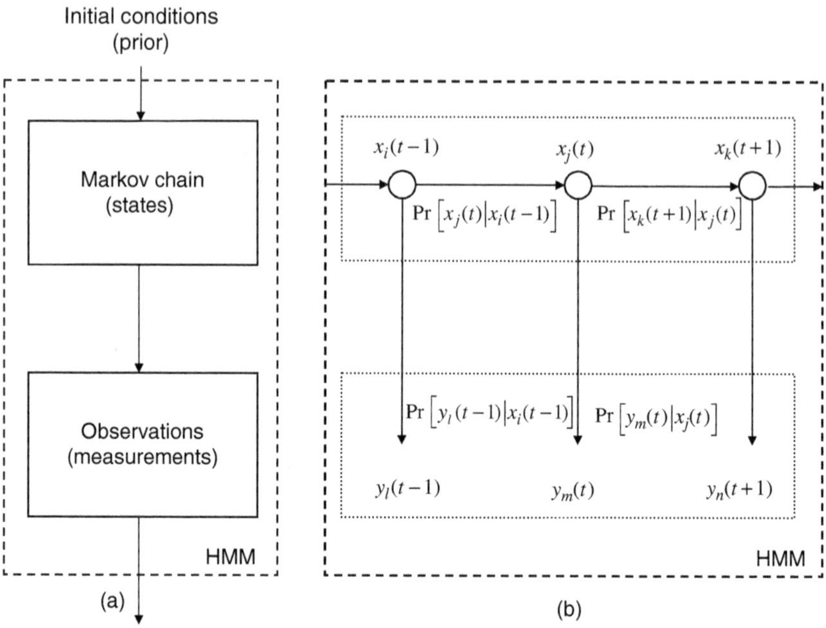

FIGURE 9.2 Hidden Markov model structure: (a) Markov chain (states) and observations (measurements); (b) Markov chain with state-transition probabilities and observations with measurement probabilities (likelihoods).

(observation space) with $\mathcal{Y} = \{\mathcal{Y}_1, \ldots, \mathcal{Y}_{N_y}\}$. We define the corresponding "discrete" *observation probability* (likelihood) by[1]

$$c_{k\ell}(t,t) := \Pr(y_\ell(t)|x_k(t)) \tag{9.3}$$

Again for the homogeneous case, $c_{k\ell}(t,t) \to c_{k\ell}$ and $c_{k\ell} \geq 0$ and $\sum_{\ell=1}^{N_y} c_{k\ell} = 1$. As in the chain, we have the associated $N_y \times N_x$ *observation probability* matrix given by $C = [c_{k\ell}]$ for $k = 1, \ldots, N_x$; $\ell = 1, \ldots, N_y$ with $C \in R^{N_y \times N_x}$. The final ingredient to characterize the HMM is the prior or initial probability distribution given by $\Pr(x(0) = \mathcal{X}_i(0))$; $i = 1, \ldots, N_x$ which represents the initial probability of the chain.

Summarizing an HMM is specified by the model $\Sigma := \{A, C, \Pr(x_i(0))\}$ (homogeneous case) where

- The *state-transition probability* matrix is

$$A = [a_{mn}] = \Pr(x_n(t)|x_m(t-1)); \quad m, n = 1, \ldots, N_x;$$

[1] This probability expression has two subscripts to annotate the discrete state $(x_k(t))$ and the discrete measurement or observation $(y_\ell(t))$. Most references assume a continuous observation and use the notation $c_k(y(t))$ [9–12].

- The *observation probability* matrix is

$$C = [c_{k\ell}] = \Pr(y_\ell(t)|x_k(t)) \quad \text{for } k = 1, \ldots, N_x; \quad \ell = 1, \ldots, N_y; \text{ and}$$

- The *prior* probability is

$$\Pr(x_i(0)); \quad i = 1, \ldots, N_x$$

where N_x, N_y are the number of states and observations (measurements), respectively (see Fig. 9.2b).

We must realize a subtle point that in contrast to dynamic physical systems where the states and measurements can be any real value or number, the HMM states or observations can *only* assume predefined integer values, $\mathcal{X} = \{\mathcal{X}_1, \ldots, \mathcal{X}_{N_x}\}$ and $\mathcal{Y} = \{\mathcal{Y}_1, \ldots, \mathcal{Y}_{N_y}\}$—this is very important to comprehend. It is the transition and observation probabilities that drive the occurrence of an individual state and observation event, since both are merely (integer) realizations of model outputs. For example, the mapping or quantization of a "real" physical communications signal to a binary coded representation takes the form of a sequence of a 0 or 1 integer value at each time step which are mapped to the observation sequence (see Section 9.7).

It should also be noted that with the addition of the observation process it is simple to define an underlying HMM *state–space model* as

$$\begin{aligned} x(t) &= Ax(t-1) + w(t-1) \\ y(t) &= Cx(t) + v(t) \end{aligned} \tag{9.4}$$

where w, v are uncorrelated (white) sequences (noise) with x, y the usual state and measurement sequences and associated initial conditions $x(0)$, all specified by the HMM above Fig. 9.2b.

Note that the additive sequences (noise) are *not* necessarily Gaussian and therefore the linear Bayesian processor (Kalman filter) is *not* optimum in this case. However, it has been shown [13] that under some (sufficient) conditions (stationarity, etc.) an optimal minimum error variance estimator (Kalman filter) can be constructed based on a stochastic realization of an HMM. The results of this design are capable of providing reasonable estimates of the HMM states and observations. It is also important to understand that there exist state–space representations in which *both* discrete HMM models are combined with dynamic (physical) state–space systems. For instance, one prevalent form is termed *switching* models in which the discrete HMM determines which underlying dynamic model applies at a given time step. This is an approach frequently used in target tracking problems [14] to provide multiple model choices.

9.3 PROPERTIES OF THE HIDDEN MARKOV MODEL

In this section we investigate some of the underlying probabilistic properties of the discrete HMM. To no surprise it matches our Bayesian processor development in the previous chapters. After all, once placed in the state–space representation all Bayesian properties should hold. We start with the joint distribution.

The HMM is a probabilistic model of the joint collection of random variates (Y_t, X_t). Critical properties of the HMM rely on basic Bayesian methodologies and developments. Perhaps the most useful notions inherited from the Markov chain theory are the two major properties of conditional independence that are used over and over again along with Bayes' rule. That is, under an HMM there are two assumptions enabling the development of the underlying techniques:

1. The hidden variables are first-order Markov[2]: $\Pr(x(t)|X_{t-1}, Y_{t-1}) = \Pr(x(t)|x(t-1))$ (state transition); and
2. The observation is independent of other variates given the state (at time t): $\Pr(y(t)|Y_{t-1}, X_t) = \Pr(y(t)|x(t))$ (likelihood).

These properties imply that the underlying joint distribution can be expanded as

$$\begin{aligned}\Pr(Y_t, X_t) &= \Pr(y(t), Y_{t-1}, x(t), X_{t-1}) \\ &= \Pr(y(t), x(t)|Y_{t-1}, X_{t-1}) \times \Pr(Y_{t-1}, X_{t-1})\end{aligned} \tag{9.5}$$

Continuing to apply Bayes' rule to this expression gives

$$\begin{aligned}\Pr(Y_t, X_t) = &[\Pr(y(t)|x(t), X_{t-1}, Y_{t-1}) \times \Pr(x(t)|X_{t-1}, Y_{t-1})] \\ &\times \Pr(Y_{t-1}, X_{t-1})\end{aligned} \tag{9.6}$$

or finally

$$\Pr(Y_t, X_t) = \Pr(y(t)|x(t)) \times \Pr(x(t)|x(t-1)) \times \Pr(Y_{t-1}, X_{t-1}) \tag{9.7}$$

where we have applied the conditional independence properties of the HMM. From the chain rule and these independence properties, we can expand this expression even further to obtain

$$\Pr(Y_t, X_t) = \prod_{k=0}^{t} \Pr(y(k)|x(k)) \times \prod_{k=2}^{t} \Pr(x(k)|x(k-1)) \times \Pr(x(0)) \tag{9.8}$$

[2] Any Nth-order Markov process can be transformed to a first-order process [15].

Thus, in order to characterize an HMM we require the following probabilities:

- Prior: $\Pr(x(0))$;
- Transition: $\Pr(x(t)|x(t-1))$; and
- Likelihood: $\Pr(y(t)|x(t))$.

These quantities correspond to the classic[3] definition of an HMM [10]. The underlying Markov chain is usually assumed to be homogeneous in time with associated stochastic state-transition matrix defined before by $A = a_{mn} = \Pr(x_n(t)|x_m(t-1))\forall t$ and the observation (measurement) probability is given by $C = [c_{k\ell}] = \Pr(y_\ell(t)|x_k(t))$. The HMM parameters are usually specified by $\Sigma = (A, C, \Pr(x_i(0)))$. Here the measurements Y_t are *observed* and the states or internal variables are *hidden*. In order to *generate* (simulate) samples from the HMM, the initial "state" distribution is generated followed by the likelihood (prior → transition → likelihood). It is important to understand that each measurement sample simulated requires new state samples, that is, two synthesized measurement samples originated from two different states in the hidden Markov chain.

9.4 HMM OBSERVATION PROBABILITY: EVALUATION PROBLEM

With the HMM properties of the previous section available, we can now pose the first problem of interest. With the model known and a set of observations available, how can we evaluate the performance of the HMM to faithfully synthesize observations? One way to approach this problem is to estimate the corresponding observation probability $\Pr(Y_T)$ and use it to "validate" that the model and observations are compatible (see Fig. 9.3a). This approach is especially useful when we are to compare or "match" different models to the same observation sequence and search for that model which provides the best match. Thus, calculating the total observation probability provides a solution to the *evaluation problem* of HMM, that is,

GIVEN the observation sequence Y_T and HMM parameters Σ, **FIND** the total observation probability $\Pr(Y_T)$ for $Y_T = \{y(0), \ldots, y(T)\}$.

The *observation probability* is obtained by marginalizing (summing over) the total probability

$$\Pr(Y_t) = \sum_{X_t} \Pr(Y_t, X_t) = \sum_{X_t} \left(\prod_{k=0}^{t} \Pr(y(k)|x(k)) \times \prod_{k=2}^{t} \Pr(x(k)|x(k-1)) \times \Pr(x(0)) \right) \tag{9.9}$$

[3] Classical notation: $\pi_o \rightarrow \Pr(x(0))$; $a_{ij} \rightarrow \Pr(x_j(t)|x_i(t-1))$; and $b_i(y(t)) \rightarrow \Pr(y(t)|x_i(t))$ (continuous observation) or $b_{ij} \rightarrow \Pr(y_j(t)|x_i(t))$ (discrete observation).

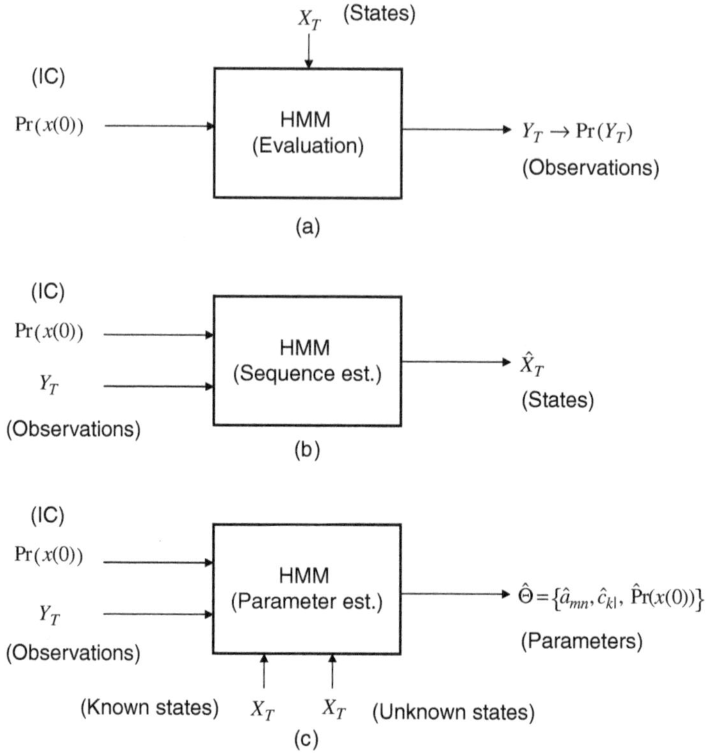

FIGURE 9.3 HMM basic problems: (a) evaluation problem; (b) sequence estimation problem; (c) parameter estimation problem.

Blindly computing this summation is a very inefficient method to estimate the desired probability. Instead, we factor the states using their first-order Markov property (conditional independence) such that

$$\Pr(Y_t) = \sum_{x(t),x(t-1)} \Pr(Y_t, x(t), x(t-1)) \tag{9.10}$$

but continuing the expansion over Y_t we have

$$\begin{aligned}\Pr(Y_t, x(t), x(t-1)) &= \Pr(Y_{t-1}, x(t-1), y(t), x(t)) \\ &= \Pr(y(t), x(t)|Y_{t-1}, x(t-1)) \times \Pr(x(t-1), Y_{t-1})\end{aligned}$$

or

$$\begin{aligned}\Pr(Y_t, x(t), x(t-1)) = \Pr(y(t)|x(t), Y_{t-1}, \ x(t-1)) \times \Pr(x(t)|Y_{t-1}, x(t-1)) \\ \times \Pr(x(t-1), Y_{t-1})\end{aligned}$$

which yields the final expression

$$\Pr(Y_t, x(t), x(t-1)) = \Pr(y(t)|x(t)) \times \Pr(x(t)|x(t-1)) \times \Pr(Y_{t-1}, x(t-1)) \quad (9.11)$$

Marginalizing, we obtain

$$\begin{aligned}\Pr(Y_t, x(t)) &= \sum_{x(t-1)} \Pr(Y_t, x(t), x(t-1)) \\ &= \sum_{x(t-1)} \Pr(y(t)|x(t)) \times \Pr(x(t)|x(t-1)) \times \Pr(Y_{t-1}, x(t-1)) \end{aligned} \quad (9.12)$$

Define the *forward operator* as $\mathcal{F}_{\mathcal{X}}(t) := \Pr(Y_t, x(t) = \mathcal{X})$, then rewriting Eq. 9.11 we have

$$\begin{aligned}\mathcal{F}_k(t) &= \Pr(Y_t, x_k(t)) \\ &= \Pr(y(t)|x_k(t)) \sum_{\ell} \Pr(x_k(t)|x_\ell(t-1)) \times \Pr(Y_{t-1}, x_\ell(t-1))\end{aligned}$$

or substituting for the previous time step, we obtain the *forward recursion* for the HMM

$$\mathcal{F}_k(t) = \Pr(y(t)|x_k(t)) \sum_{\ell} \Pr(x_k(t)|x_\ell(t-1)) \times \mathcal{F}_\ell(t-1)) \quad (9.13)$$

Now, if we assume a stationary chain, then $A = [a_{k\ell}]$, $C = [c_{k\ell}]$, and this result can be expressed in terms of transition probabilities simply as

$$\mathcal{F}_k(t) = \sum_{\ell} a_{k\ell} \times c_{k\ell} \times \mathcal{F}_\ell(t-1) \quad \text{for } c_{k\ell} = \Pr(y_\ell(T)|x_k(T)) \quad (9.14)$$

Clearly marginalizing over $x(t)$ gives the total *observation probability* as

$$\Pr(Y_T) = \sum_k \mathcal{F}_k(T) = \sum_{x_k(t)} \Pr(Y_T, x_k(t)) \quad (9.15)$$

Thus, we have the *forward recursion algorithm* for HMM that can be used to obtain the total observation probability as

- Initialize: $\mathcal{F}_k(0) = \Pr(x_k(0)) \times c_{k0}$;
- Recursion: $\mathcal{F}_k(t) = \sum_\ell a_{k\ell} \times c_{k\ell} \times \mathcal{F}_\ell(t-1)$;
- Termination: $\Pr(Y_T) = \sum_k \mathcal{F}_k(T)$.

This algorithm will be used not only to estimate the observation probability as the solution to the evaluation problem, but also to combine with another recursion to estimate model states and parameters.

Before we close this section, consider the following example of simulating an HMM.

Example 9.2

Suppose we have a discrete binary signal with the two states ($N_x = 2$) specified by $\{x_1(t) = 1,\ x_2(t) = 2\}$. The observation is also discrete with $N_y = 3$ and specified by $\{y_1(t) = 1, y_2(t) = 2, y_3(t) = 3\}$. The transition and observation probabilities are given by

$$a_{mn} = \Pr(x_n(t)|x_m(t-1)); \quad m, n = 1, 2 \quad \text{and} \quad c_{kn} = \Pr(y_k(t)|x_n(t)); \ k = 1, 2, 3$$

with $a_{mn} = \{0.6, 0.4; 0.3, 0.7\}$ and $c_{kn} = \{0.50, 0.25, 0.25; 0.35, 0.25, 0.40\}$. Construct the state-transition probability matrix A and the corresponding directed graph. The transition probability is given by

$$A = \begin{bmatrix} 0.6 & 0.4 \\ 0.3 & 0.7 \end{bmatrix}; \quad C = \begin{bmatrix} 0.50 & 0.25 & 0.25 \\ 0.35 & 0.25 & 0.40 \end{bmatrix} \tag{9.16}$$

The resulting directed graph is identical to that in Fig. 9.1. The transition probability implies that once a particular state is occupied, it is more than likely to remain in that state since $a_{11} = 0.6$ and $a_{22} = 0.7$ rather than transition to the other states, $a_{12} = 0.4$ and $a_{21} = 0.3$. So we expect the state transitions to essentially have longer time steps with fewer transitions. The observation probability on the other hand seems almost equally likely to transition with $c_{11} = 0.5$ implying that when occupying state 1 it is most probable that the observation output will be 1 with $c_{23} = 0.4$ next, that is, if in state 2 then the output 3 is most likely.

Using MATLAB a simulation was performed for 100 samples with the results shown in Fig. 9.4. The state transitions are shown in Fig. 9.4a and appear to conform the intuition afforded by the transition probability, while the observations also follow as well. The evaluation problem can be solved by estimating the corresponding total observation probability which is $\ln \Pr(Y_T) = -106$ and then making additional runs with various HMM for comparison. △△△

9.5 STATE ESTIMATION IN HMM: THE VITERBI TECHNIQUE

In this section we develop solutions to the state estimation problem for two cases of interest: (1) *individual* hidden state estimation, that is, the state estimate at a given time step; and (2) *entire* sequence or "all" time-steps hidden state estimation problem. Both these problems lead to reconstructing the entire sequence of hidden states strictly from the observations and HMM. Think of receiving a signal and being asked to retrieve or recognize the individual symbols from a coded sequence. Here the

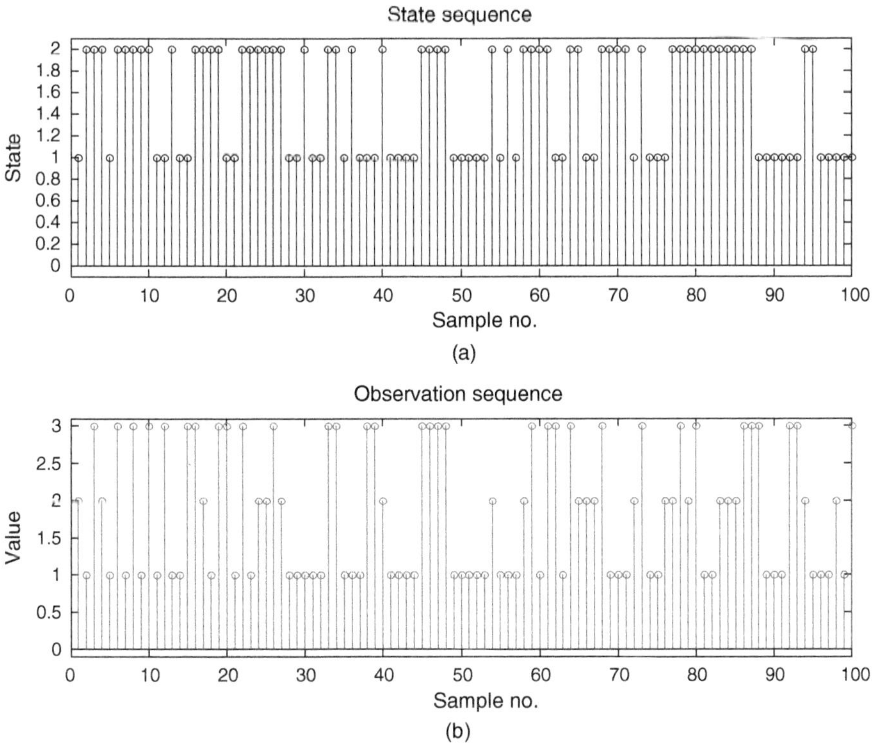

FIGURE 9.4 HMM realization of a discrete two-state Markov chain and observation: (a) hidden state simulation; (b) observation simulation. Note that both states/observations can assume only integer values governed by the transition and observation probabilities.

hidden states are the embedded sequence and the observations are the noisy digitized measurements.

9.5.1 Individual Hidden State Estimation

State estimation in HMM provides a methodology by which the hidden variables or states embedded within the HMM can be extracted from knowledge of the model parameters Σ and the noisy observations as illustrated in Fig. 9.3b. The basic problem to be solved is:

GIVEN the observation sequence, Y_T and *HMM* parameters Σ, **FIND** the "best" (*MAP*) estimate $\hat{x}_k(t)$ of the hidden state at time t based on the posterior distribution $\Pr(x_k(t)|Y_T)$, that is,

$$\hat{x}_k(t) = \arg\max_{x_k} \Pr(x_k(t)|Y_T) \quad \text{for } x_k(t) \Leftrightarrow x(t) = \mathcal{X}_k$$

The solution to this estimation problem is analogous to the forward recursion algorithm and incorporates the so-called backward algorithm, since it proceeds sequentially backwards in time (smoothing). The solution which follows is accomplished by decomposing or partitioning the total observation sequence into two subsequences: $Y_t = \{y(0), \ldots, y(t)\}$ and $Y_{t+1:T} := \{y(t+1), \ldots, y(T)\}$. To see this let us investigate the solution to the state estimation problem assuming uninformative priors

$$\Pr(x_k(t)|Y_T) = \frac{\Pr(Y_T, x_k(t))}{\Pr(Y_T)} \propto \Pr(Y_T, x_k(t)) \tag{9.17}$$

Partitioning Y_T as above we have $Y_T = \{Y_t, Y_{t+1:T}\}$ which can be used to decompose the joint distribution as

$$\Pr(Y_T, x_k(t)) = \Pr(Y_t, Y_{t+1:T}, x_k(t)) = \Pr(Y_t, x_k(t)) \times \Pr(Y_{t+1:T}|Y_t, x_k(t))$$

Applying the Markov property along with the conditional independence properties of the HMM, we have

$$\Pr(Y_T, x_k(t)) = \Pr(Y_t, x_k(t)) \times \Pr(Y_{t+1:T}|x_k(t)) \tag{9.18}$$

Defining the *backward operator* as $\mathcal{B}_k(t) := \Pr(Y_{t+1:T}|x_k(t))$ we can write the marginalization

$$\Pr(Y_{t+1:T}|x_k(t)) = \sum_{x_\ell(t+1)} \Pr(x_\ell(t+1), y(t+1), Y_{t+2:T}|x_k(t)) \tag{9.19}$$

and applying Bayes' rule

$$\begin{aligned}\Pr(Y_{t+1:T}|x_k(t)) = \sum_{x_\ell(t+1)} & \Pr(Y_{t+2:T}|x_\ell(t+1), y(t+1), x_k(t)) \\ & \times \Pr(y(t+1), x_\ell(t+1)|x_k(t))\end{aligned}$$

Now using the Markovian independence properties of the HMM and expanding the last term using Bayes' rule, we obtain

$$\begin{aligned}\Pr(Y_{t+1:T}|x_k(t)) = \sum_{x_\ell(t+1)} & \Pr(Y_{t+2:T}|x_\ell(t+1)) \\ & \times \Pr(y(t+1)|x_\ell(t+1)) \times \Pr(x_\ell(t+1)|x_k(t))\end{aligned}$$

Using the definition of the backward operator, we obtain the final *backward recursion* as

$$\mathcal{B}_k(t) = \sum_{\ell} a_{k\ell} \Pr(y(t+1)|x_\ell(t+1)) \mathcal{B}_\ell(t+1) \quad \text{for } t = T-1, T-2 \ldots, 1, 0 \tag{9.20}$$

with $\mathcal{B}_k(0) = 1 \forall k$. This relation, when coupled with the forward operator can also be used to calculate the desired posterior probability for state estimation, since

$$\Pr(Y_T, x_k(t)) = \Pr(Y_t, x_k(t)) \times \mathcal{B}_k(t) = \mathcal{F}_k(t) \times \mathcal{B}_k(t) \tag{9.21}$$

Thus, we have by marginalization that the total observation probability can be estimated by

$$\Pr(Y_T) = \sum_k \mathcal{F}_k(T) \times \mathcal{B}_k(t) \tag{9.22}$$

and therefore the *posterior distribution* is given by

$$\Pr(x_k(t)|Y_T) = \frac{\mathcal{F}_k(t) \times \mathcal{B}_k(t)}{\sum_k \mathcal{F}_k(T) \times \mathcal{B}_k(t)} \tag{9.23}$$

leading to the desired state estimate $\hat{x}_{MAP}(t)$.

With this information available, we now have the solution to the *individual* hidden state estimation problem using the *backward recursion algorithm* as

- Initialize: $\mathcal{B}_k(0) = 1 \forall k$;
- Recursion: $\mathcal{B}_k(t) = \sum_\ell a_{k\ell} \Pr(y(t+1)|x(t+1)\mathcal{B}_\ell(t+1)$;
- Termination: $\Pr(Y_T) = \sum_k \mathcal{F}_k(T) \times \mathcal{B}_k(t)$;
- Posterior: $\Pr(x_k(t)|Y_T) = \frac{\mathcal{F}_k(t) \times \mathcal{B}_k(t)}{\Pr(Y_T)}$; and
- Estimation: $\hat{x}_k(t) = \arg\max_{x_k} \Pr(x_k(t)|Y_T)$.

Together the forward–backward recursions are the *key* ingredient to estimating the HMM parameters from noisy observation data as well which will be discussed subsequently, but first we consider extending the MAP state estimation for the individual state ($\hat{x}_k(t)$) to the problem of estimating the *entire* sequence of hidden states ($\hat{X}_T$) from the observation data (Y_T).

An example follows to demonstrate the forward–backward recursion in estimating the posterior distribution.

Example 9.3

Suppose we have the discrete binary signal ($N_x = 2$) specified by $\{x_1(t) = 1, x_2(t) = 2\}$ and the discrete observation specified by $\{y_1(t) = 1,\ y_2(t) = 2,\ y_3(t) = 3\}$. Here we attempt to "decode" the message from the observations, that is, consider the binary state sequence generated by the HMM along with the corresponding the observations and we wish to extract the coded state sequence at each time step using the forward–backward approach discussed above.

We apply MATLAB (**hmmdecode**) to perform the estimation using the forward–backward approach for the 100 observation samples. Using the synthesized output

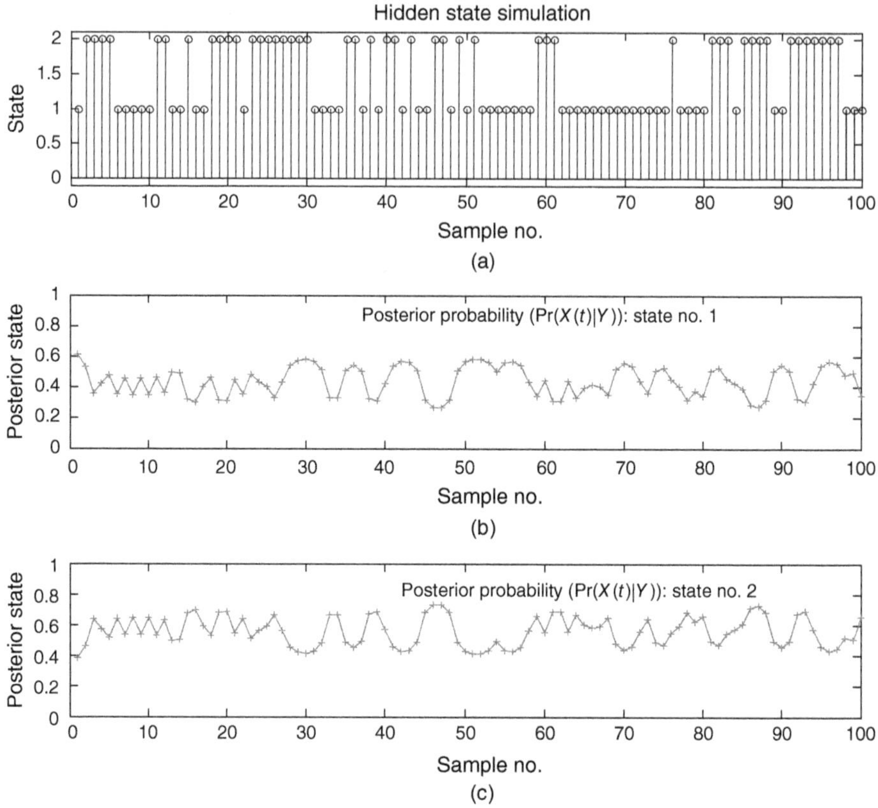

FIGURE 9.5 HMM realization of a discrete two-state Markov chain and observation: (a) hidden state sequence realization; (b) state 1 posterior probability estimate $\hat{\Pr}(x_1(t)|Y_t)$; (c) state 2 posterior probability estimate $\hat{\Pr}(x_2(t)|Y_t)$.

data and corresponding transition and observation probability matrices, we can estimate the corresponding posterior distribution for each individual state $\Pr(x_k(t)|Y_t)$ with the results shown in Fig. 9.5. The combined state transitions are shown in Fig. 9.5a and the posterior state probabilities in Figs. 9.5b and 9.5c, respectively. We can see that the estimated probabilities "match" the states reasonably well with the state transitioning according to the estimated posterior at each time step, that is, when the HMM is in state 1, the posterior probability is high relative to that of state 2 and visa versa. This completes the decoding example. Next we consider estimating the entire state sequence. △△△

9.5.2 Entire Hidden State Sequence Estimation

The maximum a posteriori estimation of the state at time t using the forward–backward recursion algorithm above can be extended to reconstruct the entire hidden state sequence which provides a more meaningful solution when attempting to extract a critical coded message from a hostile environment or accurately extracting a DNA sequence for forensic analysis. Unfortunately, estimating the *individually* "most likely" states as in the previous subsection does not imply that the *entire* sequence is estimated with minimal probability of error. Therefore, the state estimation problem must be based on jointly estimating "all" states in the sequence to obtain the optimal solution. The individual state estimates at each time step of the forward–backward algorithm minimize the error probabilities of individual states maximizing the expected number of correctly estimated states [16]. However, we are interested in estimating the *entire* state sequence, that is, we would like to solve the following problem:

GIVEN the observation sequence Y_T and *HMM* parameters Σ, **FIND** the "best" (*MAP*) estimate the sequence $\hat{X}_T$ where $\hat{X}_T = \{\hat{x}_k(0), \ldots, \hat{x}_k(T)\}$ of the entire hidden state sequence from time-step 0 to time-step T based on the posterior distribution $\Pr(X_T|Y_T)$, that is,

$$\hat{X}_T = \arg\max_{X_T} \Pr(X_T|Y_T)$$

Since we are seeking a sequential solution to this problem, we must track the estimate at time-step t. Following the development in [16], for each $x(t)$ a partial sequence of length $t+1$ is defined for each possible state; therefore, there are N_x partial sequences for each t. An alternative is to use the joint distribution, since the observation sequence is fixed in length Y_T, that is, we require

$$\hat{X}_{t-1} = \arg\max_{X_{t-1}} \Pr(X_{t-1}, x(t), Y_t) \quad \text{for } x(t) \text{ endpoint} \tag{9.24}$$

At each time step, the maximal path (X_{t-1}) problem terminating in $x(t)$ given Y_t is transformed into the maximization problem of finding the best path ending in $x(t+1)$ given Y_{t+1}. This follows directly by applying the chain rule to the joint probability distribution

$$\begin{aligned}\Pr(X_{t+1}, Y_{t+1}) &= \Pr(x(t+1), X_t, y(t+1), Y_t)\\ &= \Pr(x(t+1), y(t+1)|X_t, Y_t) \times \Pr(X_t, Y_t)\end{aligned} \tag{9.25}$$

applying Bayes' rule along with the conditional independence properties of the chain gives

$$\Pr(X_{t+1}, Y_{t+1}) = \Pr(y(t+1)|x(t+1)) \times \Pr(x(t+1)|x(t)) \times \Pr(X_t, Y_t) \tag{9.26}$$

Recursively maximizing the probability gives

$$\begin{aligned}\max_{X_t} \Pr(X_{t+1}, Y_{t+1}) &= \max_{X_t}\{\Pr(y(t+1)|x(t+1)) \times \Pr(x(t+1)|x(t)) \times \Pr(X_t, Y_t)\}\\ &= \Pr(y(t+1)|x(t+1)) \times \max_{X_t}\{\Pr(x(t+1)|x(t)) \times \Pr(X_t, Y_t)\}\end{aligned}$$

$$= \Pr(y(t+1)|x(t+1)) \times \max_{x(t)} \left\{ \Pr(x(t+1)|x(t)) \times \max_{X_{t-1}} \left\{ \Pr(X_t, Y_t) \right\} \right\}$$

Now this gives us a recursion with $\mathcal{V}(x(t)) := \max_{X_{t-1}}\{\Pr(X_t, Y_t)\}$

$$\mathcal{V}(x(t+1)) = \Pr(y(t+1)|x(t+1)) \times \arg\max_{x(t)}\{\Pr(x(t+1)|x(t)) \times \mathcal{V}(x(t))\} \tag{9.27}$$

Defining the smoothing variable as

$$\mathcal{U}(x(t)) := \max_{X_{t-1}}\{\Pr(x(t+1)|x(t)) \times \mathcal{V}(x(t))\} \tag{9.28}$$

enables us to construct the entire *state (sequence) estimation* or equivalently the *Viterbi algorithm* as [16]:

- Initialize:

$$\begin{aligned} \mathcal{V}(x(0)) &= \Pr(x(0)), \Pr(y(0)|x(0)) \\ \mathcal{U}(x(0)) &= 0 \quad \text{for } x(0) = 1, \dots, N_x \end{aligned}$$

- Recursion:

$$\begin{aligned} \mathcal{V}(x(t)) &= \Pr(y(t)|x(t)) \times \max_{x(t-1)}\{\Pr(x(t)|x(t-1)) \times \mathcal{V}(x(t-1))\} \\ \mathcal{U}(x(t)) &= \max_{x(t-1)}\{\Pr(x(t)|x(t-1)) \times \mathcal{V}(x(t-1))\} \quad \text{for } x(t) = 1, \dots, N_x; \\ t &= 2, \dots, T \end{aligned}$$

- Termination:

$$\begin{aligned} P &= \max_{x(T)}\{\mathcal{V}(x(T))\} \\ \hat{x}(T|T) &= \arg\max_{x(T)}\{\mathcal{V}(x(T))\} \end{aligned}$$

- Smoothing: $\hat{x}(t|T) = \mathcal{U}(\hat{x}(t+1)|T) \quad \text{for } t = T-1, T-2, \dots, 0$

The Viterbi algorithm uses these recursions and smoothing relations to estimate the "optimal path" and has the same order of operations as the forward algorithm discussed in the previous section. It has proved to be an extremely popular and robust algorithm to perform decoding. Consider the following example of a path estimate.

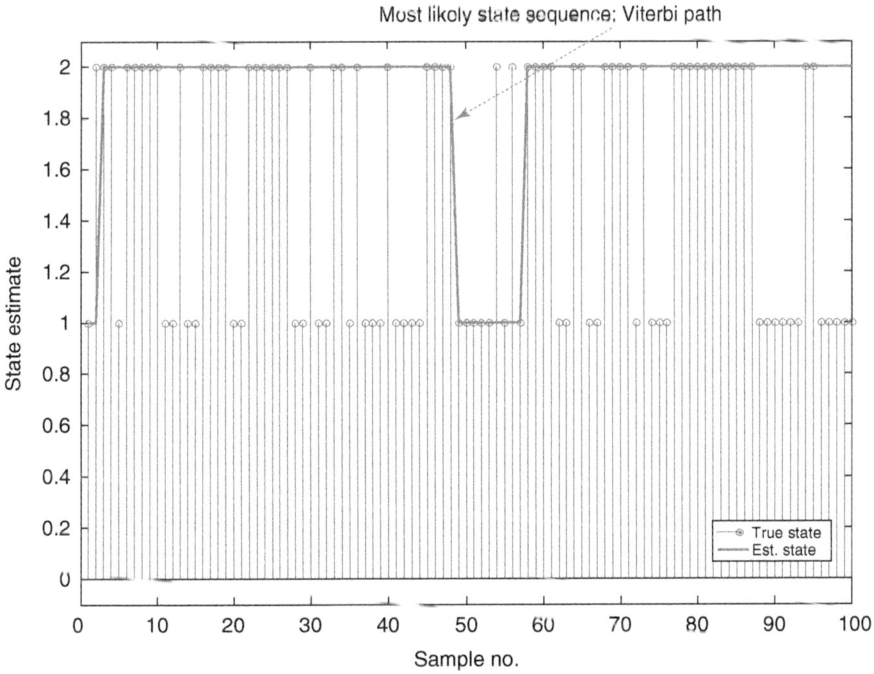

FIGURE 9.6 HMM realization of a discrete two-state Markov chain sequence and the results of the *Viterbi path* estimation with an 52% match (estimated-to-actual states).

Example 9.4

Using the discrete binary signal and observation of the previous example we would like to "decode" the entire message from the observations, that is, we wish to extract the "entire" coded sequence at each time step using the Viterbi approach discussed above.

We apply MATLAB (**hmmviterbi**) to perform the optimal entire state sequence (path) estimation using the Viterbi algorithm for the 100 observation samples. Using the synthesized output data and corresponding transition and observation probability matrices, as before, we obtain the estimation results shown in Fig. 9.6. Here we can observe the path (dark solid line) which corresponds to the sequence estimation. Note that it is a simple path but contains most of the states and leads directly to the desired result. The matching capability of this approach is captured by estimating the percentage of the time that the actual sequence agrees with the estimated. For this simulation, the estimated matches the actual sequence 52% of the time. △△△

This completes the state sequence estimation example, next we consider estimating the model parameters.

9.6 PARAMETER ESTIMATION IN HMM: THE EM/BAUM–WELCH TECHNIQUE

The most challenging problem in HMM is the development of the model in the first place. Just as in dynamic (physical) systems theory [17, 18], the system identification/parameter estimation problem is still a highly researched problem especially for nonlinear systems. The basic estimation problem consists of two major issues: (1) estimation of the underlying internal structure (interconnections, state assignments, etc.); and (2) HMM parameter estimation consisting of the transition and observation probabilities and initial conditions assuming that the *internal structure* of (1) is known a priori.

For dynamic physical systems, the identification of the internal structural model usually evolves from first principles where relations governing the phenomenology are assembled. For nonphysical systems, parametric models (with interconnections) are assumed (e.g., ARMA) and then the solution to the parameter estimation problem follows [19]. HMM are very similar from this perspective. Their structure is developed from an internal probabilistic representation that is application driven. For instance, the well-known problem in signal processing of recovering a transmitted random telegraph signal from noisy observations is representative. Here the problem is to "decode" the signal into a sequence of zeros or ones with the probability distribution of the transitions (zero-to-one) assumed known (Poisson). An HMM can be structurally developed to model this problem quite easily [2]. Another example is the decoding of DNA strings for forensic analysis. Here the same modeling principle applies to develop the internal structural model [5]. In any case developing the internal model is usually the task of the phenomenologist, whether a physical or nonphysical system, and the next step is to "fit" the parameters of this structure to the model—the primary focus of this section. Thus, we discuss the development of the parameter estimation techniques applied to estimate the parameters of an HMM. Here we assume the number of states and measurements are known along with the internal structure and the problem becomes a matter of "fitting" these well-defined parameters (transition probabilities, observation probabilities, and initial conditions) to the known model internal structure.

Parameter estimation for HMM has been a difficult and challenging problem, especially in an on-line environment [1]. The original efforts of Baum–Welch [20] have led to the general expectation-maximization (EM) algorithm (see Chapter 2) which is a very powerful iterative approach using likelihood estimation techniques to solve this problem [21, 22]. Here we briefly outline the iterative approach and then show how the algorithm is a special case of EM.

The basic HMM *parameter estimation problem* is (see Fig. 9.3c):

GIVEN a set or J-sets[4] of observation sequences $\{Y_t(j)\}; j=1, \ldots, J$ along with the underlying HMM internal structure Σ, **FIND** the "best" (MAP) estimate of the

[4] These sets are called training sets, a term that evolves from the classification/neural-net technical area.

underlying parameters $\hat{\Theta}_{MAP} := \{a_{mn}, c_{k\ell}, P(0)\}$ maximizing the posterior distribution $\Pr(\Theta|Y_t)$.

We will discuss this problem in two parts: (1) state sequence is known a priori; and (2) state sequence is *unknown* [5].

9.6.1 Parameter Estimation with State Sequence Known

When the state sequence is "known" a priori, then the parameter estimation problem is much simpler as in the case of physical systems for the design of optimal inputs for system identification [18].

We define the following posterior probability

$$\mathcal{P}_{mn}(t, t+1) := \Pr(x_m(t), x_n(t+1)|Y_T, \Theta)$$

This posterior is the probability of the joint event that a path (state sequence) passes through state m at time-step t and through state n at $t+1$ *and* the HMM generates a sequence of observations Y_T given the model parameters Θ.

To analyze this probability further we apply Bayes' rule and then partition the data as before

$$\begin{aligned}\mathcal{P}_{mn}(t, t+1) &= \frac{\Pr(x_m(t), x_n(t+1), Y_T|\Theta)}{\Pr(Y_T|\Theta)}\\ &= \frac{\Pr(x_m(t), x_n(t+1), Y_t, Y_{t+1:T}|\Theta)}{\Pr(Y_T|\Theta)}\end{aligned} \tag{9.29}$$

now applying Bayes' rule to the numerator results in

$$\begin{aligned}\Pr(x_m(t), x_n(t+1), Y_t, Y_{t+1:T}|\Theta) &= \Pr(x_n(t+1), Y_{t+1:T}|Y_t, x_m(t), \Theta)\\ &\quad\times \Pr(Y_t, x_m(t)|\Theta)\end{aligned} \tag{9.30}$$

The last term is just $\mathcal{F}_m(t)$, the forward operator (with the parameter set Θ given). Now concentrating on the remaining term of this expression, we extract $y(t+1)$ from the data term and apply Bayes' rule again to obtain

$$\begin{aligned}&\Pr(x_n(t+1), y(t+1), Y_{t+2:T}|Y_t, x_m(t), \Theta)\\ &\quad= \Pr(Y_{t+2:T}|x_n(t+1), y(t+1), Y_t, x_m(t), \Theta)\\ &\quad\quad\times \Pr(x_n(t+1), y(t+1)|Y_t, x_m(t), \Theta)\end{aligned} \tag{9.31}$$

where the last term above decomposes further to

$$\begin{aligned}\Pr(x_n(t+1), y(t+1)|Y_t, x_m(t), \Theta) &= \Pr(y(t+1)|x_n(t+1), Y_t, x_m(t), \Theta)\\ &\quad\times \Pr(x_n(t+1)|x_m(t), Y_t, \Theta)\end{aligned} \tag{9.32}$$

enabling us to simplify each of these terms individually to yield

$$\begin{aligned}\Pr(Y_{t+2:T}|x_n(t+1), y(t+1), Y_t, x_m(t), \Theta) &\rightarrow \Pr(Y_{t+2:T}|x_n(t+1), \Theta)\\ \Pr(x_n(t+1), y(t+1)|Y_t, x_m(t), \Theta) &\rightarrow \Pr(y(t+1)|x_n(t+1), \Theta)\\ &\quad \times \Pr(x_n(t+1)|x_m(t), \Theta)\end{aligned} \tag{9.33}$$

and therefore we obtain the expression

$$\begin{aligned}\mathcal{P}_{mn}(t, t+1) = {} & \Pr(Y_t, x_m(t)) \times \Pr(x_n(t+1)|x_m(t), \Theta)\\ & \times \Pr(y(t+1)|x_n(t+1), \Theta) \times \Pr(Y_{t+2:T}|x_n(t+1), \Theta)/\Pr(Y_T|\Theta)\end{aligned}$$

Finally, substituting for the known parameters, we have the desired result

$$\mathcal{P}_{mn}(t, t+1) = \frac{\mathcal{F}_m(t) \times a_{mn} \times c_{kn} \times \mathcal{B}_n(t+1)}{\Pr(Y_T|\Theta)} \tag{9.34}$$

where $\mathcal{F}_k(t)$ encompasses the *past* history ending at time t and state m while $\mathcal{B}_n(t+1)$ accounts for the path's *future* which at time $t+1$ is at state n evolving until the end. The product term $(a_{mn} \times c_{kn})$ takes into account the current activity at t with discrete observation $y_k(t+1) \rightarrow y(t+1)$.

With this term determined, we then define the *posterior distribution* from Eq. 9.23 as

$$\mathcal{O}_m(t) := \Pr(x_m(t)|Y_T) \tag{9.35}$$

which is a probability of the joint event that a path passes through state m at time-step t *and* the HMM generates a sequence of observations Y_T given the model parameters Θ.

Note that both probabilities are related, since one can be obtained through marginalization of the other

$$\mathcal{O}_m(t) = \sum_{n=1}^{N_x} \mathcal{P}_{mn}(t, t+1) \tag{9.36}$$

Summing both these quantities "across time" enables us to obtain the *expected number of times* in state m and the expected number of transitions *away* from state m for Y (see Ref. [15] for more details)

$$\sum_{t=1}^{T-1} \mathcal{O}_m(t) \tag{9.37}$$

Similarly, the expected number of *transitions* from state m to state n for Y is given by

$$\sum_{t=1}^{T-1} \mathcal{P}_{mn}(t) \tag{9.38}$$

Thus, using these expectations and counting estimation of probabilities [5], we are able to obtain the Baum–Welch estimates

$$\begin{aligned} \hat{P}_m(0) &= \mathcal{O}_m(1) \\ \hat{a}_{mn} &= \frac{\sum_{t=1}^{T-1} \mathcal{P}_{mn}(t)}{\sum_{t=1}^{T-1} \mathcal{O}_m(t)} \\ \hat{c}_{kn} &= \frac{\sum_{t=1}^{T} \mathcal{P}_{mn}(t)}{\sum_{t=1}^{T} \mathcal{O}_m(t)} \quad \text{such that } y(t) = \mathcal{Y}_k \end{aligned} \tag{9.39}$$

where

- $P_m(0)$ is the expected number of times in state $x_n(t)$ at $t = 1$;
- a_{mn} is the expected number of transitions from state $x_m(t)$ to state $x_n(t)$ over the expected transitions in state $x_m(t)$; and
- c_{kn} is the expected number of times in state $x_n(t)$ and observing $y_k(t)$ over the expected number of times in state $x_n(t)$.

Thus, when all of the paths are known (this case), it is possible to *count* the number of times each particular transition or output observation is applied in a set of training data. It has been shown that counting functions, say $N_{mn}(x(t))$ for the state transitions and $N_{kn}(y(t))$ for the output observations, provide maximum likelihood estimates for the desired model parameters [5], such that

$$\hat{a}_{mn} = \frac{N_{mn}(x(t))}{\sum_n N_{mn}(x(t))} \quad \text{and} \quad \hat{c}_{kn} = \frac{N_{kn}(y(t))}{\sum_n N_{kn}(y(t))} \tag{9.40}$$

Next we consider the unknown path case and combine the above results to establish the algorithm.

9.6.2 Parameter Estimation with State Sequence Unknown

In this section we consider the case where the state sequence is "not known" [5] and must be determined using the current parameter estimates available. When the paths are unknown for training sequences, a closed-form equation is nonexistent and therefore some type of iterative approach must be applied (e.g., EM [22–30]). The EM/Baum–Welch approach precisely solves the HMM parameter estimation problem in an iterative manner. It first estimates the counting functions $N_{mn}(x(t))$ for states and $N_{kn}(y(t))$ for observations by considering possible paths for the training sequences using current model parameters Θ and then calculates the new estimates using Eq. 9.40. The algorithm continues to iterate until the log-likelihood function $\ln \Pr(Y_T|\Theta)$ no longer increases with each iteration. Baum [20] has shown that the overall log-likelihood increases with each iteration indicating convergence to a local maximum.

More precisely, this technique estimates the counting functions $N_{mn}(x(t))$ and $N_{kn}(y(t))$ as the expected number of times each transition or output is utilized from the *given* training sequences. It also uses the identical forward/backward operators as before (see Section 9.5) using the posterior probability $\mathcal{P}_{mn}(t, t+1)$ of Eq. 9.34. From this relation, we can derive the *expected number of times* that a_{mn} is used by summing over all possible positions and over all training sequences $Y_T(j); j = 1, \ldots, J$. We can also use the training sequences to derive the *expected number of times the observation occurs* to obtain

$$\begin{aligned}
N_{mn}(x(t)) &= \sum_j \frac{1}{\Pr(Y_T|\Theta)} \sum_t \mathcal{F}_m(t,j) \times a_{mn} \times c_{kn} \times \mathcal{B}_n(t+1,j) \\
N_{k\ell}(y(t)) &= \sum_j \frac{1}{\Pr(Y_T|\Theta)} \sum_t \mathcal{F}_k(t,j) \times \mathcal{B}_k(t,j)
\end{aligned} \tag{9.41}$$

Once these expectations are estimated, the model parameters are updated as above and these new estimates are used in the counting functions. We summarize the EM/Baum–Welch algorithm as:

- Initialization: $\hat{\Theta}_0$, $N_{mn}(x(t))$ and $N_{k\ell}(y(t))$;
- Forward/backward recursions: $\mathcal{F}_k(t,j)$ and $\mathcal{B}_k(t,j)$ of Eqs. 9.12 and 9.20;
- Counting functions: $N_{mn}(x(t))$ and $N_{k\ell}(y(t))$ of Eq. 9.41;
- Parameter estimation: $\hat{\Theta} = \{a_{mn}, c_{kn}, P(x(0))\}$ of Eq. 9.40;
- Likelihood: $\Pr(Y_T|\hat{\Theta})$; and
- Termination: $\Pr(Y_T|\hat{\Theta}) < \tau$ for τa threshold.

This completes the algorithm. It should be noted that an alternative approach to searching over all paths is to use the *Viterbi paths* providing the most probable paths for all of the training sequences. However, this approach does not maximize the true likelihood. It is known that Viterbi training does *not* perform as well as the Baum–Welch, but it is still popular when applied to decoding problems.

There are also a number of techniques that practitioners use to enhance numerical performance and convergence of this technique. These include: (1) logarithmic transformation of the product probabilities to create sums [5]; (2) scaling both forward and backward operators [10]; (3) initial conditions; and (4) training data issues [23]. In closing, we note that the Baum–Welch algorithm is just a special case of the EM algorithm of Section 2.3. That is, the E-step of the EM algorithm is given by [15]

$$\begin{aligned}\text{E-step:}Q(\Theta,\hat{\Theta}_{i-1}) &= \sum_x \ln \Pr(x|Y_T,\Theta)\times\Pr(x|Y_T,\hat{\Theta}_{i-1})\\ &= \sum_x \ln \hat{P}(x(0))\Pr(Y_T,x(t)|\hat{\Theta}) + \sum_x\left(\sum_{t=1}^{T}\ln\hat{a}_{mn}(t-1,t)\right)\\ &\quad\times\Pr(Y_T,x(t)|\hat{\Theta}) + \sum_x\left(\sum_{t=1}^{T}\ln\hat{c}_{kn}(t,t)\right)\times\Pr(Y_T,x(t)|\hat{\Theta})\end{aligned} \tag{9.42}$$

where $\Pr(Y_T,x(t)|\hat{\Theta}) = P(x(0))\prod_{t=0}^{T}\hat{c}_{kn}(t,t)\times\prod_{t=1}^{T}\hat{a}_{mn}(t,t+1)$. Optimizing these terms leads precisely to the expressions in Eq. 9.39 (see Ref. [10] or [15] for details). Thus, the E-step of the EM consists of estimating the required expectations using the forward/backward recursions which completely determines $Q(\Theta,\hat{\Theta})$ and the maximum (M-step) consists of substituting these terms into the corresponding likelihood.

Example 9.5

Again using discrete binary signal and the discrete observation of the previous examples, we perform the parameter estimation of the transition and observation probabilities, first using the "known" (actual) state sequence and then generating an ensemble of training sequences ($N = 25$) to perform the EM/Baum–Welch algorithm (**hmmtrain**) with a maximum of 500 iterations and a error tolerance of 1×10^{-4}. Here, we also use the MATLAB maximum likelihood estimation with the "known" state sequence (**hmmestimate**) as well to compare performance. The resulting parameter estimates and percentage errors are:

EM/BAUM–WELCH PARAMETER ESTIMATES

$$A_{TRU} = \begin{bmatrix}0.6 & 0.4\\0.3 & 0.7\end{bmatrix}; \qquad C_{TRU} = \begin{bmatrix}0.50 & 0.25 & 0.25\\0.35 & 0.25 & 0.40\end{bmatrix}$$

$$\hat{A}_{BW} = \begin{bmatrix}0.62 & 0.38\\0.30 & 0.7\end{bmatrix}; \qquad A_{\%ERR} = \begin{bmatrix}3 & 4\\0 & 0\end{bmatrix}$$

$$\hat{C}_{BW} = \begin{bmatrix}0.51 & 0.29 & 0.19\\0.34 & 0.22 & 0.44\end{bmatrix}; \qquad C_{\%ERR} = \begin{bmatrix}3 & 18 & 23\\3 & 13 & 11\end{bmatrix}$$

Thus, the parameter estimates are quite reasonable under these conditions. Note the initial probability matrices are automatically established by this implementation in MATLAB; however, it is possible to alter them if desired.

MAXIMUM LIKELIHOOD PARAMETER ESTIMATES

$$A_{TRU} = \begin{bmatrix}0.6 & 0.4\\0.3 & 0.7\end{bmatrix}; \qquad C_{TRU} = \begin{bmatrix}0.50 & 0.25 & 0.25\\0.35 & 0.25 & 0.40\end{bmatrix}$$

$$\hat{A}_{ML} = \begin{bmatrix} 0.67 & 0.33 \\ 0.36 & 0.64 \end{bmatrix}; \qquad A_{\%ERR} = \begin{bmatrix} 12 & 18 \\ 21 & 9 \end{bmatrix}$$

$$\hat{C}_{ML} = \begin{bmatrix} 0.45 & 0.23 & 0.32 \\ 0.47 & 0.19 & 0.34 \end{bmatrix}; \qquad C_{\%ERR} = \begin{bmatrix} 9 & 9 & 28 \\ 34 & 23 & 15 \end{bmatrix}$$

Again the estimates appear quite reasonable. Longer sequences can be employed to improve the estimates even further. We see that there is a distinct advantage when the state sequence is known a priori, because the training sequences are *not* required. The Viterbi initialization was also executed on this data, but it did not perform near as well as the Baum–Welch technique. △△△

This completes the section, next we consider a case study.

9.7 CASE STUDY: TIME-REVERSAL DECODING

In this section, we consider applying the Viterbi algorithm to decode a message transmitted through a hostile environment with reverberations along with the processor and decoding algorithm. Acoustic time-reversal (T/R) communications is an application area motivated by the recent theoretical advances in T/R theory [30]. Although perceived by many in signal processing as simply an application of matched-filter theory, a T/R receiver offers an interesting solution to the communications problem for a highly reverberant channel. This case study briefly describes an acoustic communications experiment of data gathered in air and its associated signal processing. The experiment is developed to evaluate the performance of a point-to-point T/R receiver designed to extract a transmitted code information sequence propagating in a hostile, highly reverberant environment. These results are merely used to "synthesize" an HMM based on the raw/quantized acoustic measurements and then used to extract the transition and observation probabilities for simulation and evaluation. Even though this case study is based on real data, it is only chosen to illustrate the application of HMM techniques *after* data is simulated through the HMM process (evaluation).

From a signal processing perspective, T/R processing appears to be an application of matched filtering in which the output signal-to-noise ratio (SNR) is maximized. This T/R replicant is then cross-correlated with the noisy received signal to produce the optimal filtered output [31]. However, it becomes more complicated in the spatiotemporal case in which the optimal matched filter must not only match the transmitted temporal function, but also the corresponding spatiotemporal channel medium impulse response or the so-called Green's function. It has been shown that T/R techniques are applicable to spatiotemporal phenomena that satisfy a wave-type equation possessing the T/R invariance property [30]. Thus, T/R is the dynamic broadband analog of the well-known phase conjugate mirror used to focus narrowband monochromatic waves. It represents the "optimal" spatiotemporal matched filter in the sense of maximizing the output SNR. It is essentially a technique which can be used to "remove" the aberrations created by an inhomogeneous or random channel.

In communications, the T/R receiver can overcome the inherent noise created by the medium providing the enhancement required to extract the transmitted information sequence. Here we ignore the array aspects of T/R by considering only point-to-point communications. In this case study the realization of a T/R receiver is briefly discussed and applied to a noisy microphone measurement in a hostile environment. It is then used to estimate the required transition and observation matrices for eventual synthesis/analysis.

For T/R, the matched filter in additive white noise is identical to that posed above with a "known" Green's function of the medium replacing the known signal replicant [31]. The Green's function $g(r, r_o; t)$ is the result of a point-to-point communication link between a station (source) at r_o to a master station (receiver) at r. In this case, the matched-filter solution is again found by maximizing SNR_{out} leading to the solution that is satisfied with equality at some time T. If the resulting filter response is $f(t)$ then the solution is given by

$$f(t) = g(r, r_o; T - t) \tag{9.43}$$

Thus, for T/R, the optimal matched-filter solution is the time-reversed Green's function from the link station-to-master station (source-to-receiver) or visa versa. Comparing these results with the standard matched-filter solution found in the literature, the Green's function of the channel is time reversed rather than the transmitted replicant signal as in radar or sonar. Note that since T/R theory requires reciprocity [30], the result of Eq. 9.43 is valid for both transmission and reception, that is, $g(r, r_o; T - t) \leftrightarrow g(r_o, r; T - t)$. Note also that if an array is included to sample the spatial field or transmit a wave, then these results include the focus at link station (source) position r_o yielding the optimal spatiotemporal matched-filter solution $g(r_\ell, r_o; T - t)$ at sensor position r_ℓ.

So we see that in transmitting a coded signal (state sequence) through a disruptive medium the distorting effects can be mitigated by time reversing the estimated media Green's function and creating an effective receiver. The details of this mechanism are discussed in Refs. [31–33] and is beyond the scope of this case study. Here we just describe one of the variety of receiver types that can be used, once Green's function is estimated from pilot signals transmitted from transmitter (speaker) to receiver (microphone) producing $\hat{g}(r, r_o; T - t)$.

With the estimated Green's function or impulse response available, we choose to apply T/R processing on reception [31] to the noisy received data, $y(t) = g(r; t) * i(t)$ with $i(t)$ the coded information (state) sequence. On reception, the estimated Green's function is reversed and convolved with the receiver input to give

$$R(t) = z(t) * \hat{g}(r; -t) = g(r; t) * i(t) * \hat{g}(r; -t) = C_{g\hat{g}}(t) * i(t) \tag{9.44}$$

where $C_{g\hat{g}}$ is the estimated autocorrelation function of the medium possessing all of the reflection and scattering information—but modified for code signal enhancement. We show a typical T/R receiver output that was used to "synthesize" a discrete state and output sequence for transition and observation probability estimates. We show the receiver structure in Fig. 9.7a where the time-reversed Green's function is convolved with the received data and then quantized to recover the code. Actual

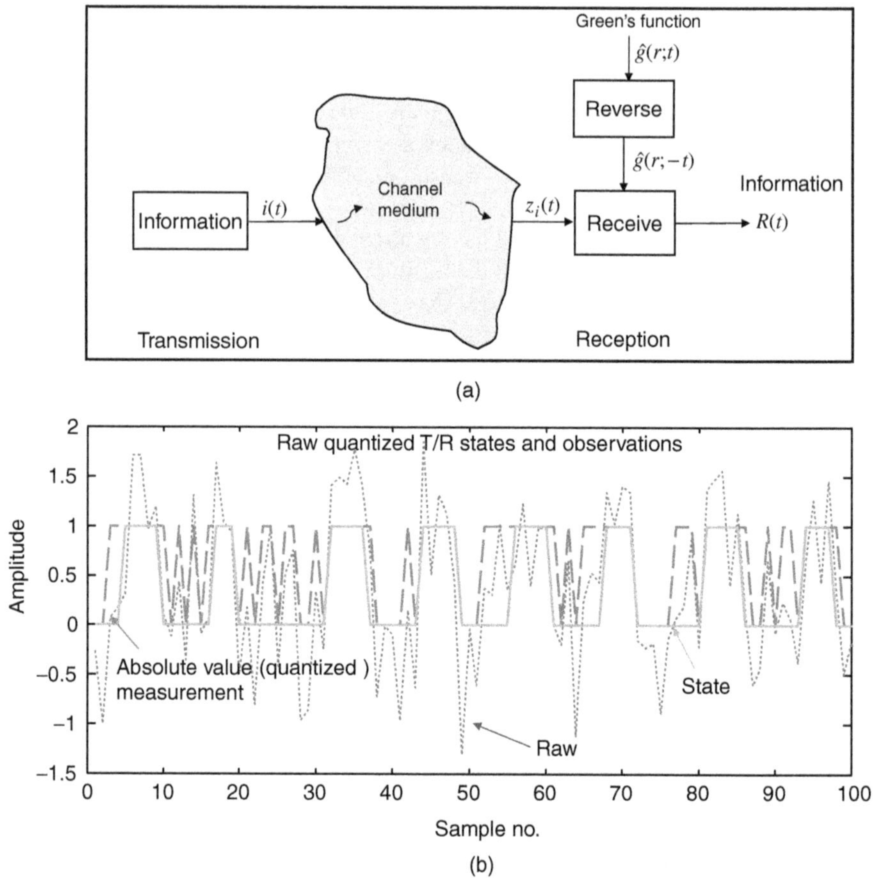

FIGURE 9.7 T/R processor acoustic microphone data: (a) T/R receiver structure; (b) raw measurement data, synthesized observation and state data input for HMM parameter estimation.

T/R processed data is shown in Fig. 9.7b along with quantized state and observation sequence extracted for illustrative purposes and eventual application of the HMM techniques.

The results of processing these quantized sequences using the EM/Baum–Welch algorithm are:

EM/BAUM–WELCH PARAMETER ESTIMATES

$$\hat{A}_{BW} = \begin{bmatrix} 0.42 & 0.58 \\ 0.28 & 0.72 \end{bmatrix}; \qquad \hat{C}_{BW} = \begin{bmatrix} 1 & 0 \\ 0.22 & 0.78 \end{bmatrix}$$

Next we used these extracted probability matrices to synthesize "realistic" T/R data for HMM processing, the results are shown in Fig. 9.8. With this available we

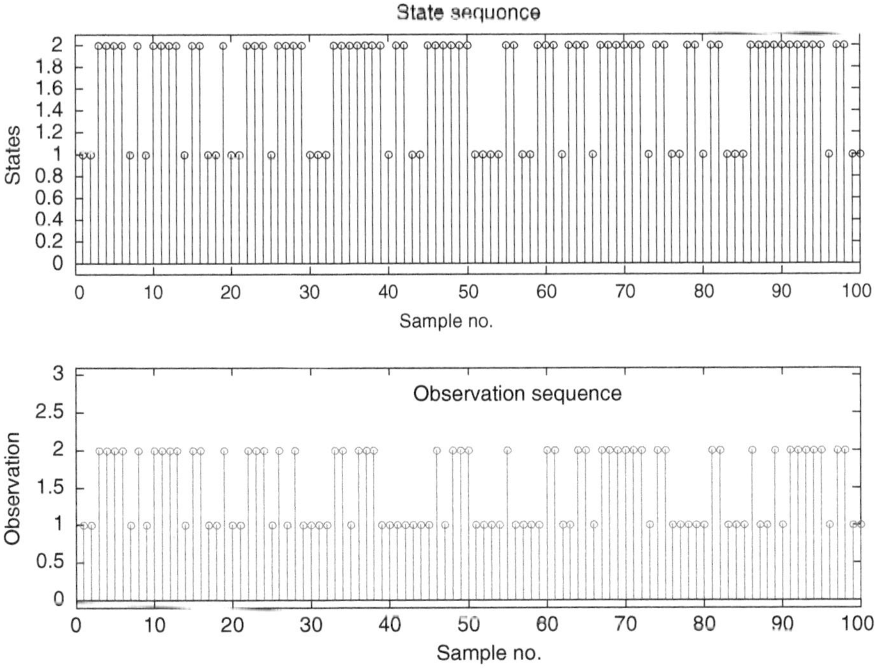

FIGURE 9.8 HMM realization of the T/R discrete Markov chain and observation: (a) hidden state sequence realization; (b) observation realization.

proceed as before and estimate the individual states as depicted by the posterior probabilities in Fig. 9.9. These results are quite reasonable as can be observed by comparing samples with the aligned probability functions at each time step. Next the entire state sequence was estimated using the "most likely" Viterbi approach and the results are illustrated in Fig. 9.10, where both the actual (synthesized time-reversed) state sequences are shown along with the Viterbi result superimposed. The agreement is quite good with an 85% matching (in time) of the actual with the estimated states.

With these synthesized probability matrices, we performed the parameter estimation approach as before to give

EM/BAUM–WELCH PARAMETER ESTIMATES

$$A_{TRU} = \begin{bmatrix} 0.42 & 0.58 \\ 0.28 & 0.72 \end{bmatrix}; \qquad C_{TRU} = \begin{bmatrix} 1 & 0 \\ 0.22 & 0.78 \end{bmatrix}$$

$$\hat{A}_{BW} = \begin{bmatrix} 0.45 & 0.55 \\ 0.27 & 0.73 \end{bmatrix}; \qquad A_{\%ERR} = \begin{bmatrix} 6 & 5 \\ 4 & 2 \end{bmatrix}$$

$$\hat{C}_{BW} = \begin{bmatrix} 1 & 0 \\ 0.24 & 0.76 \end{bmatrix}; \qquad C_{\%ERR} = \begin{bmatrix} 0 & 0 \\ 8 & 2 \end{bmatrix}$$

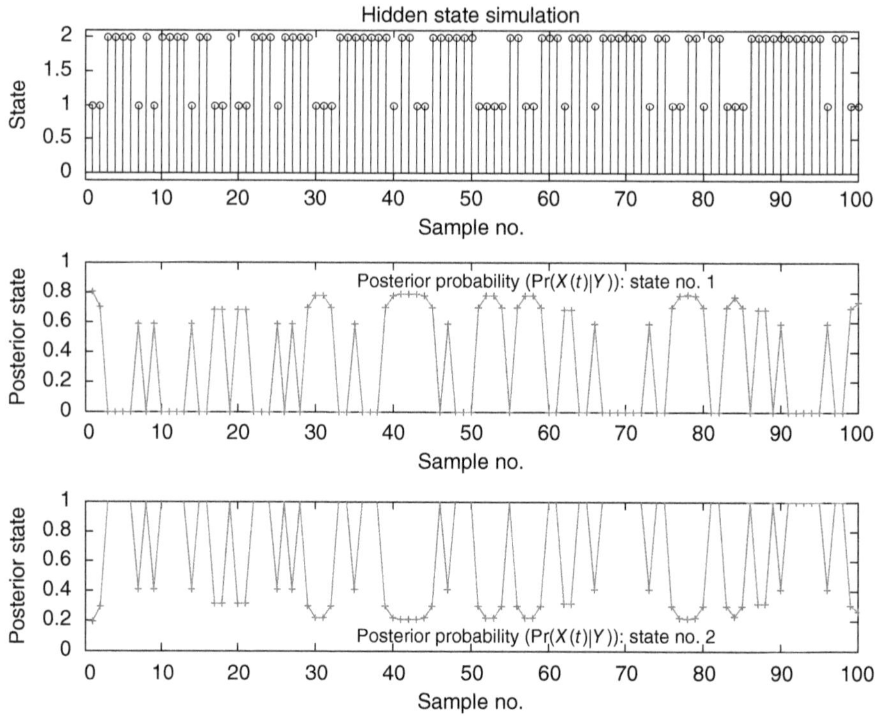

FIGURE 9.9 HMM estimation of the T/R discrete two-state Markov chain and observation: (a) hidden state sequence realization; (b) state 1 posterior probability estimate $\hat{\Pr}(x_1(t)|Y_t)$; (c) state 2 posterior probability estimate $\hat{\Pr}(x_2(t)|Y_t)$.

Thus, the parameter estimates are quite reasonable under these conditions. Note the initial probability matrices are automatically established by this implementation in MATLAB; however, it is possible to alter them if desired.

MAXIMUM LIKELIHOOD PARAMETER ESTIMATES

$$A_{TRU} = \begin{bmatrix} 0.42 & 0.58 \\ 0.28 & 0.72 \end{bmatrix}; \qquad C_{TRU} = \begin{bmatrix} 1 & 0 \\ 0.22 & 0.78 \end{bmatrix}$$

$$\hat{A}_{ML} = \begin{bmatrix} 0.42 & 0.58 \\ 0.29 & 0.71 \end{bmatrix}; \qquad A_{\%ERR} = \begin{bmatrix} 0 & 0 \\ 1 & 1 \end{bmatrix}$$

$$\hat{C}_{ML} = \begin{bmatrix} 1 & 0 \\ 0.24 & 0.76 \end{bmatrix}; \qquad C_{\%ERR} = \begin{bmatrix} 0 & 0 \\ 11 & 3 \end{bmatrix}$$

The estimates are quite reasonable. We see that there is a distinct advantage when the state sequence is known a priori, then the training sequences are not required.

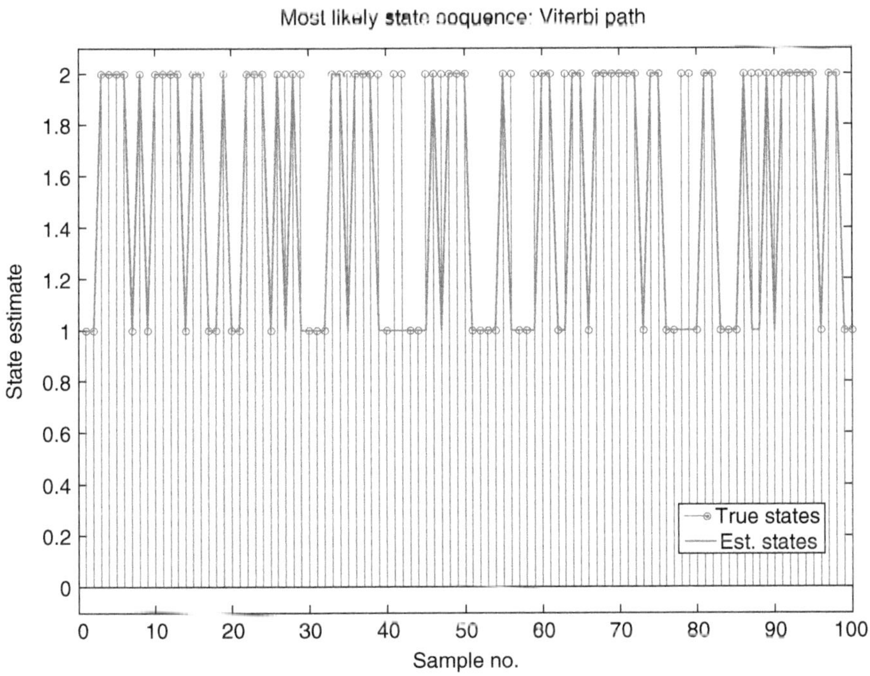

FIGURE 9.10 Entire state sequence estimation using the Viterbi algorithm illustrating the most likely path with an 82% match between the actual state sequence and the estimated.

Again the Viterbi initialization was also executed on this data but it did not perform very well. This completes the case study.

9.8 SUMMARY

In this chapter we have introduced the concept of HMM and illustrated their internal characteristics through a state–space representation. We first developed the concepts of Markov and hidden Markov chains and showed how they were related. Next, we investigated properties of the HMM illustrating how the Bayesian concepts easily transfer over to this discrete representation. We next investigated the three fundamental problems along with some variations: (1) the evaluation (simulation) problem; (2) the state estimation problem; and (3) the parameter estimation problem. A careful analysis of each led us to the popular *Viterbi* decoding technique and the specialized EM algorithm popularly called the Baum–Welch technique. We concluded with a case study to decode a transmitted coded sequence from data enhanced by a T/R processor.

MATLAB NOTES

MATLAB has a *Statistics Toolbox* that incorporates the capability to develop and process hidden Markov models (HMM) along with demonstrations and a tutorial. **hmmgenerate** synthesizes a sequence for an HMM, while the command **hmmdecode** calculates the posterior state probabilities of a given sequence (Section 9.3). The most likely (entire) state path can be estimated using the Viterbi algorithm of Section 9.5 (**hmmviterbi**). Training sequences can be used to solve the HMM parameter estimation problem (Section 9.6) while the EM/Baum–Welch technique is used to estimate the model parameters using the **hmmestimate** command. The PDF estimators include the usual histogram (**hist**) as well as the sophisticated kernel density estimator (**ksdensity**) offering a variety of kernel (window) functions (Gaussian, etc.).

There also exists NETLAB which is a free MATLAB software package that includes HMM algorithms (see Ref. [24] for details and website). The EM algorithm is well documented [20–22, 25–29] and is an integral part of each of these packages.

REFERENCES

1. O. Cappe, "Ten years of HMMs," *www.tsi.enst.fr/cappe/docs/hmmmbib*, 2001.
2. A. Papoulis and S. Pillai, *Probability, Random Variables and Stochastic Processes*, 4th Ed. (Englewood Cliffs, NJ: McGraw-Hill, 2002).
3. R. Hogg, J. McKlean, and A. Craig, *Introduction to Mathematical Statistics*, 6th Ed. (Englewood Cliffs, NJ: Prentice-Hall, 2005).
4. D. Bertsekas and J. Tsitsiklis, *Introduction to Probability* (M.I.T. Lecture Notes, Course 6.041–6.043, 2000).
5. R. Durbin, S. Eddy, A. Krough and G. Mitchison, *Biological Sequence Analysis* (Cambridge, UK: Cambridge University Press, 1990).
6. D. MacKay, *Information Theory, Inference and Learning Algorithms* (Cambridge, UK: Cambridge University Press, 2006).
7. R. Elliott, L. Aggoun, and J. Moore, *Hidden Markov Models* (New York: Springer, 1994).
8. O. Cappe, E. Moulines, and T. Ryden, *Inference in Hidden Markov Models* (New York: Springer-Verlag, 2005).
9. L. Rabiner and B. Juang, "An introduction to hidden Markov models," *IEEE ASSP Mag.*, **3**, 1, 4–16, 1986.
10. L. Rabiner, "A tutorial on hidden Markov models and selected applications in speech recognition," *Proc. IEEE*, **77**, 2, 257–286, 1989.
11. A. Poritz, "Hidden Markov models: A guided tour." *Proc. IEEE Confr. Acoustics, Speech and Sig. Proc.*, 7–13, 1988.
12. L. Rabiner and B. Juang, *Fundamentals of Speech Recognition* (Englewood Cliffs, NJ: Prentice-Hall, 1993).

13. L. White, "A comparison between optimal and Kalman filtering for hidden Markov processes," *IEEE Sig. Proc. Letters*, **5**, 5, 124–126, 1998.
14. S. Fruhwirth-Schnatter, *Finite Mixture and Markov Switching Models* (New York: Springer, 2006).
15. J. Bilmes, "What HMMs Can Do?" *Univ. Wash. Elect. Engr.*, UWEETR-2002-0003, 2002.
16. F. van der Heijden, R. Duin, D. de Ridder, and D. Tax, *Classification, Parameter Estimation and State Estimation* (Hoboken, NJ: John Wiley & Sons, Inc., 2004).
17. G. Goodwin and R. Payne, *Dynamic System Identification* (New York: Academic Press, 1976).
18. L. Ljung, *System Identification: Theory for the User* (Englewood Cliffs, NJ: Prentice-Hall, 1987).
19. J. Candy, *Model-Based Signal Processing* (Hoboken, NJ: John Wiley & Sons, Inc./IEEE Press, 2006).
20. L. Baum, T. Petrie, G. Soules, and N. Weiss, "A maximization technique occurring in the statistical analysis of probabilistic functions of Markov Chains," *Ann. Math. Statist.*, **41**, 1, 164–171, 1970.
21. G. McLachlan and T. Krishnan, *The EM Algorithm and Extensions* (Hoboken, NJ: John Wiley & Sons, Inc., 1997).
22. T. Moon, "The expectation-maximization algorithm," *IEEE Sig. Proc. Mag.*, **13**, 6, 47–60, 1996.
23. S. Theodoridis and K. Koutroumbas, *Pattern Recognition* (New York: Academic Press, 1999).
24. I. Nabney, *NETLAB Algorithms for Pattern Recognition* (New York: Springer, 2001).
25. J. Bilmes, "A gentle tutorial of the EM algorithm and its application to parameter estimation for Gaussian mixture and hidden Markov models," *Internat. Comp. Sci. Instit.*, TR-97-021, 1998.
26. A. Dempster, N. Laird, and D. Rubin, "Maximum likelihood from incomplete data via the EM algorithm," *J. R. Statist. Soc.*, **B**, **39**, 1, 1–38, 1977.
27. R. Duda and P. Hart, *Pattern Classification* (New York: John Wiley & Sons, Inc., 2000).
28. C. Wu, "On the convergence properties of the EM algorithm," *Ann. Statist.*, **11**, 1, 95–103, 1983.
29. R. Redner and H. Walker, "Mixture densities, maximum likelihood and the EM algorithm," *SIAM Rev.*, **26**, 2, 195–239, 1984.
30. M. Fink, "Time reversal in acoustics," *Contemp. Phys.*, **37**, (2), 95–109, 1996.
31. J. Candy, A. Meyer, A. Poggio, and B. Guidry, "Time reversal processing for an acoustic communications experiment in a hostile reverberant environment," *J. Acoust. Society Amer.*, **115**, (4), 1621–1631, 2004.
32. J. Candy, A. Poggio, D. Chambers, C. Robbins, and B. Guidry, "Multichannel time reversal communications a hostile reverberant environment," *J. Acoust. Society Amer.*, **118**, (4), 2339–2354, 2005.
33. J. Candy, D. Chambers, C. Robbins, B. Guidry, A. Poggio, F. Dowla, and C. Hertzog, "Wideband multichannel time-reversal acoustic communications highly reverberant environments," *J. Acoust. Society Amer.*, **120** (2), 838–851, 2006.
34. G. Kitagawa and W. Gersch, *Smoothness Priors Analysis of Time Series* (New York: Springer-Verlag, 1996).

35. A. Doucet, N. de Freitas, and N. Gordon, *Sequential Monte Carlo Methods in Practice* (New York: Springer, 2001).

PROBLEMS

9.1 Suppose we have a noisy binary channel with input (symbol) x and output (symbol) y such that $x \in \mathcal{X} = \{0, 1\}$ and $y \in \mathcal{Y} = \{0, 1\}$. The transition probability matrix is

$$A = \begin{bmatrix} \Pr(y=0|x=0) & \Pr(y=0|x=1) \\ \Pr(y=1|x=0) & \Pr(y=1|x=1) \end{bmatrix} = \begin{bmatrix} 0.85 & 0.15 \\ 0.15 & 0.85 \end{bmatrix}$$

Assume we observe the symbol $y=1$ and $x \sim \Pr(x(0)) = [0.90.1]'$, then

(*a*) What is posterior probability of x?

(*b*) Let $x=1$, what is the posterior if this is true?

(*c*) Let $x=0$, what is the posterior if this is true?

(*d*) Suppose $y=0$, what are the corresponding posteriors of x?

9.2 A fly [4] moves along a straight line in unit increments. At each time period it moves one unit to the left with probability 0.3, one unit to the right with probability 0.3 and stays in place with probability 0.4. A spider is hiding at positions 1 and m: if the fly lands there, it is captured by the spider and the process ends.

(*a*) Construct a Markov chain model, assuming the fly starts at positions $2, \ldots, m-1$.

(*b*) Sketch the corresponding directed graph.

(*c*) What is the probability of the following state evolution sequence: $\Pr(\mathcal{X}_1 = 2, \mathcal{X}_3 = 3, \mathcal{X}_4 = 4|\mathcal{X}_2)$?

9.3 Consider placing a ball in one of the N compartments at each event. Each compartment can hold multiple balls. Let $x_i; i = 0, \ldots, N$ be the state where k compartments are occupied. At the next event, the next ball can go into one of the occupied compartments with probability k/N or an empty compartment.

(*a*) Create a Markov chain for this problem.

(*b*) What is the state diagram?

(*c*) What is the transition matrix, A.

9.4 Suppose we have a discrete state–space and discrete-time measurement system (sampled data). Defining the discrete (finite) states as $x_i(t)$ and the measurements as $y(t)$, then:

(*a*) What is the state prediction probability $\Pr(x_i(t)|Y_{t-1})$?

(*b*) What is the state posterior distribution $\Pr(x_i(t)|Y_t)$?

(*c*) Sketch out the steps of the Bayesian filtering operation at times $t-1$ and t.

9.5 Autoregressive ($AR(N_a)$) models (all-pole) occupy a large part of the signal processing literature especially in speech applications [10].

(*a*) An AR model is an example of a Markov chain on a continuous space, show that the $AR(1)$ model forms a Markov chain, that is,

$$y(t) = a_1 y(t-1) + \epsilon(t), \quad \epsilon \sim \mathcal{N}(0, \sigma^2)$$

(*b*) Show an $AR(N_a)$ model also forms a Markov chain. (*Hint*: Place the AR model in state–space form).

(*c*) Does an $ARMA(N_a, N_a)$ model form a Markov chain as well?

9.6 Autoregressive ($AR(N_a)$) *switching* models also occupy a significant part of the time series literature [14]. It is a model where the mean can switch between two values, μ_0 and μ_1. It enables the time series with several regimes or local nonstationarities to be represented as

$$\begin{aligned}\Pr(y(t)|y(t-1), x(t), x(t-1)) &\sim \mathcal{N}(\mu_{x(t)} + a_1(y(t-1) - \mu_{x(t-1)})), \sigma^2) \\ \Pr(x(t)|x(t-1)) &\sim p_{x(t-1)} \mathcal{I}_{x(t-1)}(x(t)) \\ &\quad + (1 - p_{x(t-1)}) \mathcal{I}_{1-x(t-1)}(x(t))\end{aligned}$$

where the hidden state $x(t)$ takes values in $\{0, 1\}$ with initial values $x(0) = \mu_0$ and $y(0) = 0$ *and* $\mathcal{I}$ *is an indicator function.*

(*a*) What is the sampling distribution $\Pr(x(t)|x(t-1),\ x(t+1),\ y(t),\ y(t-1),\ y(t+1))$?

(*b*) Using the priors $\Pr(\mu_1 - \mu_0) \sim \mathcal{N}(0, \gamma^2)$, $\Pr(a, \sigma^2)$ with $p_0, p_1 \sim \mathcal{U}(0, 1)$ simulate the AR switching model for $N = 500$ samples. What are the final parameter estimates for $\{\mu_0, \mu_1, a, p_1, p_2\}$?

9.7 Consider a four-state Markov switching model [34, 35] with the set of discrete states given by $\mathcal{S} = \{1, 2, 3, 4\}$ representing a Markov chain with transition matrix

$$A = \begin{bmatrix} 1-3\alpha & \alpha & \alpha & \alpha \\ \beta & 1-\beta-2\gamma & \gamma & \gamma \\ \beta & \gamma & 1-\beta-2\gamma & \gamma \\ \beta & \gamma & \gamma & 1-\beta-2 \end{bmatrix}$$

The observation sequence is specified by a set of $AR(2)$ models

$$y(t) = a_{1i} y(t-1) + a_{2i} y(t-2) + \epsilon_i(t) \quad \text{for } \mathcal{S}_i;\ i = 1, \ldots, 4$$

where $\epsilon_i \sim \mathcal{N}(0, \sigma^2)$

(*a*) What is the (state) prediction probability for this model $\Pr(s(t)|Y_{t-1})$?

(*b*) What is the (state) filtering posterior probability for this model $\Pr(s(t)|Y_t)$?

(*c*) What is the likelihood for the following set of parameters describing this model $\Theta = \{\alpha, \beta, \gamma, \sigma^2, a_{1i}, a_{2i}\}; i = 1, \ldots, 4$?

(*d*) Simulate the system with the following parameters $\{\alpha, \beta, \gamma, \sigma^2\} = 0.0033, 0.016, 0.002, 0.1$ and $\{a_{1i}, a_{2i}\} = \{(1.785, -0.903), (1.344, -0.903), (1.386, -0.640), (0.800, -0.640)\}$ for $N = 1000$ samples.

9.8 Suppose we have an HMM model with discrete state–space defined by $\mathcal{X} = \{x_i(t) = \mathcal{X}_i\} i = 1, \ldots, N_x$ with state-transition matrix A. We would like to establish a parameter estimation problem given the observation sequence Y_T. Assume that the complete likelihood is given by $\Pr(y, x|\theta)$ and the density is given by $\Pr(y|x_i, \theta) = \mathcal{N}(\mu_{x(t)}, \Sigma)$ with known diagonal covariance and unknown mean. Develop the EM algorithm to estimate the parameter μ, that is,

(*a*) What is the Q-step?

(*b*) What is the M-step?

9.9 State–space models are perhaps the most important class of linear dynamic systems characterized by

$$x(t+1) = Ax(t) + w(t)$$
$$y(t) = Cx(t) + v(t)$$

Develop the EM algorithm for this class of model based on the usual Gauss–Markov assumptions.

(*a*) Suppose we are asked to estimate the A and C parameters. What is corre-sponding the E-step? (*Hint*: Use gradient techniques.)

(*b*) What is the associated M-step?

10

SEQUENTIAL BAYESIAN DETECTION

10.1 INTRODUCTION

Detection theory and more specifically sequential detection theory is closely coupled to sequential estimation techniques and is often the primary reason for constructing the estimator in the first place. Sequential techniques find application in many technical application areas such as radar, sonar (detection/tracking), biomedical (anomaly detection/localization), speech (recognition/tracking), communications (real-time/obstructed environments), the sciences (e.g., seismology (earthquakes), structures (vibrations), materials (additive manufacturing/threat detection), radiation (threat detection), etc.) and of course, a huge variety of military applications. In this chapter, we develop sequential detection techniques primarily aimed at the binary decision problem. Here, we enable the extension of these estimation methods to an entire class of problems especially when a physical model is available that can be incorporated into the algorithm leading to model-based detection. We develop the Bayesian approach to decision theory primarily aimed at a coupling of sequential Bayesian estimation to sequential decision-making.

Decision theory is rooted in hypothesis testing with the Bayesian approach providing the underlying statistical basis for its evolution. We start with the binary decision problem and develop the usual Bayesian solutions based on probability-of-error minimization leading to the well-known Bayes' risk criterion. Next, the classical Neyman–Pearson detection approach (maximize detection probability for fixed false-alarm probability) is developed and compared to the Bayesian schemes illustrating their similarity and differences. Detectors for multiple measurements,

Bayesian Signal Processing: Classical, Modern, and Particle Filtering Methods, Second Edition. James V. Candy.

multiple hypotheses, and multivariate (vector) processes are developed to provide a more pragmatic set of solutions. Next, we investigate a variety of performance criteria based on the receiver operating characteristic (ROC) curve and its variants that provide the foundations for classical analysis. Other metrics (e.g., area under curve, etc.) associated with the ROC curve are introduced and applied as well. Once these "batch schemes" are developed, we introduce the sequential approach to solving these problems in pseudo real-time evolving to sequential model-based detection. The set of model-based processing schemes are coupled to the detector providing the mechanism to solve a variety of problems in real time. Finally, we investigate an important class of problems that motivate the application of model-based detection—failure detection/classification. Here we develop the approach using the model-based detection schemes and conclude with a realistic case study.

10.2 BINARY DETECTION PROBLEM

The basic detection problem that we face is that we are given a set of noisy measurement data that has been generated by some unknown source and we are asked to "decide" whether it is present or not. This formulation is the simple radar, sonar, tumor, flaw, or threat detection problems. Thus, the fundamental decision problem is that we have a random source or target (Θ) that is known by either hypotheses ($\mathcal{H}_0$ or $\mathcal{H}_1$) generating a unique signal ($S(\Theta)$) that is observed or measured [1–3]. Both source and signal reside in the source and signal spaces. The signal is "mapped" from the signal space through some probabilistic transition mechanism ($\Pr(\mathbf{Y}|\mathbf{S}(\Theta))$) creating points in the measurement space ($\mathbf{Y}$). Finally, the measurement is mapped into an equivalent decision space, ($\mathbf{D}$) that is made up of disjoint regions of the measurement space ($\mathbf{Y} \Longleftrightarrow \mathbf{D} = \cup_i \mathbf{D}_i$) by a decision function or rule ($\mathcal{D}(\mathbf{Y}|\mathcal{H}_\ell)$). Here the measurement is mapped into subregions of the measurement space that are defined explicitly by the particular decision function (similar to clustering), that is, the decision function partitions the measurement space into a number of disjoint regions corresponding to each hypothesis. The basic decision problem is illustrated (simply) in Fig. 10.1. The true source generates one of the hypotheses as an output that initiates a signal. For instance, a code "A" might correspond to a unit positive pulse while a "B" might correspond to a unit negative pulse. The signal probabilistic transition mechanism, which knows the *true* hypotheses, separates these hypothesis from the measurements and generates points in the measurement space according to some probabilistic distribution. A decision rule (function) using the measurements is developed to decide which of the hypotheses to accept as correct. The ultimate decision is based on knowledge of the a priori probability of each hypothesis (Bayesian) and of any of the conditional probabilities inherent in the transition mechanism.

Formally, we can define the simple binary detection problem of a noisy measurement $y(t)$ consisting of a signal $s(t)$ in noise $v(t)$ or just noise as

$$
\begin{aligned}
\mathcal{H}_0 &: y(t) = v(t) && \text{[Noise]} \\
\mathcal{H}_1 &: y(t) = s(t) + v(t) && \text{[Signal+Noise]}
\end{aligned}
\tag{10.1}
$$

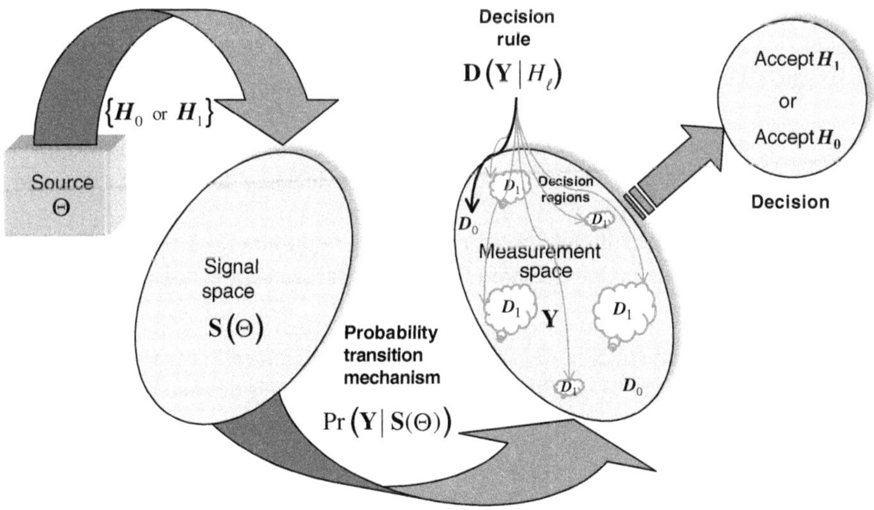

FIGURE 10.1 Fundamental decision problem: (a) source, Θ (Truth); (b) signal space, **S**(Θ); (c) measurement space, **Y** with probability transition (Pr(**Y**|**S**(Θ)); (d) decision space, **D** = $\{D_0 \cup D_1\}$ with decision rule $\mathcal{D}(\mathbf{Y}|\mathcal{H}_\ell)$.

where the subscript notation refers to the noise ("0") or alternate signal hypothesis ("1") of the test corresponding to the two respective hypotheses $(\mathcal{H}_0, \mathcal{H}_1)$ and corresponding decision regions (D_0, D_1) that partition the observation or measurement space.

Perhaps the simplest approach is to solve the "single" measurement problem in order to illustrate the approach that is taken in classically solving this type of binary decision problem. The rudiments of decision theory follow directly from this problem.

10.2.1 Classical Detection

In order to solve the classical binary detection problem, we require the following ingredients: the underlying probabilistic transition mechanisms under each hypothesis $\Pr(y(t)|\mathcal{H}_\ell); \ell = 0, 1$; the prior probabilities $\Pr(\mathcal{H}_\ell); \ell = 0, 1$; a decision rule $\mathcal{D}(y(t)|\mathcal{H}_\ell)); \ell = 0, 1$ and some manner to determine a threshold for the test. That is,

$$\mathcal{D}\big(y(t)\big) \underset{\mathcal{H}_0}{\overset{\mathcal{H}_1}{\gtrless}} \tau \tag{10.2}$$

which implies that if the decision function is greater than the threshold τ, accept hypothesis $\mathcal{H}_1$ that the signal is present or if it is less than the threshold, accept the null hypothesis $\mathcal{H}_0$ that the signal is not present.

Detectors are not perfect by any means, so to be realistic, we will require some manner to capture the errors that can be made. Therefore, we define four probabilities associated with our decision problem:

P_{DET}	a signal detection declared when the measurement is a signal (correct)
P_{FA}	a false alarm when a signal detection is declared and the measurement is noise (wrong)
P_{Miss}	a noise detection declared when the measurement is a signal (miss)
P_{Reject}	a noise detection declared when the measurement is noise (correct)

Based on our probabilistic transition mechanism (assumed known), we are able to calculate these probabilities. The *probability of detection* is defined by

$$P_{\text{DET}}(\tau) = \int_{\tau:D_1}^{\infty} \Pr\big(\mathcal{D}(y(t))|\mathcal{H}_1\big)d\mathcal{D} \tag{10.3}$$

with D_1 a partition of the measurement space. The *probability of false alarm* is defined by

$$P_{\text{FA}}(\tau) = \int_{\tau:D_1}^{\infty} \Pr\big(\mathcal{D}(y(t))|\mathcal{H}_0\big)d\mathcal{D} \tag{10.4}$$

Note that this integral is performed over the region D_1 leading to false alarms as illustrated in Fig. 10.2. Also related to these two probabilities is the *probability of a miss* given by

$$P_{\text{Miss}}(\tau) = \int_{-\infty}^{\tau:D_0} \Pr\big(\mathcal{D}(y(t))|\mathcal{H}_1\big)d\mathcal{D} = 1 - P_{\text{DET}}(\tau) \tag{10.5}$$

There are errors associated with each "wrong" decision. Suppose the decision space is partitioned into two regions (binary: signal/noise problem) with region D_0 corresponding to the noise ($\mathcal{H}_0$) and region D_1 corresponding to the signal ($\mathcal{H}_1$). There are two possible mechanisms in which an error can occur: (1) either a decision $\mathcal{D}(y(t))$ falls into region D_1 and the measurement is noise or (2) the measurement is a signal and falls into D_0. Since these events are mutually exclusive, we can define the *total probability-of-error* as

$$P_{\text{Error}} := \Pr\big(\mathcal{D} \in D_1, D_0\big) + \Pr\big(\mathcal{D} \in D_0, D_1\big) \tag{10.6}$$

applying Bayes' rule $\big(\Pr(A,B) = \Pr(A|B) \times \Pr(B)\big)$ to the joint distribution gives

$$P_{\text{Error}} = \Pr\big(\mathcal{D}|\mathcal{H}_1\big) \times \Pr\big(D_0\big) + \Pr\big(\mathcal{D}|\mathcal{H}_0\big) \times \Pr\big(D_1\big) \tag{10.7}$$

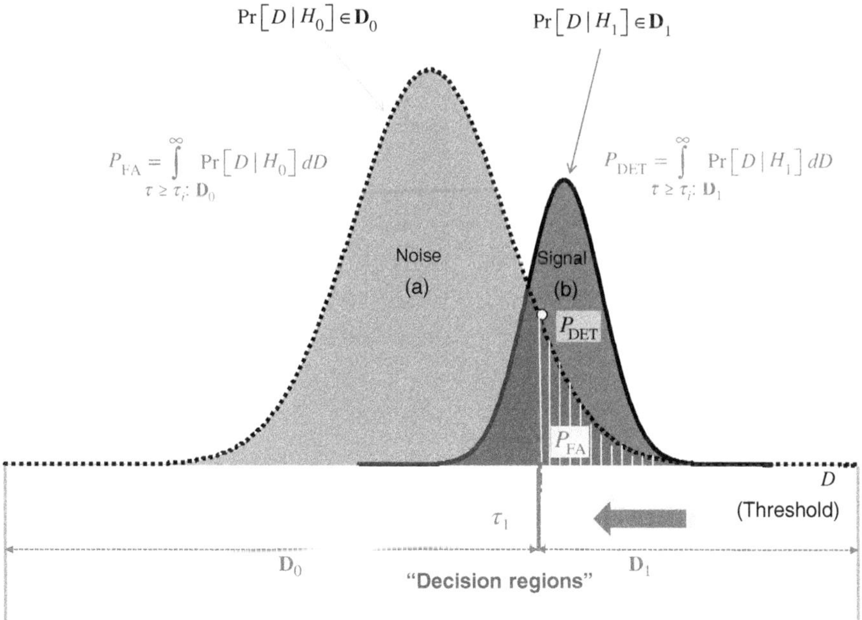

FIGURE 10.2 Decision function probability distributions with defined decision regions: (a) noise probability distribution $\Pr(\mathcal{D}(y(t))|\mathcal{H}_0)$; (b) signal probability distribution $\Pr(\mathcal{D}(y(t))|\mathcal{H}_1)$.

where $\Pr(D_0)$, $\Pr(D_1)$ are the prior probabilities (prevalence [4]) associated with each hypothesis and of course $\mathcal{H}_\ell$ is the hypothesis associated with its corresponding decision region $\mathcal{D}_\ell$.

We now have all of the ingredients we need to develop the detector except the most crucial—the decision function. There are a wide variety of decision functions that have evolved and they are typically based on a particular decision criterion. We start with possibly the most intuitive—the maximum likelihood decision criterion that is based on the likelihood probabilities under each hypothesis $\Pr(y(t)|\mathcal{H}_\ell)$; $\ell = 0, 1$. We decide on the "most likely" hypothesis based on its likelihood, that is, the *likelihood decision function* is [5]

$$\mathcal{D}(y(t)) = \begin{cases} \mathcal{D}(y(t)|\mathcal{H}_0)), & \text{if } \Pr(y(t)|\mathcal{H}_0) > \Pr(y(t)|\mathcal{H}_1) \\ \mathcal{D}(y(t)|\mathcal{H}_1)), & \text{if } \Pr(y(t)|\mathcal{H}_1) > \Pr(y(t)|\mathcal{H}_0) \end{cases} \tag{10.8}$$

Here we have defined the *decision regions* as

$$\begin{aligned} D_0 &= \{y(t) : \Pr(y(t)|\mathcal{H}_0) > \Pr(y(t)|\mathcal{H}_1)\} \\ D_1 &= \{y(t) : \Pr(y(t)|\mathcal{H}_0) < \Pr(y(t)|\mathcal{H}_1)\} \end{aligned} \tag{10.9}$$

So we see that the decision rule is to select the hypothesis that is most likely.

An equivalent method for representing this decision rule is to define the *likelihood-ratio* by

$$\mathcal{D}(y(t)) \longrightarrow \mathcal{L}(y(t)) := \frac{\Pr(y(t)|\mathcal{H}_1)}{\Pr(y(t)|\mathcal{H}_0)} \tag{10.10}$$

with decision regions

$$\begin{aligned} D_0 &= \{y(t) : \mathcal{L}(y(t)) < 1\} \\ D_1 &= \{y(t) : \mathcal{L}(y(t)) > 1\} \end{aligned} \tag{10.11}$$

leading to the *maximum likelihood detector* (ML)

$$\mathcal{L}(y(t)) \underset{\mathcal{H}_0}{\overset{\mathcal{H}_1}{\gtrless}} 1 \tag{10.12}$$

Consider the following example of detecting a constant signal in Gaussian noise.

Example 10.1

Suppose we have a constant signal $s(t) = \mu \ \forall \ t$ that is contaminated with additive zero-mean, Gaussian measurement noise $v(t)$ of variance R_{vv} and we would like to construct a likelihood detector to decide whether or not the signal is present. With the signal present, the measurement model is simply

$$y(t) = s(t) + v(t)$$

and the probability distributions under each hypothesis are Gaussian:

$$\begin{aligned} \Pr(y(t)|\mathcal{H}_0) &= \frac{1}{\sqrt{2\pi R_{vv}}} \exp\left\{ -\frac{1}{2R_{vv}} y^2(t) \right\} \\ \Pr(y(t)|\mathcal{H}_1) &= \frac{1}{\sqrt{2\pi R_{vv}}} \exp\left\{ -\frac{1}{2R_{vv}} (y(t)-\mu)^2 \right\} \end{aligned}$$

The likelihood decision function is

$$\mathcal{L}(y(t)) = \frac{\Pr(y(t)|\mathcal{H}_1)}{\Pr(y(t)|\mathcal{H}_0)} = \frac{\frac{1}{\sqrt{2\pi R_{vv}}} \exp\left\{ -\frac{1}{2R_{vv}} (y(t)-\mu)^2 \right\}}{\frac{1}{\sqrt{2\pi R_{vv}}} \exp\left\{ -\frac{1}{2R_{vv}} y^2(t) \right\}}$$

or

$$\mathcal{L}(y(t)) = \exp\left\{-\frac{1}{2R_{vv}}(-2y(t)\mu + \mu^2)\right\} = \exp\left\{\frac{1}{R_{vv}}\left(y(t)\mu - \frac{\mu^2}{2}\right)\right\}$$

which for a given μ and R_{vv} yield the decision rule

$$\mathcal{L}(y(t)) = \exp\left\{\frac{1}{R_{vv}}\left(y(t)\mu - \frac{\mu^2}{2}\right)\right\} \underset{\mathcal{H}_0}{\overset{\mathcal{H}_1}{\gtrless}} 1$$

△△△

10.2.2 Bayesian Detection

The Bayesian approach to the binary detection problem revolves around the probability of various hypotheses given the data, while the classical approach revolves around the probability of the data given the specific hypothesis. As before, similar to the development of the maximum likelihood detector, we select the maximum a posteriori (MAP) criterion for our Bayesian decision function specified by

$$\mathcal{D}(y(t)) = \begin{cases} \mathcal{D}(y(t)|\mathcal{H}_0), & \text{if } \Pr(\mathcal{H}_0|y(t)) > \Pr(\mathcal{H}_1|y(t)) \\ \mathcal{D}(y(t)|\mathcal{H}_1), & \text{if } \Pr(\mathcal{H}_1|y(t)) > \Pr(\mathcal{H}_0|y(t)) \end{cases} \tag{10.13}$$

Here we have defined the *decision regions* as

$$\begin{aligned} D_0 &= \{y(t) : \Pr(\mathcal{H}_0|y(t)) > \Pr(\mathcal{H}_1|y(t))\} \\ D_1 &= \{y(t) : \Pr(\mathcal{H}_0|y(t)) < \Pr(\mathcal{H}_1|y(t))\} \end{aligned} \tag{10.14}$$

As before, an equivalent method for representing this decision rule is to define the *likelihood-ratio* by the ratio of posterior distributions

$$\mathcal{D}(y(t)) \longrightarrow \mathcal{L}(y(t)) := \frac{\Pr(\mathcal{H}_1|y(t))}{\Pr(\mathcal{H}_0|y(t))} = \frac{\Pr(y(t)|\mathcal{H}_1) \times \Pr(\mathcal{H}_1)}{\Pr(y(t)|\mathcal{H}_0) \times \Pr(\mathcal{H}_0)} \tag{10.15}$$

where we have applied Bayes' rule (likelihood × prior). Therefore, the decision regions become

$$\begin{aligned} D_0 &= \left\{y(t) : \mathcal{L}(y(t)) < \frac{\Pr(\mathcal{H}_0)}{\Pr(\mathcal{H}_1)}\right\} \\ D_1 &= \left\{y(t) : \mathcal{L}(y(t)) > \frac{\Pr(\mathcal{H}_0)}{\Pr(\mathcal{H}_1)}\right\} \end{aligned} \tag{10.16}$$

leading to the MAP detector

$$\mathcal{L}\big(y(t)\big) \underset{\mathcal{H}_0}{\overset{\mathcal{H}_1}{\gtrless}} \frac{\Pr(\mathcal{H}_0)}{\Pr(\mathcal{H}_1)} \tag{10.17}$$

So we note that in the Bayesian case, the detector is based on the conditional and prior probabilities as mentioned previously. Let us return to the Gaussian example to illustrate the differences between the classical likelihood and MAP detection schemes.

Example 10.2

We have a constant signal $s(t) = \mu \; \forall \; t$ contaminated with additive zero-mean, Gaussian measurement noise $v(t)$ of variance R_{vv} and we would like to construct a MAP detector to decide whether or not the signal is present. The measurement model is

$$y(t) = s(t) + v(t)$$

and the probability distributions under each hypothesis are Gaussian as before in the maximum likelihood case, but now we also have a *prior probability* $\mathcal{P}$ to incorporate so that

$$\Pr(\mathcal{H}_1) = \mathcal{P}$$
$$\Pr(\mathcal{H}_0) = 1 - \mathcal{P}$$

The MAP decision function (as in the previous example) is given by

$$\mathcal{L}(y(t)) = \exp\left\{\frac{1}{R_{vv}}(y(t)\mu - \frac{\mu^2}{2})\right\} \underset{\mathcal{H}_0}{\overset{\mathcal{H}_1}{\gtrless}} \frac{1-\mathcal{P}}{\mathcal{P}}$$

for a given μ and R_{vv}. △△△

Before we leave this discussion of the binary detection problem, let us consider one extension that of unknown parameters in either or both signal and noise models that leads us to the so-called composite hypothesis test.

10.2.3 Composite Binary Detection

In many applications the signal and noise uncertainties are unknown and possibly random, but can be parameterized in some form such as an unknown signal amplitude with a random phase or a distribution with unknown mean and variance. Binary detection problems in which the hypotheses do not have any unknown or random parameters are termed *simple* hypotheses, while if they do possess unknown/random parameters they are termed *composite*. In this case the decision problem is formulated with a parameter vector, say $\boldsymbol{\Theta}$, which in the classical case can be considered deterministic but unknown or in the random case characterized by its distribution $\Pr(\boldsymbol{\Theta})$. Let us return to the binary problem as before and incorporate the unknown vector of parameters.

Formally, we can define the composite binary detection problem of a noisy measurement $y(t)$ consisting of a signal $s(t;\theta_s)$ in noise $v(t;\theta_v)$ or just noise as

$$\begin{aligned} \mathcal{H}_0 &: y(t) = v(t;\theta_v) && \text{[Noise+Uncertainties]} \\ \mathcal{H}_1 &: y(t) = s(t;\theta_s) + v(t;\theta_v) && \text{[Signal+Noise+Uncertainties]} \end{aligned} \tag{10.18}$$

where the unknown/random parameters for the signal and noise are represented by θ_s and θ_v and can be captured functionally in the parameter vector $\boldsymbol{\Theta}$.

We start with the *likelihood-ratio* decision function from the Bayesian perspective and assume that $\boldsymbol{\Theta}$ is random with prior distribution $\Pr(\boldsymbol{\Theta})$, then we have the joint distribution

$$\mathcal{L}(y(t);\boldsymbol{\Theta}) := \frac{\Pr(y(t),\boldsymbol{\Theta}|\mathcal{H}_1)}{\Pr(y(t),\boldsymbol{\Theta}|\mathcal{H}_0)} = \frac{\Pr(y(t)|\boldsymbol{\Theta},\mathcal{H}_1)\times\Pr(\boldsymbol{\Theta}|\mathcal{H}_1)}{\Pr(y(t)|\boldsymbol{\Theta},\mathcal{H}_0)\times\Pr(\boldsymbol{\Theta}|\mathcal{H}_0)} \tag{10.19}$$

where we have applied Bayes' rule (likelihood × prior) as before.

For the case where we assume that the parameters are continuous and *random* with a prior, then the *composite likelihood-ratio* is determined in terms of the corresponding marginal conditional distributions as

$$\mathcal{L}(y(t);\boldsymbol{\Theta}) := \frac{\int_{-\infty}^{\infty}\Pr(y(t),\boldsymbol{\Theta}|\mathcal{H}_1)d\boldsymbol{\Theta}}{\int_{-\infty}^{\infty}\Pr(y(t),\boldsymbol{\Theta}|\mathcal{H}_0)d\boldsymbol{\Theta}} = \frac{\int_{-\infty}^{\infty}\Pr(y(t)|\boldsymbol{\Theta},\mathcal{H}_1)\times\Pr(\boldsymbol{\Theta}|\mathcal{H}_1)d\boldsymbol{\Theta}}{\int_{-\infty}^{\infty}\Pr(y(t)|\boldsymbol{\Theta},\mathcal{H}_0)\times\Pr(\boldsymbol{\Theta}|\mathcal{H}_0)d\boldsymbol{\Theta}} \tag{10.20}$$

That is, by integrating over $\boldsymbol{\Theta}$, the likelihood-ratio is determined and the composite hypothesis test is now treated just as a simple test.

However, if the parameters are assumed non-random but unknown or random without a known distribution, then the so-called *generalized likelihood-ratio test* (GLRT) evolves. Here an estimate of the parameter $\hat{\boldsymbol{\Theta}}$ under each hypothesis is performed *first* and then the simple hypothesis test is performed, that is, we have

$$\mathcal{L}(y(t);\hat{\boldsymbol{\Theta}}) := \frac{\underset{\boldsymbol{\Theta}}{\arg\max}\ \Pr(y(t),\boldsymbol{\Theta}|\mathcal{H}_1)}{\underset{\boldsymbol{\Theta}}{\arg\max}\ \Pr(y(t),\boldsymbol{\Theta}|\mathcal{H}_0)} \tag{10.21}$$

Let us consider the previous example and now incorporate an unknown parameter θ into the detection problem [5].

Example 10.3

We have a random signal $s(t) = \theta$ contaminated with additive zero-mean, Gaussian measurement noise $v(t)$ of known variance R_{vv} and we would like to construct both MAP and maximum likelihood detectors to decide whether or not the signal is present. The measurement model is

$$y(t) = s(t; \theta) + v(t)$$

and the probability distributions under each hypothesis are Gaussian and have a prior probability $\Pr(\mathcal{H}_1) = \mathcal{P}$ as before and a prior parametric distribution given by

$$\Pr(\theta|\mathcal{H}_1) = \begin{cases} 1 & \text{if } 1 \leq \theta < 2 \\ 0 & \text{otherwise} \end{cases}$$

The composite MAP of Eq. 10.20 is therefore

$$\mathcal{L}\big(y(t); \theta\big) = \frac{\frac{1}{\sqrt{2\pi R_{vv}}} \int_1^2 \exp\left\{ -\frac{1}{2R_{vv}}(y(t) - \theta)^2 \right\} d\theta}{\frac{1}{\sqrt{2\pi R_{vv}}} \exp\left\{ -\frac{1}{2R_{vv}} y^2(t) \right\}}$$

or

$$\mathcal{L}(y(t); \theta) = \frac{\int_1^2 \exp\left\{ -\frac{1}{2R_{vv}}(y(t) - \theta)^2 \right\} d\theta}{\exp\left\{ -\frac{1}{2R_{vv}} y^2(t) \right\}} \overset{\mathcal{H}_1}{\underset{\mathcal{H}_0}{\gtrless}} \frac{1 - \mathcal{P}}{\mathcal{P}}$$

for a specified R_{vv}.

Next, the maximum likelihood decision function is specified by first obtaining the maximum likelihood parameter estimator over each hypothesis. For this problem, we need only to maximize the conditional probability under $\mathcal{H}_1$ which is obtained by differentiating this distribution, setting the result to zero and solving to obtain

$$\hat{\Theta}_{\text{ML}} = \underset{\boldsymbol{\Theta}}{\arg\max} \Pr(y(t), \boldsymbol{\Theta}|\mathcal{H}_1) = y(t)$$

and substituting into the conditional distribution, we obtain

$$\Pr(y(t), \hat{\boldsymbol{\Theta}}|\mathcal{H}_1)\Big|_{\Theta = \hat{\Theta}_{\text{ML}}} = \frac{1}{\sqrt{2\pi R_{vv}}} \exp\left\{ -\frac{1}{2R_{vv}}(y(t) - \hat{\Theta}_{\text{ML}})^2 \right\} = \frac{1}{\sqrt{2\pi R_{vv}}}$$

$$\mathcal{L}(y(t); \Theta) = \exp\left\{ \frac{1}{2R_{vv}} y^2(t) \right\} \overset{\mathcal{H}_1}{\underset{\mathcal{H}_0}{\gtrless}} 1$$

△△△

This completes the discussion of the classical and Bayesian approaches to decision theory, next we extend these results leading to a variety of the popular detection techniques.

10.3 DECISION CRITERIA

In detection theory, each decision function defines a particular and unique scheme which can range from *ad hoc* techniques like "flipping a coin" to a sophisticated and complex mapping of the measurement space into (hopefully) disjoint decision regions. In this section, we concentrate on various decision criteria and show how they lead to a variety of so-called "optimal" decision functions.

10.3.1 Probability-of-Error Criterion

We start with the total *probability-of-error* usually termed "Bayes' error" and develop the Bayesian solution to minimize this error. Recall that the total probability-of-error with decision regions under each hypotheses defined by D_0 and D_1 is given by [5]

$$P_{\text{Error}} = \Pr(D_1|\mathcal{H}_0) \times \Pr(\mathcal{H}_0) + \Pr(D_0|\mathcal{H}_1) \times \Pr(\mathcal{H}_1) \tag{10.22}$$

and since the priors are mutually exclusive, we have

$$\Pr(\mathcal{H}_0) + \Pr(\mathcal{H}_1) = 1 \tag{10.23}$$

We can express the $\Pr(D_0|\mathcal{H}_1)$ as

$$\Pr(D_0|\mathcal{H}_1) = \int_{D_0} \Pr(y(t)|\mathcal{H}_1)dy + \int_{D_1} \Pr(y(t)|\mathcal{H}_1)dy - \int_{D_1} \Pr(y(t)|\mathcal{H}_1)dy \tag{10.24}$$

or

$$\Pr(D_0|\mathcal{H}_1) = 1 - \int_{D_1} \Pr(y(t)|\mathcal{H}_1)dy = 1 - \Pr(D_1|\mathcal{H}_1) \tag{10.25}$$

since the sum of the first two terms is unity. We can also express

$$\Pr(D_1|\mathcal{H}_0) = 1 - \int_{D_1} \Pr(y(t)|\mathcal{H}_0)dy \tag{10.26}$$

as a function of D_1. With these relations in mind, the probability-of-error of Eq. 10.22 can be expressed as

$$\begin{aligned} P_{\text{Error}} &= \Pr(\mathcal{H}_0)\int_{D_1}\Pr(y(t)|\mathcal{H}_0)dy + \Pr(\mathcal{H}_1)\left[1 - \int_{D_1}\Pr(y(t)|\mathcal{H}_1)dy\right] \\ &= \Pr(\mathcal{H}_1) + \int_{D_1}\left[\Pr(\mathcal{H}_0)\Pr(y(t)|\mathcal{H}_0) - \Pr(\mathcal{H}_1)\Pr(y(t)|\mathcal{H}_1)\right]dy \end{aligned} \quad (10.27)$$

Minimizing this expression is accomplished by selecting the decision region D_1 such that the second term in braces is negative, that is,

$$D_1 = \left\{y(t) : \left[\Pr(\mathcal{H}_0)\Pr(y(t)|\mathcal{H}_0) - \Pr(\mathcal{H}_1)\Pr(y(t)|\mathcal{H}_1)\right] < 0\right\} \quad (10.28)$$

leading to region $D_0 > 0$. Therefore, in terms of the likelihood-ratio, we have the equivalent decision region

$$D_1 = \left\{y(t) : \mathcal{L}(y(t)) > \frac{\Pr(\mathcal{H}_1)}{\Pr(\mathcal{H}_0)}\right\} \quad (10.29)$$

and therefore the decision rule for the minimum *probability-of-error* (P-E) detector is

$$\mathcal{L}(y(t)) \underset{\mathcal{H}_0}{\overset{\mathcal{H}_1}{\gtrless}} \frac{\Pr(\mathcal{H}_0)}{\Pr(\mathcal{H}_1)} \quad (10.30)$$

Comparing this result to the MAP detector, we see that they are *identical*, that is,

$$\mathcal{L}_{P-E}(y(t)) \Longleftrightarrow \mathcal{L}_{\text{MAP}}(y(t))$$

10.3.2 Bayes Risk Criterion

We generalize our previous results for the MAP detector to incorporate a set of "cost functions" that enable the designer to penalize errors that might be quite serious for a particular application. For instance, a high miss probability in a stock transaction may cost profit, but a miss of a tumor in medicine may cost a life.

We define the following *costs* in the *binary* detection problem associated with the types of errors that can evolve

C_{00}——cost of accepting $\mathcal{H}_0$ when $\mathcal{H}_0$ is true (right)
C_{01}——cost of accepting $\mathcal{H}_0$ when $\mathcal{H}_1$ is true (wrong)

C_{10}——cost of accepting $\mathcal{H}_1$ when $\mathcal{H}_0$ is true (wrong)
C_{11}—cost of accepting $\mathcal{H}_1$ when $\mathcal{H}_1$ is true (right)

These costs are basically meaningless unless we know their a priori probabilities; therefore, we define the *Bayes' risk* (B-R) as the "expected" cost for these cost as

$$\mathfrak{R} = \mathrm{E}\{\mathbf{C}\} = \sum_{i,j} C_{ij} \times \Pr(\mathcal{H}_i|\mathcal{H}_j) \times \Pr(\mathcal{H}_{i\,\mathrm{or}\,j}) \tag{10.31}$$

or expanding

$$\begin{aligned}\mathfrak{R} &= C_{00}\Pr(\mathcal{H}_0|\mathcal{H}_0) \times \Pr(\mathcal{H}_0) + C_{01}\Pr(\mathcal{H}_0|\mathcal{H}_1) \times \Pr(\mathcal{H}_1)\\ &\quad + C_{10}\Pr(\mathcal{H}_1|\mathcal{H}_0) \times \Pr(\mathcal{H}_0) + C_{11}\Pr(\mathcal{H}_1|\mathcal{H}_1) \times \Pr(\mathcal{H}_1)\end{aligned}$$

Grouping this expression in terms of each decision region, we have

$$\begin{aligned}\mathfrak{R} &= \left[C_{00}\Pr(\mathcal{H}_0|\mathcal{H}_0)\Pr(\mathcal{H}_0) + C_{01}\Pr(\mathcal{H}_0|\mathcal{H}_1)\Pr(\mathcal{H}_1)\right]_{D_0}\\ &\quad + \left[C_{10}\Pr(\mathcal{H}_1|\mathcal{H}_0)\Pr(\mathcal{H}_0) + C_{11}\Pr(\mathcal{H}_1|\mathcal{H}_1)\Pr(\mathcal{H}_1)\right]_{D_1}\end{aligned}$$

Since the decision function maps the measurements into disjoint regions ($\mathbf{Y} \Longleftrightarrow \mathbf{D} = \cup_i \mathbf{D}_i$), we have that each of these terms can be specified directly within the measurement space as

$$\begin{aligned}\mathfrak{R} &= \left[\int_{D_0} C_{00}\Pr(y(t)|\mathcal{H}_0)\Pr(\mathcal{H}_0) + C_{01}\Pr(y(t)|\mathcal{H}_1)\Pr(\mathcal{H}_1)dy\right]\\ &\quad + \left[\int_{D_1} C_{10}\Pr(y(t)|\mathcal{H}_0)\Pr(\mathcal{H}_0) + C_{11}\Pr(y(t)|\mathcal{H}_1)\Pr(\mathcal{H}_1)dy\right]\end{aligned}$$

To minimize Bayes' risk with respect to the decision regions $\mathbf{D}_0$ and $\mathbf{D}_1$, we select hypothesis $\mathcal{H}_1$ for those values of $y(t)$ that yield a greater value for the second integral, that is,

$$\begin{aligned}&C_{10}\Pr(y(t)|\mathcal{H}_0)\Pr(\mathcal{H}_0) + C_{11}\Pr(y(t)|\mathcal{H}_1)\Pr(\mathcal{H}_1)\\ &\qquad > C_{00}\Pr(y(t)|\mathcal{H}_0)\Pr(\mathcal{H}_0) + C_{01}\Pr(y(t)|\mathcal{H}_1)\Pr(\mathcal{H}_1)\end{aligned}$$

hypothesis $\mathcal{H}_0$ is selected when the inequality is reversed; therefore, combining like-terms, we have

$$(C_{10} - C_{00})\Pr(\mathcal{H}_0)\Pr(y(t)|\mathcal{H}_0) \underset{>}{<} (C_{01} - C_{11})\Pr(\mathcal{H}_1)\Pr(y(t)|\mathcal{H}_1) \tag{10.32}$$

creating the *likelihood-ratio* as before,

$$\mathcal{L}\big(y(t)\big) = \frac{\Pr(y(t)|\mathcal{H}_1)}{\Pr(y(t)|\mathcal{H}_0)}$$

we obtain the *Bayes' risk* detector as

$$\mathcal{L}\big(y(t)\big) \underset{\mathcal{H}_0}{\overset{\mathcal{H}_1}{\gtrless}} \frac{\Pr(\mathcal{H}_0)\big(C_{10} - C_{00}\big)}{\Pr(\mathcal{H}_1)\big(C_{01} - C_{11}\big)} \tag{10.33}$$

Compare this to Eq. 10.17 (and Example 10.2) and we see (as before) that both schemes "minimize the probability-of-error" and are identical if we let $C_{00} = C_{11} = 0$ and $C_{01} = C_{10} = 1$. The advantage of the Bayes' risk scheme is that we *penalize* our errors heavily depending on the severity of the error relative to our application. Next, we consider a different approach to the binary detection problem in which we focus on the detection probability.

10.3.3 Neyman–Pearson Criterion

In decision problems lacking in any a priori information such as prior probabilities or costs where it is challenging to construct any reasonable guesses, the Neyman–Pearson detector has evolved as a predominant choice under these conditions. This scheme evolves directly from an optimization perspective and has motivated a large number of "optimization-based" approaches to solve the binary detection problem [6]. The Neyman–Pearson (N-P) approach constrains the false-alarm probability while maximizing the detection probability, since it is not possible to simultaneously maximize detection and minimize false alarms. It is considered a *frequentist* rather than a statistical or equivalently *Bayesian* approach [6–8]; however, it is applied quite heavily in engineering applications.

Formally, the *Neyman–Pearson decision criterion* can be stated as "fix the probability of false-alarm P_{FA} at probability p and then maximize the probability of detection P_{DET}". That is,

$$\arg\max_{D_1} P_{\text{DET}} \text{ subject to } P_{\text{FA}} = p$$

Following the usual Lagrange optimization approach with Lagrange multiplier λ, we have the augmented decision criterion in which we would like to select the decision region $\mathbf{D}_1$ to maximize [5]

$$\mathcal{J}_{\text{N-P}} = \Pr(\mathcal{H}_1|\mathcal{H}_1) - \lambda \times \big[\Pr(\mathcal{H}_1|\mathcal{H}_0) - p\big] \tag{10.34}$$

Substituting the conditional distributions, we have

$$\mathcal{J}_{\text{N-P}} = \int_{D_1} \Pr(y(t)|\mathcal{H}_1)dy - \lambda \times \int_{D_1} \left[\Pr(y(t)|\mathcal{H}_0) - p\right]dy \tag{10.35}$$

We see that the maximization problem is now unconstrained but with the Lagrange multiplier λ parameterizing D_1. It is selected to satisfy the constraint such that the last term of Eq. 10.34 is zero and therefore the detection probability $\Pr(\mathcal{H}_1|\mathcal{H}_1)$ is maximized.

Grouping terms in Eq. 10.35 we have

$$\mathcal{J}_{\text{N-P}} = \int_{D_1} \left[\Pr(y(t)|\mathcal{H}_1) - \lambda\Pr(y(t)|\mathcal{H}_0)\right]dy + \lambda p \tag{10.36}$$

To maximize this expression by the choice of D_1, we see the last term is fixed (assuming λ is positive) and the first term is positive for

$$D_1 = \left\{y(t) : \left[\Pr(y(t)|\mathcal{H}_1) - \lambda\Pr(y(t)|\mathcal{H}_0)\right] > 0\right\} \tag{10.37}$$

subject to the constraint that

$$\Pr(\mathcal{H}_1|\mathcal{H}_0) = \int_{D_1} \Pr(y(t)|\mathcal{H}_0)dy = p \tag{10.38}$$

The fundamental approach is to calculate $\Pr(\mathcal{H}_1|\mathcal{H}_0)$ as a function of λ and find the value of λ such that $\Pr(\mathcal{H}_1|\mathcal{H}_0) = p$.

Therefore, in terms of likelihood-ratio we have that

$$\mathcal{L}\big(y(t)\big) = \frac{\Pr(y(t)|\mathcal{H}_1)}{\Pr(y(t)|\mathcal{H}_0)}$$

we obtain the *Neyman–Pearson* detector as

$$\mathcal{L}\big(y(t)\big) \underset{\mathcal{H}_0}{\overset{\mathcal{H}_1}{\gtrless}} \lambda \tag{10.39}$$

We again note that similar to the techniques we have discussed, the decision functions are identical in form testing a *likelihood-ratio* against a threshold. It is the thresholds that differ in each approach: ML is a unit threshold, MAP is a ratio of priors, B- R is a ratio of priors with cost constraints and N-P is λ constrained by a fixed false-alarm probability p. To illustrate the N-P scheme, consider our usual Gaussian example as before.

Example 10.4

Again we have a constant signal $s(t) = \mu \; \forall \; t$ that is contaminated with additive zero-mean, Gaussian measurement noise $v(t)$ of variance R_{vv} and we would like to construct an N-P detector to decide whether or not the signal is present under the constraint that the false-alarm probability must be β.

Recall from Example 10.1 that the Neyman–Pearson (likelihood) decision function is given by

$$\mathcal{L}(y(t)) = \exp\left\{\frac{1}{R_{vv}}\left(y(t)\mu - \frac{\mu^2}{2}\right)\right\} \underset{\mathcal{H}_0}{\overset{\mathcal{H}_1}{\gtrless}} \lambda$$

To determine λ, we must solve the integral equation for

$$\Pr(\mathcal{H}_1|\mathcal{H}_0) = \int_{D_1} \Pr(y(t)|\mathcal{H}_0)dy = \int_{\lambda}^{\infty} \frac{1}{\sqrt{2\pi R_{vv}}} \exp\left\{-\frac{1}{2R_{vv}} y^2(t)\right\}dy = \beta$$

which for the Gaussian case can easily be determined, once β is specified. △△△

This completes the section on the variety of decision criteria that we will consider for our problems; however, we have only considered a *single measurement* at time t, $y(t)$—how do we handle multiple or a "batch" of measurements?

It should also be noted that there are a wide variety of optimization-based decision criteria (e.g., minimax, local optimal) that exist, but are beyond the scope of this text (see Refs. [3,5–9] for more detailed texts).

10.3.4 Multiple (Batch) Measurements

In most pragmatic problems, multiple measurements are required to make a reasonable decision with any confidence at all. There are some nuances that evolve when considering a batch of data. To see this, define the batch by the set of measurements $Y_N = \{y(0), y(1), \ldots, y(N)\}$ and now the hypothesis test is identical to that of Eq. 10.18 with the fact that the samples are indexed from $t = 0, 1, 2, \ldots, N$. Statistically, the measurements are *jointly* distributed and if they are also considered *independent*, then

$$\begin{aligned}\Pr\left(Y_N|\mathcal{H}_\ell\right) &= \Pr\left(y(0), y(1), \ldots, y(N)|\mathcal{H}_\ell\right) \\ &= \Pr\left(y(0)|\mathcal{H}_\ell\right) \times \Pr\left(y(1)|\mathcal{H}_\ell\right) \times, \ldots, \times \Pr\left(y(N)|\mathcal{H}_\ell\right) \quad (10.40)\end{aligned}$$

So we see from the detection perspective that the decision functions now incorpo rate joint rather than a single measurement distribution, that is, the *likelihood-ratio* for multiple measurements is simply

$$\mathcal{L}(Y_N) = \frac{\Pr(Y_N|\mathcal{H}_1)}{\Pr(Y_N|\mathcal{H}_0)} = \frac{\Pr(y(0), y(1), \ldots, y(N)|\mathcal{H}_1)}{\Pr(y(0), y(1), \ldots, y(N)|\mathcal{H}_0)} \tag{10.41}$$

If each measurement is independent, then the likelihood-ratio becomes

$$\mathcal{L}(Y_N) = \frac{\Pr(Y_N|\mathcal{H}_1)}{\Pr(Y_N|\mathcal{H}_0)} = \frac{\prod_{t=1}^{N}\Pr(y(t)|\mathcal{H}_1)}{\prod_{t=1}^{N}\Pr(y(t)|\mathcal{H}_0)} \tag{10.42}$$

therefore, thresholds for each of the previous detection schemes remain the same since the distributions just change to joint, that is,

$$\begin{aligned}
\mathcal{L}_{\text{ML}}(y(t)) &\longrightarrow \mathcal{L}_{\text{ML}}(Y_N)\\
\mathcal{L}_{\text{MAP}}(y(t)) &\longrightarrow \mathcal{L}_{\text{MAP}}(Y_N)\\
\mathcal{L}_{\text{B-R}}(y(t)) &\longrightarrow \mathcal{L}_{\text{B-R}}(Y_N)\\
\mathcal{L}_{\text{N-P}}(y(t)) &\longrightarrow \mathcal{L}_{\text{N-P}}(Y_N)
\end{aligned}$$

To illustrate this point, let us return to our Gaussian example problem and recalculate the likelihood-ratio for the batch measurement problem.

Example 10.5

Suppose we have a constant signal $s(t) = \mu \ \forall\ t$ that is contaminated with additive zero-mean, Gaussian measurement noise $v(t)$ of variance R_{vv} and we would like to construct the likelihood-ratio for a variety of detectors to decide whether or not the signal is present. With the signal present, the measurement model is simply

$$y(t) = s(t) + v(t); \quad \text{for } t = 1, \ldots, N$$

and the probability distributions for each individual measurement is Gaussian

$$\begin{aligned}
\Pr(y(t)|\mathcal{H}_0) &= \frac{1}{\sqrt{2\pi R_{vv}}}\exp\left\{-\frac{1}{2R_{vv}}y^2(t)\right\}\\
\Pr(y(t)|\mathcal{H}_1) &= \frac{1}{\sqrt{2\pi R_{vv}}}\exp\left\{-\frac{1}{2R_{vv}}(y(t)-\mu)^2\right\}
\end{aligned}$$

Assuming that the measurements are independent, then the joint distribution under each hypothesis is given by

$$\Pr\big(Y_N|\mathcal{H}_\ell\big) = \prod_{t=1}^{N} \Pr\big(y(t)|\mathcal{H}_\ell\big); \text{for } \ell = 0, 1$$

The likelihood decision function is

$$\mathcal{L}\big(Y_N\big) = \frac{\Pr(Y_N|\mathcal{H}_1)}{\Pr(Y_N|\mathcal{H}_0)} = \frac{\frac{1}{\sqrt{2\pi R_{vv}}}\prod_{t=1}^{N} \exp\left\{ - \frac{1}{2R_{vv}}(y(t) - \mu)^2 \right\}}{\frac{1}{\sqrt{2\pi R_{vv}}}\prod_{t=1}^{N} \exp\left\{ - \frac{1}{2R_{vv}} y^2(t) \right\}}$$

or

$$\mathcal{L}(Y_N) = \prod_{t=1}^{N} \exp\left\{ - \frac{1}{2R_{vv}}(-2y(t)\mu + \mu^2) \right\} = \prod_{t=1}^{N} \exp\left\{ \frac{1}{R_{vv}} \left(y(t)\mu - \frac{\mu^2}{2} \right) \right\}$$

which for a given μ and R_{vv} yield the decision rule

$$\mathcal{L}(Y_N) = \prod_{t=1}^{N} \exp\left\{ \frac{1}{R_{vv}} \left(y(t)\mu - \frac{\mu^2}{2} \right) \right\} \underset{\mathcal{H}_0}{\overset{\mathcal{H}_1}{\gtrless}} \tau$$

where τ is specified by the particular detector selected. △△△

10.3.5 Multichannel Measurements

In many cases, multiple channel or equivalently multivariable measurements are made ranging from antenna arrays of radar systems, hydrophone arrays of sonar systems to geophone arrays for seismic activity as well as microphones arrays for speech application. In fact, in many cases a variety of measurements are made and combined or "fused" together to get a complete view of the measurement space. In all of these applications, a multichannel measurement system provides a set of data at a particular instant of time captured by a measurement vector $\mathbf{y}$. Assuming jointly distributed, independent *vector* measurements, the statistics and likelihood decision function for the multichannel measurement case are

$$\begin{aligned}\Pr\big(\mathbf{Y}_N|\mathcal{H}_\ell\big) &= \Pr\big(\mathbf{y}(0), \mathbf{y}(1), \ldots, \mathbf{y}(N)|\mathcal{H}_\ell\big) \\ &= \Pr\big(\mathbf{y}(0)|\mathcal{H}_\ell\big) \times \Pr\big(\mathbf{y}(1)|\mathcal{H}_\ell\big)\times, \ldots, \times\Pr\big(\mathbf{y}(N)|\mathcal{H}_\ell\big) \quad (10.43)\end{aligned}$$

where $\mathbf{y}(t) \in \mathcal{R}^{N_y \times 1}$ and the batch N-vector data set $\mathbf{Y}_N = \{\mathbf{y}(0), \mathbf{y}(1), \ldots, \mathbf{y}(N)\}$.

So we see from the detection perspective the decision functions now incorporate joint vectors, that is, the *likelihood-ratio* for multiple measurements is

$$\mathcal{L}(\mathbf{Y}_N) = \frac{\Pr(\mathbf{Y}_N|\mathcal{H}_1)}{\Pr(\mathbf{Y}_N|\mathcal{H}_0)} = \frac{\Pr(\mathbf{y}(0), \mathbf{y}(1), \ldots, \mathbf{y}(N)|\mathcal{H}_1)}{\Pr(\mathbf{y}(0), \mathbf{y}(1), \ldots, \mathbf{y}(N)|\mathcal{H}_0)} \tag{10.44}$$

With independent measurements, the likelihood-ratio becomes

$$\mathcal{L}(\mathbf{Y}_N) = \frac{\Pr(\mathbf{Y}_N|\mathcal{H}_1)}{\Pr(\mathbf{Y}_N|\mathcal{H}_0)} = \frac{\prod_{t=1}^{N} \Pr(\mathbf{y}(t)|\mathcal{H}_1)}{\prod_{t=1}^{N} \Pr(\mathbf{y}(t)|\mathcal{H}_0)} \tag{10.45}$$

Therefore, as before in the multiple measurement case, thresholds for each of the previous detection schemes remain the same since the distributions just change to joint vector distributions, that is,

$$\begin{aligned}
\mathcal{L}_{\text{ML}}(Y_N) &\longrightarrow \mathcal{L}_{\text{ML}}(\mathbf{Y}_N) \\
\mathcal{L}_{\text{MAP}}(Y_N) &\longrightarrow \mathcal{L}_{\text{MAP}}(\mathbf{Y}_N) \\
\mathcal{L}_{\text{B-R}}(Y_N) &\longrightarrow \mathcal{L}_{\text{B-R}}(\mathbf{Y}_N) \\
\mathcal{L}_{\text{N-P}}(Y_N) &\longrightarrow \mathcal{L}_{\text{N-P}}(\mathbf{Y}_N)
\end{aligned}$$

Let us return to our Gaussian example problem, construct it as multichannel and recalculate the likelihood-ratio for the batch measurement problem.

Example 10.6

Suppose, we have a vector signal $\mathbf{s}(t) = \boldsymbol{\mu} \in \mathcal{R}^{N_y \times 1}$ that is contaminated with additive zero-mean, Gaussian measurement noise $\mathbf{v}(t)$ of variance $\mathbf{R}_{vv}$ and we would like to construct the likelihood-ratio for a variety of detectors to decide whether or not the signal is present. With the signal present, the measurement model is simply

$$\mathbf{y}(t) = \mathbf{s}(t) + \mathbf{v}(t)$$

for $\mathbf{y}, \mathbf{v} \in \mathcal{R}^{N_y \times 1}$.

The distribution in vector-matrix form under each hypothesis is given by

$$\begin{aligned}
\Pr(\mathbf{Y}_t|\mathcal{H}_0) &= (2\pi)^{-\frac{N}{2}} \times |\mathbf{R}_{vv}|^{-\frac{N}{2}} \times \exp\left\{-\frac{1}{2}\mathbf{y}^T(t)\mathbf{R}_{vv}^{-1}\mathbf{y}(t)\right\} \\
\Pr(\mathbf{Y}_t|\mathcal{H}_1) &= (2\pi)^{-\frac{N}{2}} \times |\mathbf{R}_{vv}|^{-\frac{N}{2}} \times \exp\left\{-\frac{1}{2}(\mathbf{y}(t) - \boldsymbol{\mu})' \mathbf{R}_{vv}^{-1}(\mathbf{y}(t) - \boldsymbol{\mu})\right\}
\end{aligned}$$

where $\mathbf{R}_{vv} = \text{diag}(R_{vv}) \in \mathcal{R}^{N_y \times N_y}$ and $\boldsymbol{\mu} \in \mathcal{R}^{N_y \times 1}$.

Assuming that the measurements are independent, then

$$\Pr\big(\mathbf{Y}_N|\mathcal{H}_\ell\big) = \prod_{t=1}^{N} \Pr\big(\mathbf{y}(t)|\mathcal{H}_\ell\big); \text{for } \ell = 0, 1$$

The likelihood decision function is

$$\mathcal{L}\big(\mathbf{Y}_N\big) = \frac{\Pr(\mathbf{Y}_N|\mathcal{H}_1)}{\Pr(\mathbf{Y}_N|\mathcal{H}_0)} = \frac{(2\pi)^{-\frac{N}{2}} \times |\mathbf{R}_{vv}|^{-\frac{N}{2}} \prod_{t=1}^{N} \exp\Big\{ -\frac{1}{2}(\mathbf{y}(t)-\boldsymbol{\mu})^T \mathbf{R}_{vv}^{-1}(\mathbf{y}(t)-\boldsymbol{\mu})\Big\}}{(2\pi)^{-\frac{N}{2}} \times |\mathbf{R}_{vv}|^{-\frac{N}{2}} \prod_{t=1}^{N} \exp\Big\{ -\frac{1}{2}\mathbf{y}^T(t)\mathbf{R}_{vv}^{-1}\mathbf{y}(t)\Big\}}$$

or simplifying

$$\mathcal{L}(\mathbf{Y}_N) = \frac{1}{2}\prod_{t=1}^{N} \exp\Big\{\boldsymbol{\mu}^T\mathbf{R}_{vv}^{-1}\mathbf{y}(t) + \mathbf{y}^T(t)\mathbf{R}_{vv}^{-1}\boldsymbol{\mu} - \boldsymbol{\mu}^T\mathbf{R}_{vv}^{-1}\boldsymbol{\mu}\Big\} \underset{\mathcal{H}_0}{\overset{\mathcal{H}_1}{\gtrless}} \tau$$

which for a given μ and R_{vv} yield the decision rule and τ is specified by the particular detector selected. △△△

This completes the section on multivariate detection, next we consider multiple hypothesis testing.

10.3.6 Multiple Hypotheses

Up to this point we have discussed the binary decision problem, that is, given two hypotheses choose between them to make a decision. What happens when we have more than two? This is the multiple hypotheses problem which occurs frequently in many applications. For instance, in undersea operations, we may ask the question, was that a submarine or a whale or a rock or a mine? Multiple hypothesis testing or detection has evolved to be known as a "classification" problem in which a large number of texts have been written [2], [6], [10], [11].

In this section, we briefly discuss the multiple hypotheses decision problem from the Bayesian risk perspective developed for the binary problem in Section 10.3.2 (see Eq. 10.31) and generalize it by incorporating the M-multiple hypotheses directly into the risk criterion for minimization, that is, [3]

$$\mathfrak{R} = \mathrm{E}\{\mathbf{C}\} = \sum_{i=1}^{M}\sum_{j=1}^{M} \mathrm{C}_{ij} \times \Pr(\mathcal{H}_j) \times \Pr(\mathcal{H}_i|\mathcal{H}_j) \tag{10.46}$$

where the measurement is $y \in Y_i$ or equivalently $y \in \mathbf{D}_i$, then $\mathcal{H}_j$ is selected such that

$$\Pr(\mathcal{H}_i|\mathcal{H}_j) = \int_{\mathbf{D_i}} \Pr(y(t)|\mathcal{H}_j)dy \tag{10.47}$$

and

C_{ij}———cost of accepting $\mathcal{H}_i$ when $\mathcal{H}_j$ is true
$\Pr(\mathcal{H}_j)$———probability of $\mathcal{H}_j$ with $\mathcal{H}_j$ true
$\Pr(y(t)|\mathcal{H}_j)$——conditional probability of the measurement given that $\mathcal{H}_j$ is true
$\mathbf{D}$———measurement space divided into disjoint sets such that $\sum_{i=0}^{M-1} \mathbf{D_i}$

After substituting Eq. 10.47 into Eq. 10.46, the average risk becomes

$$\Re - \sum_{i=1}^{M} \int_{\mathbf{D_i}} \sum_{j=1}^{M} C_{ij} \times \Pr(\mathcal{H}_j) \times \Pr(y(t)|\mathcal{H}_j)dy \tag{10.48}$$

Recall that, as in the binary case, the Bayes' criterion is to select the decision regions $\mathbf{D}_i; i = 1, \ldots, M$ such that the average risk $\Re$ is minimized. Therefore, this can be achieved by selecting $y \in Y_i$ or equivalently $y \in \mathbf{D}_i$ if (see Ref. [5] for proof)

$$\sum_{k=1}^{M} C_{jk} \times \Pr(\mathcal{H}_k) \times \Pr(y(t)|\mathcal{H}_k) < \sum_{k=1}^{M} C_{mk} \times \Pr(\mathcal{H}_k) \times \Pr(y(t)|\mathcal{H}_k) \;\forall\, j \neq m \tag{10.49}$$

Now, if we let these costs be

$$C_{ij} = \begin{cases} 0, & \text{if } i = j \\ 1, & \text{if } i \neq j \end{cases} \tag{10.50}$$

which essentially implies that the errors are of equal importance, then the risk is the probability-of-error P_{Error} and Eq. 10.49 simplifies. That is, we accept $\mathcal{H}_k$, if

$$\Pr(\mathcal{H}_k) \times \Pr(y(t)|\mathcal{H}_k) > \Pr(\mathcal{H}_m) \times \Pr(y(t)|\mathcal{H}_m) \;\forall\, m \neq k \tag{10.51}$$

Applying Bayes' rule we have the posteriors and therefore the equivalent decision rule consists of accepting $\mathcal{H}_k$ if

$$\Pr(\mathcal{H}_k)|y(t)) > \Pr(\mathcal{H}_m|y(t)) \;\forall\, m \neq k \tag{10.52}$$

which implies that for the M-ary hypotheses, we must calculate each of the a posteriori probabilities and *select the largest.*

These results also can be extended to the composite hypotheses case as well. There the required probabilities for each hypothesis are given by

$$\int_{-\infty}^{\infty} \Pr(\mathcal{H}_i|y(t),\Theta) \times \Pr(\Theta|y(t))d\Theta \text{ for } i = 0, 1, \ldots, M-1 \qquad (10.53)$$

and accept hypothesis $\Pr(\mathcal{H}_i)$ for i the *largest* value of Eq. 10.53.

Consider the previous "batch" example of 10.14 and extend it to the multiple hypotheses case.

Example 10.7

Suppose that we have a set of three constant signals $s_i(t) = \mu_i; i = 1, \ldots, 3$ that are contaminated with additive zero-mean, Gaussian measurement noise $v(t)$ of variance R_{vv} and we would like to construct a decision function to decide which particular signal is present. That is, we would like to decide which hypothesis is true

$$\begin{aligned} \mathcal{H}_1 &: y(t) = s_1(t) + v(t) && \text{[Signal no. 1]} \\ \mathcal{H}_2 &: y(t) = s_2(t) + v(t) && \text{[Signal no. 2]} \\ \mathcal{H}_3 &: y(t) = s_3(t) + v(t) && \text{[Signal no. 3]} \end{aligned}$$

The measurement model for this problem is simply

$$y(t) = s_i(t) + v(t); \quad \text{for } t = 1, \ldots, N; i = 1, \ldots, 3$$

and the probability distributions for each individual measurement is Gaussian

$$\Pr(y(t)|\mathcal{H}_i) = \frac{1}{\sqrt{2\pi R_{vv}}} \exp\left\{ -\frac{1}{2R_{vv}}(y(t) - \mu_i)^2 \right\}; i = 1, \ldots, 3$$

Assuming that the measurements are independent, then the joint distribution under each hypothesis is given by

$$\Pr\left(Y_N|\mathcal{H}_i\right) = \prod_{t=1}^{N} \Pr\left(y(t)|\mathcal{H}_i\right); \text{for } i = 1, 2, 3$$

and taking logarithms, we have

$$\ln \Pr\left(Y_N|\mathcal{H}_i\right) = N \ln\left(\frac{1}{\sqrt{2\pi R_{vv}}}\right) - \frac{1}{2R_{vv}} \sum_{t=1}^{N} (y(t) - \mu_i)^2; i = 1, \ldots, 3$$

Further assuming equal priors $\Pr(\mathcal{H}_1) = \Pr(\mathcal{H}_2) = \Pr(\mathcal{H}_3)$, then the decision function of Eq. 10.51 requires that we select the largest probability, that is

$$\ln \Pr(y(t)|\mathcal{H}_i) > \ln \Pr(y(t)|\mathcal{H}_m)\ \forall\ m \neq i$$

or substituting

$$N \ln\left(\frac{1}{\sqrt{2\pi R_{vv}}}\right) - \frac{1}{2R_{vv}} \sum_{t=1}^{N} (y(t) - \mu_i)^2 > N \ln\left(\frac{1}{\sqrt{2\pi R_{vv}}}\right) - \frac{1}{2R_{vv}} \sum_{t=1}^{N} (y(t) - \mu_m)^2$$

Removing the constant terms, cancelling like-terms, and multiplying by the minus sign changes the inequality leading to

$$\sum_{t=1}^{N} (y(t) - \mu_i)^2 < \sum_{t=1}^{N} (y(t) - \mu_m)^2\ \forall\ m \neq i$$

and a "minimum", then the decision function becomes

$$\mathcal{D}_i(Y_N) = \min_i \left(\sum_{t=1}^{N} (y(t) - \mu_i)^2 \right); i = 1, \ldots, 3$$

which achieves its minimum for the largest value of μ_i. △△△

This completes the section on multiple hypotheses, next we consider techniques to evaluate detection performance.

10.4 PERFORMANCE METRICS

Detection theory is a fundamental tool in decision analysis [1], [10], [12]. However, many decision functions both formal and informal evolve in a wide variety of applications. The fundamental detection problem of interest is simply *given* a set of data and a decision function, *decide* whether or not a signal is present.

Detection (binary), signal or non-signal, and classification (signal type) methods form the basis of signal detection processors including modern machine-learning algorithms. The major question that arises when investigating these methods [1–20] is their underlying performance on problems of interest. There exists a variety of metrics that can be applied to evaluate algorithm performance ranging from confusion matrices to sophisticated statistical hypothesis tests [3], but perhaps the most basic and most robust method is the calculation of the ROC curve. The ROC curve is simply a graph of detection ($P_{\rm DET}$) versus false-alarm ($P_{\rm FA}$) probabilities parameterized by threshold τ. This particular metric has evolved from the analysis of radar systems during World War II [2], [13] as a critical tool for diagnostic testing in medicine, to pattern recognition in forensics, and a wide variety of other applications [1]. There are many individual metrics that can be extracted from an ROC curve including sensitivity, specificity, cost/benefit analysis along with a set of specific features like area under the curve (AUC) and minimum probability-of-error (MinE) [1]. All of these problems have one requirement in common—they must be analyzed in some uniform manner so that their detection performance can be evaluated. This requirement leads directly to the ROC curve, since it provides all of the fundamental information from which most other metrics are derived (e.g., area under ROC curve, AUC) [21–25].

10.4.1 Receiver Operating Characteristic (ROC) Curves

A typical ROC curve is shown in Fig. 10.3 where we observe that the "ROC space" is defined by the 1×1 square region in the (P_{FA},P_{DET})-plane. The graph is monotonically increasing from $(0, 0)$ to $(1, 1)$. Detection performance can range from complete alternative (negative) hypothesis at $(0, 0)$ to complete hypothesis (positive) detection at $(1, 1)$ with "perfect detection" at $(P_{FA}, P_{DET}) = (0, 1)$ and

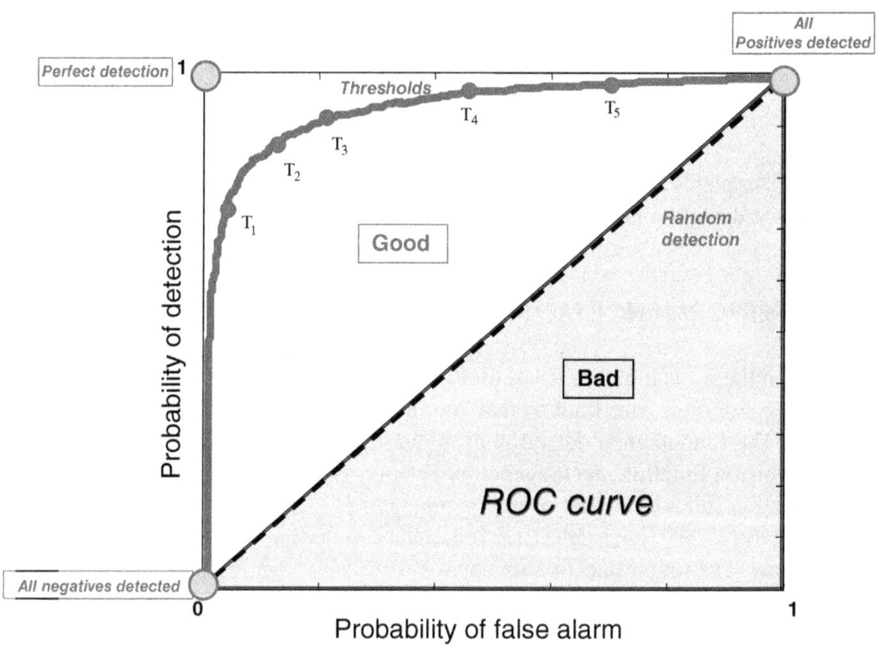

FIGURE 10.3 Receiver operating characteristic (ROC) curve: detection (P_{DET}) versus false-alarm probability (P_{FA}) for selected thresholds (τ) indicating various performance metrics including: (a) all negative ($(P_{FA},P_{DET}) = (0, 0)$) detection; (b) all positive ($(P_{FA},P_{DET}) = (1, 1)$) detection; (c) perfect ($(P_{FA},P_{DET}) = (0, 1)$) detection; (d) random (coin toss) detection ($P_{DET} = P_{FA}$).

"random detection" (e.g., coin toss) along the cross-diagonal or 45^0-line from $(0, 0)$ to $(1, 1)$, that is, $P_{\rm DET} = P_{\rm FA}$. Curves lying above this line are considered a representation of "good" detection performance (i.e., better than a coin toss), while those lying below the line are considered "bad" detection performance (i.e., worse than a coin toss). Each (operating) point along the ROC curve is parameterized by a threshold value,$\tau_n; n = 1, \ldots, N$ defining a particular expected *operating point* of the detector under analysis, that is, $(P_{\rm FA}(\tau_n), P_{\rm DET}(\tau_n))$ at τ_n. ROC curves are a function of detection and false-alarm probability density (or mass) functions determined by sweeping the threshold through the decision PDFs and calculating the underlying area overlaps [1]. We shall discuss this in more detail shortly. ROC curves are generated from "known" explicit decision functions based on the particular problem (under investigation) statistics. For example, Gaussian decision functions lead to explicit error function calculations that can be calculated analytically [1]. Unfortunately, most decision function probability distributions (not necessarily data distributions) are unknown or too complex to evaluate directly; therefore, we must resort to numerical techniques (e.g., series approximations) to calculate the ROCs or resort to "brute" force techniques, if possible, by generating large ensemble realizations and applying counting methods [11]. Even though it appears to be a straightforward calculation, ROC curves still possess a bit of "mysticism" because we rarely find simple interrelations between $P_{\rm DET}$ and $P_{\rm FA}$ such as a random detector $P_{\rm DET}(\tau_n) = P_{\rm FA}(\tau_n) \; \forall \; n$. For instance, the well-known *Neyman–Pearson* detection [1] algorithm fixes $P_{\rm FA}$ and then maximizes $P_{\rm DET}$ for a specified threshold (τ_n). Typically, we know the decision function employed in the detector can perform the required integration (analytically or numerically) to determine the detection and false-alarm probabilities enabling us to generate various points along the curve at each threshold, tracing out the curve. So we see that the ROC curve, though simple in concept, can present a challenge to calculate depending on the underlying problem scenario and availability of good measurements and known or unknown decision distributions.

In this chapter, we discuss the basic theory required to comprehend (intuitively and mathematically) the ROC and its inherent features that enable us to ascertain the performance of detection algorithms. We first develop the theory and from it develop a variety of metrics that can be extracted from the ROC to increase our understanding of the embedded information. We apply the analysis to examples and use the ROC to compare performance of a variety of detection techniques.

ROC Curve Generation Next, let us investigate just how an ROC curve is generated, first, intuitively (pictorially) and then numerically. Consider the two (binary problem) decision distribution functions shown in Fig. 10.4—the disturbance and the signal. From the figure we observe that their means (distribution peaks) differ providing an offset and they overlap one another. It is this overlapping that dictates the shape of the ROC and the performance of the detector. If we allow the decision variable (threshold) to assume various values and sweep it from $-\infty$ to $+\infty$, an ROC curve is traced out. That is, for each threshold value τ_n, a vertical line drawn from $\mathcal{D} = \tau_n$

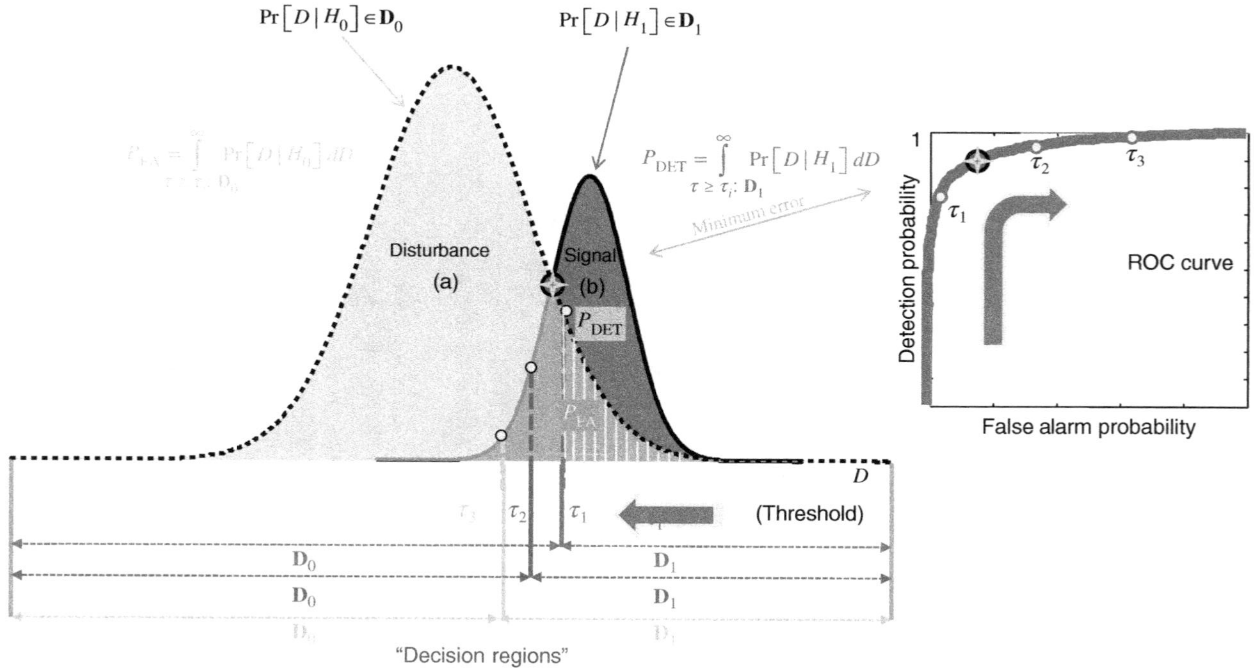

FIGURE 10.4 Decision function probability distributions for selected thresholds (τ_n) and defined decision regions: (a) disturbance probability distribution $\Pr(\mathcal{D}(y(t))|\mathcal{H}_0)$; (b) signal probability distribution $\Pr(\mathcal{D}(y(t))|\mathcal{H}_1)$. Note that the star is location of minimum error probability (maximum detection/minimum false-alarm and optimal decision point (ODP)).

intersects either curve associated with the underlying hypothesis $\Pr(\mathcal{D}|\mathcal{H}_\ell)$; $\ell = 0, 1$ defining the lower integration limit for both P_{DET} and P_{FA} (see Eqs. 2, 3). Performing the integration at τ_n calculates the areas under both decision distributions (see Fig. 10.4) providing a corresponding operating point $(P_{FA}(\tau_n), P_{DET}(\tau_n))$ on the ROC curve for the selected threshold. Sweeping τ_n; $n = 1, 2, \ldots$ the lower integration limits (thresholds) are varied and the corresponding areas defined by these boundary limits trace out the entire ROC curve. From Fig. 10.4 we also see that selecting the thresholds defines the new boundaries for the decision regions D_0 and D_1 as well as the corresponding areas of integration for all of the operating points. There exists a particular point of *minimum error* corresponding to the intersection of the conditional distributions (star in Fig. 10.4), providing the maximum detection and minimum false-alarm probabilities for an optimum (Bayes) detector design [1]. Note also that for a given threshold both P_{Miss} and P_{Reject} can be calculated by changing the integration limits to $-\infty$ to τ, if desired. So we see that knowledge of the decision function PDFs can provide us with some insight about functions generating the ROC curve and its construction.

Numerical calculation of the ROC curve can evolve from a variety of methods. Perhaps the simplest is through integration, either analytically (where possible) or numerically, with the particular availability of the detection and false-alarm probabilities and their known forms (analytic function or numerical values). Recall that it is necessary for both probabilities to evolve from "known" functions or data.

One particular technique that can be applied to calculate the ROC directly from known data sequences generated either through simulations or controlled experiments is the so-called *brute force method*. Once the *decision sequences* are calculated, then this method can be applied using simple counting methods. For this method, known data are generated and input to the decision function which is varied according to selected threshold values similar to sweeping the thresholds illustrated in Fig. 10.4, that is, for the decision problem, we must *decide*

$$\mathcal{D}(y(t)) \underset{\text{[Non-signal/Disturbance]}}{\overset{\text{[Signal]}}{\gtrless}} \tau_n$$

Once this decision function is obtained, then for any threshold value, we can calculate the probability of detection and the probability of false alarm. These are obtained empirically by estimating the probability of detection as the ratio of the number of correct signal decisions declared for that threshold with the signal present over the total number of signal samples. The false-alarm probability is the ratio of the number of signal decisions declared for that threshold without the signal present over the total number of non-signal or disturbance samples. These estimated probabilities provide a single point on the ROC curve at the selected threshold. Varying the threshold and performing the same calculation for both detection and false-alarm probabilities generates the entire ROC curve. That is, we calculate the detection and false-alarm probabilities based on the decisions made for each realization such that

$$P_{\rm DET}(n) = \frac{\text{No. of detections (Signal present)}}{\text{Total no. of signal realizations } (K)};\ P_{\rm FA}(n) = \frac{\text{No. of detections (Signal NOT present)}}{\text{Total no. of non-signal realizations } (K)}$$

$$\text{with realizations } \{\mathcal{D}(y(t))\}\ ; t = 1, \ldots, K \text{ and } @\ \tau_n \text{ for } n = 1, \ldots, N$$

enabling the generation of a single point $(P_{\rm FA}(\tau_n), P_{\rm DET}(\tau_n))$ on the ROC curve for the specified threshold value (τ_n).

Summarizing, we can now see that the selection of the decision threshold clearly specifies an operating point of the detector by its location on the ROC curve. It is the overlap of the decision function distributions (conditional) that determine the corresponding detection and false-alarm probabilities. For instance, if there is *no* overlap then perfect detection can be achieved ((0,1) on ROC). Therefore, by sweeping through the thresholds, loci of operating points are defined tracing out the entire ROC curve.

The steps required to generate the ROC curve from "known" data are

ROC generation: Brute force method

1. *Generate* two ensembles of system realizations (or signal/non-signal) through simulations or controlled experiments;
2. For each realization, *calculate* the corresponding decision function and compare its value to the threshold;
3. *Estimate* the detection and false-alarm probabilities at the specified threshold (above);
4. Continue to choose new thresholds and compare the decision function while *accumulating* the *counts* of detection and false-alarm probabilities generating the ROC curve.

So we see that the dynamic ROC curve can easily be used to evaluate individual *detector* performance as well as compare techniques.

Average ROC (AROC) One of the questions with an ROC curve is whether or not it is actually the "true" curve. Perhaps a better way to pose the problem is to ask what is the best estimate of an ROC curve from synthesized or actual measurement data? One way to approach this problem is to ask for the estimate, but also include a measure of precision (standard deviation) along with it.

One technique is to create an ensemble of ROC s which can be accomplished in a variety of ways such as through (1) simulation; (2) controlled experimental data; or (3) sectioning of actual data. When data are sparse and a simulation is not possible

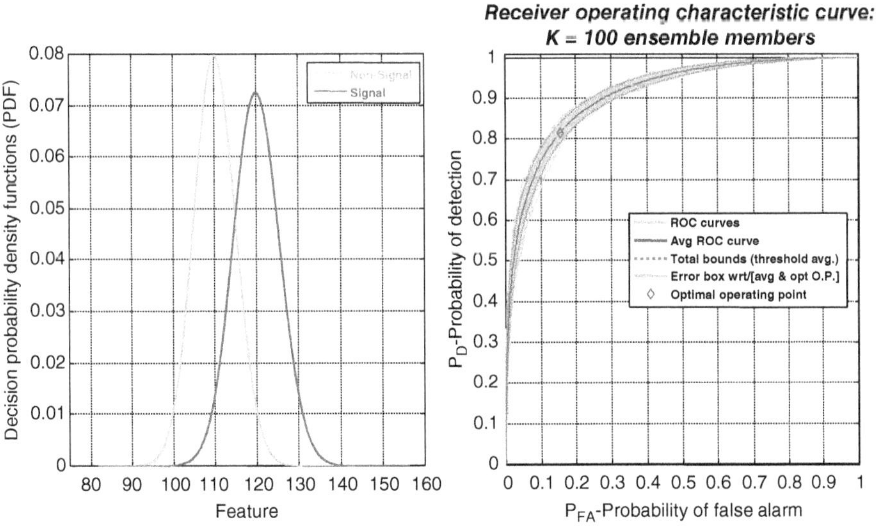

FIGURE 10.5 Threshold averaging of ROC curves: (a) example decision PDFs; (b) ROC ensemble (100 members) results with average (solid) and $\pm 2\sigma$ bounds (dots). The ODP (diamond) and 2σ–by–1σ uncertainty box is also shown (zoom box).

then "bootstrap" methods can be applied to generate the known (signal or non-signal) data to generate the ROC curves [26].

A simple method that can easily be applied would be to generate an *ensemble* of realizations (when possible) and create an ensemble average along with its corresponding confidence bounds to assess the quality of the ROC curve estimates as long as the thresholds are the same for each member [20]. For a Gaussian problem, we show an ensemble of 100 curves with the mean estimate $\overline{\mathbf{R}}$ along with the $\pm 2\overline{\sigma}$ bounds in Fig. 10.5. It must be noted that simple averaging or so-called *vertical* averaging [27] assumes that *all* of the threshold values for both detection and false-alarm probabilities are identical (aligned) for each member. If the thresholds are not available and we only have values of detection and false-alarm probabilities, then the value of the P_{FA} is fixed and the values of the P_{DET} for each ensemble value are interpolated and then averaged.

Another method of averaging is called *threshold averaging* where the detection, false-alarm, and threshold values are available for "each" realization of the ROC curve. Here the thresholds are chosen and the corresponding detection probabilities selected (or interpolated) to provide a threshold and false-alarm probability with the resulting L-ensemble of ROC curves averaged. One way that is employed is to: (1) find the global minimum/maximum thresholds for each ensemble member, and (2) generate a set of thresholds based on these values using them to generate all ROC curves in the ensemble. This way the ROCs are forced to "threshold align" and then averaging across the ensemble is accomplished. This is the *method of choice*, when possible, that is, if we generate an ensemble of ROC curves,

$\mathcal{R}_\ell(P_{\rm DET}(\tau_n), {\rm P}_{\rm FA}(\tau_n))$; $\ell = 1, \ldots, L$, $\forall\ n$, then the ensemble *average* ROC, $\overline{\mathcal{R}}$ is given by

$$\overline{\mathcal{R}}(P_{\rm DET}, P_{\rm FA}) = \frac{1}{L}\sum_{\ell=1}^{L} \mathcal{R}_\ell(P_{\rm DET}(\tau_n), P_{\rm FA}(\tau_n));\quad \forall\ n \tag{10.54}$$

with the ensemble standard deviation given by

$$\overline{\sigma}(P_{\rm DET}, P_{\rm FA}) = \sqrt{\frac{1}{L}\sum_{\ell=1}^{L}(\mathcal{R}_\ell(P_{\rm DET}(\tau_n), P_{\rm FA}(\tau_n)) - \overline{\mathcal{R}}(P_{\rm DET}, P_{\rm FA}))^2} \tag{10.55}$$

Example 10.8

As an example using a set of Gaussian distributions, we perform simple vertical averaging on an ensemble of 100 members (gray lines). The average (solid line) ROCs shown in Fig. 10.5 along with the corresponding confidence limits or bounds (dots). So we see that we have an estimated ROC with its associated precision metric indicating the uncertainty in the estimate. Of course if desired, the uncertainty can be decreased by incorporating more and more realizations (ensemble members) of the ROC ensuring tighter and tighter bounds and therefore a more precise estimate of the ROC curve. Also shown in the figure (zoom box) is the "optimal decision (threshold) point" (ODP) and its corresponding 2σ–by–1σ uncertainty box. This ODP value is the *best* operating point for the detector. We shall discuss this in more detail in the next section. △△△

Next, let us consider a variety of metrics that can be derived and extracted from the ROC curve.

ROC Properties In this section, we discuss a variety of properties that can be extracted from a ROC. Knowledge of the ROC specifies detection performance and enables us to compare various *detection* techniques. The shape and location of points along the curve can be used for performance analysis and specifications.

The basic properties of an ROC curve are well known [1], [2], [10], and can be summarized succinctly by

1. the ROC curve is monotonically increasing
2. the slope of the ROC curve is identical to its threshold value at that point

$$\tau_n = dP_{\rm DET}(\tau_n)/dP_{\rm FA}(\tau_n)$$

3. decision functions have ROC s that typically lie above the random detection curve (coin toss)
4. a *sufficient statistic* (contains all of the statistical information required) is desirable for decisions

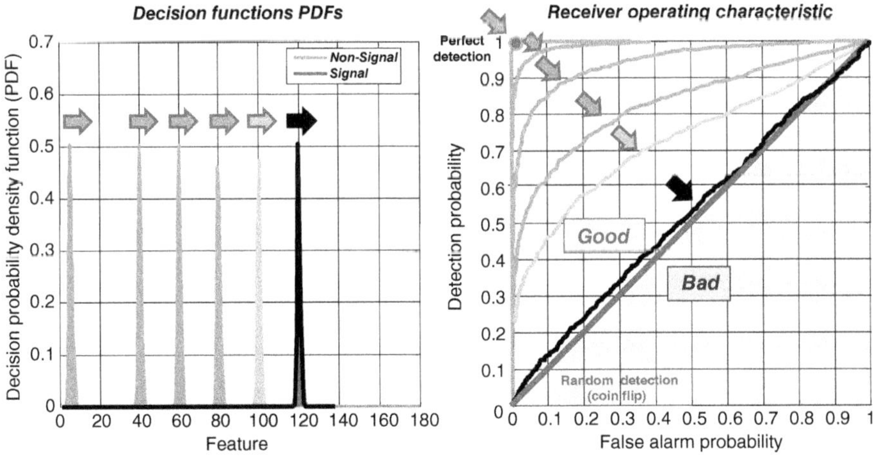

FIGURE 10.6 Family of ROC curves: as the decision function PDFs move closer to the signal PDF, the ROCs move closer to the random detector performance. All parameters of the ROC are listed below.

Besides these basic properties, there are certain metrics that can be determined from the ROC curve that are used to give a "feel" toward a single quantitative number that can be used to assess the *detection* performance such as.

Distance, d′ Consider a family of ROC curves generated from a set of individual non-signal items, assumed Gaussian, with different location parameters (means and variances) relative to a signal item as shown in Fig. 10.6. We know if the distance, say d', between the non-signal and signal distribution is large [2], then "perfect detection" can be achieved as shown in the figure by the PDF and its corresponding ROC curve [1], [12]. If the distribution is Gaussian, then the distance is simply the Euclidean distance of the ratio of individual means (M_0) to standard deviations ($\sqrt{V_0}$) relative to the signal distribution parameters ($M_1, \sqrt{V_1}$), that is,

$$d' = \left| \frac{M_1}{\sqrt{V_1}} - \frac{M_0}{\sqrt{V_0}} \right| \tag{10.56}$$

As the non-signal mean of each individual (non-signal) item increases (in the figure), the corresponding non-signal distribution moves closer and closer to the signal distribution increasing the overlap and decreasing desirable detector performance. This is illustrated in the figure by the non-signal decision function PDFs migrating closer to the signal decision function PDF and the corresponding ROC curves approaching the random detector (coin toss) performance (cross-diagonal or 45^0-line) signifying complete overlap of the distributions. From the distance metric, we gain insight into the *separability* of the signal and non-signal PDFs and we can observe performance degradation as they overlap more and more or equivalently as the distance gets smaller and smaller. A large distance metric means a large separation (smaller overlap) and higher expected detection performance. From Fig. 10.6, we see an example of this for the respective distances: $d' = 4.6, 3.2, 2.4, 1.6, 0.8, 0.01, 0.0$ with the ROC curve moving closer to the random detector (coin toss) performance as d' decreases.

1. $\text{ROC}_1 : d'_1 = 4.6, \text{AUC}_1 = 1.00, \text{MinE}_1 = 2.1;$
2. $\text{ROC}_2 : d'_2 = 3.2, \text{AUC}_2 = 0.99, \text{MinE}_2 = 8.4;$
3. $\text{ROC}_3 : d'_3 = 2.4, \text{AUC}_3 = 0.87, \text{MinE}_3 = 14.4;$
4. $\text{ROC}_4 : d'_4 = 1.6, \text{AUC}_4 = 0.74, \text{MinE}_4 = 21.5;$
5. $\text{ROC}_5 : d'_5 = 0.8, \text{AUC}_5 = 0.53, \text{MinE}_5 = 27.4;$
6. $\text{ROC}_6 : d'_6 = 0.5, \text{AUC}_6 = 0.05, \text{MinE}_6 = 29.7.$

Area Under the Curve (AUC) Another method to compare the performance of *detection* techniques is to calculate the area under the ROC curve [27–29]. In fact, it has been shown that the AUC is statistically consistent [30]. The AUC is a portion of the area of the ROC space (1×1 or unit square), its value is always between 0 and 1; however, since the 45^0-line from $(0, 0)$ to $(1, 1)$ represents the random guess, no pragmatic *detector* would have an AUC less than 0.5; therefore, AUC $\geq$ 0.5. The AUC has an important statistical property, that is, the AUC of a *detector* is equivalent to the probability that the classifier will *rank* a randomly selected signal ($\mathcal{H}_1$) measurement "higher" than a randomly selected non-signal ($\mathcal{H}_0$) measurement [27] implying that it is very sensitive to detecting signals. In practice, the AUC performs very well and is frequently employed as a general metric of detection performance. It is calculated numerically by simple trapezoidal integration as

$$\text{AUC} = \sum_n P_{\text{DET}}(\tau_n)\Delta P_{\text{FA}}(\tau_n) + \frac{1}{2}\Delta P_{\text{DET}}(\tau_n) \times \Delta P_{\text{FA}}(\tau_n) \tag{10.57}$$

$$\text{where} \quad \Delta P_{\text{DET}}(\tau_n) = -(P_{\text{DET}}(\tau_n) - P_{\text{DET}}(\tau_{n-1}))$$

$$\Delta P_{\text{FA}}(\tau_n) = (P_{\text{FA}}(\tau_n) - P_{\text{FA}}(\tau_{n-1}))$$

Referring to Fig. 10.6, the area under each ROC curve is given by

$$\text{AUC} = 1.00, 0.99, 0.96, 0.87, 0.74, 0.53, 0.50$$

clearly demonstrating the robustness of this metric and its ability to predict overall performance. Thus, the larger the AUC or equivalently the closer its value is to *unity* (perfect detection), the better the expected detection performance.

Minimum Probability-of-Error The minimum (attainable) probability-of-error or equivalently Bayesian error corresponds to the intersection point of the decision probabilities as shown as the "star" in Fig. 10.4. It evolves directly from the Gaussian distribution assumption and requires the solution of an optimization problem. We would like to investigate the decision errors more closely, that is, the probability-of-error in making the decision—right or wrong. Assuming that we have a binary decision problem with the usual two hypotheses: $\mathcal{H}_0$ and $\mathcal{H}_1$, as before and the data set, $\mathbf{Y} := \{y_1, \ldots, y_K\}$; therefore, the *total probability-of-error* is

$$\Pr(\text{Error}|\mathbf{Y}) = \Pr(\mathbf{Y}, \mathcal{H}_0|\mathcal{H}_1) + \Pr(\mathbf{Y}, \mathcal{H}_1|\mathcal{H}_0) \tag{10.58}$$

or applying Bayes' rule, we have

$$\Pr(\text{Error}|\mathbf{Y}) = \int_{\mathcal{H}_1} \Pr(\mathbf{Y}|\mathcal{H}_0)\Pr(\mathcal{H}_0)d\mathbf{Y} + \int_{\mathcal{H}_0} \Pr(\mathbf{Y}|\mathcal{H}_1)\Pr(\mathcal{H}_1)d\mathbf{Y} \tag{10.59}$$

Thus, the probability-of-error is based on making the *wrong* decision, that is,

$$\Pr(\text{Error}|\text{Data}) := \Pr(\text{Error}|\mathbf{Y}) = \begin{cases} \Pr\left(\mathcal{H}_1|\mathbf{Y}\right) \text{ if we decide on } \mathcal{H}_0 & [\text{Miss}] \\ \text{or} & \\ \Pr\left(\mathcal{H}_0|\mathbf{Y}\right) \text{ if we decide on } \mathcal{H}_1 & [\text{False alarm}] \end{cases} \tag{10.60}$$

Bayesian decisions are based on the principle of selecting the hypothesis corresponding to the largest posterior probability such that

$$\begin{array}{lll} \text{If} & \Pr\left(\mathcal{H}_0|\mathbf{Y}\right) > \Pr\left(\mathcal{H}_1|\mathbf{Y}\right) & \text{decide on } \mathcal{H}_0 \\ & \text{Otherwise} & \text{decide on } \mathcal{H}_1 \end{array} \tag{10.61}$$

Under this decision function (Bayesian) the error probability becomes

$$\Pr(\text{Error}|\mathbf{Y}) = min[\Pr(\mathcal{H}_0|\mathbf{Y}), \Pr(\mathcal{H}_1|\mathbf{Y})] \leq \Pr(\mathcal{H}_0|\mathbf{Y})^{\beta} \times \Pr(\mathcal{H}_1|\mathbf{Y})^{1-\beta}; 0 \leq \beta \leq 1 \tag{10.62}$$

A general error integral can be upper bounded [1] using this inequality to give

$$\Pr(\text{Error}|\mathcal{U}) \leq P^{\beta}_{\mathbf{Y}}(\mathcal{H}_0|\mathbf{Y}) \times P^{1-\beta}_{\mathbf{Y}}(\mathcal{H}_1|\mathbf{Y}) \times \int \mathrm{p}^{\beta}_{\mathbf{Y}}(\mathbf{Y}|\mathcal{H}_0) \times \mathrm{p}^{1-\beta}_{\mathbf{Y}}(\mathbf{Y}|\mathcal{H}_1)\, d\mathbf{Y}; 0 \leq \beta \leq 1 \quad (10.63)$$

where $\mathrm{p}_{\mathbf{Y}}(\cdot)$ is a probability density function and $\mathrm{P}_{\mathbf{Y}}(\cdot)$ the corresponding distribution. If the conditional probabilities are Gaussian, then the integral in this expression can be simplified to

$$\int \mathrm{p}^{\beta}_{\mathbf{Y}}(\mathbf{Y}|\mathcal{H}_0) \times \mathrm{p}^{1-\beta}_{\mathbf{Y}}(\mathbf{Y}|\mathcal{H}_1)\ d\mathbf{Y} = e^{-\kappa(\beta)} \quad (10.64)$$

where $\kappa(\beta)$ is a function of the means and variances of the distributions (see Ref. [11] for details) leading to

$$\Pr(\text{Error}|\mathbf{Y}) \ \leq\ P^{\beta}_{\mathbf{Y}}(\mathcal{H}_0|\mathbf{Y}) \times P^{1-\beta}_{\mathbf{Y}}(\mathcal{H}_1|\mathbf{Y}) \times e^{-\kappa(\beta)}; 0 \leq \beta \leq 1 \quad (10.65)$$

and this is called the *Chernoff upper bound* on the error probability Pr(Error) [11]. The bound is calculated analytically or numerically by finding the value of β that minimizes $e^{-\kappa(\beta)}$ with the error calculated by substituting the β_{min} from the optimizer into the integrals of Eq. 10.64.

If the decision functions are Gaussian ($\mathbf{N}(M_0, V_0), \mathbf{N}(M_1, V_1)$), then the bound expression becomes [11]

$$\kappa(\beta) = \frac{\beta(1-\beta)}{2}\left(M_1 - M_0\right)' \left[\beta V_0 - (1-\beta)V_1\right]^{-1} \left(M_1 - M_0\right) + 1/\ 2\ \ln \frac{\left|\beta V_0 - (1-\beta)V_1\right|}{\left|V_0\right|^{\beta}\left|V_1\right|^{1-\beta}} \quad (10.66)$$

For the Gaussian example of Fig. 10.6, we have the following sequence of bounds in percentage of error

$$\text{MinE}\ (\%)\ =\ 2.1\%, 8.4\%, 14.4\%, 21.5\%, 27.4\%, 29.7\%$$

which occurs because of the overlapping in the tails of the Gaussian distributions.

Confusion Matrix Another set of metrics that can be applied to the decision problem has been developed by diagnosticians (medicine, finance, etc.) and is based on the so-called confusion matrix which is shown in Fig. 10.7 [31]. For a binary (signal/non-signal) decision problem, it is a 2×2 matrix with critical ROC statistics extracted at a particular operating point on the ROC curve. On the diagonals of the matrix, we have the detection and rejection probabilities at a selected threshold τ_n

(a)

Confusion matrix			
	Truth:		
Hypothesized choices:	P_{true}	N_{true}	
P'_{est}	TP (No. of signal detections declared—signal present)	FP (No. of signal detections declared—signal Not present)	P'_{total} = TP+FP
N'_{est}	FN (No. of non-signal detections declared—non-signal Not present)	TN (No. of non-signal detections declared—non-signal present)	N'_{total} = TN+FN
Totals:	$\mathbf{P}_{total}$ = TP+FN	$\mathbf{N}_{total}$ = FP+TN	N = TP + FN + FP+TN

(b)

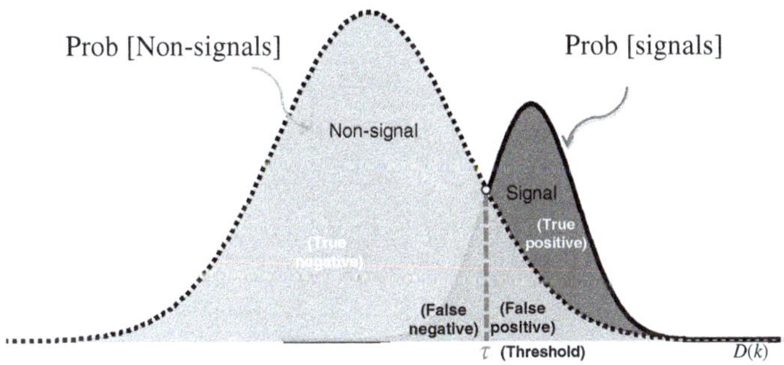

FIGURE 10.7 Confusion matrix for binary decision problem: (a) matrix entries: Diagonals are detection (TP) and rejection (TN) probabilities; off-diagonals are false-alarm (FP) and miss (FN) probabilities, totals are columns and row sums, rows headers are "truth" and columns are "estimates" (guesses or predictions); (b) decision probabilities distinguishing regions mapped to matrix at a given threshold location.

with (slight notational change: $\tau_n \to n$) $P_{\text{DET}}(n), P_{\text{Reject}}(n)$ and on the off-diagonals, we have the corresponding false-alarm and miss probabilities $P_{\text{FA}}(n), P_{\text{Miss}}(n)$.

A common jargon in diagnostics maps these probabilities into

$$P_{\text{DET}}(n) \Rightarrow P_{\text{TP}} \quad --- \quad \text{True positive rate}$$

$$P_{\text{Reject}}(n) \Rightarrow P_{\text{TN}} \quad --- \quad \text{True negative rate}$$

$$P_{\text{FA}}(n) \Rightarrow P_{\text{FP}} \quad --- \quad \text{False positive rate}$$

$$P_{\text{Miss}}(n) \Rightarrow P_{\text{FN}} \quad --- \quad \text{False negative rate}$$

From these distribution values extracted from the ROC curve, we can calculate other meaningful statistics.

We define a positive instance or event as a "signal" and a negative event as a "non-signal." The true positives (TP) are the number of correct detections declared and the corresponding true positive rate (tpr) is defined by

$$P_{\rm DET}(n) = {\rm tpr} = \frac{\rm TP}{P} = \frac{\text{No. of correct threat detections}}{\text{Total no. of threat realizations}} \tag{10.67}$$

corresponding to the detection probability at τ_n.

Likewise, the false positives (FP) are the number of detections declared when the signal is *not* present and the corresponding false positive rate (fpr) is

$$P_{\rm FA}(n) = {\rm fpr} = \frac{\rm FP}{N} = \frac{\text{No. of incorrect threat detections}}{\text{Total no. of non-threat realizations}} \tag{10.68}$$

corresponding to the false-alarm probability at τ_n.

The true negatives (TN) are the number of true non-signals declared and the true negative rate (tnr) is given by

$$P_{\rm Reject}(n) = {\rm tnr} = \frac{\rm TN}{N} = \frac{\text{No. of correct non-threat detections}}{\text{Total no. of non-threat realizations}} \tag{10.69}$$

corresponding to the rejection probability at τ_n.

Finally, the false negatives (FN) are the number of non-detections declared, then the non-signal is *not* present and the false negative rate (fnr) is defined to be

$$P_{\rm Miss}(n) = {\rm fnr} = \frac{\rm FN}{P} = \frac{\text{No. of incorrect non-threat detections}}{\text{Total no. of threat realizations}} \tag{10.70}$$

corresponding to the miss probability at τ_n.

Examining the confusion matrix further, we have that the column summations ($P_{\rm Total}, N_{\rm Total}$) are the frequencies of occurrence of the truth (actual) items (signal/non-signal) and the row summations ($P'_{\rm Total}, N'_{\rm Total}$) are the frequencies of the choices (estimates) of the item. The total sample size $N_{\rm All}$ is the overall sum of all the occurrences.

From these values extracted at a given operating point on the ROC curve, we can calculate a variety of useful metrics such as

1. Accuracy (ACC) is defined as the total number of correct decisions (signals and non-signals) divided by the total possible correct, that is,

$$\text{ACC} = \frac{\text{Total no. of correct decisions}}{\text{Total no. of possible correct}} = \frac{\rm TP + TN}{P + N} \tag{10.71}$$

2. Precision is usually defined as the "positive (signal)" predictive value (PPV) or the "negative (non-signal)" predictive value (NPV), since they offer an

indication of how well the detector can predict signals or non-signals and is given by

$$\text{PPV} = \frac{\text{No. of threat decisions}}{\text{Total threat decisions}} = \frac{\text{TP}}{\text{TP} + \text{FP}}$$

$$\text{NPV} = \frac{\text{No. ofnonthreat decisions}}{\text{Total non-threat decisions}} = \frac{\text{TN}}{\text{TN} + \text{FN}}$$

1. Specificity: SPEC = $\text{P}_{\text{Reject}} = \text{tnr} = 1 - \text{P}_{\text{FA}} = 1 - \text{fpr}$
2. Sensitivity: SENS = $\text{tpr} = \text{P}_{\text{DET}}$

It should be noted that ACC can be a misleading metric that should be used with caution [21]. This completes the section on metrics derived from ROC curves.

Optimum Decision (Threshold) Point (ODP) After obtaining the ROC curve from a particular *detector* or equivalently a *detection algorithm*, it is natural to ask the question: What is the "best" threshold τ to set and its corresponding operating point $(P_{\text{FA}}, P_{\text{DET}})$? That is, what is the best tradeoff between cost (P_{FA}) and benefit (P_{DET}) for the particular detection scheme? In order to answer this question, we can cast the problem into one of Bayes' risk by first defining the various costs of making a decision, defining the criterion in terms of these costs and determining the threshold value (operating point) that minimizes this risk. Thus, in this section we develop the relations to calculate the optimum decision (threshold) point (ODP) required to choose the best operating point from the estimated or average ROC curve. The Bayes' risk criterion for this *detector* has the following costs (weights) associated with decision making

- C_{00}——cost of accepting $\mathcal{H}_0$ when $\mathcal{H}_0$ is true [Rejection]
- C_{01}——cost of accepting $\mathcal{H}_0$ when $\mathcal{H}_1$ is true [Miss]
- C_{10}——cost of accepting $\mathcal{H}_1$ when $\mathcal{H}_0$ is true [False alarm]
- C_{11}——cost of accepting $\mathcal{H}_1$ when $\mathcal{H}_1$ is true [Detection]

Also required are the prior probabilities associated with the underlying hypotheses $\Pr(\mathcal{H}_1)$ (signal) and $\Pr(\mathcal{H}_0)$ (non-signal). With this information, we define the *Bayes' risk criterion* [3], [30] as before true

$$\begin{aligned}\Re = {} & C_{00}\Pr\big(\text{accept } \mathcal{H}_0, \mathcal{H}_0 \text{ true}\big) + C_{01}\Pr\big(\text{accept } \mathcal{H}_0, \mathcal{H}_1 \text{ true}\big) \\ & + C_{10}\Pr\big(\text{accept } \mathcal{H}_1, \mathcal{H}_0 \text{ true}\big) + C_{11}\Pr\big(\text{accept } \mathcal{H}_1, \mathcal{H}_1 \text{ true}\big) \quad (10.72)\end{aligned}$$

Applying Bayes' rule $(\Pr(A, B) = \Pr(A|B) \times \Pr(B))$ to this expression, we obtain

$$\begin{aligned}\Re = {} & C_{00}\Pr(\mathcal{H}_0) \times \Pr\big(\text{accept } \mathcal{H}_0|\mathcal{H}_0 \text{ true}\big) + C_{01}\Pr(\mathcal{H}_0) \times \Pr\big(\text{accept } \mathcal{H}_0|\mathcal{H}_1 \text{ true}\big) \\ & + C_{10}\Pr(\mathcal{H}_1) \times \Pr\big(\text{accept } \mathcal{H}_1|\mathcal{H}_0 \text{ true}\big) + C_{11}\Pr(\mathcal{H}_1) \times \Pr\big(\text{accept } \mathcal{H}_1|\mathcal{H}_1 \text{ true}\big)\end{aligned}$$

Recognizing the last term in each summand as known probabilities, we have

$$\begin{aligned}\mathfrak{R} &= C_{00}\Pr(\mathcal{H}_0) \times P_{\text{Reject}} + C_{01}\Pr(\mathcal{H}_0) \times P_{\text{Miss}} \\ &+ C_{10}\Pr(\mathcal{H}_1) \times P_{\text{FA}} + C_{11}\Pr(\mathcal{H}_1) \times P_{\text{DET}}\end{aligned}$$

Substituting for P_{Reject} and P_{Miss} in terms of false-alarm and detection probabilities and gathering like-terms, we obtain the risk as

$$\begin{aligned}\mathfrak{R} &= C_{00}\Pr(\mathcal{H}_0) + C_{01}\Pr(\mathcal{H}_1) + (C_{10} - C_{00})\Pr(\mathcal{H}_0) \\ &\times P_{\text{FA}} + (C_{11} - C_{01})\Pr(\mathcal{H}_1) \times P_{\text{DET}}\end{aligned} \tag{10.74}$$

This expression is minimized by differentiating the risk criterion with respect to the false-alarm probability, setting the result to zero and solving for the threshold or equivalently the differential or slope of the ROC curve to give

$$\frac{d\mathfrak{R}}{dp_{\text{FA}}} = (C_{10} - C_{00})\Pr(\mathcal{H}_0) + (C_{11} - C_{01})\Pr(\mathcal{H}_1) \times \frac{dP_{\text{DET}}(\tau)}{dP_{\text{FA}}(\tau)} = 0 \tag{10.75}$$

where solving for the slope of the ROC with $\tau \rightarrow \tau^*$

$$\tau^* = \left.\frac{dP_{\text{DET}}(\tau)}{dP_{\text{FA}}(\tau)}\right|_{\tau=\tau^*} = \frac{(C_{10} - C_{00})\Pr(\mathcal{H}_0)}{(C_{01} - C_{11})\Pr(\mathcal{H}_1)} \tag{10.76}$$

gives the slope leading to the desired ODP. We illustrate this calculation in the following example.

Example 10.9

In this example, we return to the example illustrated in Fig. 10.5 where we show two Gaussian decision functions $\mathbf{N}(110, 5)$ and $\mathbf{N}(120, 5.5)$ where the notation $\mathcal{N}(M, V)$ is a normal distribution with mean M and variance V. An ensemble of 100-members was generated for the non-signal or disturbance and signal distributions, respectively. Using the "brute force" method described earlier, each member ROC curve was estimated and threshold averaged. The results are shown in Fig. 10.8. First we note the (average) AROC obtained using the threshold averaging method discussed earlier along with its corresponding $\pm 2\sigma$ confidence limits (bounds). We also observe the optimal decision threshold point (diamond) and its associated uncertainty $(2\sigma\text{–by–}1\sigma)$ box indicating the maximum and minimum uncertainty for both P_{DET} (vertical sides) and P_{FA} (horizontal sides). In Fig. 10.8, we see some of the other metrics (e.g., AUC, ODP) calculated as well as the inset of the ODP. These metrics are summarized in Table 10.1.

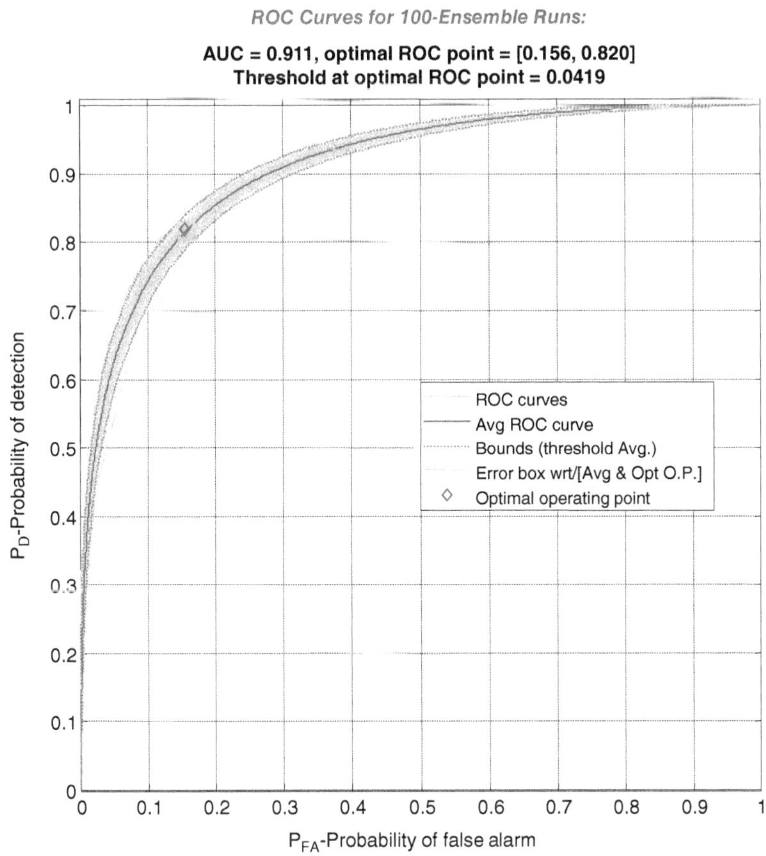

FIGURE 10.8 Final ROC curve for Gaussian decision function example including metrics along with legend with optimal decision function operating point, average (threshold averaging) and bounds.

TABLE 10.1 Results for Gaussian Decision Function Example

Ensemble Average Results	
No. of ensemble members	100
Area under curve	0.91
Distance	4
Min. error probability	14.9
Optimum decision point	[0.16,0.82]
Confusion matrix	[0.82 0.16;0.18 0.84]
Detection probability	0.82
False alarm probability	0.16
Rejection probability	0.84
Miss probability	0.18
Accuracy	0.83
Positive pred. value	0.94
Negative pred. value	0.61

△△△

In this section, we have developed the concept of performance metrics for automated detection systems. We have shown both pictorially and mathematically (briefly) how the ROC curve evolves from the detection and false-alarm probabilities. We have also shown how a variety of performance metrics can be extracted from the ROC curve computation and how some enable a "single number" point (AUC, ODP, ACC, etc.) that can be used for performance ranking the performance of *detection* systems. Next, we investigate methods that provide an alternative to "batch" detection processing—the sequential approach.

10.5 SEQUENTIAL DETECTION

In many real-time applications, a decision must be made quite rapidly. Batch schemes are exactly that—a batch of data is gathered and made available to the detector and a decision is made. For off-line applications this is acceptable, but a fighter pilot in a life/death situation must have an instantaneous indicator that a lethal weapon not a bird is approaching the aircraft. A reasonable approach to this problem of making a reliable decision with high confidence in a timely manner is to develop a sequential detection processor. At each measurement sample $\mathbf{y}(t)$, we *sequentially update* the decision function and compare it to the thresholds to perform the detection—"sample-by-sample". Here as each sample is collected producing the measurement sequence, the detector processes that measurement and attempts to "decide" whether or not it has evolved from the signal. Therefore, for each measurement the decision function is "sequentially" updated and compared to the detection thresholds obtained from the ROC curve operating point enabling a rapid decision. Once the threshold is crossed, the decision (signal or non-signal) is made and the measurement is processed; however, if not enough data are available to make the decision, then another measurement is collected. Realize, of course, that the decision function is random (see Fig. 10.9), but once either of the thresholds is crossed the detection is achieved (star in figure) and a decision is made.

We continue with the multiple channel/multiple measurement development, since other scenarios can be considered special cases of this formulation. Thus, an alternative to the batch approach is the sequential method which can be developed by expanding the likelihood-ratio for each instantaneous vector measurement to obtain

$$\mathcal{L}(\mathbf{Y}_N) = \frac{\Pr(\mathbf{Y}_N|\mathcal{H}_1)}{\Pr(\mathbf{Y}_N|\mathcal{H}_0)} = \frac{\Pr(\mathbf{y}(t_0), \mathbf{y}(t_1), \ldots, \mathbf{y}(t_N)|\mathcal{H}_1)}{\Pr(\mathbf{y}(t_0), \mathbf{y}(t_1), \ldots, \mathbf{y}(t_N)|\mathcal{H}_0)} \tag{10.77}$$

From the chain rule of probability

$$\Pr(\mathbf{Y}_N|\mathcal{H}_\ell) = \prod_{k=0}^{t} \Pr(y(t-k)|Y_{t-k-1}; \mathcal{H}_\ell); \ell = 0, 1 \tag{10.78}$$

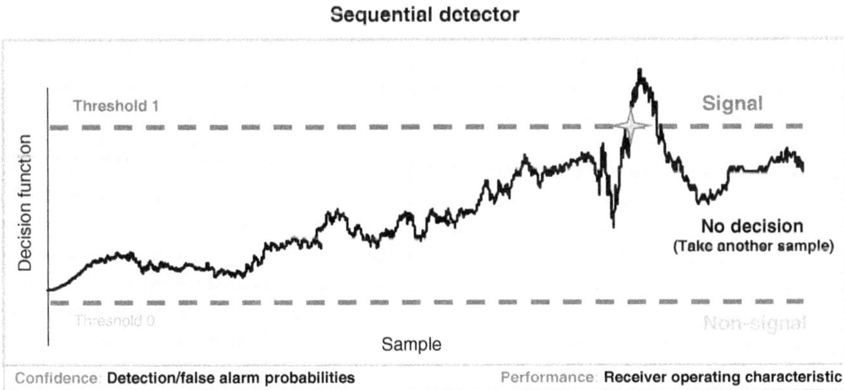

FIGURE 10.9 Conceptual implementation of the sequential detection technique. As each individual sample is extracted, the decision function is calculated and compared to thresholds to "decide" if the signal is present or not (non-signal). Quantitative performance and sequential thresholds are determined from estimated ROC curves.

and Bayes' rule [20], we have that

$$\Pr\left(\mathbf{Y}_N|\mathcal{H}_\ell\right) = \Pr\left(\mathbf{y}(N), \mathbf{Y}_{N-1}|\mathcal{H}_\ell\right) = \Pr\left(\mathbf{y}(N)|\mathbf{Y}_{N-1}, \mathcal{H}_\ell\right) \times \Pr\left(\mathbf{Y}_{N-1}|\mathcal{H}_\ell\right) \tag{10.79}$$

for $\ell = 0, 1$.

Substituting this expression into the likelihood-ratio above, replacing $t \rightarrow N$ grouping and reordering, we obtain

$$\mathcal{L}(\mathbf{Y}_t) = \frac{\Pr\left(\mathbf{Y}_{t-1}|\mathcal{H}_1\right)}{\Pr\left(\mathbf{Y}_{t-1}|\mathcal{H}_0\right)} \times \frac{\Pr\left(\mathbf{y}(t)|\mathbf{Y}_{t-1}, \mathcal{H}_1\right)}{\Pr\left(\mathbf{y}(t)|\mathbf{Y}_{t-1}, \mathcal{H}_0\right)} \tag{10.80}$$

and the recursion or equivalently *sequential likelihood-ratio* for the tth measurement follows as

$$\mathcal{L}(\mathbf{Y}_t) = \mathcal{L}(\mathbf{Y}_{t-1}) \times \frac{\Pr\left(\mathbf{y}(t)|\mathbf{Y}_{t-1}, \mathcal{H}_1\right)}{\Pr\left(\mathbf{y}(t)|\mathbf{Y}_{t-1}, \mathcal{H}_0\right)} \tag{10.81}$$

with $t = 0, \ldots, N$ and $\Pr\left(\mathbf{y}(0)|\mathbf{Y}_{-1}, \mathcal{H}_\ell\right) = \Pr\left(\mathbf{y}(0)|\mathcal{H}_\ell\right)$, the *prior* under each hypothesis.

Before we close this section, it should be noted that many problems can be characterized by distributions that are members of the *exponential family*; therefore, we can take the natural logarithm of Eq. 10.81 producing a monotonic function that does not alter the decision theoretic results. Thus, the sequential *log-likelihood-ratio* $\Lambda(\mathbf{Y}_t) = \ln \mathcal{L}(\mathbf{Y}_t)$ is defined by

$$\Lambda(\mathbf{Y}_t) = \Lambda(\mathbf{Y}_{t-1}) + \ln \Pr\left(\mathbf{y}(t)|\mathbf{Y}_{t-1}, \mathcal{H}_1\right) - \ln \Pr\left(\mathbf{y}(t)|\mathbf{Y}_{t-1}, \mathcal{H}_0\right) \tag{10.82}$$

10.5.1 Sequential Decision Theory

In this subsection, we discuss two approaches to sequential decision theory: the Bayesian approach and the Neyman–Pearson approach [3]. For the batch approach, we compared the likelihood decision function to a threshold that evolved from the specific decision criterion selected. In contrast, for the sequential decision function (above) there must be two thresholds, an upper and a lower, that vary with the number of measurements—sequentially. If the decision function exceeds the upper threshold then $\mathcal{H}_1$ is decided, while if it is below the lower then $\mathcal{H}_0$ is decided. In between the thresholds, no decision is made and another data sample must be processed.

Bayesian Approach To determine the threshold of a test, the prior probabilities of each hypothesis are required, that is, from Bayes' rule for the posterior probability at the tth step $\Pr(\mathcal{H}_\ell|t)$, we have that [8]

$$\Pr\left(\mathcal{H}_1|t\right) = \frac{\Pr\left(t|\mathcal{H}_1\right) \times \Pr(\mathcal{H}_1)}{\Pr\left(t|\mathcal{H}_0\right) \times \Pr(\mathcal{H}_0) + \Pr\left(t|\mathcal{H}_1\right) \times \Pr(\mathcal{H}_1)} \tag{10.83}$$

similarly for $\mathcal{H}_0$

$$\Pr\left(\mathcal{H}_0|t\right) = \frac{\Pr\left(t|\mathcal{H}_0\right) \times \Pr(\mathcal{H}_0)}{\Pr\left(t|\mathcal{H}_0\right) \times \Pr(\mathcal{H}_0) + \Pr\left(t|\mathcal{H}_1\right) \times \Pr(\mathcal{H}_1)} \tag{10.84}$$

where the posterior for each hypothesis satisfy the relation

$$\Pr\left(\mathcal{H}_0|t\right) + \Pr\left(\mathcal{H}_1|t\right) = 1$$

Therefore, dividing Eq. 10.83 by Eq. 10.84 we have that

$$\frac{\Pr\left(\mathcal{H}_1|t\right)}{\Pr\left(\mathcal{H}_0|t\right)} = \frac{\Pr\left(t|\mathcal{H}_1\right)}{\Pr\left(t|\mathcal{H}_0\right)} \times \frac{\Pr\left(\mathcal{H}_1\right)}{\Pr(\mathcal{H}_0)} = \mathcal{L}(t) \times \frac{\Pr\left(\mathcal{H}_1\right)}{\Pr(\mathcal{H}_0)} \tag{10.85}$$

In statistics [8], this expression relates the *posterior odds ratio* to the *prior odds ratio* through the likelihood which is called the *Bayes' factor*.

From the sequential perspective [3], we can consider the posterior at step t to be related to the posterior at the previous stage $t-1$ by

$$\begin{aligned}\Pr\left(\mathcal{H}_1;t\right) &= \Pr\left(\mathcal{H}_1|y(t-1)\right)\\ &= \frac{\Pr\left(y(t-1)|\mathcal{H}_1\right) \times \Pr(\mathcal{H}_1;t-1)}{\Pr\left(y(t-1)|\mathcal{H}_0\right) \times \Pr(\mathcal{H}_0;t-1) + \Pr\left(y(t-1)|\mathcal{H}_1\right) \times \Pr(\mathcal{H}_1;t-1)}\end{aligned} \tag{10.86}$$

and for the null hypothesis we have a similar relation using Bayes' rule for the posterior

$$\Pr(\mathcal{H}_0;t) = \Pr(\mathcal{H}_0|y(t-1))$$
$$= \frac{\Pr(y(t-1)|\mathcal{H}_0) \times \Pr(\mathcal{H}_0;t-1)}{\Pr(y(t-1)|\mathcal{H}_0) \times \Pr(\mathcal{H}_0;t-1) + \Pr(y(t-1)|\mathcal{H}_1) \times \Pr(\mathcal{H}_1;t-1)} \quad (10.87)$$

Dividing Eq. 10.86 by Eq. 10.87 we obtain

$$\frac{\Pr(\mathcal{H}_1;t)}{\Pr(\mathcal{H}_0;t)} = \mathcal{L}(y(t-1)) \times \frac{\Pr(\mathcal{H}_1;t-1)}{\Pr(\mathcal{H}_0;t-1)} \quad (10.88)$$

If we impose the independent measurement constraint, then we can easily show by induction that

$$\frac{\Pr(\mathcal{H}_1;t)}{\Pr(\mathcal{H}_0;t)} = \mathcal{L}(Y_{t-1}) \times \frac{\Pr(\mathcal{H}_1)}{\Pr(\mathcal{H}_0)} \quad (10.89)$$

completing the general Bayes' approach to sequential hypothesis testing.

Neyman–Pearson: Wald SPRT Approach The *sequential probability-ratio test* (SPRT) is a modification of the Neyman–Pearson scheme for sequential detection [32]. It follows directly from the likelihood decision function assuming that after N-samples $\mathcal{H}_1$ is accepted so that

$$\mathcal{L}(Y_N) = \frac{\Pr(Y_N|\mathcal{H}_1)}{\Pr(Y_N|\mathcal{H}_0)} = \mathcal{T}_1(N) \quad (10.90)$$

cross-multiplying and integrating both sides over D_1, we have that

$$\int_{D_1} \Pr(Y_N|\mathcal{H}_1)dY = \mathcal{T}_1(N) \times \int_{D_1} \Pr(Y_N|\mathcal{H}_0)dY \quad (10.91)$$

or in terms of detection, miss and false-alarm probabilities this relation becomes

$$P_{\text{DET}} = 1 - P_{\text{Miss}} = \mathcal{T}_1(N) \times P_{\text{FA}} \quad (10.92)$$

Alternatively, if $\mathcal{H}_0$ is accepted, then following the same approach we have

$$\mathcal{L}(Y_N) = \frac{\Pr(Y_N|\mathcal{H}_1)}{\Pr(Y_N|\mathcal{H}_0)} = \mathcal{T}_0(N) \quad (10.93)$$

and cross-multiplying and integrating both sides over D_0, we obtain

$$P_{\text{Miss}} = \mathcal{T}_0(N) \times \left(1 - P_{\text{FA}}\right) \tag{10.94}$$

Solving these relations individually, the thresholds are specified in terms of the false-alarm (P_{FA}) and miss (P_{M}) probabilities with $t \longrightarrow N$ as

$$\mathcal{T}_0(t) = \frac{P_{\text{Miss}}(t)}{1 - P_{\text{FA}}(t)}; \qquad \mathcal{T}_1(t) = \frac{1 - P_{\text{Miss}}(t)}{P_{\text{FA}}(t)} \tag{10.95}$$

These thresholds are determined from an ROC curve (detection vs. false-alarm probabilities) obtained by simulation or a controlled experiment to calculate the decision function. That is, an operating point is selected from the ROC corresponding to specific detection (or equivalently miss) and false-alarm probabilities, specifying the required thresholds which are calculated according to Eq. 10.95 for each parameter update.

Therefore, the SPRT is now completely specified by

$$\begin{array}{lll} \mathcal{L}\big(\mathbf{y}(t)\big) & > \mathcal{T}_1(t) & \text{Accept } \mathcal{H}_1 \\ \mathcal{T}_0(t) \leq \mathcal{L}\big(\mathbf{y}(t)\big) & \leq \mathcal{T}_1(t) & \text{Continue} \\ \mathcal{L}\big(\mathbf{y}(t)\big) & < \mathcal{T}_0(t) & \text{Accept } \mathcal{H}_0 \end{array} \tag{10.96}$$

It should be noted that the non-sequential Bayesian detection technique is optimum in the sense that no other hypothesis test can achieve a smaller average risk (smallest error probabilities [32]).

For the case of the sequential log-likelihood, the SPRT is given by

$$\begin{array}{lll} \Lambda\big(\mathbf{y}(t)\big) & > \ln \mathcal{T}_1(t) & \text{Accept } \mathcal{H}_1 \\ \ln \mathcal{T}_0(t) \leq \Lambda\big(\mathbf{y}(t)\big) & \leq \ln \mathcal{T}_1(t) & \text{Continue} \\ \Lambda(\big(\mathbf{y}(t)\big) & < \ln \mathcal{T}_0(t) & \text{Accept } \mathcal{H}_0 \end{array} \tag{10.97}$$

This approach was developed under the assumption that the number of measurements required to achieve this optimum is known (fixed) in advance. In relaxing this constraint, sequential detection techniques are superior to fixed non-sequential Bayesian detection in the sense that "on the average" a substantially smaller number of measurements are required to achieve the same error probabilities. The disadvantage though is that this number of measurements required to make a decision is a random number and therefore we must estimate its expected value or average value.

With this in mind, we show that the *average number of measurements* $\hat{\mathcal{N}} = E\{N|\mathcal{H}_\ell\}$ required to reach a decision can be estimated [3], [32]. Starting with the log-likelihood ratio for a batch of N measurements $\mathbf{Y}_N$, assuming each is independent and identically distributed (*i.i.d.*) then so is each $\Lambda(y(t))$ and therefore we have that

$$\Lambda(\mathbf{Y}_N) = \sum_{t=1}^{N} \Lambda(y(t)) \tag{10.98}$$

The joint expectation conditioned on each hypothesis $\mathcal{H}_\ell; \ell = 0, 1$ and N is

$$E_\Lambda\left\{\Lambda(\mathbf{Y}_N)|\mathcal{H}_\ell\right\} = E_N\left\{E_\Lambda\left\{\Lambda(\mathbf{Y}_N)|N, \mathcal{H}_\ell\right\}\right\} \tag{10.99}$$

but because of independence and performing the expectation over N, we have

$$E_\Lambda\left\{\Lambda(\mathbf{Y}_N)|\mathcal{H}_\ell\right\} = E\{N|\mathcal{H}_\ell\} \times E\left\{\Lambda(\mathbf{y}(t))|\mathcal{H}_\ell\right\} \tag{10.100}$$

and therefore we have that the *average* number of measurements required can be estimated from

$$\hat{\mathcal{N}} = E\{N|\mathcal{H}_\ell\} = \frac{E\left\{\Lambda(\mathbf{Y}_N)|\mathcal{H}_\ell\right\}}{E\left\{\Lambda(\mathbf{y}(t))|\mathcal{H}_\ell\right\}} \tag{10.101}$$

The following possible conditions *on* the sequential decision boundaries (ignoring the "continue" option)

$$\begin{aligned}
&\Lambda(\mathbf{Y}_N) = \ln \mathcal{T}_0 \text{ with } \mathcal{H}_0 \text{ true which implies } \Pr(\ln \mathcal{T}_0|\mathcal{H}_0) = 1 - P_{FA}\\
&\Lambda(\mathbf{Y}_N) = \ln \mathcal{T}_0 \text{ with } \mathcal{H}_1 \text{ true which implies } \Pr(\ln \mathcal{T}_0|\mathcal{H}_1) = P_{Miss}\\
&\Lambda(\mathbf{Y}_N) = \ln \mathcal{T}_1 \text{ with } \mathcal{H}_0 \text{ true which implies } \Pr(\ln \mathcal{T}_1|\mathcal{H}_0) = P_{FA}\\
&\Lambda(\mathbf{Y}_N) = \ln \mathcal{T}_1 \text{ with } \mathcal{H}_1 \text{ true which implies } \Pr(\ln \mathcal{T}_1|\mathcal{H}_1) = 1 - P_{Miss}
\end{aligned}$$

Combining these relations, we see that the expected log-likelihood under a given hypothesis is

$$E\{\Lambda(\mathbf{Y}_N)|\mathcal{H}_\ell\} = \begin{cases} \ln \mathcal{T}_0(1 - P_{FA}) + \ln \mathcal{T}_1 P_{FA} & \mathcal{H}_0 \text{ true}\\ \ln \mathcal{T}_0 P_{Miss} + \ln \mathcal{T}_1(1 - P_{Miss}) & \mathcal{H}_1 \text{ true} \end{cases} \tag{10.102}$$

Finally substituting into Eq. 10.101, we obtain the relations for the *average number of measurements* to make a decision using the log-likelihood decision function to perform the SPRT as

$$\hat{\mathcal{N}} = \begin{cases} \dfrac{\ln \mathcal{T}_0(1-P_{FA})+\ln \mathcal{T}_1 P_{FA}}{E\{\Lambda(\mathbf{y}(t))|\mathcal{H}_\ell\}} & \mathcal{H}_0 \text{ true}\\[2ex] \dfrac{\ln \mathcal{T}_0 P_{Miss}+\ln \mathcal{T}_1(1-P_{Miss})}{E\{\Lambda(\mathbf{y}(t))|\mathcal{H}_\ell\}} & \mathcal{H}_1 \text{ true} \end{cases} \tag{10.103}$$

Returning to the likelihood relations, we obtain similar expressions for the expected number of measurements to make a decision with the expected likelihood under a given hypothesis given by

$$E\{\mathcal{L}(\mathbf{Y}_N)|\mathcal{H}_\ell\} = \begin{cases} \mathcal{T}_0(1-P_{\mathrm{FA}}) + \mathcal{T}_1 P_{\mathrm{FA}} & \mathcal{H}_0 \text{ true} \\ \mathcal{T}_0 P_{\mathrm{Miss}} + \mathcal{T}_1(1-P_{\mathrm{Miss}}) & \mathcal{H}_1 \text{ true} \end{cases} \tag{10.104}$$

and the average number of measurement samples

$$\hat{\mathcal{N}} = \begin{cases} \frac{\mathcal{T}_0(1-P_{\mathrm{FA}})+\mathcal{T}_1 P_{\mathrm{FA}}}{E\{\mathcal{L}(\mathbf{y}(t))|\mathcal{H}_\ell\}} & \mathcal{H}_0 \text{ true} \\ \frac{\mathcal{T}_0 P_{\mathrm{Miss}}+\mathcal{T}_1(1-P_{\mathrm{Miss}})}{E\{\mathcal{L}(\mathbf{y}(t))|\mathcal{H}_\ell\}} & \mathcal{H}_1 \text{ true} \end{cases} \tag{10.105}$$

Next, we return to our previous example and develop the "sequential" solution.

Example 10.10

Suppose we have the vector signal $\mathbf{s}(t) = \boldsymbol{\mu} \in \mathcal{R}^{N_y \times 1}$ that is contaminated with additive, zero-mean Gaussian measurement noise $\mathbf{v}(t)$ of variance $\mathbf{R}_{vv}$ and we would like to construct the sequential likelihood-ratio under the same assumptions of Example 10.6.

The likelihood decision function is

$$\mathcal{L}(\mathbf{Y}_t) = \mathcal{L}(\mathbf{Y}_{t-1}) \times \frac{1}{2}\exp\left\{\boldsymbol{\mu}^T\mathbf{R}_{vv}^{-1}\mathbf{y}(t) + \mathbf{y}^T(t)\mathbf{R}_{vv}^{-1}\boldsymbol{\mu}\right\} \begin{array}{c} \mathcal{H}_1 \\ > \\ = \\ < \\ \mathcal{H}_0 \end{array} \begin{array}{c} \mathcal{T}_1 + \frac{1}{2}\boldsymbol{\mu}^T\mathbf{R}_{vv}^{-1}\boldsymbol{\mu} \\ \\ \text{No decision} \\ \\ \mathcal{T}_0 + \frac{1}{2}\boldsymbol{\mu}^T\mathbf{R}_{vv}^{-1}\boldsymbol{\mu} \end{array}$$

where we have moved all *known* functions into the threshold. For the log-likelihood we have

$$\Lambda(\mathbf{Y}_t) = \Lambda(\mathbf{Y}_{t-1}) + \frac{1}{2}\boldsymbol{\mu}^T\mathbf{R}_{vv}^{-1}\mathbf{y}(t) + \frac{1}{2}\mathbf{y}^T(t)\mathbf{R}_{vv}^{-1}\boldsymbol{\mu} \begin{array}{c} \mathcal{H}_1 \\ > \\ = \\ < \\ \mathcal{H}_0 \end{array} \begin{array}{c} \ln\mathcal{T}_1 + \frac{1}{2}\boldsymbol{\mu}^T\mathbf{R}_{vv}^{-1}\boldsymbol{\mu} \\ \\ \text{No decision} \\ \\ \ln\mathcal{T}_0 + \frac{1}{2}\boldsymbol{\mu}^T\mathbf{R}_{vv}^{-1}\boldsymbol{\mu} \end{array}$$

Since the average sample log-likelihood under each hypothesis for this problem is found by examining the thresholds,

$$\Lambda\big(y(t)|\mathcal{H}_\ell\big) = \frac{1}{2}\boldsymbol{\mu}^T\mathbf{R}_{vv}^{-1}\boldsymbol{\mu}$$

we can estimate the average number of measurements required to make a decision from Eq. 10.103

The average number of samples can be estimated from

$$\hat{\mathcal{N}} = \begin{cases} \frac{2\times\left(\ln \mathcal{T}_0(1-P_{\rm FA})+\ln \mathcal{T}_1 P_{\rm FA}\right)}{\boldsymbol{\mu}^T \mathbf{R}_{vv}^{-1}\boldsymbol{\mu}} & \mathcal{H}_0 \text{ true} \\ \frac{2\times\left(\ln \mathcal{T}_0 P_{\rm Miss}+\ln \mathcal{T}_1(1-P_{\rm Miss})\right)}{\boldsymbol{\mu}^T \mathbf{R}_{vv}^{-1}\boldsymbol{\mu}} & \mathcal{H}_1 \text{ true} \end{cases}$$

△△△

This completes the example, next we consider the sequential model-based solutions to solve the binary detection problem when a priori information about the signal is known.

10.6 MODEL-BASED SEQUENTIAL DETECTION

In this section, we arrive at the heart of this chapter—the application of the sequential detection theory to Bayesian model-based techniques. Here we show how to solve the joint signal estimation/detection problem by combining some of the sequential processors with the sequential likelihood detection approach. All of the results are based on creating the likelihood decision function and specifying the threshold (ROC curves) corresponding to the decision function selected. We start with the linear Gauss–Markov model set, discuss the nonlinear case, and then generalize to the non-Gaussian case.

10.6.1 Linear Gaussian Model-Based Processor

In this section, we develop the model-based approach to sequential detection employing the sequential likelihood-ratio as the mechanism to implement the model-based designs starting with the linear time-varying (non-stationary) problem (see Refs.[3]and [33] for more details). We begin with the development of the generic Gauss–Markov signal model defined by its *state* (signal) *vector* $\mathbf{s}(t)$, where the linear state-space process model is given by

$$\mathbf{s}(t) = A(t-1)\mathbf{s}(t-1) + B(t-1)\mathbf{u}(t-1) + \mathbf{w}(t-1) \tag{10.106}$$

with corresponding measurement model as

$$\mathbf{y}(t) = C(t)\mathbf{s}(t) + \mathbf{v}(t) \tag{10.107}$$

where $\mathbf{s}$, $\mathbf{w}$ are the N_s-dimensional signal and process noise vectors, with $\mathbf{y}$, $\mathbf{v}$ the N_y-dimensional measurement and noise vectors along with $\mathbf{u}$ the N_u-dimensional known input vector. Both $\mathbf{w}$ and $\mathbf{v}$ are zero-mean, Gaussian with respective covariances $R_{\mathbf{ww}}(t-1)$ and $R_{\mathbf{vv}}(t)$. The corresponding system, input and measurement matrices are appropriately dimensioned and given by $A(t-1)$, $B(t-1)$, and $C(t)$, respectively. The initial state vector is Gaussian with $\mathbf{s}(0) \sim \mathcal{N}(\overline{\mathbf{s}}(0), P_{\mathbf{ss}}(0))$.

Because the underlying distributions are Gaussian, we know that the optimal solution to the signal estimation problem is given by the model-based processor (Kalman filter of Section 5.3) providing the predicted conditional mean estimate $\hat{\mathbf{s}}(t|t-1) = E\{\mathbf{s}(t)|Y_{t-1})$[1] with corresponding predicted conditional (error) covariance $\tilde{P}(t|t-1)$ [33, 34].

With this signal model in hand, we can now define the binary problem to decide whether the measurement contains the signal and noise or just noise alone, that is, we are testing the hypotheses that

$$\begin{aligned}
\mathcal{H}_0 &: \quad \mathbf{y}(t) = \mathbf{v}(t) && \text{[Noise]} \\
\mathcal{H}_1 &: \quad \mathbf{y}(t) = C(t)\mathbf{s}(t) + \mathbf{v}(t) && \text{[Signal+Noise]}
\end{aligned}$$

From the underlying Gauss–Markov assumption, the sequential likelihood-ratio solution is specified by the ratio of conditional Gaussian posterior distributions $\Pr(\mathbf{y}(t)|Y_{t-1}; \mathcal{H}_i); i = 0, 1$. That is,

$$\mathcal{L}(\mathbf{y}(t)) = \frac{\Pr(\mathbf{y}(t)|Y_{t-1}; \mathcal{H}_1)}{\Pr(\mathbf{y}(t)|Y_{t-1}; \mathcal{H}_0)} \tag{10.108}$$

with the conditional probabilities distributed as

$$\begin{aligned}
\Pr(\mathbf{y}(t)|Y_{t-1}; \mathcal{H}_0) &\sim \mathcal{N}\left(\mathbf{y}(t) : 0, R_{\mathbf{vv}(t)}\right) \\
\Pr(\mathbf{y}(t)|Y_{t-1}; \mathcal{H}_1) &\sim \mathcal{N}\left(\mathbf{y}(t) : C(t)\hat{\mathbf{s}}(t|t-1), R_{\mathbf{ee}}(t)\right)
\end{aligned}$$

where the *innovations* $\mathbf{e}(t)$ and its corresponding covariance $R_{\mathbf{ee}}(t)$ are obtained as outputs of the model-based processor (Kalman filter) specified by

$$\begin{aligned}
\mathbf{e}(t) &= \mathbf{y}(t) - \hat{\mathbf{y}}(t|t-1) = \mathbf{y}(t) - C(t)\hat{\mathbf{s}}(t|t-1) \\
R_{\mathbf{ee}}(t) &= C(t)\tilde{P}(t|t-1)C'(t) + R_{\mathbf{vv}}(t)
\end{aligned} \tag{10.109}$$

Thus, under the null hypothesis we have that

$$\Pr(\mathbf{y}(t)|Y_{t-1}; \mathcal{H}_0) = \frac{1}{(2\pi)^{N_y/2}|R_{\mathbf{vv}}(t)|^{1/2}} \exp\left\{-\frac{1}{2}\mathbf{y}'(t)R_{\mathbf{vv}}^{-1}(t)\mathbf{y}(t)\right\} \tag{10.110}$$

while the conditional probability under alternate hypothesis is given by

$$\begin{aligned}
\Pr(\mathbf{y}(t)|Y_{t-1}; \mathcal{H}_1) = {}& \frac{1}{(2\pi)^{N_y/2}|R_{\mathbf{ee}}(t)|^{1/2}} \times \\
& \exp\left\{-\frac{1}{2}(\mathbf{y}(t) - \hat{\mathbf{y}}(t|t-1))'R_{ee}^{-1}(t)(\mathbf{y}(t) - \hat{\mathbf{y}}(t|t-1))\right\}
\end{aligned} \tag{10.111}$$

[1] Recall that this notation is defined in terms of predicted *conditional means* and *covariances* by $\hat{\mathbf{s}}(t|t-1)$ and $\tilde{P}(t|t-1) = \text{cov}(\tilde{\mathbf{s}}(t|t-1))$ for the *predicted state estimation error*, $\tilde{\mathbf{s}}(t|t-1) = \mathbf{s}(t) - \hat{\mathbf{s}}(t|t-1)$.

or simply

$$\Pr(\mathbf{y}(t)|Y_{t-1};\mathcal{H}_1) = \frac{1}{(2\pi)^{N_y/2}|R_{\mathbf{ee}}(t)|^{1/2}} \exp\left\{-\frac{1}{2}\mathbf{e}'(t)R_{\mathbf{ee}}^{-1}(t)\mathbf{e}(t)\right\} \quad (10.112)$$

Moving all *known* terms to the threshold, the required sequential log-likelihood of Eq. 10.82 for the linear Gauss–Markov signal model becomes

$$\Lambda(\mathbf{Y}_t) = \Lambda(\mathbf{Y}_{t-1}) - \frac{1}{2}\mathbf{e}'(t)R_{\mathbf{ee}}^{-1}(t)\mathbf{e}(t) + \frac{1}{2}\mathbf{y}'(t)R_{\mathbf{vv}}^{-1}(t)\mathbf{y}(t) \begin{array}{cc} \mathcal{H}_1 & \mathrm{T}_1 \\ > & \\ = & \text{No decision} \\ < & \\ \mathcal{H}_0 & \mathrm{T}_0 \end{array}$$

$$\mathrm{T}_\ell(t) = \ln \mathcal{T}_\ell(t) - \ln\left(\frac{1}{(2\pi)^{N_y/2}|R_{\mathbf{ee}}(t)|^{1/2}}\right) + \ln\left(\frac{1}{(2\pi)^{N_y/2}|R_{\mathbf{vv}}(t)|^{1/2}}\right); \ell = 0, 1 \quad (10.113)$$

All that remains is to specify the predicted and corrected means and covariances to implement the sequential detector. These are available as part of the model-based algorithm given by

$$\begin{aligned} \hat{\mathbf{s}}(t|t-1) &= A(t-1)\hat{\mathbf{s}}(t-1|t-1) + B(t-1)\mathbf{u}(t-1) && \text{[Prediction]} \\ \tilde{P}(t|t-1) &= A(t-1)\tilde{P}(t-1|t-1)A'(t-1) + R_{\mathbf{ww}}(t-1) && \text{[Prediction cov.]} \\ \hat{\mathbf{s}}(t|t) &= \hat{\mathbf{s}}(t|t-1) + K(t)\mathbf{e}(t) && \text{[Correction]} \\ \tilde{P}(t|t) &= [\mathbf{I} - K(t)C(t)]\,\tilde{P}(t|t-1) && \text{[Correction cov.]} \end{aligned}$$

for $K(t)$ the corresponding gain (see Ref. [33] for more details), next we consider the *RC*-circuit problem of Example 5.1 and construct a model-based sequential detector.

Example 10.11

Let us return to the *RC*-circuit of Example 5.1 where we have the Gauss–Markov model given by

$$\begin{aligned} s(t) &= 0.97s(t-1) + 100u(t-1) + w(t-1) \\ y(t) &= 2s(t) + v(t) \end{aligned} \quad (10.114)$$

$w \sim \mathcal{N}(0, 10^{-5})$, $v \sim \mathcal{N}(0, 4)$ and $s(0) \sim \mathcal{N}(2.5, 10^{-6})$.

We simulated this system as before shown in Fig. 5.3 and constructed the model-based processor resulting in Fig. 5.4.

The log-likelihood decision function is

$$\Lambda(t) = \Lambda(t-1) - \frac{e^2(t)}{2R_{ee}(t)} + \frac{y^2(t)}{8} \begin{array}{c} \mathcal{H}_1 \\ > \\ = \\ < \\ \mathcal{H}_0 \end{array} \begin{array}{c} \mathrm{T}_1 \\ \\ \text{No decision} \\ \\ \mathrm{T}_0 \end{array}$$

$$\mathrm{T}_\ell(t) = \ln \mathcal{T}_\ell(t) - \ln\left(\frac{1}{\sqrt{2\pi|R_{ee}(t)|}}\right) + \ln\left(\frac{1}{2\sqrt{2\pi}}\right); \ell = 0, 1$$

where we have moved all *known* functions into the threshold.

Recall that in order to estimate the thresholds, we first must estimate the ROC curve. The simulation was executed first for the known signal case as shown in Fig. 5.4 and then for the case of noise only. The results are shown in Fig. 10.10 where we see the estimated ROC curve in 10.10d and the log-likelihood for both cases with the corresponding thresholds in 10.10a,b along with the underlying PMFs in 10.10c. Here

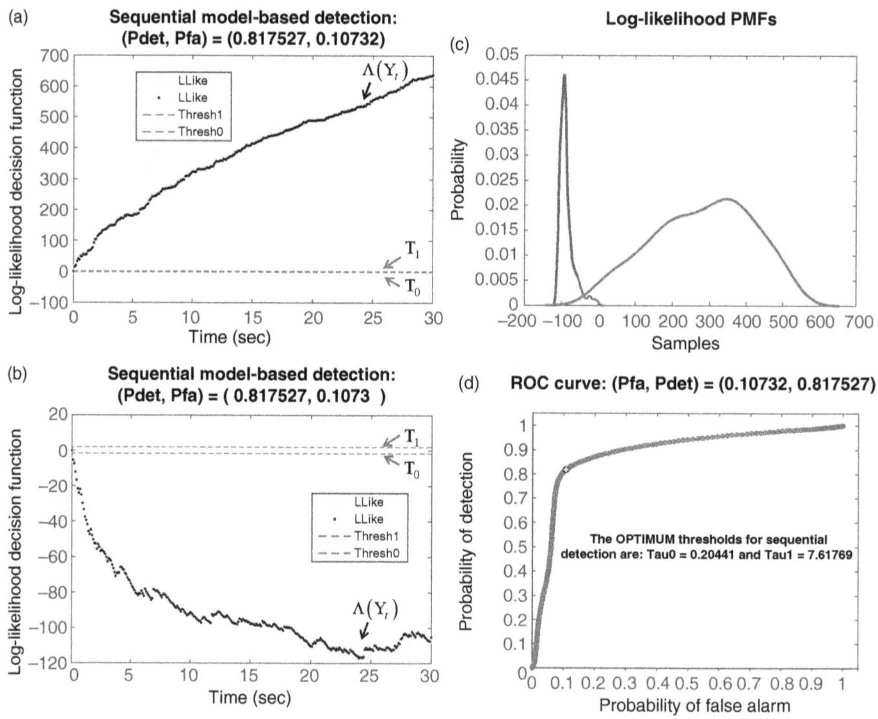

FIGURE 10.10 Sequential model-based detector design for *RC*-Circuit problem: (a) signal case: log-likelihood detection; (b) noise case: log-likelihood detection; (c) log-likelihood PMFs; (d) ROC curve.

the detection and false-alarm probabilities were $P_{\text{DET}} = 0.81$ and $P_{\text{FA}} = 0.11$ at the optimum threshold. Using Eq. 10.113, we calculate the corresponding thresholds as $\ln \mathcal{T}_0 = -1.59$ and $\ln \mathcal{T}_1 = 2.03$ leading to the thresholds $\mathrm{T}_1(t) = 1.75 + \ln \frac{1}{\sqrt{2\pi |R_{ee}(t)|}}$ and $\mathrm{T}_0(t) = -1.87 + \ln \frac{1}{\sqrt{2\pi |R_{ee}(t)|}}$. In steady-state we have that $\mathrm{T}_1 \approx 2.03$ and $\mathrm{T}_1 \approx -1.59$. △△△

This completes the example, next we consider the sequential model-based solutions to solve the binary detection problem for nonlinear processes.

10.6.2 Nonlinear Gaussian Model-Based Processor

In this section, we use the development of the linear case as a road map and develop the sequential model-based detector for the nonlinear case. We start with the nonlinear (approximate) Gauss–Markov model given by

$$
\begin{aligned}
\mathbf{s}(t) &= \mathrm{a}\left[\mathbf{s}(t-1)\right] + \mathrm{b}\left[\mathbf{u}(t-1)\right] + \mathbf{w}(t-1) \\
\mathbf{y}(t) &= \mathrm{c}\left[\mathbf{s}(t)\right] + \mathbf{v}(t)
\end{aligned} \tag{10.115}
$$

where (as before) $\mathbf{s}$, $\mathbf{w}$ are the N_s-dimensional signal and process noise vectors with $\mathbf{y}$, $\mathbf{v}$ the N_y-dimensional measurement and noise vectors along with $\mathbf{u}$ the N_u-dimensional known input vector. Both $\mathbf{w}$ and $\mathbf{v}$ are zero-mean Gaussian with respective covariances $R_{\mathbf{ww}}(t-1)$ and $R_{\mathbf{vv}}(t)$. The corresponding *nonlinear* system, input and measurement vector functions are appropriately dimensioned and given by $\mathrm{a}[\cdot]$, $\mathrm{b}[\cdot]$ and $\mathrm{c}[\cdot]$, respectively. The initial state vector is Gaussian with $\mathbf{s}(0) \sim \mathcal{N}(\overline{\mathbf{s}}(0), P_{\mathbf{ss}}(0))$.

Because the underlying distributions are assumed approximately Gaussian, we know that one possible solution to the signal estimation problem is given by the model-based processor (extended Kalman filter[2]) providing the predicted conditional mean estimate $\hat{\mathbf{s}}(t|t-1)$, with corresponding predicted conditional (error) covariance $\tilde{P}(t|t-1)$ [33].

With this nonlinear signal model in hand, we can now define the binary problem to decide whether the measurement contains the signal or just noise alone, that is, we are testing the hypotheses that

$$
\begin{aligned}
\mathcal{H}_0 &: \quad \mathbf{y}(t) = \mathbf{v}(t) && \text{[Noise]} \\
\mathcal{H}_1 &: \quad \mathbf{y}(t) = \mathrm{c}\left[\mathbf{s}(t)\right] + \mathbf{v}(t) && \text{[Signal+Noise]}
\end{aligned} \tag{10.116}
$$

Before we proceed with the sequential detection development for nonlinear, non-Gaussian problems, we have chosen to employ the first-order Taylor series representation to approximate the nonlinear vector functions. The development evolves

[2] It is well known that some of the modern estimation techniques available ([33], [35]) offer improved alternatives like the unscented Kalman filter of Chapter 6, but we choose this formulation from a pedagogical perspective, since its easily tracks the linear case developed in the previous subsection.

quite naturally from a linearized model-based processor (linearized Kalman filter) [33]. These approximations when expanded about a reference $\mathbf{s} = \mathbf{s}^*$ take the general form

$$c[\mathbf{s}(t)] \approx c\left[\mathbf{s}^*(t)\right] + \mathbf{C}\left[\mathbf{s}^*(t)\right] \times \left(\mathbf{s}(t) - \mathbf{s}^*(t)\right) + \text{H.O.T.} \tag{10.117}$$

with Jacobian matrix $\mathbf{C}[\mathbf{s}^*(t)] := \frac{\partial}{\partial \mathbf{s}} c[\mathbf{s}]\big|_{\mathbf{s}=\mathbf{s}^*(t)}$. Each of the nonlinear system, input and measurement functions are approximated in this manner using the Taylor series. For the *ad hoc* nonlinear EKF processor, the most current estimate available, $\hat{\mathbf{s}}(t|\ *)$, is used as a reference during the prediction and correction steps for the linearization, that is, the state estimates are

$$\begin{aligned} \hat{\mathbf{s}}(t|t-1) &= a\left[\hat{\mathbf{s}}(t-1|t-1)\right] + b\left[\mathbf{u}(t-1)\right] \\ \hat{\mathbf{s}}(t|t) &= \hat{\mathbf{s}}(t|t-1) + \mathbf{K}\left[\hat{\mathbf{s}}(t|t-1)\right]\mathbf{e}(t) \end{aligned} \tag{10.118}$$

where the Jacobians appear in the covariance equations.

With this in mind and from the underlying *approximate* Gauss–Markov assumption, the sequential likelihood-ratio solution is again specified by the ratio of conditional Gaussian posterior distributions $\Pr(\mathbf{y}(t)|Y_{t-1};\mathcal{H}_i); i = 0, 1$ as before. That is,

$$\begin{aligned} \Pr(\mathbf{y}(t)|Y_{t-1};\mathcal{H}_0) &\sim \mathcal{N}\left(\mathbf{y}(t) : \mathbf{0}, R_{\mathbf{vv}}(t)\right) \\ \Pr(\mathbf{y}(t)|Y_{t-1};\mathcal{H}_1) &\sim \mathcal{N}\left(\mathbf{y}(t) : c\left[\hat{\mathbf{s}}(t|t-1)\right], R_{\mathbf{ee}}(t)\right) \end{aligned}$$

where the *innovations* $\mathbf{e}(t)$ and its corresponding covariance $R_{\mathbf{ee}}(t)$ are obtained as outputs of the model-based processor (extended Kalman filter) specified by

$$\begin{aligned} \mathbf{e}(t) &= \mathbf{y}(t) - \hat{\mathbf{y}}(t|t-1) = \mathbf{y}(t) - c\left[\hat{\mathbf{s}}(t|t-1)\right] \\ R_{\mathbf{ee}}(t) &= \mathbf{C}\left[\hat{\mathbf{s}}(t|t-1)\right]\tilde{P}(t|t-1)\mathbf{C}'\left[\hat{\mathbf{s}}(t|t-1)\right] + R_{\mathbf{vv}}(t) \end{aligned} \tag{10.119}$$

Thus, under the null hypothesis we have that

$$\Pr(\mathbf{y}(t)|Y_{t-1};\mathcal{H}_0) = \frac{1}{(2\pi)^{N_y/2}|R_{\mathbf{vv}}(t)|^{1/2}} \exp\left\{-\frac{1}{2}\mathbf{y}'(t)R_{\mathbf{vv}}^{-1}(t)\mathbf{y}(t)\right\} \tag{10.120}$$

while the conditional probability under alternate hypothesis is given by

$$\Pr(\mathbf{y}(t)|Y_{t-1};\mathcal{H}_1) = \frac{1}{(2\pi)^{N_y/2}|R_{\mathbf{ee}}(t)|^{1/2}} \exp\left\{-\frac{1}{2}\mathbf{e}'(t)R_{\mathbf{ee}}^{-1}(t)\mathbf{e}(t)\right\} \tag{10.121}$$

Again moving all known terms to the threshold, the required sequential log-likelihood for the nonlinear Gauss–Markov signal model becomes

$$\Lambda(Y_t) = \Lambda(Y_{t-1}) - \frac{1}{2}\mathbf{e}'(t)R_{\mathbf{ee}}^{-1}(t)\mathbf{e}(t) + \frac{1}{2}\mathbf{y}'(t)R_{\mathbf{vv}}^{-1}(t)\mathbf{y}(t) \quad \begin{matrix} \mathcal{H}_1 & \mathrm{T}_1 \\ > & \\ = & \text{No decision} \\ < & \\ \mathcal{H}_0 & \mathrm{T}_0 \end{matrix}$$

$$\mathrm{T}_\ell(t) = \ln \mathcal{T}_\ell(t) - \ln\left(\frac{1}{(2\pi)^{N_y/2}|R_{\mathbf{ee}}(t)|^{1/2}}\right) + \ln\left(\frac{1}{(2\pi)^{N_y/2}|R_{\mathbf{vv}}(t)|^{1/2}}\right); \ell = 0, 1 \tag{10.122}$$

All that remains is to specify the predicted and corrected covariances to implement the sequential detector, since the state estimates we given in Eq. 10.118. These are available as part of the nonlinear model-based algorithm (EKF) given by

$$\tilde{P}(t|t-1) = \mathbf{A}\left[\hat{\mathbf{s}}(t|t-1)\right]\tilde{P}(t-1|t-1)\mathbf{A}'\left[\hat{\mathbf{s}}(t|t-1)\right] + R_{\mathbf{ww}}(t-1) \quad \text{[Prediction cov.]}$$

$$\tilde{P}(t|t) = \left[\mathbf{I} - \mathbf{K}\left[\hat{\mathbf{s}}(t|t-1)\right]\mathbf{C}\left[\hat{\mathbf{s}}(t|t-1)\right]\right]\tilde{P}(t|t-1) \quad \text{[Correction cov.]}$$

for $\mathbf{A}$ the system Jacobian and $\mathbf{K}\left[\hat{\mathbf{s}}(t|t-1)\right]$ the corresponding gain (see Ref. [33] for more details). This completes the development.

Example 10.12

Let us return to the nonlinear problem of Example 5.2 where we have the nonlinear Gauss–Markov model given by

$$s(t) = (1 - 0.05\triangle T)s(t-1) + 0.04\triangle T s^2(t-1) + w(t-1)$$
$$y(t) = s^2(t) + s^3(t) + v(t)$$

$w \sim \mathcal{N}(0, 10^{-6})$, $v \sim \mathcal{N}(0, 0.09)$ and $s(0) \sim \mathcal{N}(2.3, 10^{-2})$ and $\triangle T = 0.01$ sec.

We simulated this system as before and constructed the model-based processor resulting in Fig. 5.5.

The log-likelihood decision function is

$$\Lambda(t) = \Lambda(t-1) - \frac{e^2(t)}{2R_{ee}(t)} + \frac{y^2(t)}{5.6} \quad \begin{matrix} \mathcal{H}_1 & \mathrm{T}_1 \\ > & \\ = & \text{No decision} \\ < & \\ \mathcal{H}_0 & \mathrm{T}_0 \end{matrix}$$

$$\mathrm{T}_\ell(t) = \ln \mathcal{T}_\ell(t) - \ln\left(\frac{1}{\sqrt{2\pi|R_{ee}(t)|}}\right) + \ln\left(\frac{1}{2\sqrt{2\pi}}\right); \ell = 0, 1$$

where we have moved all *known* functions into the threshold.

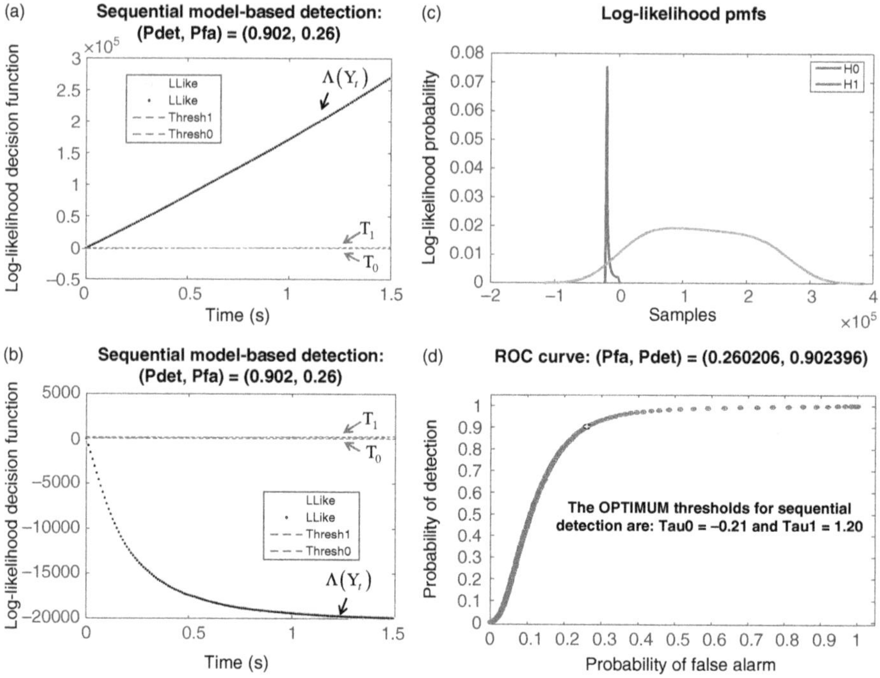

FIGURE 10.11 Sequential model-based detector design for nonlinear problem: (a) signal case: log-likelihood detection; (b) noise case: log-likelihood detection; (c) log-likelihood PMFs; (d) ROC curve. (AUC = 0.86.)

Recall that in order to estimate the thresholds, we first must estimate the ROC curve. The simulation was executed first for the known signal case as shown in Fig. 5.4 and then for the case of noise only. The results are shown in Fig. 10.11 where we see the estimated ROC curve in 10.11d and the log-likelihood for both cases with the corresponding thresholds in 10.11a,b along with the underlying PMFs in 10.11c. Here the detection and false-alarm probabilities were $P_{\rm DET} = 0.90$ and $P_{\rm FA} = 0.26$ at the optimum threshold. Using Eq. 10.113 we calculate the corresponding thresholds as $\ln \mathcal{T}_0 = -0.203$ and $\ln \mathcal{T}_1 = 1.24$ leading to the thresholds $\mathrm{T}_1(t) = 1.10 + \ln \frac{1}{\sqrt{2\pi|R_{ee}(t)|}}$ and $\mathrm{T}_0(t) = -2.17 + \ln \frac{1}{\sqrt{2\pi|R_{ee}(t)|}}$. In steady-state we have that $\mathrm{T}_1 \approx 1.20$ and $\mathrm{T}_0 \approx -0.205$. The AUC for this detector was AUC = 0.86. △△△

This completes the example, next we consider the sequential model-based solutions to solve the binary detection problem for non-Gaussian processes.

10.6.3 Non-Gaussian Model-Based Processor

Non-Gaussian processes are prevalent in many applications such as tracking, biomedical processing and financial problems [36–39]. The same basic nonlinear additive noise model of Eq. 1.101 governs the dynamics and measurements of the non-Gaussian processor although it is *not* limited to additive noise which could be multiplicative. Perhaps it is best to start from the Bayesian perspective in this case and point out that the "data" likelihood distribution (evidence of Section 2.53) is required to implement a sequential detection scheme. We require an estimate of the *data likelihood*

$$\hat{\Pr}(\mathbf{y}(t)|Y_{t-1};\mathcal{H}_\ell);\ell=0,1 \tag{10.123}$$

under each hypothesis of Eq. 1.102.

In order to obtain this estimate, we can use the particle filter processors of Chapter 7. Recall that the generic PF (Table 7.1) provides an estimate of the filtering distribution both *before* the required resampling as a non-uniformly weighted probability mass function

$$\hat{\Pr}(\mathbf{x}(t)|Y_t)=\sum_{i=1}^{N_p}\mathcal{W}_i(t)\delta\big(x(t)-x_i(t)\big) \tag{10.124}$$

where $\mathcal{W}$ is the normalized importance weight and *after* resampling ($\hat{x}_i \rightarrow x_i$) by the uniformly weighted PMF

$$\hat{\Pr}(\mathbf{x}(t)|Y_t)=\frac{1}{N_p}\sum_{i=1}^{N_p}\delta\big(x(t)-\hat{x}_i(t)\big) \tag{10.125}$$

These estimates, depending on the choice of importance distribution $q(\cdot)$ (see Section 7.3), lead to particular sets of weights. Therefore from Bayes' rule the data likelihood of Eq. 2.81

$$\Pr(y(t)|Y_{t-1})=\int \Pr\big(y(t)|x(t)\big)\times\Pr(x(t)|Y_{t-1})dx(t) \tag{10.126}$$

coupled with the assumption of a "perfect sampler" for the one-step prediction distribution

$$\hat{\Pr}(x(t)|Y_{t-1})=\frac{1}{N_p}\sum_{i=1}^{N_p}\delta\big(x(t)-x_i(t)\big)$$

gives the *estimated* data likelihood as

$$\begin{aligned}\hat{\Pr}(y(t)|Y_{t-1})&=\int \Pr(y(t)|x(t))\times\left(\frac{1}{N_p}\sum_{i=1}^{N_p}\delta\big(x(t)-x_i(t)\big)\right)dx(t)\\&=\frac{1}{N_p}\sum_{i=1}^{N_p}\int \Pr(y(t)|x(t))\times\delta\big(x(t)-x_i(t)\big)dx(t)\end{aligned} \tag{10.127}$$

or

$$\hat{\Pr}(y(t)|Y_{t-1}) \approx \frac{1}{N_p} \sum_{i=1}^{N_p} \Pr\big(y(t)|x_i(t)\big) \tag{10.128}$$

If we further assume a first-order Markov process ($Y_{t-1} \to y(t-1)$) and choose the importance distribution to be the state transition $q(x(t)|x(t-1), Y_t) \to \Pr\big(x(t)|x(t-1)\big)$, then the bootstrap PF of Table 7.2 results with the state likelihood weight, that is, $\mathcal{W}_i(t) \to \Pr\big(y(t)|x_i(t)\big)$ or Eq. 10.128 can be approximated by

$$\hat{\Pr}\big(y(t)|y(t-1)\big) = \frac{1}{N_p} \sum_{i=1}^{N_p} \Pr\big(y(t)|x_i(t)\big) \approx \frac{1}{N_p} \sum_{i=1}^{N_p} \mathcal{W}_i(t) \tag{10.129}$$

or simply defining the ensemble mean of the weight at each instant as

$$\overline{\mathcal{W}}(t) = \frac{1}{N_p} \sum_{i=1}^{N_p} \mathcal{W}_i(t) \tag{10.130}$$

we have

$$\hat{\Pr}\big(y(t)|y(t-1)\big) \approx \overline{\mathcal{W}}(t) \tag{10.131}$$

this is an a priori estimate that employs the predicted state (point) estimate $\hat{x}(t|t-1)$.

Other approaches to estimating the data likelihood can be obtained as well [40, 41]. They also evolve from Bayes' rule and the Chapman–Kolomogorov relations

$$\hat{\Pr}(y(t)|Y_{t-1}) = \int \Pr(y(t)|x(t)) \times \Pr(x(t)|Y_{t-1})\, dx(t) \tag{10.132}$$

or equivalently

$$\hat{\Pr}(y(t)|Y_{t-1}) = \int \Pr(y(t)|x(t-1)) \times \Pr(x(t-1)|Y_{t-1})\, dx(t-1) \tag{10.133}$$

Using the predicted estimate (weight) at time t, we have that

$$\hat{\Pr}(x(t)|Y_{t-1}) = \frac{1}{N_p} \sum_{i=1}^{N_p} \mathcal{W}_i(t)\delta\big(x(t) - x_i(t)\big) \tag{10.134}$$

then following the same steps as before and using the sifting property of the delta function, we have

$$\hat{\Pr}(y(t)|Y_{t-1}) \approx \frac{1}{N_p}\sum_{i=1}^{N_p} \mathcal{W}_i(t) \times \Pr\big(y(t)|x_i(t)\big) \tag{10.135}$$

while using the filtered estimate at time $t-1$, we obtain

$$\hat{\Pr}(x(t-1)|Y_{t-1}) \approx \frac{1}{N_p}\sum_{i=1}^{N_p} \mathcal{W}_i(t-1)\delta\big(x(t-1) - x_i(t-1)\big) \tag{10.136}$$

leading to

$$\hat{\Pr}(y(t)|Y_{t-1}) = \frac{1}{N_p}\sum_{i=1}^{N_p} \mathcal{W}_i(t-1) \times \Pr\big(y(t)|x_i(t-1)\big) \tag{10.137}$$

More details of these processors can be found in Ref. [36], [40] and [41]; here we will apply the bootstrap PF approximation of Eq. 10.131 to develop our sequential detectors.

For the non-Gaussian detection problem, we again calculate the likelihood-ratio decision function using the estimated data likelihood from the PF as

$$\mathcal{L}\big(y(t)\big) = \frac{\hat{\Pr}(y(t)|Y_{t-1};\mathcal{H}_1)}{\hat{\Pr}(y(t)|Y_{t-1};\mathcal{H}_0)} \tag{10.138}$$

Taking logarithms and substituting the empirical distribution estimates the sequential log-likelihood formulation results

$$\Lambda\big(y(t)\big) = \Lambda\big(y(t-1)\big) + \ln\hat{\Pr}(y(t)|Y_{t-1};\mathcal{H}_1) - \ln\hat{\Pr}(y(t)|Y_{t-1};\mathcal{H}_0) \tag{10.139}$$

Consider the nonlinear example of the previous subsection.

Example 10.13

Let us return to the problem where we again have the nonlinear model given by

$$s(t) = (1 - 0.05\triangle T)s(t-1) + 0.04\triangle T s^2(t-1) + w(t-1)$$
$$y(t) = s^2(t) + s^3(t) + v(t)$$

$w \sim \mathcal{N}(0, 10^{-6})$, $v \sim \mathcal{N}(0, 0.09)$, and $s(0) \sim \mathcal{N}(2.3, 10^{-2})$ and $\triangle T = 0.01$ sec.

The required distribution using the bootstrap PF can be approximated by

$$\ln\hat{\Pr}(y(t)|Y_{t-1};\mathcal{H}_1) \approx \ln(\overline{\mathcal{W}}(t;\mathcal{H}_1))$$

with the distribution under the null hypothesis given by

$$\ln \hat{\Pr}(y(t)|Y_{t-1};\mathcal{H}_0) = \ln\left(\frac{1}{\sqrt{2\pi \times 0.09}}\right) - y(t)R_{vv}^{-1}y(t)$$

The resulting non-Gaussian, log-likelihood decision function with threshold is given by

$$\Lambda(t) = \Lambda(t-1) + \ln\left(\overline{\mathcal{W}}(t;\mathcal{H}_1)\right) - \ln\left(\frac{1}{\sqrt{2\pi \times 0.09}}\right) + \frac{y^2(t)}{2 \times 0.09} \begin{array}{cc} \mathcal{H}_1 & \mathrm{T}_1 \\ > & \\ = & \text{No decision} \\ < & \\ \mathcal{H}_0 & \mathrm{T}_0 \end{array}$$

$\mathrm{T}_\ell(t) = \ln \mathcal{T}_\ell(t); \ell = 0, 1.$

As before, to estimate the thresholds we first estimate the ROC curve. The decision function simulation was executed first for the known signal case and then for the case of noise only. The results are shown in Fig. 10.12 where we see the

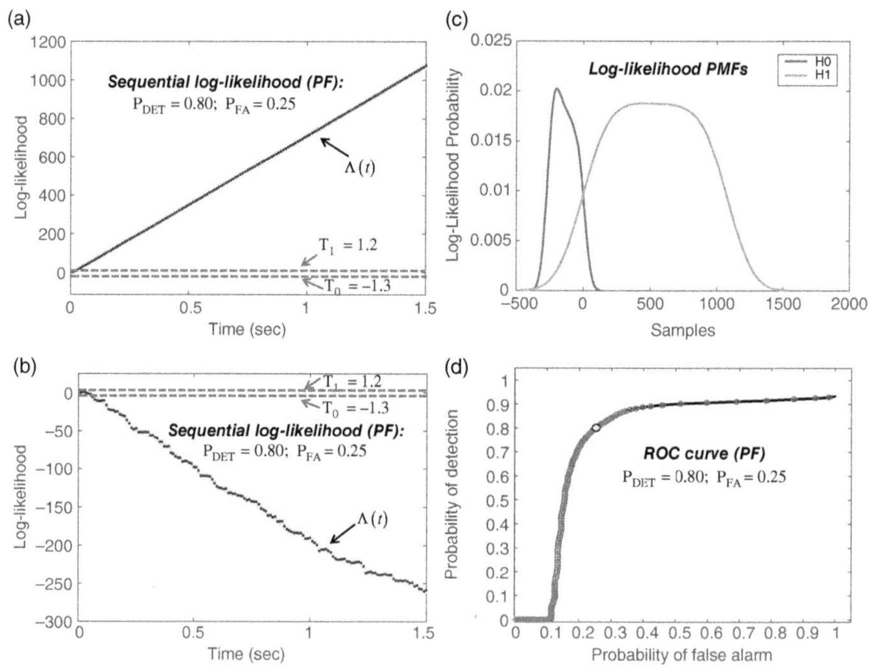

FIGURE 10.12 Sequential model-based detector design for nonlinear, non-Gaussian problem: (a) signal case: log-likelihood detection; (b) noise case: log-likelihood detection; (c) log-likelihood PMFs; (d) ROC curve. (AUC = 0.75.)

estimated ROC curve in 10.12d and the log-likelihood for both cases with the corresponding thresholds in 10.12a,b along with the underlying PMFs in 10.12c. Here the detection and false-alarm probabilities were $P_{DET} = 0.80$ and $P_{FA} = 0.25$ at the optimum threshold. Using Eq. 10.113 we calculate the corresponding thresholds as $\ln \mathcal{T}_0 = -1.33$ and $\ln \mathcal{T}_1 = 1.17$ shown in the figures. The AUC for this detector was $AUC = 0.75$. △△△

This completes the example, next we consider the change or anomaly detection problem.

10.7 MODEL-BASED CHANGE (ANOMALY) DETECTION

When a system under investigation changes its character in some way or another, this change can be reflected in a number of different ways. For instance, the dynamic response of a structure subjected to an extraneous source such as an earthquake can be severely altered or a motor running flawlessly beginning to shutter altering its performance perhaps due to wear, or an aircraft abruptly altering altitude due to a subsystem malfunction. All of these occurrences are examples that indicate the system operating normally suddenly or slowly can change its operation requiring some form of monitoring to not only detect the resulting change or anomaly, but also predict its potential occurrence and even classify its change mechanism. In this section, we briefly discuss how monitoring characteristics of a system can be used to detect changes in its performance. This is an area that has been heavily researched for quite a while, the interested reader can find much more detail in Refs. [42], [43], and [44–47]. Here we just introduce the idea of detecting a change by monitoring the "normal" response of a system and detect when that response becomes "abnormal".

Changes, from a signal processing perspective, can be captured in a number of ways: altered statistics of the process (change in mean), altered spectral response (frequency change), altered parametric response (model parametric change), other characteristics that uniquely represent the system performance. In all of these cases, it is desirable to find a feature that uniquely captures the system performance. From the model-based perspective, it is well known [33] that a system model contaminated by additive Gaussian noise when employed in a sequential model-based scheme yields a one-step *prediction error* or equivalently an *innovations* sequence that can be investigated statistically to provide an online change detection monitor.

A model-based approach to develop diagnostic techniques for change or anomaly detection is based on a well-established theoretical formalism that ensures a careful and systematic analysis of the system or classes of systems under investigation and their associated change mechanisms. Models can be developed from a first-principles mathematical representation or by utilizing sensor measurements to "fit" the model parameters to the data. In cases where the change mechanism is not well understood, the model-fitting approach offers an effective solution. In fact, in contrast to many approaches of solving this problem, the model-based approach does not require an explicit representation of the change mechanism at all to design an online monitor for initial change detection. It is based on obtaining a representation of the system during normal certification operations, or even more desirable, during quality assurance or acceptance testing. Once these normal operational characteristics of the system are known, the processor is developed based on the properties of the *innovations* sequence, the difference between the measured and predicted measurements obtained from the underlying *normal* system model. Once this condition or "difference" monitor detects a change, it can then be classified.

10.7.1 Model-Based Detection

In this section, we discuss the design of a detector to monitor the performance of the model-based processor and indicate when the model is no longer adequate or equivalently does not track the measured data. First, we briefly discuss the required theory with details to follow. Once this is accomplished, we then discuss the development of a practical processor.

Developing the model-based processor for the *change detection problem* is based on a stochastic representation of the underlying dynamical state-space system. We use a linear, continuous-discrete Gauss–Markov representation, that is, excite the system with white Gaussian noise and take into account the measurement uncertainty, we have

$$\begin{aligned}\dot{x}(t) &= Ax(t) + Bu(t) + w(t) && \text{for } w \sim \mathcal{N}(0, R_{ww}) \\ y(t_k) &= Cx(t_k) + v(t_k) && \text{for } v \sim \mathcal{N}(0, R_{vv})\end{aligned} \quad (10.140)$$

where t_k is the sample time at the output of the digitizer (analog-to-digital converter) and the initial state is assumed Gaussian $x(0) \sim \mathcal{N}(\overline{x}(0), \overline{P}(0))$. Thus, this represents a stochastic extension of the deterministic system. The system response as well as the states are contaminated with noise and uncertainties (unknown parameters); therefore, we formulate the enhancement problem as a state/response estimation problem using an embedded model developed from first principles or "fit" to the data using parameter estimation techniques. In any case, the *system enhancement problem* can be formulated succinctly as:

Given a set of noisy response sampled-measurements $\{y(t_k)\}$ and known excitations $\{u(t_k)\}$ along with the corresponding Gauss–Markov model of Section 4.5, find the best estimate of the response (and state), $\hat{y}(t_k)$ $(\hat{x}(t_k|t_k))$.

Discretizing the continuous system, the optimal sampled-data solution to this problem is well known [33] and is given (succinctly) by

$$\begin{aligned}\hat{x}(t_k|t_{k-1}) &= A\hat{x}(t_{k-1}|t_{k-1}) + Bu(t_{k-1}) \\ \hat{y}(t_k) &= C\hat{x}(t_k|t_{k-1}) \\ e(t_k) &= y(t_k) - \hat{y}(t_k) \\ \hat{x}(t_k|t_k) &= \hat{x}(t_k|t_{k-1}) + K(t_k)e(t_k)\end{aligned} \quad (10.141)$$

where $\hat{x}(t_k|t_k)$ is the state estimate at time t_k based on all of the data up to time t_k and K is the gain (Kalman) matrix obtained from a set of error covariance calculations (see Ref. [33] for details).

Thus, based on the assumed Gauss–Markov model, the optimum processor or Kalman filter has the known excitation $(u(t_k))$ and noisy measurement data $(y(t_k))$ as input and produces an estimate of the states $(\hat{x}(t_k|t_k))$, enhanced or filtered response measurement $(\hat{y}(t_k))$, and the *innovations* sequence $(e(t_k))$, as output. The statistical properties of the innovations are used to assess the performance of the processor since a *necessary and sufficient condition* for the MBP to be optimal is that the innovations sequence *must* be *zero-mean* and *white* [33]. The innovations sequence is zero-mean and white *only* when the process and measurement models reflect the true system response (and states) for a properly tuned MBP. Statistical changes in $e(t_k)$ reflect changes from the normal or expected operation; therefore, we can utilize these changes to *monitor* the performance of the model embedded in the processor.

Theoretically, it can be shown that when "model mismatch" occurs, the innovations become *non-zero* mean and are *no longer* white; therefore, we must develop a monitor to *decide* whether or not the innovations satisfy the required properties for optimality [33]. Assuming a normal (no changes) system initially, the MBP is tuned to produce a zero-mean, white innovations establishing the *normality* operating condition. When something changes in the system, the underlying normal model will no longer "match" the measured response data indicating a change from normal condition to what we define as *abnormal* or an anomaly alerting us to a potential change; therefore, we declare that a change has occurred and attempt to classify its type. This is the underlying principle of the monitor/detector or more properly "change detector."

10.7.2 Optimal Innovations Detection

To formally pose the sequential change detection problem, we appeal to classical detection theory as before. We are to test the binary hypothesis that the innovations sequence is zero-mean and white

$$\begin{aligned} \mathcal{H}_0 &: \quad e(t_k) \sim \mathcal{N}\left(0, R_{ee}(t_k)\right) && \text{[White]} \\ \mathcal{H}_1 &: \quad e(t_k) \sim \mathcal{N}\left(\overline{\mu}_e, \overline{R}_{ee}(t_k)\right) && \text{[Non-white]} \end{aligned}$$

which is a statistical test for the zero-mean and whiteness of the innovations sequence. Note that we assume that we *know* the model error and how to calculate $\overline{\mu}_e(t_k)$ and $\overline{R}_{ee}(t_k)$, a priori.

The optimal solution to this binary detection problem is based on applying the *Neyman–Pearson theorem* leading to the usual likelihood-ratio and is given by the ratio of probabilities for the *sequential innovations detector*, that is,

$$\mathcal{L}[E_{t_N}] = \frac{\Pr(E_{t_N}|\mathcal{H}_1)}{\Pr(E_{t_N}|\mathcal{H}_0)} \quad \begin{array}{c} \mathcal{H}_1 \\ > \\ < \\ \mathcal{H}_0 \end{array} \quad \tau$$

where $E_{t_N} := \{e(t_0), e(t_1), \ldots, e(t_N)\}$ or expanding

$$\mathcal{L}[E_{t_N}] = \frac{\Pr(e(t_0), e(t_1), \ldots, e(t_N)|\mathcal{H}_1)}{\Pr(e(t_0), e(t_1), \ldots, e(t_N)|\mathcal{H}_0)}$$

but from the chain rule of probability [33], we have that

$$\Pr(E_{t_k}|\mathcal{H}_i) = \Pr(e(t_k)|E_{t_{k-1}}; \mathcal{H}_i) \times \cdots \times \Pr(e(t_1)|e(t_0); \mathcal{H}_i) \times \Pr(e(t_0)|\mathcal{H}_i)$$

which can be expressed succinctly using Bayes' rule as

$$\Pr(E_{t_k}|\mathcal{H}_i) = \Pr(e(t_k), E_{t_{k-1}}|\mathcal{H}_i) = \Pr(e(t_k)|E_{t_{k-1}}; \mathcal{H}_i) \times \Pr(E_{t_{k-1}}|\mathcal{H}_i)$$

Substituting these expressions (replacing t_N by t) into the likelihood-ratio (above) and grouping, we obtain

$$\mathcal{L}[E_{t_k}] = \left[\frac{\Pr(E_{t_{k-1}}|\mathcal{H}_1)}{\Pr(E_{t_{k-1}}|\mathcal{H}_0)}\right] \times \frac{\Pr(e(t_k)|E_{t_{k-1}}; \mathcal{H}_1)}{\Pr(e(t_k)|E_{t_{k-1}}; \mathcal{H}_0)}$$

and therefore, we have the recursion or equivalently sequential likelihood as

$$\mathcal{L}[E_{t_k}] = \mathcal{L}[E_{t_{k-1}}] \times \frac{\Pr(e(t_k)|E_{t_{k-1}}; \mathcal{H}_1)}{\Pr(e(t_k)|E_{t_{k-1}}; \mathcal{H}_0)}$$

Taking logarithms, we obtain the relation for the *sequential log-likelihood* as before

$$\lambda_e(t_k) = \ln \mathcal{L}[E_{t_k}] = \lambda_e(t_{k-1}) + \ln \Pr(e(t_k)|E_{t_{k-1}}, \mathcal{H}_1) - \ln \Pr(e(t_k)|E_{t_{k-1}} \mathcal{H}_0)$$

The corresponding Wald or *sequential probability ratio test* (SPRT) [32] is then given by

$$\begin{array}{rccl} & \lambda_e(t_k) & \geq \ln \tau_1 & \text{Accept } \mathcal{H}_1 \\ \ln \tau_0 & \leq \lambda_e(t_k) & \leq \ln \tau_1 & \text{Continue} \\ & \lambda_e(t_k) & \leq \ln \tau_0 & \text{Accept } \mathcal{H}_0 \end{array}$$

and the underlying conditional Gaussian distributions are given by

$$\Pr(e(t_k)|E_{t_{k-1}}, \mathcal{H}_0) = (2\pi)^{-N/2}|\mathbf{R}_{ee}(t_k)|^{-1/2} \exp\left(-\frac{1}{2}e'(t_k)\mathbf{R}_{ee}^{-1}(t_k)e(t_k)\right)$$

and

$$\Pr(e(t_k)|E_{t_{k-1}}, \mathcal{H}_1) = (2\pi)^{-N/2}|\bar{\mathbf{R}}_{ee}(t_k)|^{-1/2}$$
$$\exp\left(-\frac{1}{2}(e(t_k) - \bar{\mu}_e(t_k))'\bar{\mathbf{R}}_{ee}^{-1}(t_k)(e(t_k) - \bar{\mu}_e(t_k))\right)$$

If we include the determinants in the thresholds, we have the modified decision function

$$\Lambda_e(t_k) = \Lambda_e(t_{k-1}) + \frac{1}{2}e'(t_k)\mathbf{R}_{ee}^{-1}(t_k)e(t_k) - \frac{1}{2}(e(t_k) - \bar{\mu}_e(t_k))'\bar{\mathbf{R}}_{ee}^{-1}(t_k)(e(t_k) - \bar{\mu}_e(t_k))$$

which yields the new test

$$\begin{array}{lll} & \Lambda_e(t_k) \geq \Upsilon_1 & \text{Accept } \mathcal{H}_1 \\ \Upsilon_0 \leq & \Lambda_e(t_k) \leq \Upsilon_1 & \text{Continue} \\ & \Lambda_e(t_k) \leq \Upsilon_0 & \text{Accept } \mathcal{H}_0 \end{array}$$

where

$$\begin{aligned} \Upsilon_1 &= \ln \tau_1 + \frac{1}{2}\ln|\mathbf{R}_{ee}(t_k)|^{-1/2} - \frac{1}{2}|\bar{\mathbf{R}}_{ee}(t_k)|^{-1/2} \\ \Upsilon_0 &= \ln \tau_0 + \frac{1}{2}\ln|\mathbf{R}_{ee}(t_k)|^{-1/2} - \frac{1}{2}|\bar{\mathbf{R}}_{ee}(t_k)|^{-1/2} \end{aligned} \tag{10.142}$$

The implementation of this "optimal" change detector/monitor presents a basic problem, such as how to obtain estimates of the innovations (abnormal) mean and covariance required, but it does illustrate a potential optimal solution to the model monitoring problem. As mentioned, the sequential innovations detector requires a priori knowledge of the actual model "mismatch" and structurally how it enters the process model to obtain $[\bar{\mu}_e(t_k)), \bar{\mathbf{R}}_{ee}(t_k)]$ for the monitor.

10.7.3 Practical Model-Based Change Detection

Because of the inability to characterize statistical estimates of the innovations under abnormal conditions, we must develop more pragmatic statistical methods for change detection that incorporate this underlying knowledge. Here we present the techniques already developed in Section 5.7 for "tuning" the optimal MBP. We briefly review these developments.

Note that there exist other prediction error models (e.g., AR, ARMA, ARMAX) that still require the optimality conditions (zero-mean, white) to validate their performance. So the following statistical tests are *not* just limited to the Kalman-based state-space processors, but can be applied to all one-step prediction error techniques. Recall that a *necessary and sufficient condition* for an MBP to be optimal is that the innovation sequence is zero-mean and white. Note also that these optimality conditions are not necessarily true for output error models. We review the zero-mean, whiteness, and weighted sum-squared residual tests as detectors (hypothesis tests) themselves in light of this chapter.

Zero-Mean Detection In some cases, if the innovations are assumed ergodic, then a composite hypothesis test can be constructed using sample statistical estimators for the required statistics [33]. When the processor is "tuned", it provides an optimal or minimum (error) variance estimate of the state. The innovations sequence, which was instrumental in deriving the processor, also provides the starting point to check the MBP operation.

If we assume that the innovation sequence is ergodic and Gaussian, then we can use the sample mean as the test statistic to estimate μ_e, the population mean [33]. The sample mean for the ith *component* of e_i is given by

$$\hat{\mu}_e(i) = \frac{1}{N} \sum_{t=1}^{N} e_i(t_k) \text{ for } i = 1, \ldots, N_y \tag{10.143}$$

where $\hat{\mu}_e(i) \sim \mathcal{N}(\mu_e, R_{ee}(i)/N)$ and N is the number of data samples. We perform a statistical hypothesis test to "decide" if the innovation mean is null [33]. We test that the mean of the ith component of the innovation vector $e_i(t_k)$ is

$$\begin{aligned} \mathcal{H}_0 &: \quad \mu_e(i) = 0 \\ \mathcal{H}_1 &: \quad \mu_e(i) \neq 0 \end{aligned}$$

As our test statistic we use the sample mean. At the α-significance level, the probability of rejecting the null hypothesis $\mathcal{H}_0$ is given by

$$\Pr\left(\left| \frac{\hat{\mu}_e(i) - \mu_e(i)}{\sqrt{R_{ee}(i)/N}} > \frac{\tau_i - \mu_e(i)}{\sqrt{R_{ee}(i)/N}} \right| \right) = \alpha \quad \text{for } i = 1, \ldots, N_y \tag{10.144}$$

Therefore, the *zero-mean test* [33] on each component innovation e_i is given by

$$\hat{\mu}_e(i) \underset{\text{Accept } \mathcal{H}_0}{\overset{\text{Reject } \mathcal{H}_0}{\gtrless}} \tau_i \text{ for } i = 1, \ldots, N_y \tag{10.145}$$

Under the null hypothesis $\mathcal{H}_0$, each $\mu_e(i)$ is assumed zero. Therefore, at the 5 % significance level ($\alpha = 0.05$), we have that the threshold is

$$\tau_i = 1.96 \sqrt{\frac{\hat{R}_{ee}(i)}{N}} \tag{10.146}$$

where $\hat{R}_{ee}(i)$ is the *sample variance* (assuming ergodicity) estimated by

$$\hat{R}_{ee}(i) = \frac{1}{N}\sum_{t=1}^{N} e_i^2(t_k) \tag{10.147}$$

Component Whiteness Detection Under the same assumptions, we can perform a *whiteness test* [33], that is, check statistically that the innovations covariance corresponds to that of an uncorrelated (white) sequence. Assuming ergodicity of the innovations sequence, we use the sample covariance function as our test statistic with the ith component covariance and lag k are given by

$$\hat{R}_{ee}(i,k) = \frac{1}{N}\sum_{n=k+1}^{N} \left(e_i(t_n) - \hat{\mu}_e(i)\right)\left(e_i(t_{n+k}) - \hat{\mu}_e(i)\right) \tag{10.148}$$

We actually use the *normalized covariance* test statistic

$$\hat{\rho}_{ee}(i,k) = \frac{\hat{R}_{ee}(i,k)}{\hat{R}_{ee}(i)} \tag{10.149}$$

Asymptotically, it can be shown that

$$\hat{\rho}_{ee}(i,k) \sim \mathcal{N}(0, 1/N)$$

Therefore, the 95% confidence interval estimate is

$$I_{\rho_{ee}} = \hat{\rho}_{ee}(i,k) \pm \frac{1.96}{\sqrt{N}} \text{ for } N > 30 \tag{10.150}$$

Hence, under the null hypothesis, 95% of the $\hat{\rho}_{ee}(i,k)$ values must lie within this confidence interval to accept $\mathcal{H}_0$. That is, for each *component* innovation sequence to be considered statistically white. The *whiteness test* of Eq. 10.150 is very useful to detect modeling inaccuracies from individual component innovations. However, for complex systems with a large number of measurement channels, it becomes computationally burdensome to investigate each innovation component wise especially under the limiting ergodicity assumptions.

Vector Whiteness Detection A statistic containing *all* of the innovation information is the *weighted sum-squared residual* (WSSR). It aggregates all of the innovation *vector* information over some finite window of length M. It can be shown that the WSSR is related to a maximum-likelihood estimate of the normalized innovations variance. The WSSR test statistic is given by

$$\rho_e(\ell) := \sum_{k=\ell-M+1}^{\ell} e(t_k)' R_{ee}^{-1}(t_k) e(t_k) \text{ for } \ell \geq M \qquad e \in \mathcal{R}^{N_e \times 1} \tag{10.151}$$

The WSSR *hypothesis test* is based on

$$\begin{array}{lll} \mathcal{H}_0 : & \rho_e(\ell) & \text{[White]} \\ \mathcal{H}_1 : & \rho_e(\ell) & \text{[Non-white]} \end{array}$$

and is given by

$$\rho_e(\ell) \mathop{\gtrless}_{\mathcal{H}_0}^{\mathcal{H}_1} \tau \tag{10.152}$$

Under the null hypothesis, the WSSR is chi-squared distributed, $\rho_e(\ell) \sim \chi^2(N_e M)$. However, for $N_e M > 30$, $\rho(\ell)$ is approximately Gaussian $\mathcal{N}(N_e M, 2N_e M)$ (see Ref. [48] for more details). At the α-significance level, the probability of rejecting the null hypothesis is given by

$$\Pr\left(\left|\frac{\rho(\ell) - N_e M}{\sqrt{2N_e M}} > \frac{\tau - N_e M}{\sqrt{2N_e M}}\right|\right) = \alpha \tag{10.153}$$

For a level of significance of $\alpha = 0.05$, we have

$$\tau = N_e M + 1.96\sqrt{2N_e M} \tag{10.154}$$

Thus, we see that the WSSR can be considered a "whiteness test" of the innovations *vector* over a finite window of length N. Note that since $\left[\{e(t_k)\}, \{R_{ee}(t_k)\}\right]$ are obtained from the state-space MBP algorithm directly, they can be used for both *stationary* as well as *nonstationary* processes. In fact, in practice for a large number of measurement components, the WSSR is used to "tune" the filter and then the component innovations are individually extracted and checked for zero-mean/whiteness to detect changes. Note also that the adjustable parameter of the WSSR statistic is the window length M, which essentially controls the width of the window sliding through the innovations sequence. Therefore, even if we cannot implement the optimal innovations detector, we can still perform a set of pragmatic statistical hypothesis tests to investigate the condition of the underlying system and classify its change.

Consider the following example to illustrate the pragmatic approach.

Example 10.14

Let us return to the *RC*-circuit of Example 5.1 where we have the Gauss–Markov model given by

$$s(t) = 0.97s(t-1) + 100u(t-1) + w(t-1)$$
$$y(t) = 2s(t) + v(t)$$

$w \sim \mathcal{N}(0, 10^{-5})$, $v \sim \mathcal{N}(0, 4)$, and $s(0) \sim \mathcal{N}(2.5, 10^{-6})$.

We simulated this system as before shown in Fig. 5.3 and constructed the model-based processor resulting in Fig. 5.4. For the abnormal case we modified the system dynamics model by decreasing it by 10%, that is, $A \rightarrow A - 0.1 \times A$. The resulting system model changed and the various pragmatic tests were used to monitor the response and detect any anomalies. The results are shown in Fig. 10.13. In the

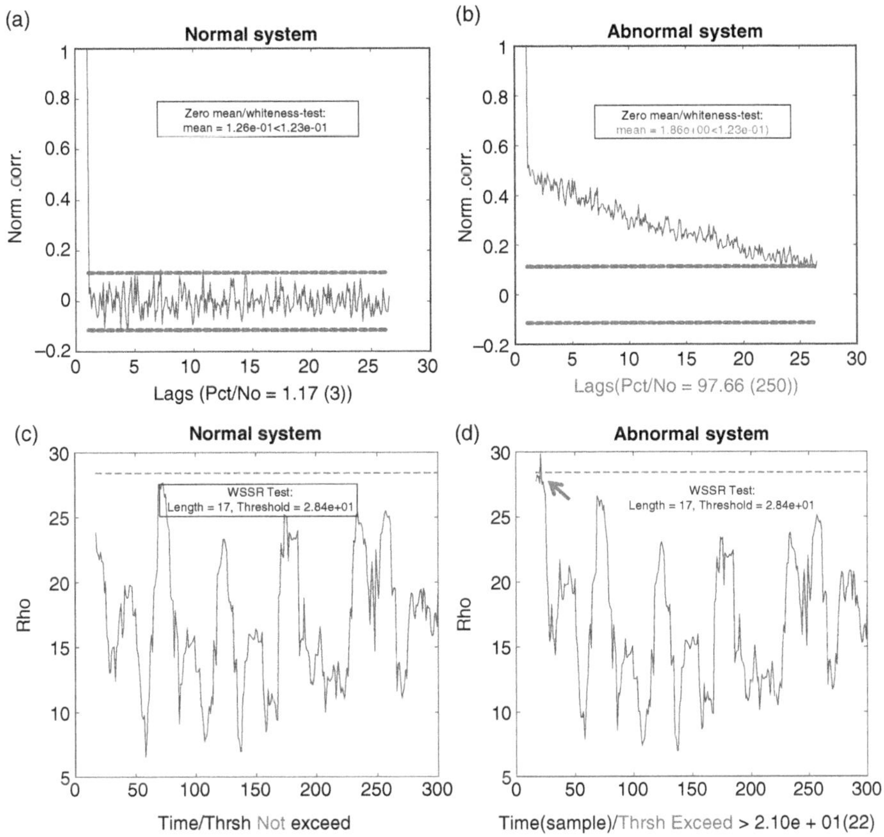

FIGURE 10.13 Sequential model-based detector design for *RC*-circuit problem: (a) Normal case: zero-mean/whiteness (0.13 ≈ 0.13 and 1.2% out) detection; (b) Abnormal case: zero-mean/whiteness (1.9 > 0.13 and 98% out) detection; (c) Normal case: WSSR detection (not exceeded); (d) Abnormal case: WSSR detection (exceeded).

figure, the zero-mean and whiteness detectors clearly indicate an anomaly with the mean significantly exceeding its bound $1.9 > 0.13$ and the percent of samples for the whiteness greatly exceeding 5% at 98% as illustrated in (a) and (b). The WSSR statistic is also able to pick up the "change" in response with the threshold "exceeded" as shown in (c) and (d). This completes the example. If there was a transient or "abrupt" change the WSSR would be able to discern and locate it, while there is a possibility that the zero-mean/whiteness test would be insensitive to it. We demonstrate this in the case study to follow. △△△

10.8 CASE STUDY: REENTRY VEHICLE CHANGE DETECTION

In this section, we apply the pragmatic detectors to tracking anomalies of a space vehicle as it re-enters the atmosphere. Here, we employ an autoregressive moving average (ARMA) model to capture the system dynamics as compared to the state-space models discussed previously. The tracking of space vehicles from launch to splash down and extracting critical information about the vehicle dynamics is a problem of high concern whether it be a Space Shuttle launched to deliver a new communications satellite or a reentry vehicle launched to analyze its flight performance. In any case, the basic question of interest is how to extract the vehicle dynamics from noisy and incomplete trajectory information gathered by various independent radar sensors and detect anomalies.

We start with a signal processing model that captures some of the simple vehicle dynamics in terms of its precession and nutation signature that give an indication of the dynamics. Consider the following signal processing representation of the reentry vehicle (RV) response received at the radar. Suppose we model the dynamics of the RV signal as [49],

$$\begin{aligned} s(t) &= A_p \sin(2\pi f_p t + \phi_p) \left[1 + \sum_{k=1}^{N_v} A_v(k) \sin(2\pi f_v(k) t + \phi_v(k))\right] \\ z(t) &= s(t) + c(t) + v(t) \end{aligned} \tag{10.155}$$

where $z(t)$ is the measured radar return; $s(t)$ is the RV signal model including precession and nutation v; $c(t)$ is the trajectory (trend); $v(t)$ is Gaussian random noise $\mathcal{N}(0, \sigma_v^2)$; A_p, A_v are the precession and nutation amplitudes; f_p, f_v are the precession and nutation frequencies; and ϕ_p, ϕ_v are the precession and nutation phase.

We develop a recursive-in-time approach to instantaneous spectral estimation enabling us to produce the desired spectrogram (amplitude vs. time vs. frequency) for event detection. Suppose we parametrically model the enhanced measurement data by a time-frequency representation specified by an ARMA model. This model has the general difference equation form given by

$$A(q^{-1}, t)y(t) = C(q^{-1}, t)e(t) \tag{10.156}$$

for the enhanced measurement $y(t)$, contaminated with zero-mean, white Gaussian noise $e \sim \mathcal{N}(0, \sigma_{ee}^2)$, with the corresponding instantaneous polynomials at the instant t defined by

$$A(q^{-1}, t) = 1 + a_1(t)q^{-1} + \cdots + a_{N_a}(t)q^{-N_a}$$
$$C(q^{-1}, t) = c_o + c_1(t)q^{-1} + \cdots + c_{N_c}(t)q^{-N_c} \tag{10.157}$$

Here, the backward shift or delay operator is defined by $q^{-i}y(t) := y(t-i)$ and therefore, we can write Eq. 10.156 simply as

$$y(t) = -\sum_{k=1}^{N_a} a_k(t)y(t-k) + \sum_{k=0}^{N_c} c_k(t)e(t-k) \tag{10.158}$$

If we transform the difference equation using the discrete Fourier transform (DFT), then we obtain the *instantaneous transfer function* (ignoring stochastic aspect)

$$H(e^{j2\pi f}, t) = \frac{Y(e^{j2\pi f}, t)}{E(e^{j2\pi f}, t)} = \frac{C(e^{j2\pi f}, t)}{A(e^{j2\pi f}, t)} \tag{10.159}$$

or more appropriately the corresponding *instantaneous power spectrum* defined by

$$S(f, t) := \left|H(e^{j2\pi f}, t)\right|^2 = \left|\frac{C(e^{j2\pi f}, t)}{A(e^{j2\pi f}, t)}\right|^2 \tag{10.160}$$

So we see that if we use a model-based parametric ARMA(N_a,N_c) representation of the enhanced measurement signal and transform it to the spectral domain, then we can obtain the instantaneous spectral estimate. We are primarily interested in estimating the spectrum at each time instant, we chose to apply a recursive-in-time or one-step prediction error technique—the recursive prediction error method (RPEM) (see Ref. [33] for details). Using this approach, we expect the spectrogram will enable us to extract the dominant precession spectral peak as it temporally evolves as well as the nutational side band frequencies. After some significant pre-processing, the enhanced measurement is available for spectrogram estimation [49].

Recursive-in-time parameter estimation algorithms take on the following generic form:

$$\underline{\hat{\Theta}}(t+1) = \underline{\hat{\Theta}}(t) + \underline{K}(t)e(t) \qquad \text{[Parameter update]}$$

$$e(t) = y(t) - \hat{y}(t) = y(t) - \underline{\varphi}'(t)\underline{\hat{\Theta}}(t) \qquad \text{[Prediction error]}$$

$$(10.161)$$

$$\underline{\varphi}(t) \equiv \left[y(t-1) \quad \cdots \quad y(t-N_a-1) \mid \hat{e}(t) \quad \cdots \quad \hat{e}(t-N_c)\right],$$
$$\underline{\hat{\Theta}}(t) \equiv \left[-\hat{a}_1(t-1) \quad \cdots \quad -\hat{a}_{N_a}(t-N_a-1) \mid \hat{c}_o(t) \quad \cdots \quad \hat{c}_{N_c}(t-N_c)\right]'$$

with $\underline{K}(t)$ the gain or weighting vector and the "hat" symbol defining the best (minimum error variance) estimate at the specified time. There are also many variations and forms of this basic recursion [33], but here we limit our application to the recursive prediction error method (RPEM) based on a local Gauss–Newton optimization method. We show the instantaneous spectrogram estimation for an RV simulation (to follow) in Fig. 10.15.

Event Detection We developed a set of event detectors based on the instantaneous ARMA signal models and recursive algorithms discussed above. The underlying idea motivating this type of detector is based on the statistical properties of the prediction error or innovations sequence which is (simply) the difference between the measurement and its prediction using the ARMA model parameters. When the model and its parameters "track" the data (after initialization), then any "abrupt" change in parameter estimates or data will be reflected directly in the prediction error sequence. The pragmatic detectors test the sequence for statistically significant changes indicating an event or possible phase change. The main point to realize is that since we are using a recursive-in-time estimator, the prediction error represents "how well the ARMA model fits the data at each time instant". If it is an optimal fit, then the prediction error sequence should be zero-mean and statistically white (uncorrelated) and the vector detectors of the previous section evolve naturally from this property.

When using these statistical tests care must be taken. If the models are nonlinear or nonstationary, then the usual whiteness/zero-mean tests, that is, testing that 95% of the sample (normalized) prediction error correlations lie within the bounds and the corresponding zero-mean test both rely on quasi-stationary assumptions and sample statistics to estimate the required correlations. However, it can be argued heuristically that when the ARMA estimator is tuned, the nonstationarities are being tracked by the processor even in the nonlinear case and therefore, the prediction errors should be covariance stationary.

When data are nonstationary, then a more reliable statistic to use is the WSSR which is a measure of the overall global estimation performance of the processor again determining the "whiteness" of the prediction error sequence. It essentially aggregates all of the information available in the prediction error and tests whiteness by requiring that the decision function $\rho(\ell)$ lies below the specified threshold to be deemed statistically white. If the WSSR statistic does lie beneath the calculated threshold, then theoretically, the estimator is tuned and said to converge or in our problem there is no abrupt phase change. Here the N-sample window is designed to slide through the prediction error data and estimate its whiteness. Even in the worst case where these estimators may not prove to be completely consistent, the processor (when tuned) predicts the nonstationary prediction error covariance $\sigma_{ee}^2(k)$ enabling a simple (varying with ℓ) confidence interval to be constructed and used for testing. Thus, overall performance of the processor and therefore the *event detector* can be assessed by analyzing the statistical properties of the prediction errors.

The final detector is *ad hoc*, but based on the premise above that abrupt changes in the data will be reflected by abrupt changes in the parameters of the ARMA model. The *parametric change detector* is obtained by averaging over all of the recursive parameter estimates and searching for significant peaks to detect events. It is defined by

$$\underline{\mu}_a(t_k) = \frac{1}{N_a + N_c} \sum_{i=1}^{N_a+N_c} \underline{\alpha}_i(t_k)$$

$$\underline{\mu}_a(t_k) \underset{\mathcal{H}_o}{\overset{\mathcal{H}_1}{\gtrless}} \tau \tag{10.162}$$

This completes our discussion of the model-based signal processing approach taken to analyze the radar measurements, extract the required RV dynamics and detect abrupt events.

10.8.1 Simulation Results

For this application, we synthesize the response of a ballistic vehicle that has been launched and measured with a radar tracking system as it re-enters the earth's atmosphere. The vehicle dynamic motion, assumed stable, is characterized by it's precession frequency and the small angular frequencies called nutations that it undergoes during reentry. We also synthesize some events that are simulated as simple phase changes in the temporal response. For our simulation, we assumed a precession frequency of 0.235 Hz and nutation frequencies ranging between 1.6 ± 0.24 Hz, 2.4 ± 0.24 Hz, 3.2 ± 0.24 Hz, and 4.0 ± 0.24 Hz. In the spectrum, these nutation frequencies appear in the side bands, since the signal model is essentially an amplitude modulated signal. So for this data we extract the precession frequency and the eight sidebands. We also synthesized three (3) events at 550, 590, and 640 sec. We simulated the data with additive zero-mean, Gaussian noise with a variance of $SNR = 20$ dB for 1024 samples and processed the data using the ARMA model given by ARMA(35,25). The synthesized RV radar data are shown in Fig. 10.14 along with a motion compensation pre-processing step.

We ran the instantaneous spectrogram estimator using the RPEM over the ionosphere-free enhanced measurements and the results are shown in Fig. 10.15 below. Here we see that the spectrogram image clearly displays the dominant precession spectral peak (0.235 Hz) as it temporally evolves as well as the nutational side band frequencies. Note with this high SNR the synthesized "phase changes" are also clearly observable as indicated by the arrows. The measurement is time aligned and shown below the spectrogram as well. This example demonstrates the effectiveness of the adaptive ARMA model spectrogram MBP on synthetic data.

We executed these detectors over synthesized data sets in the ionosphere-free case with the results shown in Fig. 10.16. We observe that in this case all three of the detectors perform well in detecting the changes in nutational phase. In *a*, the WSSR or phase change detector shows three peaks above the threshold at the correct event

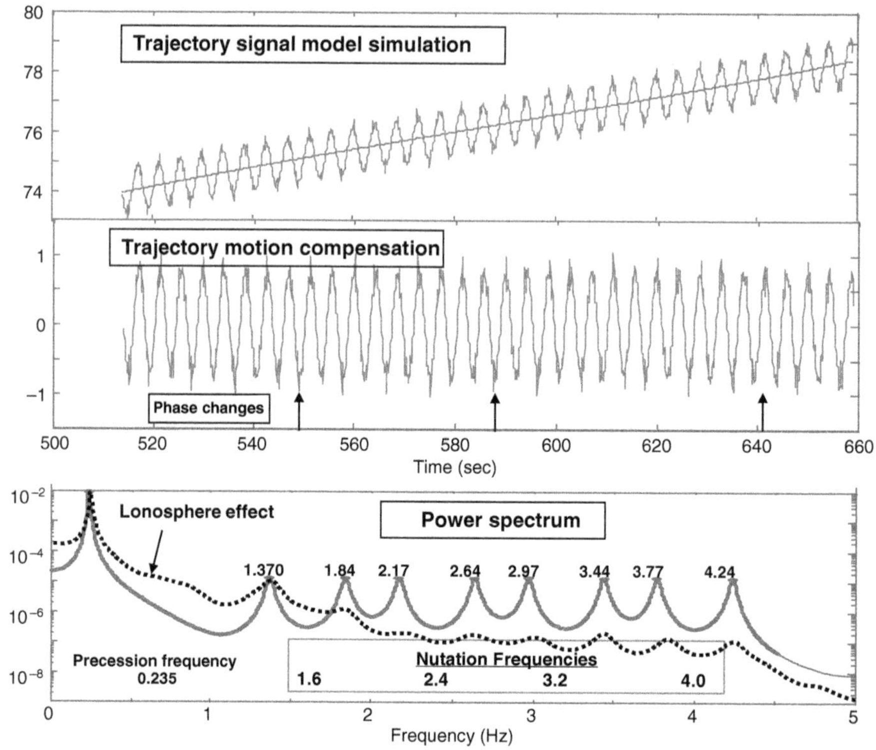

FIGURE 10.14 Trajectory simulation and tracking: (a) simulation; (b) motion compensation; (c) spectral estimate with ionospheric effects (dotted).

times (550, 580, 630 sec) as do the parametric change (shown in b, c) and whiteness *d* detectors. This completes our discussion of the estimation and monitoring of the RV dynamics to detect abrupt events.

10.9 SUMMARY

In this chapter, we developed the fundamental decision or detection problem. Starting with the classical binary hypothesis problem, we introduced the notion of a likelihood-ratio decision function that was carried throughout investigating the Bayesian approach to detector design for a single measurement both when the underlying signals and noise are completely known and unknown leading to composite hypothesis testing and the so-called generalized likelihood-ratio test. Next, a variety of detection approaches were developed based on the most popular decision criteria: probability-of-error, Bayes' risk, and the well-known Neyman–Pearson criterion. Multiple or batch measurements were then considered and shown to lead to similar results for each scheme just modifying the likelihood-ratio in each case. The

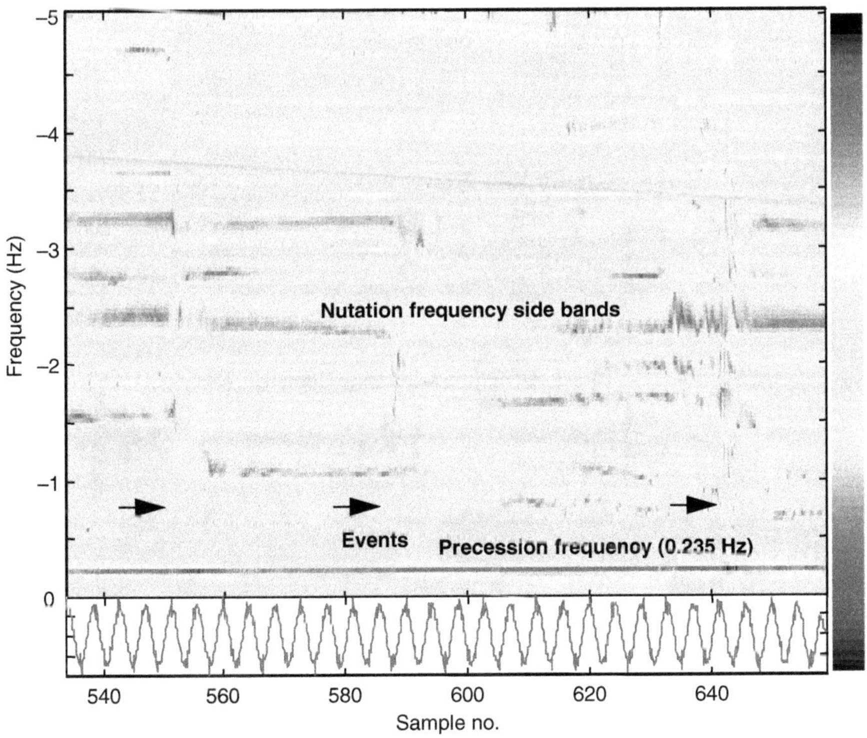

FIGURE 10.15 Adaptive ARMA Spectrogram Estimation for Event Detection: RPEM algorithm for ARMA(35,25). Events detected are shown by the arrows.

multichannel measurement case then followed again leading to similar results. Finally, the multiple hypothesis case was developed from the Bayesian probability-of-error perspective. Performance metrics, critical tools in detector analysis, were developed in an effort to assess and compare detectors. Starting with the ROC curve other metrics evolved (e.g., AUC) enabling a "quick" performance analysis. Development of an ROC curve was discussed along with various approaches leading to the average ROC curve with accompanying statistics, when ensemble data are created or are available. Other metrics followed: distance, MinE, optimum decision point, and the confusion matrices. Besides being available for performance analysis and comparing various detection schemes, the ROC curve is *necessary* to establish an acceptable threshold to set the detector for desired performance once the operating point (P_{FA}, P_{DET}) is selected.

With all of this information at hand, the heart of the chapter followed—*sequential detection*. Not only was this approach developed from the batch likelihood-ratio but the underlying theory leading to the average number of samples required for detection followed. Next, model-based detection evolved with developments for linear, nonlinear, and non-Gaussian model sets. Here it was shown how each of the sequential

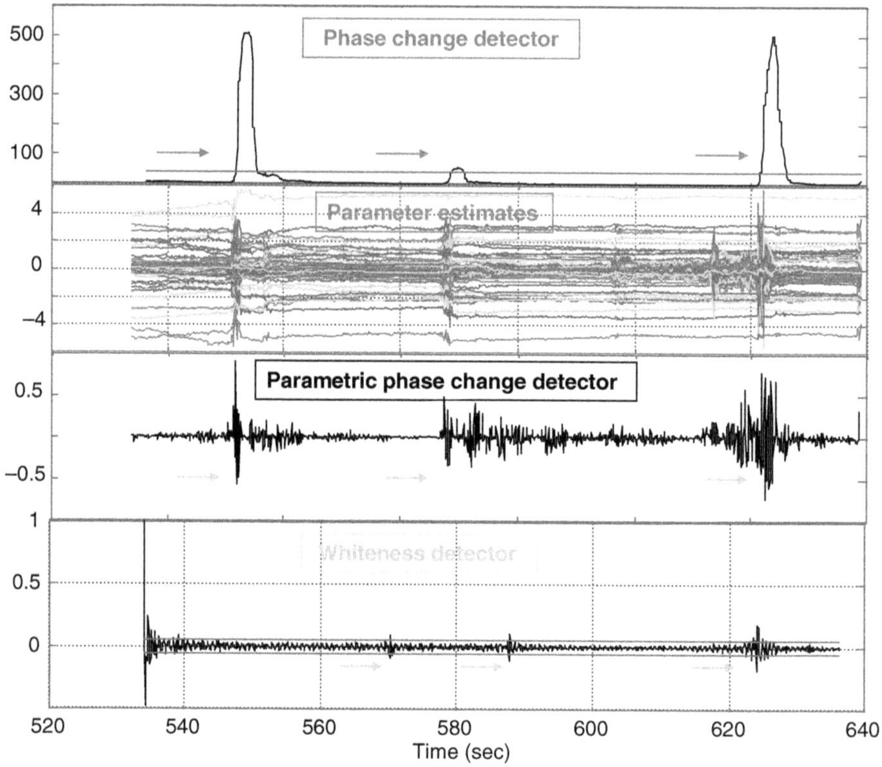

FIGURE 10.16 Event detection of enhanced measurement: (a) phase change detection ($T = 25$, Threshold = 38.8); (b) parameter estimates, ARMA(32,25); (c) parametric change detector; (d) whiteness event detector (4.4% out).

Bayesian (signal) model-based processors coupled with the newly developed sequential (likelihood-ratio) detectors could provide a powerful solution to a wide variety of decision problems. The idea of detecting a *change* or *anomaly* of a system from normal operation was exploited using this approach leading to the so-called family of *innovations-based* detectors. Finally, a case study followed showing the applicability of the model-based designs to anomaly detection.

MATLAB NOTES

MATLAB is a command oriented vector-matrix package with a simple yet effective command language featuring a wide variety of embedded *C* language constructs making it ideal for signal processing applications and graphics. MATLAB has a *Statistics Toolbox* that incorporates a large suite of PDFs and CDFs as well as "inverse" CDF functions ideal for simulation-based algorithms. The **mhsample** command incorporates the Metropolis, Metropolis–Hastings, and Metropolis independence samplers in a single command while the Gibbs sampling approach is adequately represented by the more efficient slice sampler (**slicesample**). There are even specific "tools" for sampling as well as the inverse CDF method captured in the **randsample** command. PDF estimators include the usual histogram (**hist**) as well as the sophisticated kernel density estimator (**ksdensity**) offering a variety of kernel (window) functions (Gaussian, etc.). Perhaps the most relevant command for detection is the (**perfcurve**) command which enables the development of receiver operating characteristic (ROC) curves along with the majority of metrics, discussed such as the confusion matrix, optimum threshold, detection and false-alarm probability, etc. MATLAB also offers a sophisticated classification tool, the "Bayesian learner" for classification or multiple hypothesis testing problems.

In terms of statistical testing for particle filtering diagnostics MATLAB offers the chi-square "goodness-of-fit" test **chi2gof** as well as the Kolmogorov–Smirnov distribution test **kstest**. Residuals can be tested for whiteness using the Durbin–Watson test statistic **dwtest** while "normality" is easily checked using the **normplot** command indicating the closeness of the test distribution to a Gaussian. Other statistics are also evaluated using the **mean**, **moment**, **skewness**, **std**, **var** and **kurtosis** commands. Type *help stats* in MATLAB to get more details or go to the MathWorks website.

REFERENCES

1. S. Theodoridis and K. Koutroumbas, *Pattern Recognition* (New York: Academic Press, 1998).
2. H. Van Trees, *Detection, Estimation and Modulation Theory, pt. 1*, (New York: John Wiley & Sons, Inc., 1968).
3. A. Sage and J. Melsa, *Estimation Theory with Applications to Communications and Control* (New York: McGraw-Hill, 1971).
4. D. Middleton and R. Esposito, "New results in the theory of simultaneous optimum detection and estimation of signals in noise," *Problemy Peredachi Informatsii*, **6**, 3, 3–20, 1970.
5. J. Melsa and D. Cohn, *Detection and Estimation Theory* (New York: McGraw-Hill, 1978).
6. K. Murphy, *Machine Learning* (Cambridge, MA: MIT Press, 2014).
7. A. Gelman, J. Carlin, H. Stern, and D. Rubin, *Bayesian Data Analysis* (New York: Chapman & Hall, 2004).
8. S. Press, *Subjective and Objective Bayesian Statistics* (Hoboken, NJ: John Wiley & Sons, Inc., 2003).
9. S. Kay, *Fundamentals of Statistical Signal Processing: Detection Theory* (Englewood Cliffs, NJ: Prentice-Hall, 1998).
10. R. Duda, P. Hart, and D. Stork, *Pattern Classification*, 2nd Ed. (Hoboken, NJ: John Wiley & Sons, Inc., 2001).
11. C. M. Bishop, *Pattern Recognition and Machine Learning* (New York: Springer, 2006).

12. K. Fukunaga, *Statistical Pattern Recognition* (New York: Academic Press, 1990).
13. W. Peterson, T. Birdsall, and W. Fox, "The theory of signal detectability," Transactions of the IRE Professional Group on Information Theory, PGIT-4, 171–212, 1957.
14. C. Therrien, *Decision, Estimation, and Classification: An Introduction to Pattern Recognition and Related Topics* (New York: John Wiley & Sons, Inc., 1989).
15. J. Hancock and P. Wintz, *Signal Detection Theory* (New York: McGraw-Hill, 1966).
16. D. Middleton, *Introduction to Statistical Communication Theory* (New York: McGraw-Hill, 1960).
17. D. Middleton and R. Esposito, "Simultaneous optimum detection and estimation of signals in noise," *IEEE Transactions on Information Theory*, IT-14, 3, 434–444, 1968.
18. E. Sullivan and D. Middleton, "Estimation and detection issues in matched-field processing," *IEEE Trans. Oceanic Engr.*, **18**, 3, 156–167, 1993.
19. J. Candy and D. Middleton, Joint detection-estimation: a model-based solution, LLNL Report, UCRL-JC-120219, 1995.
20. A. Papoulis, *Probability, Random Variables and Stochastic Processes* (New York: McGraw-Hill, 1965).
21. C. E. Metz, "Basic principles of ROC analysis," *Semin Nucl. Med.*, **VIII**, 4, 1978.
22. T. Fawcett, "ROC graphs: notes and practical considerations for researchers," Technical Report, (Palo Alto, CA: HP Laboratories, 2004).
23. T. Fawcett, "An introduction to ROC analysis," *Pattern Recogn. Letters*, **27**, 861–874, 2006.
24. H. Chernoff, *Sequential Analysis and Optimal Design* (Philadelphia, PA: SIAM, 1972).
25. A. Wald, *Sequential Analysis* (New York: Dover Publications, 1947.
26. A. Zoubir and D. Iskander, *Bootstrap Techniques for Signal Processing* (Cambridge, UK: Cambridge University Press, 2004).
27. A. Bradley, "AUC: a statistically consistent and more discriminating measure than accuracy," *Pattern Recogn.*, **30**, 7, 1145–1159, 1997.
28. J. Swets,"Indices of discrimination or diagnostic accuracy: their ROCs and implied models," *Psycho. Bulletin*, **99**, 1, 100–117, 1986.
29. K. Horsch, M. L. Giger, and C. Metz, "Prevalence scaling: application to an intelligent workstation for the diagnosis of breast cancer," *Acad. Radiol.*, **15**, 1446–1457, 2008.
30. R. Irwin and T. Irwin, "A principled approach to setting optimal diagnostic thresholds: where ROC and indifference curves meet," *Europ. J. Inter. Med.*, 230–234, 2011.
31. C. Ling, J. Huang, and H. Zhang, "The use of the area under the ROC curve in the evaluation of machine learning algorithms," in Proceedings of 16th Canadian Confr. Art. Intell., Springer, 2003.
32. A. Wald, *Sequential Analysis* (New York: Dover Publications, 1973).
33. J. Candy, *Model-Based Signal Processing* (Hoboken, NJ: Wiley/IEEE Press, 2006).
34. A. Jazwinski, *Stochastic Processes and Filtering Theory* (New York: Academic Press, 1970).
35. J. Candy, *Bayesian Signal Processing, Classical, Modern and Particle Filtering* (Hoboken, New Jersey: Wiley/IEEE Press, 2009).
36. P. Li and V. Kadirkamanathan, "Particle methods based likelihood-ratio approach to fault diagnosis in nonlinear stochastic systems," *IEEE Trans. Syst., Man and Cyber. Pt. C*, **31**, 3, 337–342, 2001.

37. C. Nemeth, P. Fearnhead, and L. Mihaylova, "Sequential Monte Carlo methods for state and parameter estimation in abruptly changing environments," *IEEE Trans. Signal Proc.*, **62**, 5, 1245–1255, 2014.
38. P. Djuric, J. Kotecha, J. Zhang, Y. Huang, T. Ghirmai, M. Bugallo, and J. Miguez, "Particle filtering," *IEEE Signal Proc. Mag.*, **20**, 5, 19–38, 2003.
39. S. Kassam, *Signal Detection in Non-Gaussian Noise* (New York: Springer-Verlag, 1988).
40. A. Doucet, S. Godsill, and C. Andrieu, "On sequential Monte Carlo sampling methods for Bayesian filtering," *Statist. Comput.*, **10**, 176–183, 2000.
41. C. Andrieu, A. Doucet, S. Singh, and V. Tadic, "Particle methods for change detection, system identification, and control," Proceedings of the IEEE, **92**, 3, 423–438, 2004.
42. R. Hogg and A. Craig, *Introduction to Mathematical Statistics* (New York: MacMillan, 1970).
43. L. Scharf, *Statistical Signal Processing: Detection, Estimation, and Time Series Analysis* (Reading, MA: Addison-Wesley, 1990).
44. M. Basseville and I. Nikiforov, *Detection of Abrupt Changes: Theory and Application* (Englewood Cliffs, NJ: Prentice-Hall, 1993).
45. F. Gustafasson, *Adaptive Filtering and Change Detection* (Hoboken, New Jersey: John Wiley & Sons, 2000).
46. R. Mehra and J. Peschon, "An innovations approach to fault detection and diagnosis in dynamic systems," *Automatica*, **7**, 637–640, 1971.
47. A. Willsky, "A survey of failure detection in dynamic systems," *Automatica*, **12**, 601 611, 1976.
48. B. Anderson and J. Moore, *Optimum Filtering* (New Jersey: Prentice-Hall, 1979).
49. J. Candy, "Processing of reentry vehicle signatures from radar tracking data," LLNL Report, UCRL-130433, 1998.

PROBLEMS

10.1 We are given a sequence of counting measurements $\{y_k\}$ and asked to design a set of different detectors based on a random signal that is Poisson distributed $s \sim \mathcal{P}(\lambda_s)$ and contaminated with additive Poisson noise given by

$$y_k = s_k + v_k \text{ for } v \sim \mathcal{P}(\lambda_v)$$

Design the following set of detectors to detect the random signal from a *single* measurement.

(*a*) Maximum likelihood detector.

(*b*) Maximum a posteriori detector with priors $\Pr(\mathcal{H}_0)$ and $\Pr(\mathcal{H}_1)$. How does this detector compare to the one in (a)? Under what conditions are they identical?

(*c*) Bayes' risk detector with costs: $C_{ii} = \kappa_{ii}$ and $C_{ij} = \kappa_{ij}$. How does this detector compare to the one in (b)? Under what conditions are they identical?

(*d*) Neyman–Pearson detector with threshold τ.

(*e*) Repeat this problem for *multiple* measurements.

10.2 We have measured a parameter vector $\mathbf{\Theta}$ and would like to decide if

$$\mathcal{H}_0 : \mathbf{\Theta} = \mathbf{\Theta}_0$$
$$\mathcal{H}_1 : \mathbf{\Theta} = \mathbf{\Theta}_1$$

where each vector is multivariate Gaussian distributed such that $\mathcal{N}(\mathbf{m_0}, \mathbf{R_0})$ and $\mathcal{N}(\mathbf{m_1}, \mathbf{R_1})$.

(*a*) Develop the maximum likelihood, maximum a posteriori (equal priors) Bayes' risk ($C_{ii} = 0, C_{ij} = 1$); and Neyman–Pearson detectors with threshold τ to decide which vector is selected from a measurement Θ.

(*b*) What if $\mathbf{m_1} = \mathbf{m_2} = \mathbf{m}$. What are the new decision functions?

(*c*) What if $\mathbf{R_1} = \mathbf{R_2} = \mathbf{R}$. What are the new decision functions now?

10.3 A person is attempting to estimate the *DC* voltage level in a battery with a "cheap" voltmeter and the needle continues to fluctuate wildly. He wants to detect whether the battery is completely dead (0 V) or alive (greater than 0 V). Initially, the person takes a *single* "guess" (measurement) to decide. Assuming a signal in additive zero-mean, Gaussian noise measurement model with variance σ, develop the corresponding decision problem.

(*a*) Set up the hypothesis test for *d*, the *DC* voltage.

(*b*) Develop the maximum likelihood, maximum a posteriori (equal priors), Bayes' risk ($C_{ii} = 0, C_{ij} = 1$), and Neyman–Pearson detectors with threshold τ to decide whether or not to replace the battery.

(*c*) Calculate the detection, miss and false-alarm probabilities under these assumptions.

(*d*) Frustrated, the person now take a sequence of guesses (multiple measurements) in order to make the decision. What are the new decision functions? Calculate the corresponding *miss* and *false-alarm* probabilities. What is the total probability-of-error?

(*e*) Still not satisfied with the results, the person decides that the level is actually random and can be considered Gaussian such that $d \sim \mathcal{N}(\overline{d}, \sigma_d^2)$, calculate the detection, miss and false-alarm probabilities under these assumptions for the Neyman–Pearson detector for a single measurement.

10.4 Consider multivariate Gaussian distributions under binary hypotheses: $\mathcal{N}(\mathbf{m}_\ell, \mathbf{R}_\ell); \ell = 0, 1$ with the following means and covariances:

$$\mathbf{m}_0 = [0 \quad 0]^T \text{ and } \mathbf{m}_1 = [0 \quad 2]^T$$

with corresponding covariance matrices

$$\mathbf{R}_0 = \text{diag}(1.00 \quad 0.25) \text{ and } \mathbf{R}_0 = \text{diag}(0.25 \quad 1.00)$$

(*a*) Design a Neyman–Pearson detector for threshold τ.

(*b*) What type curves are the decision boundaries?

(*c*) Sketch the decision region with these boundaries.

10.5 Suppose we are given the hypothesis to test as

$$\mathcal{H}_0 : y(t) = v(t) \text{ for } v \sim \mathcal{N}(0, 1)$$
$$\mathcal{H}_1 : y(t) = \mu + v(t)$$

(*a*) Develop the maximum likelihood, maximum a posterior (equal priors), Bayes' risk ($C_{ii} = 0, C_{ij} = 1$), and Neyman–Pearson detectors with threshold τ to decide on the presence or absence of the signal ($\mu = 1$) from a single measurement.

(*b*) Calculate the detection and false-alarm probabilities.

(*c*) For a fixed false-alarm probability of $\Pr(\mathcal{D}|\mathcal{H}_0) = 0.25$, calculate the threshold for each detector.

(*d*) Sketch (or calculate) the PDFs and ROC curve for this problem and annotate the critical regions.

(*e*) Consider incorporating multiple measurements $\{y(t)\}; t = 1, \ldots, N$, how do these detectors change? What are the new decision functions for each?

(*f*) Sketch the PDFs and ROC curve for this case and annotate the critical regions.

10.6 Calculate the total probability-of-error for the previous problem (c), if the priors are: $\Pr(\mathcal{H}_0) = 0.25$ and $\Pr(\mathcal{H}_1) = 0.75$.

10.7 We have an exponentially distributed measurement and would like to decide if it is from an $\mathcal{E}(1)$ or $\mathcal{E}(\lambda)$, that is, a binary decision problem.

(*a*) Determine the Neyman–Pearson detector for a threshold of τ.

(*b*) Determine the ROC curve for this problem.

(*c*) Calculate the slope of the ROC curve. What is its significance?

10.8 Similar to the above binary decision problem, suppose we have the hypotheses distributed: $\Pr(y|\mathcal{H}_0) = 0.5 \exp\{-|y|\}$ and $\Pr(y|\mathcal{H}_1) = \exp\{-2|y|\}$ with corresponding costs given by $C_{ii} = 0, C_{01} = 1, C_{10} = 2$, and priors $\Pr(\mathcal{H}_0) = 0.25$ and $\Pr(H_1) = 0.75$.

(a) Calculate the detection and false-alarm probabilities.

(b) Calculate the Bayes' risk.

10.9 Consider the following multiple hypothesis decision problem

$$\mathcal{H}_0 : y(t) = v(t) \text{ for } v \sim \mathcal{N}(0, R_{vv})$$
$$\mathcal{H}_1 : y(t) = \mu + v(t)$$
$$\mathcal{H}_2 : y(t) = -\mu + v(t)$$

(*a*) With μ assumed *known* and positive and the hypotheses equally likely with costs $C_{ii} = 0, C_{ij} = 1$, develop the MAP detector for this problem.

(*b*) Sketch the PDFs. What are the decision regions?

10.10 For the binary decision problem under the following hypotheses

$$\mathcal{H}_0 : \theta(t) = \theta_0(t) \text{ for } v \sim \mathcal{N}(m_{\theta_0}, R_{\theta_0\theta_0})$$
$$\mathcal{H}_1 : \theta(t) = \theta_1(t) \text{ for } v \sim \mathcal{N}(m_{\theta_1}, R_{\theta_1\theta_1})$$

use the following statistics to simulate Gaussian sequences under each hypothesis with $m_{\theta_0} = 110, R_{\theta_0\theta_0} = 5.0$ and $m_{\theta_1} = 120, R_{\theta_1\theta_1} = 5.5$. Using the simulated data, calculate the corresponding Neyman–Pearson decision functions under each hypothesis, then

(*a*) Generate the ROC curve (you can use the **perfcurve** of MATLAB).

(*b*) Calculate the performance metrics: AUC, PPV, NPV, MinE, ODP with threshold.

(*c*) Calculate the confusion matrix.

(*d*) Calculate the ACC, PPV, NPV. Compare your results with those of Table 1 of Example 10.8.

10.11 Given the following decision problem

$$\mathcal{H}_0 : y(t) = v(t) \text{ for } v \sim \mathcal{N}(0, R_{vv})$$
$$\mathcal{H}_1 : y(t) = \theta(t) + v(t)$$

(*a*) Assume $\theta(t)$ is deterministic, but *unknown*. Develop the generalized likelihood-ratio test (GLRT) for threshold τ and a *single* measurement $y(t)$.

(*b*) Assume $\theta(t)$ is *random* with $\theta \sim \mathcal{N}(0, R_{\theta\theta})$, develop the Neyman–Pearson (N-P) detector for this case.

(*c*) Suppose we now have a set of measurements $\{y(t)\}; t = 1, \ldots, N$ and assume $\theta(t)$ is represented by a random-walk with variance $R_{w_\theta w_\theta}$. What is the MAP estimate of $\theta \rightarrow \hat{\theta}_{\text{MAP}}$? How does this change the GLRT (explain)?

(*d*) Again assume $\theta(t)$ is deterministic, but unknown. Develop the *sequential* N-P detector assuming independent measurements. Determine the average number of samples required for the termination of the test assuming that the operating point obtained from an ROC curve is: $P_{\text{DET}} = 0.95$ and $P_{\text{FA}} = 0.10$.

10.12 Suppose we have a *known* pulse-like signal in zero-mean, Gaussian noise. Like a radar or sonar system, we transmit the pulse $p(t)$ into the environment and would like to decide whether it is present or not, that is we test the hypotheses

$$\mathcal{H}_0 : y(t) = v(t) \text{ for } v \sim \mathcal{N}(0, R_{vv})$$
$$\mathcal{H}_1 : y(t) = p(t) + v(t)$$

(*a*) Assuming we have a set of measurements $\{y(t)\}; t = 1, \ldots, N$ available, develop the N-P detector with threshold τ. What is the resulting decision function?

(*b*) Suppose we interpret the pulse as a finite impulse response (all-zero) filter with weights

$$W(t) = p(N - t - 1); t = 0, 1, \ldots, N - 1$$

Using the convolution relationship $z(t) = W(t) * y(t)$, show that this is equivalent to the decision function (*matched-filter*) in (a).

(*c*) Determine the performance of the matched-filter for this problem, that is, estimate its ROC curve.

10.13 Suppose we are given the measurement vector $\mathbf{y}(t) := [y(0)y(1) \cdots y(N-1)]$ and we would like to test the hypotheses

$$\mathcal{H}_0 \cdot \mathbf{y}(t) = \mathbf{v}(t) \text{ for } \mathbf{v} \sim \mathcal{N}(0, \mathbf{R}_{vv})$$
$$\mathcal{H}_1 : \mathbf{y}(t) = \mathbf{s}(t) + \mathbf{v}(t)$$

where $\mathbf{y}, \mathbf{s}, \mathbf{v} \in \mathcal{R}^{N \times 1}$, $\mathbf{s}$ *known*.

(*a*) Develop the Neyman–Pearson detector with threshold τ. What is the resulting decision function?

(*b*) Suppose the noise is uncorrelated, that is, $\mathbf{R}_{vv} = \sigma_v^2 I$ is diagonal with equal variances. For the N-P detector, how does this decision function differ from that of the previous example.

(*c*) Suppose the noise is uncorrelated, but with unequal variances, $\mathbf{R}_{vv} = \sigma_v^2(n)I$ for $n = 0, \ldots, N-1$. What is the new decision function for the N-P detector?

10.14 We would like to develop a simple scheme to estimate a change in the statistics of a given signal $s(t)$ in real-time, but we can only measure the signal that is contaminated in noise

$$y(t) = s(t) + v(t) \text{ for } v \sim \mathcal{N}(0, \sigma_v^2)$$

If the signal is assumed deterministic, then

(*a*) Develop a sequential estimate of the signal along with a sequential estimate of its statistics (*Hint*: calculate the sequential mean/variance estimators).

(*b*) With these statistics in hand construct a sequential test to detect a change (*Hint*: dynamic confidence interval).

(*c*) Can this problem be placed in an equivalent hypothesis testing framework? What is it?

(*d*) Assume the signal is a sinusoid that changes its phase at the midpoint of its measured time interval for various SNRs (SNR: signal energy/noise variance), then simulate this signal and construct the sequential detector.

10.15 Design a sequential "reference" detector using the Neyman–Pearson criterion that is capable of deciding between a Gauss–Markov (state-space) model defined by $\Sigma := \{A, C, \hat{x}(0), \hat{P}(0), R_{ww}, R_{vv}\}$ and a known reference model $\overline{\Sigma} := \{\overline{A}, \overline{C}, \overline{x}(0), \overline{P}(0), R_{vv}\}$.

The *reference model* is given by

$$x(t) = \overline{A}\overline{x}(t-1)$$
$$y(t) = \overline{C}\overline{x}(t) + v(t) \qquad \text{for } v \sim \mathcal{N}(0, R_{vv})$$

and

$$x(t) = Ax(t-1) + w(t) \quad \text{for } w \sim \mathcal{N}(0, R_{ww})$$
$$y(t) = Cx(t) + v(t) \qquad \text{for } v \sim \mathcal{N}(0, R_{vv})$$

(*a*) Determine the decision function and the thresholds τ_0 and τ_1.

(*b*) Could this general formulation be applied as a *change* detector? Explain.

10.16 For the *RC*-circuit problem of Example 5.1, assume that we only know the component values within 10% precision.

(*a*) Simulate the true circuit. Construct a model-based solution and perform the zero-mean/whiteness along with the WSSR tests to ensure a tuned processor.

(*b*) Let the resistor value be $R \pm 0.1R$ and simulate its response. Apply the "normal" model-based processor in (a) and perform the same statistical tests. How do they compare?

(*c*) Develop a Neyman–Pearson sequential detector for this problem and observe the decision function. Is it sensitive enough to detect this parametric change?

(*d*) Repeat these simulations for a better component $R \pm 0.05R$. Is the change detected?

10.17 Suppose our estimates of the trajectory from our nominal guesses of the parameters in Example 5.2 are off by 10%, that is, for the nonlinear system given by

$$x(t) = (1 - \theta_1 \triangle T)x(t-1) + \theta_2 \triangle T)x^2(t-1) + w(t-1); \text{ for } w \sim \mathcal{N}(0, 1 \times 10^{-6})$$
$$y(t) = x^2(t) + x^3(t) + v(t) \text{ for } v \sim \mathcal{N}(0, 0.09)$$

Investigate the possibility of detecting changes using the zero-mean/whiteness and WSSR tests for $\theta_1 \pm 0.1\theta_1$ and then $\theta_2 \pm 0.1\theta_2$.

(*a*) For this nonlinear problem, construct a model-based processor (EKF, UKF, or PF) and "tune" it for the true parameters by performing the zero-mean/whiteness along with the WSSR tests to ensure a tuned processor.

(*b*) Let $\theta_1 \pm 0.1\theta_1$ and observe the MBP performance? Discuss the results.

(*c*) Let $\theta_2 \pm 0.1\theta_2$ and observe the MBP performance? Discuss the results.

(*d*) Develop an N-P sequential detector for this problem and observe the decision function. Does it detect the change? Compare this performance to the zero-mean/whiteness and WSSR tests.

11

BAYESIAN PROCESSORS FOR PHYSICS-BASED APPLICATIONS

In this chapter, we develop a set of Bayesian signal processing applications based on the underlying physical phenomenology generating the measured process. The complexity of the process and the desire for enhanced signals drive the design primarily indicated by the process model. We motivate each application, succinctly develop the process and measurement models, develop the BSP, and analyze its performance. More details on any of the designs can be obtained from the references for the interested reader. The main objective is to demonstrate the applicability of the Bayesian approach to a variety of interesting applications and analyze their performance.

11.1 OPTIMAL POSITION ESTIMATION FOR THE AUTOMATIC ALIGNMENT

The alignment of high-energy laser beams for fusion experiments demands high precision and accuracy of the underlying positioning algorithms whether it be for actuator control or monitoring the beam line for potential anomalies. This section discusses the development of online, optimal position estimators in the form of Bayesian processors to achieve the desired results. Here, we discuss the modeling, development, implementation, and processing of Bayesian model-based processors applied to both simulated and measured beam line data.

Bayesian Signal Processing: Classical, Modern, and Particle Filtering Methods, Second Edition. James V. Candy.

11.1.1 Background

Alignment of a high-power beam is a complex and critical process requiring precise and accurate measurements. Misalignment of such a beam could easily destroy costly optics causing a deleterious disruption of an entire experiment. The alignment of large operative, short pulse, laser systems is a significant and costly endeavor dating back to the late sixties and early seventies [1]. Contemporary imaging systems employ high-resolution video cameras to image and accurate position control systems to align the beam. This approach estimates the current beam position from the image, adjusts mirrors relative to an accurate reference measurement of physical beam center, and minimizes their difference or deviation [2–9]. However, even with these sophisticated measurements, the beam position estimate is still a function of the inherent beam noise caused by the internal beam line gas turbulence as well as instrumentation noise [10]. Here, we introduce the idea of post-processing these uncertain measurements using advanced signal processing techniques to "enhance" the raw data and "detect" any beam line anomalies during laser system operations [11].

High-power, tightly focused laser beams are required to achieve successful ignition and therefore fusion at the Lawrence Livermore National Laboratory (LLNL) National Ignition Facility (NIF) [12]. These beams must simultaneously focus precisely on a nanoscale target capsule to succeed. Therefore, there are a large number of alignment measurements that must be performed along the NIF beam line to assure that the pointing and alignment control system centers the beam in order to provide the maximum energy on the fusion target located in the associated chamber [12–14]. An automatic alignment (AA) system was designed and implemented to assure successful deployment of the high energy beam in each of the 192 beam lines. However, since a variety of techniques are provided to perform the alignment, there is a quantifiable uncertainty associated with each technique that may or may not meet the desired accuracy and precision specifications associated at each control point. Therefore, there is a need for a post-processing technique, which accepts as input an uncertain position measurement and provides as output an improved position estimate (see Fig. 11.1). As illustrated in the figure, the measured image position estimate is compared to the reference image estimate providing an position error that is input to the mirror control system that moves the associated mirrors correcting the beam position.

Perhaps the most challenging of all beam line measurements are those made on the KDP (potassium dihydrogen phosphate) crystals. These crystals are critical elements used to double or triple the frequency of the laser beam as it passes providing a shorter wavelength. The higher frequency determined from the target physics enables laser plasma interactions for fusion. The NIF final optics assembly matches lenses to this particular frequency producing the beam that is tightly focused on the target capsule required to achieve fusion ignition. In order for the KDP crystals to optimally double or triple the laser operating frequency, they must be precisely positioned at the appropriate angle. This is one of the critical tasks of the AA system. The KDP measurement consists of using a charge coupled device (CCD) imaging camera, which produces a noisy back reflection image of a diagnostic alignment beam. The

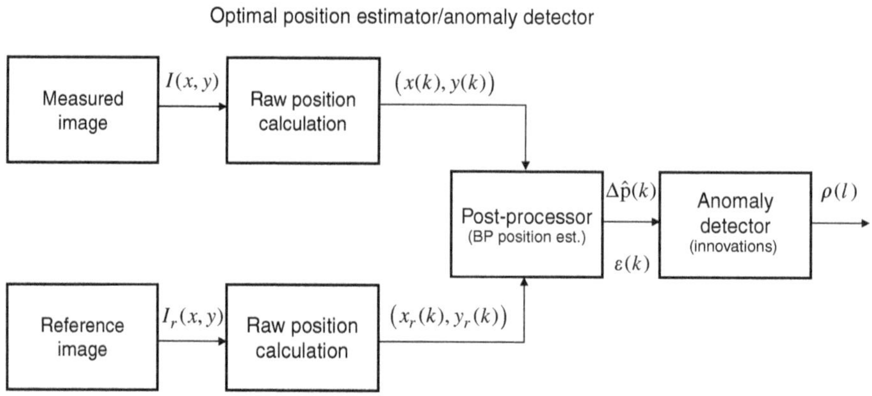

FIGURE 11.1 Optimal position estimation with anomaly detector: Bayesian post-processing and innovations-based anomaly detection.

noise is caused by the camera itself as well as uncertainties that are due to small pointing errors made during the measurement. A sophisticated two-dimensional (2D) phase-only, matched-filter [15] algorithm was developed to provide the initial raw position estimates.

The image acquired from the CCD imaging camera produces both noisy measurement and reference images. A precise reference image is used to provide the desired fiducial that is used by the alignment system. Corrections to align the measured image with the reference is accomplished using the dedicated control loops that adjust pointing mirror stepping motors until the deviations between both reference and measured positions are within acceptable limits [14]. Ultimately, the goal is to make this difference zero assuring proper beam alignment. Fig. 11.2 shows a sequence of measured KDP images with position estimates (O) along with the corresponding reference image measurement (+). Note how the control loop adjusts the KDP measurements (O) until the centroid position converges to the reference position (+). The objective, therefore, of the control loop is to adjust the beam mirrors such that both measured and reference positions completely overlap (zero deviation) as shown in the bottom of this figure. Thus, the smaller the XY-deviations, the closer the beam is to the centerline reference assuring a tightly focused, high-energy beam on target—the goal of the alignment system. Should, for some reason, an anomaly develops in any beam line, it may not be possible to align for a particular shot. The timely detection of beam line anomalies are necessary to avoid future problems, which, if left unmitigated, could result in less than optimum performance of the laser. In this work, we show how a Bayesian processor (BP) can be used to detect anomalies, online, during the calibration phase of a laser shot.

Thus, we discuss the feasibility of applying a Bayesian processor as an online post-processing technique to improve the final position estimates provided by a variety of estimators and detect anomalies in a high-energy laser beam line [11]. These estimators are used to align high-powered laser beams for experiments at NIF. We

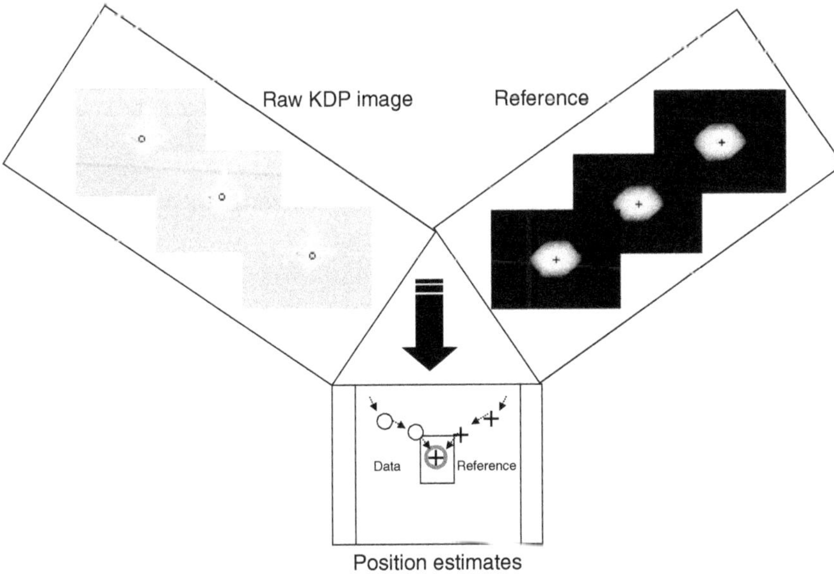

FIGURE 11.2 Raw KDP crystal back-reflection image and reference position estimates.

first motivate a stochastic model of the overall process and measurement system. Next, we discuss the underlying theory for both position estimation and anomaly detection. With the theory in hand, we develop the processor for the KDP application through ensemble statistics, simulation, and application to real measurement and reference data.

11.1.2 Stochastic Modeling of Position Measurements

The typical beam position estimator is accomplished by calculating the centroid of the measured images [16]. The point is that these measured positions are derived from the associated noisy CCD images. We define the true centroid positions by the position vector $\mathbf{p}(k) := [\mathbf{x}(k)|\mathbf{y}(k)]'$. where $\mathbf{p} \in R^{N_p \times 1}$ and k is the sample time. Since quite a number of images are acquired daily, a large database consisting of position estimates are available, say, $P(t) = \{\mathbf{p}(k); k = 1, \ldots, K\}$ and the position estimates are to be updated continuously. Since we know that the images are contaminated with noise and uncertainty, a more reasonable position measurement model is given by

$$\mathbf{z}(k) = C\mathbf{p}(k) + \mathbf{v}(k) \tag{11.1}$$

where $\mathbf{z}, \mathbf{v} \in \mathbf{R}^{N_z \times 1}, C \in \mathbf{R}^{N_z \times N_p}$, and $\mathbf{v} \sim N(0, \mathbf{R}_{vv})$, that is, the measurement noise is assumed zero-mean, multivariate Gaussian with covariance matrix, $\mathbf{R}_{vv} \in R^{N_z \times N_z}$.

We also note that since the CCD camera uses the identical beam line to measure both images, we model the position estimates as piecewise constants contaminated with beam line noise besides that noise contributed by the measurement systems. This process noise can be considered fluctuations caused by the inherent system optical transfer functions and turbulence caused by the argon gas-filled housing during the laser beam propagation (boiling noise). Therefore, we assume that the contaminated position measurement is represented as

$$\dot{\mathbf{p}}(t) = \mathbf{0} + \mathbf{w}(t) \quad \text{with} \quad \mathbf{p}(0) = [x(0)|y(0)]' \tag{11.2}$$

in continuous time, but if we discretize over the k-sample images, then using first differences we obtain

$$\dot{\mathbf{p}}(t) = \frac{\mathbf{p}(t_{k+1}) - \mathbf{p}(t_k)}{\Delta t_k} = \mathbf{w}(t_k) \quad \text{for} \quad \Delta t_k := t_{k+1} - t_k \tag{11.3}$$

Substituting into Eq. 11.2, we obtain the following Gauss–Markov position model

$$\mathbf{p}(t_{k+1}) = \mathbf{p}(t_k) + \Delta_{t_k}\mathbf{w}(t_k) \tag{11.4}$$

where $\mathbf{p}, \mathbf{w} \in \mathbf{R}^{N_p \times 1}$ and $\mathbf{w} \sim \mathbf{N}(0, \mathbf{R}_{ww})$ along with the accompanying measurement model of Eq. 11.1. This completes the basic description of the underlying Gauss–Markov model. We note in passing that for a complete description of the general model including means, $\mathbf{m}_p(k), \mathbf{m}_z(k)$ with their associated covariances see Ref. [11] for more details.

Since we have both final measured and reference position estimates, we re-define a more convenient position vector by grouping the positions as

$$\mathbf{p}(k) := \begin{bmatrix} x(k) \\ y(k) \\ --- \\ x_r(k) \\ y_r(k) \end{bmatrix} \quad \text{and} \quad \mathbf{w}(k) := \begin{bmatrix} w_x(k) \\ w_y(k) \\ --- \\ w_{x_r}(k) \\ w_{y_r}(k) \end{bmatrix}$$

where $\mathbf{p}(0) \sim \mathrm{N}\left(\bar{\mathbf{p}}(0), \bar{\mathrm{R}}_{\mathrm{pp}}(0)\right)$ are the respective measurement and reference positions coordinates (pixels) with the corresponding process noise covariance matrix given by

$$\mathbf{R}_{ww} = \begin{bmatrix} \bar{\mathbf{R}}_{ww} & | & 0 \\ - & - & - \\ 0 & | & \bar{\mathbf{R}}_{w_r w_r} \end{bmatrix}$$

since each of the measured and reference images are uncorrelated.

This type of formulation enables us to estimate the process noise covariances independently for each image as well as characterize their uncertainties individually.

Since the control loop jointly uses both the reference and measured images to produce its final centroid position estimates, we model the measurement matrix as the deviation (difference) between these data contaminated with independent measurement noise; that is, our measurement model of Eq. 11.1 becomes

$$\mathbf{z}(k) = \begin{bmatrix} 1 & 0 & -1 & 0 \\ 0 & 1 & 0 & -1 \end{bmatrix} \mathbf{p}(k) + \mathbf{v}(k)$$

with measurement covariance matrix

$$\mathbf{R}_{vv} = \begin{bmatrix} \sigma^2_{vv} & | & 0 \\ - & - & - \\ 0 & | & \sigma^2_{v_r v_r} \end{bmatrix}$$

and $\sigma^2_{vv} = \sigma^2_{xx} + \sigma^2_{yy}$ and $\sigma^2_{v_r v_r} = \sigma^2_{x_r x_r} + \sigma^2_{y_r y_r}$. Note that the deviations are defined by

$$\Delta\mathbf{p}(k) := C\mathbf{p}(k) = \begin{bmatrix} \Delta x(k) \\ \Delta y(k) \end{bmatrix} = \begin{bmatrix} x(k) - x_r(k) \\ y(k) - y_r(k) \end{bmatrix}$$

This completes the section on position (uncertainty) modeling; next, we consider the development of the optimal position estimator.

11.1.3 Bayesian Position Estimation and Detection

It is well known that the optimal solution to the position estimation problem under the Gauss–Markov model assumption is provided by the Bayesian model-based (state–space) processor or Kalman filter of Chapter 5. This solution can be considered a predictor-corrector design in which the processor uses the model (random walk) to predict in absence of a measurement and then corrects the estimate when the measurement becomes available. Under the assumption of a perfect model, that is, the model embedded in the process exactly matches the process, the BP is capable of achieving the optimal minimum (error) variance estimate under the Gaussian assumptions.

For the position estimation problem, the BP takes on the following predictor-update form where $t_k \rightarrow k$:

- Prediction: $\hat{\mathbf{p}}(k+1|k) = \hat{\mathbf{p}}(k|k)$
- Innovation: $\varepsilon(k+1) = \mathbf{z}(k+1) - \hat{\mathbf{z}}(k+1|k) = \mathbf{z}(k+1) - \mathbf{C}\hat{\mathbf{p}}(k+1|k)$
- Update: $\hat{\mathbf{p}}(k+1|k+1) = \hat{\mathbf{p}}(k+1|k) + \mathbf{G}(k+1)\varepsilon(k+1)$
- Gain: $\mathbf{G}(k+1) = \tilde{\mathbf{P}}(k+1|k)\mathbf{C}^T\mathbf{R}^{-1}_{\varepsilon\varepsilon}(k+1)$

for $\hat{\mathbf{p}}(k+1|k) := E\{\mathbf{p}(k)|\mathbf{p}(k-1), \ldots, \mathbf{p}(1)\}$ and $\hat{\mathbf{z}}(k+1|k)$, the underlying conditional means, and $\mathbf{G}(k+1) \in \mathbf{R}^{N_p \times N_z}$ is the corresponding gain matrix calculated from the error covariance matrix $\tilde{\mathbf{P}}(k+1|k) = \mathrm{Cov}\,(\tilde{\mathbf{p}}(k+1|k))$ with position error

$\tilde{\mathbf{p}}(k+1|k) := \mathbf{p}(k+1) - \hat{\mathbf{p}}(k+1|k)$. The innovations covariance matrix is defined by $\mathbf{R}_{\epsilon\epsilon}(k+1)$.

Recall that for a Gauss–Markov representation a necessary and sufficient condition for the BP to be optimal is that the innovations sequence be zero-mean and white (uncorrelated)—conditions that we test during processor design. It is possible to exploit this property of the innovations to detect anomalies in the system revealing potential problems in the beam line in pseudo real-time—a large asset when attempting to automate the alignment system. We apply two detection techniques to monitor the position estimates from the daily measurements: zero-mean/whiteness detector and the WSSRdetector (see Section 5.7 for details). Both of these techniques rely on the assumption that the position deviations in the loop should change little during the measurement cycles. The underlying idea in anomaly detection is that the processors are "tuned" during calibration to operate in an optimal manner (innovations zero-mean/white). However, when an anomaly occurs, the innovations no longer maintain their optimal statistical properties leading to a "change from normal" and an anomaly is declared. With this detection, a decision must be made to either classify the anomaly or perform some other action. The main point to realize is that since we are using a recursive-in-time BP, the innovations represent "how well the position model and its underlying statistics represent the raw data." If it is an optimal fit, then the innovations should be zero-mean and statistically white (uncorrelated) and the detectors evolve naturally.

This completes the description of the BP and the associated anomaly detectors; next, we consider a simulation of the data and processor to predict the expected performance.

11.1.4 Application: Beam Line Data

In this section, we discuss the application of optimal position estimators based on daily historical position estimates (database) provided by the KDP algorithm. The data are a record of 48 days of final positions output from both the accurate reference system (reference data) and the estimated final position output from the KDP back-reflection data coupled to a phase-only, matched filter [15] imaging algorithm. The motivation for applying the Bayesian approach to this data set is to obtain a more accurate and precise estimate of error deviations from the reference characterizing the overall control loop performance in that position of the beam line. Theoretically, the position deviations between the reference and the KDP estimates should ideally be zero, but because of the beam line noise and variations, CCD camera limitations and control system tolerances, this is not the case. Instead, the overall error statistics are used to bound the performance and assure that they remain within design specification.

11.1.4.1 BP Design In the design of the BP, the usual procedure is to (1) develop the required models; (2) simulate a set of position data characterized by any of the a-priori information available; (3) develop the minimum error variance design; and (4) apply the processor to the available data set evaluating its performance. We developed the basic model set assuming a Gauss–Markov structure as in the previous subsection. These paramcters of this model were estimated from the statistics of a large ensemble (> 5000) of images and performance statistics of the KDP centroiding algorithm. We used this information to construct the initial simulation of the BP to "match" with the historical data available (48 days) and the corresponding ensemble statistics.

Thus, our approach is to first perform a simulation of the measurement process using estimated statistics from the data and then apply it to the actual data. During the simulation phase, we are able to analyze the performance of the BP and assure ourselves that all of the models and statistics are correct. We expect to obtain the minimum variance estimates; if not achieved, then it is usually an implementation issue. Once the simulation and model adjustments have been made, we apply the processor to the measured data.

11.1.4.2 Simulation Now that we have developed the models and have some estimates from the ensemble statistics of the database, we are now able to perform a Gauss–Markov simulation to assess the feasibility of the BP. We simulated a set of data based on the following parameters using the mean XY-position estimates, that is, $x_i = \mu_{x_i} \pm 1.96\sigma_i$ for both the measured and reference data:

$$\mathbf{x}(0) = [344 \pm 8,\ 270 \pm 13.8,\ 344 \pm 7.8,\ 270 \pm 14.4]'$$

We used the mean values as the initial position estimates with low error variances (1×10^{-6}). We chose an uncorrelated measurement (deviation) noise covariance matrix as: $\mathbf{R}_{vv} = \text{diag}\,[0.0076 \quad 0.0078]$. The process noise covariance selected is

$$\mathbf{R}_{ww} = \begin{bmatrix} \bar{\mathbf{R}}_{ww}(1) & 0 \\ 0 & \bar{\mathbf{R}}_{ww}(2) \end{bmatrix} \quad \text{for} \quad \bar{\mathbf{R}}_{ww}(1) = \begin{bmatrix} 50.0 & -26.5 \\ -26.5 & 50.0 \end{bmatrix};$$

$$\bar{\mathbf{R}}_{ww}(2) = \begin{bmatrix} 50.0 & -37.7 \\ -37.7 & 50.0 \end{bmatrix}$$

The synthesized positions have small variations due to the process noise, but are essentially constants (mean values). The simulated noisy DX-, DY-deviation measurements (dotted line) including the process and measurement noise are shown in Fig. 11.3. Next, we executed the BP over this data set and the position deviations (solid line) are also shown in the figure. Here we see that much of the raw measurement and process noise have been removed by the processor and only the deviation errors are shown. Recall that they should be close to zero for the alignment control system to be operating efficiently.

To validate the results, we investigate the minimum variance design procedure. We investigate the statistics of the innovations sequence and check that they are zero-mean and white for optimality. Both innovations sequences are zero-mean: (0.30 < 3.5;0.95 < 3.3) and white: (4% out; 0% out); with the WSSR decision function lying below the threshold for a window of size N: (threshold = 69.6, $N - 25$). These statistics indicate that the minimum variance position deviation estimates have been

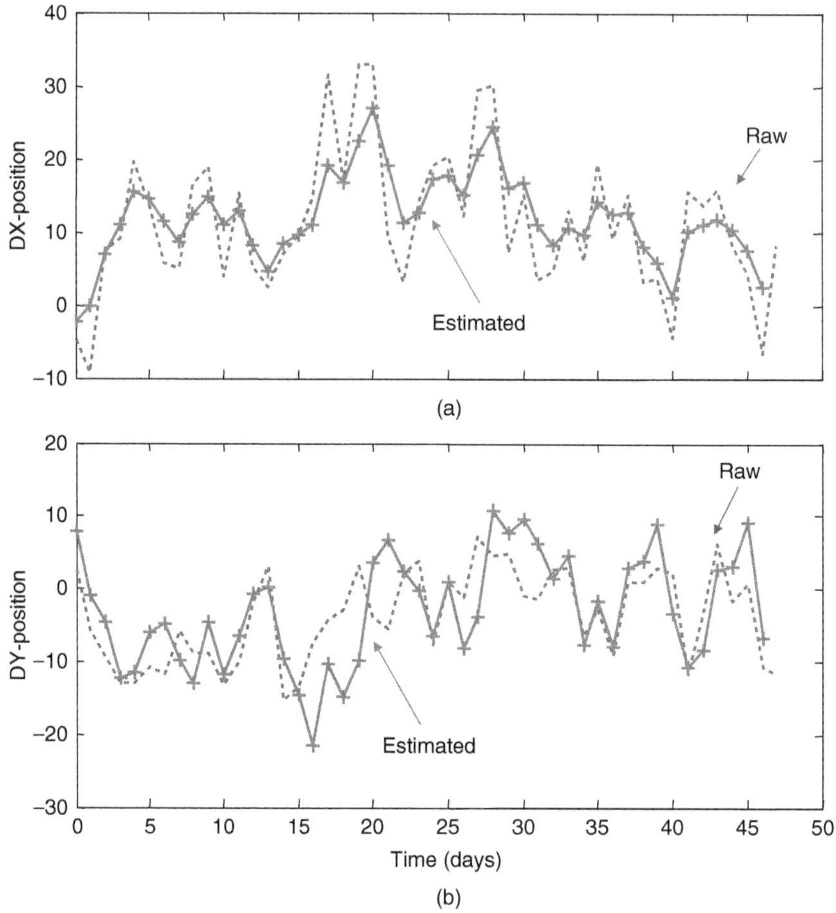

FIGURE 11.3 Simulated XY-position deviations: raw (dotted) and estimated (solid). (a) X-deviation. (b) Y-deviation.

achieved. This result is expected, since the models used in the BP are identical to those in the Gauss–Markov simulator. However, the results can be interpreted as the "best" (minimum variance) one could hope to do under these circumstances. Next, we apply the BP to the actual deviation data.

11.1.5 Results: Beam Line (KDP Deviation) Data

In order to process the KDP deviation data, we must develop parameters for the underlying model embedded in the processor. We start with guesses of the noise (process and measurement) statistics from the simulator and then adjust them accordingly. For instance, if the innovation sequence lies outside its predicted bounds implying that the measurement noise variance is too small, then it can be adjusted to satisfy

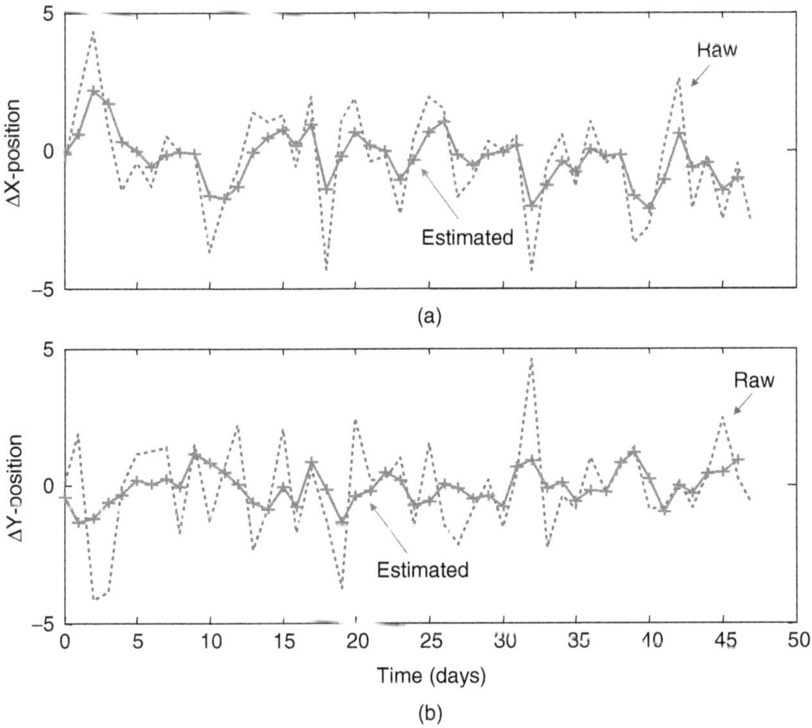

FIGURE 11.4 Actual KDP XY-deviations and estimated XY-positions: raw measured data (dashed); estimated data (solid). (a) △X-position. (b) △Y-position.

this constraint and "match" the data. The process noise is actually quite difficult to select, since it is directly proportional to the BP gain. In essence, once all of the other model parameters are reasonably adjusted, they are then held fixed and the process noise covariance is varied to achieve the "best" possible (minimum error variance) innovations statistics (zero-mean/white, WSSR below threshold).

The results of the BP design for the deviation data are shown in Figs. 11.4 and 11.5. We see the raw KDP position deviation estimates along with the optimal processor results over the 48-day period in Fig. 11.4. It is clear that the processor is tracking the trends in the data while reducing the noise or equivalently enhancing the SNR. To confirm this, we observe the three-sigma error ellipsoid plots of Fig. 11.5 where we observe that much of the uncertainty has been removed and the estimated deviations are clearly clustered (centered) around the (0,0) position and have a much smaller ellipsoid (better precision) than the raw data. Again, to confirm the optimality of the processor, we check the zero-mean-whiteness/ WSSR statistics which give the following results: zero-mean: $0.08 < 0.80$; $0.05 < 0.71$; approximately white: 4% out, 8% out; and WSSR decision function lies below the threshold for a window of size N: threshold $= 69.6, N = 25$. These statistics again indicate that the minimum variance design has been achieved and the processor, along with its associated statistics, is

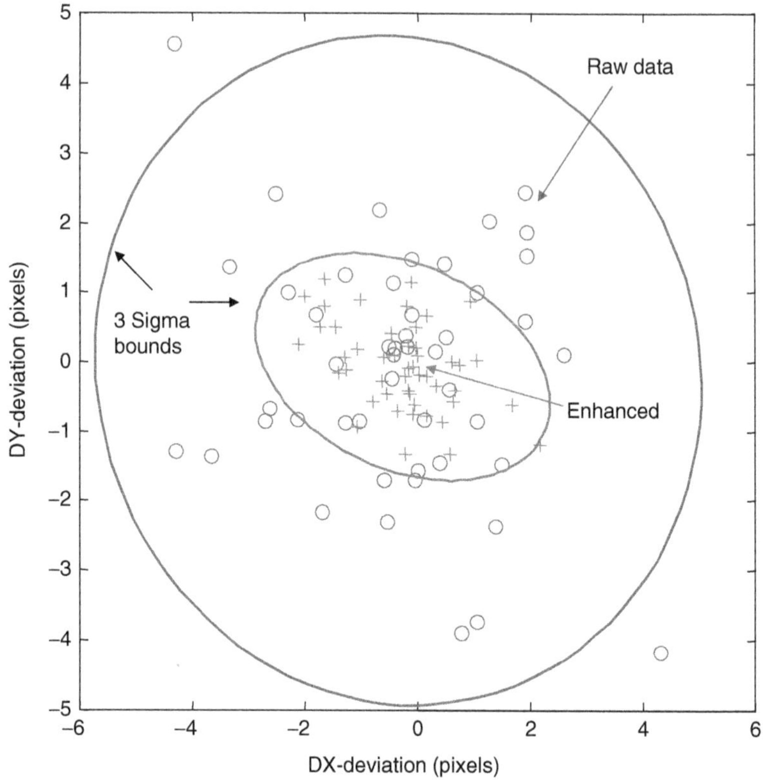

FIGURE 11.5 Three-sigma error ellipsoid for KDP XY-position deviations from simulated data: raw deviations (O) and estimated (+) with less outliers.

valid and optimal. With this in mind, we can also observe the performance of the BP in estimating (filtering) both measured KDP and reference positions. The results are shown in Fig. 11.6. Here, we see the estimated (filtered) positions observing that they are random (conditional mean) and vary slightly from the initial means indicating that the process noise is not too severe in this case. This completes the BP design for the position estimation problem. Next, we investigate the feasibility of using this approach for anomaly detection.

11.1.6 Results: Anomaly Detection

We use the actual KDP measurement data as before to assess the feasibility of applying the BP as an effective means to detect anomalies (off-normal) that could occur in the beam line. Here, we assume that the BP has been "tuned" to normal operations for the particular beam line and optics. Thus, the innovations are zero-mean, white, and the WSSR lies below the threshold for optimal design. To "simulate" an off-normal condition (in terms of position estimates), we increase the amplitude of both X and

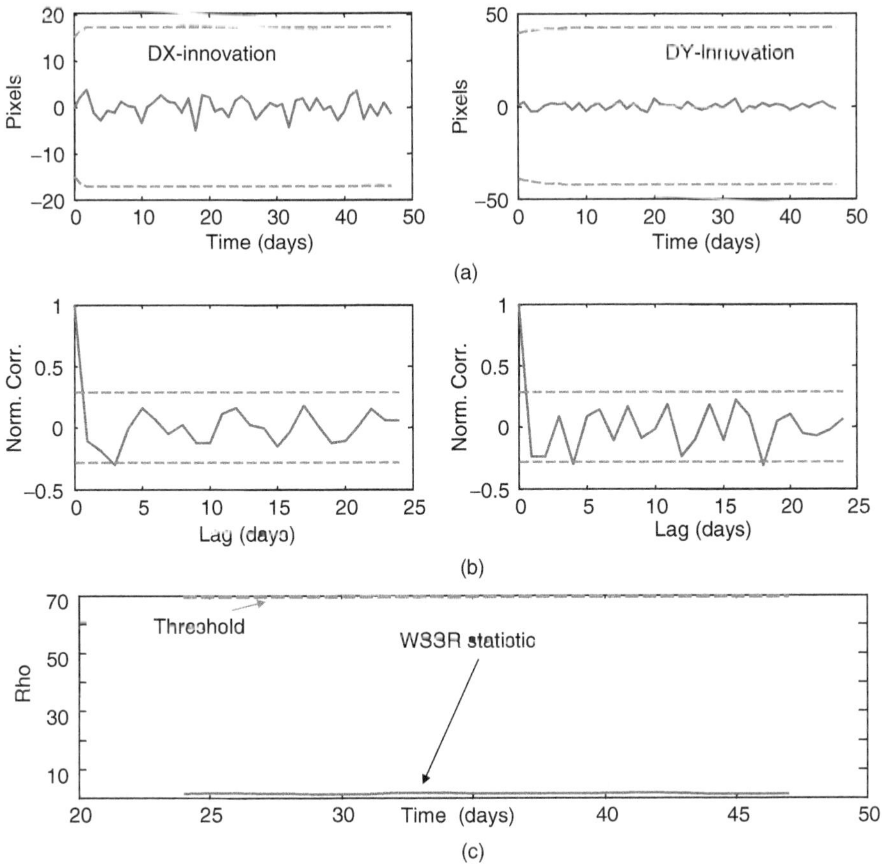

FIGURE 11.6 BP design for KDP XY-position deviation data: (a) innovations for X-deviation and Y-deviation; (b) optimality tests: zero-mean (0.08 < 0.80; 0.05 < 0.71) and whiteness (4% out; 8% out); (c) WSSR test (threshold = 69.6, $N = 25$).

Y deviations fivefold in the raw measurement data during the period of 29–35 days. This can be thought of as corresponding to a pulse-like transient in either or both of the measurement and/or reference position data. We applied the optimal (for no anomaly) BP to these data to determine whether or not it could detect and track the unknown transient anomaly.

The results of the effect on the deviation measurements are shown in Fig. 11.7. We see the normal measurement and then the transient "jump" within the prescribed time period. It is interesting to note that there is a corresponding disturbance in the estimated (filtered) measurement indicating that there is enough of a disruption at this SNR to cause the BP to track it. However, the BP is not able to track the transient extremely well the expected results. Next, we observe the deviation innovation sequences output by the BP in Fig. 11.8. We again observe the transient anomaly in both sequences, since the position estimates do not track it well enough. The

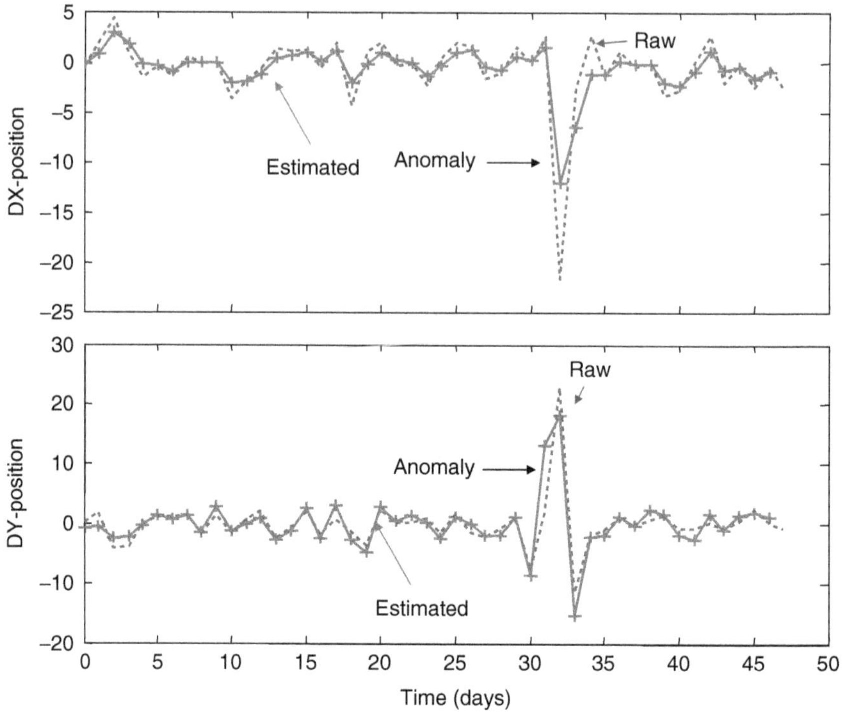

FIGURE 11.7 Raw KDP data XY-deviations with simulated anomaly: raw (dotted) and estimated (solid).

zero-mean/whiteness tests seem a bit too insensitive to the rapid change (six samples) and do not dramatically detect it even the whiteness tests do not indicate a non-white sequence (0% out; 4% out). On the other hand, the WSSR test clearly detects "change from normal" caused by the simulated beam line anomalies almost instantaneously indicating a feasible solution. This is again expected since the WSSR statistic ($\rho(\ell)$) can be tuned to transient disruption by selecting the appropriate window length. For this problem, we chose: $N = 5$; threshold $= 18$. To verify the performance of the zero-mean/whiteness test, we observe the estimated positions (states) in Fig. 11.9. Here, we see that the BP actually responds to such a high amplitude level change in the anomaly transient. The processor could easily be "tuned" to ignore the change of the states by decreasing the process noise variance. Finally, the scatter plots for this case are shown in Fig. 11.10: one for the optimal solution (no anomaly) and one for the solution with the anomaly. The size difference of the error ellipsoids in both cases is obvious due to the added uncertainty of the modeled anomaly. The three-sigma error ellipsoid for the case with an anomaly suspiciously indicates that something is different, since the estimated deviation uncertainty (ellipsoid) is actually about the same size or larger than that of the raw data. This completes the study of applying a model-based anomaly detector.

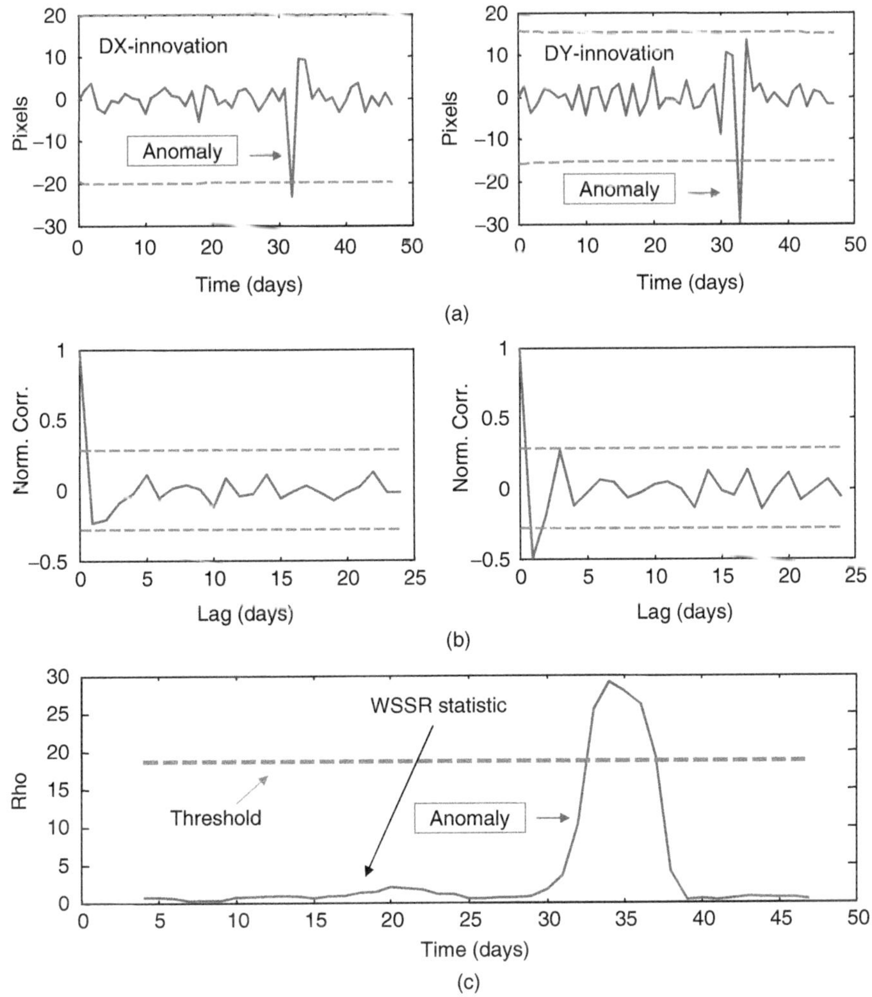

FIGURE 11.8 BP detection with simulated anomaly: (a) DX and DY innovations with anomaly; (b) optimality tests: zero-mean ($0.07 < 1.7$; $0.05 < 2.3$); whiteness (0% out; 4% out); (c) WSSR test (threshold = 18; $N = 5$).

11.2 SEQUENTIAL DETECTION OF BROADBAND OCEAN ACOUSTIC SOURCES

Acoustic sources found in the ocean environment are spatially complex and broadband due to temperature related sound-speed variations and dispersion that complicate the analysis of received acoustic data considerably. A model-based approach is developed to detect a broadband source in a shallow ocean environment characterized by a normal-mode propagation model. Here, we develop the Bayesian solution to the

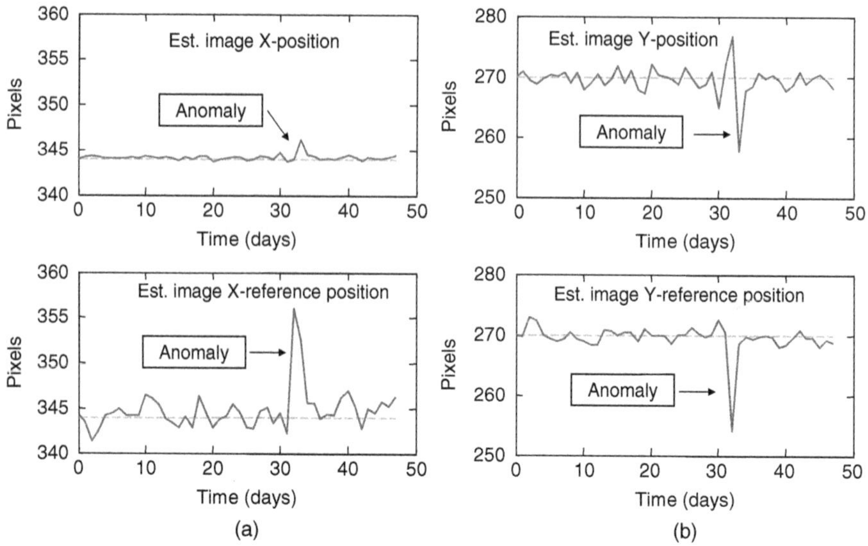

FIGURE 11.9 Estimated XY-positions with anomaly: (a) measured position estimates; (b) reference position estimates.

source detection problem while simultaneously enhancing both the modal functions and pressure-field measurements.

11.2.1 Background

Sound propagation in the shallow ocean initiated by sources significantly complicates the analysis of received acoustic data especially when they are broadband. Temperature changes resulting in sound-speed variations can cause highly dispersive propagation resulting in a statistically nonstationary signal. Highly correlated propagation channels obscure the problem of extracting target signals from disturbances and noise even further. Noise and inherent uncertainties weather ambient, distant shipping, and wind blown surface generated disrupt this uncertain environment primarily because propagation is through the same acoustic channel. When propagating in the shallow ocean, these source characteristics complicate the analysis of received acoustic data considerably—especially in littoral regions providing an important challenge for signal processing [17–23].

Employing a vertical hydrophone sensor array, the improved SNR of broadband acoustic pressure-field measurements using a multichannel Bayesian technique to estimate or filter the noisy measurement data is discussed. Here, the approach is developed for the broadband source using a normal-mode propagation model. Propagation theory predicts that a different modal structure evolves for each spectral line; therefore, it is not surprising that the multichannel solution to this problem results in a scheme that requires a bank of processors—each employing its own underlying modal structure for the narrow frequency band that it operates.

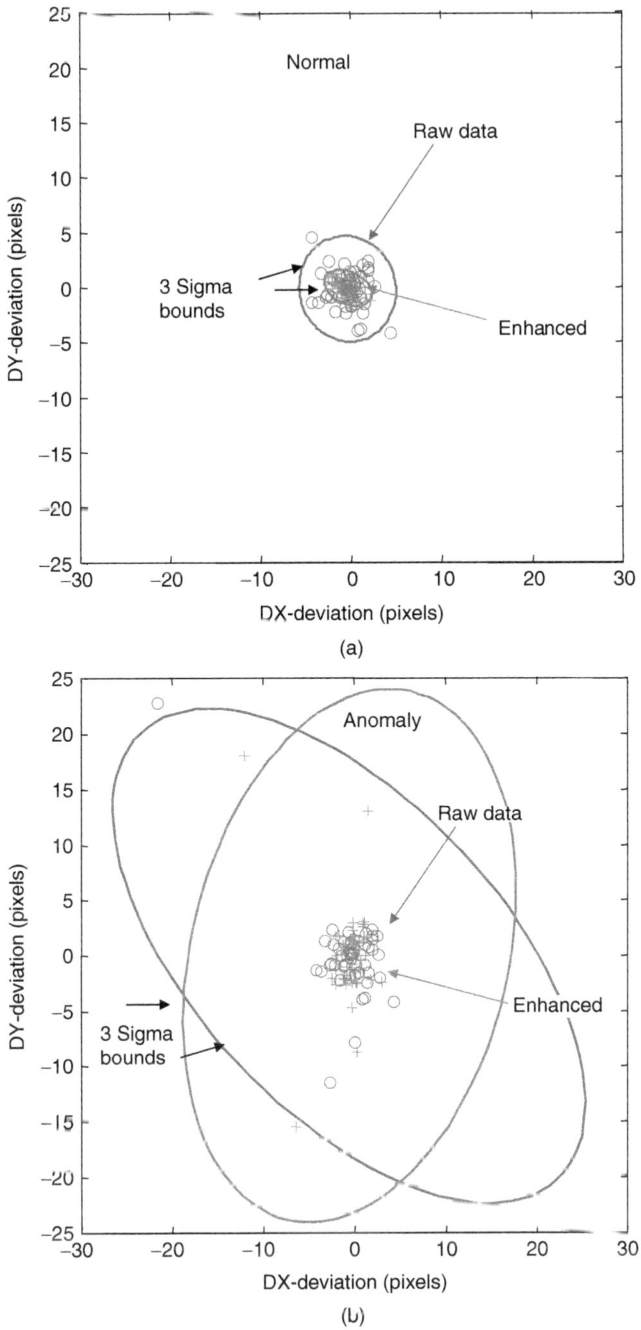

FIGURE 11.10 Three-sigma error ellipsoid and estimated XY-position deviations: (a) measured data with no anomaly; (b) measured data with anomaly. O, measured data, +, enhanced data.

The methodology employed is based on a state–space representation of the normal-mode propagation model [24]. Following Refs. [25–27], the Bayesian solution of using state–space forward propagators is discussed. It is shown that each state–space processor is naturally decoupled in modal space and recombined in the measurement space to provide improved signal level estimates. That is, the estimated or equivalently "filtered" modal/pressure-field data with improved SNR evolves providing a solution to the signal/measurement *enhancement problem* and subsequent broadband *source detection problem.*

In this study, we develop a sequential Bayesian solution to the signal enhancement/detection problem. First, we estimate both modal functions and pressure-field measurements from noisy, uncertain, broadband acoustic signals in a shallow ocean environment using a Markov chain Monte Carlo (MCMC) sampling approach (see Chapter 7). We evaluate the processor performance by applying the Kullback–Leibler information metric along with the classical "sanity" tests. These techniques lead to a nonparametric representation of the posterior modal (state/measurement) distributions and their associated statistics that are then used to develop a sequential source detection technique. We perform ensemble (100-members) testing to bound the Bayesian estimates and then develop a sequential solution to the broadband source detection problem. To evaluate detection performance, we estimate the receiver operating characteristic (ROC) curve of Section 10.4.

11.2.2 Broadband State-Space Ocean Acoustic Propagators

In this section, we discuss the development of a broadband propagator eventually employed in a Bayesian scheme to enhance noisy pressure-field measurements from a vertical array of hydrophone sensors. First, we briefly discuss the propagator from normal-mode theory following the Green's function approach [28] and then extend it using a state–space representation to develop a forward propagation scheme for eventual use in Bayesian processor design.

It is well known [28–30] in ocean acoustics that the pressure-field *solution* to the Helmholtz equation under the appropriate assumptions can be expressed as the sum of normal modes

$$p(r_s, z, t) = \sum_{m=1}^{M} a\mathcal{H}_o\left(\kappa_r(m)r_s\right)\phi_m(z_s)\phi_m(z)e^{i\omega_o t} \tag{11.5}$$

where p is the acoustic pressure-field; a is the source amplitude; $\mathcal{H}_o$ is the zeroth-order Hankel function; ϕ_m is the mth modal function evaluated at z and source depth z_s; $\kappa_r(m)$ is the horizontal wave number associated with the mth mode; ω_o is the temporal source frequency, and r_s is the horizontal range of the source. The wave numbers satisfy the corresponding *dispersion relation*

$$\kappa^2 = \frac{\omega^2}{c^2(z)} = \kappa_r^2(m) + \kappa_z^2(m), \qquad m = 1, \ldots, M \tag{11.6}$$

where the depth-dependent κ_z is the vertical wave number and c the corresponding depth-dependent sound-speed profile. Taking the temporal Fourier transform of the pressure-field, we obtain

$$p(r_s, z, \omega) = \sum_{m=1}^{M} a\mathcal{H}_o\left(\kappa_r(m)r_s\right)\phi_m(z_s)\phi_m(z)\delta(\omega - \omega_o) \tag{11.7}$$

indicating a narrowband solution or equivalently a line at ω_o in the temporal frequency domain.

This modal representation can be extended to include a broadband source $s(t)$ with corresponding spectrum $S_s(\omega)$. In this case, the ocean medium is specified by its *Green's* function (impulse response) which can be expressed in terms of the inherent normal modes spanning the water column

$$\mathcal{G}(r, z, \omega) = \sum_{m} \mathcal{H}_o\left(\kappa_r(m)r\right)\phi_m(z_s)\phi_m(z) \tag{11.8}$$

and therefore the resulting pressure-field in the temporal frequency domain is given by

$$p(r_s, z, \omega) = \mathcal{G}(r_s, z, \omega)S(\omega) \tag{11.9}$$

which equivalently results from the temporal convolution of $\mathcal{G}(r, z, t)$ and $s(t)$.

Suppose we decompose the continuous source spectrum into a sampled or discrete spectrum using a periodic impulse (frequency) sampler, then it follows that

$$S_s(\omega) = \triangle\omega\sum_{q} S(\omega)\delta(\omega - \omega_q) = \triangle\omega\sum_{q} S(\omega_q)\delta(\omega - \omega_q) \tag{11.10}$$

from the sampling properties of Fourier transforms. Therefore, a broadband signal spectrum can be decomposed into a set of narrowband components assuming an impulse sampled spectrum.

Utilizing this property in Eq. 11.10 and extracting just the qth source frequency, we have

$$p(r_s, z, \omega_q) = \mathcal{G}(r_s, z, \omega)S_s(\omega_q) \tag{11.11}$$

where $S_s(\omega_q)$ is interpreted as a single narrowband impulse at ω_q with amplitude, $a_q = \triangle\omega|S(\omega_q)|$. Suppose that the broadband source is assumed to be bandlimited and sampled $S_s(\omega), \omega_1 \leq \omega \leq \omega_Q$ then

$$S_s(\omega) = \triangle\omega\sum_{q=1}^{Q} S(\omega_q)\delta(\omega - \omega_q) \tag{11.12}$$

Thus, the normal-mode solution to the Helmholtz equation for the broadband source problem can be decomposed into a series of narrowband solutions, that is,

$$p(r, z, \omega_q) = \sum_{m=1}^{M_q} a_q \mathcal{H}_o\left(\kappa_r(m, q)r\right) \phi_m(z_s, \omega_q)\phi_m(z, \omega_q) \tag{11.13}$$

where M_q is the number of modes spanning the water column at frequency ω_q with the corresponding dispersion relation satisfying

$$\kappa_r^2(m, q) = \frac{\omega_q^2}{c^2(z)} - \kappa_z^2(m, q), \qquad m = 1, \dots, M_q; q = 1, \dots, Q \tag{11.14}$$

and the pressure-field at depth z and time t can be expressed in terms of the bilateral inverse (discrete) Fourier transform as

$$p(r, z, t) = \int_{-\infty}^{\infty} p(r, z, \omega)e^{j\omega t} d\omega \approx \frac{1}{Q} \sum_{q=-(Q-1)}^{Q-1} p(r, z, \omega_q)e^{j\omega_q t} \tag{11.15}$$

Suppose we further assume an L-element vertical sensor array, then $z \rightarrow z_\ell$, $\ell = 1, \dots, L$ and therefore, the pressure-field at the array for the qth temporal frequency of Eq. 11.13 becomes

$$p(r_s, z_\ell, \omega_q) = \sum_{m=1}^{M_q} \beta_m(r_s, z_s, \omega_q)\phi_m(z_\ell, \omega_q) \tag{11.16}$$

where $\beta_m(r_s, z_s, \omega_q) := a_q \mathcal{H}_o\left(\kappa_r(m, q)r_s\right) \phi_m(z_s, \omega_q)$ is the mth-modal coefficient at the qth temporal frequency.

In terms of these models, it has been shown [26] that the *broadband state-space propagator* can be expressed as

$$\begin{aligned} \frac{d}{dz}\mathbf{\Phi}(z, \Omega) &= \mathbf{A}(z, \Omega)\mathbf{\Phi}(z, \Omega) \\ p(z_\ell, \Omega) &= \mathbf{C}(r_s, z_s, \Omega)\mathbf{\Phi}(z_\ell, \Omega) \end{aligned} \tag{11.17}$$

where $\mathbf{\Phi}(z, \Omega) \in \mathcal{R}^{2M\times 1}$, $\mathbf{A}(z, \Omega) \in \mathcal{R}^{2M\times 2M}$, $\mathbf{C}(r_s, z_s, \Omega) \in \mathcal{C}^{1\times 2M}$ with $M = \sum_{q=1}^{Q} M_q$ and $\Omega := \{\omega_q\}; q = 1, \dots, Q$ is the entire set of discrete temporal frequencies.

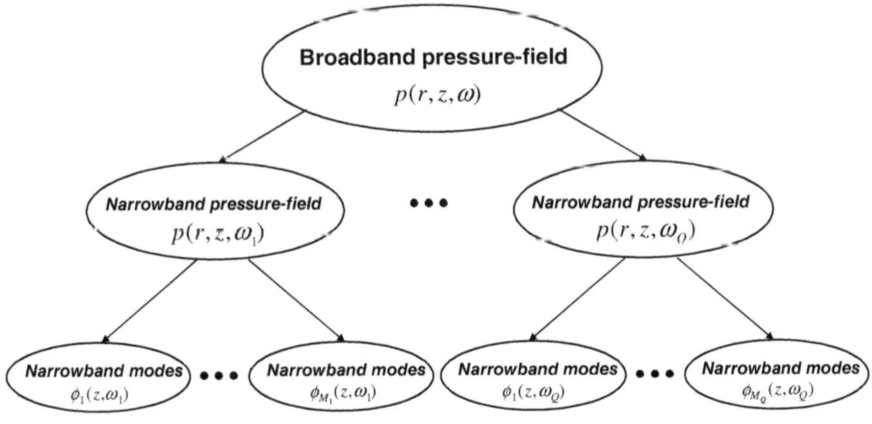

FIGURE 11.11 Decomposition of broadband signal into narrowband components based on normal-mode propagation representation.

The internal structure of this overall processor admits the following decomposition:

$$\frac{d}{dz}\mathbf{\Phi}(z,\Omega) = \begin{bmatrix} \mathbf{A}(\mathbf{z},\omega_0) & O & \cdots & O \\ O & \mathbf{A}(z,\omega_1) & \cdots & O \\ \vdots & & \ddots & \vdots \\ O & O & \cdots & \mathbf{A}(z,\omega_{Q-1}) \end{bmatrix} \begin{bmatrix} \mathbf{\Phi}(z,\omega_0) \\ \mathbf{\Phi}(z,\omega_1) \\ \vdots \\ \mathbf{\Phi}(z,\omega_{Q-1}) \end{bmatrix} \quad (11.18)$$

and $\mathbf{A}(z,\omega_q) = \text{diag}\left[\mathbf{A}_1(z,\omega_q)\cdots\mathbf{A}_{M_q}(z,\omega_q)\right] \in \mathbf{R}^{2M_q \times 2M_q}$ and

$$\mathbf{A}_m(z,\omega_q) = \begin{bmatrix} 0 & 1 \\ -\kappa_z^2(m,q) & 0 \end{bmatrix}, \qquad m = 1,\ldots,M_q \quad (11.19)$$

with the pressure-field sensor measurement model given by (see Fig. 11.11).

$$p(z_\ell,\Omega) = \left[\frac{1}{Q}\mathbf{C}_1^T(r_s,z_s,\omega_0)e^{j\omega_0 t}\cdots\frac{1}{Q}\mathbf{C}_Q^T(r_s,z_s,\omega_{Q-1})e^{j\omega_{Q-1}t}\right]\begin{bmatrix} \mathbf{\Phi}(z_\ell,\omega_0) \\ \vdots \\ \mathbf{\Phi}(z_\ell,\omega_{Q-1}) \end{bmatrix} \quad (11.20)$$

This deterministic model can be extended to a stochastic Gauss–Markov representation [11] given by

$$\begin{aligned} \frac{d}{dz}\mathbf{\Phi}(z,\Omega) &= \mathbf{A}(z,\Omega)\mathbf{\Phi}(z,\Omega) + \mathbf{w}(z,\Omega) \\ p(z_\ell,\Omega) &= \mathbf{C}(r_s,z_s,\Omega)\mathbf{\Phi}(z_\ell,\Omega) + v(z,\Omega) \end{aligned} \quad (11.21)$$

where $\mathbf{w}$, v are additive, zero-mean Gaussian noise sources with respective spectral covariance matrices $\mathbf{R}_{ww}(z,\Omega)$, and $R_{vv}(z,\Omega)$.

11.2.3 Discrete Normal-Mode State-Space Representation

Since our array spatially samples the pressure-field discretizing depth, we discretize the differential state equations of Eq. 11.21 in depth. Using a *central difference* approach for improved numerical stability, we have

$$\frac{d^2\phi_m(z,\omega_q)}{dz^2} \approx \frac{\phi_m(z_\ell,\omega_q) - 2\phi_m(z_{\ell-1},\omega_q) + \phi_m(z_{\ell-2},\omega_q)}{\triangle z_\ell^2}$$

for $\triangle z_\ell := z_\ell - z_{\ell-1}$. Substituting this approximation into the modal relations gives

$$\phi_m(z_\ell,\omega_q) - 2\phi_m(z_{\ell-1},\omega_q) + \phi_m(z_{\ell-2},\omega_q) + \triangle z_\ell^2\kappa_z^2(m,q)\phi_m(z_{\ell-1},\omega_q) = 0$$

where $m = 1,\ldots,M_q$ and z_ℓ is the location of the ℓth sensor. Defining the discrete broadband modal state vector as $\phi_m(z_\ell,\omega_q) := [\phi_m(z_{\ell-2},\omega_q)|\phi_m(z_{\ell-1},\omega_q)]^T$, we obtain the following set of difference equations for the mth mode at the qth frequency

$$\begin{aligned}\phi_{m1}(z_\ell,\omega_q) &= \phi_{m2}(z_{\ell-1},\omega_q)\\ \phi_{m2}(z_\ell,\omega_q) &= -\phi_{m1}(z_{\ell-1},\omega_q) + \left(2 - \triangle z_\ell^2\kappa_z^2(m,q)\right)\phi_{m2}(z_{\ell-1},\omega_q)\end{aligned}$$

with each of the corresponding modal A-submatrices given by

$$\mathbf{A}_m(z,\omega_q) = \begin{bmatrix} 0 & 1 \\ -1 & 2 - \triangle z_\ell^2\kappa_z^2(m,q)\end{bmatrix}; \quad m = 1,\ldots,M_q$$

11.2.4 Broadband Bayesian Processor

For our pressure-field/modal function estimation problem, we define the underlying broadband pressure-field/modal function *signal enhancement problem* as:

GIVEN a set of noisy broadband pressure-field measurements, $P_\ell := \{\mathbf{p}(z_1,\Omega),\ldots,\mathbf{p}(z_\ell,\Omega)\}$ and the underlying stochastic model of Eq. 11.21, FIND an estimate of the posterior distribution, $\hat{\Pr}\left[\mathbf{\Phi}(z_\ell,\Omega)|P_\ell\right]$ and infer the corresponding enhanced estimates of the broadband modal functions, $\hat{\mathbf{\Phi}}(z_\ell,\Omega)$ and pressure-field, $\hat{\mathbf{p}}(z_\ell,\Omega)$.

The solution to this problem can be obtained by estimating the joint *a posteriori* distribution applying Bayes' theorem, that is, the joint probability of the broadband modal functions at the ℓth depth based on the set of pressure-field measurements up to z_ℓ is given by

$$\Pr\left[\boldsymbol{\Phi}_\ell | P_\ell\right] = \frac{\Pr\left[P_\ell | \boldsymbol{\Phi}_\ell\right] \times \Pr\left[\boldsymbol{\Phi}_\ell\right]}{\Pr\left[P_\ell\right]} \tag{11.22}$$

where $\boldsymbol{\Phi}_\ell := \{\boldsymbol{\Phi}(z_1, \Omega), \ldots, \boldsymbol{\Phi}(z_\ell, \Omega)\}$ is the set of broadband modal functions (states) and P_ℓ is the set of broadband measurements up to and including $p(z_\ell, \Omega)$.

Following Ref. [26], the measurement *likelihood* distribution, that is, the joint set of measurements up to and including the ℓth based on the set of broadband modal functions at depth z_ℓ can be factored using Bayes' rule as

$$\Pr\left[P_\ell | \boldsymbol{\Phi}_\ell\right] = \Pr\left[\mathbf{p}(z_\ell, \Omega) | \boldsymbol{\Phi}(z_\ell, \Omega)\right] \times \Pr\left[P_{\ell-1} | \boldsymbol{\Phi}_{\ell-1}\right] \tag{11.23}$$

under the assumed independence of $\mathbf{p}(z_\ell, \Omega)$ with $P_{\ell-1}$ and $\boldsymbol{\Phi}_{\ell-1}$, and $\boldsymbol{\Phi}(z_\ell, \Omega)$ with $P_{\ell-1}$.

Similarly, the joint *prior* modal distribution at depth z_ℓ can also be factored to give the (modal) *transition* distribution

$$\Pr\left[\boldsymbol{\Phi}_\ell\right] = \Pr\left[\boldsymbol{\Phi}(z_\ell, \Omega), \boldsymbol{\Phi}_{\ell-1}\right] = \Pr\left[\boldsymbol{\Phi}(z_\ell, \Omega) | \boldsymbol{\Phi}(z_{\ell-1}, \Omega)\right] \times \Pr\left[\boldsymbol{\Phi}_{\ell-1}\right]$$

under the first-order Markov assumption, while that for the joint *evidence* is factored as

$$\Pr\left[\mathbf{P}_\ell\right] = \Pr\left[\mathbf{p}(z_\ell, \Omega), \mathbf{P}_{\ell-1}\right] = \Pr\left[\mathbf{p}(z_\ell, \Omega) | \mathbf{P}(z_{\ell-1}, \Omega)\right] \times \Pr\left[\mathbf{P}_{\ell-1}\right]$$

Substituting these relations into Eq. 11.22, cancelling like terms, assuming conditional independence and recognizing the Bayes' relation for the previous step $\Pr\left[\boldsymbol{\Phi}_{\ell-1} | P_{\ell-1}\right]$, we obtain the sequential Bayes' expression for the joint posterior density of the set of broadband modal functions based on the pressure-field measurements as

$$\Pr\left[\boldsymbol{\Phi}_\ell | P_\ell\right] = \left(\frac{\Pr\left[\mathbf{p}(z_\ell, \Omega) | \boldsymbol{\Phi}(z_\ell, \Omega)\right] \times \Pr\left[\boldsymbol{\Phi}(z_\ell, \Omega) | \boldsymbol{\Phi}(z_{\ell-1}, \Omega)\right]}{\Pr\left[\mathbf{p}(z_\ell, \Omega) | P_{\ell-1}\right]}\right) \times \Pr\left[\boldsymbol{\Phi}_{\ell-1} | P_{\ell-1}\right] \tag{11.24}$$

consisting of the *measurement* likelihood, *modal* transition prior and evidence (for details see Section 2.5.3).

11.2.5 Broadband Particle Filters

One approach to estimate the required posterior distribution of Eq. 11.21 from noisy broadband measurements is to develop the so-called particle filter (PF) [31]. A *particle filter* is a completely different approach to nonlinear filtering that is capable of characterizing multi-modal distributions. In fact, it might be easier to think of the PF as a histogram or kernel density-like estimator in the sense that it is an estimated empirical probability mass function (PMF) that approximates the desired posterior distribution such that inferences can be performed and statistics extracted directly. Here, the idea is a change in thinking where we attempt to develop an empirical estimation of the posterior distribution following a purely Bayesian approach using Monte Carlo (MC) sampling theory as its enabling foundation. As one might expect, the computational burden of the PF is much higher than that of other processors, since it must provide an estimate of the underlying state posterior distribution state by state at each z_ℓ-step along with the fact that the number of samples to characterize the distribution is equal to the number of particles (N_p). The estimated *empirical posterior distribution* for the broadband signal enhancement problem is given by

$$\hat{\Pr}[\mathbf{\Phi}(z_\ell, \Omega)|P_\ell] = \sum_{i=1}^{N_p} \mathcal{W}_i(z_\ell, \Omega) \times \delta\big(\mathbf{\Phi}(z_\ell, \Omega) - \mathbf{\Phi}_i(z_\ell, \Omega)\big) \tag{11.25}$$

where the weights and posterior distribution estimated by the PF are given by

P_ℓ is the set of batch pressure-field measurements,
$\mathcal{W}_i(z_\ell, \Omega)$ is the ith broadband weight (normalized) at depth z_ℓ
$\mathbf{\Phi}_i(z_\ell, \Omega)$ is the ith particle of the broadband modal function at depth z_ℓ

Once the underlying posterior is available, estimates of important statistics can be extracted directly. For instance, the maximum a posteriori (MAP) estimate is found by locating a particular particle $\mathbf{\Phi}_i(z_\ell, \Omega)$ corresponding to the maximum of the PMF at that depth, that is,

$$\hat{\mathbf{\Phi}}_{\text{MAP}}(z_\ell, \Omega) = \max_i \hat{\Pr}\left[\mathbf{\Phi}_i(z_\ell, \Omega)|P_\ell\right] \tag{11.26}$$

while the conditional mean (CM) or equivalently the minimum mean-squared error (MMSE) estimate is estimated by

$$\hat{\mathbf{\Phi}}_{\text{CM}}(z_\ell, \Omega) \approx \frac{1}{N_p} \sum_{i=1}^{N_p} \mathcal{W}_i(z_\ell, \Omega) \times \mathbf{\Phi}_i(z_\ell, \Omega) \tag{11.27}$$

The Bayesian solution for this problem is based on determining the associated broadband weighting function that leads to the estimated empirical posterior distribution. A sampling or equivalently importance distribution $\text{I}(\mathbf{\Phi}(z_\ell, \Omega)|P_\ell)$ is *selected* first, then the weight (unnormalized) is determined by the ratio of the desired posterior to this choice (see Section 7.5.1)

$$W(z_\ell, \Omega) := \frac{\Pr[\mathbf{\Phi}(z_\ell, \Omega)|P_\ell]}{\mathrm{I}(\mathbf{\Phi}(z_\ell, \Omega)|P_\ell)}$$

which can be expanded using Bayes' rule of Eq. 11.22 to provide the "sequential" generic weight

$$W(z_\ell, \Omega) := W(z_{\ell-1}, \Omega) \times \frac{\Pr[p(z_\ell, \Omega)|\mathbf{\Phi}(z_\ell, \Omega)] \times \Pr[\mathbf{\Phi}(z_\ell, \Omega)|\mathbf{\Phi}(z_{\ell-1}, \Omega)]}{\mathrm{I}(\mathbf{\Phi}(z_\ell, \Omega)|P_\ell)} \tag{11.28}$$

where the numerator is the product of the *likelihood* and the state *transition* probability.

11.2.6 Broadband Bootstrap Particle Filter

There are a variety of PF algorithms available based on the choice of the importance distribution [26]. Perhaps the simplest is the *bootstrap* technique. The PF design for our problem using the bootstrap approach employs the state transition probability as its importance distribution, that is,

$$\mathrm{I}(\Phi(z_\ell, \Omega)|P_\ell) := \Pr[\Phi(z_\ell, \Omega)|\Phi(z_{\ell-1}, \Omega)] \tag{11.29}$$

Substituting this expression into Eq. 11.28 leads to the weighting function

$$\mathcal{W}(z_\ell, \Omega) = \mathcal{W}(z_{\ell-1}, \Omega) \times \Pr[p(z_\ell, \Omega)|\Phi(z_\ell, \Omega)] \tag{11.30}$$

which is simply the *likelihood distribution.*

For the bootstrap implementation, we draw samples from the state *transition* distribution using the dynamic modal model of Eq. 11.21 driven by the process uncertainty $\mathbf{w}_i(z_{\ell-1}, \Omega)$ to generate the set of particles, $\{\mathbf{\Phi}_i(z_\ell, \Omega)\}$ for each $i = 1, \ldots, N_p$.

The *likelihood*, on the other hand, is determined from the pressure-field measurement model for each mode giving the likelihood (assuming Gaussian measurement noise) as

$$\Pr[p(z_\ell, \Omega)|\mathbf{\Phi}_i(z_\ell, \Omega)] = \frac{1}{\sqrt{2\pi R_{vv}(\Omega)}} \times \exp\left\{-\frac{1}{2R_{vv}(\Omega)}(p(z_\ell, \Omega) - \mathbf{C}(r_s, z_s, \Omega)\mathbf{\Phi}_i(z_\ell, \Omega))^2\right\} \tag{11.31}$$

Thus, we can estimate the posterior distribution using a sequential Monte Carlo approach and construct a *bootstrap particle filter* of Section 7.5 using the following steps:

- Initialize: $\boldsymbol{\Phi}_i(z_0,\Omega), \mathbf{w}_i(z_0,\Omega) \sim \mathcal{N}\left(0, \mathbf{R}_{ww}(z_0,\Omega)\right)$, $v(z_0,\Omega) \sim \mathcal{N}\left(0, \mathbf{R}_{vv}\,(z_0,\Omega)\right), W_i(z_0,\Omega) = 1/N_p; i = 1,\ldots,N_p$
- State transition: $\boldsymbol{\Phi}_i(z_\ell,\Omega) = \mathbf{A}(z_{\ell-1},\Omega)\boldsymbol{\Phi}_i(z_{\ell-1},\Omega) + \mathbf{w}_i(z_{\ell-1},\Omega)$
- Likelihood probability: $\Pr[p(z_\ell,\Omega)|\boldsymbol{\Phi}_i(z_\ell,\Omega)]$ of Eq. 11.31
- Weights: $W_i(z_\ell,\Omega) = W_i(z_{\ell-1},\Omega) \times \Pr[p(z_\ell,\Omega)|\boldsymbol{\Phi}_i(z_\ell,\Omega)]$
- Normalize: $\mathcal{W}_i(z_\ell,\Omega) = \frac{W_i(z_\ell,\Omega)}{\sum_{i=1}^{N_p} W_i(z_\ell,\Omega)}$
- Resample: $\begin{cases} \hat{\boldsymbol{\Phi}}_i(z_\ell,\Omega) \Rightarrow \boldsymbol{\Phi}_i(z_\ell,\Omega) & N_{eff} \le N_\tau \quad \text{[Resample]} \\ & N_{eff} > N_\tau \quad \text{[No resample]} \end{cases}$
- Posterior: $\hat{\Pr}[\boldsymbol{\Phi}_i(z_\ell,\Omega)|P_\ell] = \sum_{i=1}^{N_p} \mathcal{W}_i(z_\ell,\Omega)\delta(\boldsymbol{\Phi}(z_\ell,\Omega) - \hat{\boldsymbol{\Phi}}_i(z_\ell,\Omega))$
- MAP estimate: $\hat{\boldsymbol{\Phi}}_{\text{MAP}}(z,\Omega) = \max_i \hat{\Pr}[\boldsymbol{\Phi}_i(z_\ell,\Omega)|P_\ell]$
- CM estimate: $\hat{\boldsymbol{\Phi}}_{CM}(z,\Omega) = \frac{1}{N_p}\sum_{i=1}^{N_p} \mathcal{W}_i(z_\ell,\Omega) \times \boldsymbol{\Phi}_i(z_\ell,\Omega)$

We see that after initialization, the broadband modal model is used to generate the particles during the *prediction* step. These particles, $\boldsymbol{\Phi}_i(z_\ell,\Omega)$, are then incorporated into the likelihood distribution providing the *update* step. Recall that resampling of the particles (see Section 7.4) is usually required to prevent degeneracy of the associated weights which increase in variance at each depth making it impossible to avoid the degradation. *Resampling* consists of processing the particles with their associated weights duplicating those of large weights (probabilities) and discarding those of small weights. In this way, only those particles of highest probabilities (importance) are retained enabling a coalescence at the peaks of the resulting posterior distribution while mitigating the degradation. A measure based on the coefficient of variation, the *effective* particle sample size given by

$$N_{eff} = \frac{1}{\sum_{i=1}^{N_p} W_i^2(z_\ell,\Omega)} \tag{11.32}$$

is the underlying *decision statistic* such that when its value is less than a predetermined threshold resampling is performed.

We should also note from the structure of the PF algorithm that each mode (two states) is driven by the modal process noise/uncertainty vector $\mathbf{w}_i(z_\ell)$. This process, coupled with the underlying block diagonal structure of $\mathbf{A}(z_\ell,\Omega)$, enables a *parallel* construct of the overall processor that is eventually combined by the measurement model of the likelihood distribution. Here, in contrast to the approximate (unimodal) model-based (Kalman) scheme of Section 5.3, *no* approximations are required to obtain this overall block-diagonal structure and parallel implementation.

11.2.7 Bayesian Performance Metrics

Particle filters are basically developed to provide estimates of the underlying posterior distributions for the problem under investigation in order to extract meaningful statistics. Compared to the usual model-based (Kalman) processors (see Section 5.7) which utilize their inherent "optimality" property (zero-mean, uncorrelated residuals) to assess performance, the estimated posterior of the PF must be compared to the true distribution of the underlying process it is to estimate. One way to achieve this comparison as developed in Section 7.6, is to utilize the associated *divergence statistic* of the Kullback–Leibler information as a metric.

Our interest lies in comparing two probability distributions to determine "how close" they are to one another. Even though, the $\mathcal{I}_{\rm KL}$ does quantify the *difference* between the true and estimated distribution, it is not a distance metric to answer this question due to its lack of symmetry. However, the *Kullback divergence* (KD) defined by

$$
\begin{aligned}
&\mathcal{J}_{\rm KD}\left(\Pr[\boldsymbol{\Phi}_i(z_\ell,\Omega)|P_\ell];\hat{\Pr}[\boldsymbol{\Phi}_i(z_\ell,\Omega)|P_\ell]\right)\\
&\quad=\mathcal{I}_{\rm KL}\left(\Pr[\boldsymbol{\Phi}_i(z_\ell,\Omega)|P_\ell];\hat{\Pr}[\boldsymbol{\Phi}_i(z_\ell,\Omega)|P_\ell]\right)\\
&\quad+\mathcal{I}_{\rm KL}\left(\hat{\Pr}[\boldsymbol{\Phi}_i(z_\ell,\Omega)|P_\ell];\Pr[\boldsymbol{\Phi}_i(z_\ell,\Omega)|P_\ell]\right) \qquad (11.33)
\end{aligned}
$$

is a *distance metric* between distributions indicating "how close" one distribution is to the other or from our perspective, "how well does it approximate" the posterior. Thus, the KD is a very important metric that can be applied to assess the performance of Bayesian processors providing a metric between the true and estimated posterior distributions that we apply to our broadband problem.

11.2.8 Sequential Detection

In this section, we develop a sequential detector that incorporates the particle filter to solve the basic problem of detecting a broadband source from noisy pressure field measurements. For our decision problem, we define the underlying *broadband source detection problem* as:

GIVEN a set of noisy broadband pressure-field measurements, $\{P_\ell\}; \ell = 1,\dots,L$ with the underlying stochastic model of Eq. 11.21 and the "data likelihood" distribution, $\Pr[p(z_\ell,\Omega)|P_{\ell-1}]$, DECIDE whether or not the broadband source is present where $P_\ell := \{\mathbf{p}(z_1,\Omega),\dots,\mathbf{p}(z_\ell,\Omega)\}$.

This is a binary decision problem; therefore, we are to test the hypothesis of whether the noisy pressure-field measurements have evolved from a broadband source or not. The corresponding binary hypothesis test is

$$\begin{aligned}
&\mathcal{H}_0 : p(z_\ell, \Omega) = \upsilon(z_\ell, \Omega) && \text{[Noise]}\\
&\mathcal{H}_1 : p(z_\ell, \Omega) = \mathbf{C}(r_s, z_s, \Omega)\mathbf{\Phi}_i(z_\ell, \Omega) + \upsilon(z_\ell, \Omega) && \text{[Source]}
\end{aligned} \tag{11.34}$$

An optimal solution to this decision problem is based on applying the *Neyman–Pearson theorem* leading to the likelihood ratio [32, 33] defined by

$$\mathcal{L}(P_L) := \frac{\Pr[P_L|\mathcal{H}_1]}{\Pr[P_L|\mathcal{H}_0]} \quad \begin{matrix}\mathcal{H}_1\\ >\\ <\\ \mathcal{H}_0\end{matrix} \ \mathcal{T} \tag{11.35}$$

where $P_L = \{p(z_1, \Omega), \ldots, p(z_L, \Omega)\}$ with threshold $\mathcal{T}$.

Following Section 10.5, expanding the likelihood-ratio for each depth, we obtain

$$\mathcal{L}(P_L) = \frac{\Pr[p(z_1, \Omega), \ldots, p(z_L, \Omega)|\mathcal{H}_1]}{\Pr[p(z_1, \Omega), \ldots, p(z_L, \Omega)|\mathcal{H}_0]} \tag{11.36}$$

As before, from the chain rule of probability and Bayes' theorem [34], we have

$$\Pr[P_L|\mathcal{H}_i] = \Pr[p(z_L, \Omega), P_{L-1}|\mathcal{H}_i] = \Pr[p(z_L, \Omega)|P_{L-1}, \mathcal{H}_i] \times \Pr[P_{L-1}|\mathcal{H}_i] \tag{11.37}$$

Substituting these expressions into the likelihood ratio above, replacing $L \to \ell$, and grouping, we obtain the sequential form of likelihood ratio as [33]

$$\mathcal{L}\big(p(z_\ell, \Omega)\big) = \mathcal{L}\big(p(z_{\ell-1}, \Omega)\big) \times \frac{\Pr[p(z_\ell, \Omega)|p(z_{\ell-1}, \Omega), \mathcal{H}_1]}{\Pr[p(z_\ell, \Omega)|p(z_{\ell-1}, \Omega), \mathcal{H}_0]} \tag{11.38}$$

Taking logarithms simplifies the computations; therefore, we define $\Lambda\big(p(z_\ell, \Omega)\big) := \ln \mathcal{L}\big(p(z_\ell, \Omega)\big)$ to obtain the *sequential log-likelihood-ratio* decision function for the *broadband source detection problem* as

$$\begin{aligned}
\Lambda\big(p(z_\ell, \Omega)\big) &= \Lambda\big(p(z_{\ell-1}, \Omega)\big) + \ln \Pr[p(z_\ell, \Omega)|p(z_{\ell-1}, \Omega), \mathcal{H}_1]\\
&\quad - \ln \Pr[p(z_\ell, \Omega)|p(z_{\ell-1}, \Omega), \mathcal{H}_0]
\end{aligned} \tag{11.39}$$

To complete this development, we require thresholds that lead to the Wald *sequential probability-ratio test* [33] of Section 10.5

$$\begin{array}{lll} \Lambda(p(z_\ell,\Omega)) & > \ln \mathcal{T}_1(\ell) & \text{Accept } \mathcal{H}_1 \\ \ln \mathcal{T}_0(\ell) \le \Lambda(p(z_\ell,\Omega)) & \le \ln \mathcal{T}_1(\ell) & \text{Continue} \\ \Lambda(p(z_\ell,\Omega)) & < \ln \mathcal{T}_0(\ell) & \text{Accept } \mathcal{H}_0 \end{array}$$

These thresholds can be obtained from an operating point (P_{FA},P_{DET}) on the associated ROC curve obtained through simulation or experimental data and are specified by

$$\mathcal{T}_0(\ell) = \frac{1 - P_{DET}(\ell)}{P_{FA}(\ell)} \qquad \mathcal{T}_1(\ell) = \frac{P_{DET}(\ell)}{P_{FA}(\ell)} \tag{11.40}$$

For this problem, the particle filter is used to estimate the desired *data likelihood* distribution based on the nonparametric estimate of the posterior distribution, that is, we require an estimate of the

$$\hat{\Pr}[p(z_\ell,\Omega)|p(z_{\ell-1},\Omega); \mathcal{H}_\ell)]; \ell = 0, 1 \tag{11.41}$$

under each hypothesis of Eq. 11.34.

Recall from Section 10.5 that assuming a first-order Markov process $\left(P_{\ell-1} \to p(z_{\ell-1},\Omega)\right)$ and choosing the importance distribution to be the state transition $\mathrm{I}(\mathbf{\Phi}(z_\ell,\Omega)|\mathbf{\Phi}(z_{\ell-1},\Omega), P_{\ell-1}) \to \Pr\left[\mathbf{\Phi}(z_\ell,\Omega)|\mathbf{\Phi}(z_{\ell-1},\Omega)\right]$, then the bootstrap PF results with weight $\mathcal{W}_i(z_\ell,\Omega) \to \Pr\left[p(z_\ell,\Omega)|\mathbf{\Phi}_i(z_\ell,\Omega)\right]$. Therefore, the data likelihood can be approximated by

$$\begin{aligned} \hat{\Pr}\left[p(z_\ell,\Omega)|p(z_{\ell-1},\Omega)\right] &= \frac{1}{N_p}\sum_{i=1}^{N_p} \Pr\left[p(z_\ell,\Omega)|\mathbf{\Phi}_i(z_\ell,\Omega)\right] \\ &\approx \frac{1}{N_p}\sum_{i=1}^{N_p} \mathcal{W}_i(z_\ell,\Omega) =: \overline{\mathcal{W}}(z_\ell,\Omega) \end{aligned} \tag{11.42}$$

where $\overline{\mathcal{W}}$ is the ensemble mean of the weight at each depth [35, 36].

Thus, the corresponding *sequential log-likelihood decision function* for the broadband source detection problem is

$$\Lambda(p(z_\ell,\Omega)) = \Lambda(p(z_{\ell-1},\Omega)) + \ln \overline{\mathcal{W}}(z_\ell,\Omega;\mathcal{H}_1) - \ln \overline{\mathcal{W}}(z_\ell,\Omega;\mathcal{H}_0) \tag{11.43}$$

This completes the development of the sequential detector that is applied to our simulated data.

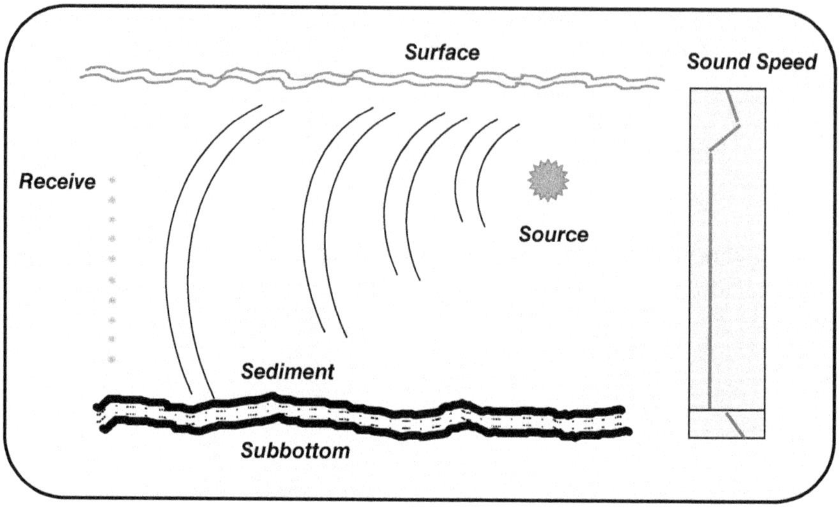

Ocean:	**flat bottom, three layers, 100 m water column, 2.5 m sediment, 100 element vertical array**
Source:	**50–100 Hz source at 10 km range and 50 m depth**

FIGURE 11.12 Shallow ocean environment problem: channel (100 m) with broadband (50 Hz) source located at range $r_s = 10\,\text{km}$ and depth $z_s = 50\,\text{m}$.

11.2.9 Broadband BSP Design

In this section, we discuss the application of the Bayesian processor to data synthesized by a broadband normal-mode model using the state–space forward propagator and the particle filter along with the sequential log-likelihood detector.

Let us consider a basic shallow water channel depicted in Fig. 11.12. We assume a flat bottom, range-independent three-layer environment with a channel depth of 100 m, a sediment depth of 2.5 m, and a subbottom. A vertical line array of 100 sensors with spacing of $\triangle z = 1\,\text{m}$ spans the entire water column and a broadband source of unit amplitude and 50 Hz bandwidth ranging from 50 to 100 Hz in 10 Hz increments is located at a depth of 50 m and a range of 10 km from the array. The sound-speed profile in the water column and the sediment are sketched in the figure and specified along with the other problem parameters. SNAP, a normal-mode propagation simulator [37] is applied to solve this shallow water problem and executed over the set of discrete temporal source frequencies. This boundary value problem was solved using SNAP and the results at each narrowband frequency are obtained. We note that as the temporal frequency increases, the *number* of modes increases thereby increasing the corresponding order of the state–space. Details of the problem parameters are given in Ref. [26].

The parameters obtained from SNAP are now used to construct the broadband state–space and measurement models of the previous subsection. Here, we use the set of horizontal wave numbers, $\{\kappa(m, q)\}, m = 1, \ldots, M_q; q = 1, \ldots, Q$, and sound speed

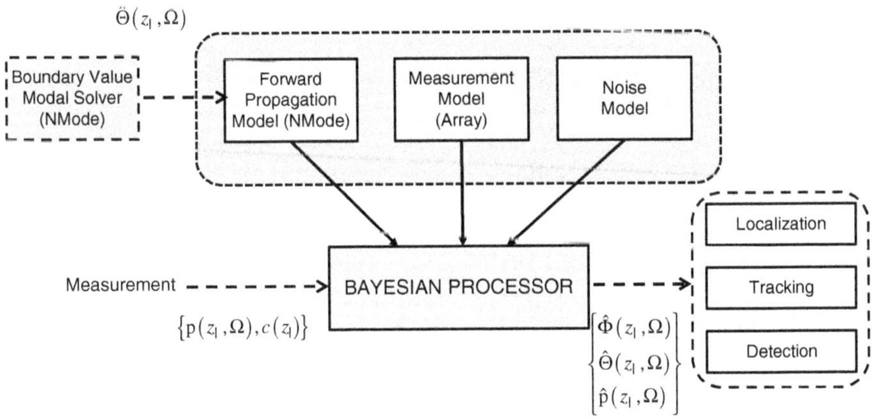

FIGURE 11.13 Bayesian processor design: boundary value solver provides initial state-space parameters for modal forward propagation, measurement, and noise/uncertainty models (state-space forward propagator) and data provide the inputs of pressure-field and sound speed. Bayesian processor outputs can be incorporated into potential ocean acoustic applications such as localization, tracking, and detection.

$\{c(z_\ell)\}$ to implement the state models along with the corresponding modal function values $\{\phi_{m1}(z_s, \omega_q)\}$, as well as the Hankel functions $\{\mathcal{H}_o(\kappa(m, q)r_s)\}$ to construct the measurement models. The final set of parameters for our simulation are the modal and measurement noise covariance matrices required by the Gauss–Markov model (see Ref. [38] for more details).

It is important to realize that the state–space "forward" propagators do *not* offer an alternative solution to the Helmholtz equation (not to be confused with a marching method), but rather use the parameters *from* the boundary value solution to obtain a set of initial conditions/parameters for the propagator construction. Even the adaptive model-based processors still utilize the boundary value solutions to "initialize" the processing [24] (see Fig. 11.13).

With this information available, a 100-member ensemble of stochastic simulations was performed at $\text{SNR}_{\text{in}} = 10\,\text{dB}$ (modes) and $\text{SNR}_{\text{out}} = 0\,\text{dB}$ (pressure-field) based on the SNAP parameters of Ref. [26].

11.2.9.1 Classical Performance Metrics Initially, we develop the processor for a single realization and perform some of the classical statistical tests to ensure the validity of the embedded model. In statistical signal processing, a standard statistical test to determine whether or not a processor is operating properly is whether the residual sequence (difference between the measurement and the processor "predicted" measurement) has a small mean-squared error (MSE). Specifically for sequential algorithms, the one-step prediction error (or residual) has the property that the prediction error, or equivalently the residual sequence, should be zero-mean and uncorrelated (or white). Pragmatically, residuals that satisfy these properties ensure that the embedded model-based processor has "removed" all correlation (of the model) from the data. The remaining measurement (residual) is uncorrelated or equivalently provides no further information about the signal (or model). Indeed this is the case, since the zero-mean test confirms that the residual mean to be smaller than the calculated bound ($8.9 \times 10^{-6} < 0.245$) and uncorrelated with only 1.6% of the correlation samples lying outside of the boundaries (5% or more are considered correlated) for our 125-*particle design*.

Since our data are nonstationary, the *weighted sum-squared residual* (WSSR) is a more reliable statistic to use to determine the "whiteness" of the residual (innovation) sequence [38]. Recall (see Section 5.7) that the WSSR uses this sequence to test whiteness by requiring that the constructed decision function lies below a threshold. If the WSSR statistic does lie beneath the calculated threshold, then theoretically, the estimator is considered tuned and the embedded model deemed appropriate. For this application, the WSSR statistic also confirms the whiteness with its decision function lying below the threshold of 81.5 with a 60-sample sliding window.

11.2.9.2 Ensemble Statistics In order to evaluate the processor performance, we construct an ensemble of 100-member realizations and calculate the ensemble statistics. The following results are based on ensemble average estimates, that is, all of the subsequent estimator results have been averaged over the 100-realizations to provide a more meaningful representation of "what could be expected" from the processor when implemented.

The results of the Bayesian design are shown in Fig. 11.14 where we see the enhanced pressure-field (MAP) and the true measurement (mean) as well as the

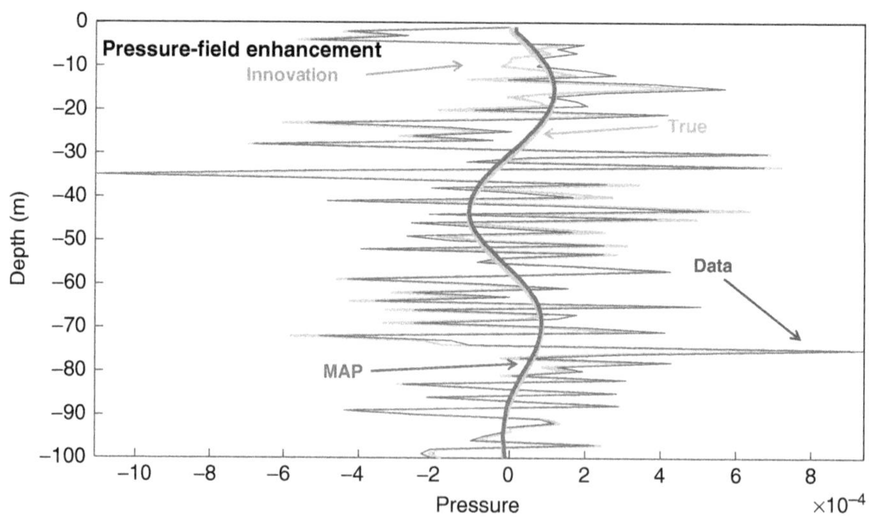

FIGURE 11.14 Broadband shallow water environment pressure-field enhancement: raw data (SNR = 0dB), particle filter (125-particles) estimates (MAP, CM) and residual errors (innovations).

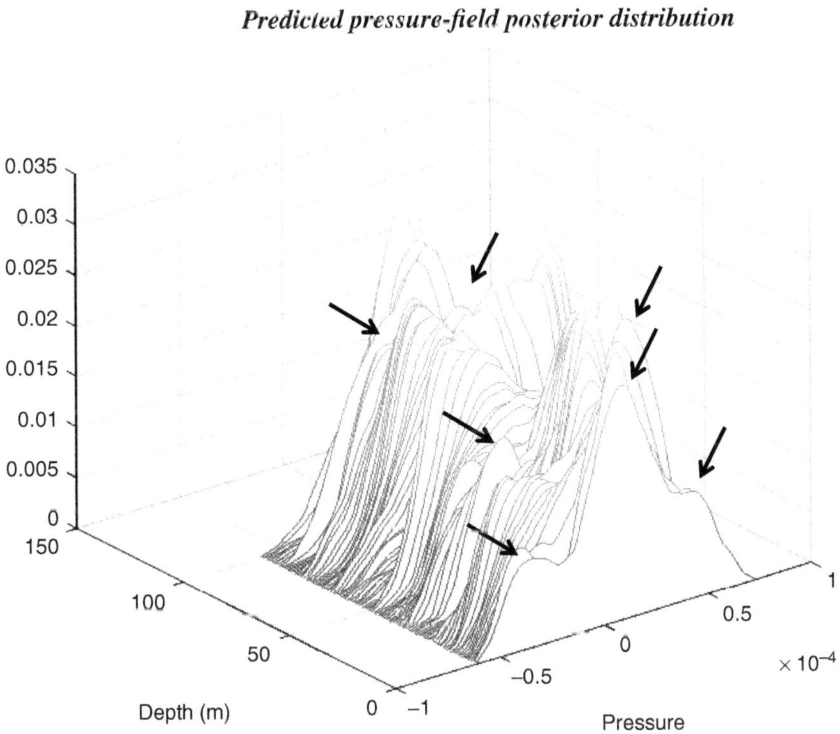

FIGURE 11.15 Predicted pressure-field multi-modal posterior distribution (125-particles) surface (depth vs. pressure vs. probability) indicating a multi-modal surface.

raw data and corresponding innovations sequence as a function of depth. Note from Eq. 11.21 that the modal estimates $\hat{\mathbf{\Phi}}(z_\ell, \omega_q)$, along with the measurement model $C^T(r_s, z_s, \omega_q)$, are used to construct the enhanced pressure-field $\hat{\mathbf{p}}(z_\ell, \omega_q)$ at each temporal frequency. To complete the performance analysis, we observe the posterior pressure-field distribution predicted by the PF at each depth in Fig. 11.15. Clearly this distribution is not unimodal, but both MAP/CM inferences "track" the mean (true) pressure-field quite well. We have also estimated the KD statistic which is shown in Fig. 11.19 a which will be discussed subsequently.

The estimated modal functions (ensemble averaged) extracted from the noisy pressure-field measurements are shown in Figs. 11.16 and 11.17—bounding the estimates we could hope to achieve for this type of data. We use the *annotation*: (Frequency/Mode No.), that is, $M50 - 1$ is the first modal function at 50 Hz frequency. Here we observe that the MAP and CM estimates inferred from the predicted posterior modal distributions essentially overlay one another and "track" the true (noise-free) functions very well in spite of the uncertainties in the synthesized modal data. In Fig. 11.16, we observe the enhanced modal function estimates (ensemble averaged) and uncertain data (DATA) corresponding to the two modes at 50 Hz, three at 60 Hz and

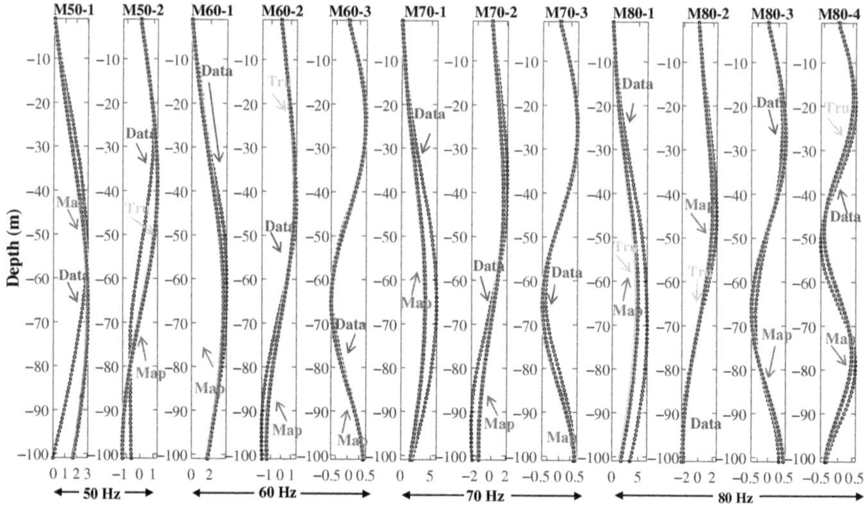

FIGURE 11.16 Broadband shallow water environment mode tracking: particle filter (125-particles) modal function estimation (MAP, CM) with true (True) and uncertain modes (Data) at 50–80 Hz frequencies.

70 Hz, and four at 80 Hz. In Fig. 11.17, we see the ensemble estimates for both 90 Hz (four modes) and 100 Hz (five modes). The RMS MSE estimates for each of the modal functions are shown in Table 11.1 where we observe small modal function tracking errors.

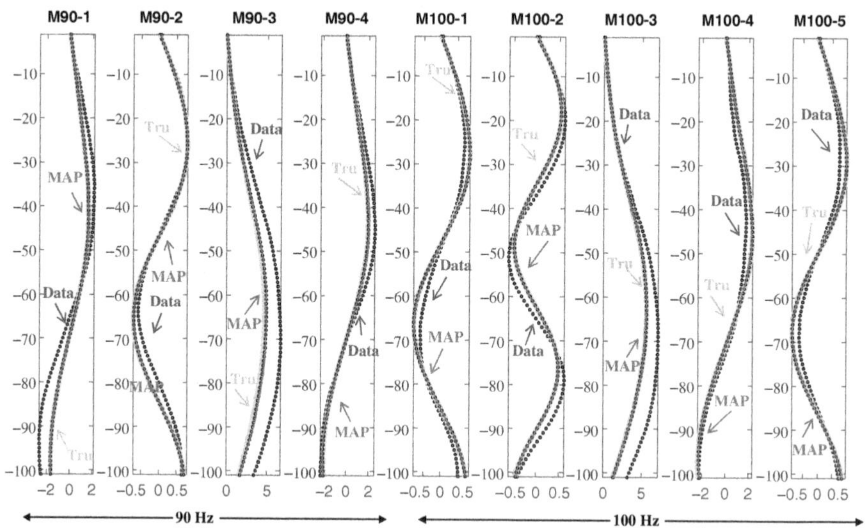

FIGURE 11.17 Broadband shallow water environment mode tracking: particle filter (125-particles) modal function (21) estimation (MAP, CM) with true (True) and uncertain modes (Data) at 90–100 Hz frequencies.

TABLE 11.1 Modal Parameters and Tracking Estimation

		Particle Filter Performance			
Frequency (Hz)	Mode No.	Modal Coeff.	Wave No.	MSE	KLD
50	1	0.122	0.207386	0.03244469	0.02333190
	2	−0.070	0.202489	0.06390081	0.00042730
60	1	0.125	0.249339	0.02057981	0.01498970
	2	−0.063	0.245106	0.02262004	0.00480695
	3	−0.097	0.237639	0.00725442	0.00045842
70	1	0.127	0.291255	0.1173259	0.01173030
	2	−0.057	0.287522	0.07374703	0.00414254
	3	−0.108	0.280907	0.01009431	0.00070735
80	1	0.129	0.333144	0.1376377	0.01020890
	2	−0.052	0.329802	0.03716592	0.00605280
	3	−0.113	0.323913	0.01584901	0.00058421
	4	0.087	0.315526	0.01425319	0.00067141
90	1	0.130	0.375015	0.1104416	0.00699367
	2	−0.047	0.371985	0.02928122	0.00476581
	3	−0.117	0.366685	0.01980555	0.00059496
	4	0.084	0.358997	0.00782307	0.00031112
100	1	0.131	0.416871	0.3176212	0.00648074
	2	−0.043	0.414098	0.06316895	0.00449904
	3	−0.119	0.409278	0.01011795	0.00079068
	4	0.079	0.402279	0.01544138	0.00031464
	5	0.081	0.393399	0.00450539	0.00019501

Thus, we see that modal function tracking of the broadband data presents another aspect of the PF estimates. We show the posterior PMF surfaces (depth vs. modal amplitude vs. probability) for a single realization of mode: $M50 - 1$ in Fig. 11.18. The surface indicates a multi-modal posterior PMF at various depth slices for this nonstationary process. This completes the modal analysis over the ensemble; next, we investigate the information theoretical results.

11.2.9.3 Performance Metrics In this section, we investigate the performance of the broadband PF over the 100-member ensemble of synthetic pressure-field measurements using the information metrics discussed in Section 7.6. The KD divergence of Eq. 11.33 provides us with a metric to determine "how close" the estimated posterior is to the true distribution. Since we have simulations available, we can generate deterministic or noise-free modal/pressure-field data that can be used to estimate the true distributions. The ensemble empirical PMF distributions provided by the PF can then be compared with the truth using the KD divergence criterion with approximate equality specified by values *close to zero*. As an illustration, we choose time slices of the pressure-field and modal function and estimate these PMFs for each as shown in Fig. 11.19 along with their corresponding KL metrics. We used a kernel density estimator (smoothed histogram) from each of the random sequence

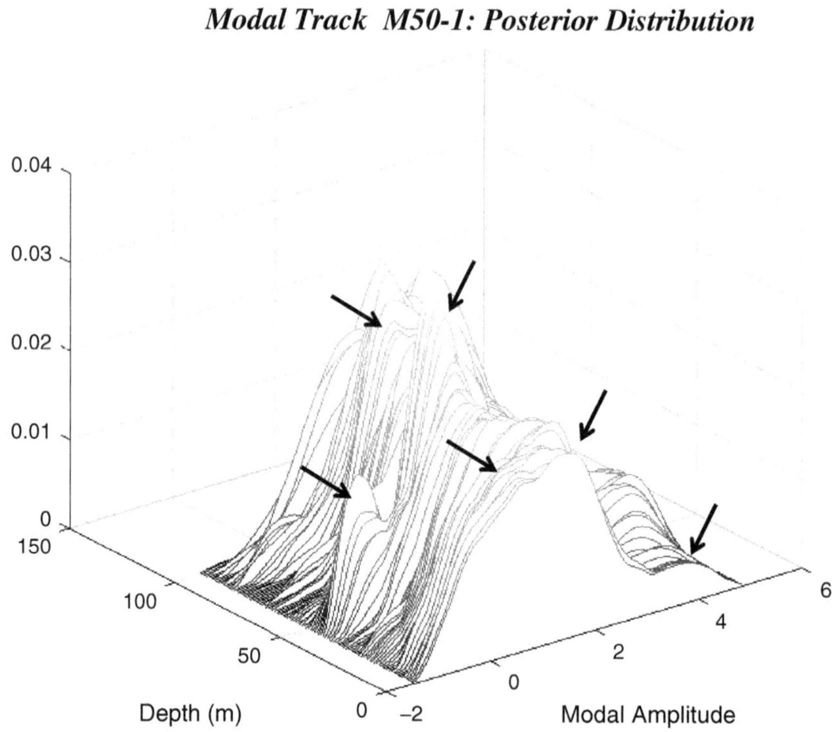

FIGURE 11.18 Estimated multi-modal (arrows) posterior modal distribution (125-particles) surface (depth vs. amplitude vs. probability) for mode: $M50 - 1$.

realizations for the raw synthesized pressure-field data, the true, and MAP and CM estimates.

For the pressure-field shown in Fig. 11.19a, it is clear that the broadband enhanced estimates are quite good compared to the noisy data as evidenced by estimated PMFs and respective the KL divergences of 0.25 (Data), 0.002 (MAP) and 0.0003 (CM). The $M50 - 1$ mode also confirms a converged posterior with KL divergences of 0.070 (Data), 0.021 (MAP), and 0.097 (CM).

Since the modal process is nonstationary in space (depth z), the multi-modal posterior varies at each discrete depth; therefore, we perform the above analysis on all 21 (42-state) modal functions and the pressure-field at each depth. We can employ the KL divergence metric, calculate it at each depth (PMF) providing a set of "KD trajectories" for the measurement and each mode summarizing the Bayesian processor overall performance. Note that the *perfect* PMF would yield a "zero" (trajectory) at every depth slice. Thus these trajectories estimate the KD for each PMF depth slice compressing the information into a single number with summary statistics (mean/median) to follow. The results of the KD divergence metric are shown in Fig. 11.20 comparing each of the multi-modal PMFs to the truth. We include the median

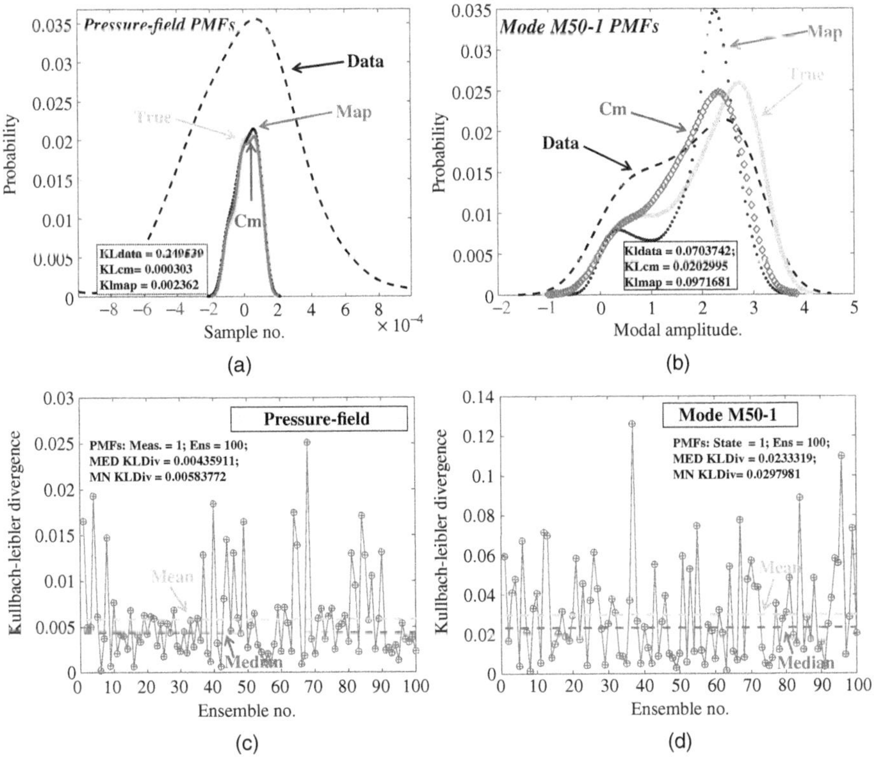

FIGURE 11.19 Multi-modal PMF data, true, MAP, CM estimates: (a) pressure-field (KD = 0.0003); (b) mode: $M50-1$ (KD = 0.02); (c) KL-divergence: pressure-field; (d) KL-divergence: mode: $M50-1$ (125-particles).

estimates as a single metric performance indicator as well as the corresponding MSE estimates in Table 11.1 for each mode.

Next, we apply the enhanced modal/pressure-field estimator as a pre-processor to solve the broadband source detection problem using the sequential detection approach. First, we generated data sets under each hypothesis and then estimate the log-likelihood decision functions for each. From these functions, we estimated the corresponding PMFs using a kernel density estimator of Section 3.2 as shown in Fig. 11.20a. From these distributions, we estimated the corresponding ROC curve of Fig. 11.20b and obtained the optimal operating point [32] which was estimated as $(P_{FA}, P_{DET}) = (0.034, 0.998)$, that is, 99.8% detection with a 3.4% false alarm rate for these SNRs. We calculated the required thresholds of Eq. 11.40 for the sequential source detector and applied it to the noisy (0 dB) pressure-field data shown in Fig. 11.14. The results for both the source present and absent scenarios are depicted in Fig. 11.20c and 11.20d indicating a detection in (c) and a non-detection (noise) in (d). These results indicate the applicability of this solution to broadband source detection problem and characterize the processor performance for these SNRs.

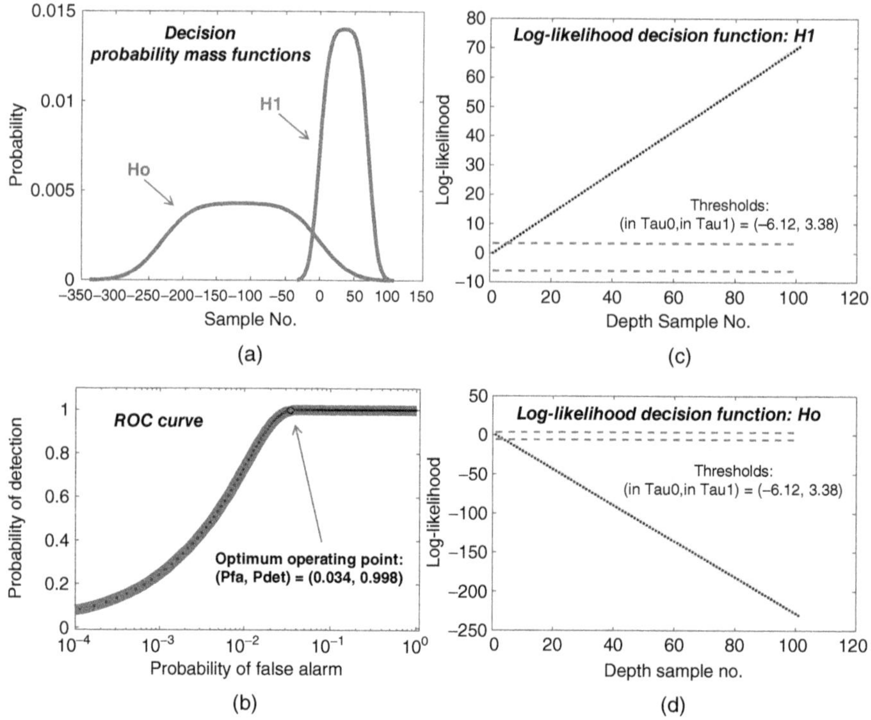

FIGURE 11.20 Sequential detection application: (a) log-likelihood PMF distributions; (b) ROC curve; (c) log-likelihood decision function under hypothesis $\mathcal{H}_1$: source detection; (d) log-likelihood decision function under hypothesis $\mathcal{H}_0$: noise.

11.2.10 Summary

In this study, we developed a sequential Bayesian solution to the broadband modal function tracking and pressure-field enhancement problem using a particle filter then coupled it to a sequential source detection scheme. We demonstrated the application of the Kullback–Leibler divergence statistic as a reliable metric to evaluate the particle filter performance for this nonstationary problem. Next, we developed a sequential detector that was capable of detecting the broadband source in noisy pressure-field measurements and evaluated its performance from the associated ROC curve. We showed the results of the design demonstrating the capability of such an approach.

11.3 BAYESIAN PROCESSING FOR BIOTHREATS

The design of a "smart" physics-based processor for microcantilever sensor arrays to detect various target species in solution based on the deflections of a functionalized array is discussed. A proof-of-concept design is demonstrated and shown to perform quite well on experimental data.

11.3.1 Background

Smart sensors with embedded processors offer unique advantages for applications that must gather large amounts of data and continuously monitor evolving conditions for potential changes (e.g., machine condition monitoring) or potential threats (e.g., biological, chemical, nuclear). Unfortunately, the usual processing techniques such as nonparametric methods like wavelets or parametric methods like autoregressive-moving average (ARMA) models do not capture the true essence of the problem physics required to extract the desired information, detect the change, or monitor the environment for threats. The underlying physical phenomenology governing the propagation physics is usually quite complex governed by nonlinearities typically characterized by nonlinear differential/difference equations. Coupling the resulting nonlinear processor to the sensor performing the measurement has not been considered a realistic possibility until now with the evolution of high-speed microcomputer chips that can easily be incorporated into the sensor design. We consider the design of an algorithm coupled to a microelectromechanical sensor (MEMS) to estimate the presence of critical materials or chemicals in solution.

Microcantilevers are powerful transducers for sensing inorganic, organic, and biological molecules, since they readily bend or deflect in the presence of a very small number of target molecules (nanomolar to femtomolar concentrations) [39] as shown in Fig. 11.21. The number of potential target chemicals is large, ranging from DNA [40] to explosives [41], implying that these sensors may be useful in defense, medicine, drug discovery, and environmental monitoring. Microcantilevers are capable of recognizing antibodies [42] and nerve agent products (hydrofluoric acid) in solution [43]. However, a major limitation of these sensors is that their

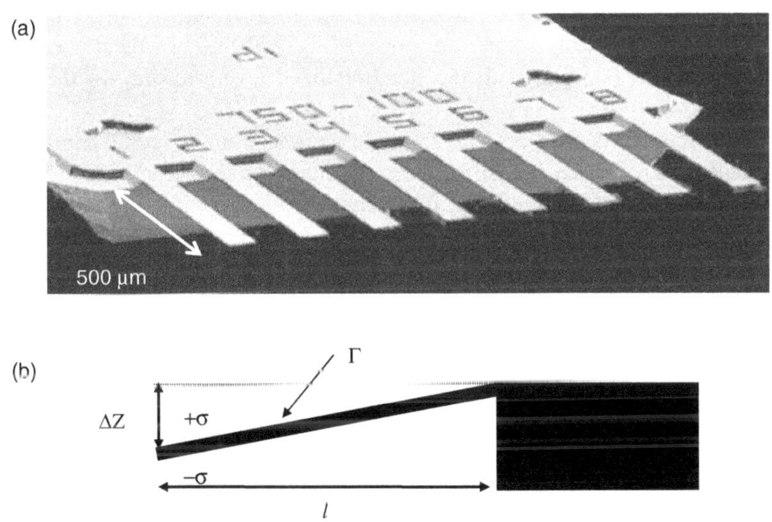

FIGURE 11.21 Micromachined cantilever array: (a) eight (8)-element lever array; (b) lever deflection.

signal-to-noise ratio (SNR) is low in many operational environments of interest. Therefore, we discuss the design of a "smart sensor" design combining the array with a physics-based processor to minimize its inherent limitations and maximize the output SNR for enhancement.

We investigate the physics-based Bayesian approach [38] to develop a multichannel processor evolving into a smart sensor for this application. This approach is essentially incorporating mathematical models of both physical phenomenology (chemistry/flow dynamics) and the measurement process (cantilever array including noise) into the processor to extract the desired information. In this way, the resulting Bayesian processor (BSP) enables the interpretation of results directly in terms of the problem physics.

We discuss the design of physics-based signal processing to micromachined cantilever measurement arrays to estimate the critical materials in solution. We briefly present the underlying physical phenomenology and reduce it to a simple model for processor development. Unknown parameters in this model are "fit" from independent experimental data. Once these parameters are estimated, we use minimum error variance techniques for the BSP design [38]. We then apply the resulting processor to experimental data, demonstrating the overall enhancement that would lead to an eventual "smart sensor" design. The resulting processor is based on nonlinear evolution equations leading to an extended Bayesian processor or classically, the extended Kalman filter, (EKF) that can be implemented in an on-line manner yielding the enhanced data as its output.

11.3.1.1 Microcantilever Sensors The dynamics of the fluids flowing over the cantilever array of Fig. 11.21 is influenced by two major factors: temperature and flow. The temperature is dependent on many variables and the dynamics are relatively slow, creating a disturbance to the cantilever sensor system. Flow in the medium associated with the array induces a stress along with the chemical forces created by the molecules bonding with the functionalized levers leading to the measured deflection.

Micromachined cantilevers can function as detection devices when one side is fabricated to be chemically distinct from the other, as shown in Fig. 11.21b. Functionalization can be accomplished, for example, by evaporating a thin (10's of nm) film of metal such as gold on the top of the chip, then immersing the cantilever chip in a "probe" chemical that will bind preferentially to the thin gold film. The lever acts as a sensor when it is exposed to a second "target" chemical that reacts with the probe, since the reaction causes a free energy change that induces stress at the cantilever surface. Differential surface stress, $\triangle\sigma$, in turn, induces a deflection of the cantilever that can be measured optically or electronically. In this section, we describe results of experiments with gold-coated cantilevers exposed to 2-mercaptoethanol, a small sulfur-terminated molecule with high affinity for gold.

When the microcantilever is immersed in a fluid and it has been functionalized to attract the target molecules, the changes in surface stress can be predicted as a function of surface loading. The evolution dynamics of this chemical interaction is captured by the well-known Langmuir kinetics in terms of a set of ordinary nonlinear differential equations. Rather than propagate these equations directly, we choose to use their solution in our processor that accounts for the adsorption–desorption kinetics. We developed an approximation to the Langmuir evolution equations based on a stirred tank reactor to estimate the target concentration as a function of time under continuous flow conditions. Experimentally, the applied chemical input signal is a constant concentration initiated by a step (function) increase at time $t_{\rm ON}$ and terminated at time $t_{\rm OFF}$ [44, 45].

Based on this representation of the process evolution physics, the dynamic (normalized) surface concentration of the interacting molecules on the surface of the cantilever, $\Gamma(t) = \frac{\tilde{\Gamma}(t)}{\tilde{\Gamma}_{MAX}}$, is given by the relations

$$\Gamma(t) = \begin{cases} 0 & t < t_{\rm ON} \\ \left(\frac{c(t)}{c(t)+k_a/k_d}\right)\{1 - \exp[-(k_a c(t) + k_d)(t - t_{ON})]\} & t_{\rm ON} \le t \le t_{\rm OFF} \\ \sqrt{\frac{1}{2k_d(t-t_{OFF})}} & t > t_{\rm OFF} \end{cases} \tag{11.44}$$

where k_a, k_d and $c(t)$ are the respective adsorption rate constant $[{\rm M}]^{-1}{\rm s}^{-1}$, desorption rate constant $({\rm cm}^{-2}{\rm M}^{-1}{\rm s}^{-1})$, and bulk concentration of the target molecules in solution (moles/liter) with $\tilde{\Gamma}_{MAX}$, the maximum surface concentration of the species of interest $({\rm cm}^{-2}{\rm M}^{-1})$.

The total free energy change of the cantilever surface, $\triangle G$, is related to the surface stress difference $\triangle\sigma$, between top and bottom side of the cantilever by:

$$\triangle\sigma(t) = \triangle G(t)\Gamma(t)/M_{\rm A} \tag{11.45}$$

where $\triangle G$ has the unit J/mole and is the change in the sum of all of the contributions to the free energy of the surface of the cantilever with $M_{\rm A}$ as Avogadro's number (constant). The differential surface stress in the cantilever induces a chemically induced deflection $\Delta z^C(t)$, using a variant of Stoney's equation [44] which implies that the deflection of the cantilever is directly proportional to the difference in surface stress (signal) on the cantilever surface relating this stress difference to the surface coverage and free energy of absorption, that is,

$$\Delta z^C(t) = \beta\Delta\sigma(t) \quad \text{for} \quad \beta = \frac{3\ell^2(1-\nu)}{E\delta^2} \tag{11.46}$$

where E is the Young's modulus, ν is the Poisson's ratio, and ℓ and δ are the cantilever length and thickness, respectively. This model can be used to predict changes in surface stress as a function of surface loading.

The measurement model is more complicated, since it is the superposition of both the chemical and temperature deflection phenomena

$$y_\ell(t) = \Delta z^C_\ell(t) + \Delta z^T(t) \quad \text{for} \quad \ell = 1, \ldots, L \tag{11.47}$$

where Δz_ℓ^C is the chemical deflection, different for different cantilevers, and $\Delta z^T(t)$ is the thermal deflection, assumed to be the same for all cantilevers. Since we have approximated the physics developing the well-founded formulation of Eq. 11.44, we know that there is uncertainty in both the measurements (noise) and the model parameters. Therefore, we cast the problem into a Gauss–Markov (*GM*) state–space framework [38] representing these uncertainties as additive Gaussian processes, that is,

$$\begin{aligned} \Delta G(t) &= \Delta G(t-1) + w(t-1) && \text{[State]} \\ y_\ell(t) &= \beta_\ell \Gamma(t;\Theta)\Delta G(t) + \Delta z^T(t) + v_\ell(t) && \text{[Measurements]} \end{aligned} \tag{11.48}$$

where $\ell = 1, \ldots, L$; $\Gamma(t;\Theta)$ is given by Eq. 11.44 with unknown model parameters defined by the vector, $\Theta = [k_a k_d \tilde{\Gamma}_{\text{MAX}}]$ and free energy (state) modeled as a random walk ($\triangle\dot{G}(t) = 0$). This representation, therefore, creates the foundation for our physics-based processor design. The process uncertainty is modeled by w and the corresponding measurement uncertainty as v, both zero-mean Gaussian with respective covariances R_{ww} and R_{vv}. With this representation in mind, the cantilever signal enhancement problem is defined as:

GIVEN a set of noisy deflection measurements $\{y_\ell(t)\}$ with known bulk concentration inputs $\{c(t)\}$ and unknown parameters Θ, FIND the best (minimum error variance) estimate of the deflection $\hat{y}(t|t-1)$, that is, the conditional mean at t based on the data up to time $t-1$.

The design of the processor for this problem is illustrated in Fig. 11.22. After the cantilever physics model is developed, it is used (1) to extract the required parameters using a physics-based parameter estimator; (2) to synthesize "data" for the initial processor designs; and (3) to enhance the noisy measurements being incorporated into the final BSP structure. In the figure, we see that the complex mathematical model of Eq. 11.44 (dashed box) is used to perform the physics-based parameter estimation using independent experimental data to extract the required parameters (adsorption/desorption rate constants and maximum concentration) as well as initially simulate data for BP design studies, once these parameters are extracted. The actual experimental data replace the synthesized and are used to validate the processor performance.

11.3.2 Parameter Estimation

The basic approach is to first estimate the model parameters Θ (off-line) from an independent set of deflection measurements, and then, incorporate them into the BSP to enhance the experimental (proof-of-concept) data. That is, we extract the critical absorption, desorption rate constants and maximum concentration parameters for each channel as: $\Theta_\ell = [k_a(\ell) k_d(\ell) \tilde{\Gamma}_{MAX}(\ell)]$. The parameter estimator employed was

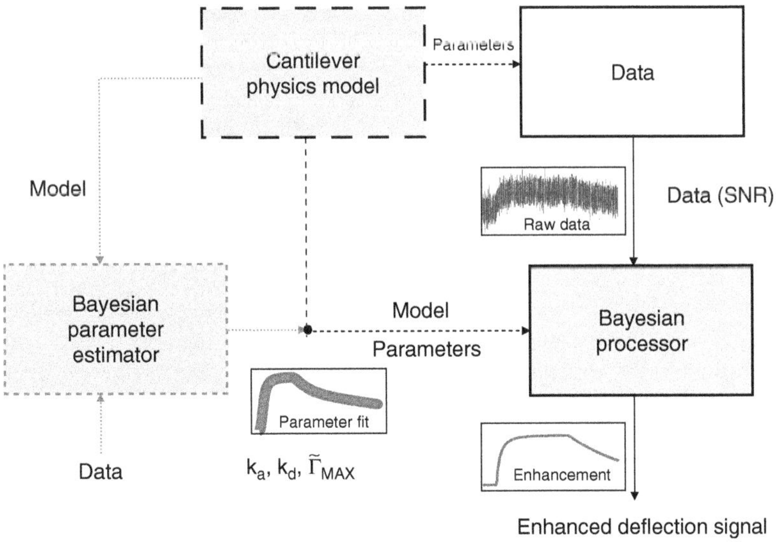

FIGURE 11.22 Physics-based approach to microcantilever "smart sensor" design: physics evolution model, parameter estimation (off-line), Bayesian processor and data enhancement.

a nonlinear least-squares solution using the Nelder–Meade polytope search algorithm [46]. This algorithm is based on minimizing

$$\min_{\Theta_\ell} J(\Theta) = \sum_{t=1}^{N_t} \epsilon_\ell^2(t;\Theta) \quad \text{for} \quad \epsilon(t;\Theta) := y_\ell(t) - \hat{y}_\ell(t;\Theta) \tag{11.49}$$

where the estimated cantilever measurement at the ℓth-lever is given by

$$\hat{y}_\ell(t;\hat{\Theta}) = \Delta\hat{z}_\ell^C(t;\hat{\Theta}) + \Delta\hat{z}^T(t) = \beta_\ell\Gamma(t;\hat{\Theta})\Delta G(t) + \Delta\hat{z}^T(t) \tag{11.50}$$

We executed this estimator on raw experimental deflection data and estimated the parameters for each lever. The extracted parameters reasonably predicted the filtered cantilever response and the resulting error was uncorrelated as discussed below.

11.3.3 Bayesian Processor Design

Next, using these estimated parameters, we developed BSP for the multichannel deflection data. The GM model was used to synthesize the multichannel data and provide known truth data to "tune" or adjust the processor noise covariance matrices that provide the "knobs" for BSP design (see Fig. 11.11 and Ref. [38] for details). Since the simulation model and that used in the processor are identical, optimal (minimum error variance) performance is achieved (zero-mean, uncorrelated errors) providing the starting point for application to the experimental data. The final (simplified) BSP algorithm (assuming the parameters $\hat{\Theta}$ have been estimated) is shown

TABLE 11.2 Bayesian Microcantilever Array Processor (BSP)

$\Delta\hat{G}(t\|t-1) = \Delta\hat{G}(t-1\|t-1)$	[Free energy prediction]
$\hat{\mathbf{y}}_\ell(t\|t-1) = \beta_\ell\Gamma(t;\hat{\Theta})\Delta\hat{G}(t\|t-1) + \Delta\hat{z}^T(t)$	[Deflection prediction]
$\boldsymbol{\varepsilon}_\ell(t) = \mathbf{y}_\ell(t) - \hat{\mathbf{y}}_\ell(t\|t-1)$	[Innovation or residual]
$\Delta\hat{G}(t\|t) = \Delta\hat{G}(t\|t-1) + \mathbf{k}'(t)\boldsymbol{\varepsilon}(t)$	[Free energy correction]

where $\mathbf{k}$ is the corresponding weight or gain and $\hat{\Theta}$ is the output estimate from the parameter estimator.

in Table 11.2. Here, we see that the algorithm has a classical predictor-corrector form where the prediction estimates (conditional mean) the free energy, filters the measurement, estimates the innovation, and corrects the final (filtered) free energy estimate.

11.3.4 Results

First, we took the filtered measurement signals, *averaged* them to a single measurement, fitted the parameters using an off-line optimization technique [46], and used the parameters in the Bayesian processor. The results are shown in Fig. 11.23 where

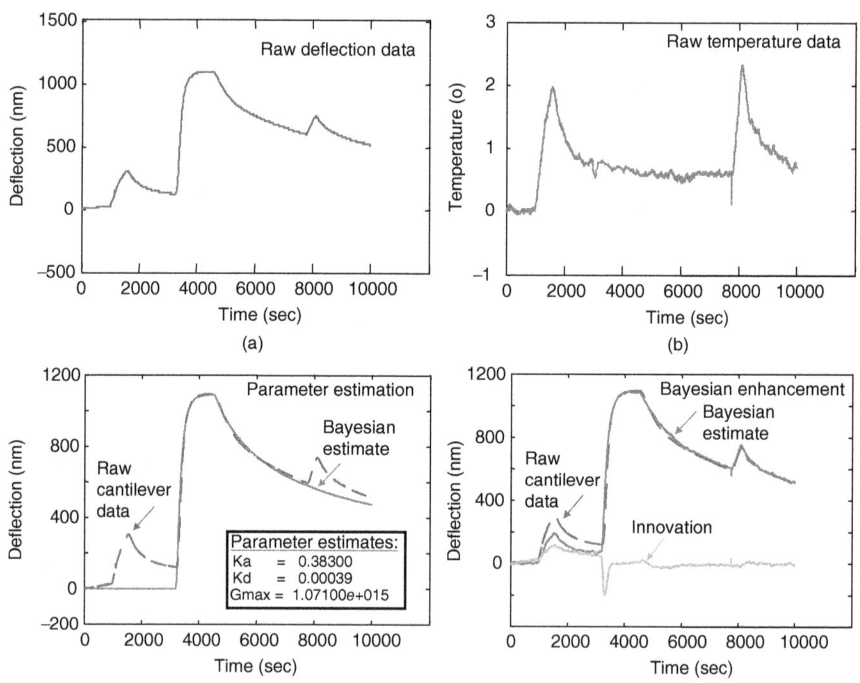

FIGURE 11.23 Bayesian processing for the "average" microcantilever sensor array signal: (a) raw deflection data; (b) raw temperature data; (c) parameter estimation; (d) Bayesian estimation (enhancement).

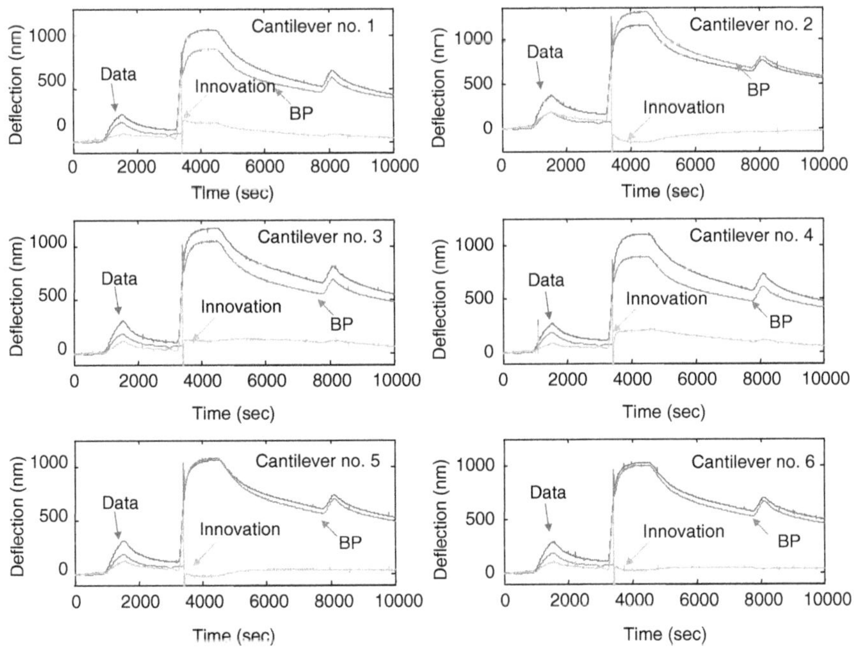

FIGURE 11.24 Bayesian processing of microcantilever experimental (proof-of-concept) data: raw data, enhanced (BP) deflection measurement and residual results for each lever.

we see the raw deflection and temperature measurements in 11.23a and 11.23b. The parameter estimator results are shown in Fig. 11.23c along with the state (signal) estimator in 11.23d. The parametric fit is quite reasonable as is the signal enhancement. However, it is clear from the innovations that the optimal processor is clearly not Gaussian, since it is not zero-mean or white. Next, we used the BP with the free energy as our piecewise constant parameter (state) and the nonlinear cantilever array model with 6 elements along with a filtered estimate of the temperature profile in the processor $\triangle\hat{z}^T(t)$. By tuning the measurement noise covariance parameters (R_{vv}), we demonstrate that the BP is capable of tracking the noisy cantilever deflection data reasonably well; however, the performance is again suboptimal, since the innovations (shown in each figure), although quite small, are not uncorrelated. The results are shown in Fig. 11.24 where we see the raw measured cantilever data, Bayesian processor estimates and the corresponding residual errors or innovations. The results are quite reasonable except for the systematic bias error in the processor output at each lever. The bias is created by out lack of knowledge of the initial concentration input and can easily be compensated (gain) at each lever as well. The dynamics appear to be captured by the model especially in cantilever 5. From the figure, we note that the dynamics of the individual levers (on-set and off-sets) are quite close to the expected. This design demonstrates that even complex physical systems can be incorporated into physics-based processors enabling the development of a "smart" sensor.

11.4 BAYESIAN PROCESSING FOR THE DETECTION OF RADIOACTIVE SOURCES

With the increase in terrorist activities throughout the world, the need to develop techniques capable of detecting radioactive sources in a timely manner is a critical requirement. The development of Bayesian processors for the detection of contraband stems from the fact that the posterior distribution is clearly multi-modal eliminating the usual Gaussian-based processors. Here, we represent a radionuclide as a unique superposition (union) of monoenergetic sources that are transported through the path for measurement and counting as illustrated in Fig. 11.25. The measured data consists of a low-count, impulsive-like, time series measurement (energy vs. time) in the form of an *event mode sequence* (EMS) obtained from the electronics. We use this representation to formulate and solve the radionuclide contraband detection problem [47–51]. That is, detection is used to *decide* whether or not a target (threat) is present.

Radionuclide source detection is a critical technology to detect the transportation of illicit radiological materials by potential terrorists. Detection of these materials is particularly difficult due to the inherent low-count emissions produced. These emissions result when sources are shielded to disguise their existence or, when being transported, are in relative motion with respect to the sensors. Some work has been accomplished on this problem [52–55]. Here, we model the source radionuclides by decomposing them uniquely as a superposition (union) of monoenergetic sources. Each γ-ray emitted is then smeared and distorted as it is transported on a path to the output of the detector for measurement and counting. We start with the "physics-based" approach to solving this suite of problems and then discuss the measurement system employed to detect γ-rays and show how the mononenergetic approach leads to a compound Poisson-driven Markov process [47] which is amplified, shaped and digitized for further processing. The processor is developed using state–space representations of the transition probability and associated likelihood and we apply it to synthesized data to evaluate its performance.

11.4.1 Physics-Based Processing Model

Gamma-ray interactions are subject to the usual physical interaction constraints of scattering and attenuation as well as uncertainties intrinsic to the detection

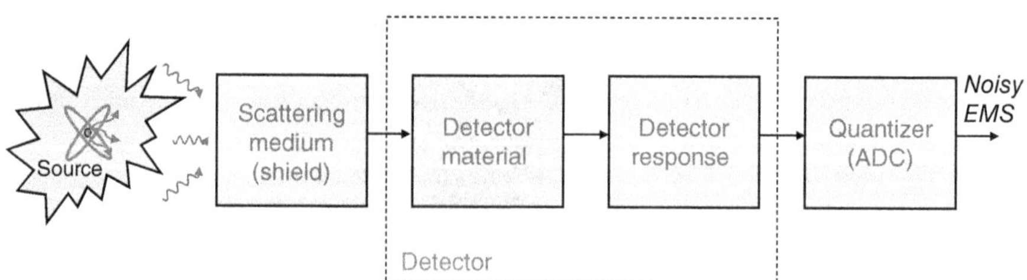

FIGURE 11.25 Gamma-ray evolution and measurement: radionuclide source (EMS), medium transport (physics), detector material interaction, detector temporal response (preamplification/pulse shaping) and A/D conversion with quantization noise.

process. Energy detectors are designed to estimate the γ-ray energy from the measured electron current. A typical detector is plagued with a variety of extraneous measurement uncertainty that creates inaccuracy and spreading of the measured current impulse (and therefore γ-ray energy). The evolution of a γ-ray as it is transported through the medium and interacts with materials, shield, and the detector is shown in Fig. 11.25.

The unique characterization of a radionuclide based on its electromagnetic emissions has been an intense area of research and development for a long period of time [48–51]. It is well known that a particular radionuclide can be uniquely characterized by the energies and relative intensities of the emitted γ-rays. Mathematically, we define the pair $[\{\epsilon_m\}, \{\lambda_m\}]$ as the respective energy (MeV) and photon detection rate of the mth γ-ray line of the radionuclide. In the absence of shielding, the detection rate λ_m is the product of the absolute decay rate per unit mass σ, the mass $\mathcal{M}$, the absolute intensity per decay α_m, a geometric factor S (typically the solid angle subtended by the detector), and a detector efficiency η_m at energy ϵ_m, that is, $\lambda_m = \eta_m \sigma \mathcal{M} \alpha_m S$.

The basic construct of a γ-ray detector starts with the material selected. We concentrate on semiconductor materials, primarily HPGe, since it provides outstanding performance enabling us to establish an upper bound for our experimental results. The detector typically outputs a list of *events*, each consisting of an event time and an amplitude (pulse height or energy). After a set time, a histogram of number of events versus pulse height (*pulse height spectrum* or PHS) is calculated. In the absence of shielding, a particular radionuclide is revealed by the presence of peaks in the PHS at the expected energies (γ-ray lines) for that nuclide. The areas of the peaks are proportional to the intensities (detection rates). The widths of the peaks are defined by the detector resolution and efficiency.

For sequential detection, we process each event in the event mode sequence individually using a model with parameters that are analogous to the features in a PHS. Since a radionuclide emits γ-rays at specific energies and rates, the sequence of emitted photons can be characterized by sets of measured energies and arrival times at the detector: $\left[\{\epsilon_m\}, \{\tau_m\}\right]$, $m = 1, \ldots, M_\epsilon$. The index m represents the *mth* γ-ray line (energy) for a radionuclide with M_ϵ lines. It is convenient to think of each line as a separate source. Then the emission from a given radionuclide is represented as a superposition of individual γ-ray sources. We will refer to this as the *monoenergetic decomposition* of the EMS for the radionuclide. Fig. 11.26 shows a graphical depiction of the monoenergetic decomposition of an EMS with three γ-ray lines.

To be more precise, define $\xi(n; \epsilon_m(n), \tau_m(n))$, a component of an EMS, as the nth measured photon arrival from the mth monoenergetic source of *energy* ϵ_m, at *arrival time* $\tau_m(n)$ with associated *detection rate* λ_m. Though the source energy may be ϵ_m, the finite resolution of the detector introduces a random component to the measured energy. Thus, the energy of the nth arrival is more accurately represented as a random variable $\epsilon_m(n)$. The resulting impulse representation for a single photon arrival is $\xi(n; \epsilon_m, \tau_m) = \epsilon_m(n)\delta(t - \tau_m(n))$. By photon arrival, we refer only to the arrival of photons that are measured by the detector and contribute to the EMS. Thus terms like *arrival time*, *interarrival time*, and *arrival rate* refer to photons that contribute to the EMS output of the detector.

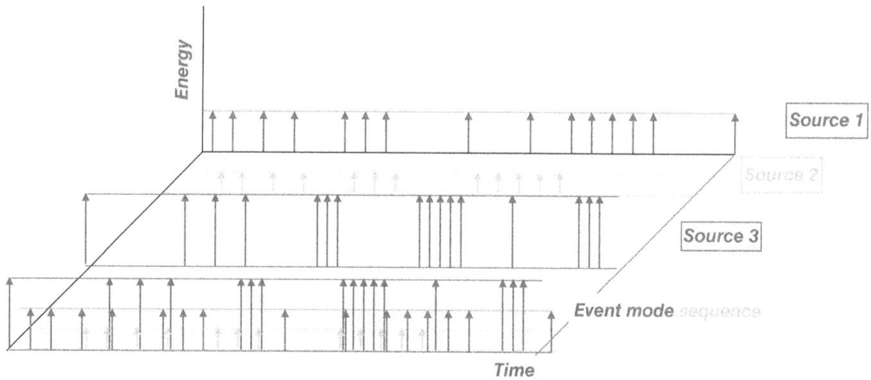

FIGURE 11.26 Monoenergetic source decomposition: individual source constituent EMS from ideal composite (superposition).

In order to define the entire emission sequence over a specified time interval, $[t_o, T)$, we introduce the set notation $\tilde{\tau}_m := \{\tau_m(1) \cdots \tau_m(N_\epsilon(m))\}$ and $\tilde{\epsilon}_m := \{\epsilon_m(1) \cdots \epsilon_m(N_\epsilon(m))\}$ with $N_\epsilon(m)$ the total number of *counts* for the mth source in the interval. Therefore, $\xi(n; \tilde{\epsilon}_m, \tilde{\tau}_m)$ represents an impulse train of random energies and arrivals up to the nth arrival (see Fig. 11.26).

The *interarrival* time is defined by $\triangle\tau_m(n) = \tau_m(n) - \tau_m(n-1)$ for $\triangle\tau_m(0) = t_o$ with the corresponding set definition (above) of $\triangle\tilde{\tau}_m(n)$. Using this definition (and with a slight abuse of notation), we can write the photon arrival as $\xi(n; \epsilon_m, \tau_m) \Rightarrow \xi(n; \epsilon_m, \triangle\tau_m) = \epsilon_m(n)\delta(t - \tau_m(n))$, where it is understood that $\tau_m(n) = \tau_m(n-1) + \triangle\tau_m(n)$. We can then rewrite the EMS in terms of interarrivals just as easily as arrival times, that is, the *monoenergetic source representation* of a radionuclide characterized by its unique set of energy/interarrival pairs $\{\epsilon_m, \triangle\tau_m\}$ is given by

$$\xi(N_\epsilon(m); \tilde{\epsilon}_m, \triangle\tilde{\tau}_m) \sum_{n=1}^{N_\epsilon(m)} \epsilon_m(n)\delta(t - \tau_m(n)) \text{ at detection rate } \lambda_m(n) \tag{11.51}$$

for t_o known.

Suppose we have a radionuclide source whose EMS is decomposed into its M_ϵ monoenergetic source components $\xi(n; \tilde{\epsilon}_m, \triangle\tilde{\tau}_m)$. The full EMS is given by the union of the components: $\cup_{m=1}^{M_\epsilon} \xi(n; \tilde{\epsilon}_m, \triangle\tilde{\tau}_m) \Leftrightarrow \sum_{m=1}^{M_\epsilon} \xi(n; \tilde{\epsilon}_m, \triangle\tilde{\tau}_m)$ where the last equivalence results from the pragmatic assumption that it is highly *improbable* that any two arrivals will overlap. Thus, it follows that a radionuclide (RN) can be represented in terms of its monoenergetic decomposition, that is, the EMS monoenergetic source representation is:

$$\mathcal{R}(N; \underline{\epsilon}, \triangle\underline{\tau}) = \sum_{m=1}^{M_\epsilon} \sum_{n=1}^{N_\epsilon(m)} \epsilon_m(n)\delta(t - \tau_m(n)) \tag{11.52}$$

for $\underline{\epsilon} := \{\tilde{\epsilon}_1, \ldots, \tilde{\epsilon}_{M_\epsilon}\}$, the complete set of energies composing $\mathcal{R}$ along with $\triangle\underline{\tau} := \{\triangle\tilde{\tau}_1, \ldots, \triangle\tilde{\tau}_{M_\epsilon}\}$, the corresponding set of interarrival times. The arrival index N is the least upper bound of the set of $N_\epsilon(m)$.

The emission of photons follows a well-defined probability structure, that is, since only one photon is emitted for each event there is a fixed probability (absolute intensity α_m) that the photon is emitted with energy ϵ_m out of M_ϵ possibilities. The probabilities for a given radionuclide are specified in its energy decay scheme [49]. Therefore, we model this decay structure by incorporating an *indicator function* defined by Refs. [34] and [56]: $\mathcal{I}_j(m) = 1m = j$ where $\mathcal{I}_j(m)$ is a random variable such that $\Pr(\mathcal{I}_j(m) = 1|\mathcal{R}(n; \underline{\epsilon}, \underline{\triangle\tau})) = \Pr(\mathcal{I}_j(m) = 1|\Xi_n) = \alpha_j$ for α_j the corresponding absolute intensity (emission probability) of the jth monoenergetic radionuclide component and $\Xi_n := \{\xi(1), \ldots, \xi(n)\}$, the set of n-EMS arrival measurements. Using the indicator function, we can write the jth monoenergetic component of the EMS as

$$\mathcal{R}_j(N; \underline{\epsilon}, \triangle\underline{\tau}) = \sum_{m=1}^{M_\epsilon} \sum_{n=1}^{N_\epsilon(m)} \mathcal{I}_j(m)\epsilon_m(n)\delta(t - \tau_m(n)) \tag{11.53}$$

where we model the energy variations as Gaussian, $\epsilon_m \sim \mathcal{N}(\overline{\epsilon}_m, \sigma^2_{\epsilon_m})$ the interarrivals as exponential, $\triangle\tau_m \sim \mathcal{E}(\alpha_m\lambda_{\triangle\tau_m}\triangle\tau_m(n))$ [49] with the emission probability as α_m.

Thus, this unique physics-based representation provides the basic signal model for subsequent processing.

11.4.2 Radionuclide Detection

Since all of the measurement data and required parameters evolve from the EMS, we are in search of a technique that enables us to "decide" when a particular target radionuclide is present or not. After the single photon is pre-processed by the acquisition system, the energy and arrival time measurements are passed onto the energy/rate discriminators to "decide" on the photon's status (accept or reject). If acceptable, the parameter estimates are sequentially updated and provided as input to the decision function for detection (identification). Thus, the decomposition implies that each unique energy/interarrival component of the target radionuclide be processed individually in a separate channel.

In order to develop a sequential processor [33,57], we must test the binary hypothesis that the measured EMS has evolved from the targeted RN of Eq. 11.53. Therefore, we specify the hypothesis test,

$$\begin{aligned} \mathcal{H}_0 &: \xi(n; \underline{\epsilon}, \triangle\underline{\tau}) = \mathcal{R}(n; \underline{\epsilon}, \triangle\underline{\tau}) + v(n) \quad \text{[Nontarget]} \\ \mathcal{H}_1 &: \xi(n; \underline{\epsilon}, \triangle\underline{\tau}) = \mathcal{R}(n; \underline{\epsilon}^t, \triangle\underline{\tau}^t) + v(n) \quad \text{[Target]} \end{aligned} \tag{11.54}$$

where $\mathcal{R}(n; \underline{\epsilon}, \triangle\underline{\tau})$ is the EMS model, $\xi(n; \underline{\epsilon}, \triangle\underline{\tau})$ is the random composite EMS contaminated with zero-mean, Gaussian measurement (instrumentation) noise,

$v \sim \mathcal{N}(0, \sigma_v^2)$ and the superscript notation "t" represents the *true* or *target* value obtained from the Tables.

The optimal solution to this binary decision problem is based on applying the *Neyman–Pearson theorem* leading to the likelihood ratio [33, 57] (see Section 10.3). Since the distributions under investigation are members of the exponential family [34], taking logarithms simplifies the computations. The *sequential log-likelihood ratio* is

$$\Lambda[\Xi_n] = \Lambda[\Xi_{n-1}] + \ln \Pr(\xi(n; \underline{\epsilon}, \triangle\underline{\tau})|\Xi_{n-1}, \mathcal{H}_1) \\ - \ln \Pr(\xi(n; \underline{\epsilon}, \triangle\underline{\tau})|\Xi_{n-1}, \mathcal{H}_0) \tag{11.55}$$

and therefore, the Wald *sequential probability-ratio test* becomes [33]

$$\begin{array}{lll} \Lambda[\Xi_n] & > \ln \mathcal{T}_1(n) & \text{Accept} \quad \mathcal{H}_1 \\ \ln \mathcal{T}_0(n) \leq \Lambda[\Xi_n] & \leq \ln \mathcal{T}_1(n) & \text{Continue} \\ \Lambda[\Xi_n] & < \ln \mathcal{T}_0(n) & \text{Accept} \quad \mathcal{H}_0 \end{array} \tag{11.56}$$

where the thresholds are specified in terms of the false alarm ($\mathrm{P_{FA}}$) and miss ($\mathrm{P_{Miss}}$) probabilities as

$$\mathcal{T}_0(n) = \frac{\mathrm{P_{Miss}}(n)}{\mathrm{P_{FA}}(n)} \qquad \mathcal{T}_1(n) = \frac{1 - \mathrm{P_{Miss}}(n)}{\mathrm{P_{FA}}(n)}$$

The thresholds are determined from a receiver operating characteristic (ROC) curve (detection versus false alarm probabilities) for each radionuclide decision function. An operating point is selected from the ROC corresponding to specific detection and false-alarm probabilities and therefore specifying the required thresholds which are calculated according to Eq. 11.56 for each radionuclide. So we see that at each photon arrival (at n), we *sequentially update* the likelihood and thresholds to perform the detection—"photon-by-photon".

To implement the sequential detection approach, the required distributions of Eq. 11.55 must be specified in order to calculate the decision function; therefore, we have

$$\Pr(\xi(n; \underline{\epsilon}, \triangle\underline{\tau})|\Xi_{n-1}, \mathcal{H}_\ell) = \\ \Pr(\mathcal{R}(n; \underline{\epsilon}, \triangle\underline{\tau})|\Xi_{n-1}, \mathcal{H}_\ell) + \Pr(v(n)|\Xi_{n-1}, \mathcal{H}_\ell) \tag{11.57}$$

for the hypotheses specified by $\mathcal{H}_\ell; \ell = 0, 1$ (see Eq. 11.54).

We start by estimating the RN posterior distribution (or its equivalent) from the model parameters and uncertain data, that is,

$$\Pr\left(\mathcal{R}(n; \underline{\epsilon}, \triangle\underline{\tau})|\Xi_{n-1}, \mathcal{H}_\ell\right) \Leftrightarrow \Pr\left(\underline{\epsilon}, \triangle\underline{\tau}, \mathcal{I}_j(m)|\Xi_{n-1}, \mathcal{H}_\ell\right)$$

Using the monoenergetic radionuclide model, we have

$$\begin{aligned}&\Pr(\xi(n;\underline{\epsilon},\triangle\underline{\tau})|\Xi_{n-1},\mathcal{H}_\ell) =\\ &\Pr(\underline{\epsilon}(n),\triangle\underline{\tau}(n),\mathcal{I}_j(m)|\Xi_{n-1},\mathcal{H}_\ell) + \Pr(v(n)|\Xi_{n-1},\mathcal{H}_\ell)\end{aligned} \tag{11.59}$$

where the argument n in the sets $\underline{\epsilon}(n)$ and $\triangle\underline{\tau}(n)$ specify all energies and interarrival times up to the nth event. Applying Bayes' rule, we obtain the decomposition

$$\begin{aligned}&\Pr(\xi(n;\underline{\epsilon},\triangle\underline{\tau})|\Xi_{n-1},\mathcal{H}_\ell) =\\ &\Pr(\triangle\underline{\tau}(n)|\underline{\epsilon}(n),\mathcal{I}_j(m),\Xi_{n-1},\mathcal{H}_\ell)\times\Pr(\underline{\epsilon}(n)|\mathcal{I}_j(m),\Xi_{n-1},\mathcal{H}_\ell)\\ &\times\Pr(\mathcal{I}_j(m)|\Xi_{n-1},\mathcal{H}_\ell) + \Pr(v(n)|\Xi_{n-1},\mathcal{H}_\ell)\end{aligned} \tag{11.60}$$

Decomposing the parameter vectors using the fact that each arrival has "no memory" and applying the chain rule of probability, we obtain the desired posterior distribution for detection

$$\begin{aligned}\Pr(\xi(n;\underline{\epsilon},\triangle\underline{\tau})|\Xi_{n-1},\mathcal{H}_\ell) = \prod_{m=1}^{M_\epsilon}\Pr(\triangle\tilde{\tau}_m(n)|\tilde{\epsilon}_m(n),\mathcal{I}_j(m),\Xi_{n-1},\mathcal{H}_\ell)\times\\ \Pr(\epsilon_m(n)|\mathcal{I}_j(m),\Xi_{n-1},\mathcal{H}_\ell)\times\Pr(\mathcal{I}_j(m)|\Xi_{n-1},\mathcal{H}_\ell) + \Pr(v(n)|\Xi_{n-1},\mathcal{H}_\ell);\ \ell=0,1\end{aligned} \tag{11.61}$$

Let us define the notation

$$\begin{aligned}\Theta_m(n;\theta) := \Pr(\triangle\tilde{\tau}_m(n)|\tilde{\epsilon}_m(n),\mathcal{I}_m(k),\Xi_{n-1},\mathcal{H}_\ell)\times\\ \Pr(\tilde{\epsilon}_m(n)|\mathcal{I}_m(k),\Xi_{n-1},\mathcal{H}_\ell)\times\Pr(\mathcal{I}_m(k)|\Xi_{n-1},\mathcal{H}_\ell)\end{aligned} \tag{11.62}$$

for θ representing the vector of unknown parameters (energy, interarrival time, emission probability). Now substituting this expression into Eq. 11.55 and *ignoring* the instrumentation noise, the *sequential log-likelihood ratio* decision function is given by

$$\Lambda[\Xi_n] = \Lambda[\Xi_{n-1}] + \sum_{m=1}^{M_\epsilon}\ln\Theta_m(n;\theta^t) - \sum_{m=1}^{M_\epsilon}\ln\Theta_m(n;\theta) \tag{11.63}$$

giving us the general form for the sequential solution. Note that this decision function incorporates not only the energy and interarrival times for each constituent monoenergetic source composing the target radionuclide but also the probability of emission which acts as a weighting function in the overall superposition enabling all of the energies to be combined in making the decision.

Next, we develop the statistical models to implement the decision function. Each individual target RN is uniquely specified by its parameter set, $\{\underline{\epsilon}^t,\triangle\underline{\tau}^t,\underline{\alpha}^t\}$ known from the Tables. Using the energy distribution and its decomposition of Eq. 11.60, the jth-monoenergetic source component selected by the indicator function $\mathcal{I}_j(m)$ is therefore

$$\Pr(\tilde{\epsilon}_j(n), \mathcal{I}_j(m)|\Xi_{n-1}, \mathcal{H}_\ell) = \alpha_j \Pr(\tilde{\epsilon}_j(n)|\Xi_{n-1}, \mathcal{H}_\ell) \tag{11.64}$$

where we have applied $\Pr(\mathcal{I}_j(m) = 1|\Xi_{n-1}) = \alpha_j$.

Each energy component is assumed Gaussian with corresponding distribution

$$\Pr(\epsilon_m(n)|\mathcal{I}_j(m), \Xi_{n-1}, \mathcal{H}_\ell) \sim \mathcal{N}(\overline{\epsilon}_j(n), \sigma^2_{\epsilon_j}) = \frac{1}{\sqrt{2\pi}\sigma_{\epsilon_j}} \exp\left\{-\frac{(\epsilon_j(n) - \overline{\epsilon}_j(n))^2}{2\sigma^2_{\epsilon_j}}\right\} \tag{11.65}$$

This models the effect of finite detector resolution and other experimental uncertainties in measuring the energy. The interarrival times $\triangle\underline{\tau}$ are exponentially distributed such that

$$\Pr(\triangle\underline{\tau}(n) \mid \underline{\epsilon}(n), \mathcal{I}_j(m), \Xi_{n-1}, \mathcal{H}_\ell) = \prod_{m=1}^{M_\epsilon} \Pr(\triangle\tau_m(n)|\epsilon_m(n), \mathcal{I}_j(m), \Xi_{n-1}, \mathcal{H}_\ell) =$$
$$\prod_{m=1}^{M_\epsilon} \alpha_m \lambda_{\triangle\tau_m} \exp\{-\alpha_m \lambda_{\triangle\tau_m} \triangle\tau_m(n)\} \tag{11.66}$$

with α_m the emission probability and detection rate $\lambda_{\triangle\tau_m} = 1/\overline{\triangle\tau_m}$, that is, the reciprocal of the mean interarrival time.

For the jth monoenergetic source component, we have

$$\Pr(\triangle\underline{\tau}(n) \mid \underline{\epsilon}(n), \mathcal{I}_j(m), \Xi_{n-1}, \mathcal{H}_\ell) \sim \mathcal{E}(\alpha_j \lambda_{\triangle\tau_j} \triangle\tau_j(n)) =$$
$$\alpha_j \lambda_{\triangle\tau_j} \exp\{-\alpha_j \lambda_{\triangle\tau_j} \triangle\tau_j(n)\} \tag{11.67}$$

Both the energy $\epsilon_j(n)$ and interarrival time $\triangle\tau_j(n)$ data are obtained from the detector arrival measurement.

Thus, for the radiation detection problem, we have under hypothesis $\mathcal{H}_0$

$$\Theta_m(n;\theta) = \frac{\alpha_m^2 \lambda_{\triangle\tau_m}}{\sqrt{2\pi}\sigma_{\epsilon_m}} \times \exp\left\{-\alpha_m \lambda_{\triangle\tau_m} \triangle\tau_m(n) - \frac{(\epsilon_m(n) - \overline{\epsilon}_m(n))^2}{2\sigma^2_{\epsilon_m}(n)}\right\} \tag{11.68}$$

and under hypothesis $\mathcal{H}_1$

$$\Theta_m(n;\theta^t) = \frac{(\alpha^t_m)^2 \lambda^t_{\triangle\tau_m}}{\sqrt{2\pi}\sigma_{\epsilon^t_m}} \times \exp\left\{-\alpha^t_m \lambda^t_{\triangle\tau_m} \triangle\tau_m(n) - \frac{(\epsilon_m(n) - \epsilon^t_m)^2}{2\sigma^2_{\epsilon^t_m}}\right\} \tag{11.69}$$

Substituting these expressions into Eq. 11.63 leads to the *sequential log-likelihood ratio* decision function

$$\Lambda[\Xi_n] = \Lambda[\Xi_{n-1}] + \sum_{m=1}^{M_\epsilon} \left[\ln\left(\frac{(\alpha_m^t)^2 \lambda^t_{\triangle\tau_m}}{\sqrt{2\pi}\sigma_{\epsilon_m^t}}\right) - \ln\left(\frac{\alpha_m^2 \lambda_{\triangle\tau_m}}{\sqrt{2\pi}\sigma_{\epsilon_m}}\right) \right.$$
$$\left. +\left(\alpha_m \lambda_{\triangle\tau_m} - \alpha_m^t \lambda^t_{\triangle\tau_m}\right)\triangle\tau_m(n) + \frac{1}{2}\left(\frac{\epsilon_m(n) - \overline{\epsilon}_m(n)}{\sigma_{\epsilon_m}}\right)^2 - \frac{1}{2}\left(\frac{\epsilon_m(n) - \epsilon_m^t(n)}{\sigma_{\epsilon_m^t}}\right)^2 \right] \tag{11.70}$$

This completes the structure of the sequential Bayesian radiation detector; we discuss the actual photon-by-photon implementation of this processor next.

11.4.3 Implementation

In the previous section, we developed the radionuclide detection approach leading to the sequential decision function. However, in order to implement the scheme, we must estimate the required parameters to reach the optimal decision. In this section, we discuss a more pragmatic approach to estimate the parameters and check their validity. The implementation of the algorithm is based on one major physics-based consideration: sequential or photon-by-photon processing leading to an individual processor (channel) for each monoenergetic source γ-ray.

There are three phases to the implementation after acquiring the individual photon energy and arrival time measurement: (1) discrimination; (2) estimation; and (3) detection as shown in Fig. 11.27a. The *discrimination* phase consists of determining the appropriate monoenergetic channel based on first testing (confidence interval) for the correct energy and second testing for a consistent interarrival time. Once this is accomplished, the *estimation* phase follows. Here, the parameters required to calculate the decision function must be estimated. Finally, in the *detection* phase with the decision function updated, sequential threshold testing proceeds where all of the individual arrival information is combined and the log-likelihood decision function updated at each arrival to detect (and identify) the targeted radionuclide.

The pragmatic implementation is accomplished in various stages: (1) photon discrimination; (2) monoenergetic parameter estimation; (3) decision function calculation and (4) threshold comparison as illustrated in Fig. 11.27b. Discrimination is performed with the "true" parameters obtained from the tables of radionuclides (energy, emission probability, and interarrival(rate)). From this information, we construct the confidence intervals to decide if the photon arrival is valid for one of the targeted radionuclide components. If so, we then move to the next step and perform the parameter estimation using a linear Kalman filter for energy (Gaussian model) and particle filter for interarrival (exponential). The emission probability is calculated by sequentially updating valid counts in the channel. With the parameter estimates available we then calculate the decision function. Finally, in order to calculate the required thresholds for the detector, we must generate an ROC curve from simulation

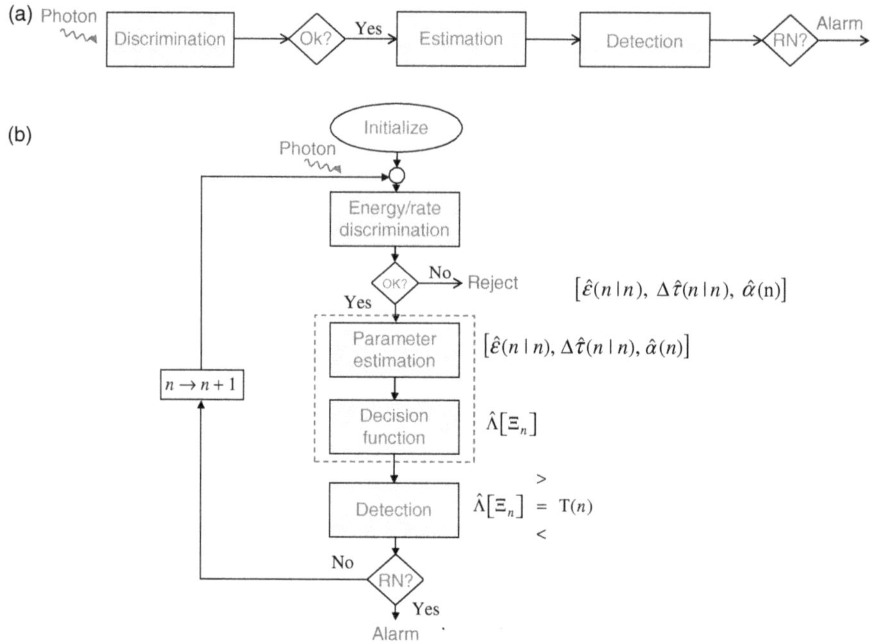

FIGURE 11.27 Implementation of radionuclide detection showing discrimination, estimation, and detection phases: (a) simplified flow of basic algorithm; (b) detailed flow diagram.

or high-fidelity data and pick an operating point specified by the desired detection and false alarm probabilities.

11.4.3.1 Discrimination Based on the parallel/distributed architecture of the detector/identifier, we first apply an energy discriminator to "decide" which channel the photon should be processed followed by a detection rate discriminator using the interarrival for verification. The discriminators use implied hypothesis testing by constructing confidence intervals about the means of the respective parameters [34]. The *energy* discriminator performs the following confidence interval test to accept or reject the photon in a channel:

$$[\epsilon^t - \kappa_\gamma \sigma_\xi \leq \epsilon_m(n) \leq \epsilon^t + \kappa_\gamma \sigma_\xi] \tag{11.71}$$

where ϵ^t is the true (channel) energy associated with the targeted radionuclide, κ_γ is the respective confidence coefficient with associated confidence level γ, and σ_ξ is the standard deviation associated with the precision of the measurement.

The interarrival measurement is tested similarly by the following confidence interval discriminator. For large n, the estimate of the mean interarrival time is approximately Gaussian, $\triangle\hat{\bar{\tau}} \sim \mathcal{N}(\triangle\tau^t, \triangle\tau^t/\sqrt{n})$.

$$[\triangle\tau^t - \kappa_{\tilde{\gamma}}\sigma_{\triangle\bar{\tau}} \leq \triangle\hat{\bar{\tau}}_m(n) \leq \triangle\tau^t + \kappa_{\tilde{\gamma}}\sigma_{\triangle\bar{\tau}}] \tag{11.72}$$

where $\triangle\tau^t$ is the true (channel) interarrival time associated with the targeted radionuclide, $\kappa_{\tilde{\gamma}}$ is the respective confidence coefficient with associated confidence level $\tilde{\gamma}$, and $\sigma_{\triangle\bar{\tau}}$ is the standard deviation of the variance of the estimated mean interarrival time $\triangle\hat{\bar{\tau}}_m$. Events with interarrival times that cause the estimated mean to deviate outside the confidence bounds are rejected.

Since the mean interarrival time is *not* necessarily the actual source interarrival, but may include scattered or background photon interarrivals, we use an estimated PHS obtained during the calibration phase to approximate the percentage (β_o) of downscattered photons to estimate $\triangle\tau^t$ [49]. β_o is based on the ratio of a fixed number of higher energy bin counts to the target energy bin count. Because of its Poisson nature, we average the interarrivals associated with the mth bin taking its reciprocal to obtain the composite (total) rate λ_{ϵ_m}. Finally, we estimate the true detection rate as

$$\lambda_m^t = (1-\beta_o)\lambda_{\epsilon_m}, \qquad \triangle\tau^t = \frac{1}{\lambda_m^t} \tag{11.73}$$

11.4.3.2 Estimation In order to implement this radionuclide detection scheme, we must estimate the underlying parameters (energy, interarrival time, emission probability) distributions to enable the calculation of the decision function. We define an overall parameter vector by

$$\begin{aligned}\theta &:= [\underline{\epsilon} \mid \underline{\tau} \mid \underline{\alpha}] = \\ &[\epsilon_1 \cdots \epsilon_{M_\epsilon} \mid \triangle\tau_1 \cdots \triangle\tau_{M_\epsilon} \mid \alpha_1 \cdots \alpha_{M_\epsilon}]\end{aligned}$$

for $\theta \in \mathcal{R}^{3M_\epsilon \times 1}$ requiring $3M_\epsilon$ parameters to specify an unknown radionuclide.

Since the measurement instrument measures both energy and arrival (interarrival) time from the EMS, it can be considered a vector measurement with components defined by

$$\underline{\xi}(n) := [\xi_\epsilon(n) \mid \xi_{\triangle\tau}(n)]' \tag{11.74}$$

After discrimination, when accepted, the raw measurement is then processed using Bayesian techniques for enhancement: the linear Kalman filter (LKF) and a PF(see Section 7.2). Both of these techniques are developed in the appendices.

11.4.3.3 Energy Estimation The energy estimator is based on the Markovian representation as

$$\begin{aligned} \epsilon_m(n) &= \epsilon_m(n-1) + w_{\epsilon_m}(n-1) \\ \xi_\epsilon(n) &= \epsilon_m(n) + v_\epsilon(n) \end{aligned} \tag{11.75}$$

with $\epsilon_m(0) \sim \mathcal{N}(\epsilon_m^t, R_{\epsilon_m^t \epsilon_m^t})$ and $w_{\epsilon_m} \sim \mathcal{N}(0, R_{w_{\epsilon_m} w_{\epsilon_m}})$; $v_\epsilon \sim \mathcal{N}(0, R_{v_\epsilon v_\epsilon})$.

This model is a linear Gauss–Markov model; therefore, the optimal Bayesian processor is the linear Kalman filter with posterior distribution specified by $\Pr(\epsilon_m(n)|\Xi_n) \sim \mathcal{N}(\hat{\epsilon}_m(n|n), \tilde{\sigma}^2_{\epsilon_m}(n|n))$ and therefore, the Kalman filter is

$$\begin{aligned} \hat{\epsilon}_m(n|n) &= \hat{\epsilon}_m(n|n-1) + K_{\epsilon_m}(n)\big(\xi_\epsilon(n) - \hat{\epsilon}_m(n|n-1)\big) \\ K_{\epsilon_m}(n) &= \tilde{\sigma}^2_{\epsilon_m}(n|n)/\sigma^2_{i_m}(n) \end{aligned} \tag{11.76}$$

where $\hat{\epsilon}_m(n|n)$ is the conditional mean estimate of the mth-energy at arrival time n based on all of the data up to n; $\tilde{\sigma}^2_{\epsilon_m}$ is the estimated error covariance $\text{cov}(\epsilon_m(n) - \hat{\epsilon}_m(n|n))$; $i_m(n)$ is the innovations sequence with covariance, $\sigma^2_{i_m}(n) = \text{cov}(i_m(n))$ and K_{ϵ_m} is the weight or gain matrix (see Section 5.3).

11.4.3.4 Interarrival Estimation From the statistics of the EMS process, we know that the interarrival times for the mth monoenergetic source are exponentially distributed with parameter $\alpha_m \lambda_{\triangle\tau_m}$ such that $\triangle\tau_m(n) \sim \mathcal{E}(\alpha_m \lambda_{\triangle\tau_m} \triangle\tau_m(n)) = \alpha_m \lambda_{\triangle\tau_m} \exp(-\alpha_m \lambda_{\triangle\tau_m} \times \triangle\tau_m(n))$ [47]. Our process model for the interarrival time is just an exponential random variable, while the underlying measurement model is this variable contaminated with exponential measurement (instrumentation) noise given by

$$\begin{aligned} \triangle\tau_m(n) &= w_{\triangle\tau_m}(n) \\ \xi_{\triangle\tau}(n) &= \triangle\tau_m(n) + v_{\triangle\tau_m}(n) \end{aligned} \tag{11.77}$$

for $w_{\triangle\tau_m} \sim \mathcal{E}(\alpha_m \lambda_{\triangle\tau_m} \triangle\tau_m(n))$ and $v_{\triangle\tau_m} \sim \mathcal{E}(\lambda_v v_{\triangle\tau_m}(n))$.

We estimate the posterior distribution using a sequential Monte Carlo approach and construct a *bootstrap particle filter* (see Section 7.5) to obtain

$$\hat{\Pr}(\triangle\tau_m(n)|\Xi_n) = \sum_{i=1}^{N_p} \mathcal{W}_i(n)\delta(\xi_{\triangle\tau}(n) - \xi_{\triangle\tau_{m_i}}(n)) \tag{11.78}$$

with the *weight* given by $W_i(n) = W_i(n-1) \times \hat{\Pr}(\xi_{\triangle\tau}(n)|\triangle\tau_{m_i}(n))$ for N_p-particles.

The MAPestimate is then obtained as:

$$\begin{aligned} \triangle\hat{\tau}_m(n|n) &= \arg\max \hat{\Pr}(\triangle\tau_m(n)|\Xi_n) \\ \hat{\lambda}_{\triangle\tau_m}(n|n) &= \frac{1}{\triangle\hat{\tau}_m(n|n)} \end{aligned} \tag{11.79}$$

11.4.3.5 *Emission Probability Estimation* We use a sequential counting technique to estimate the emission probability parameter. At each accepted arrival (after discrimination), the counting estimator for the mth-monoenergetic component of the radionuclide is given by

$$\hat{\alpha}_m(n) = \frac{N_{\epsilon_m}(n)}{M_\epsilon(n)} \tag{11.80}$$

where $N_{\epsilon_m}(n)$ is the total counts for the mth source at arrival time n or simply the mth *channel count* and $M_\epsilon(n)$ is the *total count* of all of the RN channels or monoenergetic sources at n.

11.4.4 Detection

Sequential radionuclide detection is implemented in a channel-by-channel framework. Basically, the individual distributions are calculated in *parallel* at each channel and then combined in the detector where the required parameters are replaced by their estimates. At each arrival after discrimination, the accepted channel jth photon is processed by the energy and interarrival parameter estimators ($\hat{\theta}$) providing the input to the sequential decision function (log-likelihood ratio) along with the true parameters ($\theta^t \rightarrow [\{\epsilon^t_m\}, \{\triangle\tau^t_m\}, \{\alpha^t_m\}]; m = 1, \ldots, M_\epsilon$) from the Tables. Using the relations developed for the joint parametric distributions, the following general functional form is implemented using the results of the sequential estimators to give

$$\Lambda[\Xi_n] = \Lambda[\Xi_{n-1}] + \sum_{m=1}^{M_\epsilon} \ln \Theta_m(n; \theta^t) - \sum_{m=1}^{M_\epsilon} \ln \Theta_m(n; \hat{\theta}) \tag{11.81}$$

Combining similar terms, we obtain the final sequential log-likelihood ratio radionuclide decision function specified by:

$$\begin{aligned}\Lambda[\Xi_n] = \Lambda[\Xi_{n-1}] + \sum_{m=1}^{M_\epsilon} \ln\left(\frac{(\alpha^t_m)^2 \lambda^t_{\triangle\tau_m}}{\sqrt{2\pi}\sigma_{\epsilon^t_m}}\right) - \ln\left(\frac{\hat{\alpha}^2_m \hat{\lambda}_{\triangle\tau_m}(n|n)}{\sqrt{2\pi}\hat{\sigma}_{\epsilon_m}(n|n)}\right) + \\ \left(\hat{\alpha}_m \hat{\lambda}_{\triangle\tau_m}(n|n) - \alpha^t_m \lambda^t_{\triangle\tau_m}\right)\xi_{\triangle\tau}(n) + \frac{1}{2}\left(\frac{\xi_\epsilon(n) - \hat{\epsilon}_m(n|n)}{\hat{\sigma}_{\epsilon_m}(n|n)}\right)^2 \\ -\frac{1}{2}\left(\frac{\xi_\epsilon(n) - \epsilon^t_m(n)}{\sigma_{\epsilon^t_m}}\right)^2\end{aligned} \tag{11.82}$$

We observe that after successful discrimination the parameters are estimated and employed to calculate the log-likelihood function of Eq. 11.82. These parameters are estimated channel-by-channel (mth channel) and the overall decision function updated sequentially (in arrival time). Once the parameters are estimated, they are implemented in each channel log-likelihood partial calculation ($\Theta_m(n;\theta)$) and all of the partial sums are combined along with the previous (in arrival time) log likelihood to sequentially update the new log likelihood at time n. The function is then compared to the threshold to see if a detection is possible. If not, the next photon is processed and the log-likelihood updated to see if a decision can be made. This sequential radionuclide process continues until there is enough data for a statistical decision to be made.

11.4.5 Data

A proof-of-concept experiment was developed to assess the feasibility of the sequential Bayesian processor. *Three* source radionuclides were targeted in a laboratory environment contaminated with background and extraneous sources. Each target source and background was individually counted with the results combined to generate the controlled "feasibility" data set. Here the signal-to-noise ratio (SNR), background, etc. were fixed by this environment.

This set of composite radionuclide EMS "feasibility" data consisted of the three radionuclides cobalt (^{60}Co), cesium (^{137}Cs), barium (^{133}Ba) represented by 2, 1, and 5 monoenergetic sources, respectively, along with background and the extraneous source (potassium).

11.4.6 Radionuclide Detection

The sequential detection paradigm of Table 11.3 was applied to the "feasibility" data set. Based on the experimental SNR, the selected operating point (detection and false-alarm probabilities) was (98%, 2%) specifying the thresholds which were calculated according to Eq. 11.56 for each radionuclide. Here, an ROC curve was synthesized using simulated EMS and noise sequences to obtain these probabilities or operating point. After an initial calibration phase of the algorithm which consisted of "tuning" the processors on simulated and controlled data, setting initial parameters, etc., the overall results of the processing are shown in Fig. 11.28. We note four columns of data, the first column is the composite pulse-height spectrum, with the second the composite EMS with the circles representing the *discriminator* output photons. Notice that they are chosen by the discriminator confidence intervals for *both* energy and the corresponding interarrival and align with the PHS energy "lines". The third column represents the *estimated* or enhanced energy of the processed (channel) photon, while the final column is the decision function (log likelihood) for each of the targeted radionuclides.

As each photon is processed, the decision function is sequentially updated until one of the thresholds (target/nontarget) is crossed (lighter crosses in figure) declaring a threat or nonthreat. Note that barium is detected (threshold exceeded) first (0.513 sec) followed by the cesium (0.678 sec) and then cobalt (3.02 sec).

TABLE 11.3 Sequential Radiation Detection

Discrimination	
$[\epsilon^t - \kappa_\gamma \sigma_\xi \le \epsilon_m(n) \le \epsilon^t + \kappa_\gamma \sigma_\xi]$	[Energy]
$[\Delta\tau^t - \kappa_{\tilde\gamma} \sigma_{\Delta\tilde\tau} \le \Delta\hat{\tilde\tau}_m(n) \le \Delta\tau^t + \kappa_{\tilde\gamma} \sigma_{\Delta\tilde\tau}]$	[Interarrival]
Estimation	
$\hat\epsilon_m(n\|n) = \hat\epsilon_m(n\|n-1) + K_{\epsilon_m} i_m(n)$	[Energy]
$\Delta\hat\tau_m(n\|n) = \arg\max \hat{\Pr}(\Delta\tau_m(n)\|\Xi_n)$	[Interarrival]
$\hat\lambda_{\Delta\tau_m}(n\|n) = \dfrac{1}{\Delta\hat\tau_m(n\|n)}$	[Detection rate]
$\hat\alpha_m(n) = \dfrac{N_{\epsilon_m}(n)}{M_\epsilon(n))}$	[Emission probability]
$\Lambda[\Xi_n] = \Lambda[\Xi_{n-1}] + \ln \Theta_m(n;\theta^t) - \ln \Theta_m(n;\hat\theta)$	[Decn]
Detection	
$\Lambda[\Xi_n] \begin{array}{c} \mathcal{H}_1 \\ \ge \\ = \\ \le \\ \mathcal{H}_0 \end{array} \begin{array}{c} \mathcal{T}_1 \\ \text{Continue} \\ \mathcal{T}_0 \end{array}$	[Log-likelihood]

11.4.7 Summary

We have demonstrated that a sequential Bayesian detector can be developed to provide a feasible solution to the radiation detection problem by defining a target radionuclide(s) and its monoenergetic decomposition model evolving from the underlying transport physics of the photon and measurement process. Starting with the decomposition, we developed a probabilistic framework to theoretically define the problem. Under certain assumed distributions, a particular realization of the process was successfully developed and applied to experimental data demonstrating its overall feasibility. The key idea was to process the data, photon by photon, rejecting any extraneous nontargeted radionuclide measurements and process only those photons that correspond to the targeted (threat) radionuclide(s). Thus, the Bayesian approach provides a unique processor capturing all available physics information photon by photon.

11.5 SEQUENTIAL THREAT DETECTION: AN X-RAY PHYSICS-BASED APPROACH

The timely and accurate detection of threat contraband especially for ports-of-entry is an extremely critical problem of national security for many countries. The characterization of signal processing models based on X-ray transport physics is a crucial element in advanced sequential Bayesian processor designs. Incorporating the underlying statistics of X-ray interactions with materials offering a potentially unique signature of an object under investigation leads to a (stochastic) physics-based approach.

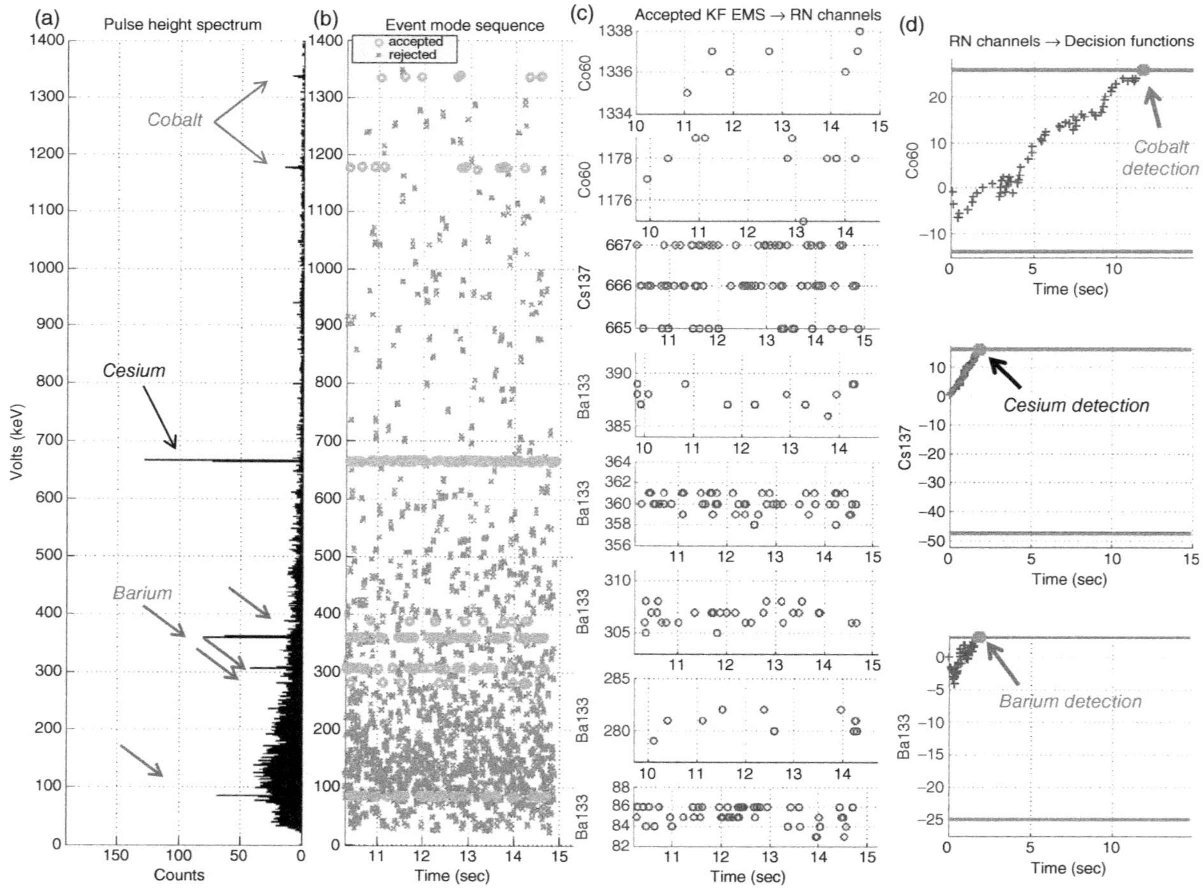

FIGURE 11.28 Sequential Bayesian detection and identification: (a) pulse-height spectrum (after calibration); (b) EMS with discrimination (circles); (c) enhanced energy estimates; (d) log-likelihood decision functions for ^{60}Co (detection time: 3.02 sec), ^{137}Cs (detection time: 0.67 sec), and ^{133}Ba (detection time: 0.51 sec) radionuclide detection/identification.

A sequential Bayesian processor is developed to provide a reliable and accurate solution with high confidence in a timely manner for this problem based on a set of synthesized object intensity data.

The basic approach is model-based incorporating the X-ray propagation physics into the state–space framework. That is, the initial polyenergetic source intensity is measured at the output of a detector array (multi-output). These data along with a-priori knowledge of the targeted threat are incorporated into a threat model with the appropriate attenuation coefficient parameters. Next, measured object intensities are estimated from the raw detector data using a linear model-based processor (MBP) or Kalman filter to compute the log-likelihood decision function. Then at each step (spatial), the intensity is predicted by the MBP and a decision (threat/nonthreat/no decision) is made.

11.5.1 Physics-Based Models

The characterization of a wealth of items that could be enclosed in a shipping container covers a wide range of features. From an X-ray interrogation perspective for threat detection, we are concerned primarily with the interaction of X-ray photons (energy) with the variety of materials enclosed within.

11.5.1.1 Background: X-ray Physics When X-rays interact with matter (materials), they undergo attenuation effects, that is, an X-ray or equivalently beam of photons passing through matter is depleted or attenuated as photons are removed through absorption and scattering [51, 58]. Here photon energy (ϵ) is transformed to heat (absorption) or scattered with its energy dissipated and redirected. Assume we have a perfectly parallel monoenergetic (single energy) beam of X-rays with intensity $\mathcal{I}(\epsilon, x)$ irradiating a uniformly thick, single-material absorber as shown in Fig. 11.29, then the transmitted (output) intensity is governed by

$$\frac{d}{dx}\mathcal{I}(\epsilon, x) = -\mu(\epsilon, \mathcal{Z}, \rho)\mathcal{I}(\epsilon, x) \tag{11.83}$$

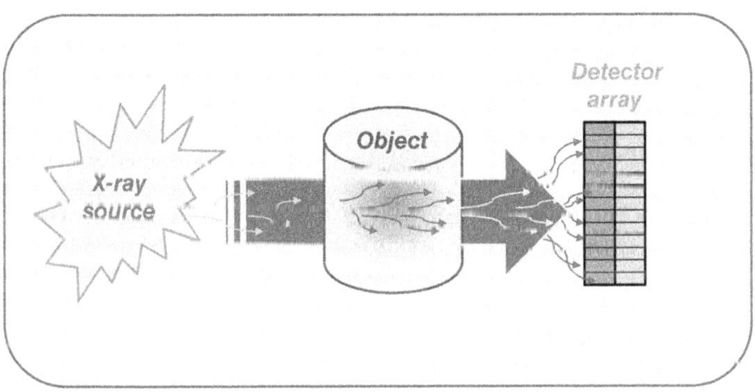

FIGURE 11.29 X-ray transmission physics: polyenergetic source, object, detector array measurements, and sequential Bayesian processing for threat detection.

where $\mu(\epsilon, \mathcal{Z}, \rho)$ is the attenuation coefficient (cm^{-1}) at energy ϵ (keV) for a "single" material of mass density ρ (gcm^{-3}) with atomic number $\mathcal{Z}$ along the X-direction of path length $\triangle x = x - x_o$ (cm) with $\mathcal{I}(\epsilon_o, x_o)$ is the initial intensity. Solving this equation leads to the *Lambert–Beer Law* [51, 58] given by

$$\mathcal{I}(\epsilon, x) = \mathcal{I}(\epsilon_o, x_o) \times \exp\{-\mu(\epsilon, \mathcal{Z}, \rho)(x - x_o)\} \tag{11.84}$$

This relation can be extended further to incorporate composite materials ($(\mathcal{Z}, \rho) \longrightarrow (\mathcal{Z}_m, \rho_m); m = 1, \ldots, M$) and multiple source energies ($S(\epsilon_n); n = 1, \ldots, N$) giving the general expression

$$\mathcal{I}(S(\epsilon_n), x) = \mathcal{I}(S(\epsilon_o), x_o) \times \exp\left\{ - \int \left(\sum_{m=1}^{M} \mu_m(S(\epsilon_n), \mathcal{Z}_m \rho_m) \triangle x \right) d\epsilon \right\} \tag{11.85}$$

for $n = 1, \ldots, N$.

From these results, we see that the attenuation coefficients contain all of the critical physical information required to uniquely characterize the material under irradiation by the source X-rays.

11.5.1.2 Physics-Based State-Space Models

For our problem, we return to the original first-order differential transmission relation of Eq. 11.83 and realize that it is *already* in the well-known state–space form [38]. We consider this model a "lumped model" where the object is characterized by a set of average attenuation coefficients. Assuming that a polyenergetic source consisting of a set of discrete monoenergetic energies, $\epsilon_n; n = 1, \ldots, N$, irradiates the material (object), Beer's law is now a *set* of first-order differential transport equations, that is,

$$\frac{d}{dx}\mathcal{I}(\epsilon_n, x) = -\mu(\epsilon_n, \mathcal{Z}, \rho)\mathcal{I}(\epsilon_n, x); n = 1, \ldots, N \tag{11.86}$$

with incident intensities given by $\mathcal{I}(\epsilon_{no}, x_{no})$ identifying each monoenergetic source component. Expanding over n, we obtain the equivalent *state–space transmission model* given by

$$\frac{d}{dx}\begin{bmatrix} \mathcal{I}(\epsilon_1, x) \\ \vdots \\ \mathcal{I}(\epsilon_N, x) \end{bmatrix} = \begin{bmatrix} -\mu(\epsilon_1, \mathcal{Z}, \rho) & & O \\ & \ddots & \\ O & & -\mu(\epsilon_N, \mathcal{Z}, \rho) \end{bmatrix} \begin{bmatrix} \mathcal{I}(\epsilon_1, x) \\ \vdots \\ \mathcal{I}(\epsilon_N, x) \end{bmatrix}$$

which can be expressed simply in vector-matrix form as

$$\frac{d}{dx}\underline{\mathcal{I}}(\epsilon, x) = A(\epsilon, \mathcal{Z}, \rho)\underline{\mathcal{I}}(\epsilon, x) \tag{11.87}$$

where we have the state intensity vector $\underline{\mathcal{I}} \in \mathcal{R}^{N\times 1}$, the system (physics) matrix, $A \in \mathcal{R}^{N\times N}$ with initial (incident) intensity vector defined by

$$\underline{\mathcal{I}}(\epsilon_o, x_o) := \left[\mathcal{I}(\epsilon_{1o}, x_{1o}), \ldots, \mathcal{I}(\epsilon_{No}, x_{No})\right]^T$$

and ^{T} the matrix transpose operator. It is well known that the solution to the linear shift invariant (A not a function of $\mathcal{I}$) state equations is given by the *state transition matrix* as [38]

$$\mathcal{T}(\epsilon, x) = e^{A(\epsilon,\mathcal{Z},\rho)(x-x_o)} \tag{11.88}$$

with $\mathcal{T} \in \mathcal{R}^{N\times N}$ yielding Beer's law for a polyenergetic source constructed as a composition of monoenergetic sources, that is,

$$\underline{\mathcal{I}}(\epsilon, x) = \mathcal{T}(\epsilon, x)\underline{\mathcal{I}}(\epsilon_o, x) = e^{A(\epsilon,\mathcal{Z},\rho)(x-x_o)}\underline{\mathcal{I}}(\epsilon_o, x_o) \tag{11.89}$$

This representation can be considered a "lumped model" where the object is characterized by a set of average attenuation coefficients $\{\mu(\epsilon_n, \mathcal{Z}, \rho)\}n = 1, \ldots, N$, one at each energy. Thus, energetic photons interacting with the object (material) are attenuated according to each associated attenuation coefficient. An alternate representation incorporates the discretization of the object into voxels through CT-reconstruction followed by expressing transport as a set of coupled differential/difference equations [38]. State–space methods can be applied in either of these representations. Here, we concentrate on the *lumped model*, since it is the simplest to comprehend.

Next, let us consider an X-ray beam or equivalently ray path consisting of polyenergetic photons irradiating the material and measured by a detector element, the ℓth. We choose to use a scintillation measurement system composed of a crystal converting the X-ray radiation to visible light photons coupled to a photo diode converting the photon energy to electrical charge/current. With this measurement system, the detector is related to the photon intensity through a set of energy-dependent conversion constants, that is, for the ℓth detector element and a set of N-energies, we have

$$d_\ell(\epsilon, x) = \underline{c}_\ell^T(\epsilon, \Delta\epsilon)\underline{\mathcal{I}}_\ell(\epsilon, x) = \sum_{n=1}^{N} c_\ell(\epsilon_n)\mathcal{I}_\ell(\epsilon_n, x) \tag{11.90}$$

where $\underline{c}_\ell, \underline{\mathcal{I}}_\ell \in \mathcal{R}^{N\times 1}$, and $\underline{c}_\ell$ is an energy-dependent vector of scintillation scale factors converting photon energy to charge (current).

For acquisition speed, an array of L-elements is normally applied, that is, expanding over ℓ, we obtain the *detector measurement model* as

$$\begin{bmatrix} d_1(\epsilon, x) \\ --- \\ \vdots \\ --- \\ d_L(\epsilon, x) \end{bmatrix} = \begin{bmatrix} \underline{c}_1^T(\epsilon, \triangle\epsilon) \\ --- \\ \vdots \\ --- \\ \underline{c}_L^T(\epsilon, \triangle\epsilon) \end{bmatrix} \underline{\mathcal{I}}(\epsilon, x) \tag{11.91}$$

which can also be expressed simply as

$$\underline{\mathbf{d}}(\epsilon, x) = \mathbf{C}(\epsilon, \triangle\epsilon)\underline{\mathcal{I}}(\epsilon, x) \tag{11.92}$$

where we have the detector measurement vector $\underline{\mathbf{d}} \in \mathcal{R}^{L\times 1}$, the measurement system matrix $\mathbf{C} \in \mathcal{R}^{L\times N}$, and the corresponding state intensity vector $\underline{\mathcal{I}} \in \mathcal{R}^{N\times 1}$.

Thus, the overall physics-based state–space description (deterministic) of the X-ray transmission and detector system is summarized by combining the relations of Eqs. 11.87 and 11.93 for *each* ray and detector element as

$$\frac{d}{dx}\underline{\mathcal{I}}(\epsilon, x) = \mathbf{A}(\epsilon; \theta)\underline{\mathcal{I}}(\epsilon, x)$$

$$\underline{\mathbf{d}}(\epsilon, x) = \mathbf{C}(\epsilon, \triangle\epsilon)\underline{\mathcal{I}}(\epsilon, x) \tag{11.93}$$

where the N-state intensity vector by $\underline{\mathcal{I}}$, the $N \times N$ system matrix is defined by $\mathbf{A}(\epsilon; \theta) := \text{diag}\left[-\mu(\epsilon_1; \theta), \ldots, -\mu(\epsilon_N; \theta)\right]$ and the *material parameter set* by $\theta := \{\mathcal{Z}, \rho\}$.

We summarize the development of the state–space representation in Fig. 11.30. Here, we see the evolution of the photon intensity proportional to the detector output current. Initially, the polyenergetic source is represented by the composition of a

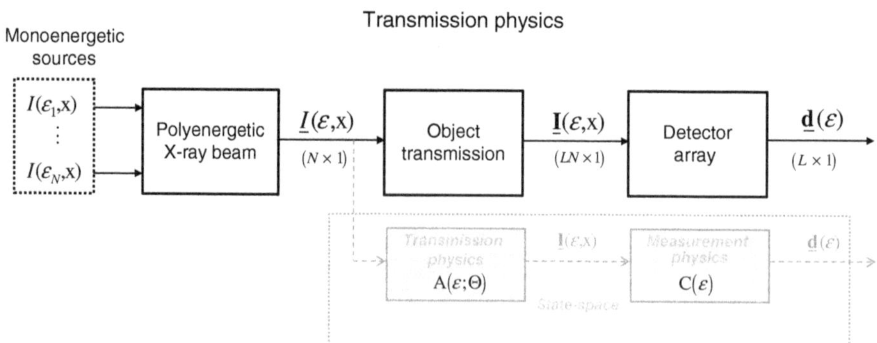

FIGURE 11.30 X-ray transmission: transmission physics with "lumped" state–space model (dashed).

set of monoenergetic sources ($\underline{\mathcal{I}}(\epsilon, x)$) irradiating the object. The transmitted output ($\underline{\mathcal{I}}(\epsilon, x)$) is collected by the detector array, integrated, producing a set of output currents proportional to the transmitted photon energies ($\underline{\mathbf{d}}(\epsilon, x)$) used as an input to image the object attenuation.

11.5.2 X-ray State-Space Simulation

In this section, we briefly discuss the development of a state–space simulator capable of synthesizing X-ray detector outputs for generating noisy detector data. The approach involves discretizing the solution using the state transition matrix which is a matrix exponential [38] and generating data for an object.

The basic *deterministic* state–space model of Eq. 11.93 can easily be extended to a *stochastic* representation assuming additive, zero-mean, Gaussian noise resulting in the *Gauss–Markov* state–space model either given by the differential equation of Eq. 11.93 or equivalently in terms of the state transition matrix of Eq. 11.89. The system of equations may be discretized to obtain a *sampled data* representation [38] resulting in the following *discrete Gauss–Markov model* given by

$$\underline{\mathcal{I}}(\epsilon, x_k) = \mathcal{T}(\epsilon, \triangle x_k)\underline{\mathcal{I}}(\epsilon, x_{k-1}) + \triangle x_k \mathcal{T}(\epsilon, \triangle x_k)\underline{\mathbf{w}}(x_{k-1}) \tag{11.94}$$

with corresponding measurement (detector array) model

$$\underline{\mathbf{d}}(\epsilon, x_k) = \mathbf{C}(e, \triangle\epsilon)\underline{\mathcal{I}}(\epsilon, x_{k-1}) + \underline{v}(x_k) \tag{11.95}$$

for $N_{\mathcal{I}}$-intensity (state) and white, Gaussian process uncertainty/noise vectors, $\underline{\mathcal{I}}$, $\underline{\mathbf{w}}$ with $w \sim \mathcal{N}(0, R_{ww})$. Here $\underline{\mathbf{d}}$ is the N_d-measurement (detector array) with corresponding white, Gaussian uncertainty/noise (electronic) vector $\underline{v}$ for $v \sim \mathcal{N}(0, R_{vv})$ along with the corresponding $N_d \times N_{\mathcal{I}}$ measurement matrix $\mathbf{C}$. The incremental path length is specified by $\triangle x_k := x_k - x_{k-1}$ with $(N_{\mathcal{I}} \times N_{\mathcal{I}})$ state transition matrix $\mathcal{T}$,

$$\mathcal{T}(\epsilon, \triangle x_k) = \exp\{\mathbf{A}(\epsilon; \theta)\triangle x_k\} \tag{11.96}$$

for $\mathbf{A}(\epsilon; \theta) = \text{diag}\left[-\mu(\epsilon_1; \theta), \ldots, -\mu(\epsilon_{N_d}; \theta)\right]$ the material attenuation coefficients and set of intrinsic material properties $\{\theta\}$. In addition to the intensity and measurement models, the Gauss–Markov representation also propagates the associated statistics (state/measurements) means and variances which are used to construct uncertainty bounds around the synthesized data, that is, the *mean state intensity* is given by the relation

$$\underline{\mathcal{M}}_{\mathcal{I}}(\epsilon, x_k) = \mathcal{T}(\epsilon, \triangle x_k)\underline{\mathcal{M}}_{\mathcal{I}}(\epsilon, x_{k-1}) \tag{11.97}$$

along with the *mean detector measurement*

$$\underline{\mathcal{M}}_{\mathbf{d}}(\epsilon, x_k) = \mathbf{C}(\epsilon, \triangle\epsilon)\underline{\mathcal{M}}_{\mathcal{I}}(\epsilon, x_{k-1}) \tag{11.98}$$

The corresponding state and measurement *covariances* are given by

$$\mathcal{P}_{\mathcal{I}}(x_k) = \mathcal{T}(\epsilon, \triangle x_k)\mathcal{P}_{\mathcal{I}}(x_{k-1})\mathcal{T}^T(\epsilon, \triangle x_k) + \mathbf{R}_{ww}(x_{k-1}) \tag{11.99}$$

$$\mathcal{R}_{\mathbf{dd}}(x_k) = \mathbf{C}(\epsilon, \triangle \epsilon)\mathcal{P}_{\mathcal{I}}(x_{k-1})\mathbf{C}^T(\epsilon, \triangle \epsilon) + \mathbf{R}_{vv}(x_k) \tag{11.100}$$

To summarize this "lumped model" representation, the discrete Gauss–Markov model characterizes the photon transport through a homogeneous object irradiated by a polyenergetic source with uncertainties approximated by additive zero mean, white Gaussian sequences both in intensity (process) and detector intensity measurements (measurement). Next, we develop the simulator.

In this problem, we assume that an object or item has been isolated and extracted from a CT-image resulting in an attenuation coefficient sequence (ACS). From this sequence, a set of average attenuation coefficients is obtained at *each* known source energy and represented by $\{\mu(\epsilon_n, \theta\}n = 1, \ldots, N$. We developed a simple transport simulator based on the lumped model and the state–space framework to synthesize a threat object composed of ammonium nitrate (NH_4NO_3) material characterized uniquely by its set of energy-dependent attenuation coefficients obtained from the Tables [51] based on a polyenergetic source and detector bands at 30–160 keV incremented every 10 keV. The transport is calculated every $\triangle x = 100$ mm. We assume that an X-ray beam (ray) at 0° passes through the object (no clutter) and is measured by an eight-element scintillation detector array. We also assume that uncertainties propagate and are captured by the model discussed above with intensity (process) uncertainty of covariance $R_{ww} = 25$ and detector measurement noise of covariance $R_{vv} = 10^7$. The scanner volume can be considered a 1 cubic meter spherical volume. The problem parameters are summarized in Table 11.4. We have $N_{\mathcal{I}} = 14$ states (and

TABLE 11.4 Transport Physics for Ammonium Nitrate

Energy (keV)	Atten. Coefficients
30	0.95388
40	0.68456
50	0.58017
60	0.52641
70	0.49307
80	0.47019
90	0.45164
100	0.43706
110	0.42389
120	0.41288
130	0.40292
140	0.39344
150	0.38511
160	0.37761

average attenuation coefficients) with $N_d = 8$ elements in the detector array. The simulation was performed for a single beam (ray) propagating through the object at a path length of 100 cm (scan volume) with the object of cross-section 6 cm located 30 cm from the source.

The photons are at the initial intensity $\mathcal{I}(\epsilon_o, x_o)$ and then drop off exponentially based on their respective attenuation coefficients and detector measurements as they interact with the material (ammonium nitrate) with the path length incremented in $\triangle x = 0.1$ cm steps. Note that in reality the detector array provides only a single snapshot of integrated intensity at a given x-location providing a measured projection [58]. These lumped model simulations will be used as input to the sequential detector algorithm.

11.5.3 Sequential Threat Detection

Detection of threats residing in a variety of containers is essentially a binary detection problem. Here, we investigate a solution to this decision problem to detect the presence of threat materials using X-ray radiation as the primary modality (e.g., pot scanners) to irradiate each object and assess its status—threat or nonthreat. Therefore, we define the problem in terms of a binary hypothesis test. From the scanner, we use detector outputs directly available from each of the array elements; that is, the binary threat detection problem is based on testing the hypotheses

$$\begin{aligned}
\mathcal{H}_0 &: \underline{\mathbf{d}}(\epsilon, x_k) = \mathbf{C}(\epsilon, \triangle\epsilon)\underline{\mathcal{I}}(\epsilon, x_k; \theta) + \underline{\mathbf{v}}(x_k) \quad \text{[Nonthreat]} \\
\mathcal{H}_1 &: \underline{\mathbf{d}}(\epsilon, x_k) = \mathbf{C}(\epsilon, \triangle\epsilon)\underline{\mathcal{I}}(\epsilon, x_k; \theta_t) + \underline{\mathbf{v}}(x_k) \quad \text{[Threat]}
\end{aligned} \tag{11.101}$$

where

$$\begin{aligned}
&\underline{\mathbf{d}} \text{ is the } N_d - \text{element detector array} \\
&\underline{\mathbf{C}} \text{ is the } (N_d \times N_{\mathcal{I}}) - \text{energy measurement matrix} \\
&\underline{\mathcal{I}} \text{ is the } N_{\mathcal{I}} - \text{energy dependent intensities (states)} \\
&\underline{\mathbf{v}} \text{ is the } N_d - \text{white Gaussian uncertainty/noise vector} \\
&\underline{\theta} \text{ is the set of material properties (object and target)} \\
&x_k \text{ is the length along the ray}
\end{aligned} \tag{11.102}$$

The optimal solution to this decision problem evolves directly from the Neyman–Pearson theorem [33,57] maximizing the detection probability for a fixed false-alarm probability resulting in the likelihood-ratio decision function defined by

$$\mathcal{L}[\mathcal{D}_K] := \frac{\Pr[\mathcal{D}_K|\mathcal{H}_1]}{\Pr[\mathcal{D}_K|\mathcal{H}_0]} \tag{11.103}$$

where $\mathcal{D}_K := \{\underline{\mathbf{d}}(\epsilon, x_1), \ldots, \underline{\mathbf{d}}(\epsilon, x_K)\}$ with the hypothesis test defined by

$$\mathcal{L}[\mathcal{D}_K] \underset{\mathcal{H}_0}{\overset{\mathcal{H}_1}{\gtrless}} T \tag{11.104}$$

for threshold T. This expression implies a "batch" decision; that is, we gather the K-detector array measurements, calculate the likelihood of Eq. 11.103 over the entire batch of data and compare it to the threshold to make the decision.

Expanding the likelihood ratio for each detector output, we obtain

$$\mathcal{L}[\mathcal{D}_K] = \frac{\Pr[\underline{\mathbf{d}}(\epsilon, x_1), \ldots, \underline{\mathbf{d}}(\epsilon, x_K)|\mathcal{H}_1]}{\Pr[\underline{\mathbf{d}}(\epsilon, x_1), \ldots, \underline{\mathbf{d}}(\epsilon, x_K)|\mathcal{H}_0]} \tag{11.105}$$

From the Bayes' rule [34], we have

$$\Pr[\mathcal{D}_K|\mathcal{H}_\ell] = \Pr[\underline{\mathbf{d}}(\epsilon, x_1), \ldots, \underline{\mathbf{d}}(\epsilon, x_K)|\mathcal{H}_\ell] = \Pr[\mathbf{d}(\epsilon, x_K), \mathcal{D}_{K-1}|\mathcal{H}_\ell] \tag{11.106}$$

and

$$\Pr[\underline{\mathbf{d}}(\epsilon, x_K), \mathcal{D}_{K-1}|\mathcal{H}_\ell] = \Pr[\underline{\mathbf{d}}(\epsilon, x_k)|\mathcal{D}_{K-1}, \mathcal{H}_\ell] \times \Pr[\mathcal{D}_{K-1}|\mathcal{H}_\ell]; \ell = 0, 1 \tag{11.107}$$

Substituting these expressions into the likelihood ratio above, replacing $K \to k$ and grouping, we obtain the recursion or equivalently *sequential likelihood-ratio* for the kth detector output as follows [33]

$$\mathcal{L}[\mathcal{D}_k] = \mathcal{L}[\mathcal{D}_{k-1}] \times \frac{\Pr[\underline{\mathbf{d}}(\epsilon, x_k)|\mathcal{D}_{k-1}, \mathcal{H}_1]}{\Pr[\underline{\mathbf{d}}(\epsilon, x_k)|\mathcal{D}_{k-1}, \mathcal{H}_0]}; \quad k = 0, 1, \ldots, K \tag{11.108}$$

with $\Pr[\underline{\mathbf{d}}(\epsilon, x_0)|\mathcal{D}_{-1}, \mathcal{H}_\ell] = \Pr[\underline{\mathbf{d}}(\epsilon, x_1)|\mathcal{H}_\ell]$, the *prior* under each hypothesis.

Anticipating the exponential family of distributions [34], we take logarithms to obtain the sequential *log likelihood ratio* (decision function) $\Lambda[\mathcal{D}_k]$ as

$$\Lambda[\mathcal{D}_k] = \Lambda[\mathcal{D}_{k-1}] + \ln \Pr[\underline{\mathbf{d}}(\epsilon, x_k)|\mathcal{D}_{k-1}, H_1] - \ln \Pr[\underline{\mathbf{d}}(\epsilon, x_k)|\mathcal{D}_{k-1}, H_0] \tag{11.109}$$

Therefore, the Wald *sequential probability-ratio test* is [33]

$$\begin{array}{lllll} & & \Lambda[\mathcal{D}_k] & > \ln T_1(k) & \text{Accept} \quad H_1 \\ \ln T_0(k) & \leq & \Lambda[\mathcal{D}_k] & \leq \ln T_1(k) & \text{Continue} \\ & & \Lambda[\mathcal{D}_k] & < \ln T_0(k) & \text{Accept} \quad H_0 \end{array} \tag{11.110}$$

where the thresholds are specified in terms of the false alarm (P_{FA}) and miss (P_{Miss}) probabilities (from ROC curve) as

$$T_0(k) = \frac{P_{Miss}(k)}{P_{FA}(k)}; \qquad T_1(k) = \frac{1 - P_{Miss}(k)}{P_{FA}(k)} \tag{11.111}$$

The sequential detector updates the decision function at each step along with the upper and lower decision thresholds based on the desired operating point (from an ROC curve).

For our X-ray material detection problem, we must specify the required probabilities for each hypothesis which implies that we must incorporate the underlying transport physics (Beer's law) developed previously. Incorporating uncertainty and noise into the representation, we choose the *Gauss–Markov* representation of Section 4.5 as

$$\underline{\mathcal{I}}(\epsilon, x_k) = \mathcal{T}(\epsilon, \triangle x_k)\underline{\mathcal{I}}(\epsilon, x_{k-1}) + \triangle x_k \mathcal{T}(\epsilon, \triangle x_k)\underline{\mathbf{w}}(x_{k-1}) \tag{11.112}$$

with corresponding measurement (detector array) model

$$\underline{\mathbf{d}}(\epsilon, x_k) = \mathbf{C}(e, \triangle\epsilon)\underline{\mathcal{I}}(\epsilon, x_{k-1}) + \underline{v}(x_k) \tag{11.113}$$

with $N_{\mathcal{I}}$-intensity (state) and white, Gaussian process uncertainty/noise vectors, $\underline{\mathcal{I}}$, $\underline{\mathbf{w}}$ with $w \sim \mathcal{N}(0, R_{ww})$. Here, $\underline{\mathbf{d}}$ the N_d-measurement (detector array) with corresponding white, Gaussian uncertainty/noise (electronic) vector $\underline{\mathbf{v}}$ for $v \sim \mathcal{N}(0, \mathbf{R}_{vv})$ along with the corresponding $N_d \times N_{\mathcal{I}}$ measurement matrix $\mathbf{C}$. The path length is specified by $\triangle x_k := x_k - x_{k-1}$ with $(N_{\mathcal{I}} \times N_{\mathcal{I}})$ state transition matrix $\mathcal{T}$,

$$\mathcal{T}(\epsilon, \triangle x_k) = e^{\mathbf{A}(\epsilon;\theta)\triangle x_k} \tag{11.114}$$

for $\mathbf{A}(\epsilon;\theta) = \text{diag}\left[-\mu(\epsilon_1;\theta), \ldots, -\mu(\epsilon_N;\theta)\right]$ the material attenuation coefficients and set of intrinsic material properties $\{\theta\}$.

With this representation of X-ray transport, we specify the binary decision problem as

$$\begin{aligned} \mathcal{H}_0 &: \underline{\mathbf{d}}(\epsilon, x_k) = \mathbf{C}(\epsilon, \triangle\epsilon)\hat{\underline{\mathcal{I}}}(\epsilon, x_k; \theta) + \underline{\mathbf{v}}(x_k) && \text{[Nonthreat]} \\ \mathcal{H}_1 &: \underline{\mathbf{d}}(\epsilon, x_k) = \mathbf{C}(\epsilon, \triangle\epsilon)\underline{\mathcal{I}}(\epsilon, x_k; \theta_t) + \underline{\mathbf{v}}(x_k) && \text{[Threat]} \end{aligned} \tag{11.115}$$

by the Gaussian distributions

$$\Pr[\underline{\mathbf{d}}(\epsilon, x_k)|\mathcal{D}_{k-1}; H_0] = \mathcal{N}\left(\hat{\underline{\mathbf{d}}}(x_k|x_{k-1}), \mathbf{R}_{ee}\right)$$

$$\Pr[\underline{\mathbf{d}}(\epsilon, x_k)|\mathcal{D}_{k-1}; H_1] = \mathcal{N}\left(\underline{\mathbf{d}}(\epsilon, x_k; \theta_t), \mathbf{R}_{dd}\right) \tag{11.116}$$

Taking logarithms of these distributions and inserting them into the log-likelihood expression of Eq. 11.109, we obtain

$$\Lambda[\mathcal{D}_k] = \Lambda[\mathcal{D}_{k-1}] - \ln\left((2\pi)^{-N_d/2}|\mathbf{R}_{dd}|^{-1/2}\right)$$
$$-\frac{1}{2}\left(\underline{\mathbf{d}}(\epsilon, x_k) - \underline{\mathbf{d}}(\epsilon, x_k; \theta_t)\right)^T \mathbf{R}_{dd}^{-1}\left(\underline{\mathbf{d}}(\epsilon, x_k) - \underline{\mathbf{d}}(\epsilon, x_k; \theta_t)\right) + \ln\left((2\pi)^{-N_d/2}|\mathbf{R}_{ee}|^{-1/2}\right)$$
$$+\frac{1}{2}\left(\underline{\mathbf{d}}(\epsilon, x_k) - \hat{\underline{\mathbf{d}}}(x_k|x_{k-1})\right)^T \mathbf{R}_{ee}^{-1}\left(\underline{\mathbf{d}}(\epsilon, x_k) - \hat{\underline{\mathbf{d}}}(x_k|x_{k-1})\right) \tag{11.117}$$

Since the underlying physics-based models are assumed Gauss–Markov, we know that the optimal solution to the linear estimation of the detector array measurement is the conditional mean with sequential solution provided by the well-known Kalman filter of Section 5.4. Therefore, we have $\hat{\underline{\mathbf{d}}}(x_k|x_{k-1}) = E\{\underline{\mathbf{d}}(\epsilon, x_k)|\mathcal{D}_{k-1}\}$ the conditional mean at path length x_k based on past path lengths up to x_{k-1} along with conditional covariance $R_{ee}(x_k) = \text{cov}(\underline{\mathbf{e}}(x_k))$—the innovations covariance provided by the sequential algorithm, that is,

$$\begin{aligned}
\hat{\underline{\mathcal{I}}}_\epsilon(x_k|x_{k-1}) &= \mathcal{T}(\epsilon, \triangle x_k; \theta)\hat{\underline{\mathcal{I}}}_\epsilon(x_{k-1}|x_{k-1}) && \text{[Pred. state]}\\
\hat{\underline{\mathbf{d}}}(x_k|x_{k-1}) &= \mathbf{C}(\epsilon, \triangle\epsilon)\hat{\underline{\mathcal{I}}}_\epsilon(x_k|x_{k-1}) && \text{[Pred. measurement]}\\
\tilde{\mathbf{P}}_{\mathcal{I}}(x_k|x_{k-1}) &= \mathcal{T}(\epsilon, \triangle x_k; \theta)\tilde{\mathbf{P}}_{\mathcal{I}}(x_{k-1}|x_{k-1})\mathcal{T}^T(\epsilon, \triangle x_k; \theta) + \mathbf{R}_{ww}(x_{k-1}) && \text{[Pred. cov.]}\\
\underline{\mathbf{e}}(x_k) &= \underline{\mathbf{d}}(\epsilon, x_k) - \hat{\underline{\mathbf{d}}}(x_k|x_{k-1})) && \text{[Innovations]}\\
\mathbf{R}_{ee}(x_k) &= \mathbf{C}(\epsilon, \triangle\epsilon)\tilde{\mathbf{P}}_{\mathcal{I}}(x_k|x_{k-1})\mathbf{C}^T(\epsilon, \triangle\epsilon) + \mathbf{R}_{vv}(x_k) && \text{[Innovations cov.]}\\
K(x_k) &= \tilde{\mathbf{P}}_{\mathcal{I}}(x_k|x_{k-1})\mathbf{C}^T(\epsilon, \triangle\epsilon)\mathbf{R}_{ee}^{1}(x_k) && \text{[Gain]}\\
\hat{\underline{\mathcal{I}}}_\epsilon(x_k|x_k) &= \mathcal{T}(\epsilon, \triangle x_k; \theta)\hat{\underline{\mathcal{I}}}_\epsilon(x_k|x_{k-1}) + K(x_k)\underline{\mathbf{e}}(x_k) && \text{[Updated state]}
\end{aligned} \tag{11.118}$$

Here, we ignore the update terms in the algorithm (for details see Ref. [38]).

An expression for the "target" or "true" distribution can be specified using the true (target) parameters from the available tables [51] for

$$\underline{\mathcal{M}}_{\mathcal{I}}(\epsilon, x_k) = \mathcal{T}(\epsilon, \triangle x_k)\underline{\mathcal{M}}_{\mathcal{I}}(\epsilon, x_k) \tag{11.119}$$

along with the *mean detector measurement*

$$\underline{\mathcal{M}}_{\mathbf{d}}(\epsilon, x_k) = \mathbf{C}(\epsilon, \triangle\epsilon)\underline{\mathcal{M}}_{\mathcal{I}}(\epsilon, x_{k-1}) \tag{11.120}$$

The corresponding state and measurement mean and covariances are given by Eqs. 11.97–11.100, that is,

$$\begin{aligned}
\underline{\mathbf{d}}(\epsilon, x_k; \theta_t) &\leftrightarrow \underline{\mathcal{M}}_{\mathbf{d}}(\epsilon, x_k)\\
\mathbf{R}_{dd}(x_k) &\leftrightarrow \mathbf{C}(\epsilon, \triangle\epsilon)\mathcal{P}_{\mathcal{I}}(x_{k-1})\mathbf{C}^T(\epsilon, \triangle\epsilon) + \mathbf{R}_{vv}(x_k)
\end{aligned} \tag{11.121}$$

Substituting these expressions into the *log-likelihood decision function*, we obtain

$$\Lambda[\mathcal{D}_k] = \Lambda[\mathcal{D}_{k-1}] + \kappa - \frac{1}{2}\left(\underline{\mathbf{d}}(\epsilon, x_k) - \underline{\mathcal{M}_{\mathbf{d}}}(\epsilon, x_k)\right)^T \mathbf{R}_{dd}^{-1}(x_k)$$
$$\left(\mathbf{d}(\epsilon, x_k) - \underline{\mathcal{M}_{\mathbf{d}}}(\epsilon, x_k)\right) + \frac{1}{2}\left(\underline{\mathbf{e}}^T(x_k)\mathbf{R}_{ee}^{-1}(x_k)\underline{\mathbf{e}}(x_k)\right) \quad (11.122)$$

where $\kappa := \frac{1}{2}(\ln|\mathbf{R}_{ee}(x_k)| - \ln|\mathbf{R}_{dd}(x_k)|)$.

Following the simulation, we developed the physics-based processor to enhance the intensity states and detector measurements. In Fig. 11.31, we observe the intensity (state) estimates clearly demonstrating the capability of the processor along with the enhanced measurements shown in Fig. 11.32.

Finally to demonstrate the sequential Bayesian detection algorithm using the enhanced data provided by the Kalman filter (above), we applied it to the synthesized data with a lower signal-to-noise ratio (SNR) with the resulting output shown in Fig. 11.33. We note the log-likelihood decision function and the associated bounds assuming a detection probability of 95% and a false alarm probability of 5% which are used to calculate the upper and lower thresholds. The results for this case are expected with the log-likelihood function increasing until the detection threshold is crossed.

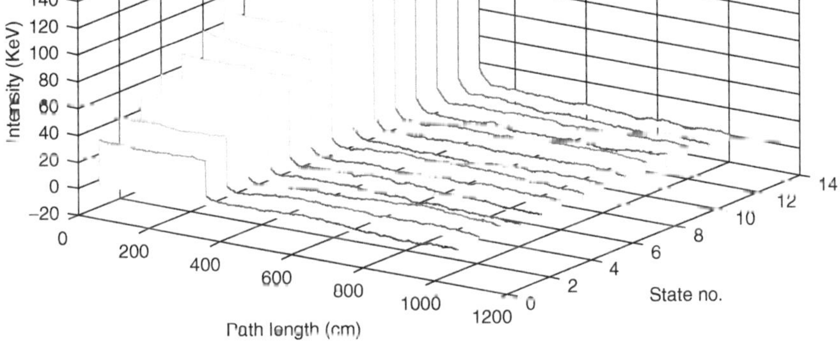

FIGURE 11.31 Physics-based processor results: intensity (state conditional mean) estimation.

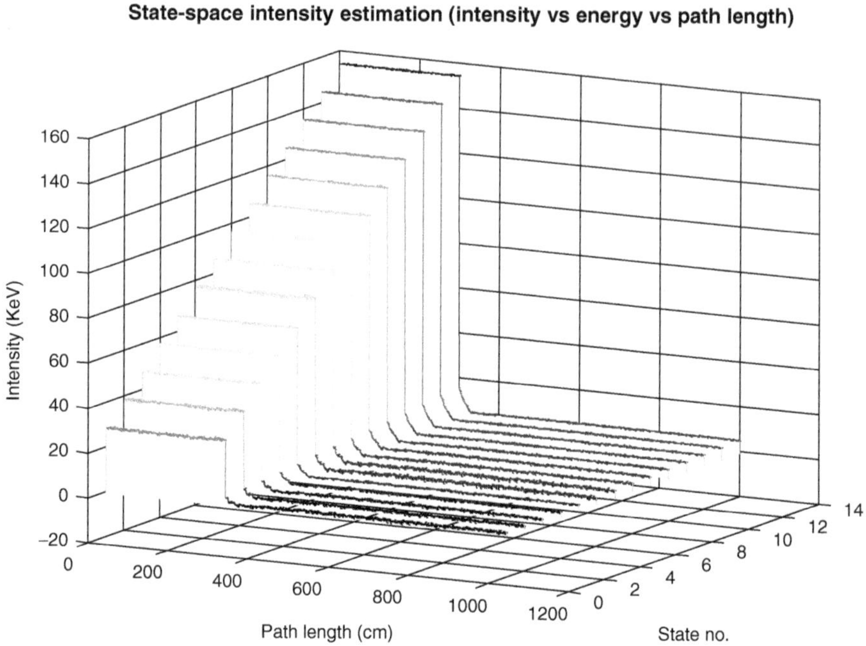

FIGURE 11.32 Physics-based processor results: detector measurement estimation.

11.5.4 Summary

We have demonstrated that a physics-based sequential Bayesian processor can be developed to provide a feasible solution to the X-ray threat contraband detection problem by defining a target X-ray object. Sequential techniques based on the state–space formulation [38] were developed demonstrating their ability to capture the X-ray transport physics for a *lumped* transport model representation. The lumped approach was further investigated and an X-ray transport synthesizer was developed including uncertainties resulting in a Gauss–Markov state–space formulation. Synthesized data for an ammonium nitrate explosive threat was then employed to investigate the performance of a sequential Bayesian detector incorporating a physics-based processor (Kalman filter) for intensity measurement enhancement. The resulting log-likelihood detection technique was capable of performing a detection under the assumed detection and false alarm probabilities (see Ref. [59] for more details).

11.6 ADAPTIVE PROCESSING FOR SHALLOW OCEAN APPLICATIONS

The shallow ocean is an ever-changing environment primarily due to temperature variations in its upper layers directly affecting sound propagation throughout. The need to develop processors capable of tracking these changes implies a stochastic as well as an environmentally adaptive design. Bayesian techniques have evolved

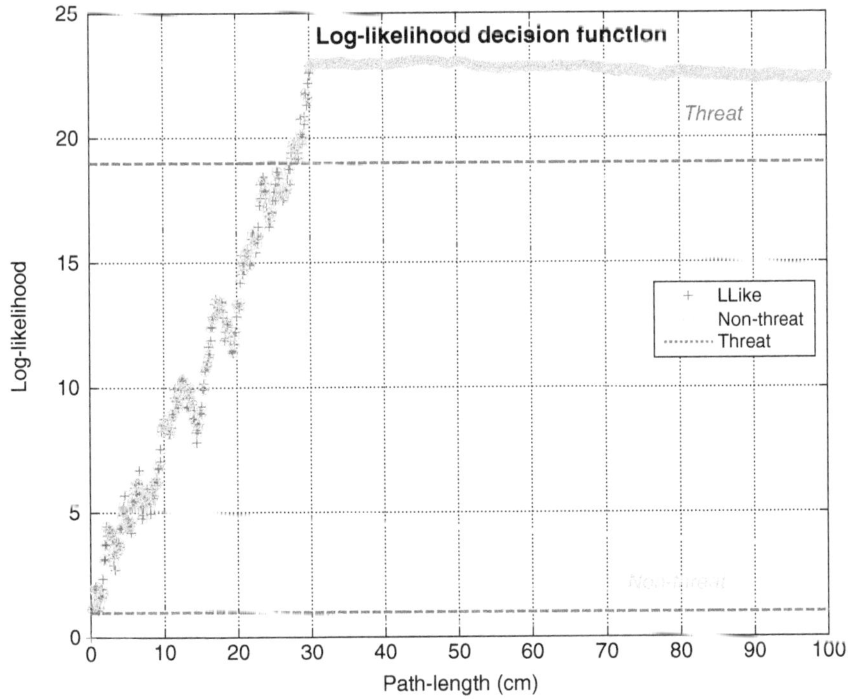

FIGURE 11.33 Physics-based processor threat detection for Gaussian case: log-likelihood-ratio decision function and thresholds for an assumed detection and false alarm probabilities of 95% and 5%, respectively.

to enable a class of processors capable of performing in such an uncertain, non-stationary, non-Gaussian, variable shallow ocean environment. Here, we develop a sequential Bayesian processor capable of providing a joint solution to the modal function tracking and environmental adaptivity problem. The focus is on the development of both a particle filter and an unscented Kalman filter capable of providing reasonable performance for this problem. These processors are applied to hydrophone measurements obtained from a vertical array. The adaptivity problem is attacked by allowing the modal coefficients and/or wavenumbers to be *jointly estimated* from the noisy measurement data along with tracking of the modal functions while simultaneously enhancing the noisy pressure-field measurements.

11.6.1 State-Space Propagator

For the ocean acoustic modal function enhancement problem, we assume a horizontally stratified ocean of depth h with a *known* horizontal source range r_s and depth z_s and that the acoustic energy from a point source can be modeled as a trapped wave governed by the Helmholtz equation [28, 60]. The standard separation of variables technique and removing the time dependence leads to a set of ordinary differential equations; that is, we obtain a "depth-only" representation of the wave equation which is an eigenvalue equation in z with

$$\frac{d^2}{dz^2}\phi_m(z) + \kappa_z^2(m)\phi_m(z) = 0, \quad m = 1, \ldots, M \tag{11.123}$$

whose eigensolutions $\{\phi_m(z)\}$ are the so-called *modal functions* and κ_z is the vertical wavenumber in the z-direction. These solutions depend on the sound- speed profile $c(z)$ and the boundary conditions at the surface and bottom as well as the corresponding *dispersion* relation given by

$$\kappa^2 = \frac{\omega^2}{c^2(z)} = \kappa_r^2(m) + \kappa_z^2(m), \quad m = 1, \ldots, M \tag{11.124}$$

where $\kappa_r(m)$ is the horizontal wavenumber (constant) associated with the mth mode in the r direction and ω is the harmonic source frequency.

By assuming a known horizontal source range a priori, we obtain a range solution given by the Hankel function $H_0(\kappa_r r_s)$ enabling the pressure-field to be represented by

$$p(r_s, z) = \sum_{m=1}^{M} \beta_m(r_s, z_s)\phi_m(z) \tag{11.125}$$

where p is the acoustic pressure; ϕ_m is the mth modal function with the modal coefficient defined by

$$\beta_m(r_s, z_s) := qH_0(\kappa_r r_s)\phi_m(z_s) \tag{11.126}$$

for (r_s, z_s) the source position and q its amplitude.

11.6.1.1 State-Space Model The depth-only eigen equation can be transformed to state–space form by defining the state vector of the mth mode as

$$\underline{\phi}_m(z) := \begin{bmatrix} \phi_m(z) \\ \frac{d}{dz}\phi_m(z) \end{bmatrix} = \begin{bmatrix} \phi_{m1}(z) \\ \phi_{m2}(z) \end{bmatrix} \tag{11.127}$$

leading to the state (vector) equation:

$$\frac{d}{dz}\underline{\phi}_m(z) = \mathbf{A}_m(z)\underline{\phi}_m(z) \tag{11.128}$$

for

$$\mathbf{A}_m(z) = \begin{bmatrix} 0 & 1 \\ -\kappa_z^2(m) & 0 \end{bmatrix} \tag{11.129}$$

Assuming that the ocean acoustic noise can be characterized by additive uncertainties, we can extend this deterministic state equation for the M modes; that is, $\Phi(z) := [\underline{\phi}_1(z)| \cdots |\underline{\phi}_M(z)]^T$ leading to the following $2M$-dimensional Gauss–Markov representation of the model:

$$\frac{d}{dz}\mathbf{\Phi}(z) = \mathbf{A}(z)\mathbf{\Phi}(z) + \mathbf{w}(z) \tag{11.130}$$

where $\mathbf{w}(z) = [w_1 w_2 \ldots w_{2M}]^T$ is additive, zero-mean random noise. The system matrix $\mathbf{A}(z)$ is

$$\mathbf{A}(z) = \begin{bmatrix} \mathbf{A_1(z)} & \cdots & 0 \\ \vdots & \ddots & \vdots \\ 0 & \cdots & \mathbf{A}_M(z) \end{bmatrix} \tag{11.131}$$

with the overall state vector given by

$$\mathbf{\Phi}(z) = [\phi_{11}\phi_{12}|\phi_{21}\phi_{22}| \ldots |\phi_{M1}\phi_{M2}]^T \tag{11.132}$$

This representation leads to the *measurement* equations, which we can write as

$$p(r_s, z) = \mathbf{C}^T(r_s, z_s)\mathbf{\Phi}(z) + v(z) \tag{11.133}$$

where

$$\mathbf{C}^T(r_s, z_s) = \left[\beta_1(r_s, z_s)0| \cdots |\beta_M(r_s, z_s)0\right. \tag{11.134}$$

The random noise terms $\mathbf{w}(z)$ and $v(z)$ can be assumed Gaussian and zero-mean with respective covariance matrices $\mathbf{R}_{ww}$ and R_{vv}. The measurement noise (v) can be used to represent the "lumped" effects of near-field acoustic noise field, flow noise on the hydrophone, and electronic noise. The modal noise ($\mathbf{w}$) can be used to represent the "lumped" uncertainty of sound-speed errors, distant shipping noise, errors in the boundary conditions, sea state effects and ocean inhomogeneities that propagate through the ocean acoustic system dynamics (normal-mode model).

Since the array spatially samples the pressure-field discretizing depth, we choose to discretize the differential state equations using a *central difference* approach for improved numerical stability; that is, assuming uniformly spaced hydrophones, from Eq. 11.123 we have

$$\frac{d^2}{dz^2}\phi_m \approx \frac{\phi_m(z_\ell) - 2\phi_m(z_{\ell-1}) + \phi_m(z_{\ell-2})}{\triangle z_\ell^2} \tag{11.135}$$

for $\triangle z_\ell := z_\ell - z_{\ell-1}$.

Applying this approximation to Eq. 11.123 gives

$$\phi_m(z_\ell) - 2\phi_m(z_{\ell-1}) + \phi_m(z_{\ell-2}) + \triangle z_\ell^2 \kappa_z^2(m)\phi_m(z_{\ell-1}) = 0$$

where z_ℓ is the location of the ℓth sensor. Defining the discrete modal state vector as $\boldsymbol{\phi}_m(z_\ell) := [\phi_m(z_{\ell-2})|\phi_m(z_{\ell-1})]^T$, we obtain the following set of difference equations for the mth mode

$$\begin{aligned}\boldsymbol{\phi}_{m1}(z_\ell) &= \boldsymbol{\phi}_{m2}(z_{\ell-1})\\ \boldsymbol{\phi}_{m2}(z_\ell) &= -\boldsymbol{\phi}_{m1}(z_{\ell-1}) + (2 - \triangle z_\ell^2 \kappa_z^2(m))\boldsymbol{\phi}_{m2}(z_{\ell-1})\end{aligned} \tag{11.136}$$

with each of the corresponding A-submatrices now given by

$$A_m(z_\ell) = \begin{bmatrix} 0 & 1 \\ -1 & 2 - \triangle z_\ell^2 \kappa_z^2(m) \end{bmatrix}; \quad m = 1, \ldots, M \tag{11.137}$$

and

$$\kappa_z^2(m) = \left(\frac{\omega^2}{c^2(z)}\right) - \kappa_r^2(m) \tag{11.138}$$

Substituting this model and combining *all* of the modes as in Eq. 11.131, the following overall *Gauss–Markov* representation of our normal-mode process and measurement is

$$\boldsymbol{\Phi}(z_\ell) = \mathbf{A}(z_\ell)\boldsymbol{\Phi}(z_{\ell-1}) + \mathbf{w}(z_\ell)$$

$$p(r_s, z_\ell) = \mathbf{C}^T(r_s, z_s)\boldsymbol{\Phi}(z_\ell) + v(z_\ell) \tag{11.139}$$

and $\boldsymbol{\Phi}, \mathbf{w} \in \mathcal{R}^{2M\times 1}$, $p, v \in \mathcal{R}^{1\times 1}$ for $\mathbf{w} \sim \mathcal{N}(0, R_{ww})$, $v \sim \mathcal{N}(0, R_{vv})$ with $\boldsymbol{\Phi}(z_\ell) \sim \mathcal{N}(\overline{\boldsymbol{\Phi}}(z_0), \overline{P}(z_0))$, $\mathbf{A} \in \mathcal{R}^{2M\times 2M}$, $C^T \in \mathcal{R}^{1\times 2M}$ and with $\sim$ meaning "distributed as".

This completes the normal-mode representation of the shallow ocean in state–space form; next, we consider *augmenting* this model with unknown parameters to create a parametrically adaptive processor.

11.6.1.2 Augmented State-Space Models

The "parametrically adaptive" processor evolves from the normal-mode representation by defining parameter sets of interest. Variations in the ocean can be reflected, parametrically, in a number of ways. For instance, sound-speed variations are related to temperature changes especially in a shallow ocean environment directly impacting the corresponding dispersion relation of Eq. 11.124 that can be parametrically captured by the horizontal wavenumber. Besides the wavenumbers, modal variations can be reflected through the measured pressure-field relations of Eq. 11.125 that can be parametrically captured by the modal coefficients of Eq. 11.126. Therefore, we choose to use the modal coefficients as well as the horizontal wavenumbers (individually) as the parameters of interest in adapting to the changing shallow ocean environment.

Case(i): Modal Coefficients

The modal coefficients of Eq. 11.126 can be used to capture modal function variations. In this case, we define the unknown *parameter vector* as

$$\theta_m(r_s, z_s) := \beta_m(r_s, z_s); \quad m = 1, \ldots, M$$

and a new "augmented" state vector for the mth mode as

$$\Phi_m(z_\ell; \theta_m) := \Phi_m(z_\ell) = [\boldsymbol{\phi}_{m1}(z_\ell)\boldsymbol{\phi}_{m2}(z_\ell)|\theta_m(z_\ell)]^T$$

With this choice of parameters (modal coefficients), the *augmented* state equations for the mth mode becomes

$$\begin{aligned}\boldsymbol{\phi}_{m1}(z_\ell) &= \boldsymbol{\phi}_{m2}(z_{\ell-1}) + w_{m1}(z_{\ell-1})\\ \boldsymbol{\phi}_{m2}(z_\ell) &= -\boldsymbol{\phi}_{m1}(z_{\ell-1}) + (2 - \triangle z_\ell^2\kappa_z^2(m))\boldsymbol{\phi}_{m2}(z_{\ell-1}) + w_{m2}(z_{\ell-1})\\ \boldsymbol{\theta}_m(z_\ell) &= \boldsymbol{\theta}_m(z_{\ell-1}) + \triangle z_\ell w_{\theta_m}(z_{\ell-1})\end{aligned} \quad (11.140)$$

where we have selected a discrete *random walk* model ($\dot{\theta}_m(z) = w_{\theta_m}(z)$) based on first differences to capture the variations of the modal coefficients with additive, zero-mean, Gaussian noise of covariance $R_{w_{\theta_m}w_{\theta_m}}$.

Note that when we augment the unknown parameters into the state vector to construct the *parametrically adaptive* processor, we assume that they are random (walks) with our pre-computed initial values specified (initial conditions or means) and their corresponding covariances used to bound their uncertainty (2σ confidence bounds).

More succinctly, for the mth mode, we can write

$$\Phi_m(z_\ell) = A_m(z_{\ell-1}; \theta)\Phi_m(z_{\ell-1}) + \mathbf{w}_m(z_{\ell-1}) \quad (11.141)$$

or expanding

$$\begin{bmatrix}\phi_m(z_\ell)\\ ---\\ \theta_m(z_\ell)\end{bmatrix} = \begin{bmatrix}A_m(z_{\ell-1}) & | & 0\\ - & - & -\\ 0 & | & 1\end{bmatrix}\begin{bmatrix}\phi_m(z_{\ell-1})\\ ---\\ \theta_m(z_{\ell-1})\end{bmatrix} + \begin{bmatrix}W_{\phi_m}(z_{\ell-1})\\ ---\\ W_{\theta_m}(z_{\ell-1})\end{bmatrix} \quad (11.142)$$

where $W_{\phi_m} \sim \mathcal{N}(0, R_{W_{\phi_m}W_{\phi_m}})$, $W_{\theta_m} \sim \mathcal{N}(0, R_{W_{\theta_m}W_{\theta_m}})$, $\phi_m(0) \sim \mathcal{N}(\overline{\phi}_m(0), R_{\phi_m\phi_m})$, $\theta_m(0) \sim \mathcal{N}(\overline{\theta}_m(0), R_{\theta_m\theta_m})$.

The corresponding *nonlinear* measurement model is given by

$$p(r_s, z_\ell) = \sum_{m=1}^{M} \theta_m(z_\ell)\phi_m(z_\ell) + v(z_\ell); \quad \ell = 1, \ldots, L \tag{11.143}$$

with dispersion (sound-speed)

$$c(z_\ell) = \frac{\omega}{\sqrt{\kappa_z^2(m) + \kappa_r^2(m)}}; \quad m = 1, \ldots, M; \ell = 1, \ldots, L \tag{11.144}$$

To complete this representation, we combine all of the modes and unknown parameters and therefore the state transition is characterized by the underlying augmented state–space model as

$$\mathbf{\Phi}(z_\ell) = \mathbf{A}(z_{\ell-1}; \Theta)\mathbf{\Phi}(z_{\ell-1}) + \mathbf{w}(z_{\ell-1})$$

and the measurement, on the other hand, is determined from the *nonlinear* pressure-field measurement model

$$p(r_s, z_\ell) = \mathbf{c}\Big[\mathbf{\Phi}(z_\ell); \Theta\Big] + v(z_\ell) \tag{11.145}$$

Note that for this case, the pressure-field is *nonlinear* in the states (modal functions) and parameters (modal coefficients), since they are multiplicands and therefore lead to non-Gaussian measurements.

Case(ii): Horizontal Wavenumbers

The horizontal wavenumbers of Eq. 11.124 can be used to capture sound-speed (temperature) variations. For this case, we define the unknown *parameter vector* as

$$\theta_m(z) := \kappa_r(m); \quad m = 1, \ldots, M$$

and a new "augmented" state vector as

$$\Phi_m(z_\ell; \theta_m) := \Phi_m(z_\ell) = [\boldsymbol{\phi}_{m1}(z_\ell)\boldsymbol{\phi}_{m2}(z_\ell)|\theta_m(z_\ell)]^T$$

With this choice of parameters (horizontal wavenumber), the augmented state equations for the mth mode become

$$\begin{aligned}
\boldsymbol{\phi}_{m1}(z_\ell) &= \boldsymbol{\phi}_{m2}(z_{\ell-1}) + w_{m1}(z_{\ell-1}) \\
\boldsymbol{\phi}_{m2}(z_\ell) &= -\boldsymbol{\phi}_{m1}(z_{\ell-1}) + \left(2 - \Delta z_\ell^2\left(\frac{\omega^2}{c^2(z_\ell)} - \theta_m^2(z_{\ell-1})\right)\right)\boldsymbol{\phi}_{m2}(z_{\ell-1}) + w_{m2}(z_{\ell-1}) \\
\boldsymbol{\theta}_m(z_\ell) &= \boldsymbol{\theta}_m(z_{\ell-1}) + \Delta z_\ell w_{\theta_m}(z_{\ell-1})
\end{aligned} \tag{11.146}$$

where we have again selected a discrete *random walk* model $\left(\dot{\theta}_m(z) = w_{\theta_m}(z)\right)$ to capture the variations of the horizontal wavenumber with additive, zero-mean, Gaussian noise of covariance $R_{w_{\theta_m} w_{\theta_m}}$. Note that even though we know that theoretically the horizontal wavenumbers are constant for each mode, we incorporate this stochastic representation due to the uncertainty inherent in the measurements and the parametric model itself.

More succinctly, for the mth mode, we can write

$$\Phi_m(z_\ell) = A_m(z_{\ell-1};\theta)\Phi_m(z_{\ell-1}) + w_m(z_{\ell-1}) \tag{11.147}$$

for

$$A_m(z_{\ell-1};\theta) = \begin{bmatrix} 0 & 1 & | & 0 \\ -1 & 2 - \triangle z_\ell^2\left(\frac{\omega^2}{c^2(z_\ell)} - \theta_m^2(z_{\ell-1})\right) & | & 0 \\ - & - & - & \\ 0 & 0 & | & 1 \end{bmatrix}$$

The corresponding measurement model is given by

$$p(r_s, z_\ell) = \sum_{m=1}^{M} \beta_m\left(r_s, z_s; \theta_m(z_\ell)\right)\phi_m(z_\ell) + v(z_\ell); \quad \ell = 1, \ldots, L \tag{11.148}$$

with

$$\beta_m(r_s, z_s; \theta_m) := qH_0(\theta_m(z_\ell)r_s)\phi_m(z_s) \tag{11.149}$$

and dispersion (sound-speed)

$$c(z_\ell;\theta_m) = \frac{\omega}{\sqrt{\kappa_z^2(m) + \theta_m^2(z_\ell)}}; \quad m = 1, \ldots, M; \ell = 1, \ldots, L \tag{11.150}$$

Here the "combined" augmented model for this case leads to both a *nonlinear state and measurement space*, that is,

$$\begin{aligned} \mathbf{\Phi}(z_\ell;\Theta) &= \mathbf{a}\left[\mathbf{\Phi}(z_{\ell-1};\Theta)\right] + \mathbf{w}(z_{\ell-1}) \\ p(r_s, z_\ell) &= c\left[\mathbf{\Phi}(z_\ell;\Theta)\right] + v(z_\ell) \end{aligned} \tag{11.151}$$

In this case, both the propagator and the pressure-field measurements are nonlinear functions of the states (modes) and unknown parameters (wavenumbers). Note that the modal coefficients are also direct functions of the estimated wavenumbers and are adapted simultaneously. Therefore, this processor is clearly non-Gaussian similar to the previous case.

It should be noted that the initial model parameters are obtained from the prior solution of the boundary value problem typically developed as part of the experimental design process and/or after the experiment has been executed. Here, the initial "guesses" at modal coefficients and modal functions themselves are calculated based on the experimental conditions such as frequencies, current-temperature-density (CTD), archival sound-speed profiles (SSP), boundary conditions, horizontal wavenumber estimators (e.g., see Refs. [61] and [62] for more details) to provide the input to the normal-mode boundary-value problem (BVP) solutions (SNAP [37], KRAKEN [63], SAFARI [64]) yielding the required parameters. These parameters are then input to the state–space, measurement, and noise/uncertainty models.

11.6.2 Processors

In this section, we briefly discuss the processors for our shallow oceanic problem with details available in Sections 6.4 and 7.5. The basic adaptive problem can be defined in terms of the mathematical models as:

GIVEN, $\left[\{p(r_s, z_\ell)\}, \{c(z_\ell)\}\right]$, a set of noisy pressure-field and sound-speed measurements varying in depth along with the underlying state–space model of Eqs. 11.147, 11.148, and 11.150 with unknown parameters $\{\theta(z_\ell)\}$, FIND the "best" (minimum error variance) estimates (joint) of the modal functions and parameters, that is, $\{\hat{\phi}_m(z_\ell)\}, \{\hat{\theta}_m(z_\ell)\}; m = 1, \ldots, M$ and measurements $\{\hat{p}(r_s, z_\ell)\}$.

The solution to this problem lies in the *joint state/parameter estimation* problem, that is, defining the *augmented state vector*,

$$\boldsymbol{\Phi}(z_\ell; \boldsymbol{\Theta}) := \begin{bmatrix} \boldsymbol{\Phi}(z_\ell) \\ --- \\ \boldsymbol{\Theta} \end{bmatrix}$$

and starting with the joint distribution applying Bayes' theorem, we obtain

$$\Pr[\boldsymbol{\Phi}(z_\ell; \boldsymbol{\Theta})|P_\ell] = \left(\frac{\Pr\left[p(r_s, z_\ell)|\boldsymbol{\Phi}(z_\ell; \boldsymbol{\Theta})\right] \times \Pr\left[\boldsymbol{\Phi}(z_\ell; \boldsymbol{\Theta})|\boldsymbol{\Phi}(z_{\ell-1}; \boldsymbol{\Theta})\right]}{\Pr\left[p(r_s, z_\ell)|P_{\ell-1}\right]} \right) \times \Pr[\boldsymbol{\Phi}(z_{\ell-1}; \boldsymbol{\Theta})|P_{\ell-1}] \tag{11.152}$$

where we have assumed conditional independence and defined the set of measurements as $P_\ell := \{p(r_s, z_1), \ldots, p(r_s, z_\ell)\}$.

Define the *joint weighting function* in terms of the likelihood, transition, and evidence as (see Section 2.5 for details)

$$W(z_\ell; \boldsymbol{\Theta}) := \left(\frac{\Pr\left[p(r_s, z_\ell)|\boldsymbol{\Phi}(z_\ell; \boldsymbol{\Theta})\right] \times \Pr\left[\boldsymbol{\Phi}(z_\ell; \boldsymbol{\Theta})|\boldsymbol{\Phi}(z_{\ell-1}; \boldsymbol{\Theta})\right]}{\Pr\left[p(r_s, z_\ell)|P_{\ell-1}\right]} \right) \tag{11.153}$$

yielding the *sequential* Bayesian posterior distribution as

$$\Pr[\boldsymbol{\Phi}(z_\ell;\boldsymbol{\Theta})|P_\ell] = W(z_\ell;\boldsymbol{\Theta}) \times \Pr[\boldsymbol{\Phi}(z_{\ell-1};\boldsymbol{\Theta})|P_{\ell-1}] \tag{11.154}$$

In this study, we are interested in investigating the applicability of the PFof Section 7.5 and the unscented Kalman filter (UKF) of Section 6.3 with the goal of analyzing their performance on pressure-field data obtained from the well-known Hudson Canyon experiments performed on the New Jersey shelf [61, 62]. Recall that the PF is a sequential MCMC Bayesian processor capable of providing reasonable performance for a multi-modal (multiple peaked) distribution problem estimating a nonparametric representation of the posterior distribution. On the other hand, the UKF is a processor capable of representing any unimodal (single-peaked) distribution using a statistical linearization technique based on sigma points that deterministically characterize the posterior.

Here, we are concerned with the *joint estimation problem* consisting of setting a prior for θ and augmenting the state vector to solve the problem as defined above thereby converting the parameter estimation problem to one of optimal filtering. Thus, the particle filter estimates the weights required to specify the posterior distribution, empirically, that is,

$$\hat{\Pr}[\boldsymbol{\Phi}(z_\ell;\boldsymbol{\Theta})|P_\ell] \approx \frac{1}{N_p}\sum_{i=1}^{N_p} \hat{W}_i(z_\ell;\boldsymbol{\Theta}) \times \delta\left(\boldsymbol{\Phi}(z_\ell;\boldsymbol{\Theta}) - \boldsymbol{\Phi}_i(z_\ell;\boldsymbol{\Theta})\right) \tag{11.155}$$

The approach we chose for our problem is to estimate these weights based on the concept of importance sampling (see Section 3.6). Recall that *importance sampling* is a technique to compute statistics with respect to one distribution using random samples drawn from another. It is a method of simulating samples from a proposal or sampling (importance) distribution to be used to approximate a targeted distribution (joint posterior) by appropriate weighting. For this choice, the weighting function is defined by

$$\mathcal{W}(z_\ell;\boldsymbol{\Theta}) := \frac{\Pr[\boldsymbol{\Phi}(z_\ell;\boldsymbol{\Theta})|P_\ell]}{q[\boldsymbol{\Phi}(z_\ell;\boldsymbol{\Theta})|P_\ell]} \tag{11.156}$$

where $q[\cdot]$ is the proposed sampling or *importance distribution*.

For the "sequential" case we have that the weighting function becomes

$$\mathcal{W}(z_\ell;\boldsymbol{\Theta}) \propto \left(\frac{\Pr\left[p(r_s,z_\ell)|\boldsymbol{\Phi}(z_\ell;\boldsymbol{\Theta})\right] \times \Pr\left[\boldsymbol{\Phi}(z_\ell;\boldsymbol{\Theta})|\boldsymbol{\Phi}(z_{\ell-1};\boldsymbol{\Theta})\right]}{q[\boldsymbol{\Phi}(z_\ell;\boldsymbol{\Theta})|\boldsymbol{\Phi}(z_{\ell-1};\boldsymbol{\Theta}),P_\ell]}\right) \times \mathcal{W}(z_{\ell-1};\boldsymbol{\Theta}) \tag{11.157}$$

where $\propto$ means *proportional to* up to a normalizing constant (evidence).

There are a variety of PF algorithms available, each evolving by a particular choice of the sampling or importance distribution, but the simplest is the *bootstrap* technique which we apply to our problem. Here, the importance distribution is selected as the transition prior, that is,

$$q\left[\mathbf{\Phi}(z_\ell;\mathbf{\Theta})|\mathbf{\Phi}(z_{\ell-1};\mathbf{\Theta}),P_\ell)\right] \longrightarrow \Pr\left[\mathbf{\Phi}(z_\ell;\mathbf{\Theta})|\mathbf{\Phi}(z_{\ell-1};\mathbf{\Theta})\right] \quad (11.158)$$

and substituting into Eq. 11.157, we obtain

$$\mathcal{W}(z_\ell;\mathbf{\Theta}) = \Pr\left[p(r_s,z_\ell)|\mathbf{\Phi}(z_\ell;\mathbf{\Theta})\right] \times \mathcal{W}(z_{\ell-1};\mathbf{\Theta}) \quad (11.159)$$

Thus, we see that once the underlying posterior is available, the estimates of important statistics can be inferred directly. For instance, the MAPestimate is simply found by locating a particular particle $\hat{\phi}_i(z_\ell)$ corresponding to the maximum of the PMF, that is, $\hat{\mathbf{\Phi}}_i(z_\ell;\mathbf{\Theta})_{MAP}$, while the conditional mean estimate is calculated by integrating the posterior to give $\hat{\mathbf{\Phi}}_i(z_\ell;\mathbf{\Theta})_{CM}$.

For the bootstrap implementation, we need only draw noise samples from the state and parameter distributions and use the dynamic models above (normal-mode/random walk) in Eq. 11.147 to generate the set of particles $\{\mathbf{\Phi}_i(z_\ell;\mathbf{\Theta})\} \longrightarrow \{\mathbf{\Phi}_i(z_\ell),\mathbf{\Theta}_i(z_\ell)\}$ for $i = 1,\ldots,N_p$. That is, both sets of particles are generated from the augmented models (linear/nonlinear) for each individual case (adaptive modal coefficients or adaptive wavenumbers) from

$$\mathbf{\Phi}_i(z_\ell;\Theta) = \begin{cases} \mathbf{A}(z_{\ell-1})\mathbf{\Phi}_i(z_{\ell-1}) + \mathbf{w}_i(z_{\ell-1}) & \text{(Case i: modal coefficients)} \\ \mathbf{a}\left[\mathbf{\Phi}_i(z_{\ell-1};\Theta)\right] + \mathbf{w}_i(z_{\ell-1}) & \text{(Case ii: wavenumbers)} \end{cases} \quad (11.160)$$

while the likelihood is determined from the nonlinear pressure-field measurement model

$$p(r_s,z_\ell) = c\left[\mathbf{\Phi}_i(z_\ell;\Theta)\right] + v(z_\ell) \quad (11.161)$$

Assuming additive Gaussian noise the likelihood probability is given by

$$\Pr[p(r_s,z_\ell)|\mathbf{\Phi}_i(z_\ell)] = \frac{1}{\sqrt{2\pi R_{vv}}} \times \exp\left\{-\frac{1}{2R_{vv}}\left(p(r_s,z_\ell) - c\left[\mathbf{\Phi}_i(z_\ell;\Theta)\right]\right)^2\right\} \quad (11.162)$$

Thus, we estimate the posterior distribution using a sequential Monte Carlo approach and construct a *bootstrap particle filter* using the following steps:

- Initialize: $\mathbf{\Phi}_i(0)$, $\mathbf{w}_i \sim \mathcal{N}(0,\mathbf{R}_{ww})$, $W_i(0) = 1/N_p; i = 1,\ldots,N_p$;

- State transition: $\boldsymbol{\Phi}_i(z_\ell;\Theta) = \begin{cases} \mathbf{A}(z_{\ell-1})\boldsymbol{\Phi}_i(z_{\ell-1}) + \mathbf{w}_i(z_{\ell-1}) & \text{(Case i)} \\ \mathbf{a}\left[\boldsymbol{\Phi}_i(z_{\ell-1};\Theta)\right] + \mathbf{w}_i(z_{\ell-1}) & \text{(Case ii)}; \end{cases}$
- Likelihood probability: $\Pr[p(r_s, z_\ell)|\boldsymbol{\Phi}_i(z_\ell)]$ of Eq. 11.162;
- Weights: $W_i(z_\ell;\boldsymbol{\Theta}) = W_i(z_{\ell-1};\boldsymbol{\Theta}) \times \Pr[p(r_s, z_\ell)|\boldsymbol{\Phi}_i(z_\ell)]$;
- Normalize: $\mathcal{W}_i(z_\ell;\boldsymbol{\Theta}) = \frac{W_i(z_\ell;\boldsymbol{\Theta})}{\sum_{i=1}^{N_p} W_i(z_\ell;\boldsymbol{\Theta})}$;
- Resample: $\tilde{\boldsymbol{\Phi}}_i(z_\ell;\boldsymbol{\Theta}) \Rightarrow \boldsymbol{\Phi}_i(z_\ell;\boldsymbol{\Theta})$;
- Posterior: $\hat{\Pr}[\boldsymbol{\Phi}(z_\ell;\boldsymbol{\Theta})|P_\ell] = \sum_{i=1}^{N_p} \mathcal{W}_i(z_\ell : \boldsymbol{\Theta}) \times \delta\left(\boldsymbol{\Phi}(z_\ell;\boldsymbol{\Theta}) - \boldsymbol{\Phi}_i(z_\ell;\boldsymbol{\Theta})\right)$; and
- MAP estimate: $\hat{\boldsymbol{\Phi}}_i(z_\ell;\boldsymbol{\Theta})_{\text{MAP}} = \max_i \hat{\Pr}[\boldsymbol{\Phi}_i(z_\ell;\boldsymbol{\Theta})|P_\ell]$;
- MMSE (CM) estimate: $\hat{\boldsymbol{\Phi}}_i(z_\ell;\boldsymbol{\Theta})_{\text{MMSE}} = \frac{1}{N_p}\sum_{i=1}^{N_p} \mathcal{W}_i(z_\ell;\boldsymbol{\Theta}) \times \boldsymbol{\Phi}_i(z_\ell;\boldsymbol{\Theta})$.

The particle filter (bootstrap) algorithm of Section 7.5 is illustrated with the prediction and update steps along with a resampling algorithm to provide convergence.

11.6.3 Model-Based Ocean Acoustic Processing

In this section, we discuss the development of the propagators for the Hudson Canyon experiment performed in 1988 in the Atlantic with the primary goal of investigating acoustic propagation (transmission and attenuation) using continuous wave data [61, 62]. The Hudson Canyon is located off the coast of New Jersey in the area of the Atlantic Margin Coring project borehole 6010. The seismic and coring data are combined with sediment properties measured at that site. Excellent agreement was determined between the model and data indicating a well-known, well-documented shallow water experiment with bottom interaction and yielding ideal data sets for investigating the applicability of an MBP to measured ocean acoustic data. The experiment was performed at low frequencies (50–600 Hz) in shallow water of 73 m depth during a period of calm sea state. A calibrated acoustic source was towed at roughly 36 m depth along the 73-m isobath radially to distances of 4–26 km. The ship speed was between 2 and 4 knots. The fixed vertical hydrophone array consisted of 24 phones spaced 2.5 m apart extending from the seafloor up to a depth of about 14 m below the surface. The CTD and SSP measurements were made at regular intervals and the data were collected under carefully controlled conditions in the ocean environment. The normalized horizontal wavenumber spectrum for a 50 Hz temporal frequency is dominated by five modes occurring at wavenumbers between 0.14 and 0.21 m^{-1} with relative amplitudes increasing with increased wavenumber. A SNAP [37] simulation was performed and the results agree quite closely, indicating a well-understood ocean environment.

In order to construct the state–space propagator, we require the set of parameters which were obtained from the experimental measurements and processing (wavenumber spectra). The horizontal wavenumber spectra were estimated using synthetic aperture processing [61]. Eight temporal frequencies were employed: four on the inbounds (75 Hz, 275 Hz, 575 Hz, 600 Hz) and four on the outbound (50 Hz, 175 Hz, 375 Hz, 425 Hz). In this application, we will confine our investigation to the 50 Hz case, which is well documented, and to horizontal ranges from 0.5 to 4 km. The raw measured data was processed (sampled, corrected, filtered, etc.) and supplied for this investigation. For this investigation, we used a single snapshot of the pressure-field across the vertical array.

11.6.3.1 Adaptive PF Design: Modal Coefficients The design and development of the environmentally adaptive PF proceeds through the following steps: (1) preprocessing the raw experimental data; (2) solving the BVP [37] to obtain initial parameter sets for each temporal frequency (e.g., modal coefficients, wavenumbers, initial conditions.); (3) state–space forward propagator simulation of synthetic data for PF analysis/design; (4) application to measured data; and (5) PF performance analysis.

Preprocessing of the measured pressure-field data follows the usual pattern of filtering, outlier removal and Fourier-transforming to obtain the complex pressure-field as a function of depth along the array. These data along with experimental conditions (frequencies, current-temperature-density (CTD), sound-speed profiles (SSP), boundary conditions, horizontal wavenumber estimators (see Ref. [61] for details) provide the input to the normal mode BVP solutions (SNAP [37]) yielding the output parameters. These parameters are then used as input to the state–space forward propagator developed previously.

The state–space propagator is then used to develop a set of synthetic pressure-field data with higher resolution than the original raw data (e.g., 46-element array rather than 23-element at half-wave inter-element spacing). This set represents the "truth" data that can be investigated when "tuning" the PF (e.g., number of particles, covariances). Once tuned, the processors are applied directly to the measured pressure-field data (23-elements) after re-adjusting some of the processor parameters (covariances). Here, the performance metrics are estimated and processor performance analyzed. Since each run of the PF is a random realization, that is, the process noise inputs are random, an ensemble of results are estimated with its statistics presented. In this way, we can achieve a detailed analysis of the processor performance prior to fielding and operational version. We constrain our discussion results to processing the noisy experimental pressure-field measurements.

We performed a series of "tuning" runs for both the UKF and PF. We primarily adjusted the process noise covariance matrix ($\mathbf{R}_{ww}$) for each of the modal functions and then executed a 100-member ensemble of realizations using these parameters. The particle filter was designed with the same parameters and 1500-particles were used to characterize the posterior PMF at each depth. Resampling (see Section 7.4) was applied at every iteration of the PF to avoid any potential degradation.

First, we investigate the enhancement capabilities of the PF in estimating the pressure-field over a 100-member ensemble shown in Fig. 11.34. The resulting figures show the averaged PF estimates. We observe the raw data (Data) as well as both MAP

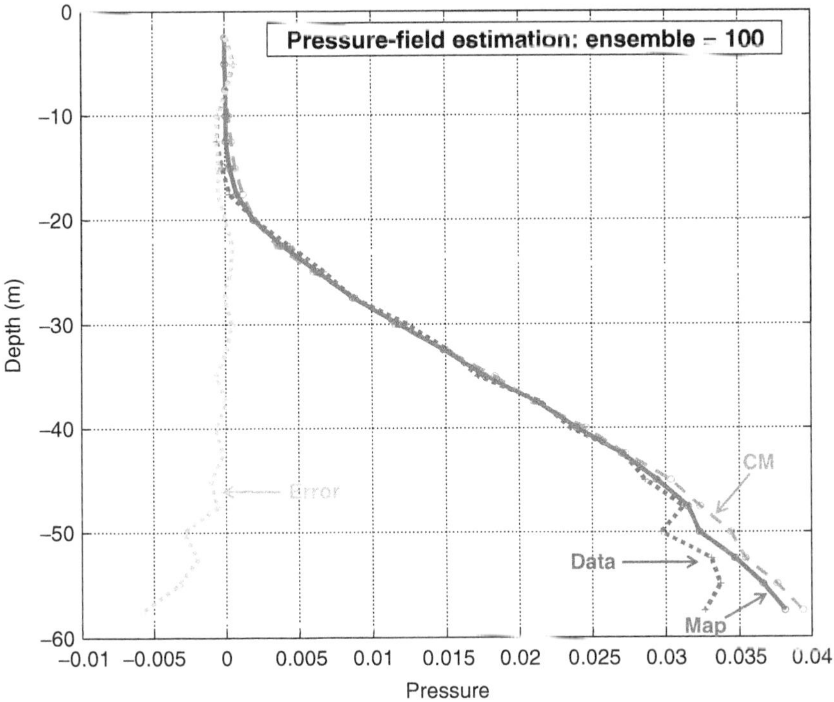

FIGURE 11.34 Raw/enhanced pressure-field (Data) data from the Hudson Canyon experiment using particle filter estimators with adaptive modal coefficients: maximum a-posteriori (MAP), conditional mean (CM) and the corresponding innovations (Error) sequence.

and CM estimates. Both estimators are capable of tracking the field quite well and even filter the erratic measurements near the bottom of the channel. The innovations or residuals (Error) are also shown in the figure. Both estimators are capable of tracking and enhancing the pressure-field. Using classical performance (sanity tests) metrics on the innovations sequence (Error), the zero-mean/whiteness tests, both processors satisfy the criteria of unbiasedness (Z-M: $6.2 \times 10^{-4} < 4.9 \times 10^{-1}$) and uncorrelated innovations, that is, less than 5% exceeding the bound (6.3%). The WSSRtest of Section 5.7 is also applied with satisfactory results, that is, *no* samples exceed the threshold indicating a functionally "tuned" processor. The UKF processor also produced reasonable results for the enhanced pressure-field (not shown).

Ensemble mode tracking results are shown in Fig. 11.35 for each of the modal function estimators, the PF (MAP/CM) and the UKF. In Fig. 11.35, we observe the performance of the PF which appears to track the modes quite well compared to the UKF. It is interesting to note that the modal coefficient estimates are constantly being adapted (adjusted) by the processor throughout the runs attesting to the nonstationary nature of the ocean statistics as illustrated in Fig. 11.36.

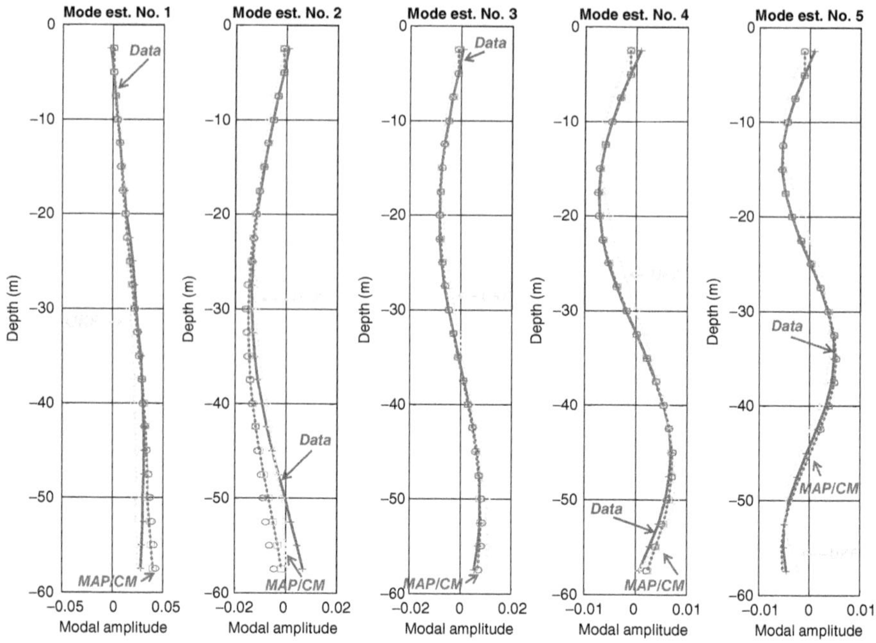

FIGURE 11.35 Modal function tracking with adaptive modal coefficient estimation: raw experimental data (Data), unscented Kalman filter (UKF), maximum a-poster (MAP) (circles) and conditional mean (CM) (squares) particle filters.

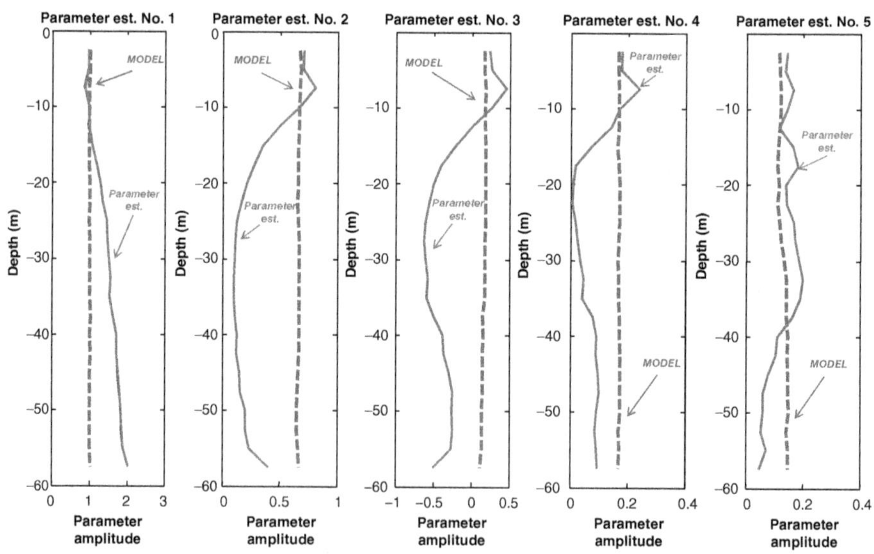

FIGURE 11.36 Adaptive modal coefficient parameter estimation data (Model) from the Hudson Canyon experiment using the MAP particle filter (Parameter est.).

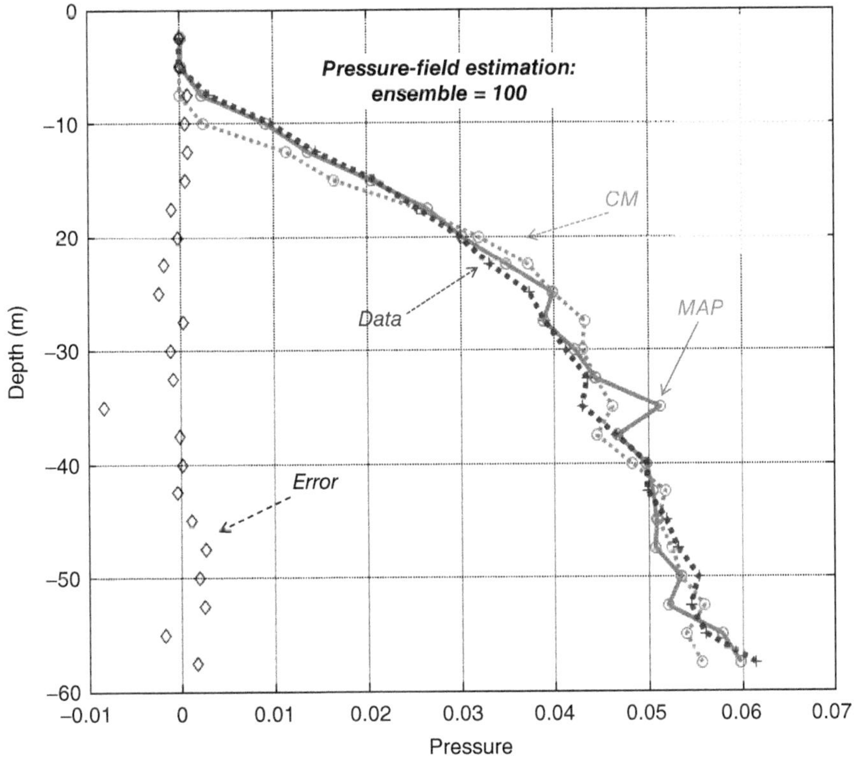

FIGURE 11.37 Raw pressure-field data/enhanced data (Data) from the Hudson Canyon experiment for a 23-element hydrophone vertical array using particle filter estimators with adaptive wavenumber processing: maximum a-posteriori (MAP), conditional mean (CM), and the corresponding innovations (Error) sequence.

This completes the analysis of the Hudson Canyon experimental data for the adaptive (modal coefficient) PF processing performance.

11.6.3.2 Adaptive PF Design: Wavenumbers As before in the modal coefficient case, we investigate the enhancement capabilities of the PF in estimating the pressure-field over a 100-member ensemble shown in Fig. 11.37. Using 1500-particles, we see the raw hydrophone data (dashed line) from the experiment as well as both MAP estimates (circles) and CM estimates (dotted line with circles). Both estimators appear to track the field quite well (true (mean) solution in dashes). The corresponding innovations (residual) sequence is also shown (diamonds). Classically, both estimators produced satisfactory zero-mean/statistical whiteness tests as well as the WSSR tests indicating a "tuned" processor [38].

The ensemble mode tracking results are shown in Fig. 11.38 for each of the modal function estimators, the PF (MAP/CM) and the UKF. In Fig. 11.38, we observe

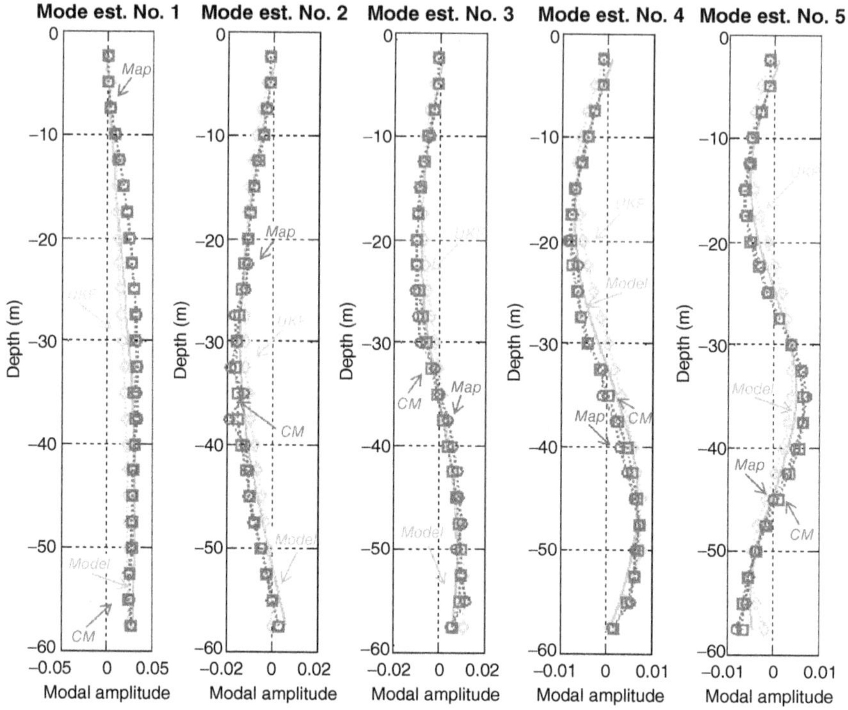

FIGURE 11.38 Modal function tracking for adaptive wavenumber estimation: Hudson Canyon data (Model) of a 23-element array, unscented Kalman filter (UKF), maximum a-posteriori (MAP) and conditional mean (CM) (squares) particle filters.

that the performance of the PF appears to track the modes quite well and better than the UKF. The root-mean-squared (modal tracking) error for each mode is quite reasonable on the order of 10^{-5} again confirming their performance. It is interesting to note that the wavenumber estimates are constantly being adapted (adjusted) by the processor throughout the runs attesting to the nonstationary nature of the ocean statistics. The ensemble average wavenumber estimates are very reasonable: (0.206, 0.197, 0.181, 0.173, 0.142; (True) 0.208, 0.199, 0.183, 0.175, 0.142. The PF and CM ensemble estimates are very close to the true values adapting to the changing ocean environment yet still preserving wavenumber values on the average.

We also illustrate the multi-modal aspect of the oceanic data by observing the modal function posterior probability PDF estimates for mode 5 illustrated in Fig. 11.39. It is clear from the plots that for each depth multiple peaks appear in the posterior estimates. The wavenumber *PDF* estimate corresponding to mode 5 is shown in Fig. 11.40. Again, we note the multiple, well-defined peaks in the posterior distribution leading to the MAP parameter estimate.

This completes the analysis of the synthesized Hudson Canyon experiment and the PF processing performance.

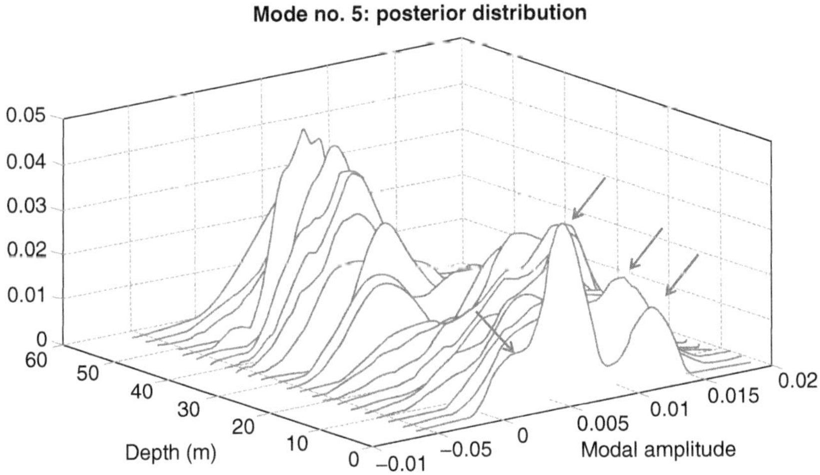

FIGURE 11.39 Probability mass function (PMF) posterior estimation (mode 5) surface for Hudson Canyon 23-element array data (particle vs. time vs. probability).

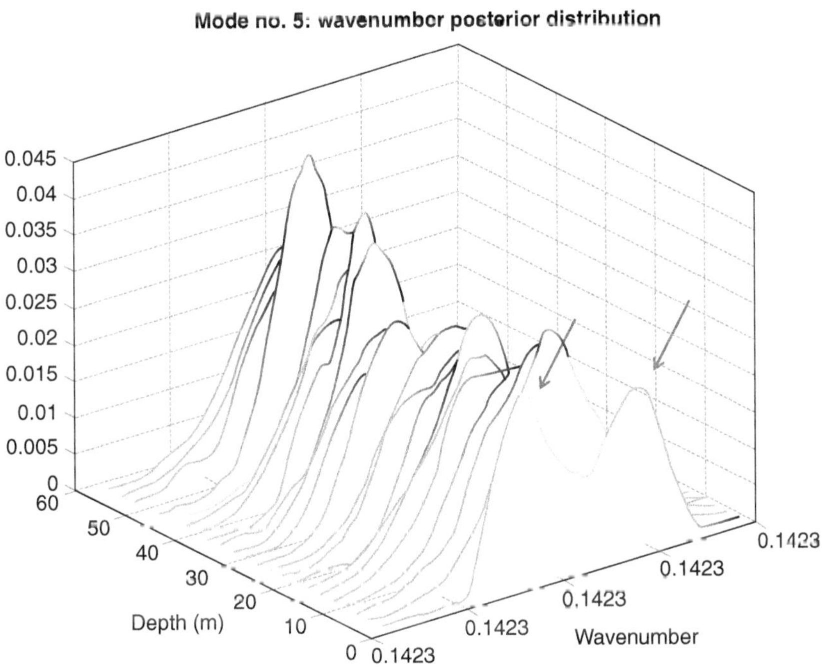

FIGURE 11.40 Probability mass function (PMF) posterior estimation (wavenumber 5) surface for Hudson Canyon 23-element array data (particle vs. time vs. probability).

11.6.4 Summary

In this study, we have discussed the development of environmentally adaptive processors capable of tracking modes and enhancing the raw pressure-field measurements obtained from a vertical hydrophone array in shallow water. The parametric adaption was based on simultaneously estimating either the modal coefficients or the horizontal wavenumbers along with the modes and pressure-field as the environmental parameters of interest. These wavenumber parameters were more challenging from a processor design perspective because of their increased sensitivity to environmental change compared to the modal coefficients. We chose a Bayesian sequential design because of the varying nature of the shallow ocean and applied a normal-mode model in state–space form to create a forward propagator. The algorithms applied were the unscented Kalman filter and the particle filter, both modern approaches applied to this problem (see Chapters 6 and 7). We compared their performance and found slightly better results of the PF over a 100-member ensemble. These results with more detail can be found in a recent paper [65].

REFERENCES

1. D. Speck, E. Bliss, J. Glaze, J. Herris, F. Holloway, J. Hun, B. Johnson, D. Kuizenga, R. Ozarski, H. Patton, P. Ruppert, G. Suski, C. Swift, and C. Thompson, "The Shiva laser-fusion facility," *IEEE J. Quantum Electr.*, **17**, 9, 1599–1619, 1981.
2. J. Liu, H. Furuhashi, A. Torii, R. Sharma, V. Chitnis, B. Singh, J. Yamada, and Y. Uchida, "Automatic mask alignment in the theta direction using moir sensors," *Nanotechnology*, **6**, 135–138, 1995.
3. W. Blum, H. Kroha, and P. Widmann, "A novel laser-alignment system for tracking detectors using transparent silicon strip sensors," *IEEE Trans. Nuclear Sci.*, **43**, 3, 1194–1199, 1996.
4. G. Seward, J. Leszcynski, and E. Mulhern, "Rapid alignment of laser beams within optical systems," *Opt. Eng.*, **36**, 5, 1414–1420, 1997.
5. A. Adolph, A. Boscheron, A. Dulac, and E. Journot, "Final optics design for the megajoule laser," *SPIE Proc.*, **3492**, 1-2, 44–50, 1999.
6. W. He, Q. Chen, R. Xu, Z. Peng, H. Yang, C. Zhu, and J. Zhao, "Image transfer based automatic alignment technique for laser-fusion facility," *Acta Opt. Sin.*, **19**, 9, 1279–1283, 1999.
7. S. Roth, S. Schael, and G. Schmidt, "A test of the laser alignment system ALMY at the TTF-FEL," *Nucl. Instr. Methods Phys. Res. A*, **475**, 537–544, 2001.
8. N. Fleurot, A. Adolf, M. Andre, J. Bruneau, C. Cavailler, M. Novaro, P. Philippe, F. Kovacs, B. Le Garrec, J. Di Nicola, and J. Leidinger, "The ligne dintegration laser (LIL): construction status and first 1-w early results," *Proc. SPIE*, **4948**, 418–424, 2003.
9. D. Liu, J. Zhu, R. Zhu, and D. Fan, "Laser beam automatic alignment in multipass amplifier," *Opt. Eng.*, **43**, 9, 2066–2070, 2004.
10. R. Zacharias, N. Beer, E. S. Bliss, S. Burkhart, S. Cohen, S. Button, R. Van Atta, S. Winters, J. Salmon, M. Latta, C. Stoiz, D. Pigg, and T. Arnold, "Alignment and

wavefront control systems of the National Ignition Facility," *Opt. Eng.*, **43**, 12, 2873–2884, 2004.

11. J. Candy, W. Mcclay, A. Awwal, and S. Ferguson, "Optimal position estimation for the automatic alignment of a high energy laser beam," *J. Opt. Soc. Amer.*, **22**, 7, 1348–1356, 2005.
12. E. Moses, "The national ignition facility comes to life," Science & Technology Review, LLNL Report, pp. 4–14, September 2003.
13. F. Holderner, E. Ables, E. Bliss, S. Boege, R. Boyd, C. Chocol, D. Davis, R. Demaret, R. English, C. Laumann, J. Miller, and S. Thomas, "Beam control and diagnostic functions in the NIF transport spatial filter," *Proc. SPIE*, **3047**, 692–699, 1997.
14. E. Bliss, F. Holderner, J. Salmon, and J. Severyn, "Beam control and laser diagnostic systems," LLNL Report, UCRL-LR-105821-99-1, pp. 79–97, 1999.
15. A. Awwal, W. Mcclay, W. Ferguson, J. Candy, T. Salmon, and P. Wegner, "Composite amplitude modulated phase-only filter based detection and tracking of the back-reflection of KDP images," in Photonic Devices and Algorithms for Computing VI, Proceedings of SPIE 5556, 2004.
16. K. Castleman, *Digital Image Processing* (Prentice-Hall, 1979).
17. A. Parvulescu, "Signal detection in a multipath medium by MESS processing," *J. Acoust. Soc. Am.*, **29**, 223–228, 1965.
18. E. Westwood, "Broadband matched-field source localization," *J. Acoust. Soc. Am.*, **91**, 5, 2777–2789, 1992.
19. C. Clay, "Optimum time domain signal transmission and source localization in a waveguide," *J. Acoust. Soc. Am.*, **81**, 660–664, 1987.
20. S. Li and C. Clay, "Optimum time domain signal transmission and source localization in a waveguide: experiments in an ideal wedge waveguide," *J. Acoust. Soc. Am.*, **82**, 1409–1417, 1987.
21. R. Brienzo and W. Hodgkiss, "Broadband matched-field processing," *J. Acoust. Soc. Am.*, **94**, 1409–1417, 1994.
22. A. Baggeroer, W. Kuperman, and H. Schmidt, "Matched-field processing: source localization in correlated noise as an optimum parameter estimation problem," *J. Acoust. Soc. Am.*, **83**, 2, 571–587, 1988.
23. T. Yang, "Broadband source localization and signature estimation," *J. Acoust. Soc. Am*, **93**, (4), 1797–1806, 1993.
24. J. Candy and E. Sullivan, "Ocean acoustic signal processing: a model-based approach," *J. Acoust. Soc. Am.*, **92**, 12, 3185–3201, 1992.
25. E. Sullivan, *Model-Based Processing for Underwater Acoustic Arrays* (New York: Springer, 2015).
26. J. Candy and E. Sullivan, "Broadband model-based processing for shallow ocean environments," *J. Acoust. Soc. Am.*, **104**, 1, 275–287, 1998.
27. J. Candy, "Broadband processing in a noisy shallow ocean environment: a particle filtering approach," *IEEE J. Oceanic Engr.*, **41**, 1, 1-16, 2016.
28. C. Clay and H. Medwin, *Acoustical Oceanography* (New York: John Wiley & Sons, Inc., 1977).
29. F. Jensen, W. Kuperman, M. Porter, and H. Schmidt, *Computational Ocean Acoustics* (New York: AIP Press, 1994).

30. A. Robinson and D. Lee, *Oceanography and Acoustics* (New York: AIP Press, 1994).
31. M. Arulampalam, S. Maskell, N. Gordon, and T. Clapp, "A tutorial on particle filters for online nonlinear/non-gaussian Bayesian tracking," *IEEE Trans. Sig. Process.*, **50**, 2, 174–188, 2002.
32. H. Van Trees, *Detection, Estimation and Modulation Theory*, Part 1 (New York: John Wiley & Sons, Inc., 1968).
33. A. Wald, *Sequential Analysis* (New York: John Wiley & Sons, 1947) (Reprint by Dover Publications, 1973).
34. A. Papoulis and S. Pillai, *Probability, Random Variables and Stochastic Processes*, 4th Ed. (New York: McGraw-Hill, 2002).
35. C. Andrieu, A. Doucet, S. Singh, and V. Tadic, "Particle methods for change detection, system identification and control," *Proc. IEEE*, **92**, 3, 423–438, 2004.
36. P. Li and V. Kadirkamanathan, "Particle methods based likelihood-ratio approach to fault diagnosis in nonlinear stochastic systems," *IEEE Trans. Syst. Man Cyber. C*, **31**, 3, 337–342, 2001.
37. F. Jensen and M. Ferla, "SNAP: the SACLANTCEN normal-mode acoustic propagation model," SACLANTCEN Report, SM-121, SACLANT Undersea Research Centre, La Spezia, Italy, 1982.
38. J. Candy, *Model-Based Signal Processing* (Hoboken, NJ: Wiley/IEEE Press, 2006).
39. Y. Tang, J. Fang, X. Xu, H. Ji, G. Brown, and T. Thundat, "Detection of femtomolar concentrations of HF using SiO_2 microcantilever," *Anal. Chem.*, **76**, 2478–2481, 2004.
40. M. Su, S. Li, and V. Dravid, "Microcantilever resonance-based DNA detection with nanoparticle probes," *Appl. Phys. Lett.*, **82**, 3562–3564, 2003.
41. G. Muralidharan, A. Wig, L. Pinnaduwage, D. Hedden, T. Thundat, and R. Lareau, "Absorption-desorption characteristics of explosive vapors investigated with microcantilevers," *Ultramicroscopy*, **97**, 433–439, 2003.
42. R. Raiteri, M. Grattarola, H. J. Butt, and P. Skladal, "Micromechanical cantilever-based biosensors," *Sens. Actuators B Chem.*, **79**, 115–126, 2001.
43. N. Lavrik, C. Tipple, M. Sepaniak, and P. Datskos, "Gold nano-structures for transduction of biomolecular interactions into micrometer scale movements," *Biomed. Microdevices*, **3**, 35–44, 2001.
44. J. Tringe, D. Clague, J. Candy, A. Simensky, C. Lee, R. Rudd, and A. Burnham, "A model-based processor design for smart microsensor arrays," *IEEE J. MEMS Syst.*, **15**, 5, 1379–1391, 2006.
45. J. Candy, D. Clague, and J. Tringe, "Model-based processing of microcantilever sensor arrays," *IEEE Signal Process. Magaz.*, **24**, 1, 125–126, 2007.
46. P. Gill, W. Murray, and M. Wright, *Practical Optimization* (New York: Academic Press, 1981).
47. D. Snyder and M. Miller, *Random Point Process in Time and Space* (New York: Springer-Verlag, 1991).
48. R. Evans, *The Atomic Nucleus* (New York: McGraw-Hill, 1985).
49. G. F. Knoll, *Radiation Detection and Measurement*, 3rd Ed. (Hoboken, NJ: John Wiley & Sons, Inc., 2000).
50. G. Gilmore and J. Hemingway, *Practical Gamma-Ray Spectrometry* (Hoboken, NJ: John Wiley & Sons, Inc., 2003).

51. N. Carron, *An Introduction to the Passage of Energetic Particles through Matter* (New York: Taylor & Francis Group, 2007).
52. S. Gulam Razul, W. Fitzgerald, and C. Andrieu, "Bayesian model selection and parameter estimation of nuclear emission spectra using RJMCMC," *Nucl. Instrum. Methods Phys. Res. A*, **497**, 492–510, 2003.
53. E. Barat and J. Heikkinen, "An algorithm for nonparametric Bayesian estimation of a Poisson intensity," *Comput. Statist.*, **12**, 3, 385–402, 1997.
54. J. Candy, E. Breitfeller, B. Guidry, D. Manatt, K. Sale, D. Chambers, M. Axelrod, and A. Meyer, "Physics-based detection of radioactive contraband: a sequential Bayesian approach," *IEEE Trans. Nucl. Sci.*, **56**, 6, 3694–3711, 2009.
55. J. Candy, D. Chambers, E. Breitfeller, B. Guidry, K. Sale, M. Axelrod, and A. Meyer, "Threat detection of radioactive contraband incorporating Compton scattering physics: a model-based processing approach," *IEEE Trans. Nucl. Sci.*, **58**, 1, 214–230, 2011.
56. J. Liu, *Monte Carlo Strategies in Scientific Computing* (New York: Springer-Verlag, 2001).
57. D. Middleton, *An Introduction to Statistical Communication Theory* (New York: McGraw-Hill, 1960).
58. A. Kak and M. Slaney, *Principles of Computerized Tomographic Imaging* (New York: IEEE Press, 1988).
59. J. Candy, "Physics-based X-ray processing: a feasibility study," LLNL Report, UCID-TR-12345, 2012.
60. C. Yardim, Z-H. Michalopoulou, and P. Gerstoft, "An overview of sequential Bayesian filtering in ocean acoustics," *IEEE J. Oceanic Eng.*, **36**, 1, 73–91, 2011.
61. W. Carey, J. Doutt, R. Evans, and L. Dillman, "Shallow watertransmission measurements taken on the New Jersey continental shelf," *IEEE J. Oceanic Eng.*, **20**, (4), 321–336, 1995.
62. A. Rogers, Y. Yamamoto, and W. Carey, "Experimental investigation of sediment effect on acoustic wave propagation in shallow water," *J. Acoust. Soc. Am.*, **93**, 1747–1761, 1993.
63. M. Porter, "The KRAKEN normal mode program," Report SM-245, SACLANTCEN, Italy, pp. 183, 1991.
64. H. Schmidt, "SAFARI: Seismo-acoustic fast field algorithm for range independent environments," Report SM-245, SACLANTCEN, Italy, pp. 142, 1987.
65. J. Candy, "Environmentally adaptive processing for shallow ocean applications: a sequential Bayesian approach," *J. Acoust. Soc. Am.*, **138**, 3, 1268–1281, 2015.

Appendix

PROBABILITY AND STATISTICS OVERVIEW

A.1 PROBABILITY THEORY

Defining a sample space (outcomes) Ω, a field (events) B, and a probability function (on a class of events) Pr, we can construct an *experiment* as the triple $\{\Omega, B, \Pr\}$.

Example A.1

Consider the experiment $\{\Omega, B, \Pr\}$ of tossing a fair coin, then we see that

$$\begin{array}{lll} \text{Sample space:} & \Omega & = \{H, T\} \\ \text{Events:} & B & = \{0, \{H\}, \{T\}\} \\ \text{Probability:} & \Pr(H) & = p \\ & \Pr(T) & = 1 - p \end{array}$$

△△△

With the idea of a sample space, probability function, and experiment in mind, we can now start to define the concept of a discrete random signal more precisely. We define a discrete *random variable* as a real function whose value is determined by the outcome of an experiment. It assigns a real number to each point of a sample space Ω, which consists of all the possible outcomes of the experiment. A random variable X and its realization x are written as

$$X(\omega) = x \quad \text{for } \omega \epsilon \Omega \tag{A.1}$$

Consider the following example of a simple experiment.

Bayesian Signal Processing: Classical, Modern, and Particle Filtering Methods, Second Edition. James V. Candy.

Example A.2

We are asked to analyze the experiment of flipping a fair coin, then the sample space consists of a head or tail as possible outcomes, that is,

$$\Omega = \{0\ H\ T\} \Longrightarrow X(\omega) = x$$
$$\omega = \{H, T\}$$

If we assign a 1 for a head and 0 for a tail, then the random variable X performs the mapping of

$$X(\omega = H) = x(H) = 1$$
$$X(\omega = T) = x(T) = 0$$

where $x(.)$ is called the sample value or realization of the random variable X. △△△

A *probability mass function* is defined in terms of the random variable, that is,

$$\mathrm{P}_X(x_i) = \Pr(X(\omega_i) = x_i) \tag{A.2}$$

and the *probability distribution function* is defined by

$$\mathrm{F}_X(x_i) = \Pr(X(\omega_i) \le x_i) \tag{A.3}$$

These are related by

$$\mathrm{P}_X(x_i) = \sum_i \mathrm{F}_X(x_i)\delta(x - x_i) \tag{A.4}$$
$$\mathrm{F}_X(x_i) = \sum_i \mathrm{P}_X(x_i)\mu(x - x_i) \tag{A.5}$$

where δ and μ are the unit impulse and step functions, respectively.

It is easy to show that the distribution function is a monotonically increasing function (see Papoulis [1] for details) satisfying the following properties:

$$\lim_{x_i \to -\infty} \mathrm{F}_X(x_i) = 0$$

and

$$\lim_{x_i \to \infty} \mathrm{F}_X(x_i) = 1$$

These properties can be used to show that the mass function satisfies

$$\sum_i \mathrm{P}_X(x_i) = 1$$

Either the distribution or the probability mass function completely describes the properties of a random variable. Given either of these functions, we can calculate probabilities that the random variable takes on values in any set of events on the real line.

To complete our coin tossing example, if we define the probability of a head occurring as p, then we can calculate the distribution and mass functions as shown in the following example.

Example A.3

Consider the coin tossing experiment and calculate the corresponding mass and distribution functions. From the previous example, we have

Sample space:	Ω	$= \{H, T\}$
Events:	B	$= \{0, \{\mathrm{H}\}, \{\mathrm{T}\}\}$
Probability:	$\mathrm{P}_X(x_1 = H)$	$= p$
	$\mathrm{P}_X(x_0 = T)$	$= 1 - p$
Random variable:	$X(\omega_1 = H)$	$= x_1 = 1$
	$X(\omega_2 = T)$	$= x_2 = 0$
Distribution:	$\mathrm{F}_X(x_i)$	$= \begin{cases} 1 & x_i \geq 1 \\ 1 - p & 0 \leq x_i \leq 1 \\ 0 & x_i < 0 \end{cases}$

The mass and distribution functions for this example are shown in Fig. A.1. Note that the sum of the mass function value must be 1, and that the maximum value of the distribution function is 1 satisfying the properties mentioned previously. △△△

If we extend the idea that a random variable is now a function of time as well, then we can define a stochastic process as discussed in Chapter 2. More formerly, a random or *stochastic process* is a two-dimensional function of t and ω:

$$X(t, \omega) \quad \omega \epsilon \Omega, \quad t \epsilon T \tag{A.6}$$

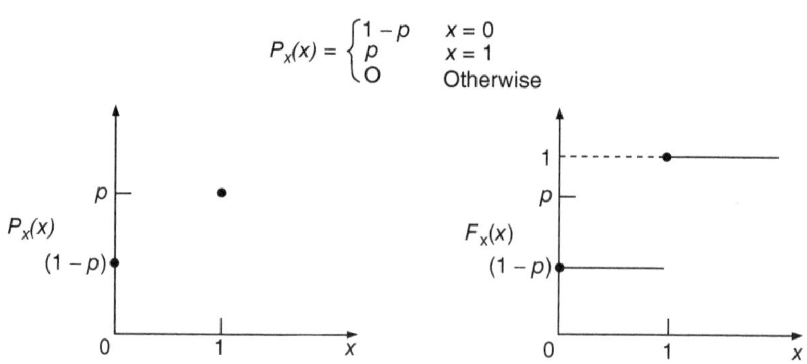

FIGURE A.1 Probability mass and distribution functions for the coin-tossing experiment.

where T is a set of index parameters (continuous or discrete) and Ω is the sample space.

We list some of the major theorems in probability theory and refer the reader to more detailed texts [1, 2].

Univariate:	$\Pr(X)$	$= \mathrm{P}_X(x)$
Bivariate:	$\Pr(X, Y)$	$= \mathrm{P}_{XY}(x, y)$
Marginal:	$\Pr(X)$	$= \sum_y \mathrm{P}_{XY}(x, y)$
Independent:	$\Pr(X, Y)$	$= \mathrm{P}_X(x) \times \mathrm{P}_Y(y)$
Conditional:	$\Pr(X\|Y)$	$= \mathrm{P}_{XY}(x, y)/\mathrm{P}_Y(y)$
Chain Rule:	$\Pr(X, Y, Z)$	$= \Pr(X\|Y, Z) \times \Pr(Y\|Z) \times \Pr(Z)$

For a random variable, we can define basic statistics in terms of the probability mass function. The *expected value* or *mean* of a random variable X, is given by

$$m_x = E\{X\}$$

and is considered the typical or representative value of a given set of data. For this reason, the mean is called a measure of central tendency. The degree to which numerical data tend to spread about the expected value is usually measured by the *variance* or equivalently, *auto-covariance* given by

$$R_{xx} = E\{(X - m_x)^2\}$$

The basic statistical measures are called *ensemble statistics* because they are measured across the ensemble ($i = 1, 2, \ldots$) of data values, that is, the expectation is always assumed over an ensemble of realizations. We summarize these statistics in terms of their mass function as

Expected value: $m_x =$	$E\{X\}$	$= \sum_i X_i \mathrm{P}_X(x_i)$
Nth-moment:	$E\{X^n\}$	$= \sum_i X_i^n \mathrm{P}_X(x_i)$
Nth-moment about mean:	$E\{(X - m_x)^n\}$	$= \sum_i (X_i - m_x)^n \mathrm{P}_X(x_i)$
Mean-squared ($N = 2$):	$E\{X^2\}$	$= \sum_i X_i^2 \mathrm{P}_X(x_i)$
Variance: $R_{xx} =$	$E\{(X_i - m_x)^2\}$	$= \sum_i (X_i - m_x)^2 \mathrm{P}_X(x_i)$
Covariance:	R_{xy}	$= E\{(X_i - m_x)(Y_j - m_y)\}$
Standard deviation:	σ_{xx}	$= \sqrt{R_{xx}}$
Conditional mean:	$E\{X\|Y\}$	$= \sum_i X_i \mathrm{P}(X_i\|Y)$
Joint conditional mean:	$E\{X\|Y, Z\}$	$= \sum_i X_i \mathrm{P}(X_i\|Y, Z)$
Conditional variance:	$R_{x\|y}$	$= E\{(X - E\{X\|Y\})^2\|Y\}$

These basic statistics possess various properties that enable them to be useful for analyzing operations on random variables, some of the more important[1] are

[1] Recall that independence states that the joint mass function can be factored, $\Pr(x, y) = \Pr(x) \times \Pr(y)$, which leads to these properties.

Linearity:	$E\{ax+b\}$	$= aE\{x\}+b=am_x+b$
Independence:	$E\{xy\}$	$= E\{x\}E\{y\}$
Variance:	$R_{xx}(ax+b)$	$= a^2R_{xx}$
Covariance		
Uncorrelated:	$E\{xy\}$	$= E\{x\}E\{y\} \quad \{R_{xy}=0\}$
Orthogonal:	$E\{xy\}$	$= 0$

Note that the expected value operation implies that for stochastic processes, these basic statistics are calculated *across* the ensemble. For example, if we want to calculate the mean of a process, that is,

$$m_x(t) = E\{X(t, \omega_i) = x_i(t)\}$$

We simply take the values of $t=0, 1, \ldots$ and calculate the mean for each value of time across ($i=1, 2, \ldots$) the ensemble. Dealing with stochastic processes is similar to dealing with random variables, except that we must account for the time indices (see Chapter 2 for more details).

Next let us define some concepts about the probabilistic information contained in a random variable. We define the (self) *information* contained in the occurrence of the random variable $X(\omega_i)=x_i$ as

$$I(x_i) = -\log_b \mathrm{P}_X(x_i) \tag{A.7}$$

where b is the base of the logarithm which results in different units for information measures (base $=2 \rightarrow$ bits), and the *entropy* or *average information* of $X(\omega_i)$ as

$$H(x_i) = -E\{I(x_i)\} = \sum_i \mathrm{P}_X(x_i) \log_b \mathrm{P}_X(x_i) \tag{A.8}$$

Consider the case where there is more than one random variable. Then, we define the *joint* mass and distribution functions of an N-dimensional random variable as

$$\mathrm{P}_X(x_1, \ldots, x_N), \quad \mathrm{F}_X(x_1, \ldots, x_N)$$

All of the basic statistical definitions remain as before, except that we replace the scalar with the joint functions. Clearly, if we think of a stochastic process as a sequence of ordered random variables, then we are dealing with joint probability functions, that is, a collection of time-indexed random variables. Suppose we have two random variables x_1 and x_2, and we know that the latter has already assumed a particular value, the we can define the *conditional* probability mass function of x_1 given that $X(\omega_2)=x_2$ has occurred by

$$\Pr(x_1 \mid x_2) := \mathrm{P}_X(X(\omega_1) \mid X(\omega_2) = x_2) \tag{A.9}$$

and it can be shown from basic probabilistic axioms (see Papoulis [1]) that

$$\Pr(x_1 \mid x_2) = \frac{\Pr(x_1, x_2)}{\Pr(x_2)} \tag{A.10}$$

Note also that this expression can also be written as

$$\Pr(x_1, x_2) = \Pr(x_2 \mid x_1)\Pr(x_1) \tag{A.11}$$

Substituting this equation into Eq. A.10 gives *Bayes' rule*, that is,

$$\Pr(x_1 \mid x_2) = \Pr(x_2 \mid x_1)\frac{\Pr(x_1)}{\Pr(x_2)} \tag{A.12}$$

If we use the definition of joint mass function and substitute into the previous definitions, then we can obtain the *probabilistic chain rule* [1–3],

$$\Pr(x_1, \ldots, x_N) = \Pr(x_1 \mid x_2, \ldots, x_N)\Pr(x_2 \mid x_3, \ldots, x_N) \ldots \Pr(x_{N-1} \mid x_N)\Pr(x_N) \tag{A.13}$$

Along with these definitions follows the idea of conditional expectation, that is,

$$E\{x_i \mid x_j\} = \sum_i X_i \Pr(x_i \mid x_j) \tag{A.14}$$

With the conditional expectation defined, we list some of their basic properties:

1. $E_x\{X|Y\} = E\{X\}$, if X and Y are independent
2. $E\{X\} = E_y\{E\{X|Y\}\}$
3. $E_x\{g(y)X|Y\} = g(y)E\{X|Y\}$
4. $E_{x,y}\{g(Y)X\} = E_y\{g(Y)E\{X|Y\}\}$
5. $E_x\{c|Y\} = c$
6. $E_x\{g(Y)|Y\} = g(Y)$
7. $E_{x,y}\{cX + dY|Z\} = cE\{X|Z\} + dE\{Y|Z\}$

The concepts of information and entropy can also be extended to the case of more than one random variable. We define the *mutual information* between two random variables x_i and x_j as

$$I(x_i; x_j) = \log_b \frac{P_X(x_i \mid x_j)}{P_X(x_i)} \tag{A.15}$$

and the *average mutual information* between $X(\omega_i)$ and $X(\omega_j)$ as

$$I(X_i;X_j) = E_{x_i x_j}\{I(x_i,x_j)\} = \sum_i \sum_j \mathrm{P}_X(x_i,x_j) I(x_i,x_j) \tag{A.16}$$

which leads to the definition of *joint entropy* as

$$H(X_i;X_j) = -\sum_i \sum_j \mathrm{P}_X(x_i,x_j) \log_b \mathrm{P}_X(x_i,x_j) \tag{A.17}$$

This completes the section on probability theory; next, let us consider an important multivariable distribution and its properties.

A.2 GAUSSIAN RANDOM VECTORS

In this section, we consider the multivariable Gaussian distribution used heavily in this text to characterize Gaussian random vectors, that is, $\mathbf{z} \sim \mathcal{N}(\mathbf{m}_z, \mathbf{R}_{zz})$ where $\mathbf{z} \in \mathcal{R}^{N_z \times 1}$ and defined by

$$\Pr(\mathbf{z}) = (2\pi)^{-N_z/2} |\mathbf{R}_{zz}|^{-1/2} \exp\left(-\frac{1}{2}(\mathbf{z}-\mathbf{m}_z)'\mathbf{R}_{zz}^{-1}(\mathbf{z}-\mathbf{m}_z)\right) \tag{A.18}$$

where the vector mean and covariance are defined by

$$\mathbf{m}_z := E\{\mathbf{z}\} \quad \text{and} \quad \mathbf{R}_{zz} = \mathrm{Cov}(\mathbf{z}) := E\{(\mathbf{z}-\mathbf{m}_z)(\mathbf{z}-\mathbf{m}_z)'\}$$

Certain properties of the Gaussian vectors are useful such as

- *Linear transformation*: Linear transformations of Gaussian variables are Gaussian; that is, if $\mathbf{z} \sim \mathcal{N}(\mathbf{m}_z, \mathbf{R}_{zz})$ and $\mathbf{y} = \mathbf{A}\mathbf{z} + \mathbf{b}$, then

$$Y \sim \mathcal{N}(\mathbf{A}\mathbf{m}_z + \mathbf{b}, \mathbf{A}\mathbf{R}_{zz}\mathbf{A}') \tag{A.19}$$

- *Uncorrelated Gaussian vectors*: Uncorrelated Gaussian vectors are independent.
- *Sums of Gaussian variables*: Sums of independent gaussian vectors yield Gaussian distributed vectors with mean and variance equal to the sums of the respective means and variances.
- *Conditional Gaussian vectors*: Conditional Gaussian vectors are Gaussian distributed; that is, if $\mathbf{x}$ and $\mathbf{y}$ are jointly Gaussian, with

$$\mathbf{m}_z = E\left\{\begin{matrix}\mathbf{x}\\ \mathbf{y}\end{matrix}\right\} = \begin{bmatrix}\mathbf{m}_\mathbf{x}\\ \mathbf{m}_y\end{bmatrix}; \quad \text{and} \quad \mathbf{R}_{zz} = \mathrm{Cov}(\mathbf{z}) = \begin{bmatrix}\mathbf{R}_{\mathbf{xx}} & \mathbf{R}_{xy}\\ \mathbf{R}_{yx} & \mathbf{R}_{yy}\end{bmatrix}$$

then the conditional distribution for $\mathbf{x}$ and $\mathbf{y}$ is also Gaussian with conditional mean and covariance given by

$$\mathbf{m}_{\mathbf{x}|\mathbf{y}} = \mathbf{m}_x + \mathbf{R}_{xy}\mathbf{R}_{yy}^{-1}(\mathbf{y} - \mathbf{m}_y)$$
$$\mathbf{R}_{\mathbf{x}|\mathbf{y}} = \mathbf{R}_{xx} - \mathbf{R}_{xy}\mathbf{R}_{yy}^{-1}\mathbf{R}_{yx}$$

and the vectors $\mathbf{x} - E\{\mathbf{x}|\mathbf{y}\}$ and $\mathbf{y}$ are independent.

- *Gaussian conditional means*: Let $\mathbf{x}$, $\mathbf{y}$, and $\mathbf{z}$ be jointly distributed Gaussian random vectors and let $\mathbf{y}$ and $\mathbf{z}$ be *independent*, then

$$E\{\mathbf{x}|\mathbf{y},\mathbf{z}\} = E\{\mathbf{x}|\mathbf{y}\} + E\{\mathbf{y}|\mathbf{z}\} - \mathbf{m}_x$$

A.3 UNCORRELATED TRANSFORMATION: GAUSSIAN RANDOM VECTORS

Suppose we have a Gaussian random vector, $\mathbf{x} \sim \mathcal{N}(\mathbf{m}_x, \mathbf{R}_{xx})$ and we would like to transform it to a normalized Gaussian random vector with the mean $\mathbf{m}_x$ removed so that $\mathbf{z} \sim \mathcal{N}(0, \mathbf{I})$. Assume that the mean has been removed ($\mathbf{z} \rightarrow \mathbf{z} - \mathbf{m}_x$), then there exists a nonsingular transformation $\mathbf{T}$ such that $\mathbf{z} = \mathbf{T}\mathbf{x}$, and therefore

$$\mathbf{R}_{zz} = \text{Cov}(\mathbf{z}) = \text{Cov}((\mathbf{T}\mathbf{x})(\mathbf{T}\mathbf{x})') = \mathbf{T}\mathbf{R}_{xx}\mathbf{T}' = \mathbf{I} \quad \text{(A.20)}$$

Thus we must find a transformation that satisfies the relation

$$\mathbf{R}_{zz} = \mathbf{I} = \mathbf{T}\mathbf{R}_{xx}\mathbf{T}' \quad \text{(A.21)}$$

Since $\mathbf{R}_{xx}$ is a positive semi-definite, symmetric matrix it can always be factored into matrix square roots ($\mathbf{R} = \mathbf{U}\mathbf{U}' = \mathbf{R}_{xx}^{1/2}\mathbf{R}_{xx}^{T/2}$) using a Cholesky decomposition [4]; therefore, Eq. A.21 implies

$$\mathbf{R}_{zz} = \mathbf{I} = (\mathbf{T}\mathbf{U})\mathbf{U}'\mathbf{T}' \quad \text{(A.22)}$$

or simply that

$$\mathbf{T} = \mathbf{U}^{-1} = \mathbf{R}_{xx}^{-1/2} \quad \text{(inverse matrix square root)} \quad \text{(A.23)}$$

and therefore

$$\mathbf{R}_{zz} = (\mathbf{R}_{xx}^{-1/2}\mathbf{U})\mathbf{U}'\mathbf{R}_{xx}^{-T/2} = \mathbf{R}_{xx}^{-1/2}\mathbf{R}_{xx}\mathbf{R}_{xx}^{-T/2} = \mathbf{I} \quad \text{(A.24)}$$

the desired result.

This discussion completes the introductory concepts of probability and random variables which is extended to include stochastic processes in Chapter 2 and throughout the text.

REFERENCES

1. A. Papoulis and S. Pillai, *Probability, Random Variables and Stochastic Processes* (New York: McGraw-Hill, 2002).
2. R. Hogg, J. McKlean, and A. Craig, *Introduction to Mathematical Statistics* (Englewood Cliffs, NJ: Prentice-Hall, 2005).
3. A. Jazwinski, *Stochastic Processes and Filtering Theory* (New York: Academic Press, 1970).
4. G. Bierman, *Factorization Methods for Discrete Sequential Estimation* (New York: Academic Press, 1977).

INDEX

Wiley Series on
Adaptive and Cognitive Dynamic Systems

Editor: Simon Haykin

Adali and Haykin · Adaptive Signal Processing: Next Generation Solutions
Beckerman · Adaptive Cooperative Systems
Candy · Model-Based Signal Processing
Candy · Bayesian Signal Processing: Classical, Modern, and Particle Filtering Methods, Second Edition
Chen, Haykin, Eggermont, and Becker · Correlative Learning: A Basis for Brain and Adaptive Systems
Chen and Gu · Control-Oriented System Identification: An H_∞ Approach
Cherkassky and Mulier · Learning from Data: Concepts, Theory, and Methods
Cirrincione and Cirrincione · Neural-Based Orthogonal Data Fitting: The EXIN Neural Networks
Costa and Haykin · Multiple-Input Multiple-Output Channel Models: Theory and Practice
Diamantaras and Kung · Principal Component Neural Networks: Theory and Applications
Haykin · Unsupervised Adaptive Filtering: Blind Source Separation
Haykin · Unsupervised Adaptive Filtering: Blind Deconvolution
Haykin and Liu · Handbook on Array Processing and Sensor Networks
Haykin and Puthussarypady · Chaotic Dynamics of Sea Clutter
Haykin and Widrow · Least Mean-Square Adaptive Filters
Hossain, Le, and Niyato · Radio Resource Management in Multi-Tier Cellular Wireless Networks
Hrycej · Neurocontrol: Towards an Industrial Control Methodology
Hyvürinen, Karhunen, and Oja · Independent Component Analysis
Kristić, Kanellakopoulos, and Kokotović · Nonlinear and Adaptive Control Design
Mann · Intelligent Image Processing
Nikias and Shao · Signal Processing with Alpha-Stable Distributions and Applications
Passino and Burgess · Stability Analysis of Discrete Event Systems
Sánchez-Peña and Sznaier · Robust Systems: Theory and Applications
Sandberg, Lo, Fancourt, Principe, Katagairi, and Haykin · Nonlinear Dynamical Systems: Feedforward Neural Network Perspectives
Sellathurai and Haykin · Space-Time Layered Information Processing for Wireless Communications
Spooner, Maggiore, Ordóñez, and Passino · Stable Adaptive Control and Estimation for Nonlinear Systems: Neural and Fuzzy Approximator Techniques
Tao · Adaptive Control Design and Analysis
Tao and Kokotović · Adaptive Control of Systems with Actuator and Sensor Nonlinearities
Tsoukalas and Uhrig · Fuzzy and Neural Approaches in Engineering
Van Hulle · Faithful Representations and Topographic Maps: From Distortion- to Information-Based Self-Organization
Vapnik · Statistical Learning Theory
Werbos · The Roots of Backpropagation: From Ordered Derivatives to Neural Networks and Political Forecasting
Yee and Haykin · Regularized Radial Bias Function Networks: Theory and Applications

www.ingramcontent.com/pod-product-compliance
Lightning Source LLC
Chambersburg PA
CBHW071307180926
51613CB00001B/146
9781119125457